The Growth of Cities

Wherever possible, the articles in these volumes have been reproduced as originally published using facsimile reproduction, inclusive of footnotes and pagination to facilitate ease of reference.

For a list of all Edward Elgar published titles visit our site on the World Wide Web at www.e-elgar.com

The Growth of Cities

Edited by

Zoltan J. Acs

University Professor
School of Public Policy, George Mason University, USA
and Research Scientist
Max Planck Institute of Economics, Jena, Germany

An Elgar Reference Collection
Cheltenham, UK • Northampton, MA, USA

Published by
Edward Elgar Publishing Limited
Glensanda House
Montpellier Parade
Cheltenham
Glos GL50 1UA
UK

Edward Elgar Publishing, Inc.
136 West Street
Suite 202
Northampton
Massachusetts 01060
USA

A catalogue record for this book is available from the British Library

ISBN-13: 978 1 84376 640 7
ISBN-10: 1 84376 640 X

Printed and bound in Great Britain by MPG Books Ltd, Bodmin, Cornwall.

Contents

Acknowledgements

The editor and publishers wish to thank the authors and the following publishers who have kindly given permission for the use of copyright material.

American Economic Association for articles: Antonio Ciccone and Robert E. Hall (1996), 'Productivity and the Density of Economic Activity', *American Economic Review*, **86** (1), March, 54–70; David B. Audretsch and Maryann P. Feldman (1996), 'R&D Spillovers and the Geography of Innovation and Production', *American Economic Review*, **86** (3), June, 630–40.

Blackwell Publishing Ltd for articles: Paul Krugman (1993), 'First Nature, Second Nature, and Metropolitan Location', *Journal of Regional Science*, **33** (2), May, 129–44; Claudio Michelacci (2003), 'Low Returns in R&D due to the Lack of Entrepreneurial Skills', *Economic Journal*, **113** (484), January, 207–25.

Elsevier for articles: Robert E. Lucas, Jr. (1988), 'On the Mechanics of Economic Development', *Journal of Monetary Economics*, **22**, 3–42; James E. Rauch (1993), 'Productivity Gains from Geographic Concentration of Human Capital: Evidence from the Cities', *Journal of Urban Economics*, **34**, 380–400; Edward L. Glaeser, José A. Scheinkman and Andrei Shleifer (1995), 'Economic Growth in a Cross-Section of Cities', *Journal of Monetary Economics*, **36**, 117–43; Curtis J. Simon and Clark Nardinelli (1996), 'The Talk of the Town: Human Capital, Information, and the Growth of English Cities, 1861 to 1961', *Explorations in Economic History*, **33**, 384–413; Luc Anselin, Attila Varga and Zoltan Acs (1997), 'Local Geographic Spillovers between University Research and High Technology Innovations', *Journal of Urban Economics*, **42**, 422–48; Maryann P. Feldman and David B. Audretsch (1999), 'Innovation in Cities: Science-Based Diversity, Specialization and Localized Competition', *European Economic Review*, **43**, 409–29; Claus Steinle and Holger Schiele (2002), 'When do Industries Cluster? A Proposal on How to Assess an Industry's Propensity to Concentrate at a Single Region or Nation', *Research Policy*, **31**, 849–58; Zoltan J. Acs and Catherine Armington (2004), 'The Impact of Geographic Differences in Human Capital on Service Firm Formation Rates', *Journal of Urban Economics*, **56** (2), September, 244–78.

Harvard Business School Publishing for article: Michael E. Porter (1998), 'Clusters and the New Economics of Competition', *Harvard Business Review*, November–December, 77–90.

MIT Press Journals and the President and Fellows of Harvard College and the Massachusetts Institute of Technology for article: Adam B. Jaffe, Manuel Trajtenberg and Rebecca Henderson (1993), 'Geographic Localization of Knowledge Spillovers as Evidenced by Patent Citations', *Quarterly Journal of Economics*, **108** (3), August, 577–98.

Oxford University Press for excerpt and article: Edward L. Glaeser (2000), 'The New Economics of Urban and Regional Growth', in Gordon L. Clark, Maryann P. Feldman and Meric S. Gertler (eds), *Oxford Handbook of Economic Geography*, Chapter 5, 83–98; Timothy Bresnahan, Alfonso Gambardella and AnnaLee Saxenian (2001), '"Old Economy" Inputs for "New Economy" Outcomes: Cluster Formation in the New Silicon Valleys', *Industrial and Corporate Change*, **10** (4), December, 835–60.

Princeton University Press and National Bureau of Economic Research for excerpt: Kenneth J. Arrow (1962), 'Economic Welfare and the Allocation of Resources For Invention', in *The Rate and Direction of Inventive Activity: Economic and Social Factors*, 609–26.

Sage Publications, Inc. for articles: Paul Krugman (1996), 'Urban Concentration: The Role of Increasing Returns and Transport Costs', *International Regional Science Review*, **19** (1&2), 5–30; Philip Cooke (2002), 'Biotechnology Clusters as Regional, Sectoral Innovation Systems', *International Regional Science Review*, **25** (1), January, 8–37.

Springer Science and Business Media for article: William J. Baumol (2004), 'Entrepreneurial Enterprises, Large Established Firms and Other Components of the Free-Market Growth Machine', *Small Business Economics*, **23** (1), August, 9–21.

Springer-Verlag GmbH & Co KG for article: Peter Nijkamp and Jacques Poot (1998), 'Spatial Perspectives on New Theories of Economic Growth', *Annals of Regional Science*, **32** (1), 7–37.

Taylor and Francis Ltd (http://www.tandf.co.uk/journals) for article: Sam Youl Lee, Richard Florida and Zoltan J. Acs (2004), 'Creativity and Entrepreneurship: A Regional Analysis of New Firm Formation', *Regional Studies*, **38** (8), November, 879–91.

University of Chicago Press for articles: Paul M. Romer (1986), 'Increasing Returns and Long-Run Growth', *Journal of Political Economy*, **94** (5), October, 1002–37; Paul M. Romer (1990), 'Endogenous Technological Change', *Journal of Political Economy*, **98** (5, Part 2), S71–S102; Paul Krugman (1991), 'Increasing Returns and Economic Geography', *Journal of Political Economy*, **99** (3), June, 483–99; Edward L. Glaeser, Hedi D. Kallal, José A. Scheinkman and Andrei Shleifer (1992), 'Growth in Cities', *Journal of Political Economy*, **100** (6), December, 1126–52; Vernon Henderson, Ari Kuncoro and Matt Turner (1995), 'Industrial Development in Cities', *Journal of Political Economy*, **103** (5), October, 1067–90.

Every effort has been made to trace all the copyright holders but if any have been inadvertently overlooked the publishers will be pleased to make the necessary arrangement at the first opportunity.

In addition the publishers wish to thank the Marshall Library of Economics, University of Cambridge, UK, the Library at the University of Warwick, UK, and the Library of Indiana University at Bloomington, USA, for their assistance in obtaining these articles.

Introduction

Zoltan J. Acs

This volume calls for a closer collaboration between economics and regional science, and suggests we study the growth of cities through a new lens – the creation and incubation of new knowledge. While the lens has many facets, there are three theoretical fields that can help us focus. First, the new growth theory suggests that knowledge is the key driving force behind economic growth and rising standards of living (Romer, Chapter 6 this volume). Second, the new economic geography suggests that regions and not countries are the real units of economic analysis (Krugman, Chapter 3) because as economies become less constrained by national frontiers they become geographically more specialized. Third, the new economics of innovation requires a systemic understanding of the institutional framework (Nelson, 1993). Innovation is a matter of producing new knowledge or combining existing knowledge in new ways. Many different kinds of actors and agents in the systems of innovation are involved in the process. Many of these interconnections can be found at the regional level as firms forge networks, clusters and specialized markets (Acs, 2002).

If we are to understand why some regions grow and others stagnate we need to understand the interactions among economic growth, economic geography and the economics of innovation (Acs and Varga, 2002, 2005). There are three fundamental questions that need to be answered. First, why and when does economic activity become concentrated in a few regions leaving others relatively underdeveloped? Second, what role does technological change play in regional economic activity? Third, how does innovation occur, and what are the key processes and institutions involved? To answer these questions three separate and distinct literatures need to be examined: the new economic geography, the new growth theory, and the new economics of innovation. Each one of the above three approaches has its strengths and weaknesses that can be integrated to develop an appropriate model of technology-led regional economic development.

The new economic geography answers the question of why economic activity concentrates in certain regions but not in others, but it leaves out innovation and economic growth. The contribution of Krugman's theory on economic concentration is not in its elements but in the way the system was put together. It has already been well known in economic geography and regional economics that decreasing transportation costs, economies scale or increasing demand favor agglomeration (Isserman, 1996). However, the way Krugman put these elements together in a general equilibrium model is novel. The model provides a case for the treatment of spatial issues in the way economists are accustomed to. The model provides a technique to analyze geographical concentration of economic activities as being induced by some initial combination of basic parameters. However, the model in its current form does not seem to be suitable for modeling technology-led regional economic growth for at least two reasons. First, Krugman's definite insistence of avoiding modeling the role of technological externalities in regional economic growth prevents the model from being applicable in innovation-led regional development since spillovers and innovation networks and clusters are in the core of this type of development, as exemplified in the literature on innovation and clusters. Second, while the model is very strong in working out the characterization of specific initial combinations of parameters favoring geographical concentration, it is weak in actually modeling the growth process. This research in addition to the articles in this volume has lead to two important works, Fujita, Krugman and Venables (1999) and Fujita and Francois Thisse (2002).

The new growth theory explains the causes of economic growth but leaves out regional considerations and ignores the key processes and institutions involved in innovation (Acs and Armington, 2006). A principal assumption in the theory of endogenous technological change is that for creating new sets of technological knowledge, the total stock of knowledge is freely accessible for anyone engaged in research. However, this assumption is not verified in the growing literature of geographic knowledge spillovers. New knowledge (potentially leading to either product or process innovations) is usually in such a tacit form that its accessibility is bounded by geographic proximity and or by the nature and extent of the interactions among actors of an innovation system. Similar to the case of relaxing the neoclassical assumption of equal availability of technological opportunities in all countries of the world, a relaxation of the assumption that knowledge is evenly distributed across space within countries seems also to be necessary. The nonexcludable part of the total stock of knowledge seems rather to be correctly classified if it is assumed to have two portions: a perfectly accessible part consisting of already established knowledge elements (obtainable via scientific publications, patent applications, etc.) and a novel tacit element, accessible by interactions among actors in the innovation system. While the first part is available without restrictions, accessibility of the second one is bounded by the nature of interactions among actors in a system of innovation or clusters, and so on.

The new economics of innovation, while explaining the institutional arrangements in the innovation process, leaves out economic growth. The system approach is a conceptual framework that many scholars and policy makers consider useful for the analysis of innovation. Although the systems of innovation approach is not considered a formal and established theory, its development has been influenced by different theories of innovation such as interactive learning theories, evolutionary theories, cluster theory and entrepreneurship. In recent years, efforts have been make to percolate general theoretical and empirical observations from this literature into a conceptual framework capable of guiding policy and loosely organized around the idea of a system of innovation. Important elements of the process of innovation tend to become regional rather than national. Regional networks of firms are creating new forms of learning and production. These changes are important and challenge the tradition roles of national systems of innovation.

I Analytical Overview

Part I provides readers with an introduction to the literature on the growth of cities. It introduces readers to two classes of literature, one focusing on economics and the other focusing on regional science. Chapter 1 by Edward Glaeser presents the new economics of urban and regional growth and is interested in three factors: the role of human capital in regional growth, the role of specialization versus diversity and the role of entrepreneurship and culture. For a larger treatment see Huriot and Thisse (2002). Finally the chapter touches on the issue of regional convergence.

Chapter 2 by Peter Nijkamp and Jacques Poot takes a strictly spatial perspective on the new theories of economic growth. Given that the new growth theory does not consider spatial perspectives this essay provides the reader in regional science a way in which to access the importance of this literature from a strictly special literature. For those interested in how the growth of cities fits into a larger urban dynamic see Capello and Nijkamp (2004).

II The New Economic Geography

Part II provides an introduction to the new economic geography. Paul Krugman, prominent in international trade theories, had turned to economic geography issues after recognizing that regions and not countries are the real units of economic analysis. Economic activities are not evenly distributed across space. An issue having central importance in economics is the explanation of why economic activities concentrate in certain areas while other areas remain relatively underdeveloped. In his basic theory Krugman presents a general equilibrium model that explains spatial concentration of economic activities with the interrelations of three parameters: increasing returns, transport cost and demand for manufacturing goods. According to the model, each of these parameters should pass certain threshold values before any kind of geographic concentration emerges. Similar to models in the new trade theory and the new growth theory, it builds extensively on the monopolistic competition models of Dixit and Stiglitz (1977).

III The New Growth Theory

Part III introduces the reader to the new growth theory of Romer and Lucas. The new growth theory identified knowledge as a key variable in explaining economic growth. Human capital in these models is an unobservable force or magnitude. According to Lucas this is no different than having a society imbued with the Protestant Ethic. However, one aspect of this particular aggregate model is the focus on the external effects of human capital. This is quite different from the internal effects that human capital models focused on before. Of course the external effects of human capital are also unobservable. Lucas suggests that there are individual interactions that are central to individual productivity. These include many of the activities that are found in cities. On purely economic grounds there is no reason that cities would not fly apart. As Lucas (Chapter 8, p. 39), points out, 'What can people be paying Manhattan or downtown Chicago rents for, if not for being near other people?'

In the Lucas model the amount of human capital in a society affects the productivity of each worker, apart from the worker's individual human capital. The acquisition of human capital by individual workers generates external effects, or spillovers of knowledge. The spillovers increase with the concentration of human capital. It is plausible that such external economies are strongest within a city, because the costs of acquiring the exchanging information – particularly small bits of information – are lower in cities. If human capital generates external economies, economies whose industries use human capital more intensively should grow faster, *ceteris paribus*.

IV Knowledge Spillover in Cities

Part IV has three chapters on knowledge spillovers. Adam Jaffe (1989) published the first paper on knowledge spillovers at the spatial level. The concept of knowledge spillovers leads to several theoretical issues. First, for analysis of endogenous growth, cities and their broader integrated economic areas provide much more suitable units than states or nations. The local economic areas centered on primary cities tend to function as open economies, with a tremendous internal mobility of capital, labor and ideas. These city-based economic areas are much more homogeneous units than those defined by the political boundaries of states, and they frequently cross state borders. National boundaries that bar mobility and national policies that encourage industrial diversification reduce the gains from factor mobility. These forces complicate analysis with cross-national samples. Cities allow us to look at units of economic growth without these concerns.

Second, not all types of industrial structure can promote knowledge spillovers equally. Identifying these structures is a major research effort. Third, knowledge spillovers do not appear to be constant over time, and affect mature and young industries differently. The empirical and theoretical literature suggests that knowledge spillovers are more important in the early states of the industry life cycle, when young firms flourish. One potential interpretation is that cities that are endogenously growing may have higher levels of entrepreneurial activity.

V Human Capital and the Growth of Cities

One implication of the Lucas model is that individuals do not capture all of the benefits from their ownership of human capital. Given the existence of human capital externalities, economically identical workers will tend to earn higher wages in human capital rich, rather than in human capital poor countries. The problem with inferring causation is that different countries are at different levels of development. This can be overcome by looking at cities within a country. In a country that has a well developed communications system the cost of capital and the level of disembodied technological knowledge will presumably be the same within its borders.

How important is human capital to the growth of cities? Studying English cities in 1861–1911 Simon and Nardinelli (Chapter 12) tell a story about city growth involving people exchanging information in certain occupations and industries. They focus on business professionals. They found that, in addition to such concentration sources of growth as ports, textiles, and other manufacturing, human capital played an important role in the long run. The providers of information did not include a large proportion of city population, but their presence may have made the difference between fat growth and stagnation. The knowledge produced by face-to-face meetings of people with high levels of human capital, rather than raw materials or large-scale manufacturing may well be the driving force behind the typical city's long-run growth.

Glaeser, Scheinkman and Schleifer (Chapter 13) examine the relationship between economic growth and initial years of schooling for a cross section of 203 US cities between 1960 and 1990. They present results on initial education and subsequent population growth. The initial median years of schooling exert a positive and significant influence on the subsequent population growth, using standard controls. A closer inspection of schooling shows that the percentage of the population with 12–15 years of schooling (high school graduates or some college) is more important than the percentage of the population with over 16 years of schooling (college graduates). This result suggests the importance of a well-educated labor force, not just of the top of the education distribution. The results for income and population growth are similar: initial years of schooling raise income, manufacturing lowers income and unemployment lowers income.

They conclude that the growth of cities is similar to that of countries. Since cities' differences are not created by different savings rates or different labor force endowments, the higher education levels influence later growth not through savings but through influencing the growth of technology. James E. Rauch (Chapter 14) finds additional support for the Lucas thesis suggesting that one additional year of Standard Metropolitan Statistical Areas (SMSA) average education can be expected to raise total factor productivity by 2.8%. Antonio Ciccone and Robert Hall examine the role of density in Chapter 15.

VI Diversity vs. Specialization in Cities

In the debate between specialization and diversity, the Marshal–Arrow–Romer (MAR) externality concerns knowledge spillovers between firms in the same industry. Knowledge accumulated and innovations produced by one firm tend to help other firms' technologies, or improvement of products, processes, or marketing, without appropriate compensation. The Jacobs–Bairoch–Rosenberg (JBR) model of externalities stresses knowledge spillovers across industries. They posit that the crucial externality in economic areas is cross-fertilization of ideas across different lines of business. Finally, several theories including Porter and Jacobs suggest that local competition rather than monopoly promotes economic growth.

While JBR, unlike MAR, assume that competition leads to economic growth, how competition is achieved and opportunity explored is left unexplained. In other words, the dynamic process by which local competition is achieved remains a black box. In Porter, local competition accelerates imitation and improvements on the original innovator's idea. This has two divergent effects – although such competition reduces the returns to the innovator, it also increases pressure to innovate in order to remain competitive. Porter believes that the second effect is by far the more important. Porter's model focuses on a set of factor conditions that he calls the 'diamond,' which includes demand conditions, related and supporting industries, and firm strategy, structure and rivalry. Regions are most likely to succeed in industries or industry segments where the 'diamond,' a term used to refer to the determinants of the system, is most favorable. The 'diamond' is a naturally reinforcing system in which new business formation is one of the key components of rivalry and competition.

There is a process linking spillovers to growth, and entrepreneurship that may be the vital ingredient in the process by which externalities generate economic growth, both within and across sectors. No matter how richly endowed an economic environment is with intellectual, social, human and financial resources, some person has to organize resources to pursue market opportunities.

VII Entrepreneurship and the Growth of Cities

One of the problems with the new growth theory is that while the theory emphasizes the role of human capital it does not explain how knowledge spillovers take place or what mechanism is at play. In other words, they do not answer the question, 'What type of activities do agents pursue that leads to faster growth?' Jovanovic and Rob (1989) develop a model where individual agents augment their knowledge through pairwise meetings at which they exchange ideas. In each time period, each individual seeking to augment his knowledge meets an agent chosen randomly from a distribution of agents. The higher the average level of human capital of the agents the more 'luck' the agents will have with their meetings and the more rapid will be the diffusion and growth of knowledge spillovers.

Part VII suggests that entrepreneurship is an important vehicle by which new knowledge is transformed into economic knowledge. If a city does not have enough entrepreneurs a given stock of knowledge may not generate the same growth as an environment rich in entrepreneurs. Chapter 19 by Claudio Michelacci suggests that low returns to research and investment may be due to a lack of entrepreneurial activity. In Chapter 20 Acs and Armington examine the impact of geographic differences in human capital on firm formation rates. They examine the relationship between the local levels of human capital and firm formation rates and find that formation rates differ with the share of adults with college degrees, especially for industries that normally require college-educated founders. They also differ strongly with the local concentration of existing establishments in the same sector especially for industries service non-local markets, suggesting that an important mechanism is the spillover of relevant knowledge.

Chapter 21 by Lee, Florida and Acs examines the role of creativity and entrepreneurship. This chapter explores whether connections exist among regional social characteristics, human capital and new firm formation. It agues that social diversity and creativity have a positive relationship with new firm formation. Building on the contribution of urbanism of Jane Jacobs (1969), Richard Florida (2002) showed that social diversity and human capital have positive and significant relationships with regional innovation production measured by per capita patent production. While it is well known that regional human capital stock positively affects new firm formation rates, little attention has been paid to the interaction between social diversity, human capital and entrepreneurship. It is argued that low barriers of entry into regional labor market and diverse culture facilitate the influx of a particular kind of human capital that promotes innovation and accelerates information flow, leading to the higher rate of new firm formation.

VIII Clusters and the Growth of Cities

In the traditional economy self-sufficiency by firms was viewed as a sign of strength. In the new economy it is a sign of weakness, with its emphasis on firms' interconnections to factor inputs, suppliers, competitors and customers (Porter, 1990). Clusters offer three key competitive advantages over vertical integration in single firms. First, productivity is enhanced by lower transaction costs and non-traded interdependencies. Second, innovation is increased which is dependent upon interactive knowledge exchange between varieties of knowledge actors. Third, new business formation is massively assisted by the mentoring, role model provision, learning, and communication gains that arise from operating in a cluster setting (Acs and Storey, 2004).

In Chapter 22, Timothy Bresnahan, Alfonso Gambardella and AnnaLee Saxenian discuss the sources of success in regional clusters of entrepreneurship and innovation. The study examines the forces that create new clusters of entrepreneurial led growth. They suggest that in addition to education policies one driver of economic growth is the low opportunity cost of human capital. When human capital has high opportunity cost because of alternative productive use this is hardly an issue for public policy. The question is far more serious when the opportunity costs for local human capital comes from artificially high wages in relatively unproductive jobs.

Chapter 23 by Philip Cooke draws on the Regional Innovation System (REGIS) project of the European Union to offer a theoretical framework for why clusters are important for the growth of cities. Using the biotechnology industry as a case study in three countries – USA, UK, Germany – Cooke identifies two key issues hindering the building of innovative regional clusters in Europe. Probably the key limitation on regional initiatives for advanced technology is funding for basic and applied research, since most regional administrations do not have remotely enough of such capital. The second limitation is venture capital and other sources of risk capital for the commercialization stages of biotechnology. He concludes that the regional level becomes the most important for the evolution of clusters including the concentration of critical research mass. Chapter 24 by Steinle and Schiele examines how to assess an industry's propensity to concentrate at a single region or nation.

IX Competition and Public Policy

One of the most important questions in the whole growth of cities literature is the role of competition. Two views on the subject suggest that either competition or monopoly is more important for economic growth. In a classic paper Arrow (Chapter 25) suggested that it was monopoly that promoted ecomic growth. However, Porter (Chapter 26) has suggested that it is competition that drives economic growth. Finally Baumol (Chapter 27) examines the interaction of entrepreneurship and competition for growth.

References

Acs, Zoltan J. (2002), *Innovation and the Growth of Cities*, Cheltenham, UK and Northampton, MA: Edward Elgar.

Acs, Zoltan J. and Catherine Armington (2006), *Entrepreneurship, Geography and American Economic Growth*, Cambridge: Cambridge University Press.

Acs, Zoltan J. and David J. Storey (2004), 'Introduction: Entrepreneurship and Economic Development', special issue on 'Entrepreneurship and regional development', *Regional Studies*, **38** (8), 871–8.

Acs, Zoltan J. and Attila Varga (2002), 'Geography, endogenous growth and innovation', in special issue: 'Regional Innovation Systems', *International Regional Science Review*, **25** (1), 132–48.

Acs, Zoltan J. and Attila Varga (2005), 'Entrepreneurship, agglomeration and technological change', *Small Business Economics*, **24** (2), 323–34.

Capello, Roberta and Peter Nijkamp (2004), *Urban Dynamics and Growth*, Amsterdam: Elsevier.

Dixit, Avinash K. and Joseph E. Stiglitz (1977), 'Monopolistic competition and optimum product diversity', *American Economic Review*, **67** (3), 297–308.

Florida, Richard (2002), *The Rise of the Creative Class*, New York: Basic Books.

Fujita, Masahisa and Jacques Francois Thisse (2002), *Economics of Agglomeration*, Cambridge: Cambridge University Press.

Fujita, Masahisa, Paul Krugman and Anthony J. Venables (1999), *The Spatial Economy*, Cambridge, MA: The MIT Press.

Huriot, Jean-Marie and Jacques-Francois Thisse (2002), *Economics of Cities*, Cambridge: Cambridge University Press.

Isserman, Andrew M. (1996), 'It's obvious, it's wrong, and anyway they said it years ago? Paul Krugman on large cities', *International Regional Science Review*, **19** (1&2), 37–48.

Jacobs, Jane (1969), *The Economy of Cities*, New York: Random House.

Jaffe, Adam (1989), 'Real effects of academic research', *American Economic Review*, **79**, 957–70.

Jovanovic, Boyan and Rafael Rob (1989), 'The Growth and Diffusion of Knowledge', *Review of Economic Studies*, **56**, 569–82.

Nelson, Richard (1993), *National Innovation Systems*, New York: Oxford University Press.

Porter, Michael (1990), *The Competitive Advantage of Nations*, New York: The Free Press.

Part I
Analytical Overview

[1]

The New Economics of Urban and Regional Growth

Edward L. Glaeser

Introduction

THE location of people and activities determines the nature and quality of our lives. Urban economists and regional scientists share this central conviction about the importance of space. This chapter presents an introduction to recent research by economists on urban and regional growth. While a primary goal of this volume is better coordination between geographers and regional scientists, in this discussion I review only works by economists, excluding the many superb papers by regional scientists in this area.

The new economics of urban and regional growth merges traditional urban economics with the new economic growth theory. The pioneering work in urban and regional growth provided basic methodological insights and an important set of stylized facts about classic urban issues, such as the interdependency between cities and metropolitan areas (Kain and Neidercorn 1963), the role of transport costs (Hoover 1937), and input price differences (Carlton 1983). The new economic growth theory suggests that cities should be understood as centers of idea creation and transmission. If this is so, then cities will grow when they are producing new ideas or when their role as intellectual centers is increasing in importance.

Advances in economic growth theory in the 1980s emphasized the role of technology, intellectual spillovers, and human capital externalities (Romer 1986; Lucas 1988). Knowledge spillovers solved the technical problem in economic theory of reconciling increasing returns (which are generally needed to generate endogenous growth) with competitive markets. This work suggests that if the global economy is endogenously growing (which is hard to deny) and if we believe in truly competitive markets (which is more controversial), then it almost follows that intellectual spillovers are a critical feature of the economic landscape. When economists began to look for intellectual spillovers, cities presented the clearest examples (see Lucas 1988); Jane Jacobs's (1969) anecdotes of technological innovations such as the invention of the bra are among the most vivid examples of intellectual spillovers.

This reasoning generated a mini-renaissance in urban economics and a reappraisal of which agglomeration economies matter. Economists since Alfred Marshall have argued that cities facilitate the flow of ideas. However, most urban research focuses on the role that density plays in reducing transport costs between suppliers and customers (Krugman 1991). However, Dumais *et al.* (1997) show that manufacturing firms in the USA since 1970 have not based their location decisions on the presence of suppliers and customers. Instead, firms locate near other firms that use the same types of workers. As new facts suggest that traditional urban models no longer explain cities and as a new body of theory emphasizes the flows of ideas, a new urban economics is taking place. In this body of research, cities are thought of as informational entities that exist to speed the flow of learning and knowledge. In this review, I examine this new literature on urban and regional growth.

Urban Growth and Agglomeration Economics

The fundamental question of urban economics is why cities exist. Elsewhere (Glaeser 1998), I argue that cities exist to eliminate transport costs for people, goods, and ideas. The urban advantage in eliminating transport costs for goods captures the classic manufacturing cities of the nineteenth century and is formalized in Krugman (1991), but this advantage is decaying over time as transports costs decline and as manufacturing leaves the city.

Models of labor market pooling or the division of labor emphasize the role of cities in facilitating the movement of people. For example, in Marshall's labor market pooling model firms group together because workers want to be able to switch firms if their employer has an idiosyncratic downturn or if the firm—worker match is bad. Indeed, in Dumais *et al.* (1997) the primary variable determining which industries locate near one another is their use of the same type of workers.

Cities also facilitate the flow of ideas between individuals and firms. In dense, urban environments proximity enables workers to acquire human capital by imitating a rich array of role models and learning by seeing. Alternatively, the flow of ideas may increase the rate of technological innovation and may lead dense cities to have a faster rate of new product innovation. If there is a greater variety of new ideas in cities, then these ideas may show up in new firms and better production processes.

This third force predicts that workers and firms will crowd together in dense areas to learn ideas from one another. This force also predicts that areas that are better at producing new ideas will grow more quickly. Cities which produce more ideas will grow because innovators may be temporarily, geographically fixed. According to this view, innovators acquire more ideas in cities and they first implement those ideas in their cities. Eventually, the idea may become old hat and the innovator may move to low cost areas, but at least temporarily the good ideas are implemented in the city. Alternatively, individuals may crowd into cities in the hope of acquiring a new idea. However, this will show up more as a static effect (the overall size of the city) rather than a dynamic effect (the growth of the city).

The Economics of Local Growth

Economic research begins with individual optimization—this is the defining characteristic of economics as a social science. In the urban and regional context, this means that the economic approach to urban growth begins with individuals choosing locations. Most commonly, economists assume that the world can be described by a spatial equilibrium where all individuals live in their preferred locations. Among homogeneous people, a spatial equilibrium implies that people's levels of welfare are constant across space. This assumption about constant utility levels across space underpins most economic work on cities and regions going back to Alonso (1964).

Non-economists tend to find this a little absurd, and indeed few economists take the assumption literally. We assume a spatial equilibrium because sensible alternatives are mind-numbingly complex, and economists cannot perform empirical (or theoretical) work that is not based on a clear model of individual choice. Indeed, macroeconomists in the 1960s and 1970s made predictions without considering individual decision-making; this led to some of the worst mistakes of postwar economic policy.

The concept of a spatial equilibrium is our guide to empirical work. For example, without a spatial equilibrium model, a research might try to measure the effects of city-specific attributes on productivity by examining whether these attributes raise profits or wages or productivity. But in equilibrium, firms will crowd near a city-specific productivity attribute until there are no more profits (and possibly no extra productivity) associated with that location-specific amenity. In equilibrium, there should not be any real differences in wages or profits across space. The empirical importance of this reasoning is that wage or profit (or even productivity) differences across space may not tell us anything about location-specific productivity.

Instead, the existence of a spatial equilibrium means that location-specific productivity effects will produce a greater density of firms and workers, and higher land prices. If a location has a high level of productivity, crowds of firms and workers will come to that place and drive up prices. The following formalization (based on Glaeser *et al.* 1995) formalizes this point. I assume that all firms produce goods with a price, which is fixed and equal to one. All firms have access to an identical constant returns to scale production function $Af(K,L,P) = AK^{\alpha}L^{\beta}P^{\gamma}$ where K is capital, L is labor, and P is the amount of land and where $1 = \alpha + \beta + \gamma$. The national market fixes the prices of capital and labor. The price of land is 'r' and there are $\bar{P}$ total units of land in the city. The wage outside equals w so workers in the city must be paid that as well. In this framework, cities have finite populations only because of the use of land in the production process. In reality, commuting times, pollution, and congestion can also put limits on the growth of cities.

The variable A captures the degree of local productivity and this is our key variable of interest. This variable captures the aspects of the city which are particular to the city. For example, A would include city and time-specific knowledge about techniques of production or the location-specific production amenities. Indeed, our primary interest tends to be estimating the determinants of A and the determinants of changes in A.

We can think of A as having a component which is a function of city-level characteristics and a city-specific component, that is, $A = \bar{A}e^{\Sigma\delta X}$, where X represents city-level characteristics (the level of human capital, the presence of a port), δ represents the coefficients on these characteristics, and $\bar{A}$ represents the base city-level productivity. Solving the model shows $K = A^{\frac{1}{\gamma}}P(\beta/w)^{\frac{\beta}{\gamma}}(\alpha/c)^{\frac{1-\beta}{\gamma}}$, $L = A^{\frac{1}{\gamma}}P(\beta/w)^{\frac{1-\alpha}{\gamma}}(\alpha/c)^{\frac{\alpha}{\gamma}}$ and $r = (1-\alpha-\beta)A^{\frac{\alpha+\beta}{\gamma}}(\beta/w)^{\frac{\beta}{\gamma}}(\alpha/c)^{\frac{\alpha}{\gamma}}$. Thus improvements in location-specific amenities (A) will show up in greater amounts of capital, greater amounts of labor, and higher property values.

In this framework, wages tell us nothing about local productivity. If we assume that workers need to buy land to live or if more crowded cities are less pleasant, then this will no longer be true. As the level of A rises, crowding and land prices will rise and workers will need to be paid more to get them to live in the city. Under those conditions, wages can be used as measures of local productivity (see Glaeser *et al.* 1995; Rappaport 1998). However, wage regressions are difficult to interpret because wages should decrease as cities become more pleasant.

Standard growth regressions tend to focus on the amount of labor (population is usually quite close to labor) in the city rather than capital or land prices, primarily because of data issues. Thus regressions use a version of the formula for L above and transform it so that:

$$Ln\left(\frac{L_{t+1}}{L_t}\right) - \Omega = \frac{1}{\gamma}Ln\left(\frac{A_{t+1}}{A_t}\right)$$

$$(1) \qquad = \frac{1}{\gamma}\left[Ln\left(\frac{\bar{A}_{t+1}}{\bar{A}_t}\right) + \sum(\delta_{t+1} - \delta_t)X_t + \sum\delta_{t+1}(X_{t+1} - X_t)\right],$$

where Ω is a term that is constant across cities. If we assume that city-specific productivity growth is a function of initial characteristics, that is, $Ln\left(\frac{A_{t+1}}{A_t}\right) = \sum \lambda X_t + \varepsilon_t$, then (1) can be rewritten as:

$$(1') \quad Ln\left(\frac{L_{t+1}}{L_t}\right) = \Omega + \frac{1}{\gamma}\sum(\lambda + \delta_{t+1} - \delta_t)X_t + \frac{1}{\gamma}\sum \delta_{t+1}(X_{t+1} - X_t) + \frac{\varepsilon_t}{\gamma}$$

Equation (1′) is the fundamental growth regression. Changes in employment are regressed on a vector of city-specific characteristics.[1] Generally, terms relating to the changes in the X variables are ignored. Formally this requires assuming that changes in the X variables are orthogonal with respect to the initial values. In practice the decision to ignore these terms comes about because of the endogeneity of these change variables. For example, if urban growth is related to increases in the level of human capital which might happen if migrants are more skilled, then controlling for the growth in the level of human capital might completely bias the other coefficients. Rappaport (1998) runs growth regressions with median housing values (as a proxy for land values); these regressions rely on an equality similar to equation (1).

This framework allows us to interpret changes in the labor force in the city and shows that an initial variable can have an effect on the growth of labor in two very distinct ways.

First, a variable may actually increase the growth rate of A. This occurs if the coefficient λ is positive. An example might be that the level of urban diversity leads to more ideas being produced in the city over time and the city becoming a more productive place.

Secondly, a variable might directly determine A and the relationship between that variable and A has become more important over time. This occurs if $\delta_{t+1} > \delta_t$. Thus, there might be a connection between human capital and increases in the level of A, not because human capital causes increases in A but rather because human capital provides a positive productive amenity for firms in each period, and the importance of human capital has risen over time.[2]

These two interpretations are linked with dynamic and static externalities respectively. Dynamic externalities occur when a city-specific feature, such as the level of diversity or the level of human capital, increases the speed with which new ideas are produced and increases the level of growth of productivity

[1] One alternative to the classic growth regression taken by Eaton and Eckstein (1997), (1′), is to follow Quah (1993) and partition the data into subgroups (e.g. cities with less than 50,000 inhabitants, cities with between 50,000 and 60,000) and then estimate a discrete transition model where the probability of moving from one group to another is estimated.

[2] A variable X might also influence population growth if initial levels of X are correlated with changes in X.

in the city. Static externalities occur when a city-specific feature increases city-specific productivity. Thus human capital may increase the level of employment growth without dynamic externalities if the importance of human capital in generating static externalities has risen over time.

Some attempts to relax the spatial equilibrium assumption have occurred. Glaeser *et al.* (1995) examines the possibility that there are costs associated with rapid adjustment to city-specific productivity shocks and suggested a simple way of allowing a slightly more flexible framework. Rappaport (1998) is much more ambitious and considers a full dynamic adjustment pattern where local shocks only work their way through the system over time. In general, slower adjustment tends to make income growth a more plausible dependent variable, as slow adjustment means that shocks to labor demand will increase wages temporarily as the labor supply adjusts. Of course, in these cases the results from looking at employment should move in the same direction as the results using wages or income. Glaeser *et al.* (1995) show that income growth and population growth tend to move together over the past thirty years across cities.

The Unit of Observation

Within the USA, units of observation can be divided into population-based units, which attempt to capture the total number of workers or population within a geographic area, and location-industry units, which capture the number of workers in a particular industry within a particular location. Population units come from the census and include census tracts, cities, counties, metropolitan areas, and states. In principle, most data sets focus on where people live, which is good if we care about consumption amenities but less ideal for measuring the effects of production amenities. Measures of where people work also exist (County Business Patterns or the Economic Census) and these are better for measuring the effect of location-specific productivity. One advantage of using large enough geographic units is that they do a better job of ensuring that people both live and work within their boundaries.

The city has the advantage of being somewhat smaller (which is valuable if we believe that intellectual spillovers are operating in small, dense areas) and being closely tied to the traditional central business district (CBD). Cities are also the natural unit of analysis in cases where politics are an important independent variable. On the negative side, city boundaries are often quite artificial and their small size means that they may be excessively influenced by neighboring political units (i.e. the growth of Cambridge is certainly strongly influenced by the growth of Boston). In principle, some of this correlation across space can be handled by including initial characteristics of the city in question and of neighboring cities.

Alternatively, metropolitan statistical areas (multi-county units generally surrounding a large city) are less artificial geographic units that are meant to include an entire labor market. These multi-county units' main problems are that they may be too large for some purposes and that they change size over time (this is a problem for cities as well, but the changes are generally less extreme). For example, the heterogeneity within a large metropolitan area may be staggeringly large.

To capture increased suburbanization and interconnection, the census bureau has often increased the size of metropolitan statistical areas to include more counties. In that case, the econometrician needs to make a decision about using initial county definitions, or end-point definitions, or allowing the changes to be incorporated in the regressions. In general, initial definitions tend to be the safest procedure. Within metropolitan areas, there is also the debate over whether to use Primary Metropolitan Statistical Areas or Consolidated Metropolitan Statistical Areas (PMSAs or CMSAs), where CMSAs are larger.

A third possible approach is to use counties. These are larger than cities and often parts of metropolitan areas. Counties are natural for studies, which focus on the entire USA. The only problem with county-level regressions is the possible interdependence between counties that are part of the same metropolitan area. States are also a possible unit of analysis for regional analysis (as in Blanchard and Katz 1992), and while the time patterns of states are extremely interesting they are less appropriate for tests of intellectual spillovers since they are so large.

A final possible population unit of analysis is the census tract, which is a tiny sub-city unit of observation (generally between 3,000 and 5,000 inhabitants). These units are interesting for examining determinants of changing demand for neighborhood characteristics. They are less helpful for looking at issues of productivity.

Another approach is to examine city-industries (which may also be metropolitan area-industries or state-industries or county-industries). These are defined as an industry in a particular geographic unit. For example, steel in the Pittsburgh area would be a city-industry. Information on these variables comes from the Economic Census of the US and County Business Patterns. The advantage of looking at city-industries is that we can control for general patterns of industrial growth in the country as a whole quite easily. We can also examine the interaction between industry variables and city variable (i.e. are more human capital-oriented industries attracted to high human capital cities). Glaeser *et al.* (1992) use city-industries, as do Henderson *et al.* (1995) and Glendon (1998).

Within this class, it is also possible to examine variables such as new plant formation. This is a subcomponent of employment growth that may capture the innovation aspects that are particularly important to tests of intellectual spillovers. Dumais *et al.* (1997) examine the determinants of the emergence of new plants. A similar interpretation can be given to the emergence of new products which is examined by Feldman and Audretsch (1999). These can also be thought of as growth regressions with particularly fine-tuned dependent variables. There is a great deal of added precision gained by looking at this type of variable.

As a final comment, it is worthwhile thinking about the relevant time period for growth regressions. Periods of less than ten years tend to have data problems in many cases. Furthermore, there is presumably a non-trivial amount of year to year variation that is not related to growth rates. Even with these high frequency shocks, we can use shorter time periods as long as the econometrician uses area-specific random effects not area-specific fixed effects. Area-specific fixed effects mean that the regression is identifying itself from very high frequency changes which is probably undesirable if the econometrician is hoping the determine the causes of long-term growth.

Facts and Controversies about Urban and Regional Growth

In the last section, I reviewed the methodological issues involved in estimating the determinants of urban growth. In this section, I discuss some basic stylized facts and their interpretation.

Local Human Capital

A significant number of papers confirm the connection between the initial level of human capital in an area and the later growth of that area. Regardless of whether human capital is measured as years of schooling, the percentage with high school degrees, the percentage of college educated, or a measure of education based on the occupational mix, there is a strong, steady connection between growth and initial skills in the area. Glaeser (1994) presents some of the first evidence on this and shows that the connection between skills and growth appears to be getting stronger over time. For example, the connection between initial skills and later growth is stronger over the 1970–90 period than over the 1950–70 period.

Glaeser *et al.* (1995) document this link between human capital and growth more thoroughly using metropolitan areas and find that initial human capital predicts population, employment, and income growth. This work finds that the percentage completing high school appears to be a more powerful variable for the year 1960. Evidence presented by Glaeser (1994) suggests that college education may have become more important in recent years. Rappaport (1998) examines a much larger set of counties and documents that the human capital–growth connection appears to hold with this much larger sample over the post-1970 period. Mody and Wang (1997) show that the connection between human capital and growth even appears in modern China.

In two important essays, Simon and Nardinelli (1996, 1998) examine the connection between human capital and city growth over a much longer time period in the USA and Great Britain. In the USA, they find that a measure of human capital (based on the occupational mix of the cities in 1880) predicts city growth for all subsequent decades. This suggests that long-running historical factors that influence the level of human capital exert a long-standing influence on city growth. The work on urban growth has not yet followed the tendency of the economics profession to focus overwhelmingly on exogenous sources of variation and this will surely be one of the changes in the urban growth literature over the next ten years. However, since the Simon and Nardinelli data relies upon such a long-term measure of human capital, this can be seen as an early attempt to create an instrument for the level of skills in the city. Glendon (1998) confirms their evidence and shows that the primary effect of human capital is not just due to the fact that high human capital cities have high human capital industries which have tended to grow faster over this century.

As suggested by regression (1′) there are two natural interpretations of the connection between human capital and later urban growth. First, the presence of skills in the metropolitan area may increase new idea production and the growth rate of the city-specific productivity level. High skilled people in high skilled industries may come up with more new ideas. Secondly, skills may create a static externality, which has grown more important over time. Rauch (1993) documents the importance of human capital spillovers by showing that nominal wages and housing prices together rise in cities with the average level of human capital (holding the worker's own level of human capital constant). For example, if one learns skills from one's neighbors (as modeled by Glaeser 1998) and if skills have become more valuable over time, then the advantages from being around skilled people may have risen over time. However, some doubt is cast on this explanation since the connection between growth rates and local human capital levels occurs in the middle of the century when the returns to skill apparently fell.

Concentration versus Diversity

Jacobs (1969) argues that new ideas are formed by combining older ideas (see also Weitzman 1998 for a formalization). Cities allow for the mixing of many different industries and occupations. In this cauldron, ideas from different areas get combined and growth occurs. A natural implication of this line of reasoning is that more diverse cities will grow faster than concentrated cities. Alternatively, industrial concentration may be more important. If new ideas are useful only within an industry, then diversity is wasteful. If learning by doing is extremely important for growth, then geographic specialization in a particular form of production may allow the knowledge learned by doing to be used by other forms.

One way to test this hypothesis is to ask whether an initial concentration of the industry later attracts more industry. Initial concentration is measured in two distinct ways. First, initial concentration is measured by the raw employment level of the city-industry. Secondly, initial concentration is measured by the employment level in the city-industry relative to the city size or relative to the size of the industry in the USA. For the second measure, Glaeser *et al.* (1992) use *Share of City Employment in Industry/Share of US Employment in Industry* which captures the degree to which the industry is overconcentrated in the city relative to the industry's concentration in the USA as a whole.

A second test asks whether the level of industrial diversity outside the city-industry in question increases growth. A primitive means of measuring this diversity is to use a four or five industry concentration measure (the share of total city employment involved in the five largest industries). Alternatively (and probably preferably), Henderson *et al.* (1995) use a Herfindahl index across industries.

The connection between initial employment and later employment growth is pretty uniformly negative (Glaeser *et al.* 1992; Henderson *et al.* 1995; Dumais *et al.* 1997). Naturally, this needs to happen if we have a stationary distribution of city-industry employment. However, this indicates that there are limits to the advantages of concentration in fomenting later growth.

Results on concentration and diversity are more mixed. For example, Glaeser *et al.* (1992) find that concentration has negative effects and that diversity is positive. Henderson *et al.* (1995) find that concentration is positive and diversity has no clear effect. However, the negative effect of initial employment in the Henderson *et al.* work is much stronger than the negative effect of raw employment in Glaeser *et al.* (1992). Thus, in some respects the Henderson *et al.* (1992) paper finds even less evidence for the presence of gains from concentration. Miracky (1994) looks at a complete set of industries in the 1970–90 period, and finds few positive or negative benefits of diversity across all specifications. He does find some negative evidence against concentration.[3]

For the moment, the role of concentration and diversity does not seem to have been resolved by the literature. Different time periods and different samples give different results which suggest that there is no universal truth on this topic. My hope is that by investigating the actual hard evidence on innovations, we will be able to assess the relative importance of idea combinations and the role of diversity and concentration.

Cultures of Entrepreneurship

Benjamin Chinitz (1961) argues that the differences between then-successful New York (and then-declining Pittsburgh came about in part because of New

[3] Garcia-Mila and McGuire (1993) also find that industrial mix is important for growth.

York's many small firms and culture of entrepreneurship. Porter (1990) also argues that local competition provides the right incentives for innovation and new ideas. Alternatively, some models suggest that local monopolists invest more in new ideas (since competitors will only steal each other's ideas). A series of urban growth papers have attempted to examine the role of local competition.

The first key question is how to measure competition. A natural measure of the amount of competition is just to measure the number of firms in the city-industry (because of data issues, occasionally plants are substituted for firms). In order to distinguish competition from scale, the number of firms is sometimes divided by the size of the industry in the area (measured by the number of workers or the amount of output). The resulting variable can be interpreted as the number of firms per worker or per unit output, which seems like a natural competition measure. However, this measure is also the inverse of the average firm size so anything that we attribute to competition may actually be a function of smaller firms. Glaeser *et al.* (1992) use the number of firms per worker in the city-industry and then divide this quantity by the number of firms per worker in the industry in the USA as a whole.

The facts about these competition measures are unequivocal. The positive effect of the competition variable is the strongest result in Glaeser *et al.* (1992). Miracky (1994) confirms the remarkable strength of this variable throughout the entire country for the post-1970 period. Every piece of research in this area that I am aware of finds a positive effect of competition on later growth.

While the facts appear to be clear, the interpretation is not. One interpretation of these results is that competition foments intellectual growth. Alternatively, cities that are exogenously growing may have a lot of new plants and firms that are also small. Miracky (1994) suggests that the competition measure captures the place the city-industry has in the industry life cycle (i.e. young growing industries have small firms). Further work will be needed to clarify these issues.

Government and Redistribution

The literature on the connection between government policy and country-level economic growth is vast and has strong conclusions suggesting that government-created distortions can badly hurt growth. The literature on government and local growth is much smaller, which is somewhat appropriate given the small differences in government policies across localities within the USA relative to the differences across countries. There is a literature on industrial location suggesting that pro-business policies (such as right-to-work laws) generally attract firms. The masterpiece of this literature is Holmes (1998) which documents the importance of pro-business policy (e.g. right-to-work laws) in attracting industry using within as well as across state variation. The key insight is that there should be a real shock on the borders of states if policies matter and Holmes finds just such a discontinuity.

The literature connecting growth and policies is much smaller. Glaeser *et al.* (1995) include a large vector of city-level political variables in our growth regressions. We find only two political variables that predict urban growth: the amount of debt per capita of the city in 1960 and spending on sanitation. We interpret the debt effect as the result of reverse causality. Cities expecting large growth borrow because they know that they will be able to repay that borrowing in the future. The sanitation effect is provocative, but hardly an overwhelming stylized fact. Rauch (1995) also shows that the municipal reforms in the early part of this century have little effect on growth.

Rappaport (1998) finds much stronger effects of government policies for the period starting in 1970. His results differ from ours both because of the later time period and because he examines a much larger range of counties. Both the size of government and the extent to which government focuses on redistribution tend to depress later population growth of the region. Vigdor (1998) shows that exogenous reductions in the size of government caused by Proposition $2^1/_2$ increase the growth of the tax base in particular areas. One explanation for the different sets of results is that transfer policies attract lower human capital people and these people were more negatively associated with city growth during the later time period because the importance of skills had risen.

These results suggest that localities can stymie their own growth with sufficiently negative policies. While local redistribution is probably bad for growth, there is not a clear message about what policies are good for growth. Using the results on human capital from above, it seems likely that policies which will increase the quality of the local human capital will increase the speed of urban growth. This stands as another area of research where we are just beginning to understand the determinants of growth.

A final connection between politics and urban growth is the role of higher levels of governments. While we do not know what policies a city can follow to increase its own growth, we do know that if states and the country as a whole favor particular cities they will grow. Eaton and Eckstein (1997) document the special growth patterns of Paris and Tokyo which are associated with the political factors affecting those cities. Ades and Glaeser (1995) document the political factors, which generate urban primacy across countries. DeLong and Shleifer (1993) show that before the industrial revolution city growth is associated with mercantile, not royal power.

The Convergence Debate

A final critical area of local growth is the debate on whether initial levels of income or population are related negatively to later growth in these variables. Barro and Sala-I-Martin (1991) showed a strong negative correlation between initial income and income growth across states within the USA and across countries. There are many possible interpretations of this finding ranging from slow-moving migration following labor demand shocks to less developed states imitating the technologies of more advanced states. Glaeser *et al.* (1995) show that holding regions constant there is little convergence in income levels across cities between 1960 and 1990.

As argued earlier, population or employment levels are more sensible measures for most local growth regressions than income levels. With population levels there is relatively little evidence of strong convergence at the metropolitan area level. Eaton and Eckstein (1997) show that city population growth in Japan and France follows a parallel pattern where there is neither divergence nor convergence. Glaeser *et al.* (1995) find that within the USA there is some convergence at the city level, but little convergence at the metropolitan area level. It is not surprising that there is more convergence at the city level since limits on land area will tend to bind more tightly for these smaller population units. In a particularly comprehensive study, Beeson *et al.* (1998) show that between 1840 and 1990, population convergence at the county level occurs only for the densest counties. More generally, they find no convergence and even some evidence of divergence.

The pattern where city growth rates are independent of the initial size of the city ties quite closely to the well-known fact of Zipf's law. Zipf's law says that the second largest city is one-half the size of the largest city and the third largest city is one-third the size of the largest city, and so on. Gabaix (1999) shows that Gibrat's law (city growth rates are independent of initial city sizes) implies Zipf's law. Thus, since Zipf's law is one of the most well-documented facts in social science, we should not be surprised that Gibrat's law also appears to hold fairly pervasively.

A final fact that bears on this debate is the stunning persistence of growth rates of population at the state and city level. Blanchard and Katz (1992) show that there is a clear correlation over time in the growth rates of states. Glaeser *et al.* (1995) show the same thing for cities. This fact can be interpreted as evidence for migration responding only slowly to location-specific shocks. Alternatively, it can mean that some places are steadily more productive in producing new ideas and new technologies. In either case, the evidence speaks against a strong form of convergence where gains in one period are offset by relative losses in the next period.

Conclusion

Cities appear to grow faster when they have higher skilled workers and when they have more competitive industrial structures. There is little clear evidence on the role of diversity versus competition. Big cities tend to grow at roughly the same rate as small cities. Big local governments that redistribute highly to the poor tend to reduce local growth. These are useful facts but they do not thoroughly answer the central questions posed at the beginning of this section. The role of human capital supports the role of intellectual spillovers, but there are certainly other interpretations. The other facts are compatible with an idea transmission view of cities, but also with other views.

There are two major avenues for future research. First, we need more attention to be paid to exogenous sources of variation (instrumental variables. All explanatory variables have the problem that they are exogenous and quite potentially correlated with other variables that may truly be driving growth. The second primary avenue for future research is to try and measure the localization of ideas more effectively. Jaffe *et al.* (1993) is the paradigmatic paper in this area which uses patent citations to document that ideas move only slowly over time. Audretsch and Feldmann (1996) also use an innovative data set measuring new product innovations to inform us about idea production in cities. New variables and data sets such as these will be invaluable in understanding the processes that drive urban growth.

References

Ades, A., and Glaeser, E. 1995. Trade and circuses: explaining urban giants. *Quarterly Journal of Economics*, 110/1: 195–258.

Alonso, W. 1964. *Location and Land Use: Toward a General Theory of Land Rent.* Cambridge, Mass.: Harvard University Press.

Audretsch, D. B., and Feldmann, M. P. 1996. Knowledge spillovers and the geography of innovation and production. *American Economic Review*, 86: 630–40.

Barro, R., and Sala-I-Martin, X. 1991. Convergence across states and regions. *Brooking Papers on Economic Activity*, 1.

Beeson, P., DeJong, D., and Troesken, W. Forthcoming. Population growth in U.S. counties, 1840–1990. *Regional Science and Urban Economics.*

Blanchard, O., and Katz, L. 1992. Regional evolutions. *Brookings Papers on Economic Activity*, 1: 1–75.

Carlton, D. 1983. The location and employment choices of new firms: an econometric model with discrete and continuous endogenous variables. *Review of Economics and Statistics*, 65/3: 440–9.

Chinitz, B. 1961. Contrasts in agglomeration: New York and Pittsburgh. *American Economic Review*, 71: 279–89.

DeLong, B., and Shleifer, A. 1993. Princes and merchants: city growth in preindustrial europe. *Journal of Law and Economics*, 36/2: 671–702.

Dumais, G., Ellison, G., and Glaeser, E. 1997. Geographic concentration as a dynamic process. *NBER Working Paper*, No. 6270.

Eaton, J., and Eckstein, Z. 1997. Cities and growth: theory and evidence from France and Japan. *Regional Science and Urban Economics*, 27/4–5: 443–74.

Feldman, M. P., and Audretsch D. B. 1999. Innovation in cities: science-based diversity, specialisation and localised competition. *European Economic Review*, 43: 409–29.

Gabaix, X. 1999. Zipf's law for cities: an explanation. *Quarterly Journal of Economics*, 114/3: 739–68.

Garcia-Mila, T., and McGuire, T. 1993. Industrial mix as a factor in the growth and variability of states' economies. *Regional Science and Urban Economics*, 23/6: 731–48.

Gaspar, J., and Glaeser, E. 1998. Information technology and the future of cities. *Journal of Urban Economics*, 43/1: 136–56.

Glaeser, E. 1994. Cities, information and economic growth. *Cityscape*, 1: 9–48.

—— 1998. Are cities dying? *Journal of Economic Perspectives*, 12/2: 139–60.

—— Kallal, H., Scheinkman, J., and Shleifer, A. 1992. Growth in cities. *Journal of Political Economy*, 100/6: 1126–52.

—— Scheinkman, J., and Shleifer, A. 1995. Economic growth in a cross-section of cities. *Journal of Monetary Economics*, 36/1: 117–43.

Glendon, S. 1998. Urban life cycles. Harvard University mimeograph.

Henderson, J. V., Kuncoro, A., and Turner, M. 1995. Industrial development in cities. *Journal of Political Economy*, 103/5: 1067–90.

Holmes, T. 1998. The effect of state policies on the location of manufacturing: evidence from state borders. *Journal of Political Economy*, 106/4: 667–705.

Hoover, E. 1937. *Location Theory and the Shoe and Leather Industries*. Cambridge, Mass.: Harvard University Press.

Jacobs, J. 1969. *The Economy of Cities*. New York: Random House.

Jaffe, A., Trajtenberg, M., and Henderson, R. 1993. Geographic localization of knowledge spillovers as evidenced by patent citations. *Quarterly Journal of Economics*, 108: 577–98.

Kain, J., and Neidercorn, D. 1963. An econometric model of metropolitan development. *Regional Science Association Papers*, 11.

Krugman, P. 1991. Increasing returns and economic geography. *Journal of Political Economy*, 99: 483–99.

Lucas, R. 1988. On the mechanics of economic development. *Journal of Monetary Economics*, 22: 3–42.

Miracky, W. 1994. Technological spillovers, the product cycle and regional growth. Massachusetts Institute of Technology dissertation.

Mody, A., and Wang, F. 1997. Explaining industrial growth in coastal China: economic reforms . . . and what else? *World Bank Economic Review*, 11/2: 293–325.

Porter, M. 1990. *The Competitive Advantage of Nations*. New York: Free Press.

Quah, D. 1993. Empirical cross-section dynamics in economic growth. *European Economic Review*, 37: 426–34.

Rappaport, J. 1998. Local economic growth. Harvard University mimeograph.

Rauch, J. 1993. Productivity gains from geographic concentration of human capital: evidence from the cities. *Journal of Urban Economics*, 34: 380–400.

—— 1995. Bureaucracy, infrastructure, and economic growth: evidence from U.S. cities during the progressive era. *American Economic Review*, 85/4: 968–79.

Romer, P. 1986. Increasing returns and long run growth. *Journal of Political Economy*, 94: 1002–37.

SIMON, C., and NARDINELLI, C. 1996. The talk of the town: human capital, information and the growth of English Cities, 1861–1961. *Explorations in Economic History*, 33/3: 384–413.

—— —— 1998. Human capital and city growth in the U.S. Mimeograph.

VIGDOR, J. 1998. Was Tiebout wrong? Evidence from Proposition 2½ in Massachusetts. Harvard University mimeograph.

WEITZMAN, M. 1998. Recombinant growth. *Quarterly Journal of Economics*, 113/2: 331–60.

[2]

Ann Reg Sci (1998) 32:7–37

The Annals of
Regional Science

Spatial perspectives on new theories of economic growth*

Peter Nijkamp[1], **Jacques Poot**[2]

[1] Department of Spatial Economics, Free University Amsterdam, De Boelelaan 1105, 1081 HV Amsterdam, The Netherlands
(Tel.: +31-20-4446094; Fax: +31-20-4446004; e-mail: pnijkamp@econ.vu.nl)
[2] School of Economics and Finance, Victoria University of Wellington, P.O. Box 600, Wellington, New Zealand
(Tel.: +64-4-4721000; Fax: +64-4-4955014; e-mail: jacques.poot@vuw.ac.nz)

Abstract. A new wave of interest in long-run economic growth emerged since the late 1980s. This paper uses a simple model to illustrate how technological change can be endogenised in macroeconomic theories of growth and then surveys how – through factor mobility, the diffusion of innovations and trade – spatial interdependence in a system of regions can influence technological change and growth. Endogenous technological change generates in our illustrative model long-run steady-state growth in a closed economy. However, it turns out that the dynamic impact of spatial interdependence depends on the specification of the model. Spatial convergence, a steady state with persisting spatial differences in growth rates and unstable growth are all theoretically possible. Issues relating to the role of aggregate demand and policy also receive attention. There is much scope for further theoretical and empirical work on endogenous growth in a spatial-economic context, while a better integration of micro and macro level approaches is also desirable.

1. Introduction

The spatial-economic landscape exhibits a panorama of changing hills and valleys of welfare levels. Why growth rates differ between nations, or regions, is a fundamental question which has intrigued economists ever since Adam Smith's (1776) "Inquiry into the nature and causes of the wealth of nations". Research on the subject, however, has not been at a steady pace during the past two centuries. Instead, there have been several waves of in-

* This is a revised version of a paper presented at the international seminar on Endogenous Growth and the Space Economy, 19–20 December 1996, Tinbergen Institute. Amsterdam. This paper was written while Poot was Visiting Professor at the Institute of Policy and Planning Sciences of the University of Tsukuba, Japan. The hospitality of the Institute is gratefully acknowledged.

terest. Temporal and spatial differences in the standard of living were important issues for classical economists such as Adam Smith, Malthus (1798) and Ricardo (1817). Except for Marx's (1867) alternative interpretation of the driving forces in long-term capitalist accumulation, further major theorising did not occur until the 20th century. During the first half of this century, Schumpeter (1934) laid the foundations for recent insights into the role of technological change and entrepreneurial competition in long-run development, while Harrod (1939) and, independently, Domar (1946) studied the growth of a Keynesian economy. In their model, the long run growth path would be likely to exhibit either growing unemployment or accelerating inflation.

The second half of this century has seen two major waves of interest in the macroeconomics of growth. The first of these commenced with the seminal articles by Solow (1956) and Swan (1956), whose neoclassical growth model provided a more plausible description of the long-term path of the economy than the Harrod-Domar model. Yet the major weakness of the standard neoclassical model, and of the wave of theoretical contributions which built upon it, was that it did not provide an explanation for the actual "engines" of long-run growth in income per worker, although its predictions were consistent with several stylised facts of economic development.

The reason for this deficiency was not that economists were ignorant about the causes of spatial or temporal differences in long-run growth rates but, instead, that causes of productivity improvements such as innovation, economies of scale and learning-by-doing had effects on the economy which violated the assumptions of perfect competition and constant returns to scale. As Krugman (1995) argued, the theoretical tools were not yet available in the 1950s to study such phenomena of increasing returns and imperfect competition within the accepted axiomatic neoclassical framework. Consequently, policy makers requiring advice on how to improve national and regional growth rates needed to look elsewhere and the fields of development economics and regional economics emerged respectively as more pragmatic responses to fill this vacuum. Thus, theoretical and empirical models of regional growth could build upon the idea of growth poles (Perroux 1955) or cumulative causation (Myrdal 1957; Kaldor 1970).

When several North-American macroeconomists returned to the issue of long-run growth in the mid 1980s, now with new tools to formulate equilibrium models with increasing returns and imperfectly competitive sectors, a revival of the field emerged starting with the influential articles by Paul Romer (1986) and Robert Lucas (1988). This literature describes macroeconomic outcomes in terms of microfoundations such as an intertemporal optimisation of consumption by rational and forward-looking households. However, the essence of this idea itself goes back to Ramsey (1928) and was earlier applied to growth modelling by Cass (1965) and Koopmans (1965).

Parallel to the development of the new theoretical explanations, a vast literature on the empirics of growth emerged, commencing with the contribution of Kormendi and Meguire (1985). Here the objective was not only to identify causes of growth, but also to find out whether growth rates across countries or regions converge or diverge over time.

The importance of increasing returns, externalities and imperfect competition for an understanding of the dynamics of economic development by means of the "New Growth Theories" has also led to two related research paradigms. One of these is the "New International Economics", which provides a reformulation of the theories of trade and trade policy (see e.g. Krugman 1988). The other is the "New Economic Geography", which attempts to explain the spatial distribution of economic activity, both in terms of urban systems and in terms of regional development (see e.g. Krugman 1991; Fujita et al. 1995).

The wave of "New Growth Theories" and its empirical counterpart has now reached a stage of maturity whose substantive contribution has already been assessed in special issues of major journals and general surveys (e.g. Ehrlich 1990; Stern 1991; Jones and Stokey 1992; Verspagen 1992; Romer 1994; Mankiw 1995; Jones and Manuelli 1997). There is also some dissent, e.g. Scott (1989) provides an alternative approach which rejects the use of the neoclassical production function. The empirical work on growth has benefited from new comparative data bases for a wide range of countries in the world (e.g. Summers and Heston 1991; International Monetary Fund, various years; World Bank, various years). However, some of the conclusions regarding the determinants of growth drawn from such data appear as yet rather fragile (Levine and Renelt 1992; Mankiw 1995). Theory and empirics are brought together skillfully in the textbook by Barro and Sala-i-Martin (1995) which provides an advanced, but accessible, treatment of all the major issues in theoretical models of economic growth and the related empirical investigations.

Thus, the present paper does not purport to provide yet another broad survey. Instead, we will focus only on the spatial aspects of the neoclassical and the new growth models. A distinction is made between differences in growth rates due to spatial variations in parameters which influence growth in closed economies and causes of differences in growth rates between open economies. The predictions of the models are also compared with the main conclusions of the well-established literature on regional economic development. Finally, implications for regional policies are identified briefly.

Many of the neoclassical and new growth models describe accumulation in closed economies. Explanations for spatial differences in growth rates between such economies must necessarily derive from differences in parameters, initial conditions or other exogenous variables. Closed economy explanations of differences in growth rates are discussed in Sect. 2.

At the regional level, there is spatial interaction in terms of trade, capital flows, migration, diffusion of technological innovation and information exchanges. Thus, the closed economy models can provide at best a very limited understanding of regional growth. Section 3 considers the implications of introducing factor mobility, trade, economic integration and innovation diffusion into models of growth. It will become clear that the extension of the new growth models to the case of open economies is not yet fully satisfactory, and much work remains to be done in this area (Barro and Sala-i-Martin 1995, p. 128).

Section 4 focuses on the transitional dynamics in closed or open economies. An understanding of the transitional dynamics will enable predictions to be made about convergence or divergence. An assessment of the sometimes contradictory empirical evidence on this matter is provided.

Section 5 compares the new theories of growth with the conventional theories of regional development. While some of these have built upon neoclassical theories (as surveyed in e.g. McCombie 1988a), the post-Keynesian perspective which puts more emphasis on the role of demand in the economy and Kaldor's formulation of technical change (Kaldor 1957), had tended to be more popular (McCombie 1988b). The contribution, and potential, of evolutionary economics for an understanding of the regional growth process is also briefly addressed in this context. The final section provides some general conclusions and suggestions for further research in the field of spatial modelling of economic growth.

2. Why growth rates differ between closed economies

The process of economic growth, by which we mean the growth in real income per person in an economy, can be described in a simple way by the neoclassical model formulated by Robert Solow (1956) and independently by Trevor Swan (1956). Their work remains important because many of the new growth models can generate steady-state long-run growth paths which resemble those of the Solow-Swan model, with the only difference being that the new models now provide endogenous explanations for aspects of the Solow-Swan model which were assumed to be constant and given, such as the production function, the rate of technological change, the propensity to save and the population growth rate. Moreover, some of the stylised facts of development of economies are consistent with the predictions of the Solow-Swan model, provided capital accumulation in the model is interpreted as including human capital accumulation through education and training (e.g. Mankiw 1995; Bal and Nijkamp 1997).

Thus, it is useful to commence with reviewing briefly the key features of the Solow-Swan model. Consider a closed economy with competitive markets and a constant returns technology. At date t, labour supply is $L(t)$. The exogenously given rate of growth of $L(t)$ is n. Real production $Y(t)$ is assumed to result from combining inputs according to

$$Y(t) = F(K(t)\,,\; L(t)e^{xt}) \qquad (1)$$

where $K(t)$ is the stock of capital at time t and e^{xt} represents the effect of exogenous labour-augmenting technical progress. The model neglects labour-leisure choices and assumes full employment. Population and labour force are therefore equivalent concepts and both grow at rate n. Equation (1) can be rewritten as

$$\hat{y} = f(\hat{k}) \tag{2}$$

where the symbol ^ denotes a quantity per effective unit of labour $L(t)e^{xt}$. We shall assume that $f(.)$ has the usual "well-behaved" properties, formalised in the Inada (1963) conditions. If the rate of depreciation of capital is a fraction δ of the stock, net investment is given by

$$\dot{K} = Y - C - \delta K \tag{3}$$

where $\cdot$ denotes a derivative with respect to time and C is the level of consumption. Under these assumptions, we can derive the following equation of motion for the amount of capital per effective unit of labour $\hat{k}$:

$$\dot{\hat{k}} = f(\hat{k}) - \hat{c} - (n + x + \delta)\hat{k} \,. \tag{4}$$

In the Solow-Swan model, the tradeoff between current and future consumption is not explicitly considered. Households simply consume a constant fraction of income, which implies that there is a constant savings rate

$$s = \frac{f(\hat{k}) - \hat{c}}{f(\hat{k})} \,. \tag{5}$$

Substituting (5) into (4) gives the fundamental growth equation:

$$\dot{\hat{k}} = sf(\hat{k}) - (n + x + \delta)\hat{k} \,. \tag{6}$$

It is straightforward to show that given the assumptions made, any initial resources $K(0)$ and $L(0)$ and the dynamics described by (6), the economy will converge to a balanced, or steady-state, path asymptotically. In the steady state, the quantities $\hat{y}$, $\hat{k}$ and $\hat{c}$ do not change and the steady-state value of $\hat{k}$, $\hat{k}^*$ is found by setting the left-hand side of (6) equal to zero. Income, capital and consumption per capita each grow in the steady state at the rate of technological progress, x. The absolute quantities Y, K and C grow at the rate $x + n$.

This simple model suggests that if countries or regions have access to the same technology (the same production function and the same rate of technical change x), income per head must in the long run grow in each one at the same rate. During the transition to the steady state, the less capital endowed economy will have a lower income per head and grow faster. In the steady state, income per head will be higher in the economies with lower population growth and lower depreciation of capital.

The model as outlined above is only consistent with some of the stylised facts about the growth of nations and regions. Mankiw (1995, p. 277) summarises the predictions of the model as follows:

1. In the long run, the economy approaches a steady state that is independent of initial conditions.
2. The steady-state level of income depends on the rate of saving and population growth.
3. The steady-state rate of growth of income per head depends only on the rate of technological progress.
4. In the steady state, the capital-to-income ratio is constant.
5. In the steady state, the marginal product of capital is constant and the marginal product of labour grows at the rate of technological change.

Predictions 2, 4 and 5 among these are broadly supported by the empirical evidence. However, predictions 1 and 3 are more contentious and have led to a large empirical literature, which we review in Sect. 4.

The oldest way to endogenise one of the aspects of the Solow-Swan model is to make savings behaviour endogenous. Following Barro and Sala-i-Martin (1995), we refer to this model as the Ramsey model since the original idea of optimal savings behaviour was developed by Ramsey (1928). Cass (1965) showed that from any starting-point optimal capital accumulation converges to the balanced Solow-Swan growth path. The advantage of the Ramsey approach is that it permits a description of the economy in terms of the rational optimising behaviour of individual households and firms, which is now the cornerstone of modern macroeconomics. It can be shown that the Solow-Swan model predictions are consistent with those of a model with intertemporal optimisation in an Arrow-Debreu competitive equilibrium framework (e.g. Romer 1989).

In the Ramsey model, households seek to maximise lifetime utility[1] given by

$$W = \int_{t=0}^{\infty} u(c)e^{nt}e^{-\rho t}dt \tag{7}$$

where $c = C/L$ and ρ is the constant rate of time preference. Note that household utility rather than individual utility is in the welfare criterion since the utility of each person's consumption is multiplied by household membership which grows at rate n. Assuming that the utility function has the form

$$u(c) = \frac{c^{1-\theta} - 1}{1 - \theta}, \tag{8}$$

[1] Note that the usual "infinite horizon" assumption has been introduced for simplicity. Alternatively, economic growth models can use the overlapping generations approach, first formulated by Maurice Allais (1947) and subsequently in the seminal paper by Paul Samuelson (1958).

marginal utility $u'(c)$ has the constant elasticity $-\theta$ with respect to c.[2] To find the consumption path $c(t)$ which maximises (7) subject to (6) is a standard dynamic optimisation problem, which can be solved by Pontryagin's maximum principle of optimal control (Pontryagin et al. 1962). It can be shown that on the optimal time path for consumption

$$\tilde{c} = [f'(\hat{k}) - \delta - \rho]/\theta \tag{9}$$

where $\tilde{}$ refers to a rate of growth, i.e. $\dot{c}/c = .\tilde{c}$ (see e.g. Barro and Sala-i-Martin 1995, chapter 2). The long-run rate of return to capital is $f'(\hat{k}^*)$ where $\hat{k}^*$is the steady-state effective capital intensity found by setting the rate of growth in per capita consumption in (9) equal to x. Hence,

$$f'(\hat{k}^*) = \delta + \rho + \theta x\,. \tag{10}$$

Some explicit formulae for the level and growth rate of income and the optimal propensity to save can be easily derived with the use of a Cobb-Douglas production function. If a denotes the share of profits in income, this production function is given by

$$f(\hat{k}) = A\hat{k}^a\,. \tag{11}$$

It is then straightforward to derive an equation for the time path of real income per capita once the steady-state has been reached:

$$y(t) = A^{\frac{1}{1-a}}\left[\frac{a}{\delta + \rho + \theta x}\right]^{\frac{a}{1-a}} e^{xt}\,. \tag{12}$$

The optimal propensity to save in the steady-state is again constant, as in the Solow-Swan model, and is equal to

$$s^* = \frac{(n + x + \delta)\hat{k}^*}{f(\hat{k}^*)} = \frac{a(n + x + \delta)}{(\delta + \rho + \theta x)}\,. \tag{13}$$

It can be seen from (13) that a low rate of time preference ρ and a high intertemporal elasticity of substitution (i.e. small θ) increase s^*. This demonstrates the well-known prediction of the neoclassical model that a thrifty society will in the long run be wealthier than an impatient one, but does not grow faster. Note also that population growth no longer affects real income per head in Eq. (12). A faster growing population will simply find it optimal to save more (see Eq. (13)).

---

[2] The reciprocal of θ is called the intertemporal elasticity of substitution σ. A special case occurs when the intertemporal elasticity of substitution is one. In this case $\theta = \sigma = 1$ and $u(c) = \ln c$. Consequently, a 1% postponement of consumption from now until later raises marginal utility in the current period by 1%.

Both the Solow-Swan and the Ramsey model provide only two reasons for differences in growth rates between regions or countries. The first is that the rate of technological change x differs between economies. Reasons for such differences are not explained. Secondly, economies may not yet be on the steady-state growth path. In this case, poor economies with a low quantity of capital per head would grow faster than rich economies. This is referred to as the *absolute convergence* property of the model.

Barro and Sala-i-Martin (1995) suggest parameter values which will make the Ramsey model generate both a plausible long-run growth rate, but also a slow convergence to this growth rate which is consistent with empirical observations (see also Sect. 4). The slow convergence can only be explained with an interpretation of capital as a broad concept which also includes human capital accumulation. With this interpretation, the share of capital in income (combined with returns to education) is perhaps 0.75. Thus, $\alpha = 0.75$. Other plausible parameter values for developed economies are: the natural growth rate $n = 0.01$, the rate of technical change $x = 0.02$; the rate of depreciation $\delta = 0.05$, the rate of time preference $\rho = 0.02$ and the elasticity of marginal utility $\theta = 3$. Substituting these values in (12) suggests an optimal propensity to save of 45% (including investments in education and training).

The purpose of many of the models of growth which have been developed during the last decade is to provide an endogenous explanation for the natural growth rate n, or the rate of technical change x, or both.[3] We will focus here primarily on the latter issue, but we refer to Barro and Becker (1989) for a model which explains the natural growth rate n in a closed economy with exogenous technical change. In such an economy, the population growth rate and the rate of technical change are inversely related. Their model is an extension of the Ramsey model, namely through the introduction of a dynastic utility function: parents care about the utility attained by their children when reaching adulthood and by subsequent generations.

Becker et al. (1990) describe a closed-economy model in which technical change and population growth are both endogenous (with productivity growth driven by education). In this case, multiple equilibria emerge with the possibility of an economy becoming trapped in a "low income growth with high fertility" steady state. This is an example of a growth model in which the steady state is sensitive to initial conditions, which is a general feature of increasing returns and other "positive-feedback loop" models (see also Arthur 1994). Thus, historical endowments and "luck" may be critical determinants of differentials in growth which we may observe between countries or regions in such models.

A "poverty trap" can also be generated by the original Solow-Swan model by simply assuming that the average product of capital $f(k)/k$ is initially declining, then increasing and finally again declining with increasing

[3] Another possibility is to make the internal rate of time preference, i.e. the discount rate, endogenous. See Drugeon (1996) for the impact of growing impatience on long-run growth.

values of k (see Barro and Sala-i-Martin 1995, pp. 49–52). The possibility of a poverty trap is a popular idea in the development literature, see for example the "Big Push" model of Lewis (1954).

The process of labour-augmenting technological change, which proceeds at a constant rate x in the Solow-Swan model, is the simplest formalization of many phenomena which may lead to long-run productivity improvements. In macroeconomic models with only one good, technological change represents a growth of knowledge. There are four ways of accumulating knowledge: research, schooling, learning-by-doing and training. Alternatively, if a model permits more than one good, technological change may be due to an increase in the variety of intermediate inputs or consumer goods. Furthermore, the quality of goods may improve for a given variety. Finally, increasing returns may result from economies of scale.

Endogenous growth models have been formulated for each of these situations. For example, Lucas (1988) describes one model in which education generates external benefits and another model in which productivity improves through learning by doing. The idea that experience spills over to other producers goes back to Arrow (1962) and is also the engine of growth in Romer's (1986) model.

Other papers explicitly describe the activities of the R&D sector which generates new knowhow, see e.g. Romer (1990) and Aghion and Howitt (1992). Osano (1992) points out that basic research may have a stronger impact on the growth rate than applied research. He proceeds therefore to formulate a growth model with two research sectors. Schmitz (1989) points to the importance of entrepreneurial imitation, while Stokey (1988) studies the effects of the introduction of new goods on long-run growth. In this context, Jovanovic (1995) notes that the emphasis on generating new knowledge is overemphasised in the new growth models. A significant proportion of resources in an economy is devoted to adopting existing recent technologies rather than inventing new ones and firms frequently adopt "dominated" technologies. Thus, new growth models which explicitly consider the adoption decision are warranted. A recent issue in this context is also the optimal timing of adoption of a new technology (Choi 1994). Early adoption may create a competitive edge, but may be costly. Late adoption may be less risky, but may also generate less benefits (see also Koski and Nijkamp 1996).

A key issue is that technological inputs create spillovers due to the fact that they are non-rival goods. New inventions are produced at a high cost for the first unit but subsequent units (e.g. photocopies) can be produced at virtually zero cost. This generates nonconvexities in production, even if such goods are partially excludable (i.e. appropriable) through patents and if adoption is costly. The technological spillover phenomenon is better captured by human capital accumulation or the introduction of new goods rather than by physical capital accumulation.

In many of the models with endogenous technological change, the characteristics of the dynamic competitive equilibrium can be traced by setting up an optimal control problem similar to the Ramsey model. Where a steady-state exists, the presence of an externality, e.g. through R&D, creates a divergence between the private and social rates of return and the competitive equilibrium may not be Pareto-optimal.

To highlight some common features of many endogenous growth models we will now formulate a simple model of endogenous technological change in which the existence and properties of the steady-state are readily established without having to explicitly solve the underlying dynamic optimisation problem.[4] This will also simplify the discussion of growth in the open economy in the next section.

As noted earlier, technological change is most conveniently interpreted as a labour-augmenting process. Thus, if N measures the effective labour input, $N = LT$, where L is the quantity of workers and T is an index of the average quality of labour input, which depends on the stock of knowledge and practices. In the Solow-Swan model, T grows at the exogenous rate x. Here we relax this assumption. Central to the current view about the process of technological innovation is that a change in T requires a production process with real resource inputs, a multi-product output, its own technology, market structure, spatial differentiation and, indeed, its own changing technology (e.g. Dosi 1988). Hence we shall assume that a change in T is generated by the following process of knowledge creation:

$$\dot{T} = H\left(\frac{R}{L}, T\right) \tag{14}$$

where R/L is expenditure per worker on activities such as education, training, R&D etc.[5] Thus, the change in T is positively related to the intensity of the effort devoted to the enhancement of labour quality as well as the current level of labour quality. This function is assumed to be homogeneous of degree one, twice differentiable and concave. Both the public sector and private sector in the economy carry out knowledge-creating activities, funded through taxes and retained profits respectively. For simplicity, we lump these activities here together and assume that a fraction τ of national income is allocated to this process of technical change. Hence,

$$R = \tau Y . \tag{15}$$

As in the case of the accumulation of physical capital, a trade-off arises in that a large value of τ reduces current consumption, but yields a higher level of output in the future. Using (14) and (15) we can derive that

$$\tilde{T} = H\left(\frac{R}{LT}, 1\right) = H(\tau\hat{y}, 1) \equiv h(\tau f(\hat{k})) \tag{16}$$

[4] Solow (1994, p. 49) has in fact argued that the intertemporally-optimising representative agent formulation in itself has had little to add to the insights of the New Growth literature.

[5] This equation is a generalisation of a model of endogenous technical change proposed by Conlisk (1967), who assumed that dT/dt would be a linear function of Y/L and T.

Households now maximise lifetime utility according to Eq. (7) as before, but consumption per capita c at any time cannot exceed $f(k) - \dot{k} - \tau f(k) - \delta k$. There are now two decision variables, the propensity to save and the propensity to allocate resources to technical change, and if a steady-state growth path exists these propensities will both be constant on the steady-state path. Consequently, income per head will still grow at a constant rate in the steady state, as in the standard neoclassical model, but (16) shows that this rate is now a function of both τ (the proportion of resources devoted to education, innovation etc.), and $\hat{k}$ (the effective capital intensity). If we assume that the labour input L again grows at an exogenous rate, n, we can derive a "fundamental growth" equation similar to Eq. (6) for the Solow-Swan model. For given s and τ, the path of the effective capital intensity $\hat{k}$ is given by

$$\dot{\hat{k}} = sf(\hat{k}) - [n + h(\tau f(\hat{k})) + \delta]\hat{k}\,. \tag{17}$$

The long-run equilibrium level of the effective capital intensity is given by $\hat{k}^*$ for which $\dot{\hat{k}} = 0$. Under the specified conditions, such an equilibrium exists and is stable (see Nijkamp and Poot 1993a).

The merit of this simple model of endogenous technical change is that, in contrast with the Solow-Swan or Ramsey model, it identifies various sources of differences in growth rates between countries and regions. Firstly, the model shows that thriftiness is good for an economy: if, e.g., a removal of tax distortions raises the propensity to save, the per capita growth rate becomes permanently higher.[6] Similarly, if capital is depreciated faster or the labour force grows faster, the rate of growth of output per capita decreases. Finally, if the proportion of income devoted to the production of technical change increases, the rate of growth of output per capita increases. These results can be readily derived from considering the effect on the per capita growth rate $h(\tau f(\hat{k}^*))$ when changing a parameter in (17) and setting $\dot{\hat{k}} = 0$.

If government can influence the parameters of the growth model (savings behaviour, R&D, population growth), the endogenous growth model also points to ways in which government could permanently raise the long-run growth rate. It is also possible to formulate endogenous growth models in which the activities of government become explicit in the form of provision of infrastructure, pure public goods, the protection of property rights and taxation policies. For example, Barro (1990) considers a model in which the total of government purchases becomes an additional production factor (external effect) in the production function of private firms.

[6] One issue, which we do not address in this paper, is the effect of savings behaviour in models of heterogeneous capital. The introduction of heterogeneous capital may lead to paradoxical results, see e.g. Garegnani (1970) and Hatta (1976). The issue was raised in the context of the "Cambridge-Cambridge" debate regarding the meaning of the concept of capital and the notion of an aggregate production function (see e.g. Jones 1975 for an introduction into this debate).

The above model generates constant long-run growth, but endogenous growth models may also generate ever-increasing growth rates due to increasing returns. An example can be found in Romer (1986), who justifies such a model by the observation that in the very long run (over several centuries) worldwide labour-productivity growth has been accelerating (see also Kremer 1993), although it is equally true that during the last forty years productivity growth exhibited a downward trend among developed countries (Romer, 1989). Nijkamp and Poot (1993b) formulate a model of increasing returns in which such ever-increasing growth is eventually checked by technological, social and economic capacity constraints.

Another plausible break on ever-increasing growth is a deterioration of the natural environment. On the one hand we may find that damage to the environment or an increasing scarcity of natural resources restricts the economy's ability to generate innovations which drive the growth in income per head (Barbier 1996). On the other hand, it is likely that a society becomes more interested in environmental preservation at higher incomes. A growing preference for a clean environment may then become incompatible with growth when the production technology exhibits increasing returns to scale (Yoshida 1995). However, if pollution abatement technologies themselves benefit from endogenous technological change, more resources devoted to such activities may lower output growth in the short run, but increase it in the long run (Bovenberg and Smulders 1996). This phenomenon is also referred to as the environmental Kutznets curve.

3. Long-run growth tendencies in open economies

Our analysis has been so far confined to the case of the closed economy. The growing importance of trade, capital flows, a diffusion of product and process innovations, and migration at the interregional and international levels, suggests that spatial interactions need to be explicitly considered, both in terms of their direct effects on growth and their effects on technological change. In this section we address these issues by considering, in this order: factor mobility, diffusion and trade.

3.1 Factor mobility

If interregional differences in preferences, factor endowments or technology generate interregional differences in returns to production factors, a reallocation of production factors may be expected. The impact of this reallocation depends on the assumed characteristics of the model of the interacting economies. In standard neoclassical analyses in which spatial factor price differentials are due to differences in factor endowments, factor movements have an equilibrating effect. However, with differences in preferences or technology, factor movements may lead to divergence in which all mobile resources are eventually attracted to one region (see e.g. Nana and Poot 1996).

This contrast is also evident in the attempts to date to formulate the neoclassical growth model in an open economy setting. For example, it is often considered plausible to assume perfect capital mobility and immobile labour. Barro and Sala-i-Martin (1995, chapter 3) find that in these circumstances a multi-country Ramsey model generates several paradoxical conclusions. Firstly, the speed of convergence to the steady state is infinite. Secondly, consumption in all but the most patient country tends to zero and assets in these countries become negative. Finally, the most patient country eventually owns all resources and consumes all output.

In practice, capital is far from being perfectly mobile and, specifically, capital does not seem to flow from rich to poor countries to the extent that neoclassical models would predict. Similarly, human capital may migrate from places where it is scarce to places where it is abundant (a process sometimes referred to as the "braindrain"), rather than vice versa (Lucas 1988). These observations may be explained by asymmetric information, imperfect credit markets or labour markets, or adjustment costs for investment (Barro and Sala-i-Martin 1995; Gordon and Bovenberg 1996). Alternatively, the incentives for capital mobility may be exaggerated, because spatial real risk-adjusted rate of return differentials may in fact be small due to significant differences between countries in human capital accumulation, the external benefits of human capital, capital market imperfections and political uncertainties (Lucas 1990; Mankiw 1995).

However, as long as a production factor moves in the "right" direction (i.e. to where its price is higher) it has in the neoclassical growth models the tendency to speed up convergence to the steady-state. As an example, we illustrate the role of factor mobility in the endogenous growth models by explicitly considering labour migration.

Separating the effect of "natural" growth and migration, the change in labour supply is given

$$\dot{L} = nL + M \tag{18}$$

in which net migration M may be assumed to be given positively related to the effective capital intensity $\hat{k}$ since the real wage is $f(\hat{k}) - \hat{k}f'(\hat{k})$, which increases with higher values of $\hat{k}$ (see also Barro and Sala-i-Martin 1995, p. 288). Hence

$$M = m(\hat{k})L \tag{19}$$

in which $m(\hat{k})$ is the migration rate, which is an increasing function of $\hat{k}$. Migrants can also bring capital with them, so that the change in the domestic capital stock is now given by:

$$\dot{K} = sF(K, \hat{L}) - \delta K + \phi M \tag{20}$$

where ϕ is the value of capital per migrant (which could be human capital when K is interpreted in a broad sense). When we combine (18)–(20) with our earlier model of endogenous technical change, the following dynamic equation for the effective capital intensity emerges:

$$\dot{\hat{k}} = sf(\hat{k}) - [h(\tau f(\hat{k})) + \delta + n + m(\hat{k})]\hat{k} + m(\hat{k})\hat{\phi} \ . \tag{21}$$

As in the previous models, Eq. (21) can be used to identify which factors influence the long-run growth rate. It is clear that in addition to policies with influence R&D, immigration policies can now also affect the long-run growth rate. For example, if immigration controls are relaxed, the $m(\hat{k})$ function may shift upward. This would lead to a lower steady-state effective capital intensity $\hat{k}^*$ and it would also lower the long-run growth rate in per capita income.

The conclusion that immigration lowers the growth rate requires the assumption that there is no change in the amount of capital which immigrants bring with them. If a new immigration policy targets specifically highly skilled workers, this raises $\hat{\phi}$ and leads to an increase in $\hat{k}^*$, so that in this case the long-run growth rate will increase.

In addition, migrants may directly influence productivity growth in various ways. If migrants provide new ideas and encourage investment which embodies new technologies, there are dynamic gains from inward migration not captured in (21). The empirical literature suggests that there are indeed dynamic gains from migration, at least in the traditional immigrant receiving countries of North America and Australasia (see also Gorter et al. forthcoming). This literature suggests that net immigration in these developed countries has raised per capita incomes. The regional literature also suggests that migrants on balance move in the "right" direction, but that this reallocation does not reduce interregional disparities (e.g. Van Dijk et al. 1989). These observations would be consistent with a "cumulative causation" process rather than neoclassical convergence. However, recent work by Persson (1994) and Cashin and Loayza (1995) finds empirical evidence for migration aiding convergence among Swedish regions and between South Pacific countries respectively.

A final point to note with respect to growth models with endogenous migration is that the migration function in Eq. (19) is a macro function, which does not explicitly consider the micro-level behaviour of individual households. For example, migration will be a function of the gain from migration in terms of wages or amenities, but this gain may depend on interregional differences in the rate of technical change rather than just the amount of capital per worker. Moreover, if the productivity growth rate between regions or countries differs, mobile labour may have the same effect as mobile capital which we discussed earlier, namely the final equilibrium may be a corner solution in which all population is concentrated in one region. To avoid such an extreme, it is plausible to introduce some capacity constraint either in terms of technological bottlenecks or congestion of a natural resource (see also Nijkamp and Poot 1993b). Braun (1993) formulates several growth models with endogenous migration.

3.2 Diffusion

Diffusion analysis has become an important field of research in industrial economics. At the micro level, it does not only focus on the distribution and adoption of new technologies (see Brown 1981; Soete and Turner 1984), but also on business services and networks related to technological transformations (Cappellin 1989; Bertuglia et al. 1997). In most diffusion studies the S-shaped (or logistic) curve forms a central component (Davies 1979; Morrill et al. 1988). Both the adoption time and the adoption rate can be pictured in this curve. The precise shape of the S-curve can then be explained from firm size, market structure, profitability of innovations etc. (Kamien and Schwartz 1982).

An important negative role can be played in this context by barriers to information transfer in a multi-region system (Giaoutzi and Nijkamp 1988). Without the right local conditions, the adoption of diffused technological innovations may also not be effective. For example when adoption requires a skilled work force, a low level of human capital accumulation will slow down technological change. Kubo and Kim (1996) find evidence of a strong complementarity between imported technology and human capital in a study of growth in Japan and Korea.

Several authors have proposed long-run growth models which incorporate the diffusion of technological change, for example Krugman (1979), Grossman and Helpman (1991) and Barro and Sala-i-Martin (1995, chapter 8).

For simplicity, we do not consider logistic diffusion here. The simplest way to investigate the effect of diffusion in our dynamical system is to replace the equation for productivity growth (16) by

$$\tilde{T} = h(\tau f(\hat{k})) + dh(\tau^f f(\hat{k}^f)) \tag{22}$$

where d is a diffusion parameter and the superscript f refers to a second country or region. This parameter is likely to vary over time and space. In the general case, the dynamics of a multi-regional system with endogenous technological change and innovation diffusion can be quite complex, but the properties of the system can be studied by means of simulation results (see Nijkamp et al. 1991).

When d is taken to be a constant parameter, it is straightforward to show that diffusion is compatible with a steady state in which both regions could grow at different rates. This result is obtained by substituting the endogenous rate of technical change (22) into (17) or (21) and by writing down a similar equation for the second country. The equilibrium effective capital intensities $\hat{k}^*$ and $\hat{k}^{f*}$ can then be found as the steady-state solution to the resulting two simultaneous differential equations (see Nijkamp and Poot 1993a).

As $\hat{k}^*$ and $\hat{k}^{f*}$ need not be identical, there can be in this model of technological change and diffusion a difference in steady-state growth rates between regions. Regions will then diverge in terms of a growing absolute difference in real income per head. If the steady-state growth rates differ, because the equilibrium capital intensities differ, there will be a persistent, and constant, difference in the rate of return on capital and an increasing real wage gap unless migration and capital movements (in opposite directions) are significant enough to reduce the factor price gaps.

If the diffusion parameter is very large, "overshooting" may take place and the system would then be characterized by saddle-path stability (see Nijkamp and Poot 1993a). Thus, ever increasing growth rates or growth rates reducing to zero are then a possibility.[7]

In conclusion, factor mobility and a plausible rate of diffusion have in this model the usual equilibrating effect of bringing capital intensities closer. In the circumstances in which a steady state exists, it is easy to identify the benefits of diffusion: compared with the situation of autarky, the equilibrium effective capital intensity is lower, the rate of return to capital is higher and income per capita grows at a faster rate.

The model discussed above treats both countries identically, i.e. diffusion takes place in both directions. As the adoption of imported technology usually also requires resources, firms must compare at the margin the cost of adopting imported technology with the cost of their own R&D activities. Barro and Sala-i-Martin (1995, p. 276) show that it is likely that a leader-follower situation will emerge in which one country eventually allocates its entire product and process development budget to imitation of foreign ideas while the R&D sector in the other generates all new ideas. Switches of roles may take place in the long run (see also Choi 1994).

Similar results regarding the choice between imported technology and R&D activity along the optimal growth path were obtained by Nijkamp and Poot (1993b). During the "take off" phase of the growth path – when real incomes are still low – productivity growth is driven by importing new technology. At the second stage, a domestic R&D sector may develop, which leads to increasing returns in the economy. However, the R&D sector eventually matures when growth becomes limited by natural resource constraints or other bottlenecks. The issue of an absorptive capacity for technological change in a growth model is also addressed by Keller (1996).

3.3 Trade

The Solow-Swan model, Ramsey model and the endogenous growth models referred to so far focus only on the supply side of the economy, i.e. the production factors and the level of technology. Within the standard neoclassical framework, trade has in fact no role to play beyond speeding up convergence (through Heckscher-Ohlin resource reallocation effects) and determining the steady-state world prices. The long-run growth rate itself remains purely a function of technological change and this process is not influenced by trade in the standard models.

---

[7] This was confirmed by simulations with a three-region model of endogenous technical change and spatial diffusion. A wide range of outcomes could emerge (see Nijkamp et al. 1991).

For example, the traditional neoclassical trade-and-growth model (Oniki and Uzawa 1965), suggests that two trading regions (with an identical rate of growth of labour supply) would, under standard conditions, move towards a long-run balanced growth path. The two regions grow on this path at identical rates and the pattern of specialisation is determined by the equilibrium factor intensities, i.e. the regions would produce relatively more of the good which uses the abundant production factor more intensively.

The restrictive assumptions of equal labour force growth and identical technological change are in fact required to ensure a stable long-run steady state in which relative prices (and therefore the terms of trade) are constant. Nijkamp and Poot (1993a) show that when the Oniki and Uzawa trade model is extended to include the model of technological change described in the previous section, a steady state is unlikely to exist and the terms of trade will continue to change. A qualitatively similar result is obtained by Lucas (1988) in a model of learning by doing and trade.

The description of technological change as a process of imperfect competition and increasing returns through increases in the variety of goods or improvements in product quality lead naturally to the question what the impact of trade would be under these more realistic assumptions of the new growth models. Grossman and Helpman (1991) provide an extensive discussion of the links between innovation, trade and growth in the open economy. Several new "trade and growth" models have been formulated. Rivera-Batiz and Romer (1991) show that when R&D activity is driven by the monopoly rights of producers to sell new intermediate goods, economic integration of these types of economies (assuming they are structurally identical) can raise the world growth rate. Thus there are both static and dynamic gains from trade in this type of model. However, Rivera-Batiz and Xie (1993) show that when the countries have different sizes and diverging resource endowments, economic integration will lower the growth rate of a country with a high (autarky) growth rate, while it will raise the growth rate of a country with a low (autarky) growth rate. Devereux and Lapham (1993) find that the post-integration equilibria in these types of trade and growth models may be unstable. Moreover, a specialisation based on comparative advantage leads to a sub-optimal investment in R&D activities by resource-rich economies (Grossman and Helpman 1994).

The Oniki and Uzawa trade- and growth-model assumed that the labour-augmenting technical change affects both the consumption and the investment goods sectors equally. In contrast, the new growth models describe how labour productivity improvements can vary between sectors, or how a trade advantage is generated by product innovations such as is described in the product cycle theory (Vernon 1966; Krugman 1979). Alternatively, it may be the level of activity in specific sectors which provides a "learning by doing" spillover benefit for the whole economy. In this case it is straightforward to show that an increase in the supply of the resource used intensively in the knowledge-generating sector speeds up growth (Grossman and Helpman 1990a). Similarly, the market allocation of resources to this sector is suboptimal because firms do not take a spillover benefit into account. Not surprisingly, the presence of a positive externality implies that subsidising the R&D sector improves welfare.

In a multi-regional context, the capture of spillover benefits from other regions increases growth, but what matters now from the policy perspective is which of the regions has a comparative advantage in the R&D sector. If subsidies are given to regions which are better at manufacturing rather than innovating, the overall growth rate may decline.

A typical model of comparative advantage and long-run growth is described by Grossman and Helpman (1990b). In their model, there are three sectors: an R&D sector, which produces blueprints for new products and also generates increases in the stock of knowledge, an intermediate goods sector and a final consumption goods sector. Resources devoted to R&D raise the number of available varieties of differentiated inputs in final production and this in turn raises total factor productivity. If this model is applied to two regions, each which fixed primary resources, a steady-state growth rate can be computed and its sensitivity to policies analysed. For example, a small R&D subsidy in both regions increases in this model the rate of growth, while a national trade policy that switches spending toward the consumer good produced by the region with a comparative advantage in R&D will cause long-run growth rates to decline. Diffusion can also have a significant effect on the long-run growth rate in this type of model. Moreover, it is also possible to study environmental externalities and environmental policies in this context (Elbasha and Roe 1996).

The increasing returns due to economies of scale or technical change in many of these models generates a sensitivity to initial conditions. It is intuitively clear that "uneven development" is a necessary outcome of such a situation: an initial discrepancy in capital-labour ratios between regions will be reinforced over time. Trade specialisation may also generate such uneven development. Kugman (1981) provides a well known example of this situation. Krugman assumed that two products, an agricultural good and a manufactured good, can be produced by means of Ricardian production techniques, with increasing external economies of scale. Such external economies are of course often empirically indistinguishable from technical change. In either case, the technical coefficients representing the input requirements per unit of output decline as the capital stock increases. In this situation the region with the larger initial capital stock has the higher profit rate and, if all profits are saved, generates the fastest capital accumulation. The result is an ever-increasing divergence between the regions, which only ends when a boundary of some kind has been reached. Krugman assumed this to be a limit to labour supply. Kubo (1995) formulated an extension of Krugman's model in which there is an interregional externality in the form of spillovers of knowledge or other benefits of regional agglomeration. In this case, a range of stable or unstable development patterns may emerge with the actual outcome dependent on the values of the parameters.

Production factor growth and commodity trade may also reinforce each other through technical change. Lucas (1988) suggested that a difference in human capital accumulation is one of the main causes of a difference in growth rates between regions or countries. Different goods have different potentials for human capital growth through on-the-job training or through learning-by-doing. Consequently, the comparative advantage which determines which goods get produced also determines the rate of growth in human capital (and therefore technical change). Lucas' (1988) model of trade and growth has features similar to Krugman's (1981) model, although the increase in the efficiency of the Ricardian production technology in the former is due to human capital accumulation through learning by doing, rather than economies of scale through physical capital accumulation. Nonetheless, if two goods are produced which are "good" substitutes (i.e. they have a substitution elasticity greater than one), there will be a tendency for complete specialisation with the direction of specialisation determined by the initial conditions.

Many of the "new growth theory", "new economic geography" and "new international economics" models have in common a possibility of multiple stable or unstable equilibria and a sensitivity to initial conditions. Consequently, they point to a role for policy to ensure that initial conditions on the growth path are generated which take the possibility of a technological comparative advantage into account. For example, to ensure that more resources are devoted to the good with a high learning-by-doing propensity, an industrial policy of "picking winners" would appear helpful in the Lucas (1988) model. The introduction of trade in this framework also generates complete specialisation. Over time, the terms of trade change continuously to reinforce the pattern of comparative advantage. Provided the goods are good substitutes, regions which produce the good which enjoys a faster technical change will continue to have a higher growth rate, resulting in a persisting change in the terms of trade. Thus, this dynamic trade model suggests again a persistent pattern of uneven development. Markusen (1996) recently emphasised that industrial policy consequently favours the highly urbanised high-income regions and that a top-down regional policy continues to be desirable to avoid increasing inequities between regions.

There is of course a fairly long tradition of emphasising uneven development in the regional growth literature, such as expounded, for example, in Myrdal's (1957) cumulative causation theory. For a recent contribution of the link between cumulative causation and trade, see Venables (1996).

A challenge in modelling regional growth with cumulative causation effects is to be able to endogenise changes in the position of individual regions in this growth continuum. Possibilities for such growth switches would include – on the demand side – the introduction of different income elasticities for different classes of goods; and on the supply side the continuing introduction of new goods, with learning potentials declining with the amount produced. Such factors could continuously shake up the existing pattern of specialisation and explain why, for example, the rapid growth in Newly Industrialised Economies has been associated with a growth of exports in products initially not produced in these countries.

4. Transitional dynamics: convergence or divergence

Parallel to the development of the new models of growth which we discussed in the previous two sections, a related empirical literature has burgeoned. However, this literature has not attempted as yet to verify empirically whether the new sophisticated models provide an adequate description of cross-section or intertemporal differences in growth rates. It is often difficult to derive from the new theories estimable equations, for example because the behaviour away from the steady state is theoretically uncertain. Also, some of the variables of the new models such as knowledge are hard to measure (Mankiw 1995).

Because it is possible to describe exactly in the Solow-Swan model how an economy develops over time from any initial position, and therefore how the transition to the steady state will be made, much of the literature on transitional dynamics and convergence is based on this traditional model (e.g. King and Rebelo 1993). Indeed, some research suggests that the empirical evidence on convergence is not consistent with theories in which trend growth rates differ across economies endogenously (Evans and Karras 1996).

There are two notions of convergence. *Weak* convergence, also called β-convergence, takes place when low income regions or countries grow faster than high-income ones, all else being equal. *Strong* convergence, also called σ-convergence, takes place when the standard deviation of the distribution of income across regions or countries declines. β-Convergence is a necessary but not sufficient condition for σ-convergence (Sala-i-Martin 1996). In simple terms, β is the slope coefficient from a regression of the growth of real income on the logarithm of its level. Convergence to the steady state would imply a negative coefficient. This type of regression can be carried out both with time-series data or with cross-section data. In the former case, the specification would be

$$\log(y_t/y_{t-1}) = a - (1 - e^{-\beta})\log(y_{t-1}) + u_t \tag{23}$$

for a time series of observations $t = 1, 2, ..., T$ while in the case of a cross section of regions or countries $i = 1, 2, ..., I$ the regression equation is

$$(1/T)\log(y_{iT}/y_{i0}) = a - [(1 - e^{-\beta T})/T]\log(y_{i0}) + u_{i0,T} \tag{24}$$

where 0 and T now refer to two points in time over which growth is computed. When the estimated Eqs. (23) and (24) provide a good fit, this would suggest so-called absolute convergence. In practice, there are many structural differences between countries or regions which would influence a cross-section equation or shocks to the growth process which would affect the time series equation. Consequently, research in this area usually proceeds to estimate (23) and (24) with a range of additional variables added and a significant β in this situation is referred to as conditional β convergence. Some data sets may in fact permit estimation with pooling of cross-section and time-series data (e.g. Knight et al. 1993).

Barro and Sala-i-Martin (1995) show how the value of β is related to the parameters of the Solow-Swan model with a Cobb-Douglas production function for a growth process that is actually described by this model. They find that $\beta = (1 - a)(x + n + \delta)$. Using the parameter values given in Sect. 2, we get $\beta = (1 - 0.75)(0.02 + 0.01 + 0.05) = 0.02$. This suggests that 2% of the gap between the current income and the steady state is reduced each year.

There is indeed some evidence of convergence at this rate. This evidence is stronger for regions than for countries and, within a cross section of countries, stronger for similar nations than for a broader cross section. For example, Barro and Sala-i-Martin (1992) and Sala-i-Martin (1996) find convergence across the states of the USA, prefectures of Japan and five European countries. Similarly, Andres et al. (1996) find convergence among OECD countries. Persson (1994) finds convergence across Swedish counties. Even Quah (1996), who is generally sceptical of claims of strong convergence in economic growth, finds evidence of such convergence across US states. Ben-David (1996) finds that convergence is stronger among countries that have strong trading relationships.

The speed of convergence is similar in many empirical studies, namely about 2% per year and therefore only consistent with the Solow-Swan model when capital is interpreted as including human capital (so that the share of capital in income is as high as 75%). Using more conventional parameter values, Mankiw (1995) finds that the Solow model suggests convergence at a speed of 4% per year, i.e. faster than is actually the case, although Cashin and Loayza (1995) find a speed of convergence at this rate after controlling for the effect of net international migration in a sample of nine South Pacific countries.

A by-product of the regression equations specified above is that the additional explanatory variables may lead to some insight into the causes of differences in growth rates. For example, Barro and Sala-i-Martin (1992) explain the persisting differences in steady-state growth rates in rather ad hoc fashion by school enrolment rates and government consumption expenditure (excluding education and defence). A cross-section of countries shows in their research a lesser tendency to convergence. Similarly, Mankiw et al. (1992) show that the textbook Solow-Swan model needs to be augmented to make the model useful in explaining differences in income per capita across countries. They find that introducing human capital accumulation explicitly (measured by secondary school enrolment rates) has the same type of positive effect on income per head as the savings ratio s^* in the standard Solow model. The augmented model provides some evidence of inter-country convergence, although again at a slow rate.

Regional openness and interconnectedness may of course be responsible for the somewhat more convincing patterns of convergence observed at the regional level than in cross-country comparisons. For example, Barro and Sala-i-Martin (1992) point to diffusion of technological change having the potential of generating convergence even if the marginal product of capital is not declining. Nonetheless, despite well known historical evidence of convergence of incomes across states of the USA (e.g. Easterlin 1960), there are fairly lengthy periods during which one can observe divergence[8] and the evidence that factor mobility operates as an equilibrating process is also rather inconclusive.[9] Richer models are needed to explain such observations.

A vast literature has emerged during the last decade on which factors, in addition to the convergence effect, explain differences in growth rates between regions or countries. A review of this literature is beyond the scope of the present paper, but see inter alia Kormendi and Meguire (1985), Baumol (1986), De Long (1988), Dowrick and Nguyen (1989), Barro (1991) and Barro and Sala-i-Martin (1992). However, Levine and Renelt (1992) find that many of the regression models may suffer from specification errors and the results are not very robust. Moreover, Evans (1995) shows that the usual approach of estimation with OLS is in this context inappropriate.

In recent years, several studies have also attempted to focus on specific variables which are often considered to be important from the growth perspective. For example, De Long and Summers (1991) find that equipment investment can raise growth rates, and more strongly than investment in structures. Perroti (1996) finds that a lower income inequality raises growth rates, while Barro (1996) finds that political freedom has a positive effect on growth. Finally, Devarajan et al. (1996) show that the composition of public expenditure can have an important effect on growth. Specifically, governments of developing countries appear to have allocated relatively too many funds to public capital expenditures, at the expense of current expenditures.

5. Alternative approaches

Although the Solow-Swan model was also extensively used for the purpose of studying the growth of regions, of which Borts and Stein (1964) is a well-known example, dissatisfaction with this model led to a search for alternatives. This dissatisfaction was due to several reasons. Firstly, the neoclassical model predicted convergence, which contradicted with many case studies of uneven development. Secondly, the theory did not provide any guidance regarding policy instruments which could help to raise the growth rate. Thirdly, the macro focus of the neoclassical model is inadequate to explain any regional differences in growth rates which may be due to significant differences in sectoral composition. In this context it is useful to note that Bernard and Jones (1996) found recently, using data on 14 OECD countries, that there is a lack of convergence in manufacturing while aggre-

[8] The periods 1840–1880, 1920–1940 and 1970–1980 are periods of interregional divergence in the USA. See also Carlino and Mills (1996).

[9] See for example the discussion in Armstrong and Taylor (1985, pp. 118–121).

gate convergence results may be driven by services. Fourthly, the neoclassical model in a regional setting only permits the study of regions as point economies, with a possible interaction through factor flows, but it does not address the distribution of economic activity over continuous space.

Consequently, many alternative approaches were proposed during the 1960s and 1970s, such as the export base model, econometric models, (multi-regional) input-output and computable general equilibrium models; multisector development planning models, cumulative causation dynamic models etc. (see e.g. Richardson 1973). Some of the current endeavours to obtain a precise micro-level description of all sectors and the requirements for general equilibrium could already be found in the optimal space-time development models of regional science literature (e.g. Isard et al. 1979). However, the questions of the existence, uniqueness and stability of optimal solutions could not yet be addressed at the time.

The rather secondary role of trade in the neoclassical model led to a significant popularity of Keynesian modelling with a heavy emphasis on demand considerations. For example, output growth in a region is driven in the well-known Kaldor-Dixon-Thirlwall model by relative competitiveness and income growth outside the region (Dixon and Thirlwall 1975). Supply-side factors play in such an export-led growth model only a role in terms of the effects of cost inflation and productivity on relative competitiveness, with the latter effect being generated by means of Verdoorn's law (Verdoorn 1949). This model explains differences in equilibrium growth rates between regions in terms of differences in price and income elasticities in the demand for exports and differences in rates of autonomous productivity growth.

In more formal terms, output growth is assumed to be export-led:

$$\tilde{Y} = \omega \tilde{X} \tag{25}$$

where $\tilde{}$ refers again to a rate of growth and X to the volume of exports. The export demand function has constant price and income elasticities

$$\tilde{X} = -\eta \tilde{p} + \xi \tilde{p}^f + \pi \tilde{Y}^f \ . \tag{26}$$

Price inflation results from fixed mark-up pricing on production costs, which in turn depend on unit wage costs w and labour productivity. Thus, in rate of change terms

$$\tilde{p} = \tilde{w} - \tilde{y} \ . \tag{27}$$

Central to this growth model is that labour productivity is partly dependent on growth of output itself, i.e. Verdoorn's Law:

$$\tilde{y} = \kappa + \lambda \tilde{Y} \ . \tag{28}$$

This equation is the result of inductive macroeconomic research and increasing returns may be the driving force behind it. However, the key weakness of this approach is that it does not explain which processes at the micro-level cause the positive feedback effect from output growth on productivity growth. Nonetheless, an extensive literature exists regarding the empirical evidence for the Verdoorn equation (reviewed in e.g. Bairam 1987). This literature suggests that the observed relationship may be the result of simultaneous responses in output and labour markets to changes in demand, combined with the effects of economies of scale and technical progress. Naturally, a simultaneous equation approach is required for empirical estimation of the parameters in (28). By and large, the empirical evidence suggests that λ is positive.

The reduced form of the model (25)–(28) is readily computed and suggests a constant rate of growth of income per worker:

$$\tilde{y} = \kappa + \frac{\lambda\omega[-\eta(\tilde{w} - \kappa) + \xi\tilde{p}^f + \pi\tilde{Y}^f]}{1 - \lambda\omega\eta} \ . \tag{29}$$

This model has unrealistic implications if it is considered in an explicit two region situation in which income growth in either region affects growth in the other region through trade between them. It is fairly straightforward to compute the reduced form for the per capita income growth rates in both regions, but dependent on the choice of parameters, these growth rates could obviously differ and would suggest a persisting trade imbalance (Nijkamp and Poot 1993a). Krugman (1989) noted that long-run balance of payments equilibrium in such a regional growth-and-trade framework necessitates a strict relationship between differences in growth rates between regions on the one hand and income elasticities of the demand for exports and imports on the other.[10] The Kaldor-Dixon-Thirlwall model is itself not informative about the processes which would ensure that the growth rates which this model generates would be consistent with long-run balance of payments equilibrium.

For example, if technical change proceeds at a different pace in two regions, growth in the more innovative region could be hampered by lower demand for its output from the less innovative, and less competitive, region. Indeed, if the Verdoorn effect is strong enough, a situation of "immiserising" growth may be generated in which a detrimental shock in the trading partner's economy (e.g. a rapid growth in nominal wages) is more than compensated by an, on balance, negative effect on the local economy.

The model discussed above does not take into account explicitly the possibility of factor flows between the regions, nor the diffusion and adoption of technological advances. These phenomena cannot be readily introduced here. For example, net migration would respond to the difference in

[10] Interestingly, the latter condition appears indeed consistent with international trade data, i.e. countries which grow fast tend to experience a high income elasticity of the demand for their exports, while the income elasticity of their demand for imports is low.

growth rates in per capita incomes, but the latter are again likely to be themselves affected by net migration. Moreover, production capacity limits are assumed unimportant here. In essence, the model describes the properties of a demand-driven steady-state growth path rather than full dynamics.[11] Yet it does make explicit that an exogenous shock to trade can have a long-term impact on the equilibrium growth rate, although our discussion suggests that the introduction of simple explicit feedback effects (here aggregate demand and relative competitiveness) can strongly modify the behaviour which may be expected in the absence of such effects.[12] Recently, Targetti and Foti (1997) showed that a blending of the ideas of neoclassical conditional convergence, Kaldorian technological progress and export-led demand growth can be fruitful for cross-section econometric explanations of country growth rates, although their results are quite sensitive to the choice of groups of countries.

The various approaches discussed in this paper have in common the interpretation of economic growth as a moving macroeconomic equilibrium in which the underlying processes can be described in rather mechanical terms. An alternative approach is to consider regional growth as the result of (un)stable evolution in a system of competing regions. Evolutionary economics has tried to examine the space-time trajectories of dynamic complex systems on the basis of biological metaphors (see for an excellent overview Nelson 1995). Notions like predator-prey, resilience, fragility and shocks are used to map out the various types of dynamics that may emerge once a system of open economies with indigenous externalities is described by nonlinear complex (synergetic) dynamics (see Nijkamp and Reggiani 1997). In such cases, a wide spectrum of dynamic behaviour may emerge, in particular if the parameters defining the architecture of a system are also becoming time-dependent in the long-run. Evolutionary theories of economic growth have already existed for some time (see, for example, the overview of Nelson and Winter 1974). Yet there is still a great potential for further in-depth study of economic growth and dynamics in a multiregional system by means of evolutionary economics.

6. Conclusions

In this paper we have surveyed the literature on the new models of economic growth which have been developed during the last decade. We have contrasted these with the earlier theories and with the common alternative approaches of studying regional development. Despite their macro-level orientation, the strength of the new models is that they give better insight into

[11] It is possible to introduce lags in the behavioural equations. Dixon and Thirlwall (1975) showed that the introduction of one-period lags in the export demand function still generates convergence to the equilibrium growth rate for plausible values of the elasticities.

[12] This is a general conclusion for models of interdependent regions. See also, for example, the models which have been developed by Frenkel and Razin (1987) to describe the effects of fiscal policies and monetary conditions on equilibrium output in a "two-region world".

the economic dynamics in an open system, specifically with respect to spatial interdependencies in the form of trade, factor mobility and innovation diffusion.

Many new models have been proposed in the literature. Such models capture one or more of the important features of development: sectoral composition, human and physical capital accumulation, natural endowments, economies of scale, trade, technological innovation and diffusion, factor mobility, government policies and market imperfections. However, the design of a coherent and unified framework appears to be far from easy. Moreover, most of the new theories require further extensive empirical scrutiny. In such empirical work it will be important to distinguish between transitional dynamics and long-run steady-state tendencies.

Much of the current literature adopts a macro perspective, albeit with carefully specified micro foundations. A common problem in empirical macroeconomics is that the macro data are sometimes unable to permit the researcher to conclusively choose between competing theories. For example, this problem has reduced what we can learn for policy formulation from the large literature on cross-country and cross-region growth regressions (see also e.g. Mankiw 1995).

Both the export-led growth model and the general equilibrium models considered in this paper had the ability to generate persisting differences in long-run growth rates in the presence of some spatial interdependency, provided there were barriers to other types of flows. Moreover, some models with increasing returns due to endogenous technical change suggest that there is a tendency for a highly interdependent system to be unstable, with a likelihood of "uneven development". While the new growth models offer interesting and appropriate foundation stones for a thorough analysis of the evolutionary patterns of a multi-regional system, it is obvious that much work in this area remains to be done. In particular, a better integration of micro- and macro-level approaches is required.

For example, the locational aspects of R&D creation, diffusion and adoption deserves much closer attention. To some extent, this issue is comparable to the infrastructure debate as presented, among others, in Biehl et al. (1986) and Nijkamp (1986). Production theories may be used to assess the implications of a favourable infrastructure in particular regions with respect to differential competitiveness. Endogenous growth initiatives may also be relevant in this context. The lumpiness of infrastructure means that the regional benefits may only be expected in the long run, while investments are to be made in the short run. Thus, financial resources have to be set aside which may have an uncertain future return, in particular since the demand responses to supply of public infrastructure are difficult to gauge. The same applies to unforeseen impacts of economic integration (the European market, NAFTA, etc.). Therefore, issues on the demand side, such as household behaviour, impediments to trade and the institutional structure, should not be ignored either.

In our context, a regional dynamisation of a production function, accompanied by a technological diffusion function with parameters dependent on information barriers on the one hand and competitive behaviour on the other, would provide a promising starting-point. Changing trade patterns, factor flows and public policies can then be incorporated to identify the long-run growth tendencies of the regions in the system.

References

Aghion P, Howitt P (1992) A model of growth through creative destruction. Econometrica 60(2):323–351

Allais M (1947) Économie et interêt. Imprimerie National, Paris

Andres J, Domenech R, Molinas C (1996) Macroeconomic performance and convergence in OECD countries. European Economic Review 40(9):1683–1704

Armstrong H, Taylor J (1985) Regional economics and policy. Philip Allen, Oxford

Arrow KJ (1962) The economic implications of learning by doing. Review of Economic Studies 29 (June):155–173

Arthur WB (1994) Increasing returns and path dependence in the economy. University of Michigan Press, Ann Arbor

Bairam EI (1987) The Verdoorn law, returns to scale and industrial growth: a review of the literature. Australian Economic Papers 26 (June):20–42

Bal F, Nijkamp P (1997) Exogenous and endogenous spatial growth models. TI Discussion Paper 97-022/3, Tinbergen Institute, Amsterdam

Barbier EB (1996) Endogenous growth and natural resource scarcity. Working Paper 45.96, Fondazione Eni Enrico Mattei, Milano, Italy

Barro RJ (1990) Government spending in a simple model of endogenous growth. Journal of Political Economy 98(5.2):S103–125

Barro RJ (1991) Economic growth in a cross section of countries. Quarterly Journal of Economics 106:407–443

Barro RJ (1996) Determinants of growth: a cross-country empirical study. NBER Working Paper No. 5698

Barro RJ, Becker GS (1989) Fertility choice in a model of economic growth. Econometrica 57(2):481–501

Barro RJ, Sala-i-Martin X (1992) Convergence. Journal of Political Economy 100(2):223–251

Barro RJ, Sala-i-Martin X (1995) Economic growth. McGraw-Hill, New York

Baumol WJ (1986) Productivity growth, convergence, and welfare: what the long run data show. American Economic Review 76(5):1072–1085

Becker GS, Murphy KM, Tamura RF (1990) Human capital, fertility, and economic growth. Journal of Political Economy 98(5.2):S12–S37

Ben-David D (1996) Trade and convergence among countries. Journal of International Economics 40:279–298

Bernard AB, Jones CI (1997) Comparing apples to oranges: productivity convergence and measurement across industries and countries. American Economic Review 86(5):1216–1238

Bertuglia CS, Lombardo S, Nijkamp P (eds) (1997) Innovation behaviour in space and time. Springer, Berlin Heidelberg New York

Biehl D et al. (1986) The contribution of infrastructure to regional development. Commission of European Communities, Brussels

Borts GH, Stein JL (1964) Economic growth in a free market. Columbia University Press, New York

Bovenberg AL, Smulders SA (1996) Transitional impacts of environmental policy in an endogenous growth model. International Economic Review 37(4):861–893

Braun J (1993) Essays on economic growth and migration. PhD Dissertation, Harvard University

Brown LA (1981) Innovation diffusion. Methuen, London

Cappellin R (1989) The diffusion of producer services in the urban system. In: Cappellin R, Nijkamp P (eds) Theories and policies of technological development at the local level. Gower, Aldershot

Carlino G, Mills L (1996) Convergence and the U.S. states: a time-series analysis. Journal of Regional Science 36(4):597–616
Cashin P, Loayza N (1995) Paradise lost? Growth, convergence and migration in the South Pacific. IMF Staff Papers 42(3):608–641
Cass D (1965) Optimum growth in an aggregative model of capital accumulation. Review of Economic Studies 32(3):233–240
Choi JP (1994) Irreversible choice of uncertain technologies with network externalities. Rand Journal of Economics 25:382–400
Conlisk J (1967) A modified neo-classical growth model with endogenous technical change. Southern Economic Journal 34(October):199–208
Davies S (1979) The diffusion of process innovations. Cambridge University Press, Cambridge
De Long JB (1988) Productivity growth, convergence, and welfare: comment. American Economic Review 78(5):1138–1154
De Long JB, Summers LH (1991) Equipment investment and economic growth. Quarterly Journal of Economics 106(2):445–502
Devarajan S, Swaroop V, Zou H (1996) The composition of public expenditure and economic growth. Journal of Monetary Economics 37(2):313–344
Devereux MB, Lapham BJ (1993) The stability of economic integration and endogenous growth. Discussion Paper No. 878. Institute for Economic Research, Queen's University, Canada
Dixon R, Thirlwall AP (1975) A model of regional growth-rate differences on Kaldorian lines. Oxford Economic Papers 27:201–214
Domar ED (1946) Capital expansion, rate of growth, and employment. Econometrica 14(April):137–147
Dosi G (1988) Sources, procedures, and microeconomic effects of innovation. Journal of Economic Literature 88(3):1120–1171
Dowrick S, Nguyen D (1989) OECD comparative economic growth 1950–85: catch-up and convergence. American Economic Review 79(5):1010–1030
Drugeon JP (1996) Impatience and long-run growth. Journal of Economic Dynamics and Control 20:281–313
Easterlin RA (1960) Interregional differences in per capita income, population, and total income, 1840-1950. In: Conference on Research in Income and Wealth, NBER Studies in Income and Wealth, vol 24
Ehrlich I (1990) The problem of development: introduction. Journal of Political Economy 98(5.2):S1–S11
Elbasha EH, Roe TL (1996) On endogenous growth: the implications of environmental externalities. Journal of Environmental Economics and Management 31:240–268
Evans P (1995) How to estimate growth regressions consistently. Department of Economics, Ohio State University, mimeo
Evans P, Karras G (1996) Convergence revisited. Journal of Monetary Economics 37:249–265
Frenkel JA, Razin A (1987) Fiscal policies and the world economy: an intertemporal approach. MIT Press, Cambridge, Mass.
Fujita M, Krugman P, Mori T (1995) On the evolution of hierarchical urban systems. Discussion Paper No. 419, Institute of Economic Research, Kyoto University
Garegnani P (1970) Heterogeneous capital, the production function and the theory of capital. Review of Economic Studies 37:407–436
Giaoutzi M, Nijkamp P (eds) (1988) Informatics and regional development. Gower, Aldershot
Gordon RH, Bovenberg AL (1996) Why is capital so immobile internationally? Possible explanations and implications for capital income taxation. American Economic Review 86(5):1057–1075
Gorter C, Nijkamp P, Poot J (forthcoming) Crossing borders: regional and urban perspectives on international migration. Avebury, Aldershot
Grossman GM, Helpman E (1990a) Trade, innovation and growth. American Economic Review 80(2):86–91

Grossman GM, Helpman E (1990b) Comparative advantage and long-run growth. American Economic Review 80(4):796–815
Grossman GM, Helpman E (1991) Innovation and growth in the global economy. MIT Press, Cambridge, Mass
Grossman GM, Helpman E (1994) Endogenous innovation in the theory of growth. Journal of Economic Perspectives 8(1):23–44
Harrod RF (1939) An essay in dynamic theory. Economic Journal 49 (June):14–33
Hatta T (1976) The paradox in capital theory and complementarity of inputs. Review of Economic Studies 43:127–142
Inada K (1963) On a two-sector model of economic growth: comments and a generalization. Review of Economic Studies 30(June):119–127
International Monetary Fund (various years) International Financial Statistics. International Monetary Fund, Washington, DC
Isard W, Liossatos P, Kanemoto Y, Kaniss PC (1979) Spatial dynamics and optimal space-time development. North-Holland Publishing Company, Amsterdam
Jones H (1975) An introduction to modern theories of economic growth. Nelson, London
Jones LE, Manuelli RE (1997) Endogenous growth theory: an introduction. Journal of Economic Dynamics and Control 21(1):11–22
Jones LE, Stokey NL (1992) Introduction: symposium on economic growth, theory and computations. Journal of Economic Theory 58:117–134
Jovanovic B (1995) Research, schooling, training and learning by doing in the theory of growth. Department of Economics, University of Pennsylvania, mimeo
Kaldor N (1957) A model of economic growth. Economic Journal 67:591–624
Kaldor N (1970) The case for regional policies. Scottish Journal of Political Economy 17(3):337–348
Kamien MI, Schwartz NL (1982) Market structure and innovation. Cambridge University Press, Cambridge
Keller W (1996) Absorptive capacity: on the creation and acquisition of technology in development. Journal of Development Economics 49(1):199–227
King RG, Rebelo ST (1993) Transitional dynamics and economic growth in the neoclassical model. American Economic Review 83(4):908–931
Knight M, Loayza N, Villanueva D (1993) Testing the neoclassical theory of economic growth. IMF Staff Papers 40(3):512–541
Koopmans TC (1965) On the concept of optimal economic growth. In: The econometric approach to development planning. North-Holland, Amsterdam
Kormendi R, Meguire P (1985) Macroeconomic determinants of growth: cross-country evidence. Journal of Monetary Economics 16:141–163
Koski H, Nijkamp P (1996) Timing of adoption of new communication technology. TI Discussion Paper 96-61/5, Tinbergen Institute, Amsterdam
Kremer M (1993) Population growth and technological change: one million B.C. to 1990. Quarterly Journal of Economics 108(3):681–716
Krugman P (1979) A model of innovation, technology transfer, and the world distribution of income. Journal of Political Economy 87:253–266
Krugman P (1981) Trade, accumulation and uneven development. Journal of Development Economics 8:149–161
Krugman P (ed) (1988) Strategic trade policy and the new international economics. MIT Press, Cambridge, Mass
Krugman P (1989) Income elasticities and real exchange rates. European Economic Review 33:1031–1054
Krugman P (1991) Geography and trade. MIT Press, Cambridge, Mass
Krugman P (1995) Development, geography and economic theory. MIT Press, Cambridge, Mass
Kubo Y (1995) Scale economies, regional externalities, and the possibility of uneven regional development. Journal of Regional Science 35(1):29–42
Kubo Y, Kim H (1996) Human capital, imported technology and economic growth: a comparative study of Korea and Japan. Discussion Paper 695, Institute of Policy and Planning Sciences, University of Tsukuba

Levine R, Renelt D (1992) A sensitivity analysis of cross-country growth regressions. American Economic Review 82(4):942–963
Lewis WA (1954) Economic development with unlimited supplies of labor. Manchester School of Economics and Social Studies 22 (May):139–191
Lucas RE (1988) On the mechanics of economic development. Journal of Monetary Economics 22(1):3–42
Lucas RE (1990) Why doesn't capital flow from rich to poor countries? American Economic Review 80(2):92–96
Malthus TR (1798) An essay on the principle of population, 1986 edition. W. Pickering, London
Mankiw NG (1995) The growth of nations. Brookings Papers on Economic Activity 1:275–310
Mankiw NG, Romer D, Weil DN (1992) A contribution to the empirics of economic growth. Quarterly Journal of Economics 107 (May):407–438
Markusen A (1996) Interaction between regional and industrial policies. International Regional Science Review 19(1&2):49–78
Marx K (1867) Capital Vol I, 1967 edition. International Publishers, New York
McCombie JSL (1988a) A synoptic view of regional growth and unemployment: I – the neoclassical theory. Urban Studies 25(4):267–281
McCombie JSL (1988b) A synoptic view of regional growth and unemployment: II – the post-Keynesian theory. Urban Studies 25(5):399–417
Morrill R, Gaile GL, Thrall GI (1988) Spatial diffusion. Sage, Beverley Hills
Myrdal G (1957) Economic theory and underdeveloped regions. Duckworth, London
Nana G, Poot J (1996) A study of trade liberalisation and factor mobility with a CGE model of Australia and New Zealand. Studies in Regional Science 26(2):27–52
Nelson R (1995) Recent evolutionary theorizing about economic change. Journal of Economic Literature 33(1):48–90
Nelson R, Winter S (1974) Neoclassical vs. evolutionary theories of economic growth: critique and prospectus. Economic Journal 84(336):886–905
Nijkamp P (1986) Infrastructure and regional development: a multidimensional policy analysis. Empirical Economics 11:1–21
Nijkamp P, Poot J (1993a) Technological progress and spatial dynamics: a theoretical reflection. In: Kohno H, Nijkamp P (eds) Potentials and bottlenecks of spatial economic development. Springer, Berlin Heidelberg New York
Nijkamp P, Poot J (1993b) Endogenous technological change, innovation diffusion and transitional dynamics in a nonlinear growth model. Australian Economic Papers 32(4):191–213
Nijkamp P, Poot J, Rouwendal J (1991) A nonlinear dynamic model of spatial development and R&D policy. Annals of Regional Science 25:287–302
Nijkamp P, Reggiani A (1997) The economics of complex spatial systems. Elsevier, Amsterdam
Oniki H, Uzawa H (1965) Patterns of trade and investment in a dynamic model of international trade. Review of Economic Studies 32:15–38
Osano H (1992) Basic research and applied R&D in a model of endogenous economic growth. Osaka Economic Papers 42(1/2):144–167
Perroti R (1996) Growth, income distribution and democracy; what the data say. Journal of Economic Growth (June)
Perroux F (1955) Note sur la notion de pôle de croissance'. Cahiers de l'Institut de Science Economique Appliquée, Series D No. 8
Perrson J (1994) Convergence in per capita income and migration across the Swedish counties 1906–1990. Institute for International Economic Studies, Stockholm University, mimeo
Pontryagin LS et al. (1962) The mathematical theory of optimal processes. Interscience Publishers, New York
Quah DT (1996) Empirics for economic growth and convergence. European Economic Review 40:1353–1375
Ramsey F (1928) A mathematical theory of saving. Economic Journal 38 (December):543-559
Ricardo D (1817) On the principles of political economy and taxation, 1951 edition. Cambridge University Press, Cambridge, Mass
Richardson HW (1973) Regional growth theory. Macmillan, London

Rivera-Batiz LA, Xie Danyang (1993) Integration among unequals. Regional Science and Urban Economics 23(3):337–354
Rivera-Batiz LA, Romer PM (1991) Economic integration and endogenous growth. Quarterly Journal of Economics 106(2):531–556
Romer PM (1986) Increasing returns and long-run growth. Journal of Political Economy 94(5):1002–1037
Romer PM (1989) Capital accumulation in the theory of long-run growth. In: Barro RJ (ed) Modern business cycle theory. Harvard University Press, Cambridge, Mass
Romer PM (1990) Endogenous technological change. Journal of Political Economy 98(5.2): S71–S102
Romer PM (1994) The origins of endogenous growth. Journal of Economic Perspectives 8(1):3–22
Sala-i-Martin X (1996) Regional cohesion: evidence and theories of regional growth and convergence. European Economic Review 40:1325–1352
Samuelson PA (1958) An exact consumption-loan model of interest with or without the social contrivance of money. Journal of Political Economy 66(6):467–482
Schmitz JA (1989) Imitation, entrepreneurship, and long-run growth. Journal of Political Economy 97(3):721–739
Schumpeter JA (1934) The theory of economic development. Harvard University Press, Cambridge, Mass
Scott M (1989) A new view of economic growth. Oxford University Press, Oxford
Smith A (1776) An inquiry into the nature and causes of the wealth of nations. 1937 edition, Random House, New York
Soete L, Turner R (1984) Technology diffusion and the rate of technical change. Economic Journal 94:612–623
Solow RM (1956) A contribution to the theory of economic growth. Quarterly Journal of Economics 70:65–94
Solow RM (1994) Perspectives on growth theory. Journal of Economic Perspectives 8(1):45–54
Stern N (1991) The determinants of growth. Economic Journal 101 (January):122–133
Stokey NL (1988) Learning by doing and the introduction of new goods. Journal of Political Economy 96:701–717
Summers R, Heston A (1991) The Penn World Table (Mark 5): an expanded set of international comparisons, 1950–1988. Quarterly Journal of Economics 106(2):327–368
Swan TW (1956) Economic growth and capital accumulation. Economic Record 32 (November):334–361
Targetti F, Foti A (1997) Growth and productivity: a model of cumulative growth and catching up. Cambridge Journal of Economics 21(1):27–43
Van Dijk J, Folmer H, Herzog Jr HW, Schlottmann AM (1989) Migration and labour market adjustment. Kluwer, Dordrecht
Venables AJ (1996) Trade policy, cumulative causation, and industrial development. Journal of Development Economics 49(1):179–197
Verdoorn PJ (1949) Fattori che regolano la sviluppo della produttivita del lavoro [Factors governing the growth of labour productivity]. L'Industria 1:3–10 [English translation: Thirlwall AP, Thirlwall G, Research in Population and Economics, 1979]
Vernon R (1966) International investment and international trade in the product cycle. Quarterly Journal of Economics 80:190–207
Verspagen B (1992) Endogenous innovation in neo-classical growth models: a survey. Journal of Macroeconomics 14(4):631–662
World Bank (various years) World Bank national accounts. World Bank, Washington, DC
Yoshida M (1995) Compatibility between growth and environmental preservation: threshold effects of the environment on health risks. Institute of Policy and Planning Sciences, University of Tsukuba

Part II
New Economic Geography

[3]

Increasing Returns and Economic Geography

Paul Krugman

Massachusetts Institute of Technology

This paper develops a simple model that shows how a country can endogenously become differentiated into an industrialized "core" and an agricultural "periphery." In order to realize scale economies while minimizing transport costs, manufacturing firms tend to locate in the region with larger demand, but the location of demand itself depends on the distribution of manufacturing. Emergence of a core-periphery pattern depends on transportation costs, economies of scale, and the share of manufacturing in national income.

The study of economic geography—of the location of factors of production in space—occupies a relatively small part of standard economic analysis. International trade theory, in particular, conventionally treats nations as dimensionless points (and frequently assumes zero transportation costs between countries as well). Admittedly, models descended from von Thünen (1826) play an important role in urban studies, while Hotelling-type models of locational competition get a reasonable degree of attention in industrial organization. On the whole, however, it seems fair to say that the study of economic geography plays at best a marginal role in economic theory.

On the face of it, this neglect is surprising. The facts of economic geography are surely among the most striking features of real-world economies, at least to laymen. For example, one of the most remarkable things about the United States is that in a generally sparsely populated country, much of whose land is fertile, the bulk of the population resides in a few clusters of metropolitan areas; a quarter of the inhabitants are crowded into a not especially inviting section of the East Coast. It has often been noted that nighttime satellite

[*Journal of Political Economy*, 1991, vol. 99, no. 3]

photos of Europe reveal little of political boundaries but clearly suggest a center-periphery pattern whose hub is somewhere in or near Belgium. A layman might have expected that these facts would play a key role in economic modeling. Yet the study of economic geography, at least within the economics profession, has lain largely dormant for the past generation (with a few notable exceptions, particularly Arthur [1989, 1990] and David [in press]).

The purpose of this paper is to suggest that application of models and techniques derived from theoretical industrial organization now allows a reconsideration of economic geography, that it is now time to attempt to incorporate the insights of the long but informal tradition in this area into formal models. In order to make the point, the paper develops a simple illustrative model designed to shed light on one of the key questions of location: Why and when does manufacturing become concentrated in a few regions, leaving others relatively undeveloped?

What we shall see is that it is possible to develop a very simple model of geographical concentration of manufacturing based on the interaction of economies of scale with transportation costs. This is perhaps not too surprising, given the kinds of results that have been emerging in recent literature (with Murphy, Shleifer, and Vishny [1989*a*, 1989*b*] perhaps the closest parallel). More interesting is the fact that this concentration of manufacturing in one location need not always happen and that whether it does depends in an interesting way on a few key parameters.

The paper is divided into four sections. Section I sets the stage with an informal discussion of the problem. Section II then sets out the analytical model. In Section III, I analyze the determination of short-run equilibrium and dynamics. Section IV analyzes the conditions under which concentration of manufacturing production does and does not occur.

I. Bases for Regional Divergence

There has been fairly extensive discussion over time of the nature of the externalities that lead to localization of particular industries. Indeed, Alfred Marshall's original exposition of the concept of external economies was illustrated with the example of industry localization. Most of the literature in this area follows Marshall in identifying three reasons for localization. First, the concentration of several firms in a single location offers a pooled market for workers with industry-specific skills, ensuring both a lower probability of unemployment and a lower probability of labor shortage. Second, localized industries can support the production of nontradable specialized inputs. Third, informational spillovers can give clustered firms a better production function than isolated producers. (Hoover [1948] gives a particularly clear discussion of agglomeration economies.)

These accounts of industry localization surely have considerable validity. In this paper, however, I shall offer a somewhat different approach aimed at answering a somewhat different question. Instead of asking why a particular industry is concentrated in a particular area—for example, carpets in Dalton, Georgia—I shall ask why manufacturing in general might end up concentrated in one or a few regions of a country, with the remaining regions playing the "peripheral" role of agricultural suppliers to the manufacturing "core." The proposed explanation correspondingly focuses on generalized external economies rather than those specific to a particular industry.

I shall also adopt the working assumption that the externalities that sometimes lead to emergence of a core-periphery pattern are *pecuniary* externalities associated with either demand or supply linkages rather than purely technological spillovers. In competitive general equilibrium, of course, pecuniary externalities have no welfare significance and could not lead to the kind of interesting dynamics we shall derive later. Over the past decade, however, it has become a familiar point that in the presence of imperfect competition and increasing returns, pecuniary externalities matter; for example, if one firm's actions affect the demand for the product of another firm whose price exceeds marginal cost, this is as much a "real" externality as if one firm's research and development spills over into the general knowledge pool. At the same time, by focusing on pecuniary externalities, we are able to make the analysis much more concrete than if we allowed external economies to arise in some invisible form. (This is particularly true when location is at issue: how far does a technological spillover spill?)

To understand the nature of the postulated pecuniary externalities, imagine a country in which there are two kinds of production, agriculture and manufacturing. Agricultural production is characterized both by constant returns to scale and by intensive use of immobile land. The geographical distribution of this production will therefore be determined largely by the exogenous distribution of suitable land. Manufactures, on the other hand, we may suppose to be characterized by increasing returns to scale and modest use of land.

Where will manufactures production take place? Because of economies of scale, production of each manufactured good will take place at only a limited number of sites. Other things equal, the preferred sites will be those with relatively large nearby demand, since producing near one's main market minimizes transportation costs. Other locations will then be served from these centrally located sites.

But where will demand be large? Some of the demand for manufactured goods will come from the agricultural sector; if that were the whole story, the distribution of manufacturing production would essentially form a lattice whose form was dictated by the distribution of agricultural land, as in the classic schemes of Christaller (1933) and Lösch (1940). But it is not the whole story: some of the demand for manufactures will come not from the agricultural sector but from the manufacturing sector itself.

This creates an obvious possibility for what Myrdal (1957) called "circular causation" and Arthur (1990) has called "positive feedback": manufactures production will tend to concentrate where there is a large market, but the market will be large where manufactures production is concentrated.

The circularity created by this Hirschman (1958)–type "backward linkage" may be reinforced by a "forward linkage": other things equal, it will be more desirable to live and produce near a concentration of manufacturing production because it will then be less expensive to buy the goods this central place provides.

This is not an original story; indeed, a story along roughly these lines has long been familiar to economic geographers, who emphasize the role of circular processes in the emergence of the U.S. manufacturing belt in the second half of the nineteenth century (see in particular Pred [1966] and Meyer [1983]). The main goal of this paper is to show that this story can be embodied in a simple yet rigorous model. However, before we move on to this model, it may be worth pursuing the intuitive story a little further to ask two questions: How far will the tendency toward geographical concentration proceed, and where will manufacturing production actually end up?

The answer to the first question is that it depends on the underlying parameters of the economy. The circularity that can generate manufacturing concentration will not matter too much if manufacturing employs only a small fraction of the population and hence generates only a small fraction of demand, or if a combination of weak economies of scale and high transportation costs induces suppliers of goods and services to the agricultural sector to locate very close to their markets. These criteria would have been satisfied in a prerailroad, preindustrial society, such as that of early nineteenth-century America. In such a society the bulk of the population would have been engaged in agriculture, the small manufacturing and commercial sector would not have been marked by very substantial economies of scale, and the costs of transportation would have ensured that most of the needs that could not be satisfied by rural production would be satisfied by small towns serving local market areas.

But now let the society spend a higher fraction of income on nonagricultural goods and services; let the factory system and eventually mass production emerge, and with them economies of large-scale production; and let canals, railroads, and finally automobiles lower transportation costs. Then the tie of production to the distribution of land will be broken. A region with a relatively large nonrural population will be an attractive place to produce both because of the large local market and because of the availability of the goods and services produced there. This will attract still more population, at the expense of regions with smaller initial production, and the process will feed on itself until the whole of the nonrural population is concentrated in a few regions.

This not entirely imaginary history suggests that small changes in the parameters of the economy may have large effects on its qualitative behavior. That is, when some index that takes into account transportation costs, economies of scale, and the share of nonagricultural goods in expenditure crosses a critical threshold, population will start to concentrate and regions to diverge; once started, this process will feed on itself.

The story also suggests that the details of the geography that emerges—which regions end up with the population—depend sensitively on initial conditions. If one region has slightly more population than another when, say, transportation costs fall below some critical level, that region ends up gaining population at the other's expense; had the distribution of population at that critical moment been only slightly different, the roles of the regions might have been reversed.

This is about as far as an informal story can take us. The next step is to develop as simple a formal model as possible to see whether the story just told can be given a more rigorous formulation.

II. A Two-Region Model

We consider a model of two regions. In this model there are assumed to be two kinds of production: agriculture, which is a constant-returns sector tied to the land, and manufactures, an increasing-returns sector that can be located in either region.

The model, like many of the models in both the new trade and the new growth literature, is a variant on the monopolistic competition framework initially proposed by Dixit and Stiglitz (1977). This framework, while admittedly special, is remarkably powerful in its ability to yield simple intuition-building treatments of seemingly intractable issues.

All individuals in this economy, then, are assumed to share a utility function of the form

$$U = C_M^{\mu} C_A^{1-\mu}, \tag{1}$$

where C_A is consumption of the agricultural good and C_M is consumption of a manufactures aggregate. Given equation (1), of course, manufactures will always receive a share μ of expenditure; this share is one of the key parameters that will determine whether regions converge or diverge.

The manufactures aggregate C_M is defined by

$$C_M = \left[\sum_{i=1}^{N} c_i^{(\sigma - 1)/\sigma} \right]^{\sigma/(\sigma - 1)}, \tag{2}$$

where N is the large number of potential products and $\sigma > 1$ is the elasticity of substitution among the products. The elasticity σ is the second parameter determining the character of equilibrium in the model.

There are two regions in the economy and two factors of production in each region. Following the simplification suggested in Krugman (1981), each factor is assumed specific to one sector. Peasants produce agricultural goods; without loss of generality we suppose that the unit labor requirement is one. The peasant population is assumed completely immobile between regions, with a given peasant supply $(1 - \mu)/2$ in each region. Workers may move between the regions; we let L_1 and L_2 be the worker supply in regions 1 and 2, respectively, and require only that the total add up to the overall number of workers μ:[1]

$$L_1 + L_2 = \mu. \tag{3}$$

The production of an individual manufactured good i involves a fixed cost and a constant marginal cost, giving rise to economies of scale:

$$L_{Mi} = \alpha + \beta x_i, \tag{4}$$

where L_{Mi} is the labor used in producing i and x_i is the good's output.

We turn next to the structure of transportation costs between the two regions. Two strong assumptions will be made for tractability. First, *transportation of agricultural output will be assumed to be costless.*[2]

[1] This choice of units ensures that the wage rate of workers equals that of peasants in long-run equilibrium.

[2] The reason for this assumption is that since agricultural products are assumed to be homogeneous, each region is either exporting or importing them, never both. But

The effect of this assumption is to ensure that the price of agricultural output and, hence, the earnings of each peasant are the same in both regions. We shall use this common agricultural price/wage rate as numeraire. Second, transportation costs for manufactured goods will be assumed to take Samuelson's "iceberg" form, in which transport costs are incurred in the good transported. Specifically, of each unit of manufactures shipped from one region to the other, only a fraction $\tau < 1$ arrives. This fraction τ, which is an inverse index of transportation costs, is the final parameter determining whether regions converge or diverge.

We can now turn to the behavior of firms. Suppose that there are a large number of manufacturing firms, each producing a single product. Then given the definition of the manufacturing aggregate (2) and the assumption of iceberg transport costs, the elasticity of demand facing any individual firm is σ (see Krugman 1980). The profit-maximizing pricing behavior of a representative firm in region 1 is therefore to set a price equal to

$$p_1 = \left(\frac{\sigma}{\sigma - 1}\right)\beta w_1, \tag{5}$$

where w_1 is the wage rate of workers in region 1; a similar equation applies in region 2. Comparing the prices of representative products, we have

$$\frac{p_1}{p_2} = \frac{w_1}{w_2}. \tag{6}$$

If there is free entry of firms into manufacturing, profits must be driven to zero. Thus it must be true that

$$(p_1 - \beta w_1)x_1 = \alpha w_1, \tag{7}$$

which implies

$$x_1 = x_2 = \frac{\alpha(\sigma - 1)}{\beta}. \tag{8}$$

That is, output per firm is the same in each region, irrespective of wage rates, relative demand, and so forth. This has the useful implication that the number of manufactured goods produced in each region

if agricultural goods are costly to transport, this would introduce a "cliff" at the point at which the two regions have equal numbers of workers and thus at which neither had to import food. This is evidently an artifact of the two-region case: if peasants were spread uniformly across a featureless plain, there would be no discontinuity.

is proportional to the number of workers, so that

$$\frac{n_1}{n_2} = \frac{L_1}{L_2}. \tag{9}$$

It should be noted that in zero-profit equilibrium, $\sigma/(\sigma - 1)$ is the ratio of the marginal product of labor to its average product, that is, the degree of economies of scale. Thus although σ is a parameter of tastes rather than technology, it can be interpreted as an inverse index of equilibrium economies of scale.

I have now laid out the basic structure of the model. The next step is to turn to the determination of equilibrium.

III. Short-Run and Long-Run Equilibrium

This model lacks any explicit dynamics. However, it is useful to have a concept of short-run equilibrium before we turn to full equilibrium. Short-run equilibrium will be defined in a Marshallian way, as an equilibrium in which the allocation of workers between regions may be taken as given. We then suppose that workers move toward the region that offers them higher real wages, leading to either convergence between regions as they move toward equality of worker/peasant ratios or divergence as the workers all congregate in one region.

To analyze short-run equilibrium, we begin by looking at the demand within each region for products of the two regions. Let c_{11} be the consumption in region 1 of a representative region 1 product, and c_{12} be the consumption in region 1 of a representative region 2 product. The price of a local product is simply its free on board price p_1; the price of a product from the other region, however, is its transport cost–inclusive price p_2/τ. Thus the relative demand for representative products is

$$\frac{c_{11}}{c_{12}} = \left(\frac{p_1\tau}{p_2}\right)^{-\sigma} = \left(\frac{w_1\tau}{w_2}\right)^{-\sigma}. \tag{10}$$

Define z_{11} as the ratio of region 1 *expenditure* on local manufactures to that on manufactures from the other region. Two points should be noted about z. First, a 1 percent rise in the relative price of region 1 goods, while reducing the relative *quantity* sold by σ percent, will reduce the *value* by only $\sigma - 1$ percent because of the valuation effect. Second, the more goods produced in region 1, the higher their share of expenditure for any given relative price. Thus

$$z_{11} = \left(\frac{n_1}{n_2}\right)\left(\frac{p_1\tau}{p_2}\right)\left(\frac{c_{11}}{c_{12}}\right) = \left(\frac{L_1}{L_2}\right)\left(\frac{w_1\tau}{w_2}\right)^{-(\sigma-1)}. \tag{11}$$

Similarly, the ratio of region 2 spending on region 1 products to spending on local products is

$$z_{12} = \left(\frac{L_1}{L_2}\right)\left(\frac{w_1}{w_2\tau}\right)^{-(\sigma-1)}. \tag{12}$$

The total income of region 1 workers is equal to the total spending on these products in both regions. (Transportation costs are included because they are assumed to be incurred in the goods themselves.) Let Y_1 and Y_2 be the regional incomes (including the wages of peasants). Then the income of region 1 workers is

$$w_1L_1 = \mu\left[\left(\frac{z_{11}}{1+z_{11}}\right)Y_1 + \left(\frac{z_{12}}{1+z_{12}}\right)Y_2\right], \tag{13}$$

and the income of region 2 workers is

$$w_2L_2 = \mu\left[\left(\frac{1}{1+z_{11}}\right)Y_1 + \left(\frac{1}{1+z_{12}}\right)Y_2\right]. \tag{14}$$

The incomes of the two regions, however, depend on the distribution of workers and their wages. Recalling that the wage rate of peasants is the numeraire, we have

$$Y_1 = \frac{1-\mu}{2} + w_1L_1 \tag{15}$$

and

$$Y_2 = \frac{1-\mu}{2} + w_2L_2. \tag{16}$$

The set of equations (11)–(16) may be regarded as a system that determines w_1 and w_2 (as well as four other variables) given the distribution of labor between regions 1 and 2. By inspection, one can see that if $L_1 = L_2$, $w_1 = w_2$. If labor is then shifted to region 1, however, the relative wage rate w_1/w_2 can move either way. The reason is that there are two opposing effects. On one side, there is the "home market effect": other things equal, the wage rate will tend to be higher in the larger market (see Krugman 1980). On the other side, there is the extent of competition: workers in the region with the smaller manufacturing labor force will face less competition for the local peasant market than those in the more populous region. In other words, there is a trade-off between proximity to the larger market and lack of competition for the local market.

As we move from short-run to long-run equilibrium, however, a third consideration enters the picture. Workers are interested not in nominal wages but in real wages, and workers in the region with the larger population will face a lower price for manufactured goods. Let $f = L_1/\mu$, the share of the manufacturing labor force in region 1. Then the true price index of manufactured goods for consumers residing in region 1 is

$$P_1 = \left[f w_1^{-(\sigma-1)} + (1-f)\left(\frac{w_2}{\tau}\right)^{-(\sigma-1)} \right]^{-1/(\sigma-1)}; \tag{17}$$

that for consumers residing in region 2 is

$$P_2 = \left[f\left(\frac{w_1}{\tau}\right)^{-(\sigma-1)} + (1-f) w_2^{-(\sigma-1)} \right]^{-1/(\sigma-1)}. \tag{18}$$

The real wages of workers in each region are

$$\omega_1 = w_1 P_1^{-\mu} \tag{19}$$

and

$$\omega_2 = w_2 P_2^{-\mu}. \tag{20}$$

From (17) and (18), it is apparent that if wage rates in the two regions are equal, a shift of workers from region 2 to region 1 will lower the price index in region 1 and raise it in region 2 and, thus, raise real wages in region 1 relative to those in region 2. This therefore adds an additional reason for divergence.

We may now ask the crucial question: How does ω_1/ω_2 vary with f? We know by symmetry that when $f = ½$, that is, when the two regions have equal numbers of workers, they offer equal real wage rates. But is this a stable equilibrium? It will be if ω_1/ω_2 decreases with f, for in that case whenever one region has a larger work force than the other, workers will tend to migrate out of that region. In this case we shall get regional convergence. On the other hand, if ω_1/ω_2 *increases* with f, workers will tend to migrate *into* the region that already has more workers, and we shall get regional divergence.[3] As we have seen, there are two forces working toward

[3] This description of dynamics actually oversimplifies in two ways. First, it implicitly assumes that ω_1/ω_2 is a monotonic function of f, or at least that it crosses one only once. In principle, this need not be the case, and there could be several stable equilibria in which both regions have nonzero manufacturing production. I have not been able to rule this out analytically, although it turns out not to be true for the numerical example considered below. The analytical discussion in the next section simply bypasses the question. Second, a dynamic story should take expectations into account. It is possible that workers may migrate into the region that initially has fewer workers because they expect other workers to do the same. This kind of self-fulfilling prophecy can occur, however, only if adjustment is rapid and discount rates are not too high. See Krugman (1991) for an analysis.

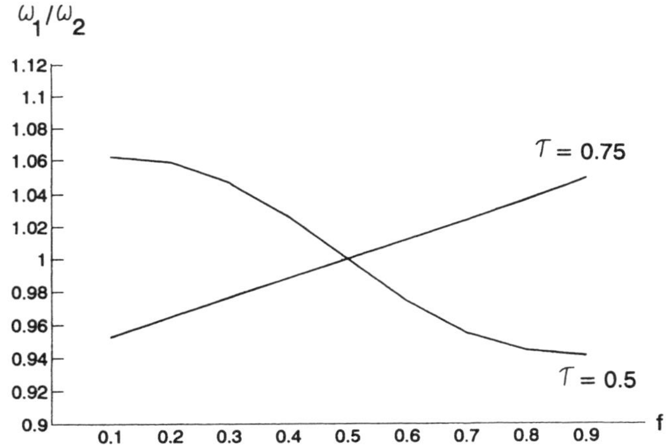

FIG. 1

divergence—the home market effect and the price index effect—and one working toward convergence, the degree of competition for the local peasant market. The question is which forces dominate.

In principle, it is possible simply to solve our model for real wages as a function of f. This is, however, difficult to do analytically. In the next section an alternative approach is used to characterize the model's behavior. For now, however, let us simply note that there are only three parameters in this model that cannot be eliminated by choice of units: the share of expenditure on manufactured goods, μ; the elasticity of substitution among products, σ; and the fraction of a good shipped that arrives, τ. The model can be quite easily solved numerically for a variety of parameters. Thus it is straightforward to show that depending on the parameter values we may have either regional convergence or regional divergence.

Figure 1 makes the point. It shows computed values of ω_1/ω_2 as a function of f in two different cases. In both cases we assume $\sigma = 4$ and $\mu = .3$. In one case, however, $\tau = .5$ (high transportation costs); in the other, $\tau = .75$ (low transportation costs). In the high-transport-cost case, the relative real wage declines as f rises. Thus in this case we would expect to see regional convergence, with the geographical distribution of the manufacturing following that of agriculture. In the low-transport-cost case, however, the slope is reversed; thus we would expect to see regional divergence.

It is possible to proceed entirely numerically from this point. If we take a somewhat different approach, however, it is possible to characterize the properties of the model analytically.

IV. Necessary Conditions for Manufacturing Concentration

Instead of asking whether an equilibrium in which workers are distributed equally between the regions is stable, this section asks whether a situation in which all workers are concentrated in one region is an equilibrium. This is not exactly the same question: as noted above, it is possible both that regional divergence might not lead to complete concentration and that there may exist stable interior equilibria even if concentration is also an equilibrium. The questions are, however, closely related, and this one is easier to answer.

Consider a situation in which all workers are concentrated in region 1 (the choice of region of course is arbitrary). Region 1 will then constitute a larger market than region 2. Since a share of total income μ is spent on manufactures and all this income goes to region 1, we have

$$\frac{Y_2}{Y_1} = \frac{1 - \mu}{1 + \mu}. \tag{21}$$

Let n be the total number of manufacturing firms; then each firm will have a *value* of sales equal to

$$V_1 = \left(\frac{\mu}{n}\right)(Y_1 + Y_2), \tag{22}$$

which is just enough to allow each firm to make zero profits.

Now we ask: Is it possible for an individual firm to commence production profitably in region 2? (I shall refer to such a hypothetical firm as a "defecting" firm.) If not, then concentration of production in region 1 is an equilibrium; if so, it is not.

In order to produce in region 2, a firm must be able to attract workers. To do so, it must compensate them for the fact that all manufactures (except its own infinitesimal contribution) must be imported; thus we must have

$$\frac{w_2}{w_1} = \left(\frac{1}{\tau}\right)^{\mu}. \tag{23}$$

Given this higher wage, the firm will charge a profit-maximizing price that is higher than that of other firms in the same proportion. We can use this fact to derive the value of the firm's sales. In region 1, the defecting firm's value of sales will be the value of sales of a representative firm times $(w_2/w_1\tau)^{-(\sigma-1)}$. In region 2, its value of sales will be that of a representative firm times $(w_2\tau/w_1)^{-(\sigma-1)}$, so the total value of the defecting firm's sales will be

$$V_2 = \left(\frac{\mu}{n}\right)\left[\left(\frac{w_2}{w_1\tau}\right)^{-(\sigma-1)} Y_1 + \left(\frac{w_2\tau}{w_1}\right)^{-(\sigma-1)} Y_2\right]. \tag{24}$$

Notice that transportation costs work to the firm's disadvantage in its sales to region 1 consumers but work to its advantage in sales to region 2 consumers (because other firms must pay them but it does not).

From (22), (23), and (24) we can (after some manipulation) derive the ratio of the value of sales by this defecting firm to the sales of firms in region 1:

$$\frac{V_2}{V_1} = \tfrac{1}{2}\tau^{\mu(\sigma-1)}[(1+\mu)\tau^{\sigma-1} + (1-\mu)\tau^{-(\sigma-1)}]. \tag{25}$$

One might think that it is profitable for a firm to defect as long as $V_2/V_1 > 1$, since firms will collect a constant fraction of any sales as a markup over marginal costs. This is not quite right, however, because fixed costs are also higher in region 2 because of the higher wage rate. So we must have $V_2/V_1 > w_2/w_1 = \tau^{-\mu}$. We must therefore define a new variable,

$$\nu = \tfrac{1}{2}\tau^{\mu\sigma}[(1+\mu)\tau^{\sigma-1} + (1-\mu)\tau^{-(\sigma-1)}]. \tag{26}$$

When $\nu < 1$, it is unprofitable for a firm to begin production in region 2 if all other manufacturing production is concentrated in region 1. Thus in this case concentration of manufactures production in one region is an equilibrium; if $\nu > 1$, it is not.

Equation (26) at first appears to be a fairly unpromising subject for analytical results. However, it yields to careful analysis.

First note what we want to do with (26). It defines a *boundary:* a set of critical parameter values that mark the division between concentration and nonconcentration. So we need to evaluate it only in the vicinity of $\nu = 1$, asking how each of the three parameters must change in order to offset a change in either of the others.

Let us begin, then, with the most straightforward of the parameters, μ. We find that

$$\frac{\partial \nu}{\partial \mu} = \nu\sigma(\ln\tau) + \tfrac{1}{2}\tau^{\sigma\mu}[\tau^{\sigma-1} - \tau^{-(\sigma-1)}] < 0. \tag{27}$$

That is, the larger the share of income spent on manufactured goods, the lower the relative sales of the defecting firm. This takes place for two reasons. First, workers demand a larger wage premium in order to move to the second region; this "forward linkage" effect is reflected

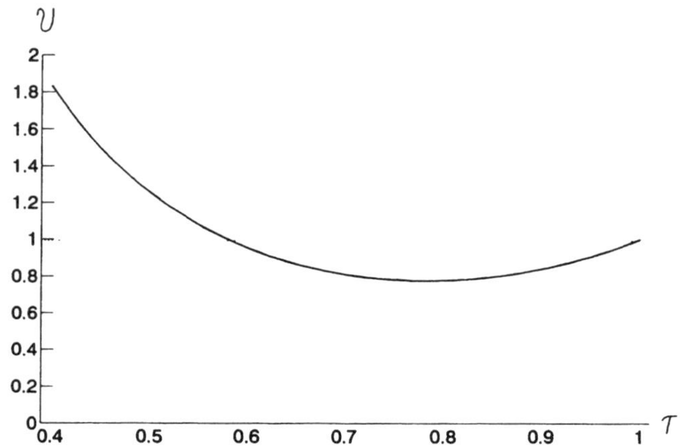

FIG. 2

in the first term. Second, the larger the share of expenditure on manufactures, the larger the relative size of the region 1 market and hence the stronger the home market effect. This "backward linkage" is reflected in the second term in (27).

Next we turn to transportation costs. From inspection of (26), we first note that when $\tau = 1$, $\nu = 1$; that is, when transport costs are zero, location is irrelevant (no surprise!). Second, we note that when τ is small, ν approaches $(1 - \mu)\tau^{1-\sigma(1-\mu)}$. Unless σ is very small or μ very large, this must exceed one for sufficiently small τ (the economics of the alternative case will be apparent shortly). Finally, we evaluate $\partial\nu/\partial\tau$:

$$\frac{\partial \nu}{\partial \tau} = \frac{\mu\sigma\nu}{\tau} + \frac{\tau^{\mu\sigma}(\sigma - 1)[(1 + \mu)\tau^{\sigma-1} - (1 - \mu)\tau^{-(\sigma-1)}]}{2\tau}. \qquad (28)$$

For τ close to one, the second term in (28) approaches $\mu(\sigma - 1) > 0$; since the first term is always positive, $\partial\nu/\partial\tau > 0$ for τ near one.

Taken together, these observations indicate a shape for ν as a function of τ that looks like figure 2 (which represents an actual calculation for $\mu = .3$, $\sigma = 4$): at low levels of τ (i.e., high transportation costs), ν exceeds one and it is profitable to defect. At some critical value of τ, ν falls below one and concentrated manufacturing is an equilibrium, and the relative value of sales then approaches one from below.

The important point from this picture is that at the critical value of τ that corresponds to the boundary between concentration and nonconcentration, $\partial\nu/\partial\tau$ is negative. That is, higher transportation costs militate against regional divergence.

We can also now interpret the case in which $\sigma(1 - \mu) < 1$, so that $\nu < 1$ even at arbitrarily low τ. This is a case in which economies of scale are so large (small σ) or the share of manufacturing in expenditure is so high (high μ) that it is unprofitable to start a firm in region 2 no matter how high transport costs are.

Finally, we calculate $\partial\nu/\partial\sigma$:

$$\begin{aligned}\frac{\partial \nu}{\partial \sigma} &= \ln(\tau)\{\mu\nu + \tfrac{1}{2}\tau^{\mu\sigma}[(1+\mu)\tau^{\sigma-1} - (1-\mu)\tau^{-(\sigma-1)}]\} \\ &= \ln(\tau)\left(\frac{\tau}{\sigma}\right)\left(\frac{\partial \nu}{\partial \tau}\right).\end{aligned} \tag{29}$$

Since we have just seen that $\partial\nu/\partial\tau$ is negative at the relevant point, this implies that $\partial\nu/\partial\sigma$ is positive. That is, a higher elasticity of substitution (which also implies smaller economies of scale in equilibrium) works against regional divergence.

The implications of these results can be seen diagrammatically. Holding σ constant, we can draw a boundary in μ, τ space. This boundary marks parameter values at which firms are just indifferent between staying in a region 1 concentration and defecting. An economy that lies inside this boundary will not develop concentrations of industry in one or the other region; an economy that lies outside the boundary will. The slope of the boundary is

$$\frac{\partial \tau}{\partial \mu} = -\frac{\partial \nu/\partial \mu}{\partial \nu/\partial \tau} < 0.$$

If we instead hold μ constant and consider changing σ, we find

$$\frac{\partial \tau}{\partial \sigma} = -\frac{\partial \nu/\partial \sigma}{\partial \nu/\partial \tau} > 0.$$

Thus an increase in σ will shift the boundary in μ, τ space outward.

Figure 3 shows calculated boundaries in μ, τ space for two values of σ, 4 and 10. The figure tells a simple story that is precisely the intuitive story given in Section I. In an economy characterized by high transportation costs, a small share of footloose manufacturing, or weak economies of scale, the distribution of manufacturing production will be determined by the distribution of the "primary stratum" of peasants. With lower transportation costs, a higher manufacturing share, or stronger economies of scale, circular causation sets in, and manufacturing will concentrate in whichever region gets a head start.

What is particularly nice about this result is that it requires no appeal to elusive concepts such as pure technological externalities: the external economies are pecuniary, arising from the desirability of selling to and buying from a region in which other producers are

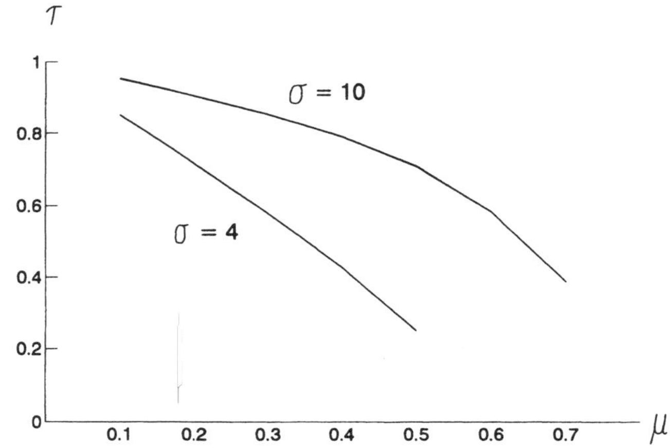

FIG. 3

concentrated. It also involves no arbitrary assumptions about the geographical extent of external economies: distance enters naturally via transportation costs, and in no other way. The behavior of the model depends on "observable" features of the tastes of individuals and the technology of firms; the interesting dynamics arise from interaction effects.

Obviously this is a vastly oversimplified model even of the core-periphery issue, and it says nothing about the localization of particular industries. The model does illustrate, however, how tools drawn from industrial organization theory can help to formalize and sharpen the insights of a much-neglected field. Thus I hope that this paper will be a stimulus to a revival of research into regional economics and economic geography.

References

Arthur, W. Brian. "Competing Technologies, Increasing Returns, and Lock-in by Historical Events." *Econ. J.* 99 (March 1989): 116–31.

———. "Positive Feedbacks in the Economy." *Scientific American* 262 (February 1990): 92–99.

Christaller, Walter. *Central Places in Southern Germany*. Jena: Fischer, 1933. English translation by Carlisle W. Baskin. London: Prentice-Hall, 1966.

David, Paul. "The Marshallian Dynamics of Industrialization: Chicago, 1850–1890." *J. Urban Econ.* (in press).

Dixit, Avinash K., and Stiglitz, Joseph E. "Monopolistic Competition and Optimum Product Diversity." *A.E.R.* 67 (June 1977): 297–308.

Hirschman, Albert O. *The Strategy of Economic Development*. New Haven, Conn.: Yale Univ. Press, 1958.

Hoover, Edgar M. *The Location of Economic Activity*. New York: McGraw-Hill, 1948.

Krugman, Paul. "Scale Economies, Product Differentiation, and the Pattern of Trade." *A.E.R.* 70 (December 1980): 950–59.

———. "Intraindustry Specialization and the Gains from Trade." *J.P.E* 89 (October 1981): 959–73.

———. "History versus Expectations." *Q.J.E.* 106 (May 1991).

Lösch, August. *The Economics of Location.* Jena: Fischer, 1940. English translation. New Haven, Conn.: Yale Univ. Press, 1954.

Murphy, Kevin M.; Shleifer, Andrei; and Vishny, Robert W. "Income Distribution, Market Size, and Industrialization." *Q.J.E.* 104 (August 1989): 537–64. (*a*)

———. "Industrialization and the Big Push." *J.P.E* 97 (October 1989): 1003–26. (*b*)

Meyer, David R. "Emergence of the American Manufacturing Belt: An Interpretation." *J. Hist. Geography* 9, no. 2 (1983): 145–74.

Myrdal, Gunnar. *Economic Theory and Under-developed Regions.* London: Duckworth, 1957.

Pred, Allan R. *The Spatial Dynamics of U.S. Urban-Industrial Growth, 1800–1914: Interpretive and Theoretical Essays.* Cambridge, Mass.: MIT Press, 1966.

von Thünen, Johann Heinrich. *The Isolated State.* Hamburg: Perthes, 1826. English translation. Oxford: Pergamon, 1966.

[4]

JOURNAL OF REGIONAL SCIENCE, VOL. 33, NO. 2, 1993, pp. 129–144

FIRST NATURE, SECOND NATURE, AND METROPOLITAN LOCATION

Paul Krugman
Department of Economics, Massachusetts Institute of Technology, Cambridge, Massachusetts 02139-1307

ABSTRACT. Economies of scale, transportation costs, and factor mobility can interact to produce agglomerations even in the absence of any pure external economies. This paper offers a monopolistic competition model of a city that serves an agricultural hinterland; unlike most analyses in location theory, the model is fully general equilibrium, but it has strong links to older concepts in geography, notably the idea of "market potential." The analysis shows that the forward and backward linkages that hold a population concentration together also allow that concentration to occur in a variety of possible sites—that is, there are multiple equilibria (indeed a continuum) for metropolitan location.

In his justly acclaimed recent book *Nature's Metropolis: Chicago and the Great West,* the historian William Cronon documents the extraordinary 19th century rise of Chicago as the central city of the American heartland. As Cronon points out, what made this rise particularly remarkable was the absence of any distinctive natural advantages of Chicago's site. The city stood on a flat plain; the river that ran through the city was barely navigable; the city's lakeside harbor was inadequate and tended to silt up. Whatever natural advantages the site did have proved transitory. Initially, Chicago seemed the natural terminus of a canal linking the watershed of the Mississippi with the Great Lakes, but when a canal was finally built it had only a few years of economic importance before being overshadowed by the railroads. Chicago's harbor on the Great Lakes was not unique, and in any case lake transportation became relatively unimportant by the 1870s as compared with rail links. Yet once Chicago had become established as a central market, as a focal point for transportation and commerce, its strength fed on itself. As Cronon puts it, the advantages that "first nature" failed to provide the city were more than made up for by the self-reinforcing advantages of "second nature": the concentration of population and production in Chicago, and the city's role as a transportation hub, provided the incentive for still more concentration of production there, and caused all roads to lead to Chicago.

Cronon's discussion suggests strongly that there were, in effect, multiple equilibria in the choice of site for the great metropolis of the American heartland—that had events played differently, another city such as Toledo or St. Louis might have come to seem just as inevitably destined for greatness as Chicago turned out to be in fact.

Received June 1992, revised and accepted November 1992.

Blackwell Publishers, 238 Main Street, Cambridge, MA 02142, USA and 108 Cowley Road, Oxford, OX4 1JF, UK

The purpose of this paper is to develop an approach to modeling urban location that shows, albeit in a highly simplified way, both why a great metropolis like Chicago might emerge to serve an agricultural hinterland and why there may be a range of potential sites for that metropolis. The paper develops a model of location in which for some parameter values there are equilibria in which the production of manufactured goods for an agricultural hinterland is concentrated at a single location. This "metropolis" could be at the agricultural region's geographical center. It need not be, however, because the metropolis is part of its own market (and also supplies part of what its own residents consume). Because of the feedback from the location of the metropolis to the geography of demand and supply, there is a range of potential metropolitan sites—although in general only locations that are not too far off the center, in a sense that the paper will make clear, will do. As Cronon would put it, the existence of the metropolis creates a "second nature" that drags the optimal location of firms with it, as long as the site is not too unreasonable.

There is, of course, an extensive existing literature in urban economics and location theory. The model differs in some important respects, however, from most of the existing formal urban models. First, unlike most theoretical models of urban systems (e.g., Henderson, 1974; and Rauch, 1991), there is no assumption of local external economies; the "centripetal" forces that hold the metropolis together are not assumed, but derived from the interaction of economies of scale at the plant level with transport costs. This gives the paper a relationship to some recent models of urban systems in which nontraded intermediate goods give rise to urban concentration (see for example, Fujita, 1988). In these models, however, the final goods produced by cities can be costlessly traded, and thus such models have nothing to say about the location of cities themselves.[1] In this paper, by contrast, there is no special class of nontraded goods. The incentive for geographically concentrated production is therefore derived rather than, as is the case in these nontraded-intermediate models, essentially assumed; and the cost of transportation of final goods constrains the possible location of the city.

The model does bear a strong resemblance to an earlier paper of my own (Krugman, 1991). The main difference here is that the model, instead of assuming that there are only two regions each of which can be regarded as a dimensionless point, examines equilibrium locations in continuous space. This puts the analysis in the tradition of the central place theory of Lösch (1940) and Christaller (1933), a tradition that in spite of enormous influence has never been the subject of fully worked-out microeconomic models. Going to a continuous representation of space requires introduction of a new method of analysis that turns out to provide a

[1]Rauch (1991) does offer a model in which there is a spatial structure of cities. This is achieved by restricting cities to a limited number of sites and giving these sites different desirability because of costs of transportation of an imported raw material. Thus the biggest city is on the site closest to the coast, and city size declines as one moves inland. This is a clever and interesting idea, but is very different from this paper both in method and in the questions asked.

rigorous justification for the commonly used geographical concept of "market potential."

The paper is in four parts. The first part sets out a basic modeling approach. The second analyzes spatial equilibrium in a one-dimensional region, in which rural population is spread along a line. The third part analyzes a two-dimensional case, in which population is distributed across a circular plain. The final section suggests some directions for extension.

1. A MODELING APPROACH

The intuition behind the approach in this paper is simple: firms that have an incentive to concentrate production at a limited number of locations prefer, other things equal, to choose locations with good access to markets; but access to markets will be good precisely where a large number of firms choose to locate. This positive feedback loop drives the formation of urban centers; it also implies that the location of such centers is not wholly determined by the underlying natural geography, that there are typically multiple locational equilibria.

In order to capture this intuition, a formal model must have three features. First, there must be some costs of transportation, so that location matters. Second, there must be some immobile factors of production, providing some form of "first nature" that constrains the possible spatial structure of the economy. Finally, there must be economies of scale in the production of at least some goods, so that there is an incentive for concentration.

The framework that will be used here is based on the two-location model in Krugman (1991), which in turn relies heavily on the monopolistic competition model of Dixit and Stiglitz (1977). We envision an economy with two sectors: constant-returns agriculture and increasing-returns manufacturing. Agricultural output is produced by geographically immobile factors, which are spread across space. Manufactures—which come in many differentiated varieties—are produced by mobile factors, which move to wherever they can achieve the highest return. Cities emerge when manufacturing firms clump together to be near the markets they provide for one another.

Specifically, we begin by assuming that everyone in the economy shares a utility function into which both the agricultural good and a manufactures aggregate enter

$$(1) \qquad U = C_M^{\mu} C_A^{1-\mu}$$

where μ is the share of manufactured goods in expenditure.

The manufactures aggregate in turn is a CES function of a large number of potential varieties, not all of which will actually be available.

$$(2) \qquad C_M = \left[\sum_i C_{Mi}^{\sigma-1/\sigma} \right]^{\sigma/\sigma-1}$$

Agricultural goods are produced by a sector-specific factor, agricultural labor. (Ideally, we would introduce land as an explicit additional factor of production. For simplicity, however, we do not; the role that should be played by land is proxied by making agricultural labor immobile[2]). They are produced under constant returns; without loss of generality we assume that the unit labor requirement is one. Manufactured goods are similarly produced by a specific factor, manufacturing labor. In the manufacturing sector, however, there are increasing returns; we introduce these by assuming that for any variety that is actually produced, there is a fixed labor input required independent of the volume of output:

$$L_{Mi} = \alpha + \beta q_{Mi} \tag{3}$$

Both kinds of labor will be assumed to be fully employed. Thus

$$L_M = \sum_i L_{Mi} \tag{4}$$

and

$$L_A = C_A \tag{5}$$

Aside from the sector-specificity of the two kinds of labor, this is essentially the Dixit-Stiglitz model. The main innovation here is to make the model spatial. This is done through two assumptions. First, agricultural labor will be assumed to be distributed across space; in the next section we will assume that it is evenly spread along a line, in the section following that it is evenly spread across a disk. Second, we assume that although agricultural goods can be costlessly transported (an assumption made purely for analytic convenience), manufactured goods are costly to transport. Specifically, we follow Samuelson's "iceberg" assumption, under which goods "melt" in transit, so that transportation costs are in effect incurred in the same goods that are shipped. The proportional rate of melting is assumed to be constant per unit of distance, implying that if a single unit of a manufactured good is shipped a distance D, the quantity that arrives is only $e^{-\tau D}$, where τ is the transport cost.

This setup is incomplete unless we specify a particular spatial structure. We may, however, quickly review several familiar features of this kind of model, well-known both from the original Dixit-Stiglitz paper and from the extensive derivative literature in the field of international trade.

First, we note that given the absence of any transportation costs for agricultural goods, all agricultural workers will receive the same nominal (although not real) wage.

Second, we note that manufacturing will have a monopolistically competitive market structure; the form of the utility function together with the assumption of "iceberg" transportation costs implies a constant elasticity of demand, and hence

[2]A model with land would introduce two complications. First, land-labor substitution would complicate the analysis considerably by requiring that we introduce land-labor substitution and a rent gradient. Second, in a model like this, in which the geographic distribution of demand plays a crucial role, we would need to worry about where the landowners live. There have certainly been rentier metropolises in history, but this paper tries to focus on different issues.

that the price of each manufactured good at the factory gate will be a constant proportional markup on the wage rate. All profits will be competed away by entry.

Third, we note that a fraction μ of total expenditure will fall on manufactured goods (including those that "melt" in transit); since profits are competed away, manufacturing workers will receive a share μ of total income, agricultural workers $1-\mu$.

Finally, we note that the elasticity of substitution between any two manufactured products is σ.

The minimalism of this framework is apparent, yet it is sufficient to generate some interesting insights into spatial equilibrium.

2. A ONE-DIMENSIONAL MODEL

We begin by considering a one-dimensional region. We assume that agricultural workers are distributed evenly along a line of unit length. What we will show is that if transport costs are not too high, there is an equilibrium in which all manufacturing is concentrated at a single point along that line. This "metropolis" could be at the region's center, but it need not be: in general there is a range of potential locations, whose width depends in an economically meaningful way on the model's parameters.

The method we will use is to posit an initial situation in which all manufacturing is concentrated at a single location, then ask whether an individual firm will find it advantageous to move to any other location. If not, then concentration of manufacturing at that location is a Nash equilibrium. (One might argue that if the metropolis is in the "wrong" place a coalition of firms will form to shift the equilibrium to a better one; I disregard that possibility in this paper).

Suppose, then, that all manufacturing is concentrated at the location x_C along the unit interval. We need to ask whether it is to the advantage of an individual firm to "defect," relocating to some other site x_A.

It turns out that the easiest way to do this is indirect. Instead of asking what the profits of such a defecting firm would be, we ask how high a real wage the defecting firm could afford to pay at x_A while still covering its fixed costs. If there is no site at which this hypothetical real wage exceeds the real wage paid at x_C, then a defecting firm will be unable to attract workers away from the city and still earn a profit; hence in this case concentration at x_C is an equilibrium.

Let w be the ratio of the *nominal* wage rate that a firm at x_A would have to pay to attract workers to that at x_C. Given the monopolistically competitive market structure, the ratio of the f.o.b. price charged by a defecting firm choosing to locate at x_A to that of a typical good manufactured at x_C will also be w. The price ratio to a consumer at location x will reflect both this f.o.b. price ratio and transport costs, which depend on the consumer's relative distance from x_A and x_C. Let p_x be this relative price to a consumer at location x: given our assumption about transport costs, it is simply

$$p_x = we^{\tau(|x_A-x| - |x_C-x|)} \tag{6}$$

Next consider the ratio of sales of a product manufactured at x_A to that of a typical good manufactured at x_C. Given the elasticity of substitution of σ, the ratio of *consumption* by a consumer at location x is

$$c = p_x^{-\sigma} \tag{7}$$

The ratio of *value* of sales to the consumer at x, however, is less sensitive to the price, because volume effects are offset by valuation effects; thus we have for the value ratio

$$s = p_x^{-(\sigma-1)} \tag{8}$$

To calculate the overall ratio of sales of a firm that defects to x_A to that of a typical product from the metropolis, we note that manufacturing workers, who account for a fraction μ of demand, are all concentrated at x_C; while agricultural workers, who account for the rest, are spread evenly along the unit interval. This implies that the overall sales ratio is

$$S = \mu[we^{\tau|x_C-x_A|}]^{1-\sigma} + (1-\mu)\int_0^1 [we^{\tau(|x_A-x|-|x_C-x|)}]^{1-\sigma}dx \tag{9}$$

or rearranging,

$$S = w^{-(\sigma-1)}[\mu[e^{-\tau|x_C-x_A|}]^{\sigma-1} + (1-\mu)\int_0^1 [e^{\tau(|x_C-x|-|x_A-x|)}]^{\sigma-1}dx] \tag{10}$$

Equation (10) determines relative sales as a function of the relative wage rate. We can now, however, turn it around to determine the maximum wage rate that the defecting firm can afford to pay by invoking the zero-profit condition. First, note that by assumption all firms at x_C are earning zero profits, with their operating surpluses just covering their fixed costs. A firm at x_A must do the same. But the operating surplus of a firm in the Dixit-Stiglitz model is proportional to its sales, while the fixed costs are incurred in manufacturing labor, which at x_A receives a relative wage w. It follows, then, that if there are to be zero profits we must have

$$S = w \tag{11}$$

Putting (10) and (11) together, we have our expression for the nominal wage rate that a firm at x_A can afford to pay

$$w = [\mu[e^{-\tau|x_C-x_A|}]^{\sigma-1} + (1-\mu)\int_0^1 [e^{\tau(|x_C-x|-|x_A-x|)}]^{\sigma-1}dx]^{1/\sigma} \tag{12}$$

To determine whether a concentration of manufactures at x_C is an equilibrium, however, we need to compare not the relative nominal wage but the relative real wage. The difference between the two comes from the fact that manufactured goods produced at x_C are part of workers' consumption basket, with a weight μ. Taking this into account, we note that the relative real wage rate is

$$\omega = we^{-\mu\tau|x_A-x_C|} \tag{13}$$

Equations (12) and (13) determine the real wage rate that a defecting firm

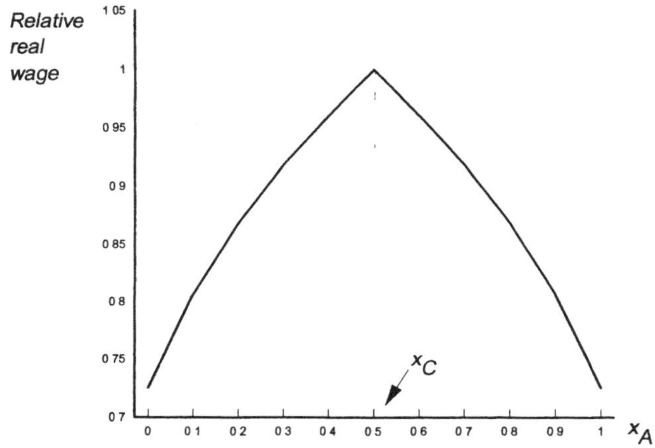

FIGURE 1: An Equilibrium Central Metropolis.

could afford to pay if it were to locate at a site x_A when all other manufacturing workers are concentrated at x_C. One way to think about (13) is that it is a kind of index of "market potential" of alternative sites. Such indices are widely used by geographers as a way to help think about plant location. In the standard calculation of market potential, the potential of a site is measured by a weighted sum of the purchasing power of all available markets, with the weights inversely proportional to distance from that site. In this case, the weights are derived from an explicit model of profit maximization, and there is also a "forward linkage" due to the role of metropolitan products in consumption. At a broad level, however, the idea is similar—a similarity that will become even more evident when we turn to the two-dimensional model of the next section.

Returning to (13), we immediately have a simple definition of an equilibrium metropolitan location. *A metropolis at x_C is an equilibrium, if, given that location, the maximum of (13) is also at x_C.*

Figure 1 illustrates such an equilibrium with the most obvious case, a metropolis located at $x_C = 0.5$, at the exact center of the line. For the purposes of this example[3] we set $\sigma = 4$, $\tau = 0.5$, $\mu = 0.2$. We see that the real wage is indeed maximized at $x_A = 0.5$, so that this is an equilibrium metropolitan location.

It might seem that a metropolis at the center of the region is always a possible equilibrium. Unfortunately, this is not quite right, because there may exist no equilibrium with only a single metropolis. If transportation costs are high enough, then even if one posits a concentration of all manufacturing at the center, firms will find it advantageous to move away from the center to get better access to the rural

[3]For the purposes of the calculation, the agricultural labor force, instead of being continuously spread along the line, was placed at 11 discrete locations (that is, at intervals of 0.1).

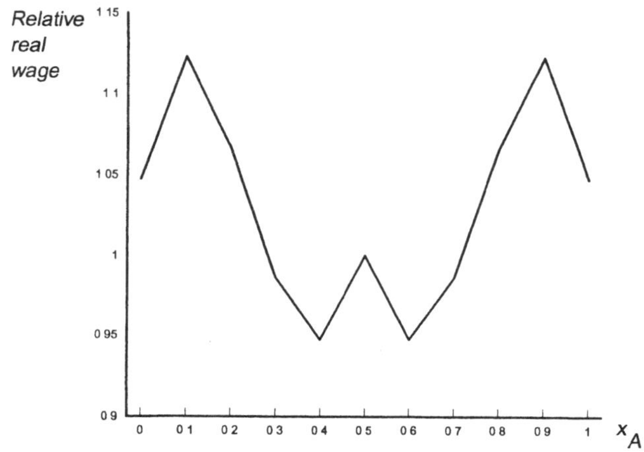

FIGURE 2: Absence of an Equilibrium Metropolis.

market. Figure 2 illustrates this point, by calculating market potential with a hypothetical metropolis at $x_C = 0.5$, but with a transport cost of $\tau = 1.5$. The central metropolis is not an equilibrium.

Presumably in the case of high transport costs equilibrium must take the form of several metropolitan centers. The lower are transport costs, the fewer and larger the metropolises can be. In this paper, however, I will not try to pursue that line of inquiry. Instead, we will simply assume that transport costs are sufficiently low that equilibria with a single urban center do exist, and restrict ourselves to examining that center's potential location.[4]

Let us now return to Figure 1. Notice that in that figure the market potential line has a "kink" at the metropolitan location. This is not an artifact of the particular numerical example. Consider what happens if we set $x_C = 0.5$, and evaluate the derivative of ω with respect to x_A for x_A just infinitesimally greater than x_C. Symmetry ensures that the derivative of the integral in (12) is zero in this case, and all the exponential terms become unity, so that this derivative to the right is simply

$$R = -\tau\mu - \frac{\sigma - 1}{\sigma}\tau\mu \tag{14}$$

If we take the derivative for x_A slightly *less* than x_C, however, we find the derivative

[4]One might expect that the ability to sustain a single metropolis would depend on the geographic extent of the region as well as on transportation costs. In this model, however, these are essentially the same thing. The size of the region (the length of the line) is normalized at unity; lengthening the line and increasing τ have exactly the same effects.

to the left to be

$$L = \tau\mu + \tau\mu\frac{\sigma - 1}{\sigma} = -R \tag{15}$$

Thus, there is a discontinuity in the slope of the market potential function.

The first term in (14) and (15) arises from the role of urban goods in consumption, introduced in (13). The second term arises from the role of urban consumers in demand [the first term in the bracketed expression in (12)]. Thus, the two terms can be seen as representing a Hirschman-type (1958) "forward linkage" and "backward linkage" respectively.[5] In other words, the kink reflects the concentration of economic mass at the metropolis.

Stepping slightly outside the model, the incentive to be at this concentration of mass makes the existence of a metropolis robust to small amounts of "noise." Suppose, for example, that some firms have a small preference to be a little left of center, while others have a small preference to be to the right. The kink in market potential at the metropolis implies that these firms will nonetheless all clump together at the same place.[6]

Returning to the model, the kink also implies that the exact center of the line is not the only possible site for the metropolis. For suppose that the metropolis lies a little bit to the left or right of the center. There will then be some incentive to move away from the metropolis toward the center. But given the kink, as long as the metropolis is not too far from the center the market potential line will still slope down in both directions. In effect, if we move the metropolis we drag the point of maximum market potential along with it.

Figure 3 illustrates this point. All parameters are the same as in Figure 1, but this time we suppose that the metropolis is located at $x_C = 0.4$, that is, somewhat left of center. Nonetheless, as we see from the figure, the point of maximum market potential is still at x_C. Thus, this is also an equilibrium metropolitan location.

Not all locations, however, are necessarily suitable for a metropolis. If the hypothetical metropolis is too far from the center, it may be advantageous for firms to move away. In Figure 4, again with the same parameters, we hypothesize a metropolis at $x_C = 0.1$. In this case, firms can afford to pay a higher real wage at sites to the right. Thus 0.1 is *not* an equilibrium metropolitan location.

It seems apparent that there is a range of potential metropolitan locations, including the center of the region but also extending some distance to either side.

[5]Forward linkages are usually interpreted to involve intermediate goods used in production rather than consumption goods. At a fundamental level, however, a forward linkage should be interpreted as a positive external economy resulting from the role of firms as producers, and this certainly meets that criterion.

[6]For example, in Cronon's case of Chicago in the 19th century, the market potential map for wheat marketing would presumably have looked different from that for slaughterhouses, which in turn would have looked different from that for lumber, and different yet again for different manufacturing industries. Nevertheless, the gravitational attraction of Chicago meant that the location of peak market potential for all of these sectors was in the same place.

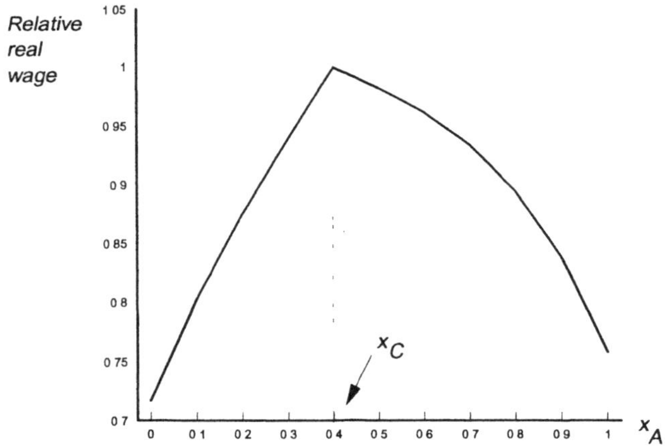

FIGURE 3: An Off-center Metropolis.

We can solve analytically for this range by applying a criterion of local stability: a necessary condition for a metropolitan site to be an equilibrium site is that given a hypothetical metropolis at that site, the market potential has a local maximum there. Given the symmetry of the problem, we need only consider locations to the left of center; obviously in that case a more desirable alternative site, if it exists, will lie to the right. So the defining criterion for the range of potential sites is: if we posit a metropolis at some $x_C < 0.5$, then $d\omega/dx_A$ for x_A slightly greater than x_C must be negative.

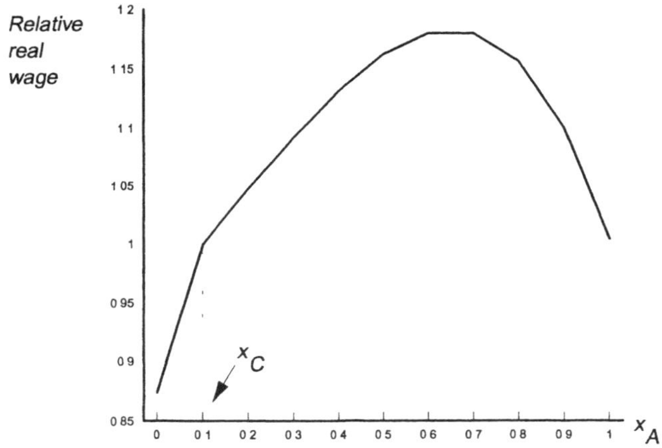

FIGURE 4: The Metropolis Cannot Be Too Far Off-center.

This local stability criterion simplifies the algebra massively: all the absolute value terms in (12) and (13) become unambiguously signed, and when the expression is evaluated in the vicinity of $x_A = x_C$ all of the exponential terms become unity. Thus letting R be the derivative of the relative real wage with respect to x_A, when x_A is just slightly greater than x_C, we have

$$R = \frac{\tau}{\sigma}[-\mu\sigma - \mu(\sigma - 1) + (1 - \mu)(\sigma - 1)(1 - 2x_C)] \tag{16}$$

Bearing in mind that for an equilibrium metropolitan location we must have $R < 0$, we note that (16) contains two negative terms and one positive. Broadly, the negative and positive effects may be seen respectively as the "centripetal" forces tending to hold a metropolis together and the "centrifugal" forces tending to pull it apart—or, as Myrdal (1957) put it, "backwash" and "spread" effects. More specifically, by examining (12) and (13) one can place direct economic interpretations on the three terms. As pointed out above, the first term in (16) comes from the role of metropolitan goods in consumption, introduced in (13); it therefore represents, in Hirschman's (1958) terms, a "forward linkage." The second term comes from the role of the metropolis as a source of demand for manufactured goods; it therefore represents a Hirschman-type "backward linkage." The third term, finally, represents the incentive to move away from the metropolis to be closer to the rural market.

The criterion $R < 0$ defines the range of potential metropolitan sites. For a metropolis to the left of center, we have

$$1 - 2x_C < \frac{\mu}{1 - \mu}\frac{2\sigma - 1}{\sigma - 1} \tag{17}$$

The larger is the right-hand side of (17), the wider the range of potential sites; if the right-hand side exceeds 1, any point on the line can accommodate a metropolis.

The width of the range depends in an economically sensible way on the parameters. It is increasing in μ, the share of manufactures in spending: this makes sense, because it is the importance of manufactures in consumption and of manufacturing workers's income in demand that gives rise to the forward and backward linkages that attract production to the metropolis. The range is decreasing in σ: the less monopoly power firms have (and hence also the smaller the degree of increasing returns in equilibrium), the less powerful the forces for metropolitan concentration.

For the case of a one-dimensional region, then, we have been able to show that if transport costs are not too high there will be an equilibrium in which all manufacturing is concentrated in a single metropolis; that this equilibrium is not unique, because there is a range of potential metropolitan sites; and that the width of this range depends in an economically meaningful way on the model's parameters. The method we used to analyze this circular logic of metropolitan concentration involves a somewhat surprising and, in my view, exciting conclusion: a city can be thought of as causing a kink in a "market potential" function, and it is precisely the existence of such a kink that in turn justifies the existence of the city.

The tradition of economic geography, however, contains a strong cartological component, in which one tries to relate theoretical concepts to actual maps. And maps, unfortunately, have two dimensions rather than one. So it is natural to ask whether the insights gained from this one-dimensional model still apply in a two-dimensional world. In the next section we show that they do.

3. A TWO-DIMENSIONAL MODEL

The two-dimensional analogue of a line is a disk. We now imagine, then, a region in which the agricultural labor force is spread evenly across a disk; we normalize the radius of the disk to unity, and let the center be $x = y = 0$. All the other assumptions are the same as in the one-dimensional model.

The analogy with the one-dimensional model immediately suggests what we are going to find. Provided that transportation costs are not too high, there will be an equilibrium with all manufacturing concentrated at a central metropolis. This metropolis could be in the geometric center of the disk, but it need not be: because changing the location of the metropolis itself changes the map of market potential, there will be a range of potential metropolitan locations. In the one-dimensional case this range was a central portion of the line segment; in the two-dimensional case it will be a central disk within the regional disk. Figure 5 illustrates schematically what we will find: the larger disk represents the region as a whole, the shaded interior disk the set of potential metropolitan locations.

The analytics of this model are very similar to the one-dimensional model, but complicated by the need to measure distances in two dimensions. Let us posit a metropolis at x_C, y_C; without loss of generality (since one can always rotate the disk), assume $y_C = 0$. We want to consider the attractiveness of an alternate location x_A, y_A. To calculate this, we need to know three distances. Let D_{AC} be the distance between the alternative location and the metropolis; we have

$$D_{AC} = \sqrt{(x_A - x_C)^2 + y_A^2} \tag{18}$$

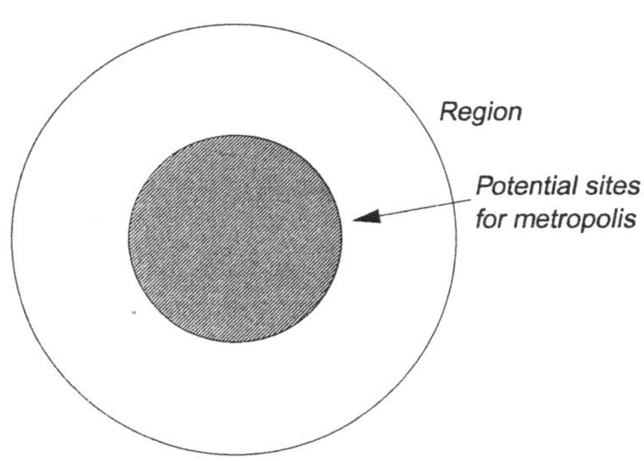

FIGURE 5: Metropolitan Location in Two Dimensions.

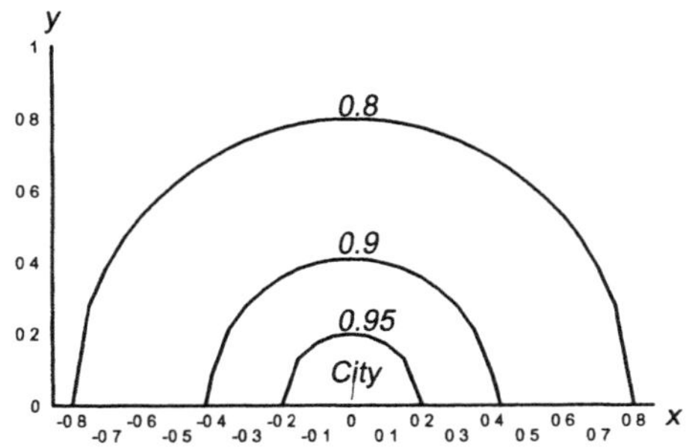

FIGURE 6: Market Potential with a Central Metropolis.

Let $D_C(x, y)$ be the distance from the metropolis to some other location (x, y)

$$D_C(x, y) = \sqrt{(x_C - x)^2 + y^2} \tag{19}$$

and let $D_A(x, y)$ be the distance from the alternate location to (x, y)

$$D_A(x, y) = \sqrt{(x - x_A)^2 + (y - y_A)^2} \tag{20}$$

By analogy with the one-dimensional case, the sales of a firm at the alternate location relative to those of one in the metropolis are

$$S = \mu[we^{-\tau D_{AC}}] + \frac{1-\mu}{\pi}\int_{-1}^{1}\int_{-\sqrt{1-x^2}}^{\sqrt{1-x^2}} [we^{\tau[D_X(x,y) - D_A(x,y)]}]^{\sigma-1} dy\, dx \tag{21}$$

The zero-profit condition once again requires that $S = w$. Thus we have

$$w = \left[\mu e^{-\tau(\sigma-1)D_{AC}} + \frac{1-\mu}{\pi}\int_{-1}^{1}\int_{-\sqrt{1-x^2}}^{\sqrt{1-x^2}} e^{(\sigma-1)\tau[D_C(x,y) - D_A(x,y)]} dy\, dx\right]^{1/\sigma} \tag{22}$$

and the relative real wage at the alternate location is

$$\omega = we^{-\mu\tau D_{AC}} \tag{23}$$

Equations (18)–(23) can be used to construct a market potential map, given the location of a hypothetical metropolis. And as in the one-dimensional case, if the location of the posited metropolis is also the point of greatest market potential, then that is an equilibrium location.

Figure 6 shows the case of a metropolis located at the geometric center of the region,[7] under the assumptions $\tau = 0.5$, $\mu = 0.3$, $\sigma = 4$. (Since the diagram is symmetric, only the upper half of the disk is shown). The market potential is

[7]As in the one-dimensional case, this example was constructed using a discrete distribution of rural population at 317 points, located at grid intervals of 0.1.

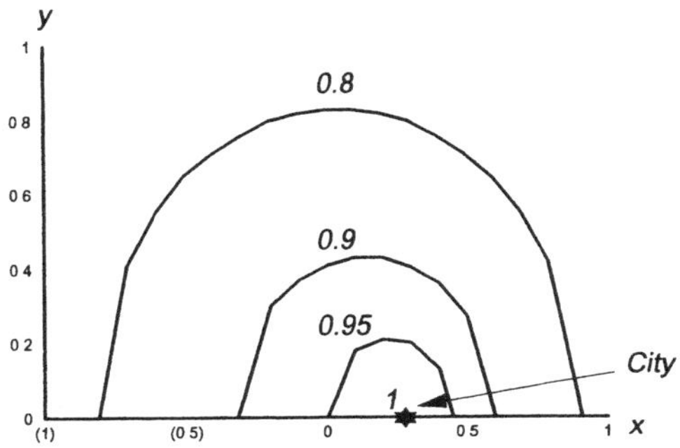

FIGURE 7: Market Potential with an Off-center Metropolis.

represented by contour lines, loci of equal real wages that are 0.95, 0.9, and 0.8 of the real wage at the metropolis; in this case, of course, the contour lines are simply circles around the metropolitan center. Since the central bulls-eye is also the point of peak market potential, a central metropolis is an equilibrium.

But a somewhat off-center metropolis may also be an equilibrium. Figure 7 shows the market potential map generated by a metropolis located at $x_C = 0.3$. The contour lines are dragged off to the east by the metropolis's economic mass; the metropolis is still the best location, and thus this proposed geography is also an equilibrium.

As in the one-dimensional model, we can analyze the determinants of the range of potential metropolitan sites by examining local stability. Suppose we posit a metropolis at $(x_C, 0)$, with $x_C > 0$. If there is a better location for a small group of workers, it will be toward the center; thus we need only consider alternative locations with $x_A < x_C$ and $y_A = 0$. For local stability, the derivative of ω with respect to x_A must be positive for x_A slightly less than x_C. Letting L represent this derivative, we have

(24)
$$L = \frac{\tau}{\sigma}\left[\mu(2\sigma - 1) + \frac{1-\mu}{\pi}(\sigma - 1)\int_{-1}^{1}\int_{-\sqrt{1-x^2}}^{\sqrt{1-x^2}}(x - x_A)[(x_A - x)^2 + y^2]^{-1/2}dy\,dx\right]$$

As in the one-dimensional model, this local derivative is simpler than the global equation, but unfortunately not quite as much so. There are two terms inside the brackets. The first term captures the forward and backward linkage effects, and is always positive. The second term is always negative: one can see this by thinking of the term as representing the weighted sum of a series of values of $x - x_A$. Now compare the weight on each positive value $x - x_A$ with that on $-(x - x_A)$; in each case the weight on the negative value is larger. So the second term in (24) captures the centrifugal forces pulling the metropolis apart.

The further the metropolis is from the geometric center of the disk, the stronger are these centrifugal forces. Differentiating with respect to $x_A = x_C$, we find

(25)

$$\frac{dL}{dx_C} = -\frac{\tau}{\sigma}\frac{1-\mu}{\pi}(\sigma - 1)\int_{-1}^{1}\int_{-\sqrt{1-x^2}}^{\sqrt{1-x^2}}\frac{y^2}{(x-x_C)^2+y^2}\sqrt{(x-x_C)^2+y^2}dy\,dx < 0$$

Thus there may, as suggested in Figure 5, be a maximum distance of the metropolis from the center. This maximum distance, and thus the area of potential metropolitan sites, depends on μ and σ. In particular, it is straightforward to show that $dL/d\mu > 0$, which implies that the radius of the range of potential metropolitan sites is increasing in μ. That is, the larger the share of income spent on manufactures, and hence the stronger the positive feedback of actual to optimal manufacturing location, the less "first nature" determines where manufacturing takes place.

4. LIMITATIONS OF THE ANALYSIS

The models developed in this paper offer a kind of "gravitational" analysis of the existence of metropolitan centers. Figure 7 in particular suggests the following metaphor: the concentration of economic mass at the metropolis bends economic space around itself, and it is precisely this curvature of the economic space that sustains the metropolitan concentration.

I would argue that this approach, in spite of the numerous special assumptions needed to yield tractable models, conveys a valuable perspective on metropolitan location. Nonetheless, the models presented here have five serious limitations.

First, the assumption that transportation of agricultural goods is costless is justified only by the (very considerable) analytical simplification it makes possible. For any realistic application of this approach to actual urban history it must be abandoned.

Second, and related, the assumption of an exogenously distributed agricultural work force is ultimately unsatisfactory. In particular, it prevents the models from accommodating von Thünen's key insight about the relationship between distance from the metropolis, land rents, and land use.

Third, the models assume that transportation cost is strictly proportional to distance. Yet in practice—and above all in Cronon's story of Chicago, which motivated this paper—increasing returns to transportation, which lead to the formation of transportation hubs, play a key role in metropolitan concentration. A preliminary effort to model this is in Krugman (1990), but no effort is made to incorporate this analysis here.

Fourth, throughout this paper the focus has been on the location of a single metropolis. A realistic analysis will have to take into account the emergence of a system of cities. At present there are two different approaches to modelling urban systems. Central place theory, deriving from Lösch (1940) and Christaller (1933) and widely used by geographers, is a powerful metaphor but lacks satisfactory microfoundations. Meanwhile, urban economists—notably Henderson (1974, 1988)—have developed models of urban systems that carefully model behavior, but which lack any spatial content. The approach in this paper is, among other things, an effort to build a bridge between these traditions, but it does not manage to get beyond the one-city case.

Finally, this paper presents only a static model. This model explains why a metropolis might persist but says nothing about the dynamics of how such centers form. Dynamic issues have been a preoccupation of much of the nonspatial urban systems literature, and will eventually have to be tackled in this framework as well.

These are, then, only preliminary models. They do, however, offer a new approach that may eventually prove able to accommodate greater realism.

REFERENCES

Christaller, Walter. 1933. *Central Places in Southern Germany.* (C. W. Baskin, trans.) London: Prentice Hall, 1966.

Cronon, William. 1991. *Nature's Metropolis: Chicago and the Great West,* New York: Norton.

Dixit, A. and Joseph Stiglitz. 1977. "Monopolistic Competition and Optimum Product Diversity," *American Economic Review,* 67, 297–308.

Fujita, Masahisa. 1988. "A Monopolistic Competition Model of Spatial Agglomeration: Differentiated Product Approach," *Regional Science and Urban Economics,* 18, 87–124.

Henderson, J. Vernon. 1974. "The Sizes and Types of Cities," *American Economic Review,* 64, 640–56.

———. 1988. *Urban Development: Theory, Fact, and Illusion,* New York: Oxford.

Hirschman, Albert O. 1958. *The Strategy of Economic Development,* New Haven: Yale University Press.

Krugman, Paul. 1990. "The Hub Effect: Or, Threeness in Interregional Trade," mimeo, Department of Economics, MIT.

———. 1991. "Increasing Returns and Economic Geography," *Journal of Political Economy,* 99, 483–499.

Lösch, August. 1940. *The Economics of Location,* (W. H. Woglom and W. F. Stolper, trans.) New Haven: Yale University Press, 1954.

Myrdal, Gunnar. 1957. *Economic Theory and Underdeveloped Regions,* London: Duckworth.

Rauch, J. 1991. "Comparative Advantage, Geographic Advantage, and the Volume of Trade," NBER Working Paper #3512.

von Thünen, Johann H. 1826. *The Isolated State.* (Peter Hall, trans.) New York: Pergamon, 1966.

[5]

©*International Regional Science Review* **19**, 1 & 2: 5–30 (1996)

'Urban Concentration: The Role of Increasing Returns and Transport Costs

Paul Krugman

Department of Economics
Stanford University
Stanford CA 94305 USA

Very large urban centers are a conspicuous feature of many developing economies, yet the subject of the size distribution of cities (as opposed to such issues as rural-urban migration) has been neglected by development economists. This article argues that some important insights into urban concentration, especially the tendency of some developing countries to have very large primate cities, can be derived from recent approaches to economic geography. Three approaches are compared: the well-established neoclassical urban systems theory, which emphasizes the tradeoff between agglomeration economies and diseconomies of city size; the new economic geography, which attempts to derive agglomeration effects from the interactions among market size, transportation costs, and increasing returns at the firm level; and a nihilistic view that cities emerge out of a random process in which there are roughly constant returns to city size. The article suggests that Washington consensus policies of reduced government intervention and trade opening may tend to reduce the size of primate cities or at least slow their relative growth.

Over the past several years there has been a broad revival of interest in issues of regional and urban development. This revival has taken two main directions. The first has focused on theoretical models of urbanization and uneven regional growth, many of them grounded in the approaches to imperfect competition and increasing returns originally developed in the "new trade" and "new growth" theories. The second, a new wave of empirical work, explores urban and regional growth patterns for clues to the nature of external economies, macroeconomic adjustment, and other aspects of the aggregate economy.

Most of this work has focused either on generic issues or on issues raised by the experience of advanced market economies like the United States. Yet arguably the issues raised by the recent work are most salient for smaller, less-wealthy countries like Mexico and Brazil.

Why might the "new economic geography" be more relevant for the developing world than for industrial countries? First, the matter is an urgent one for real-world policy. Urbanization in developing countries, and particularly the very large agglomerations such as Mexico City and São Paulo, is widely regarded as a problem. Rural-urban migration has, of course, been the subject of a vast literature in development economics, with many papers suggesting that its pace is excessive from a social point of view. Moreover, the sheer size of some cities that such migration now feeds reinforces these concerns. Although nobody can claim to have made a thorough welfare-economic study of the consequences of the emergence of huge cities in developing countries, many observers believe that something has gone wrong, that such giant cities are in some sense parasitic entities that drain vitality from their host economies—Bairoch (1988) has called these metropolises "Romes without empires"—that the environmental and social problems posed by cities with populations in the tens of millions are even greater in poor nations than in the West.

Associated with concern about urbanization and metropolitan growth is related concern about regional inequality. In many developing countries the regions that contain the big cities are also much richer per capita than other regions. The problem of core-periphery patterns within countries is not only economic and social but political as well: it is no accident that a separatist movement has emerged in Brazil's relatively rich south or that armed opposition to the central government surfaced in the bypassed southern regions of Mexico.

On the bright side, urbanization and unequal regional development may be analytically more tractable in developing than in industrial countries. The models developed in recent years, which stress the role of relatively measurable factors like economies of scale and transportation costs in determining urban growth, often seem to miss much of the story in advanced economies. For one thing, in huge economies like the United States or the European Union, static economies of scale tend to seem relatively unimportant. For another, in advanced nations that are increasingly in the business of producing information rather than tangible goods, the nature of both the external economies that induce agglomeration and the transaction costs that make distance matter becomes more and more subtle. By contrast, developing countries have much smaller internal markets. For example, although Mexico's population is one-third that of the United States, its dollar purchasing power is about the same as that of metropolitan Los Angeles. Thus, conventional scale economies remain relevant. And these countries still devote much more of their labor force and expenditure to tangible products that must be transported by road or rail.

Finally, the radical policy changes that have taken place, or may be about to take place, in some developing countries are likely to have major impacts on urban and regional development, impacts that we want to be able to predict. One need only consider the case of Mexico: the federal district in that country became dominant during a prolonged period of both import-substituting development strategy and extensive government involvement in the economy. As the country has shifted to an export-oriented trade policy, the manufacturing center of gravity has visibly shifted toward the country's northern states. Will the combining of that shift with privatization and deregulation undermine Mexico City's special role, or will other activities maintain its position?

For these reasons, then, it is natural to ask whether, and if so to what extent, the new tools of urban and regional analysis apply to developing countries. The literature on urban and regional issues in development is immense. This article explores a narrow, indeed largely technical issue: what can we learn from looking at urbanization and regional inequality in developing countries through the lens of the specific approach to economic geography that has emerged out of the new trade and growth theories? The article sketches out a minimalist new economic geography model designed to highlight the way a tension between forces of agglomeration and forces of dispersal determines city sizes. The implications of that tension are illustrated by examining a particular issue: how trade policy may affect the tendency of developing countries to have very large, primate cities. Two other factors also are explored that probably have even more important roles in determining urban structure: the centralization of government and the quality and form of transportation infrastructure.

Approaches to Urban Development

Urbanization—and uneven regional development, which is a closely related process—clearly involves a tension between the "centripetal" forces that tend to pull population and production into agglomerations and the "centrifugal" forces that tend to break such agglomerations up. The following tabulation lists the major types of centripetal and centrifugal forces that appear in various models of urban growth:

Centripetal forces

- Natural advantages of particular sites
  - Harbors, rivers, and the like
  - Central locations
- Market-size external economies
  - Access to markets (backward linkages)
  - Access to products (forward linkages)
  - Thick labor markets
- Pure external economies
  - Knowledge spillovers

Centrifugal forces

- Market-mediated forces
  - Commuting costs, urban land rent
  - Pull of dispersed resources, such as farmland

- Nonmarket forces
  - Congestion
  - Pollution

Several key distinctions among these forces are worth pointing out. Among centripetal forces there is a basic distinction between natural factors that favor a site—such as a good harbor or a central position—and external economies that are acquired and self-reinforcing advantages of a site. Among external economies there is a further key distinction between "pure" external economies, such as spillover of knowledge between nearby firms, and market-size effects, whether in the labor market or in the linkages between upstream and downstream industries.

On the side of centrifugal forces there is a similar distinction between nonmarket diseconomies (such as congestion) and factors such as land prices that are fully mediated through the market. A narrower but sometimes important distinction appears between forces that push business out of a large city, such as urban land prices, and those that pull business away, such as the existence of a dispersed rural market.

Which forces actually explain the pattern of urbanization in developing countries? The answer is, of course, all of them. Nonetheless, to say anything useful we must always rely on simplified models. The typical analytical approach therefore takes "one from column A and one from column B" and thus gets a particular story about the tension between the agglomeration and dispersion that creates an urban system. Several such approaches have achieved wide influence.

Neoclassical Urban Systems Theory

At least within the economics profession the most influential approach to urban development is probably what we might call neoclassical urban systems theory. This approach models the centripetal forces for agglomeration as pure external economies (therefore allowing the modeler to assume perfect competition)[1] and the centrifugal forces as arising from the need to commute to a central business district within each city, a need that leads to a gradient of land rents within each city. In the simplest case the tension between these forces leads to an optimal city size, though there is no guarantee that market forces will actually produce this optimal city.

This neoclassical approach has been extensively developed by Henderson (1974, 1977, 1988) and his followers, who added two important elaborations. First, Henderson pointed out that if cities are the "wrong" size,

[1] It is possible, without any real change in the structure, to derive external economies from a monopolistically competitive sector that produces nontraded inputs. See Abdel-Rahman (1988) and Rivera-Batiz (1988).

there are potential profit opportunities for a class of "city developers"; and as an empirical matter, large forward-looking private agents who seem to try to internalize external economies do play a large role in urban development in the United States. Thus, Henderson-type models adopt as a working hypothesis the assumption that competition among developers produces cities of optimal size.

Second, according to Henderson, external economies may well be industry-specific (textile plants may convey external benefits to neighboring textile plants; metalworking plants may do the same, but it is hard to see why metalworkers want textile workers nearby). On the other hand, diseconomies of commuting and land rent depend on the overall size of a city, not the size of an individual industry within that city. Thus, Henderson-type models predict the emergence of specialized cities, with each city's "export" sector producing a range of industries with mutual spillovers, and with industries that do not benefit from these spillovers seeking other locations. Because cities are specialized, this approach explains the existence of an urban system with many different types of cities; inasmuch as the optimal size of a city depends on the relative strength of external economies and city-size diseconomies, and external economies are presumably stronger in some industries than in others, cities of different types will be of different sizes. Neoclassical urban systems theory therefore offers a framework that explains the existence not only of cities but also of a system of cities of differing sizes.

Although the insights gained from this approach are impressive, it has important limitations. First, the external economies that drive agglomeration are treated for the most part as a kind of black box, making it difficult to think about what might influence their strength and thus making it hard even to start to predict how policy or other changes might affect the urban system. Second, the reliance of much of this literature on the assumption of competition between city developers, while a useful clarifying device, strains credibility when applied to huge urban areas: the Irvine Corporation may arguably have played a major role in developing a particular "edge city" within metropolitan Los Angeles, but could any private agent internalize the externalities of São Paulo? Finally, neoclassical urban systems theory is entirely nonspatial: it describes the number and types of cities, but says nothing about their locations. In the past few years an alternative approach has emerged that shares much of the framework of urban systems theory but attempts to deal with these issues.

Monopolistic Competition Theory

In this new literature, agglomeration economies are not assumed but are instead derived from the interaction among economies of scale at the plant level, transportation costs, and factor mobility. Economies of scale at the plant level inevitably imply imperfect competition; this imperfection is modeled using the same (unsatisfactory) monopolistic competition approach that has played such a large role in trade and growth theory over the past fifteen years. The "new economic geography" literature, begun in Krugman (1991a, b), bears considerable resemblance to the urban systems approach, but the black-box nature of external economies is gone, there is a spatial dimension, and the models no longer rely on the assumption of city developers who enforce optimal outcomes. In some respects, in fact, the new approach seems closer in spirit to the "cumulative process" description of urban and regional development associated with geographers such as Pred (1966).

The model described below is in this tradition, so it is worth noting the considerable limitations of this approach. Two points stand out. First, multiple-city systems are difficult to model using this approach. Where the urban systems approach easily tells a story of multiple cities of a number of different types, in monopolistically competitive spatial settings (see, for example, Krugman 1993b) multiple-city systems can at this point be modeled only with considerable difficulty, and initial efforts to get some kind of urban hierarchy have encountered surprisingly nasty problems (Fujita and Krugman 1993). Second, going from the black-box external economies of the urban systems model to the derived agglomeration effects of the monopolistic competition model may involve a degree of misplaced concreteness. We will have a seemingly clear story about linkage externalities in the manufacturing sector, but it may be that, say, informational externalities in the service sector are equally important even in developing countries. Attempts to get specific, to open up the black box, always run this risk; nonetheless, it seems greater than usual in this case.

Finally, we should point out one additional risk in both the urban systems and the monopolistic competition approaches to urban modeling: we may be trying to explain too much, engaging in a kind of Rorschach test in which we are trying to find deterministic explanations of essentially random outcomes. Although this notion does not exactly constitute a rival theory of urban systems, the idea that they are largely random creations requires at least some discussion.

Random Urban Systems

The general idea suggested by the tabulation above—that city sizes are determined by a tension between centripetal and centrifugal forces—seems to imply the conclusion that there will in any economy be a typical, equilibrium city size. In fact, one sees a whole range of city sizes. The urban systems theory explains that there are different types of cities, each with a characteristic size, and that the size distribution is actually a type distribution. Although this argument surely has some validity, it may not be a full explanation. For one thing, urban specialization is increasingly difficult to detect in advanced countries. It is a familiar point that the mix of activities within U.S. metropolitan areas has become increasingly similar since 1950, and the influential study by Glaeser and others (1992) finds, as well, that individual industries seem to grow fastest in more diverse metropolitan areas.

Moreover, the size distribution of cities is suspiciously smooth and regular. City sizes in many countries are startlingly well described by a power law of the form

$$N(S) = AS^{-\alpha} \tag{1}$$

where $N(S)$ is the number of cities that are the same size as or larger than S. Furthermore, the exponent α a is generally quite close to 1. In fact, when equation (1) is estimated for U.S. metropolitan areas, α is almost exactly 1, and it has remained close to 1 for at least a century. International evidence is not quite so strong, perhaps because of definitions of city boundaries: Rosen and Resnick (1980) show that when data for metropolitan areas rather than cities proper are used for a number of countries, α almost always moves substantially closer to 1.

Why should this matter? Because while a relationship like equation (1) is difficult to explain with an equilibrium story about determination of city size, it is quite easy to justify with a nihilistic story of the kind analyzed by Herbert Simon (Ijiri and Simon 1977). Suppose that for all practical purposes there is no equilibrium city size—that approximately constant returns to scale appear over some wide range of sizes. And suppose that cities grow through some random process, in which the expected rate of growth is independent of city size. Then as long as the random process generates a widely dispersed distribution of city sizes, that distribution will be well described by a power law like equation (1). (A possible, suggested explanation of this result is given in the appendix.)

Worse yet, such a nihilistic approach can even explain the tendency of the exponent of the power law to be close to 1. Suppose that there is some minimum viable city size, say S_0, and that the distribution of city sizes above that minimum is well described by equation (1). Then the *average* city size is

$$\bar{S} = S_0(\alpha/\alpha - 1). \tag{2}$$

In other words, α close to 1 is equivalent to the statement that the average city size is large relative to the minimum. And it is easy to imagine why this might be the case. Suppose that urban population has grown substantially over a period during which, for whatever reason, few new cities have been founded. Then the existing cities must on average grow much larger than the minimum viable size, and the estimated α will be close to 1.

This nihilistic approach raises real questions about any kind of equilibrium model of an urban system; indeed, if this interpretation is correct, there may be no optimal or equilibrium city size, simply a random process that generates population clusters of many sizes. At some level this interpretation cannot be completely right: surely city size must matter. (This is the same issue that arises in studies of the size distribution of firms, which also seems to obey power laws.) Yet the data may contain less information than we think.

On the other hand, this approach suggests that estimates of relationships like equation (1), together with related measures like "primacy," may be a useful summary indicator of the structure of a country's urban system. Primacy describes the size of the largest city relative either to total population or to some other measure, such as the population of the n largest cities. Many have studied city size distributions: Carroll (1982) provides a survey; Rosen and Resnick (1980) is a particularly clear example; Ades and Glaeser (1993) is a recent study inspired by the new economic geography literature. This literature suggests several stylized facts that may help us to think about urbanization in developing countries.

Stylized Facts

Although urban experience varies widely across nations, there seem to be four interesting empirical regularities about urban size distributions.

First, per capita income is negatively related to measures of urban concentration, whether one uses α from equation (1) or measures of primacy such as the share of the largest city in the population of the top ten. This observation confirms an impression of giant metropolitan areas in developing countries: to a large extent, of course, the developing world has big cities simply because it has so many people, but even in this light the biggest cities in these countries are disproportionately big.

Second, the concentration of urban population is closely related to the concentration of political power. Countries with federal systems, and thus geographically diffused power, have flatter distributions of city size and, in particular, smaller biggest cities than countries that do not have federal systems. Thus Tokyo, the largest city in centralized Japan, is considerably larger than New York, the biggest city of federal America, even though the United States has twice Japan's population. Australia and Canada, though developed at about the same time, have much less urban concentration than do Argentina or Chile. Dictatorships have more concentrated urban centers than do more pluralistic systems, according to Ades and Glaeser (1993).

Third, the nature of transportation infrastructure has an important effect on urban concentration. Countries in which the capital city has a uniquely central position—something that Rosen and Resnick (1980) proxy by a measure of rail density—tend, not too surprisingly, to have more populous capitals. Obviously, this effect often works in tandem with centralization of political power.

Finally, a less dramatic but still visible relationship is apparent between trade openness and urban structure. More open economies, as measured by the share of exports in gross domestic product, tend to have smaller biggest cities. (This is an other-things-equal proposition. Countries with small populations tend to be open, and also to have a large share of their population in the biggest city—consider Singapore. But countries that are more open than you would expect given their population tend to have smaller biggest cities than you would expect given their population.)[2]

At this point, then, we have described a menu of ways (far from inclusive) to think about urban systems in developing countries and have very briefly set out some stylized facts. The next step is to sketch a particular model as a basis for trying to understand those facts.

A Model of Urban Concentration

This section presents a formal model of urban concentration; the full model is presented in the appendix. As pointed out above, numerous centrifugal and centripetal forces may affect urban concentration. All of them probably play some role in practice, yet the modeler normally chooses only a few to include in any given analysis. In my own work I have generally chosen to include only the centripetal forces that arise from the interaction among economies of scale, market size, and transportation costs, that is, backward and forward linkages. Other external economies are undoubtedly at work in real urban areas, but they are omitted in the interest of keeping the models as simple as possible and of keeping a reasonable distance between assumptions and conclusions.

---

[2] Before the Rosen and Resnick study (1980) most writing on primacy assumed that export orientation would tend to *increase* primacy. The ruling image was of a primary exporting country in which the primate city would be the country's main port; the implicit argument was that the economies of scale involved in building infrastructure for exporting were larger than those involved in selling to the domestic market. One can hardly deny that this effect has existed in some times and places; the evidence that the effect runs the opposite way is not overwhelming. This kind of ambiguity arises in any attempt to summarize the richness of cross-national variation with a short list of explanatory variables.

For similar reasons we can handle only one centrifugal force at a time. It turns out to be useful to move back and forth between two different approaches. One, which is close in spirit to the neoclassical urban systems literature, involves commuting costs and land rent. The other involves the pull of a dispersed rural market. This second approach has already been described in a number of published articles, for example, Krugman (1991a, b, 1993b); thus, the formal model described here does not include this effect.

As we will see, attempting to make sense of the stylized facts described above is easiest when keeping both approaches in mind. The role of trade openness in urban concentration is most easily understood by focusing on urban land rent, while one cannot model the effects of political centralization and infrastructure without some kind of backdrop of immobile population and purchasing power.

Imagine, then, a stylized economy consisting of three locations: 0, 1, and 2. Location 0 is the "rest of the world," while 1 and 2 are two domestic locations (say, Mexico City and Monterrey). There is only one factor of production, labor. A fixed domestic supply of labor L is mobile between locations 1 and 2, but there is no international labor mobility.

In this radically oversimplified model, the issue of urban concentration reduces to just one question: how equally or unequally will the labor force be distributed between the two locations? It is, of course, a considerable stretch to relate results of this kind to the realities of multicity urban systems, but as always the hope is that despite their oversimplifications, simple models yield useful insights.

To generate diseconomies of urban concentration, we assume that in each location production must take place at a single central point. Workers, however, require land to live on. To make matters simple, we make several special assumptions. First, each worker needs a fixed living space, say, one unit of land. Second, the cities are long and narrow, so that workers are effectively spread along a line. This assumption implies that the commuting distance of the last worker in any given location is simply proportional to that location's population (as opposed to depending on the square root of population, as it would in a disk-shaped city).[3]

The diseconomies arising from the need to commute will be reflected both in land rents and in commuting costs. Workers who live in the outskirts of the town will pay no land rent but will have high commuting costs. Workers who live closer to the city center will avoid these costs, but

[3] In what are commuting costs incurred? It is easiest to assume that they are incurred only in workers' time, and that time spent commuting is time not spent working. In this case, as shown in the appendix, the net wage rate of the most remote worker in a city of population L takes the form $w(1-\gamma L)$, where w is the wage at the center.

competition will ensure that they pay an offsetting land rent. The wage net of commuting costs will decline as one moves away from the city center, but land rents will always exactly offset the differential. Thus, given any wage rate at the center, the wage net of both commuting and land rents will be a decreasing function of city size for all workers.

To explain agglomeration in the face of these diseconomies, we must introduce compensating advantages of concentration. These must arise from economies of scale. Unless economies of scale are purely external to firms, however, they must lead to imperfect competition. So we must introduce scale economies in a way that allows a tractable model of imperfect competition.

Not surprisingly, the easiest way to do this is with the familiar tricks of monopolistic competition modeling. We suppose a large number of symmetric potential products, not all actually produced. Each producer acts as a profit-maximizing monopolist, but free entry drives profits to zero. The result will be that a large concentration of population produces a large variety of differentiated products. (One might think that the average scale of production will also be larger. Unfortunately, in the Dixit-Stiglitz–type model used in the appendix, this plausible effect does not materialize: all scale gains appear in the form of variety rather than production.)

Will this advantage make such a location attractive despite high land rent and commuting costs? Only if there are costs of transacting between locations, so that a location with a large population is a good place to have access to products (a forward linkage) and to markets (a backward linkage). Thus, we next introduce transportation costs, both between domestic regions and between these regions and the rest of the world. For technical reasons involving the way that monopolistic competition must be modeled, it turns out to be extremely convenient, if silly, to assume that transport costs are incurred in the goods shipped, an assumption sometimes referred to as the iceberg assumption: if one unit of a good shipped between regions is to arrive, $\tau > 1$ units must begin the journey. The same applies to international shipments, except that the transport costs may be different.

We may think of interregional transport costs as "natural" consequences of distance (albeit affected by investments in infrastructure). The costs of transacting with the rest of the world, however, involve not only natural costs but artificial trade barriers. Thus, the level of transport costs to and from the outside world can be seen as a policy variable.

And that's it (except for the details laid out in the appendix). The interaction among economies of scale, transport costs, and labor mobility is enough to generate economies of agglomeration; the need to commute generates diseconomies of city size; and the tension between centrifugal and centripetal forces provides a framework for thinking about urban structure.

To understand how this model works, consider what would happen in the absence of foreign trade, and within that special case ask only a limited question: Under what conditions is concentration of all population in either location 1 or 2 an equilibrium? Once we have seen this case, it will be easier to understand the results when the model is opened up.

Suppose, then, that the cost of transacting with the outside world is very high, so that we can ignore the role of the rest of the world. Furthermore, consider the determination of relative real wages when almost all domestic labor is in region 1. If the real wage rate of a worker in location 2 is less than that of a worker in region 1 in this case, then concentration of all labor in region 1 is an equilibrium; otherwise it is not.

We first note that the nominal wage paid at the center of city 2 (w_2) must be less than that at the center of city 1 (w_1). The reason is that almost all output from a firm in 2 must be sold in 1 and must therefore incur transport costs. At the same time, the zero-profit output for firms is the same in each location. So goods produced at location 2 must have sufficiently lower f.o.b. prices to sell as much in 1's market as goods produced at location 1. It can then be shown that

$$w_2/w_1 = \tau^{(1-\sigma)/\sigma} < 1 \tag{3}$$

where σ is the elasticity of substitution among differentiated products.

This wage premium at location 1, which results from its dominant role as a market, essentially represents the backward linkages associated with the concentration of demand there.

Next we notice that if almost all labor is in location 1, almost all goods consumed in 2 must be imported, implying a higher price of these goods:

$$T_2/T_1 = \tau \tag{4}$$

where T_i is the price index for goods (excluding land rent) at location i.

If the wage rate is higher in 1 and the price of consumer goods lower, must not real wages be higher in 1? No—because land rent or commuting costs (or both) are higher. With almost all of the labor force L concentrated in 1, the most remote workers in 1 must commute a distance $L/2$, and all workers who live closer to the center must pay a land rent that absorbs any saving in commuting costs. Meanwhile, the small number of workers in 2 pay almost no land rent and have essentially no commuting distance. So the real wage difference turns out to be

$$w_1/w_2 = \tau^{(2\sigma-1)/\sigma}(1-\gamma L). \tag{5}$$

In this expression the first term represents the centripetal forces—the backward and forward linkages described in equations (3) and (4), which arise from the concentration of suppliers and purchasing power at location 1; the second term represents the centrifugal forces of commuting cost and land rent.[4]

Our next step is to examine the relation between trade openness and urban concentration.

Trade Openness and Urban Concentration

The previous section demonstrates how a concentration of labor in one location may be sustainable, despite the commuting and land rent diseconomies of urban size, through forward and backward linkages. Now suppose that the economy is open to international trade, albeit with some natural and perhaps artificial barriers. How does this change the story? It should be obvious that the effect is to weaken the centripetal forces while leaving the centrifugal forces as strong as before.

Consider a hypothetical primate city, a Mexico City or São Paulo, in a country with a strongly protectionist trade policy. Firms will be willing to pay a wage premium in order to locate at that center precisely because so many other firms, and thus the bulk of their market, are concentrated there. They also may be attracted by the presence of other firms producing intermediate inputs—something not explicitly represented in the model in the appendix, but similar in its effect. On the other side, workers will face high land rents or commuting costs, but these will be at least partly offset by better access to the goods and services produced in the metropolis.

But now throw this economy open to international trade. The typical firm will now sell much of its output to the world market (and perhaps get many of its intermediate inputs from that market as well). To the extent that production is for world markets rather than for the domestic market, access to the main domestic market becomes less crucial—and thus the wage premium that firms are willing to pay for a metropolitan location falls. At the same time, workers will consume more imported goods; they will therefore be less willing to accept high commuting and land costs in order to be close to the metropolitan suppliers. The result can be to make a previously sustainable metropolitan concentration unsustainable.

---

[4] In this story all workers end up concentrated in one location, earning the same wage. It thus appears to be a model of urban concentration but not of regional inequality. Suppose, however, that not all workers are mobile. Then it is apparent that a core-periphery pattern could emerge in which mobile workers agglomerate in one region, leaving behind an impoverished rump of those workers who for whatever reason cannot or will not move. Stories along the same lines can surely also be relevant to the extreme regional inequality one sees in some developing countries.

The easiest way to confirm this intuition is through numerical examples. Figures 1 and 2 show, for one set of parameters, how the qualitative behavior of our two-location model changes as the economy becomes more open (that is, as the cost of shipping goods to and from the world falls). Each figure shows how equilibrium real wage rates in the two locations vary as the share of the labor force in location 1 changes. If we assume that workers move toward whichever location offers the higher real wage rate, these figures show a picture of the economy's dynamic behavior. When the real wage differential is positive, labor moves toward location 1; when it is negative, labor moves toward location 2.

When the costs of transacting with the outside world are fairly high, so that the economy is not very open, there is an equilibrium, though unstable, in which labor is equally divided between the two locations (figure 1). If slightly more than half the labor is in location 1, that location will offer higher wages, inducing more labor to move there. This will strengthen the forward and backward linkages and induce still more labor to move there, and so on. Thus, in this closed-economy case, a cumulative process leads to a concentration of population in a single metropolis. (Obviously this result does not fully obtain in practice, but perhaps it suggests how a very large primate city is established.)

FIGURE 1
Response of Labor Force to Relative Wages under High Costs of Transacting with Outside World

If the economy is more open, we get a result like that in figure 2.[5] Now the equilibrium in which the population is equally divided between the two locations is stable, and a concentration of population in only one location is unsustainable. Thus, in this situation we tend to have two equal-size cities rather than one very large metropolis.

It is, of course, obvious that Mexican industry has been shifting its center of gravity away from Mexico City as the country has shifted toward exports. In that case, however, the explanation lies at least partly in the role of access to the U.S. border, as well as in the role of the *maquiladora* program in fostering export industry in the country's north. Our analysis suggests, however, a more generic reason why inward-looking policies may encourage the growth of primate cities, and outward-looking policies may discourage that growth; the empirical evidence described above offers at least modest support for the belief that such a generic tendency exists.

Political Centralization and Regional Inequality

While the theoretical and empirical relationship between trade policy and urban structure is a surprising, and thus gratifying, insight, it is surely not the most important reason why developing-country cities grow so large,

FIGURE 2
Response of Labor Force to Changes in Relative Wages under Relatively Low Costs of Transacting with Outside World

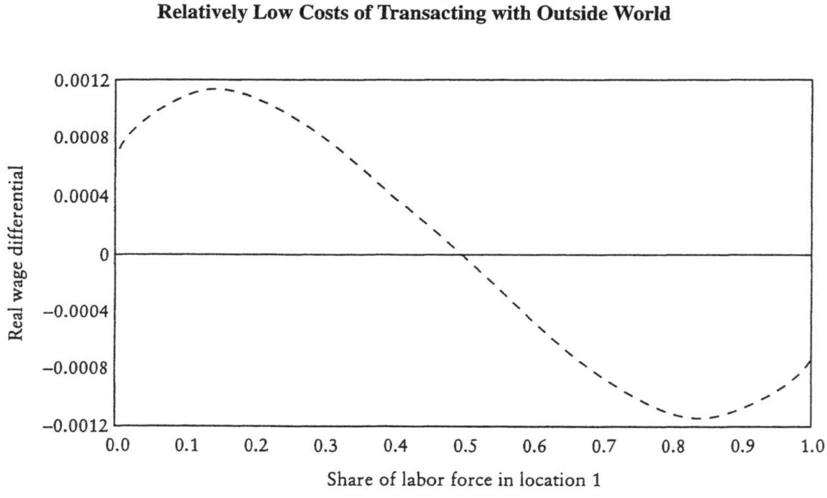

[5] Krugman and Livas Elizondo (1992) show that there may be a range of parameters in which both dispersed and concentrated stable equilibriums exist. Numerical examples suggest, however, that this range is quite narrow.

or why regional inequality is so marked in developing countries. Almost surely the most important reason is the role of political centralization.

Political centralization has effects at several levels. The most obvious is that the business of government is itself a substantial source of employment: employment in Paris is larger than it is in Frankfurt in part simply because there are so many more people working for the government, or supplying nontraded services to those who work for the government.

A more subtle source of urban concentration is the importance of access to the government, especially in highly interventionist states. In its simplest form, this is simply a result of the concentration of lobbyists. More subtly, if government policies tend to be more responsive to those close at hand (if, say, subsidies or protection to prevent strikes are more forthcoming in the capital than in the provinces), this exerts a hard-to-measure but doubtless important attraction of the capital area for business.

Economic modeling per se cannot contribute much to our understanding of these political concerns. It can, however, help us understand a further consequence of political centralization: the multiplier effects on regional concentration that can result from asymmetric government spending.

Consider a variant on the approach described in the last two sections. Put the commuting and land-rent diseconomies to the side and suppose instead that there is an immobile rural population divided between two regions. Manufacturing will be drawn to concentrate in one region by the forward and backward linkages we have already seen in action, but against this force will be the pull of the market provided by the rural population. A model along exactly these lines is worked out in Krugman (1991b). I show there that the outcome depends on the parameters. For some parameters one gets the type of result shown by the dashed curve in figure 3: the stable equilibrium is one in which manufacturing is equally divided between the two regions.

But now suppose that a government collects taxes from the rural population in both regions but spends it all in one region. Obviously, the latter region becomes the larger market, thus attracting more manufacturers. However, the forward and backward linkages that are generated attract still more manufacturing to that region, fostering a cumulative process of concentration. In figure 3 we start with an economy in which the natural state of affairs has 50 percent of manufacturing in each region. In this example a tax equal to 20 percent of rural income was collected in both regions, but spent only in region 1. The result is shown by the upward shift in the schedule relating the real wage differential to the allocation of manufacturing between the regions (the dotted curve). In this case the multiplier effects cause a concentration of approximately 85 percent of manufacturing in the favored region. The direct transfer of resources from the periphery to the core is only

FIGURE 3
Response of Manufacturing to Relative Wages

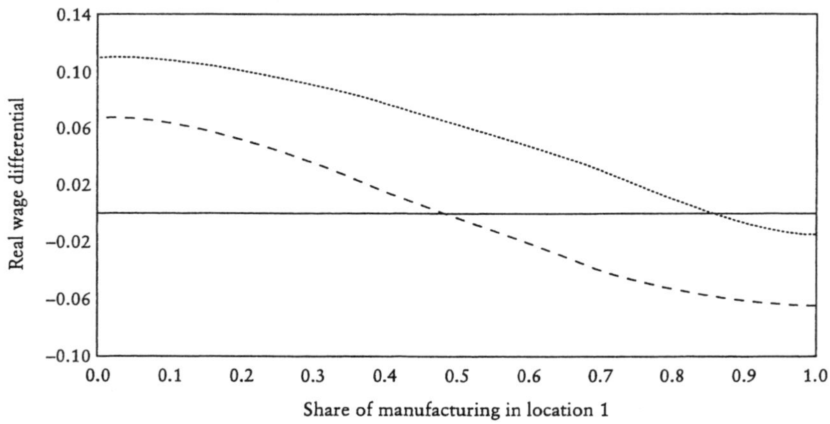

8 percent of GDP, but the end result is to raise the favored region's share of GDP (before taxes) from 50 to 74 percent.

Although it is not explicitly modeled here, there ought to be an interaction between the strength of multiplier effects producing regional concentration and the degree of openness of the economy. Locating manufacturing near the capital in order to take advantage of the market that the government and its employees provide will be much less attractive in a very open than in a very closed economy.

Transportation Infrastructure

The extent and form of a country's investments in transportation infrastructure can affect the tendency to form large urban centers in at least two ways.

First, the higher transport costs are within a country, the stronger the advantages in terms of backward and forward linkages of locating production near an established metropolitan concentration. This effect may be seen directly in equation (5), which asks whether the linkages are strong enough to sustain an established concentration in the face of the diseconomies of urban scale. In this expression, the higher are the transport costs, the more likely is the condition for sustainability to be satisfied.

The implication is that the tendency to concentrate economic activity in a single large city may be reinforced if the government neglects the transportation network. This makes intuitive sense, and corresponds to workaday perceptions about the contrast between location decisions in advanced and developing economies. In advanced economies, good transportation to markets (and good communications) is available virtually everywhere, whereas in developing countries, roads and telecommunications often peter out quickly as one moves away from the capital.

A more subtle issue involves the form of the transport system. A system that is centered on the primate city is more likely to promote concentration than one that does not favor movement of goods and services in any particular direction.

This point also seems intuitively obvious, but it may be worth sketching out how it works in formal models. Imagine, as in Krugman (1993a), a country with not two but three regions. And suppose that instead of being equal in all directions, transport costs between location 1 and both other locations are lower than those between 2 and 3, so that 1 is in effect the hub of the transport system. Then it is straightforward to show that even if all three regions offer the same size market, region 1 will be a preferred location for goods produced subject to scale economies: it offers better access to the national market than does either of the other locations. Of course, such an advantage will not usually stand alone. Typically, concentration of population and centralization of the transport system reinforce one another: transport links point toward the primate city because that is where the markets and suppliers are, and business concentration is all the greater because of the role of that city as a transport hub.

One might speculate that the apparent tendency of developing countries to have more concentrated distributions of urban size is due to an important extent to the way that their relative poverty leads to a limited transport system. In advanced countries, the volume of traffic is sufficient to ensure that good roads link even minor centers; railway lines will often provide direct connections that bypass the biggest cities.[6] In developing countries, traffic is sufficient to support good roads pointing only toward the capital, if any at all. Here, too, there is probably a political linkage—a system that centralizes political power in the capital is likely to concentrate investment in infrastructure either near it or on projects that serve it.

Policy Implications

One wants to be very careful about drawing policy implications from any discussion of urbanization and regional growth. By its nature, this is a subject that deals extensively with external economies and diseconomies; while neoclassical urban systems theory may suggest that competition

---

[6] Cronon (1990) shows that the rapid growth of Chicago in the nineteenth century came to an end largely because the growing density of the U.S. rail network made its traditional position as a rail hub increasingly unimportant.

among city developers yields optimal results, the newer literature does not contain any such suggestion. Yet the extent and even the direction of the deviations from optimality may be sensitive to the particular form of the external effects. One could in principle argue that because the growth of cities necessarily involves positive external economies, the biggest cities tend to be too small. Or one could argue that the diseconomies of congestion and pollution—or the inability of markets to internalize the benefits of creating new cities—mean that primate cities are too big. Most people have an instinctive feeling that the biggest cities are too big. I share that prejudice, but it must be said that it is only a prejudice at this point.

That said, the general moral of the models described here seems to be that a desire for cities in developing countries to be not quite so big may be fulfilled indirectly by the kinds of liberal economic policies currently favored by most international institutions for other reasons. Liberal trade policy appears likely to discourage primate city growth; so does a reduction in state intervention and a decentralization of power. Investment in better transportation infrastructure—a traditional role of government—also seems to work in the same direction.

The tentative conclusion, then, is that neoliberal policies seem likely to have the unexpected side benefit of partly alleviating any problems created by the growth of very large cities. The definite conclusion is that whatever the changes made in economic policies, their implications for urban and regional development within countries are an important, neglected issue.

References

Abdel-Rahman, H. 1988. Product differentiation, monopolistic competition, and city size. *Regional Science and Urban Economics* 18: 69–86.

Ades, A., and E. Glaeser. 1993. Trade and circuses: Explaining urban giants. Cambridge: Harvard University, Department of Economics, unpublished.

Bairoch, P. 1988. *Cities in economic development.* Chicago: University of Chicago Press.

Bak, Per. 1991. Self-organizing criticality. *Scientific American* 264 (January): 46–53.

Carroll, G. 1982. City-size distributions: What do we know after 67 years of research? *Progress in Human Geography* 6: 1–43.

Cronon, W. 1990. *Nature's metropolis: Chicago and the great West.* New York: Norton.

Fujita, M., and P. Krugman. 1993. Monopolistic competition and systems of cities. Philadelphia: University of Pennsylvania, Department of Regional Sciences, unpublished.

Glaeser, E., H. D. Kallal, J. Scheinkman, and A. Shleifer. 1992. Growth in cities. *Journal of Political Economy* 100: 1126–52.

Henderson, J. V. 1974. The sizes and types of cities. *American Economic Review* 64: 640–56.

Henderson, J. V. 1977. *Economic theory and the cities.* Oxford: Oxford University Press.

Henderson, J. V. 1988. *Urban development.* Oxford: Oxford University Press.

Ijiri, Y., and H. Simon. 1977. *Skew distributions and the sizes of business firms.* Amsterdam: North-Holland.

Krugman, P. 1991a. *Geography and trade.* Cambridge: MIT Press.

Krugman, P. 1991b. Increasing returns and economic geography. *Journal of Political Economy* 99: 483–99.

Krugman, P. 1993a. The hub effect: Or, threeness in interregional trade. In *Theory, policy, and dynamics in international trade*, eds. W. Ethier, E. Helpman, and J. P. Neary. Cambridge: Cambridge University Press.

Krugman, P. 1993b. On the number and location of cities. *European Economic Review* 37: 293–98.

Krugman, P. 1996. *The self-organizing economy.* London: Basil Blackwell.

Krugman, P., and R. Livas Elizondo. 1992. *Trade policy and the Third World metropolis.* Cambridge: National Bureau of Economic Research Working Paper 4238.

Pred, A. 1966. *The spatial dynamics of U.S. urban-industrial growth, 1800–1914.* Cambridge: MIT Press.

Rivera-Batiz, F. 1988. Increasing returns, monopolistic competition, and agglomeration economies in consumption and production. *Regional Science and Urban Economics* 18: 125–53.

Rosen, K., and M. Resnick. 1980. The size distribution of cities: An examination of the pareto law and primacy. *Journal of Urban Economics* 8: 165–86.

Simon, H. 1955. On a class of skew distribution functions. *Biometrika* 52: 425–40.

Appendix

In this appendix, I present the formal structure of the model of the determinants of urban concentration sketched out in the second section of the article and illustrated in the third. For a full description of how the model is solved, and an exploration of its properties, see Krugman and Livas Elizondo (1992).

A Formal Model of Urban Concentration

We consider an economy with three locations: 0, 1, and 2. Labor is mobile between 1 and 2, but not from the rest of world.

Each location is a linear city, populated by workers who must work in a central business district but require one unit of land to live on. Thus, if a location has a labor force L_i, the distance the last worker must commute is

$$d_j = L_j/2 . \tag{A.1}$$

We assume that commuting costs are incurred in labor: a worker is endowed with one unit of labor, but if he must commute a distance d, he arrives with a net amount of labor to sell of only

$$S = 1 - 2\gamma d . \tag{A.2}$$

These assumptions immediately allow us to describe the determination of land rent given the labor force at a location. Let w_j be the wage rate paid at the city center per unit of labor. Workers who live at the outskirts of the town will pay no land rent, but will receive a net wage of only $(1-\gamma L_j)\, w_j$ because of the time spent in commuting. Workers who live closer to the city center will receive more money, but must pay an offsetting land rent. The wage net of commuting costs declines as one moves away from the city center, but land rents always exactly offset the differential. Thus, the wage net of both commuting and land rents is $(1-\gamma L_j)\, w_j$ for all workers.

The total labor input of a location, net of commuting costs, is

$$Z_j = L_j(1 - 0.5\gamma L_j) \tag{A.3}$$

and the location's total income—*including* the income of landowners—is

$$Y_j = w_j Z_j . \tag{A.4}$$

Next, we assume that everyone in the economy shares the constant elasticity of substitution utility function

$$U = \left(\Sigma_i C_i^{\frac{\sigma-1}{\sigma}} \right)^{\frac{\sigma}{\sigma-1}} . \tag{A.5}$$

To produce any good i at location j involves a fixed as well as a variable cost:

$$Z_{ij} = \alpha + \beta Q_{ij}. \quad \text{(A.6)}$$

The properties of monopolistic competition models like this are by now very familiar. As long as many goods are produced, and as long as we make appropriate assumptions on transportation costs (see below), each producer faces an elasticity of demand equal to the elasticity of substitution, and will therefore charge a price that is a constant markup over marginal cost:

$$P_j = (\sigma/\sigma - 1)\, \beta w_j. \quad \text{(A.7)}$$

Given this pricing rule and the assumption that free entry will drive profits to zero, there is a unique zero-profit output of each product:

$$Q = (\alpha/\beta)\,(\sigma - 1). \quad \text{(A.8)}$$

And the constancy of output of each product implies that the number of goods produced at each location is simply proportional to its net labor input after commuting:

$$n_j = (Z_j/\alpha\sigma). \quad \text{(A.9)}$$

It will save notation to make two useful choices of units. First, units are chosen to make the f.o.b. price of goods produced at any given location equal to the wage rate at the region's city center. Thus:

$$P_j = w_j. \quad \text{(A.10)}$$

Second, there is no need to count goods one at a time. They can be equally well counted in batches, say, of a dozen each. To save notation, the batch size is such that

$$n_j = Z_j. \quad \text{(A.11)}$$

To preserve the constant elasticity of demand facing firms, the costs of transacting between locations must take Samuelson's "iceberg" form, in which transport costs are incurred in the goods shipped. Thus, we assume that when a unit of any good is shipped between location 1 and location 2, only $1/\tau$ units actually arrive; thus, the c.i.f. price of a good shipped from either domestic location to the other is τ times its f.o.b. price. Only a fraction $1/\rho$ of a good imported from location 0 is assumed to arrive in either location 1 or 2. For simplicity, exports are assumed to take place with zero transport costs.[1]

We take τ to represent "natural" transport costs between locations. The parameter ρ, however, is meant to be interpreted as combining natural transport costs with artificial trade barriers. It would be straightforward (and would yield similar results) in this model to introduce an explicit ad valorem tariff whose proceeds are redistributed, but here we simply imagine that any potential revenue is somehow dissipated in waste of real resources.

Given these transport costs and the utility function, we may define true consumer price indexes for manufactured goods in each location. First, let us define the shares of the three locations in the total number of products produced, which are equal to their shares of net labor input:

$$\lambda_j = \frac{n_j}{\Sigma_k n_k} = \frac{Z_j}{\Sigma_k Z_k}. \tag{A.12}$$

Let the wage rate in location 0 be the numeraire; then the true price indices are

$$T_0 = K(\lambda_0 + \lambda_1 w_1^{1-\sigma} + \lambda_2 w_2^{1-\sigma})^{\frac{1}{1-\sigma}} \tag{A.13}$$

$$T_1 = K\left[\lambda_0 \rho^{1-\sigma} + \lambda_1 w_1^{1-\sigma} + \lambda_2 (w_2 \tau)^{1-\sigma}\right]^{\frac{1}{1-\sigma}} \tag{A.14}$$

$$T_2 = K\left[\lambda_0 \rho^{1-\sigma} + \lambda_1 (w_1 \tau)^{1-\sigma} + \lambda_2 w_2^{1-\sigma}\right]^{\frac{1}{1-\sigma}} \tag{A.15}$$

where

$$K = (n_0 + n_1 + n_2)^{\frac{1}{1-\sigma}}. \tag{A.16}$$

We will take Z_0 as given. Suppose we know the allocation of labor between locations 1 and 2. Then we can determine Z_1 and Z_2. As we will see, we can then solve the model for equilibrium wage rates w_j. Labor is, however, mobile, and we will have a full equilibrium only if all domestic workers receive the same net real wage. This net real wage in location j can be defined as

$$\omega_j = w_j(1 - \gamma L_j)/T_j. \tag{A.17}$$

[1] Even though we make exports costless, an increase in ρ, which reduces imports, must necessarily decrease exports as well. The mechanism through which this happens is a rise in the prices of domestic relative to foreign output—in effect, a real overvaluation that prices domestic goods out of world markets.

A situation in which real wages are equal in the two domestic locations is an equilibrium. Such an equilibrium may, however, be unstable under any plausible adjustment story. To get some rudimentary dynamics, we impose a simple Marshallian adjustment mechanism,

$$\mathrm{d}L_1/\mathrm{d}t = -\mathrm{d}L_2/\mathrm{d}t = \delta(\omega_1 - \omega_2). \tag{A.18}$$

We have now laid out a complete formal model. It is not a model with a closed-form analytical solution. However, if one is willing to rely on numerical examples, it is straightforward to solve the equations on the computer for any given parameters and see how the wage differential depends on the allocation of labor between the two locations, thereby deriving diagrams like figures 1 to 3. As explained in the text, such pictures allow us to see how the patterns of urban or regional concentration change as the parameters change.

City Growth and Power Laws

As mentioned in the text, the size distribution of cities is startlingly well described by a power law of the form

$$N(S) = AS^{-\alpha} \tag{A.19}$$

where $N(S)$ is the number of cities with populations larger than S, and the exponent is very close to -1. (As an illustration, figure A.1 plots the log of metropolitan area rank against the log of city population for the United States in 1991.)

If the distribution of cities were continuous and there were no maximum city size, equation (A.19) would be equivalent to saying that the density of cities of size S is

$$n(S) = \alpha AS^{-\alpha-1}. \tag{A.20}$$

Now imagine that cities come only in discrete sizes, with units of, say, 10,000 people. Let k be the number of units in the population, and $n(i,k)$ be the number of cities with i units; then equation (A.19) with an exponent of -1 becomes the statement that

$$n(i,k) \approx B(k)\, i^{-2}. \tag{A.21}$$

Why should something like equation (A.21) be true? In 1955 Herbert Simon offered an ingenious explanation, which is a bit short of a formal proof. I offer here a heuristic version of Simon's argument, which is in turn less than rigorous, so it should be viewed only as a suggestive justification.

Imagine that urban growth proceeds according to the following process: new units arrive in the economy successively over time; each new unit is attached to an existing city with a probability that is proportional to the number of units already there. (In Simon's original formulation, some units

FIGURE A.1
Relation of U.S. City Rank and City Size

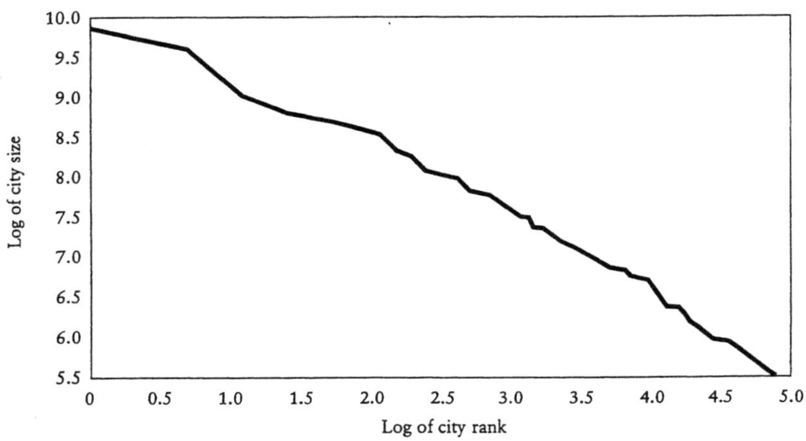

form the nuclei of new cities; I return to that issue below.) Thus, a city of size i has a probability i/k of getting the next unit.

What is the expected change in the number of cities of size i when a new unit is added? That number can change in two ways. First, a city of size $i-1$ can acquire the new unit, in which case it becomes a city of size i, adding 1 to the total. Second, a city of size i can acquire the unit, in which case it becomes a city of size $i+1$, reducing the number of i cities. It therefore follows that

$$E[\Delta n(i,k)] = \frac{(i-1)\,n(i-1,k)}{k} - \frac{in(i,k)}{k}. \tag{A.22}$$

Now comes the crucial ad hoc step. Simon asks us to imagine that the frequency distribution of city sizes approaches a steady state. This cannot be quite right, since the largest city keeps on getting bigger. But suppose that it is approximately true. Then the number of cities of size i must grow at the same rate as the population, implying

$$E[\Delta n(i,k)] = \frac{n(i,k)}{k}. \tag{A.23}$$

From equations (A.22) and (A.23) it follows that

$$\frac{n(i,k)}{n(i-1,k)} = \frac{i-1}{i+1} \tag{A.24}$$

and thus that

$$n(i,k) = \frac{i-1}{i+1}\frac{i-2}{i}\cdots\frac{1}{3}n(1,k) \tag{A.25}$$

or

$$n(i,k) = \frac{2}{i(i+1)}n(1,k) \approx 2n(1,k)\,i^{-2} \tag{A.26}$$

for large i.

That is, in the upper tail of the size distribution, equation (A.21) should be approximately true!

This derivation is a bit slippery. It can be bolstered, however, by simulation results; these show that a wide variety of stochastic growth models will produce upper tails for which equation (A.21) is very close to true. For example, in Krugman (1996) I consider a model of the following form. A city is begun by an entrepreneur who starts a business. She has two foremen, each of whom with probability p leaves to set up a new factory in the same town. Each foreman has two foremen. Suppose that the probability of defection is close to 0.5, as it must be if towns are to grow very large. Then the results are startling. In figure A.2, I started with 1,000 original businesses, and set $p = 0.49$; the figure shows the relationship between rank and size for the top 100 "cities."

There is a close affinity between Simon's work and the trendy current work on "self-organized criticality," which attempts to explain such observed power-law relationships as the Gutenberg-Richter law relating the sizes and frequencies of earthquakes (Bak 1991).

FIGURE A.2
Relation of Simulated City Rank and Size, Top 100 Cities

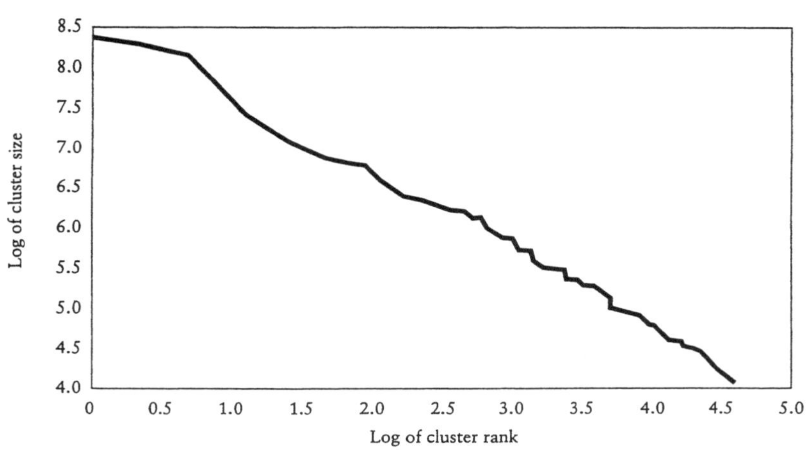

Part III
New Growth Theory

[6]

Increasing Returns and Long-Run Growth

Paul M. Romer

University of Rochester

This paper presents a fully specified model of long-run growth in which knowledge is assumed to be an input in production that has increasing marginal productivity. It is essentially a competitive equilibrium model with endogenous technological change. In contrast to models based on diminishing returns, growth rates can be increasing over time, the effects of small disturbances can be amplified by the actions of private agents, and large countries may always grow faster than small countries. Long-run evidence is offered in support of the empirical relevance of these possibilities.

I. Introduction

Because of its simplicity, the aggregate growth model analyzed by Ramsey (1928), Cass (1965), and Koopmans (1965) continues to form the basis for much of the intuition economists have about long-run growth. The rate of return on investment and the rate of growth of per capita output are expected to be decreasing functions of the level of the per capita capital stock. Over time, wage rates and capital-labor ratios across different countries are expected to converge. Consequently, initial conditions or current disturbances have no long-run effect on the level of output and consumption. For example, an exog-

This paper is based on work from my dissertation (Romer 1983). An earlier version of this paper circulated under the title "Externalities and Increasing Returns in Dynamic Competitive Analysis." At various stages I have benefited from comments by James J. Heckman, Charles M. Kahn, Robert G. King, Robert E. Lucas, Jr., Sergio Rebelo, Sherwin Rosen, José A. Scheinkman (the chairman of my thesis committee), and the referees. The usual disclaimer applies. I gratefully acknowledge the support of NSF grant no. SES-8320007 during the completion of this work.

[*Journal of Political Economy*, 1986, vol 94, no 5]

enous reduction in the stock of capital in a given country will cause prices for capital assets to increase and will therefore induce an offsetting increase in investment. In the absence of technological change, per capita output should converge to a steady-state value with no per capita growth. All these presumptions follow directly from the assumption of diminishing returns to per capita capital in the production of per capita output.

The model proposed here offers an alternative view of long-run prospects for growth. In a fully specified competitive equilibrium, per capita output can grow without bound, possibly at a rate that is monotonically increasing over time. The rate of investment and the rate of return on capital may increase rather than decrease with increases in the capital stock. The level of per capita output in different countries need not converge; growth may be persistently slower in less developed countries and may even fail to take place at all. These results do not depend on any kind of exogenously specified technical change or differences between countries. Preferences and the technology are stationary and identical. Even the size of the population can be held constant. What is crucial for all of these results is a departure from the usual assumption of diminishing returns.

While exogenous technological change is ruled out, the model here can be viewed as an equilibrium model of endogenous technological change in which long-run growth is driven primarily by the accumulation of knowledge by forward-looking, profit-maximizing agents. This focus on knowledge as the basic form of capital suggests natural changes in the formulation of the standard aggregate growth model. In contrast to physical capital that can be produced one for one from forgone output, new knowledge is assumed to be the product of a research technology that exhibits diminishing returns. That is, given the stock of knowledge at a point in time, doubling the inputs into research will not double the amount of new knowledge produced. In addition, investment in knowledge suggests a natural externality. The creation of new knowledge by one firm is assumed to have a positive external effect on the production possibilities of other firms because knowledge cannot be perfectly patented or kept secret. Most important, production of consumption goods as a function of the stock of knowledge and other inputs exhibits increasing returns; more precisely, knowledge may have an increasing marginal product. In contrast to models in which capital exhibits diminishing marginal productivity, knowledge will grow without bound. Even if all other inputs are held constant, it will not be optimal to stop at some steady state where knowledge is constant and no new research is undertaken.

These three elements—externalities, increasing returns in the production of output, and decreasing returns in the production of new knowledge—combine to produce a well-specified competitive equilibrium model of growth. Despite the presence of increasing returns, a competitive equilibrium with externalities will exist. This equilibrium is not Pareto optimal, but it is the outcome of a well-behaved positive model and is capable of explaining historical growth in the absence of government intervention. The presence of the externalities is essential for the existence of an equilibrium. Diminishing returns in the production of knowledge are required to ensure that consumption and utility do not grow too fast. But the key feature in the reversal of the standard results about growth is the assumption of increasing rather than decreasing marginal productivity of the intangible capital good knowledge.

The paper is organized as follows. Section II traces briefly the history of the idea that increasing returns are important to the explanation of long-run growth and describes some of the conceptual difficulties that impeded progress toward a formal model that relied on increasing returns. Section III presents empirical evidence in support of the model proposed here. Section IV presents a stripped-down, two-period version of the model that illustrates the tools that are used to analyze an equilibrium with externalities and increasing returns. Section V presents the analysis of the infinite-horizon, continuous-time version of the model, characterizing the social optimum and the competitive equilibrium, both with and without optimal taxes.

The primary motivation for the choice of continuous time and the restriction to a single state variable is the ease with which qualitative results can be derived using the geometry of the phase plane. In particular, once functional forms for production and preferences have been specified, useful qualitative information about the dynamics of the social optimum or the suboptimal competitive equilibrium can be extracted using simple algebra. Section VI presents several examples that illustrate the extent to which conventional presumptions about growth rates, asset prices, and cross-country comparisons may be reversed in this kind of economy.

II. Historical Origins and Relation to Earlier Work

The idea that increasing returns are central to the explanation of long-run growth is at least as old as Adam Smith's story of the pin factory. With the introduction by Alfred Marshall of the distinction between internal and external economies, it appeared that this explanation could be given a consistent, competitive equilibrium interpretation. The most prominent such attempt was made by Allyn Young in his 1928 presidential address to the Economics and Statistics section of the British Association for the Advancement of Science

(Young 1969). Subsequent economists (e.g., Hicks 1960; Kaldor 1981) have credited Young with a fundamental insight about growth, but because of the verbal nature of his argument and the difficulty of formulating explicit dynamic models, no formal model embodying that insight was developed.

Because of the technical difficulties presented by dynamic models, Marshall's concept of increasing returns that are external to a firm but internal to an industry was most widely used in static models, especially in the field of international trade. In the 1920s the logical consistency and relevance of these models began to be seriously challenged, in particular by Frank Knight, who had been a student of Young's at Cornell.[1] Subsequent work demonstrated that it is possible to construct consistent, general equilibrium models with perfect competition, increasing returns, and externalities (see, e.g., Chipman 1970). Yet Knight was at least partially correct in objecting that the concept of increasing returns that are external to the firm was vacuous, an "empty economic box" (Knight 1925). Following Smith, Marshall, and Young, most authors justified the existence of increasing returns on the basis of increasing specialization and the division of labor. It is now clear that these changes in the organization of production cannot be rigorously treated as technological externalities. Formally, increased specialization opens new markets and introduces new goods. All producers in the industry may benefit from the introduction of these goods, but they are goods, not technological externalities.[2]

Despite the objections raised by Knight, static models of increasing returns with externalities have been widely used in international trade. Typically, firm output is simply assumed to be increasing, or unit cost decreasing, in aggregate industry output. See Helpman (1984) for a recent survey. Renewed interest in dynamic models of growth driven by increasing returns was sparked in the 1960s following the publication of Arrow's (1962) paper on learning by doing. In his model, the productivity of a given firm is assumed to be an increasing function of cumulative aggregate investment for the industry. Avoiding the issues of specialization and the division of labor, Arrow argued that increasing returns arise because new knowledge is discovered as investment and production take place. The increasing returns were external to individual firms because such knowledge became publicly known.

To formalize his model, Arrow had to face two problems that arise

[1] For an account of the development of Young's ideas and of his correspondence with Knight, see Blitch (1983).

[2] For a treatment of increasing returns based on specialization, see Ethier (1982). Although the model there is essentially static, it demonstrates how specialization can be introduced in a differentiated products framework under imperfect competition.

in any optimizing model of growth in the presence of increasing returns. The first, familiar from static models, concerns the existence of a competitive equilibrium; as is now clear, if the increasing returns are external to the firm, an equilibrium can exist. The second problem, unique to dynamic optimizing models, concerns the existence of a social optimum and the finiteness of objective functions. In a standard optimizing growth model that maximizes a discounted sum or integral over an infinite horizon, the presence of increasing returns raises the possibility that feasible consumption paths may grow so fast that the objective function is not finite. An optimum can fail to exist even in the sense of an overtaking criterion. In the model of Arrow and its elaborations by Levhari (1966*a*, 1966*b*) and Sheshinski (1967), this difficulty is avoided by assuming that output as a function of capital and labor exhibits increasing returns to scale but that the marginal product of capital is diminishing given a fixed supply of labor. As a result, the rate of growth of output is limited by the rate of growth of the labor force. Interpreted as an aggregate model of growth (rather than as a model of a specific industry), this model leads to the empirically questionable implication that the rate of growth of per capita output is a monotonically increasing function of the rate of growth of the population. Like conventional models with diminishing returns, it predicts that the rate of growth in per capita consumption must go to zero in an economy with zero population growth.

The model proposed here departs from both the Ramsey-Cass-Koopmans model and the Arrow model by assuming that knowledge is a capital good with an increasing marginal product. Production of the consumption good is assumed to be globally convex, not concave, as a function of stock of knowledge when all other inputs are held constant. A finite-valued social optimum is guaranteed to exist because of diminishing returns in the research technology, which imply the existence of a maximum, technologically feasible rate of growth for knowledge. This is turn implies the existence of a maximum feasible rate of growth for per capita output. Over time, the rate of growth of output may be monotonically increasing, but it cannot exceed this upper bound.

Uzawa (1965) describes an optimizing growth model in which both intangible human capital and physical capital can be produced. In some respects, the human capital resembles knowledge as described in this paper, but Uzawa's model does not possess any form of increasing returns to scale. Instead, it considers a borderline case of constant returns to scale with linear production of human capital. In this case, unbounded growth is possible. Asymptotically, output and both types of capital grow at the same constant rate. Other optimizing models took the rate of technological change as exogenously given (e.g., Shell

1967*b*). Various descriptive models of growth with elements similar to those used here were also proposed during the 1960s (e.g., Phelps 1966; von Wiezsäcker 1966; Shell 1967*a*). Knowledge is accumulated by devoting resources to research. Production of consumption goods exhibits constant returns as a function of tangible inputs (e.g., physical capital and labor) and therefore exhibits increasing returns as a function of tangible and intangible inputs. Privately produced knowledge is in some cases assumed to be partially revealed to other agents in the economy. Because the descriptive models do not use explicit objective functions, questions of existence are generally avoided, and a full welfare analysis is not possible. Moreover, these models tend to be relatively restrictive, usually constructed so that the analysis could be carried out in terms of steady states and constant growth rate paths.

Continuous-time optimization problems with some form of increasing returns are studied in papers by Weitzman (1970), Dixit, Mirrlees, and Stern (1975), and Skiba (1978). Similar issues are considered for discrete-time models in Majumdar and Mitra (1982, 1983) and Dechert and Nishimura (1983). These papers differ from the model here primarily because they are not concerned with the existence of a competitive equilibrium. Moreover, in all these papers, the technical approach used to prove the existence of an optimum is different from that used here. They rely on either bounded instantaneous utility $U(c)$ or bounds on the degree of increasing returns in the problem; for example, the production function $f(k)$ is assumed to be such that $f(k)/k$ is bounded from above. The results here do not rely on either of these kinds of restrictions; in fact, one of the most interesting examples analyzed in Section VI violates both of these restrictions. Instead, the approach used here relies on the assumptions made concerning the research technology; the diminishing returns in research will limit the rate of growth of the state variable. A general proof that restrictions on the rate of growth of the state variable are sufficient to prove the existence of an optimum for a continuous-time maximization problem with nonconvexities is given in Romer (1986).

Because an equilibrium for the model proposed here is a competitive equilibrium with externalities, the analysis is formally similar to that used in dynamic models with more conventional kinds of externalities (e.g., Brock 1977; Hochman and Hochman 1980). It also has a close formal similarity to perfect-foresight Sidrauski models of money demand and inflation (Brock 1975) and to symmetric Nash equilibria for dynamic games (e.g., Hansen, Epple, and Roberds 1985). In each case, an equilibrium is calculated not by solving a social planning problem but rather by considering the maximization problem of an individual agent who takes as given the path of some endogenously determined aggregate variable. In the conventional analysis of externalities, the focus is generally on the social optimum and the set of taxes necessary to support it as a competitive equilibrium. While this question is addressed for this growth model, the discussion places more stress on the characterization of the competitive equilibrium without intervention since it is the most reasonable positive model of observed historical growth. One of the main contributions of this paper is to demonstrate how the analysis of this kind of suboptimal equilibrium can proceed using familiar tools like a phase plane even though the equations describing the equilibrium cannot be derived from any stationary maximization problem.

III. Motivation and Evidence

Because theories of long-run growth assume away any variation in output attributable to business cycles, it is difficult to judge the empirical success of these theories. Even if one could resolve the theoretical ambiguity about how to filter the cycles out of the data and to extract the component that growth theory seeks to explain, the longest available time series do not have enough observations to allow precise estimates of low-frequency components or long-run trends. When data aggregated into decades rather than years are used, the pattern of growth in the United States is quite variable and is apparently still influenced by cyclical movements in output (see fig. 1). Cross-country comparisons of growth rates are complicated by the difficulty of controlling for political and social variables that appear to strongly influence the growth process. With these qualifications in mind, it is useful to ask whether there is anything in the data that should cause economists to choose a model with diminishing returns, falling rates of growth, and convergence across countries rather than an alternative without these features.

Consider first the long-run trend in the growth rate of productivity or per capita gross domestic product (GDP). One revealing way to consider the long-run evidence is to distinguish at any point in time between the country that is the "leader," that is, that has the highest level of productivity, and all other countries. Growth for a country that is not a leader will reflect at least in part the process of imitation and transmission of existing knowledge, whereas the growth rate of the leader gives some indication of growth at the frontier of knowledge. Using GDP per man-hour as his measure of productivity, Maddison (1982) identifies three countries that have been leaders since 1700, the Netherlands, the United Kingdom, and the United States. Table 1 reports his estimates of the rate of growth of productivity in each country during the interval when it was the leader. When the

TABLE 1

PRODUCTIVITY GROWTH RATES FOR LEADING COUNTRIES

Lead Country	Interval	Annual Average Compound Growth Rate of GDP per Man-Hour (%)
Netherlands	1700–1785	−.07
United Kingdom	1785–1820	.5
United Kingdom	1820–90	1.4
United States	1890–1979	2.3

SOURCE.—Maddison (1982).

productivity growth rate is measured over intervals several decades long and compared over almost 3 centuries, the evidence clearly suggests that it has been increasing, not decreasing. The rate of growth of productivity increases monotonically from essentially zero growth in eighteenth-century Netherlands to 2.3 percent per year since 1890 in the United States.

Similar evidence is apparent from data for individual countries over shorter horizons. Table 2 reports growth rates in per capita GDP for the United States over five subperiods from 1800 to 1978. (The raw data used here are from Maddison [1979].) These rates also suggest a positive rather than a negative trend, but measuring growth rates over 40-year intervals hides a substantial amount of year-to-year or even decade-to-decade variation in the rate of growth. Figure 1 presents the average growth rate over the interval 1800–1839 (for which no intervening data are available) and for the subsequent 14 decades. Identifying a long-run trend in rates measured over decades is more problematical in this case, but it is straightforward to apply a simple nonparametric test for trend.

Table 3 reports the results of this kind of test for trend in the per capita rate of growth in GDP for several countries using raw data

TABLE 2

PER CAPITA GROWTH IN THE UNITED STATES

Interval	Average Annual Compound Growth Rate of Real per Capita GDP (%)
1800–1840	.58
1840–80	1.44
1880–1920	1.78
1920–60	1.68
1960–78	2.47

SOURCE.—Raw data are from Maddison (1979).

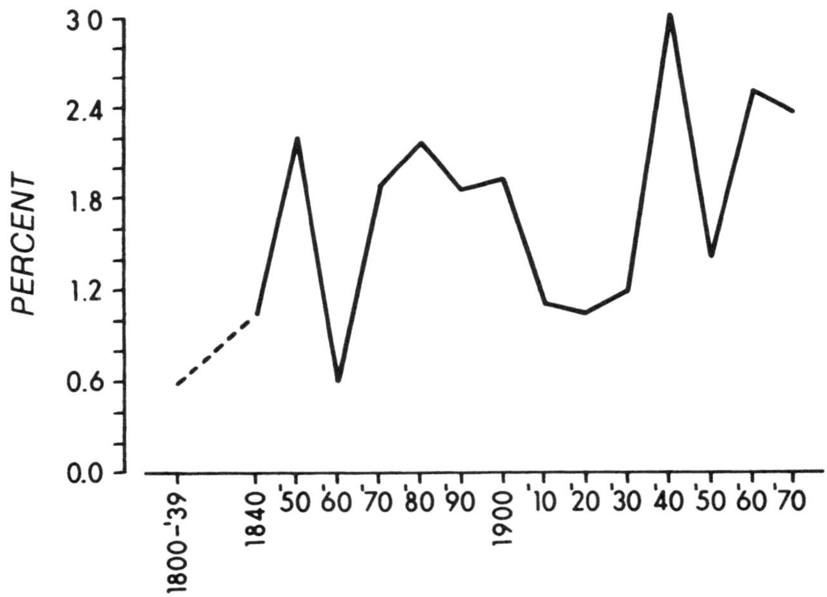

FIG. 1.—Average annual compound growth rate of per capita GDP in the United States for the interval 1800–1839 and for 14 subsequent decades. Data are taken from Maddison (1979).

from Maddison (1979). The sample includes all countries for which continuous observations on per capita GDP are available starting no later than 1870. As for the data for the United States graphed in figure 1, the growth rates used in the test for trend are measured over decades where possible. The statistic π gives the sample estimate of the probability that, for any two randomly chosen decades, the later decade has a higher growth rate.

Despite the variability evident from figure 1, the test for trend for the United States permits the rejection of the null hypothesis of a nonpositive trend at conventional significance levels. This is true even though growth over the 4 decades from 1800 to 1839 is treated as a single observation. However, rejection of the null hypothesis depends critically on the use of a sufficiently long data series. If we drop the observation on growth between 1800 and 1839, the estimate of π drops from .68 to .63 and the p-value increases from .03 to .11.[3] If we further restrict attention to the 11 decades from 1870 to 1978, π drops to .56 and the p-value increases to .29, so it is not surprising that studies that focus on the period since 1870 tend to emphasize the

[3] The p-value gives the probability of observing a value of π at least as large as the reported value under the null hypothesis that the true probability is .5.

TABLE 3

A Test for Trend in per Capita GDP Growth Rates

	Date of First Observation	Number of Observations	π	p-Value
United Kingdom	1700	20	.63	.06
France	1700	18	.69	.01
Denmark	1818	16	.70	.02
United States	1800	15	.68	.03
Germany	1850	13	.67	.06
Sweden	1861	12	.58	.25
Italy	1861	12	.76	.01
Australia	1861	12	.64	.11
Norway	1865	12	.81	.002
Japan	1870	11	.67	.07
Canada	1870	11	.64	.12

Note.—π is the sample estimate for each country of the probability that, for any two growth rates, the later one is larger. The p-value is the probability of observing a value of π at least as large as the observed value under the null hypothesis that the true probability is .5. Except in the early years when data are sparse, per capita rates of growth of GDP were measured over successive decades. (Only two observations on growth rates are available for France prior to 1820, for the United Kingdom, only two prior to 1800, for the United States, only one from 1800 to 1840.) For the calculation of the p-value, see Kendall (1962). Data are from Maddison (1979).

constancy of growth rates in the United States. Rejection does not appear to depend on the use of the rate of growth in per capita GDP rather than the rate of growth of productivity. Reliable measures of the work force prior to 1840 are not available, but using data from Kuznets (1971) for the period 1840–1960 and from the 1984 Economic Report of the President for 1960–80, one can construct a similar test for trend in the rate of growth of productivity over successive decades. The results of this test, π equal to .64 with a p-value of .10, correspond closely to those noted above for growth in per capita GDP over the similar interval, 1840–1978.

Over the entire sample of 11 countries, the estimated value for π ranges from .58 to .81, with a p-value that ranges from .25 to .002. Five out of 11 of the p-values are less than .05, permitting rejection at the 5 percent level in a one-sided test of the null hypothesis that there is a nonpositive trend in the growth rate; eight out of 11 permit rejection at the 10 percent level.

For less developed countries, no comparable long-run statistics on per capita income are available. Reynolds (1983) gives an overview of the pattern of development in such countries. Given the paucity of precise data for less developed countries, he focuses on the "turning point" at which a country first begins to exhibit a persistent upward trend in per capita income. The timing of this transition and the pace of subsequent growth are strongly influenced by the variations in the world economy. A general pattern of historically unprecedented growth for the world economy is evident starting in the last part of the 1800s and continuing to the present. This general pattern is interrupted by a significant slowdown during the years between the two world wars and by a remarkable surge from roughly 1950 to 1973. Worldwide growth since 1973 has been slow only by comparison with that surge and appears to have returned to the high rates that prevailed in the period from the late 1800s to 1914.

Although all less developed countries are affected by the worldwide economy, the effects are not uniform. For our purposes, the key observation is that those countries with more extensive prior development appear to benefit more from periods of rapid worldwide growth and suffer less during any slowdown. That is, growth rates appear to be increasing not only as a function of calendar time but also as a function of the level of development. The observation that more developed countries appear to grow relatively faster extends to a comparison of industrialized versus less developed countries as well. In the period from 1950 to 1980, when official estimates for GDP are generally available, Reynolds reports that the median rate of growth of per capita income for his sample of 41 less developed countries was 2.3 percent, "clearly below the median for the OECD countries for the same period" (p. 975).

If it is true that growth rates are not negatively correlated with the level of per capita output or capital, then there should be no tendency for the dispersion in the (logarithm of the)[4] level of per capita income to decrease over time. There should be no tendency toward convergence. This contradicts a widespread impression that convergence in this sense has been evident, especially since the Second World War. Streissler (1979) offers evidence about the source of this impression and its robustness. For each year from 1950 to 1974, he measures the variance across countries of the logarithm of the level of per capita income. In a sample of ex post industrialized countries, those countries with a level of per capita income of at least $2,700 in 1974, clear evidence of a decrease in the dispersion over time is apparent. In a sample of ex ante industrialized countries, countries with a per capita income of at least $350 in 1950, no evidence of a decrease in the variance is apparent. The first sample differs from the second because it includes Japan and excludes Argentina, Chile, Ireland, Puerto Rico, and Venezuela. As one would expect, truncating the sample at the end biases the trend toward decreasing dispersion (and

[4] Examining the dispersion in the logarithm of the level of per capita income, not dispersion in the level itself, is the correct way to test for convergence in the growth rates. If the rate of growth were constant across countries that start from different levels, the dispersion in the logarithm of the levels will stay constant, but dispersion in the levels will increase.

at the beginning toward increasing dispersion). When a sample of all possible countries is used, there is no evidence of a decrease in variance, but the interpretation of this result is complicated by the changing number of countries in the sample in each year due to data limitations.

Baumol (1985) reports similar results. When countries are grouped into industrialized, intermediate, centrally planned, and less developed economies, he argues that there is a tendency toward convergence in the level of productivity within groups, even though there is no tendency toward overall convergence. The tendency toward convergence is clear only in his group of industrialized economies, which corresponds closely to the sample of ex post industrialized countries considered by Streissler. In any case, he finds no obvious pattern in his entire sample of countries; if anything, there is a weak tendency toward divergence.[5]

The other kind of evidence that bears directly on the assumption of increasing returns in production comes from growth accounting exercises and the estimation of aggregate production functions. Economists believe that virtually all technical change is endogenous, the outcome of deliberate actions taken by economic agents. If so and if production exhibits constant returns to scale, one would expect to be able to account for the rate of growth of output in terms of the rates of growth of all inputs. The difficulty in implementing a direct test of this assertion lies in correctly measuring all the inputs to production, especially for intangible capital inputs such as knowledge. In a comprehensive attempt to account for the rates of growth in output in terms of rates of growth of all inputs, including human and nonhuman, tangible and intangible stocks of capital, Kendrick (1976) concluded that rates of growth of inputs are not sufficient to explain the rate of growth of output in the 40-year interval 1929–69. For various sectors and levels of aggregation, the rate of growth of output is 1.06–1.30 times the appropriate aggregate measure of the rate of growth for inputs. This kind of estimate is subject to substantial, unquantified uncertainty and cannot be taken as decisive support for the presence of increasing returns. But given the repeated failure of this kind of growth accounting exercise, there is no basis in the data for excluding the possibility that aggregate production functions are best described as exhibiting increasing returns.

[5] Baumol (1985) argues that the convergence he observes among the industrialized countries results from a transmission process for knowledge that takes place among the industrialized countries but does not extend to centrally planned or less developed countries. He would not agree that the apparent convergence is an artifact of an ex post choice of the industrialized countries. Since he does not treat this issue directly, it is difficult to resolve it from his data. He does admit that his groupings are "somewhat arbitrary."

IV. A Simple Two-Period Model

Even in the presence of increasing returns and externalities, calculating a social optimum is conceptually straightforward since it is equivalent to solving a maximization problem. Standard mathematical results can be used to show that a maximum exists and to characterize the solution by means of a set of necessary conditions. Despite the presence of global increasing returns, the model here does have a social optimum. The next section illustrates how it can be supported as a competitive equilibrium using a natural set of taxes and subsidies. This optimum is of theoretical and normative interest, but it cannot be a serious candidate for describing the observed long-run behavior of per capita output. To the extent that appropriate taxes and subsidies have been used at all, they are a quite recent phenomenon.

The model here also has an equilibrium in the absence of any governmental intervention. Much of the emphasis in what follows focuses on how to characterize the qualitative features of this suboptimal dynamic equilibrium. Although it is suboptimal, the competitive equilibrium does satisfy a constrained optimality criterion that can be used to simplify the analysis much as the study of the social optimization problem simplifies the analysis in standard growth models.

The use of a constrained or restricted optimization problem is not a new approach to the analysis of a suboptimal dynamic equilibrium. For example, it has been widely used in the perfect-foresight models of inflation. Nonetheless, it is useful to describe this method in some detail because previous applications do not highlight the generality of the approach and because the dynamic setting tends to obscure its basic simplicity. Hence, I start by calculating a competitive equilibrium for a greatly simplified version of the growth model.

Specifically, consider a discrete-time model of growth with two periods. Let each of S identical consumers have a twice continuously differentiable, strictly concave utility function $U(c_1, c_2)$, defined over consumption of a single output good in periods 1 and 2. Let each consumer be given an initial endowment of the output good in period 1. Suppose that production of consumption goods in period 2 is a function of the state of knowledge, denoted by k, and a set of additional factors such as physical capital, labor, and so forth, denoted by a vector $\mathbf{x}$.[6] To restrict attention to a choice problem that is essentially

[6] For most of the subsequent discussion, k will be treated as a stock of disembodied knowledge, i.e., knowledge in books. This is merely an expositional convenience and is not essential. For example, if one wants to assume that all knowledge is embodied in some kind of tangible capital such as conventional physical capital or human capital, k can be reinterpreted throughout as a composite good made up of both knowledge and the tangible capital good.

one-dimensional, assume that only the stock of knowledge can be augmented; the factors represented by $\mathbf{x}$ are available in fixed supply. To capture the basic idea that there is a trade-off between consumption today and knowledge that can be used to produce more consumption tomorrow, assume that there is a research technology that produces knowledge from forgone consumption in period 1. Because the economy here has only two periods, we need not be concerned with the problem that arises in an infinite-horizon model when consumption grows too fast and discounted utility goes to infinity. Thus we do not need diminishing returns in research to limit the rate of growth of knowledge, and we can choose a simple linear technology with units such that one unit of forgone consumption produces one unit of knowledge. A more realistic diminishing returns research technology is described in the infinite-horizon model presented in the next section.

Since newly produced private knowledge can be only partially kept secret and cannot be patented, we can represent the technology of firm i in terms of a twice continuously differentiable production function F that depends on the firm-specific inputs k_i and $\mathbf{x}_i$ and on the aggregate level of knowledge in the economy. If N is the number of firms, define this aggregate level of knowledge as $K = \Sigma_{i=1}^{N} k_i$.

The first major assumption on the production function $F(k_i, K, \mathbf{x}_i)$ is that, for any fixed value of K, F is concave as a function of k_i and $\mathbf{x}_i$. Without this assumption, a competitive equilibrium will not exist in general. Once concavity is granted, there is little loss of generality in assuming that F is homogeneous of degree one as a function of k_i and $\mathbf{x}_i$ when K is held constant; any concave function can be extended to be homogeneous of degree one by adding an additional factor to the vector $\mathbf{x}$ if necessary (Rockafellar 1970, p. 67). McKenzie (1959) refers to this additional factor as an entrepreneurial factor. It can be interpreted as an accounting device that transforms any profits into factor payments.

By the homogeneity of F in k_i and $\mathbf{x}_i$ and by the assumption that F is increasing in the aggregate stock of knowledge, K, it follows that F exhibits increasing returns to scale. For any $\psi > 1$,

$$F(\psi k_i, \psi K, \psi \mathbf{x}_i) > F(\psi k_i, K, \psi \mathbf{x}_i) = \psi F(k_i, K, \mathbf{x}_i).$$

The second major assumption strengthens this considerably. It requires that F exhibit global increasing marginal productivity of knowledge from a social point of view. That is, for any fixed $\mathbf{x}$, assume that $F(k, Nk, \mathbf{x})$, production per firm available to a dictator who can set economywide values for k, is convex in k, not concave. This strengthening of the assumption of increasing returns is what distinguishes the production function used here from the one used in the models of Arrow, Levhari, and Sheshinski.

The equilibrium for the two-period model is a standard competitive equilibrium with externalities. Each firm maximizes profits taking K, the aggregate level of knowledge, as given. Consumers supply part of their endowment of output goods and all the other factors $\mathbf{x}$ to firms in period 1. With the proceeds, they purchase output goods in period 2. Consumers and firms maximize taking prices as given. As usual, the assumption that agents treat prices and the aggregate level K as given could be rationalized in a model with a continuum of agents. Here, it is treated as the usual approximation for a large but finite number of agents. Because of the externality, all firms could benefit from a collusive agreement to invest more in research. Although this agreement would be Pareto-improving in this model, it cannot be supported for the same reasons that collusive agreements fail in models without externalities. Each firm would have an incentive to shirk, not investing its share of output in research. Even if all existing firms could be compelled to comply, for example, by an economywide merger, new entrants would still be able to free-ride and undermine the equilibrium.

Because of the assumed homogeneity of F with respect to factors that receive compensation, profits for firms will be zero and the scale and number of firms will be indeterminate. Consequently, we can simplify the notation by restricting attention to an equilibrium in which the number of firms, N, equals the number of consumers, S. Then per firm and per capita values coincide. Assuming that all firms operate at the same level of output, we can omit firm-specific subscripts.

Let $\bar{\mathbf{x}}$ denote the per capita (and per firm) endowment of the factors that cannot be augmented; let $\bar{e}$ denote the per capita endowment of the output good in period 1. To calculate an equilibrium, define a family of restricted maximization problems indexed by K:

$$
\begin{aligned}
P(K):\quad & \max_{k\in[0,\,\bar{e}]} && U(c_1, c_2) \\
& \text{subject to} && c_1 \le \bar{e} - k, \\
& && c_2 \le F(k, K, \mathbf{x}), \\
& && \mathbf{x} \le \bar{\mathbf{x}}.
\end{aligned}
$$

Since U is strictly concave and $F(k, K, \mathbf{x})$ is concave in k and $\mathbf{x}$ for each value of K, $P(K)$ will have a unique solution k for each value of K. (The solution for $\mathbf{x}$ is trivially $\bar{\mathbf{x}}$.) In general, the implied values for c_1, c_2, and k have no economic meaning. If K differs from Sk, then $F(k, K, \bar{\mathbf{x}})$ is not a feasible level of per capita consumption in period 2. Equilibrium requires that the aggregate level of knowledge that is achieved in the economy be consistent with the level that is assumed when firms make production decisions. If we define a function $\Gamma: \mathbb{R} \rightarrow \mathbb{R}$ that sends K into S times the value of k that achieves the maximum for the problem $P(K)$, this suggests fixed points of Γ as candidates for equilibria.

To see that any fixed point K^* of Γ can indeed be supported as a competitive equilibrium, observe that $P(K^*)$ is a concave maximization problem with solution $k^* = K^*/S$, $c_1^* = \bar{e} - k^*$, and $c_2^* = F(k^*, Sk^*, \bar{\mathbf{x}})$. Since it is concave, standard necessary conditions for concave problems apply. Let $\mathcal{L}$ denote a Lagrangian for $P(K^*)$ with multipliers p_1, p_2, and w:

$$\mathcal{L} = U(c_1, c_2) + p_1(\bar{e} - k - c_1) + p_2[F(k, K, \mathbf{x}) - c_2] + w(\bar{\mathbf{x}} - \mathbf{x}).$$

When an interior solution is assumed, familiar arguments show that $p_j = D_jU(c_1^*, c_2^*)$ for $j = 1, 2$, that $p_1 = p_2D_1F(k^*, Sk^*, \bar{\mathbf{x}})$, and that $w = p_2D_3F(k^*, Sk^*, \bar{\mathbf{x}})$.[7] As always, the shadow prices w and p_j can be interpreted as equilibrium prices. To see this, consider first the maximization problem of the firm: $\max_k p_2F(k, Sk^*, \mathbf{x}) - p_1k - w \cdot \mathbf{x}$. Since the firm takes both prices and the aggregate level Sk^* as given, a trivial application of the sufficient conditions for a concave maximization problem demonstrates that k^* and $\bar{\mathbf{x}}$ are optimal choices for the firm. By the homogeneity of F with respect to its first and third arguments, profits will be zero at these values. Consider next the problem of the consumer. Income to the consumer will be the value of the endowment, $I = p_1\bar{e} + w \cdot \bar{\mathbf{x}} = p_2F(k^*, Sk^*, \bar{\mathbf{x}}) + p_1(\bar{e} - k^*)$. (The second equality follows from the homogeneity of F in k and $\mathbf{x}$.) When the necessary conditions $p_j = D_jU(c_1^*, c_2^*)$ from the problem $P(K^*)$ are used, it follows immediately that c_1^* and c_2^* are solutions to the problem max $U(c_1, c_2)$ subject to the budget constraint $p_1c_1 + p_2c_2 \leq I$. Note that the marginal rate of substitution for consumers will equal the private marginal rate of transformation perceived by firms, $D_1U(c_1^*, c_2^*)/D_2U(c_1^*, c_2^*) = D_1F(k^*, Sk^*, \bar{\mathbf{x}})$. Because of the externality, this differs from the true marginal rate of transformation for the economy, $D_1F(k^*, Sk^*, \bar{\mathbf{x}}) + SD_2F(k^*, Sk^*, \bar{\mathbf{x}})$.

Arguments along these lines can be used quite generally to show that a fixed point of a mapping like Γ defined by a family of concave problems $P(K)$ can be supported as a competitive equilibrium with externalities. The necessary conditions from a version of the Kuhn-Tucker theorem generate shadow prices associated with any solution to $P(K)$. The sufficient conditions for the problems of the consumer and the firm can then be used to show that the quantities from the

[7] Here, D denotes a derivative, D_i the partial derivative with respect to the ith argument.

solution will be chosen in an equilibrium in which these prices are taken as given. Conversely, an argument similar to the usual proof of the Pareto optimality of competitive equilibrium can be used to show that any competitive equilibrium with externalities for this kind of economy will satisfy the restricted optimality condition implicit in the problem $P(K)$ (Romer 1983). That is, if K^* is an equilibrium value of aggregate knowledge, then K^*/S will solve the problem $P(K^*)$. Thus equilibria are equivalent to fixed points of the function Γ.

This allows an important simplification because it is straightforward to characterize fixed points of Γ in terms of the underlying functions U and F. Substituting the constraints from $P(K)$ into the objective and using the fact that $\mathbf{x}$ will be chosen to be $\bar{\mathbf{x}}$, define a new function $V(k, K) = U(\bar{e} - k, F(k, K, \bar{\mathbf{x}}))$. Because of the increasing marginal productivity of knowledge, V is not a concave function; but for any fixed K, it is concave in k. Then the optimal choice of k in any problem $P(K)$ is determined by the equation $D_1V(k, K) = 0$. Fixed points of Γ are then given by substituting Sk for K and solving $D_1V(k, Sk) = 0$. Given functional forms for U and F, this equation can immediately be written in explicit form. The analysis can therefore exploit a three-way equivalence between competitive equilibria with externalities, fixed points of Γ, and solutions to an explicit equation $D_1V(k, Sk) = 0$.

The key observation in this analysis is that equilibrium quantities can be characterized as the solution to a concave maximization problem. Then prices can be generated from shadow prices or multipliers for this problem. The complete statement of the problem must be sought simultaneously with its solution because the statement involves the equilibrium quantities. But since $P(K)$ is a family of concave problems, solving simultaneously for the statement of the problem and for its solution amounts to making a simple substitution in a first-order condition.

V. Infinite-Horizon Growth

A. *Description of the Model*

The analysis of the infinite-horizon growth model in continuous time proceeds exactly as in the two-period example above. Individual firms are assumed to have technologies that depend on a path $K(t)$, $t \geq 0$, for aggregate knowledge. For an arbitrary path K, we can consider an artificial planning problem $P_\infty(K)$ that maximizes the utility of a representative consumer subject to the technology implied by the path K. Assume that preferences over the single consumption good take the usual additively separable, discounted form, $\int_0^\infty U(c(t))e^{-\delta t}dt$, with $\delta >$

0. The function U is defined over the positive real numbers and can have $U(0)$ equal to a finite number or to $-\infty$, for example, when $U(c) = \ln(c)$. Following the notation from the last section, let $F(k(t), K(t), \mathbf{x}(t))$ denote the instantaneous rate of output for a firm as a function of firm-specific knowledge at time t, economywide aggregate knowledge at time t, and the level of all other inputs at t. As before, we will assume that all agents take prices as given and that firms take the aggregate path for knowledge as given.

Additional knowledge can be produced by forgoing current consumption, but the trade-off is no longer assumed to be one-for-one. By investing an amount I of forgone consumption in research, a firm with a current stock of private knowledge k induces a rate of growth $\dot{k} = G(I, k)$. The function G is assumed to be concave and homogeneous of degree one; the accumulation equation can therefore be rewritten in terms of proportional rates of growth, $\dot{k}/k = g(I/k)$, with $g(y) = G(y, 1)$. A crucial additional assumption is that g is bounded from above by a constant α. This imposes a strong form of diminishing returns in research. Given the private stock of knowledge, the marginal product of additional investment in research, Dg, falls so rapidly that g is bounded. An inessential but natural assumption is that g is bounded from below by the value $g(0) = 0$. Knowledge does not depreciate, so zero research implies zero change in k; moreover, existing knowledge cannot be converted back into consumption goods. As a normalization to fix the units of knowledge, we can specify that $Dg(0) = 1$; one unit of knowledge is the amount that would be produced by investing one unit of consumption goods at an arbitrarily slow rate.

Assume as before that factors other than knowledge are in fixed supply. This implies that physical capital, labor, and the size of the population are held constant. If labor were the only other factor in the model, exponential population growth could be allowed at the cost of additional notation; but as was emphasized in the discussion of previous models, a key distinguishing feature of this model is that population growth is not necessary for unbounded growth in per capita income. For simplicity it is left out. Allowing for accumulation of physical capital would be of more interest, but the presence of two state variables would preclude the simple geometric characterization of the dynamics that is possible in the case of one state variable. If knowledge and physical capital are assumed to be used in fixed proportions in production, the variable $k(t)$ can be interpreted as a composite capital good. (This is essentially the approach used by Arrow [1962] in the learning-by-doing model.) Given increasing marginal productivity of knowledge, increasing marginal productivity of a composite k would still be possible if the increasing marginal productivity of knowledge were sufficient to outweigh the decreasing marginal productivity associated with the physical capital.

Within the restrictions imposed by tractability and simplicity, the assumptions on the technology attempt to capture important features of actual technologies. As noted in Section II, estimated aggregate production functions do appear to exhibit some form of increasing returns to scale. Assuming that the increasing returns arise because of increasing marginal productivity of knowledge accords with the plausible conjecture that, even with fixed population and fixed physical capital, knowledge will never reach a level where its marginal product is so low that it is no longer worth the trouble it takes to do research. If the marginal product of knowledge were truly diminishing, this would imply that Newton, Darwin, and their contemporaries mined the richest veins of ideas and that scientists now must sift through the tailings and extract ideas from low-grade ore. That knowledge has an important public good characteristic is generally recognized.[8] That the production of new knowledge exhibits some form of diminishing marginal productivity at any point in time should not be controversial. For example, even though it may be possible to develop the knowledge needed to produce usable energy from nuclear fusion by devoting less than 1 percent of annual gross national product (GNP) to the research effort over a period of 20 years, it is likely that this knowledge could not be produced by next year regardless of the size of the current research effort.

B. *Existence and Characterization of a Social Optimum*

Before using necessary conditions to characterize the solutions to either the social optimization problem, denoted as PS_∞, or any of the artificial optimization problems $P_\infty(K)$, I must verify that these problems have solutions. First I state the problems precisely. Let k_0 denote the initial stock of knowledge per firm for the economy. As in the last section, I will always work with the same number of firms and consumers. Because the choice of $\mathbf{x} = \bar{\mathbf{x}}$ is trivial, I suppress this argument, writing $f(k, K) = F(k, K, \bar{\mathbf{x}})$. Also, let $\mathcal{F}(k) = f(k, Sk) = F(k, Sk, \bar{\mathbf{x}})$ denote the globally convex (per capita) production function that would be faced by a social planner. In all problems that follow, the constraint $\dot{k}(t) \geq 0$ for all $t \geq 0$ and the initial condition $k(0) = k_0$ will be understood:

[8] See, e.g., Bernstein and Nadiri (1983) for estimates from the chemical industry suggesting that spillover effects can be quite large.

$$PS_\infty: \quad \max \int_0^\infty U(c(t))e^{-\delta t}dt$$

$$\text{subject to } \frac{\dot{k}(t)}{k(t)} = g\left(\frac{\mathcal{F}(k(t)) - c(t)}{k(t)}\right);$$

$$P_\infty(K): \quad \max \int_0^\infty U(c(t))e^{-\delta t}dt$$

$$\text{subject to } \frac{\dot{k}(t)}{k(t)} = g\left(\frac{f(k(t), K(t)) - c(t)}{k(t)}\right).$$

Note that the only difference between these two problems lies in the specification of the production function. In the first case, it is convex and invariant over time. In the second, it is concave but depends on time through its dependence on the path $K(t)$. I can now state the theorem that guarantees the existence of solutions to each of these problems.

THEOREM 1. Assume that each of U, f, and g is a continuous real-valued function defined on a subset of the real line. Assume that U and g are concave. Suppose that $\mathcal{F}(k) = f(k, Sk)$ satisfies a bound $\mathcal{F}(k) \leq \mu + k^\rho$ and that $g(z)$ satisfies the bounds $0 \leq g(\mathbf{x}) \leq \alpha$ for real numbers μ, ρ, and α. Then if $\alpha\rho$ is less than the discount factor δ, PS_∞ has a finite-valued solution, and $P_\infty(K)$ has a finite-valued solution for any path $K(t)$ such that $K(t) \leq K(0)e^{\alpha t}$.

The proof, given in an appendix available on request, amounts to a check that the conditions of theorem 1 in Romer (1986) are satisfied. Note that if α is less than δ the inequality $\alpha\rho < \delta$ allows for $\rho > 1$. Thus the socially feasible production function $\mathcal{F}$ can be globally convex in k, with a marginal social product and an average social product of knowledge that increase without bound.

The analysis of the social planning problem PS_∞ in terms of a current-valued Hamiltonian and a phase plane follows along familiar lines (see, e.g., Arrow 1967; Cass and Shell 1976*a*, 1976*b*). Define $H(k, \lambda) = \max_c U(c) + \lambda\{kg([\mathcal{F}(k) - c]/k)\}$. For simplicity, assume that the functions U, f, and g are twice continuously differentiable. The first-order necessary conditions for a path $k(t)$ to be a maximum for PS_∞ are that there exists a path $\lambda(t)$ such that the system of first-order differential equations $\dot{k} = D_2H(k, \lambda)$ and $\dot{\lambda} = \delta\lambda - D_1H(k, \lambda)$ are satisfied and that the paths satisfy two boundary conditions: the initial condition on k and the transversality condition at infinity, $\lim_{t\to\infty} \lambda(t)k(t)e^{-\delta t} = 0$.[9]

[9] Proving the necessity of the transversality condition for a maximization problem that is not concave takes relatively sophisticated mathematical methods. Ekeland and

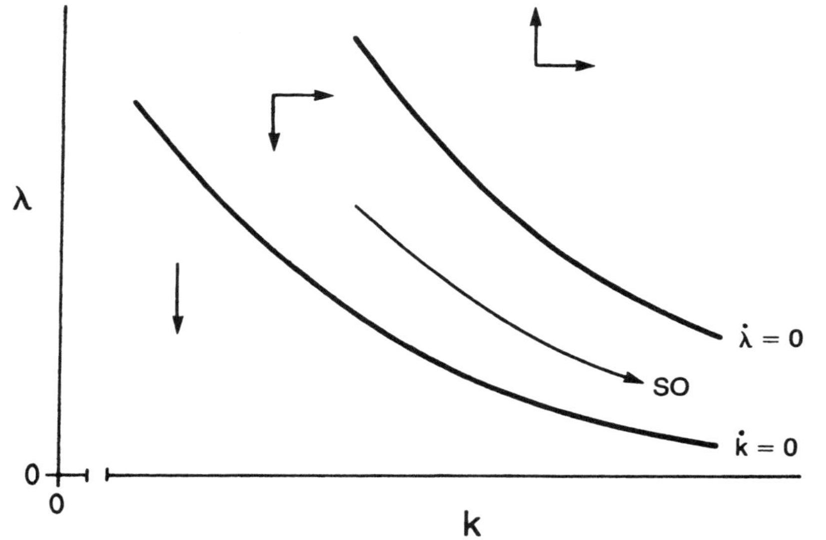

FIG. 2.—Geometry of the phase plane for a typical social optimum. Arrows indicate directions of trajectories in different sections of the plane. The rate of change of the stock of knowledge, $\dot{k}$, is zero everywhere on or below the locus denoted by $\dot{k} = 0$; *SO* denotes the socially optimal trajectory that stays everywhere between the lines $\dot{\lambda} = 0$ and $\dot{k} = 0$.

Under the assumption that $\lim_{c \to 0} DU(c) = \infty$, maximizing over c in the definition of $H(k, \lambda)$ implies that $DU(c) = \lambda Dg([\mathcal{F}(k) - c]/k)$ whenever the constraint $\dot{k} \geq 0$ is not binding; otherwise, $c = \mathcal{F}(k)$. This gives c as a function of k and λ. Substituting this expression in the equations for $\dot{k}$ and $\dot{\lambda}$ gives a system of first-order equations that depends only on k and λ.

Because of the restriction that $\dot{k}$ be nonnegative, the plane can be divided into two regions defined by $\dot{k} = 0$ and $\dot{k} \geq 0$ (see fig. 2). In a convenient abuse of the terminology, I will refer to the locus of points dividing these two regions as the $\dot{k} = 0$ locus. Along this locus, both the conditions $c = \mathcal{F}(k)$ and $DU(c) = \lambda Dg([\mathcal{F}(k) - c]/k)$ must hold. Thus the $\dot{k} = 0$ locus is defined by the equation $DU(\mathcal{F}(k)) = \lambda$. By the concavity of U, it must be a nonincreasing curve in the k-λ plane.

As usual, the equation $\dot{\lambda} = 0$ defines a simple locus in the plane. When the derivative $D_1H(k, \lambda)$ is evaluated along the $\dot{k} = 0$ locus, the equation for $\dot{\lambda}$ there can be written $\dot{\lambda}/\lambda = \delta - D\mathcal{F}(k)$. If $D\mathcal{F}$ increases without bound, there exists a value of $\hat{k}$ such that $D\mathcal{F}(k) > \delta$ for all k

Scheinkman (1983) prove the necessity of the transversality condition for nonconcave discrete-time problems. In continuous time, a proof that requires a local Lipschitz condition is given by Aubin and Clarke (1979).

larger than $\hat{k}$, and for all such k, the $\dot{\lambda} = 0$ locus lies above the $\dot{k} = 0$ locus. It may be either upward or downward sloping. If $\mathcal{F}$ were concave and satisfied the usual Inada conditions, $\dot{\lambda} = 0$ would cross $\dot{k} = 0$ from above and the resulting steady state would be stable in the usual saddle-point sense. Here, $\dot{\lambda} = 0$ may cross $\dot{k} = 0$ either from above or from below. If $D\mathcal{F}(k)$ is everywhere greater than δ, the $\dot{\lambda} = 0$ locus lies everywhere above the $\dot{k} = 0$ locus, and $\hat{k}$ can be taken to be zero. (This is the case illustrated in fig. 2.) Starting from any initial value greater than $\hat{k}$, the optimal trajectory $(\lambda(t), k(t))$, $t \geq 0$, must remain above the region where $\dot{k} = 0$. Any trajectory that crosses into this region can be shown to violate the transversality condition. Consequently, $k(t)$ grows without bound along the optimal trajectory.

This social optimum cannot be supported as a competitive equilibrium in the absence of government intervention. Any competitive firm that takes $K(t)$ as given and is faced with the social marginal products as competitive prices will choose not to remain at the optimal quantities even if it expects all other firms to do so. Each firm will face a private marginal product of knowledge (measured in terms of current output goods) equal to $D_1 f$; but the true shadow price of capital will be $D_1 f + SD_2 f > D_1 f$. Given this difference, each firm would choose to acquire less than the socially optimal amount of knowledge.

C. *Existence and Characterization of the Competitive Equilibrium*

Under a general set of conditions, this economy can be shown to have a suboptimal equilibrium in the absence of any intervention. It is completely analogous to the equilibrium for the two-period model. As in that model, it is straightforward to show that there is a three-way equivalence between competitive equilibria, fixed points of the mapping that sends a path $K(t)$ into S times the solution to $P_\infty(K)$, and solutions to an equation of the form $D_1 V(k, Sk) = 0$.[10] In the infinite-horizon case, this equation consists of a system of differential equations, which can be represented in terms of a phase plane, and a set of boundary conditions.

To derive these equations, consider the necessary conditions for the concave problem $P_\infty(K)$. Define a Hamiltonian, denoted as $\tilde{H}$ to distinguish it from the Hamiltonian H for the social planning problem PS_∞:

[10] An explicit proof of this result is given in Romer (1983). The method of proof is exactly as outlined in the two-period model. A generalized Kuhn-Tucker theorem is used to derive the necessary conditions that yield shadow prices for the maximization problems $P_\infty(K)$. Suppose K^* is a fixed point. If the consumer and the firm are faced with the shadow prices associated with $P_\infty(K^*)$, the sufficient conditions for their maximization problems are shown to be satisfied at the quantities that solve $P_\infty(K^*)$.

$$\tilde{H}(k, \lambda, K) = \max_{c} U(c) + \lambda\left[kg\left(\frac{f(k, K) - c}{k}\right)\right].$$

Then the necessary conditions for $k(t)$ to be a solution to $P_\infty(K)$ are that there exists a path $\lambda(t)$ such that $\dot{k}(t) = D_2\tilde{H}(k(t), \lambda(t), K(t))$ and $\dot{\lambda}(t) = \delta\lambda(t) - D_1\tilde{H}(k(t), \lambda(t), K(t))$ and such that the paths $k(t)$ and $\lambda(t)$ satisfy the boundary conditions $k(0) = k_0$ and $\lim_{t\to\infty} \lambda(t)k(t)e^{-\delta t} = 0$. Substituting $Sk(t)$ for $K(t)$ yields an autonomous system of differential equations, $\dot{k}(t) = D_2\tilde{H}(k(t), \lambda(t), Sk(t))$, $\dot{\lambda}(t) = \delta\lambda(t) - D_1\tilde{H}(k(t), \lambda(t), Sk(t))$, that can be characterized using the phase plane. The two boundary conditions must still hold. Any paths for $k(t)$ and $\lambda(t)$ that satisfy these equations and the boundary conditions will correspond to a competitive equilibrium, and all competitive equilibria can be characterized this way.

Before considering phase diagrams, I must show that a competitive equilibrium exists for some class of models. Standard results concerning the existence of solutions of differential equations can be used to prove that the equations for $\dot{\lambda}$ and $\dot{k}$ determine a unique trajectory through any point (k, λ) in the phase plane. The difficulty arises in showing that for any given value of k_0 there exists some value of λ_0 such that the transversality condition at infinity is satisfied along the trajectory through (k_0, λ_0). As opposed to the case in which these equations are generated by a concave maximization problem known to have a solution, there is no assurance that such a λ_0 exists.

The basic idea in the proof that such a λ_0 exists, and hence that a competitive equilibrium exists, is illustrated in example 1 from the next section. To state the general result, I need additional conditions that characterize the asymptotic behavior of the functions f and g. This is accomplished by means of an asymptotic exponent as defined by Brock and Gale (1969). Given a function $h(y)$, define the asymptotic exponent e of h as $e = \lim_{y\to\infty} \log_y|h(y)|$. Roughly speaking, $h(y)$ behaves asymptotically like the power function y^e. Also, recall that α is the maximal rate of growth for k implied by the research technology.

THEOREM 2. In addition to the assumptions of theorem 1, assume that U, f, and g are twice continuously differentiable. Assume also that $\mathcal{F}(k) = f(k, Sk)$ has an asymptotic exponent ρ such that $\rho > 1$ and $\alpha\rho < \delta$. Finally, assume that $Dg(\mathbf{x})$ has an asymptotic exponent strictly less than -1. Let $\bar{k}$ be such that $D_1 f(k, Sk) > \delta$ for all $k > \bar{k}$. Then if $k_0 > \bar{k}$, there exists a competitive equilibrium with externalities in which $c(t)$ and $k(t)$ grow without bound.

The proof is given in Romer (1983, theorem 3). The assumption on the asymptotic growth of $\mathcal{F}$ is self-explanatory. The assumption on the asymptotic exponent of Dg is sufficient to ensure the boundedness of g. The condition on $D_1 f$ will be satisfied in most cases in which $\mathcal{F}(k)$

$= f(k, Sk)$ is convex. Examples of functions satisfying these assumptions are given in the next section.

Once the conditions for the existence of a competitive equilibrium have been established, the analysis reduces once again to the study of the phase plane summarizing the information in the differential equations. In many respects, this analysis is similar to that for the social optimum for this economy. The phase plane can once again be divided into regions where $\dot{k} = 0$ and $\dot{k} > 0$. Since by definition $\mathcal{F}(k) = f(k, Sk)$, the equations for c as a function of k and λ will be identical to those in the social optimum: $DU(c) = \lambda Dg([f(k, Sk) - c]/k)$ if $\dot{k} > 0$, $c = f(k, Sk)$ if $\dot{k} = 0$. As a result, the boundary locus for the region $\dot{k} = 0$ will also be identical with that from the social optimum. The only difference arises in the equation for $\dot{\lambda}$. Although the equality $H(k, \lambda) = \tilde{H}(k, \lambda, Sk)$ does hold, the derivatives $D_1H(k, \lambda)$ and $D_1\tilde{H}(k, \lambda, Sk)$ differ. In the first case, a term involving the expression $D\mathcal{F}(k) = D_1f(k, Sk) + SD_2f(k, Sk)$ will appear. In the second case, only the first part of this expression, $D_1f(k, Sk)$, appears. Therefore, $D_1H(k, \lambda)$ is always larger than $D_1\tilde{H}(k, \lambda, Sk)$. Consequently, the $\dot{\lambda} = 0$ locus for the competitive equilibrium must lie below that for the social optimum.

As was true of the social optimum, the $\dot{\lambda} = 0$ locus can be either upward or downward sloping. If $D_1f(k, Sk) > \delta$ for all k greater than some value $\bar{k}$, the $\dot{\lambda} = 0$ locus will lie above $\dot{k} = 0$ for values of k to the right of $\bar{k}$. Then the qualitative analysis is the same as that presented for the social optimum. Starting from an initial value $k_0 > \bar{k}$, the only candidate paths for equilibria are ones that stay above the $\dot{k} = 0$ region; as before, paths that cross into this region will violate the transversality condition. A trajectory lying everywhere in the region where $\dot{k} > 0$ can fail to have $k(t)$ grow without bound only if the trajectory asymptotically approaches a critical point where $\dot{\lambda}$ and $\dot{k}$ are both zero, but no such point exists to the right of $\bar{k}$. Hence, all the trajectories that are possible candidates for an equilibrium have paths for $k(t)$ that grow without bound. The existence result in theorem 2 shows that at least one such path satisfies the transversality condition at infinity.

D. *Welfare Analysis of the Competitive Equilibrium*

The welfare analysis of the competitive equilibrium is quite simple. The intuition from simple static models with externalities or from the two-period model presented in Section III carries over intact to the dynamic model here. In the calculation of the marginal productivity of knowledge, each firm recognizes the private return to knowledge, $D_1f(k, Sk)$, but neglects the effect due to the change in the aggregate level, $SD_2f(k, Sk)$; an increase in k induces a positive external effect

$D_2f(k, Sk)$ on each of the S firms in the economy. Consequently, the amount of consumption at any point in time is too high in the competitive equilibrium and the amount of research is too low. Any intervention that shifts the allocation of current goods away from consumption and toward research will be welfare-improving. As in any model with externalities, the government can achieve Pareto improvements not available to private agents because its powers of coercion can be used to overcome problems of shirking.

If the government has access to lump-sum taxation, any number of subsidy schemes will support the social optimum. Along the paths $k^*(t)$ and $\lambda^*(t)$ from the social optimum, taxes and subsidies must be chosen so that the first partial derivative of the Hamiltonian for the competitive equilibrium with taxes equals the first partial derivative of the Hamiltonian for the social planning problem; that is, the taxes and subsidies must be chosen so that the after-tax private marginal product of knowledge is equal to the social marginal product. This can be accomplished by subsidizing holdings of k, subsidizing accumulation $\dot{k}$, or subsidizing output and taxing factors of production other than k. The simplest scheme is for the government to pay a time-varying subsidy of $\sigma_1(t)$ units of consumption goods for each unit of knowledge held by the firm. If this subsidy is chosen to be equal to the term neglected by private agents, $\sigma_1(t) = SD_2f(k^*(t), Sk^*(t))$, private and social marginal products will be equal. A subsidy $\sigma_2(t)$ paid to a firm for each unit of goods invested in research would be easier to implement but is harder to characterize. In general, solving for $\sigma_2(t)$ requires the solution of a system of differential equations that depends on the path for $k^*(t)$. In the special case in which production takes the form $f(k, K) = k^\nu K^\gamma$, the optimal subsidy can be shown to be constant, $\sigma_2 = \gamma/(\nu + \gamma)$. (This calculation is also included in the app. available on request.)

While it is clear that the social marginal product of knowledge is greater than the private marginal product in the no-intervention competitive equilibrium, this does not necessarily imply that interest rates in the socially optimal competitive equilibrium with taxes will be higher than in the suboptimal equilibrium. In each case, the real interest rate on loans made in units of output goods can be written as $r(t) = -(\dot{p}/p)$, where $p(t) = e^{-\delta t}DU(c(t))$ is the present value price for consumption goods at date t. When utility takes the constant elasticity form $U(c) = [c^{(1-\theta)} - 1]/(1 - \theta)$, this reduces to $r(t) = \delta + \theta(\dot{c}/c)$. In the linear utility case in which $\theta = 0$, r will equal δ regardless of the path for consumption and in particular will be the same in the two equilibria. This can occur even though the marginal productivity of knowledge differs because the price of knowledge in terms of consumption goods (equal to the marginal rate of transformation between knowledge and consumption goods) can vary. Holders of knowledge earn capital gains and losses as well as a direct return equal to the private marginal productivity of knowledge. In the case of linear utility, these capital gains and losses adjust so that interest rates stay the same.

This logical point notwithstanding, it is likely that interest rates will be higher in the social optimum. On average, $\dot{c}/c$ will be higher in the social optimum; higher initial rates of investment with lower initial consumption must ultimately lead to higher levels of consumption. If there is any curvature in the utility function U, so that θ is positive, interest rates in the optimum will be greater than in the no-intervention equilibrium. In contrast to the usual presumption, cost-benefit calculations in a suboptimal equilibrium should use a social rate of discount that is higher than the market rate of interest.

VI. Examples

To illustrate the range of behavior possible in this kind of model, this section examines specific functional forms for the utility function U, the production function f, and the function g describing the research technology. Because the goal is to reach qualitative conclusions with a minimum of algebra, the choice of functional form will be guided primarily by analytical convenience. For the production function, assume that f takes the form noted above, $f(k, K) = k^{\nu}K^{\gamma}$. This is convenient because it implies that the ratio of the private and social marginal products,

$$\frac{D_1 f(k, Sk)}{D_1 f(k, Sk) + SD_2 f(k, Sk)} = \frac{\nu}{\nu + \gamma},$$

is constant. Nonincreasing private marginal productivity implies that $0 < \nu \leq 1$; increasing social marginal productivity implies that $1 < \gamma + \nu$. With these parameter values, this functional form is reasonable only for large values of k. For small values of k, the private and social marginal productivity of knowledge is implausibly small; at $k = 0$, they are both zero. This causes no problem provided we take a moderately large initial k_0 as given. An analysis starting from k_0 close to zero would have to use a more complicated (and more reasonable) functional form for f.

Recall that the rate of increase of the stock of knowledge is written in the homogeneous form $\dot{k} = G(I, k) = kg(I/k)$, where I is output minus consumption. The requirements on the concave function g are the normalization $Dg(0) = 1$ and the bound $g(I/k) < \alpha$ for all I/k. An analytically simple form satisfying these requirements is $g(z) = \alpha z/(\alpha + z)$. Recalling that δ is the discount rate, note that the bound required for the existence of a social optimum as given in theorem 1 requires the additional restriction that $\alpha(\nu + \gamma) < \delta$. Given the stated parameter restrictions, it is easy to verify that f and g satisfy all the requirements of theorems 1 and 2.

A. *Example 1*

With this specification of the technology for the economy, we can readily examine the qualitative behavior of the model for logarithmic utility $U(c) = \ln(c)$. The Hamiltonian can then be written as

$$\tilde{H}(k, \lambda, K, c) = \ln(c) + \lambda k g\left(\frac{f(k, K) - c}{k}\right).$$

Along (the boundary of the region in which) $\dot{k} = 0$, $Dg(0) = 1$ implies that $c = \lambda^{-1}$, so $\dot{k} = 0$ is determined by the equation

$$\lambda = [f(k, Sk)]^{-1} = S^{-\gamma}k^{-(\nu+\gamma)}.$$

The exact form for the locus $\dot{\lambda} = 0$ is algebraically complicated, but it is straightforward to show that, for large k, $\dot{\lambda} = 0$ lies above the $\dot{k} = 0$ locus since $D_1 f(k, Sk)$ will be greater than δ. Also, if we define the curve L_1 in the phase plane by the equation $\lambda = [1/(\delta - \alpha)]k^{-1}$, the $\dot{\lambda} = 0$ locus must cross L_1 from above as indicated in figure 3. (Details are given in the app. available on request.) Thus $\dot{k} = 0$ behaves as k to the power $-(\nu + \gamma) < -1$, and $\dot{\lambda} = 0$ is eventually trapped between $\dot{k} = 0$ and a line described by k to the power -1. In figure 3, representative trajectories t_1 and t_2 together with the competitive equilibrium trajectory CE are used to indicate the direction of trajectories in the various parts of the plane instead of the usual arrows.

Because the line L_1 is of the form $\lambda = [1/(\delta - \alpha)]k^{-1}$, any trajectory that eventually remains below L_1 will satisfy the transversality condition $\lim_{t \to \infty} e^{-\delta t}k(t)\lambda(t) = 0$. Given the geometry of the phase plane, it is clear that there must exist a trajectory that always remains between the loci $\dot{\lambda} = 0$ and $\dot{k} = 0$. Given the initial value k_0, index by the value of λ all the trajectories that start at a point (k_0, λ) between the two loci. The set of λ's corresponding to trajectories that cross $\dot{\lambda} = 0$ can have no smallest value, the set of λ's that correspond to trajectories that cross $\dot{k} = 0$ can have no largest value, and the two sets must be disjoint. Thus there exists a value λ_0 such that the trajectory through (k_0, λ_0) crosses neither locus and must therefore correspond to an equilibrium.[11]

[11] This is the essence of the proof of theorem 2.

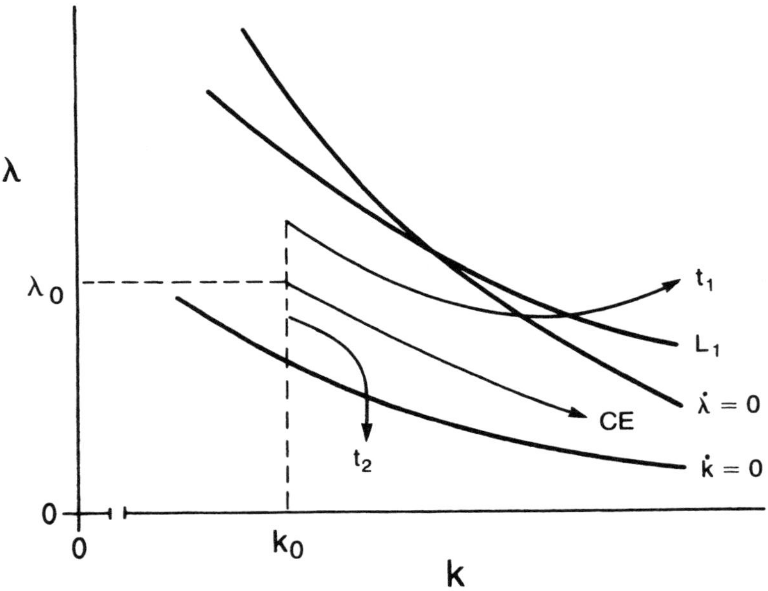

FIG. 3.—Geometry of the competitive equilibrium for example 1. The line L_1 is defined by the equation $\lambda = 1/(\delta - \alpha)k$; t_1 and t_2 denote representative trajectories in the phase plane; *CE* denotes the competitive equilibrium trajectory, which stays everywhere between the $\dot{\lambda} = 0$ and $\dot{k} = 0$ loci; λ_0 denotes the initial shadow price of knowledge corresponding to the initial stock of knowledge k_0.

In fact, the path resembles a conventional equilibrium in which the trajectory remains between the $\dot{\lambda} = 0$ and $\dot{k} = 0$ loci as it converges to a saddle point, although here it is as if the saddle point has been moved infinitely far to the right. Since the optimal trajectory cannot stop, capital grows without bound. Since the trajectory is downward sloping and since consumption is increasing in k and decreasing in λ, it is easy to see that consumption also grows without bound. Because of the difficulty of the algebra, it is not easy to describe the asymptotic rates of growth.

B. *Example 2*

Suppose now that utility is linear, $U(c) = c$. In the algebra and in the phase plane for this case, we can ignore the restriction $c \geq 0$ since it will not be binding in the region of interest. Maximizing out c from the Hamiltonian $\tilde{H}(k, \lambda, K, c) = c + \lambda k g((f - c)/k)$ implies that $c = f - \alpha k(\lambda^{.5} - 1)$. Then $f - c$ is positive (hence $\dot{k}$ is positive) if and only if $\lambda > 1$.

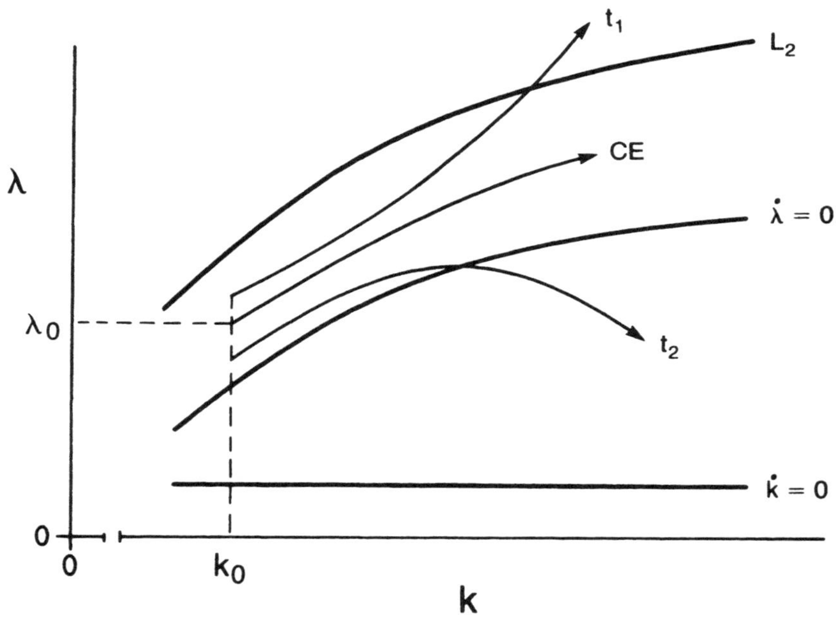

FIG. 4.—Geometry of the competitive equilibrium for example 2. The line L_2 is defined by an equation of the form $\lambda = Ak^{\nu+\gamma-1}$; t_1 and t_2 denote representative trajectories in the phase plane; *CE* denotes the competitive equilibrium trajectory that stays everywhere between L_2 and $\dot{\lambda} = 0$; λ_0 denotes the initial shadow price of knowledge.

In this example, it is possible to put tighter bounds on the behavior of the $\dot{\lambda} = 0$ locus and, more important, on the behavior of the equilibrium trajectory. As demonstrated in the appendix (available on request), $\dot{\lambda} = 0$ is upward sloping and behaves asymptotically like the power function $\lambda = Bk^{\nu+\gamma-1}$ for some constant B. For this economy, the equilibrium trajectory will lie above the $\dot{\lambda} = 0$ locus, so it is convenient to define an additional curve that will trap the equilibrium trajectory from above. For an appropriate choice of the constant A, the line L_2 defined by $\lambda = Ak^{\nu+\gamma-1}$ will lie above $\dot{\lambda} = 0$ and will have the property that trajectories must cross it from below (see fig. 4). Since trajectories must cross $\dot{\lambda} = 0$ from above, the same geometric argument as used in the last example demonstrates that there exists a trajectory that remains between these two lines. Consequently it must also behave asymptotically like $k^{\nu+\gamma-1}$. Since $k(t)$ can grow no faster than $e^{\alpha t}$, the product $\lambda(t)k(t)$ will be bounded along such a trajectory by a function of the form $e^{\alpha(\nu+\gamma)t}$. Since $\delta > (\nu + \gamma)\alpha$, this trajectory satisfies the transversality condition and corresponds to an equilibrium.

Along the equilibrium trajectory, λ behaves asymptotically like

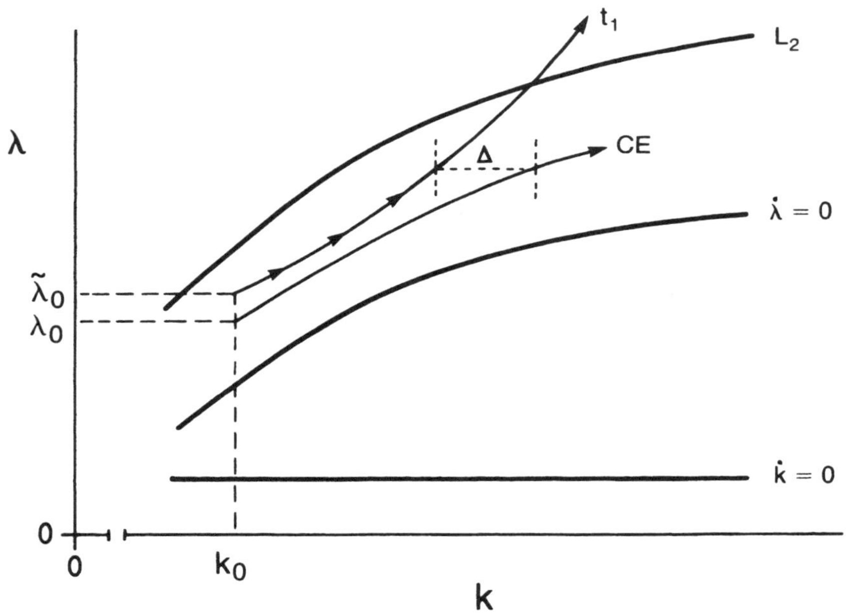

FIG. 5.—Geometry for the economy in example 2 when an exogenous increase of size Δ in the stock of knowledge is known to occur at a time $T > 0$. The equilibrium trajectory moves along t_1 until time T, at which point it is Δ units to the left of the trajectory CE. At time T, the economy jumps horizontally to CE with the change in the capital stock, but the path for $\lambda(t)$ is continuous. The equilibrium then proceeds along CE. $\tilde{\lambda}_0$ denotes the initial shadow price of knowledge in the case in which the exogenous increase will take place; λ_0 denotes the lower value that obtains in an economy in which no exogenous increase will take place.

$k^{\nu+\gamma-1}$. Given the expression noted above for c in terms of λ and k, c behaves asymptotically like $k^{\nu+\gamma} - \alpha k^{1+(.5)(\nu+\gamma-1)}$ and $I = f - c$ behaves like $k^{1+(.5)(\nu+\gamma-1)}$. Then c, I, C/k, and I/k go to infinity with k. By the assumptions on the research technology, I/k going to infinity implies that $\dot{k}/k$ approaches its upper bound α. Consequently, the percentage rate of growth of output and of consumption will be increasing, both approaching the asymptotic upper bound $\alpha(\nu + \gamma)$.

Because the equilibrium trajectory is upward sloping, this economy will exhibit different stability properties from either the conventional model or the economy with logarithmic utility described above. Figure 5 illustrates a standard exercise in which a perfect-foresight equilibrium is perturbed. Suppose that at time 0 it is known that the stock of knowledge will undergo an exogenous increase of size Δ at time T and that no other exogenous changes will occur. Usual arbitrage arguments imply that the path for any price like $\lambda(t)$ must be continuous at time T. The path followed by the equilibrium in the phase plane starts on a trajectory like t_1 such that at time T it arrives at a point exactly Δ units to the left of the trajectory CE from figure 4, which would have been the equilibrium in the absence of any exogenous change in k. As the economy evolves, it moves along t_1 then jumps Δ units to the right to the trajectory CE at time T. Since $e^{-\delta t}\lambda(t)$ can be interpreted as a time 0 market price for knowledge, a foreseen future increase in the aggregate stock of knowledge causes a time 0 increase in the price for knowledge and a consequent increase in the rate of investment in knowledge. Because of the increasing returns, the private response to an aggregate increase in the stock of knowledge will be to reinforce its effects rather than to dampen them. Since the rate of growth of the stock of knowledge is increasing in the level, this kind of disturbance causes the stock of knowledge to be larger at all future dates. Moreover, the magnitude of the difference will grow over time. Thus small current or anticipated future disturbances can potentially have large, permanent, aggregate effects.

As a comparison with the first example shows, this result requires not only that increasing returns be present but also that marginal utility not decrease too rapidly with the level of per capita consumption. If we had restricted attention to the class of bounded, constant elasticity utility functions, $[c^{(1-\theta)} + 1]/(1 - \theta)$ with $\theta > 1$, this phenomenon would not be apparent. The specific example here uses linear utility for convenience, but similar results will hold for constant elasticity utility function $[c^{(1-\theta)} - 1]/(1 - \theta)$ for values of θ close enough to zero.

C. *Example 3*

The analysis of the previous example suggests a simple multicountry model with no tendency toward convergence in the level of per capita output. Suppose each country is modeled as a separate closed economy of the type in example 2. Thus no trade in goods takes place among the different countries, and knowledge in one country has external effects only within that country. Even if all countries started out with the same initial stock of knowledge, small disturbances could create permanent differences in the level of per capita output. Since the rate of growth of the stock of knowledge is increasing over time toward an asymptotic upper bound, a smaller country s will always grow less rapidly than a larger country l. Asymptotically, the rates of growth $(\dot{k}/k)_s$ and $(\dot{k}/k)_l$ will both converge to α, but the ratios k_l/k_s and c_l/c_s will be monotonically increasing over time, and the differences $k_l(t) - k_s(t)$ and $c_l(t) - c_s(t)$ will go to infinity.

It is possible to weaken the sharp separation assumed between countries in this discussion. In particular, neither the absence of trade in consumption goods and knowledge nor the sharp restriction on the extent of the externalities is essential for the divergence noted above. As in the Arrow (1962) learning-by-doing model, suppose that all knowledge is embodied either in physical capital or as human capital. Thus k denotes a composite good composed of both knowledge and some kind of tangible capital. In this embodied form, knowledge can be freely transported between two different countries. Suppose further that the external effect of knowledge embodied in capital in place in one country extends across its border but does so with diminished intensity. For example, suppose that output of a representative firm in country 1 can be described as $f(k, K_1, K_2) = k^v(K_1^a + K_2^b)$, where k is the firm's stock of the composite good, K_1 and K_2 are the aggregates in the two countries, and the exponent a on the domestic aggregate K_1 is strictly greater than the exponent b on the foreign aggregate K_2. Production in country 2 is defined symmetrically. Then for a specific form of the research technology, Romer (1983) shows that the key restriction on the equilibrium paths Sk_1 and Sk_2 in the two countries comes from the equality of the marginal product of private knowledge imposed by the free mobility of the composite good k:

$$D_1 f(k_1, Sk_1, Sk_2) = D_1 f(k_2, Sk_2, Sk_1). \qquad (1)$$

With the functional form given above, it is easy to verify that, in addition to the symmetric solution $k_1 = k_2$, there exists an asymmetric solution. In that solution, if k_1 is larger than k_2 and growing (e.g., country 1 is industrialized and country 2 is not), the path for k_2 that satisfies this equation either can grow at a rate slower than that for country 1 or may shrink, exporting the composite good to the more developed country.[12]

This kind of steady, ongoing "capital flight" or "brain drain" does not require any fundamental difference between the two countries. They have identical technologies. If we assume that there is perfect mobility in the composite k, it can even take place when both countries start from the same initial level of k. If all agents are convinced that country 2 is destined to be the slow-growing country in an asymmetric equilibrium, a discrete amount of the composite good will jump immediately to country 1. Thereafter, the two countries will evolve according to equation (1), with country 2 growing more slowly than country 1 or possibly even shrinking.

This kind of model should not be taken too literally. A more realistic model would need to take account of other factors of production with various degrees of less than perfect mobility. Nonetheless, it does suggest that the presence of increasing returns and of multiple

[12] Details are available in an app. available from the author.

equilibria can introduce a degree of instability that is not present in conventional models. This identifies a second sense in which small disturbances can have large effects. In addition to the multiplier-type effect for a closed economy as described in the last example, a small disturbance or a small change in a policy variable such as a tax rate could conceivably have a decisive effect on which of several possible equilibria is attained.

VII. Conclusion

Recent discussions of growth have tended not to emphasize the role of increasing returns. At least in part, this reflects the absence of an empirically relevant model with increasing returns that exhibits the rigor and simplicity of the model developed by Ramsey, Cass, and Koopmans. Early attempts at such a model were seriously undermined by the loose treatment of specialization as a form of increasing returns with external effects. More recent attempts by Arrow, Levhari, and Sheshinski were limited by their dependence on exogenously specified population growth and by the implausible implication that the rate of growth of per capita income should be a monotonically increasing function of the rate of population growth. Incomplete models that took the rate of technological change as exogenously specified or that made it endogenous in a descriptive fashion could address neither welfare implications nor positive implications like the slowing of growth rates or the convergence of per capita output.

The model developed here goes part way toward filling this theoretical gap. For analytical convenience, it is limited to a case that is the polar opposite of the usual model with endogenous accumulation of physical capital and no accumulation of knowledge. But once the operation of the basic model is clear, it is straightforward to include other state variables. The implications for a model with both increasing marginal productivity of knowledge and decreasing marginal productivity of physical capital can easily be derived using the framework outlined here; however, the geometric analysis using the phase plane is impossible with more than one state variable, and numerical methods for solving dynamic equation systems must be used.[13] Since the model here can be interpreted as the special case of the two-state-variable model in which knowledge and capital are used in fixed

[13] For an example of this kind of numerical analysis in a model with a stock of knowledge and a stock of an exhaustible resource, see Romer and Sasaki (1985). As in the growth model, increasing returns associated with knowledge can reverse conventional presumptions; in particular, exhaustible resource prices can be monotonically decreasing for all time.

proportions, this kind of extension can only increase the range of possible equilibrium outcomes.

References

Arrow, Kenneth J. "The Economic Implications of Learning by Doing." *Rev. Econ. Studies* 29 (June 1962): 155–73.

———. "Applications of Control Theory to Economic Growth." In *Mathematics of the Decision Sciences,* vol. 1, edited by George B. Dantzig and Arthur F. Veinott. Providence, R.I.: American Math. Soc., 1967.

Aubin, J. P., and Clarke, F. H. "Shadow Prices and Duality for a Class of Optimal Control Problems." *SIAM J. Control and Optimization* 17 (September 1979): 567–86.

Baumol, William J. "Productivity Growth, Convergence and Welfare: What the Long Run Data Show." Research Report no. 85-27. New York: New York Univ., C. V. Starr Center, 1985.

Bernstein, J. I., and Nadiri, M. Ishaq. "Research and Development, Spillovers and Adjustment Costs: An Application of Dynamic Duality at the Firm Level." Working paper. New York: New York Univ., 1983.

Blitch, Charles P. "Allyn Young on Increasing Returns." *J. Post Keynesian Econ.* 5 (Spring 1983): 359–72.

Brock, William A. "A Simple Perfect Foresight Monetary Model." *J. Monetary Econ.* 1 (April 1975): 133–50.

———. "A Polluted Golden Age." In *Economics of Natural and Environmental Resources,* edited by Vernon L. Smith. New York: Gordon and Breach, 1977.

Brock, William A., and Gale, David. "Optimal Growth under Factor Augmenting Progress." *J. Econ. Theory* 1 (October 1969): 229–43.

Cass, David. "Optimum Growth in an Aggregative Model of Capital Accumulation." *Rev. Econ. Studies* 32 (July 1965): 233–40.

Cass, David, and Shell, Karl. "Introduction to Hamiltonian Dynamics in Economics." *J. Econ. Theory* 12 (February 1976): 1–10. (*a*)

———. "The Structure and Stability of Competitive Dynamical Systems." *J. Econ. Theory* 12 (February 1976): 31–70. (*b*)

Chipman, John S. "External Economies of Scale and Competitive Equilibrium." *Q.J.E.* 84 (August 1970): 347–85.

Dechert, W. Davis, and Nishimura, Kazuo. "A Complete Characterization of Optimal Growth Paths in an Aggregated Model with a Non-concave Production Function." *J. Econ. Theory* 31 (December 1983): 332–54.

Dixit, Avinash K.; Mirrlees, James A.; and Stern, Nicholas. "Optimum Saving with Economies of Scale." *Rev. Econ. Studies* 42 (July 1975): 303–25.

Ekeland, Ivar, and Scheinkman, José A. "Transversality Conditions for Some Infinite Horizon Discrete Time Maximization Problems." Technical Report no. 411. Stanford, Calif.: Stanford Univ., IMSSS, July 1983.

Ethier, Wilfred J. "National and International Returns to Scale in the Model Theory of International Trade." *A.E.R.* 72 (June 1982): 389–405.

Hansen, Lars Peter; Epple, Dennis; and Roberds, William. "Linear-Quadratic Duopoly Models of Resource Depletion." In *Energy, Foresight and Strategy,* edited by Thomas J. Sargent. Washington: Resources for the Future, 1985.

Helpman, Elhanan. "Increasing Returns, Imperfect Markets, and Trade Theory." In *Handbook of International Economics,* vol. 2, edited by Ronald W. Jones and Peter B. Kenen. New York: North-Holland, 1984.

Hicks, John R. "Thoughts on the Theory of Capital: The Corfu Conference." *Oxford Econ. Papers* 12 (June 1960): 123–32.
Hochman, Oded, and Hochman, Eithan. "Regeneration, Public Goods, and Economic Growth." *Econometrica* 48 (July 1980): 1233–50.
Kaldor, Nicholas. "The Role of Increasing Returns, Technical Progress and Cumulative Causation in the Theory of International Trade and Economic Growth." *Écon. Appl.* 34, no. 4 (1981): 593–617.
Kendall, M. G. *Rank Correlation Methods.* New York: Hafner, 1962.
Kendrick, John W. *The Formation and Stocks of Total Capital.* New York: Columbia Univ. Press (for N.B.E.R.), 1976.
Knight, Frank H. "On Decreasing Cost and Comparative Cost: A Rejoinder." *Q.J.E.* 39 (February 1925): 331–33.
Koopmans, Tjalling C. "On the Concept of Optimal Economic Growth." In *The Econometric Approach to Development Planning.* Amsterdam: North-Holland (for Pontificia Acad. Sci.), 1965.
Kuznets, Simon. "Notes on the Pattern of U.S. Economic Growth." In *The Reinterpretation of American Economic History,* edited by Robert W. Fogel and Stanley L. Engerman. New York: Harper and Row, 1971.
Levhari, David. "Extensions of Arrow's 'Learning by Doing.'" *Rev. Econ. Studies* 33 (April 1966): 117–31. (*a*)
———. "Further Implications of Learning by Doing." *Rev. Econ. Studies* 33 (January 1966): 31–38. (*b*)
McKenzie, Lionel W. "On the Existence of General Equilibrium for a Competitive Market." *Econometrica* 27 (January 1959): 54–71.
Maddison, Angus. "Per Capita Output in the Long Run." *Kyklos* 32, nos. 1, 2 (1979): 412–29.
———. *Phases of Capitalist Development.* New York: Oxford Univ. Press, 1982.
Majumdar, Mukul, and Mitra, Tapan. "Intertemporal Allocation with a Non-convex Technology: The Aggregative Framework." *J. Econ. Theory* 27 (June 1982): 101–36.
———. "Dynamic Optimization with a Non-convex Technology: The Case of a Linear Objective Function." *Rev. Econ. Studies* 50 (January 1983): 143–51.
Phelps, Edmund S. "Models of Technical Progress and the Golden Rule of Research." *Rev. Econ. Studies* 33 (April 1966): 133–45.
Ramsey, Frank P. "A Mathematical Theory of Saving." *Econ. J.* 38 (December 1928): 543–59.
Reynolds, Lloyd G. "The Spread of Economic Growth to the Third World: 1850–1980." *J. Econ. Literature* 21 (September 1983): 941–80.
Rockafellar, R. Tyrrell. *Convex Analysis.* Princeton, N.J.: Princeton Univ. Press, 1970.
Romer, Paul M. "Dynamic Competitive Equilibria with Externalities, Increasing Returns and Unbounded Growth." Ph.D. dissertation, Univ. Chicago, 1983.
———. "Cake Eating, Chattering and Jumps: Existence Results for Variational Problems." *Econometrica,* vol. 54 (July 1986).
Romer, Paul M., and Sasaki, Hiroo. "Monotonically Decreasing Natural Resource Prices under Perfect Foresight." Working Paper no. 19. Rochester, N.Y.: Univ. Rochester, Center Econ. Res., 1985.
Shell, Karl. "A Model of Inventive Activity and Capital Accumulation." In *Essays on the Theory of Optimal Growth,* edited by Karl Shell. Cambridge, Mass.: MIT Press, 1967. (*a*)
———. "Optimal Programs of Capital Accumulation for an Economy in Which There Is Exogenous Technological Change." In *Essays on the Theory*

of Optimal Growth, edited by Karl Shell. Cambridge, Mass.: MIT Press, 1967. (*b*)

Sheshinski, Eytan. "Optimal Accumulation with Learning by Doing." In *Essays on the Theory of Optimal Growth*, edited by Karl Shell. Cambridge, Mass.: MIT Press, 1967.

Skiba, A. K. "Optimal Growth with a Convex-Concave Production Function." *Econometrica* 46 (May 1978): 527–39.

Streissler, Erich. "Growth Models as Diffusion Processes: II. Empirical Illustrations." *Kyklos* 32, no. 3 (1979): 571–86.

Weitzman, Martin L. "Optimal Growth with Scale Economies in the Creation of Overhead Capital." *Rev. Econ. Studies* 37 (October 1970): 555–70.

von Weizsäcker, Carl C. "Tentative Notes on a Two Sector Model with Induced Technical Progress." *Rev. Econ. Studies* 33 (July 1966): 245–51.

Young, Allyn A. "Increasing Returns and Economic Progress." In *Readings in Welfare Economics*, edited by Kenneth J. Arrow and Tibor Scitovsky. Homewood, Ill.: Irwin (for American Econ. Assoc.), 1969.

Uzawa, Hirofumi. "Optimum Technical Change in an Aggregative Model of Economic Growth." *Internat. Econ. Rev.* 6 (January 1965): 18–31.

[7]

Endogenous Technological Change

Paul M. Romer

University of Chicago

Growth in this model is driven by technological change that arises from intentional investment decisions made by profit-maximizing agents. The distinguishing feature of the technology as an input is that it is neither a conventional good nor a public good; it is a nonrival, partially excludable good. Because of the nonconvexity introduced by a nonrival good, price-taking competition cannot be supported. Instead, the equilibrium is one with monopolistic competition. The main conclusions are that the stock of human capital determines the rate of growth, that too little human capital is devoted to research in equilibrium, that integration into world markets will increase growth rates, and that having a large population is not sufficient to generate growth.

I. Introduction

Output per hour worked in the United States today is 10 times as valuable as output per hour worked 100 years ago (Maddison 1982). In the 1950s, economists attributed almost all the change in output per hour worked to technological change (Abramovitz 1956; Kendrick 1956; Solow 1957). Subsequent analysis raised our estimates of

Prepared for the conference "The Problem of Economic Development: Exploring Economic Development through Free Enterprise," held at the State University of New York at Buffalo, May 1988. I have benefited from the comments of many seminar and conference participants and two discussants (Rob Vishny, Buffalo, May 1988, and Dale Jorgenson, National Bureau of Economic Research Economic Fluctuations meeting, July 1988). Discussions with Gary Becker, Karl Shell, Robert Lucas, Gene Grossman, and Elhanan Helpman were especially helpful. Research assistance was provided by Danyang Xie. The original work was supported by National Science Foundation grant SES-8618325. It was revised while I was a visitor at the Center for Advanced Study in the Behavioral Sciences and supported by NSF grant BNS87-00864.

[*Journal of Political Economy*, 1990, vol. 98, no. 5, pt. 2]

the importance of increases in the effective labor force and the effective stock of capital in generating growth in output per worker (Jorgenson, Gollop, and Fraumeni 1987), but technological change has surely been important as well. The raw materials that we use have not changed, but as a result of trial and error, experimentation, refinement, and scientific investigation, the instructions that we follow for combining raw materials have become vastly more sophisticated. One hundred years ago, all we could do to get visual stimulation from iron oxide was to use it as a pigment. Now we put it on plastic tape and use it to make videocassette recordings.

The argument presented in this paper is based on three premises. The first is that technological change—improvement in the instructions for mixing together raw materials—lies at the heart of economic growth. As a result, the model presented here resembles the Solow (1956) model with technological change. Technological change provides the incentive for continued capital accumulation, and together, capital accumulation and technological change account for much of the increase in output per hour worked.

The second premise is that technological change arises in large part because of intentional actions taken by people who respond to market incentives. Thus the model is one of endogenous rather than exogenous technological change. This does not mean that everyone who contributes to technological change is motivated by market incentives. An academic scientist who is supported by government grants may be totally insulated from them. The premise here is that market incentives nonetheless play an essential role in the process whereby new knowledge is translated into goods with practical value. Our initial understanding of electromagnetism arose from research conducted in academic institutions, but magnetic tape and home videocassette recorders resulted from attempts by private firms to earn a profit.

The third and most fundamental premise is that instructions for working with raw materials are inherently different from other economic goods. Once the cost of creating a new set of instructions has been incurred, the instructions can be used over and over again at no additional cost. Developing new and better instructions is equivalent to incurring a fixed cost. This property is taken to be the defining characteristic of technology.

Most models of aggregate growth, even those with spillovers or external effects, rely on price-taking behavior. But once these three premises are granted, it follows directly that an equilibrium with price taking cannot be supported. Section II of the paper starts by showing why this is so. It also indicates which of the premises is dropped in growth models that do depend on price-taking behavior. The argument in this section is fundamental to the motivation for the particular model of monopolistic competition that follows, but it is more general than the model itself.

In the specific model outlined in Section III, a firm incurs fixed design or research and development costs when it creates a new good. It recovers those costs by selling the new good for a price that is higher than its constant cost of production. Since there is free entry into this activity, firms earn zero profit in a present value sense.

The conclusions of the model follow directly from this specification. On the basis of results from the static theory of trade with differentiated goods (see, e.g., Helpman and Krugman 1985), one should expect that fixed costs lead to gains from increases in the size of the market and therefore to gains from trade between different countries. Perhaps the most interesting feature of the equilibrium calculated for the model constructed here is that increases in the size of the market have effects not only on the level of income and welfare but also on the rate of growth. Larger markets induce more research and faster growth.

The analysis also suggests why population is not the right measure of market size and why the presence of a large domestic market in countries such as China or India is not a substitute for trade with the rest of the world. The growth rate is increasing in the stock of human capital, but it does not depend on the total size of the labor force or the population. In a limiting case that may be relevant for historical analysis and for the poorest countries today, if the stock of human capital is too low, growth may not take place at all.

These implications of the model are taken up briefly in the final sections of the paper. Section III describes the functional forms that are used to describe the preferences and the technology for the model. It defines an equilibrium that allows for both monopolistic competition and external effects arising from knowledge spillovers. Section IV offers a brief intuitive description of a balanced growth equilibrium for the model. Section V formally characterizes the equilibrium. Section VI describes the welfare properties of the equilibrium. Section VII discusses the connection implied by the model between trade, research, and growth. Algebraic details of the derivations are placed in the Appendix.

II. Rivalry, Excludability, and Nonconvexities

Economists studying public finance have identified two fundamental attributes of any economic good: the degree to which it is rivalrous and the degree to which it is excludable (Cornes and Sandler 1986). Rivalry is a purely technological attribute. A purely rival good has the property that its use by one firm or person precludes its use by another; a purely nonrival good has the property that its use by one firm or person in no way limits its use by another. Excludability is a function of both the technology and the legal system. A good is excludable if the owner can prevent others from using it. A good such as the code for a computer program can be made excludable by means of a legal system that prohibits copying or by means of encryption and copy protection schemes.

Conventional economic goods are both rivalrous and excludable. They are privately provided and can be traded in competitive markets. By definition, public goods are both nonrival and nonexcludable. Because they are nonexcludable, they cannot be privately provided or traded in markets. Public goods can be introduced into a model of price-taking behavior by assuming the existence of a government that can levy taxes. Basic scientific research is an example of a public good that could be provided in this way and that is relevant for modeling growth.

Rivalry and excludability are closely linked because most rival goods are excludable. (A parking space in a shopping center parking lot is an example of a good that is effectively nonexcludable because the cost of enforcing excludability is too high relative to the value of the good.) The interesting case for growth theory is the set of goods that are nonrival yet excludable. The third premise cited in the Introduction implies that technology is a nonrival input. The second premise implies that technological change takes place because of the actions of self-interested individuals, so improvements in the technology must confer benefits that are at least partially excludable. The first premise therefore implies that growth is driven fundamentally by the accumulation of a partially excludable, nonrival input.

To evaluate these claims, it helps to have a specific case in mind. The example of a nonrival input used in what follows is a design for a new good. The vast majority of designs result from the research and development activities of private, profit-maximizing firms. A design is, nonetheless, nonrival. Once the design is created, it can be used as often as desired, in as many productive activities as desired.

In this sense, a design differs in a crucial way from a piece of human capital such as the ability to add. The design is nonrival but the ability to add is not. The difference arises because the ability to add is inherently tied to a physical object (a human body) whereas the design is not.[1] The ability to add is rivalrous because the person who

[1] The original version of this paper used the terms "embodied" and "disembodied" to refer to the difference between an intangible such as the ability to add, which is tied to a specific person, and an intangible such as a design, which is not. This choice of terminology is not used in this revision because embodiment has another meaning in growth theory and because the notion of rivalry already exists in the public finance literature.

possesses this ability cannot be in more than one place at the same time; nor can this person solve many problems at once. As noted above, rivalry leads to a presumption that human capital is also excludable. Thus human capital can be privately provided and traded in competitive markets. In contrast, the design is nonrival because it is independent of any physical object. It can be copied and used in as many different activities as desired.

Like any scientific concept, nonrivalry is an idealization. It is sometimes observed that a design cannot be a nonrival good because it is itself tied to the physical piece of paper or the physical computer disk on which it is stored. What is unambiguously true about a design is that the cost of replicating it with a drafter, a photocopier, or a disk drive is trivial compared to the cost of creating the design in the first place. This is not true of the ability to add. Training the second person to add is as costly as training the first. For simplicity, the arguments here will treat designs as idealized goods that are not tied to any physical good and can be costlessly replicated, but nothing hinges on whether this is literally true or merely close to being true.

Nonrivalry has two important implications for the theory of growth. First, nonrival goods can be accumulated without bound on a per capita basis, whereas a piece of human capital such as the ability to add cannot. Each person has only a finite number of years that can be spent acquiring skills. When this person dies, the skills are lost, but any nonrival good that this person produces—a scientific law; a principle of mechanical, electrical, or chemical engineering; a mathematical result; software; a patent; a mechanical drawing; or a blueprint—lives on after the person is gone. Second, treating knowledge as a nonrival good makes it possible to talk sensibly about knowledge spillovers, that is, incomplete excludability. These two features of knowledge—unbounded growth and incomplete appropriability—are features that are generally recognized as being relevant for the theory of growth. What thinking about nonrivalry shows is that these features are inextricably linked to nonconvexities.

If a nonrival input has productive value, then output cannot be a constant-returns-to-scale function of all its inputs taken together. The standard replication argument used to justify homogeneity of degree one does not apply because it is not necessary to replicate nonrival inputs. Suppose that a firm can invest 10,000 hours of engineering time to produce a design for a 20-megabyte hard disk drive for computers. Suppose that it can produce a total of 2 trillion megabytes of storage per year (i.e., 100,000 units of the drive) if it builds a $10 million factory and hires 100 workers. If it merely replicates the rival inputs—the factory and the workers—it can double its output to 4 trillion megabytes of storage per year.

Suppose that the firm could have invested 20,000 hours of engineering time in the design work instead of 10,000 hours and, by doing so, could have designed a 30-megabyte hard disk drive that could be manufactured with the same factory and workers. When the firm doubles all its inputs, it uses a 20,000-hour design, two factories, and 200 workers and produces 6 trillion megabytes of storage per year, three times the original output.

More formally, if $F(A, X)$ represents a production process that depends on rival inputs X and nonrival inputs A, then by a replication argument, it follows that $F(A, \lambda X) = \lambda F(A, X)$. This replication argument assumes that X is an exhaustive list of the rival inputs. Because the focus here is on national economies, the argument neglects integer problems that may be relevant for a small market that gets stuck between n and $n + 1$ plants. The fact that it may not be possible to actually replicate all the inputs in the list X has no bearing on this argument about the properties of $F(\cdot)$.

If A is productive as well, it follows that F cannot be a concave production function because $F(\lambda A, \lambda X) > \lambda F(A, X)$. Because of the properties of homogeneous functions, it also follows that a firm with these kinds of production possibilities could not survive as a price taker. If disk drives sold for marginal cost, annual revenue for the firm would just equal interest payments on the capital and wage payments to workers. More generally, since $F(A, X) = X \cdot (\partial F/\partial X)(A, X)$, it follows that

$$F(A, X) < A \cdot \frac{\partial F}{\partial A}(A, X) + X \cdot \frac{\partial F}{\partial X}(A, X).$$

If all inputs were paid their value marginal product, the firm would suffer losses.

This point has been made many times before (Schumpeter 1942; Arrow 1962*b*; Shell 1966, 1967, 1973; Nordhaus 1969; Wilson 1975). Previous growth models have avoided this difficulty in various ways. Solow (1956) treats A as an exogenously provided public input (i.e., an input that is both nonexcludable and nonrival). Shell (1966, 1967) treats it as a public input that is provided by the government. In each case, the factor A receives no compensation, and every individual firm is assumed to be free to exploit the entire stock of A. These models are consistent with the first premise, that technological change drives growth, and the third, that the technology is a nonrival good, but they are inconsistent with the second premise. They both deny the role that private, maximizing behavior plays in generating technological change.

In an attempt to make the evolution of A responsive to market incentives, Arrow (1962*a*) assumed that an increase in K necessarily leads to an equiproportionate increase in knowledge through "learning by doing," but he still treats knowledge as a public good. Lucas (1988) assumed in effect that it is production of human capital rather than physical capital that generates this nonrival, nonexcludable good. Both of these papers make the production of a nonrival, nonexcludable good an unintentional side effect of the production of a conventional good.

The learning-by-doing formulation has the advantage that it makes the rate of accumulation of nonrival knowledge endogenous, but it is unsatisfactory because it takes the strict proportionality between knowledge and physical capital or knowledge and education as an unexplained and exogenously given feature of the technology. It preserves the public-good character of knowledge assumed by Solow and Shell but makes it a public good that is privately provided as a side effect. Like the other public-good formulations, it rules out the possibility that firms make intentional investments in research and development.

This formulation has the additional difficulty that it is not robust. The nonrival input produced through learning by doing must be completely nonexcludable. If it were even partially excludable, Dasgupta and Stiglitz (1988) show that decentralized equilibrium with many firms would not be sustainable.

In a partial equilibrium model of an industry in which firms face upward-sloping cost curves, Shell (1973) proposed a model with price taking in which expenditure on research was compensated out of quasi rents. Griliches (1979), again in an industry setting, made this formulation more explicit. He assumed that the production function takes the form $F(A_N, A_E, X)$, where A_E represents an excludable part of the benefits of research and development and A_N represents the nonexcludable part. Since A_E is excludable, it is accumulated intentionally. The nonexcludable part A_N is created as a side effect of producing A_E. He also assumed that the function $F(\cdot)$ is homogeneous of degree one in X and A_E taken together.

In an aggregate model of growth, I made the same kind of assumption (Romer 1986). To make the dynamic analysis in this paper simple, I reduced the dynamic model to one with a single-state-variable model by assuming that the excludable good A_E that the firm produces intentionally is used in fixed proportions with physical capital. As a result, the model ends up having dynamics similar to those of Arrow's learning-by-doing model, and the mathematical equations can be interpreted equally well in terms of learning by doing that is incidental to capital production.

The advantage of the interpretation that knowledge is compensated out of quasi rents is that it allows for intentional private investments in research and development. The difficulty is that it violates the logic of the replication argument. If the input A_E is truly the result of research and development, it is a nonrival good. In this case, the function $F(\cdot)$ must be homogeneous of degree one in X alone and the Dasgupta and Stiglitz (1988) argument applies. If a nonrival input is even partially excludable, nonconvexities are present and a decentralized equilibrium cannot be sustained. Stated in terms of cost curves, the replication argument implies that long-run cost curves are horizontal when all rival inputs are treated as variable. What appear to be quasi rents are merely competitive returns to rival factors that are in fixed supply. These quasi rents cannot be used to compensate both the innovation activity and the rival fixed factors. Even if these factors are in fixed supply at the aggregate level, they presumably have alternative uses and will not be supplied to an activity if they are not paid their marginal product.

An alternative approach to growth theory with price-taking competition is to dispute the first and third premises, that technological change drives growth and that knowledge about the technology is a nonrival input. Human capital models such as those presented by King and Rebelo (1987), Jones and Manuelli (1988), Rebelo (1988), and Becker, Murphy, and Tamura (this issue) treat all forms of intangible knowledge as being analogous to human capital skills that are rivalrous and excludable. There is no nonrival input like the technology and, hence, no nonconvexities or spillovers.

The only way to accept all three premises described in the Introduction is to return to the suggestion of Schumpeter (1942) and explicitly introduce market power. Shell (1973) described a model with a single monopolist who invests in technological change, but as he recognizes, it is difficult to give an aggregate interpretation to a model dominated by a single firm. In a recent paper (Romer 1987), I presented a model with market power but also with free entry and many firms. It builds on the model of monopolistic competition in consumption goods formulated by Dixit and Stiglitz (1977), applied in a dynamic setting by Judd (1985), used in a dynamic model of trade by Grossman and Helpman (1989*c*), and extended to differentiated inputs in production by Ethier (1982). The specification used here differs from that in my earlier paper primarily because it emphasizes the importance of human capital in the research process. The earlier model showed that scale is an important determinant of the rate of growth. The analysis here shows that the correct measure of scale is not population but human capital.

III. Description of the Model

The four basic inputs in this model are capital, labor, human capital, and an index of the level of the technology. Capital is measured in units of consumption goods. Labor services L are skills such as eye-hand coordination that are available from a healthy physical body. They are measured by counts of people. As used here, human capital H is a distinct measure of the cumulative effect of activities such as formal education and on-the-job training.

The concept of human capital as years of education or training that are person specific is close to the one used in labor market contexts (e.g., Heckman 1976; Rosen 1976). It corresponds to the practice in growth accounting applications that take account of changes in the quality of the labor force due to changes in observables such as the level of education and experience (see, e.g., Gollop and Jorgenson 1980). This concept of human capital is more limited than the notion used in theoretical models of growth based on unlimited human capital accumulation such as those presented by King and Rebelo (1987), Lucas (1988), and Becker et al. (this issue). These models implicitly combine a notion of knowledge that can outlive any individual with a labor market notion of human capital that does not.

The model used here separates the rival component of knowledge, H, from the nonrival, technological component, A. Because it has an existence that is separate from that of any individual, A can grow without bound. In the specific formulation used below, each new unit of knowledge corresponds to a design for a new good, so there is no conceptual problem measuring A. It is a count of the number of designs.

The formal model of the economy has three sectors. The research sector uses human capital and the existing stock of knowledge to produce new knowledge. Specifically, it produces designs for new producer durables. An intermediate-goods sector uses the designs from the research sector together with forgone output to produce the large number of producer durables that are available for use in final-goods production at any time. In practice, one might expect research on a new design and the production of the new good to take place within the same firm, and nothing in the analysis here rules this out; design work can take place either internally or in a separate firm that sells its patent to the firm that will produce the actual good. A final-goods sector uses labor, human capital, and the set of producer durables that are available to produce final output. Output can be either consumed or saved as new capital.

To keep the dynamic analysis simple and highlight the effects of interest, several simplifying assumptions are used. The first is that the population and the supply of labor are both constant. This rules out an analysis of fertility, labor force participation, or variation in hours worked per worker. The second is that the total stock of human capital in the population is fixed and that the fraction supplied to the market is also fixed. Thus the supply of the aggregate factors L and H

is fixed. The assumption on H is made largely for technical reasons. The dynamic analysis is greatly simplified by restricting attention to equilibria with constant growth rates. In a stationary population in which people have finite lives, the only feasible constant growth rate for total years of education or experience is zero. A more complicated dynamic analysis could consider the effects of the kinds of increases in H and L that have been observed historically, but even in this kind of analysis, H must ultimately approach an upper bound.

The other simplifying assumptions are extreme assumptions on factor intensities. One has already been made implicitly. Assuming that capital can be accumulated as forgone output is equivalent to assuming that capital goods are produced in a separate sector that has the same technology as the final-output sector. Forgoing consumption is then equivalent to shifting resources from the consumption sector into the capital sector. Also, the plausible assertion that research is relatively human capital– and knowledge-intensive is translated into an extreme specification in which only knowledge and human capital are used to produce new designs or knowledge. Labor and capital do not enter at all. These kinds of restrictions will reduce the analysis of the dynamics of this system to a system of equations that can be explicitly solved by doing algebra. Presumably, a relaxation of these assumptions that preserves the factor intensity orderings used here would not change the basic dynamics of the model.

Final output Y in this model is expressed as a function of physical labor L, human capital devoted to final output H_Y, and physical capital. The unusual feature of the production technology assumed here is that it disaggregates capital into an infinite number of distinct types of producer durables. For now, let these durables be indexed by an integer i. (Soon, the index i will be assumed to be a continuous variable instead of a discrete one to avoid integer constraints.) Only a finite number of these potential inputs, the ones that have already been invented and designed, are available for use at any time. Thus if $x = \{x_i\}_{i=1}^{\infty}$ is the list of inputs used by a firm that produces final output, there is some value A such that $x_i = 0$ for all $i \geq A$. Because A changes as new producer durables are invented, it is important to be able to describe final output as a stationary function of all conceivable input lists.

In this kind of environment, a simple functional form for output is the following extension of the Cobb-Douglas production function:

$$Y(H_Y, L, x) = H_Y^{\alpha} L^{\beta} \sum_{i=1}^{\infty} x_i^{1-\alpha-\beta}. \qquad (1)$$

This production function differs from the usual production function only in its assumption about the degree to which different types of capital goods are substitutes for each other. In the conventional specification, total capital K is implicitly defined as being proportional to the sum of all the different types of capital. This definition implies that all capital goods are perfect substitutes. One additional dollar of capital in the form of a truck has the same effect on the marginal productivity of mainframe computers as an additional dollar's worth of computers. Equation (1) expresses output as an additively separable function of all the different types of capital goods so that one additional dollar of trucks has no effect on the marginal productivity of computers.

Among the various types of capital goods, one can imagine pairs of inputs that are close substitutes (trucks and trains), pairs that are complements (computers and communications networks), and many pairs that fall somewhere in between. The conventional formulation explores the case in which all durables are perfect substitutes. The model here considers the case in which all durables have additively separable effects on output. An investigation of complementarity as well as of mixtures of types of substitutability is left for future work.

Because the production function in equation (1) is homogeneous of degree one, output in the final-goods sector can be described in terms of the actions of a single, aggregate, price-taking firm. The sector that produces producer durables, however, cannot be described by a representative firm. There is a distinct firm i for each durable good i. A firm must purchase or produce a design for good i before commencing production. Once it owns the design, the firm can convert η units of final output into one durable unit of good i. As in the standard one-sector model, the formal specification here describes the sector that produces capital goods as a black box that takes final output in on one side and gives capital goods out of the other side. The correct interpretation of this formal description is that the forgone consumption is never manufactured. The resources that would have been used to produce the forgone output are used instead to manufacture capital goods. It is possible to exchange a constant number of consumption goods for each unit of capital goods if the production function used to manufacture capital goods has exactly the same functional form as the production function used to manufacture consumption goods.

Once a firm has produced a design for durable i, it can obtain an infinitely lived patent on that design. If the firm manufactures $x(i)$ units of the durable, it rents those durables to final-output firms for a rental rate $p(i)$. Since firm i will be the only seller of capital good i, it will face a downward-sloping demand curve for its good. Since the durables are assumed not to depreciate, the value of one unit of durable i is the present discounted value of the infinite stream of rental income that it generates.

There are many equivalent institutional arrangements that can support any given equilibrium. In what follows, the firm that owns the patent on the design for durable i is assumed to be the only one that manufactures it, but this is only a convenience. Whether the owner of the patent manufactures the good itself or licenses others to do so, it can extract the same monopoly profit. Design of new durables and manufacturing could take place within the same firm, but it is easier to describe the equilibrium if the research and development department is treated as a separate firm and designs are transferred for an explicit price. It is also easier to assume that the firm that buys a design and manufactures a differentiated producer durable rents its durables instead of selling them outright. In particular, this shows that there are market mechanisms that avoid the usual durable-goods-monopoly problem. Provided that the manufacturer of the durable could commit to levels of output, nothing would change if the durables were sold instead of rented. The analysis is further simplified by assuming that the durables do not depreciate. Adding depreciation would merely add a familiar term to the user cost of capital.

In parallel with the usual one-sector model and in conformity with national income accounting conventions, it is useful to define an accounting measure of total capital K as cumulative forgone output. Thus $K(t)$ evolves according to the rule

$$\dot{K}(t) = Y(t) - C(t), \tag{2}$$

where $C(t)$ denotes aggregate consumption at time t. Because it takes η units of forgone consumption to create one unit of any type of durable, this accounting measure K is related to the durable goods that are actually used in production by the rule $K = \eta \sum_{i=1}^{\infty} x_i = \eta \sum_{i=1}^{A} x_i$.

Thus H and L are fixed, and K grows by the amount of forgone consumption. It remains to specify the process for the accumulation of new designs, that is, for the growth of $A(t)$. As noted above, research output depends on the amount of human capital devoted to research. It also depends on the stock of knowledge available to a person doing research. If designs were treated as discrete indivisible objects that are not produced by a deterministic production process, the production technology for designs would have to take explicit account of both integer constraints and uncertainty. There is no doubt that both indivisibility and uncertainty are important at the micro level and over short periods of time. The simplifying assumption made here is that neither is crucial to a first-pass analysis of technological change at the aggregate level. Henceforth, the index i

for the different types of goods is treated as a continuous variable, and the sum in equation (1) is replaced by an integral:

$$Y(H_Y, L, x) = H_Y^{\alpha} L^{\beta} \int_0^{\infty} x(i)^{1-\alpha-\beta}\, di. \tag{1'}$$

(Here x must be interpreted as a function from an appropriately defined function space.) With this formal structure, the output of new designs produced by researcher j can be written as a continuous, deterministic function of the inputs applied. If the researcher possesses an amount of human capital H^j and has access to a portion A^j of the total stock of knowledge implicit in previous designs, the rate of production of new designs by researcher j will be $\delta H^j A^j$, where δ is a productivity parameter.

This formulation with a continuum of goods is close to that used by Judd (1985) in his discussion of patents. He also studies the process whereby goods are introduced as new patents are produced. The model here corresponds to Judd's discussion of the case in which patents are infinitely lived. The main differences are that Judd treats the differentiated goods as consumption goods rather than producer durables and uses a form of exogenous technological change to generate growth of productivity in the research sector. Here, growth in A by itself increases the productivity of human capital in the research sector.

Although other assumptions about secrecy and property rights could be considered, the equilibrium here is based on the assumption that anyone engaged in research has free access to the entire stock of knowledge. This is feasible because knowledge is a nonrival input. All researchers can take advantage of A at the same time. The output of researcher j is therefore $\delta H^j A$. If we sum across all people engaged in research, the aggregate stock of designs evolves according to

$$\dot{A} = \delta H_A A, \tag{3}$$

where H_A has the obvious interpretation of total human capital employed in research.

Equation (3) contains two substantive assumptions and two functional form assumptions. The first substantive assumption is that devoting more human capital to research leads to a higher rate of production of new designs. The second is that the larger the total stock of designs and knowledge is, the higher the productivity of an engineer working in the research sector will be. According to this specification, a college-educated engineer working today and one working 100 years ago have the same human capital, which is measured in terms of years of forgone participation in the labor market. The engineer working today is more productive because he or she can take advantage of all the additional knowledge accumulated as design problems were solved during the last 100 years.

The two functional form assumptions are that the output of designs is linear in each of H_A and A when the other is held constant. These assumptions, like the exclusion of the inputs L and $x(\cdot)$, are made largely for analytical convenience. Linearity in H_A is not important for the dynamic properties of the model, but weakening this assumption would require a more detailed specification of how income in the research sector is allocated to the participants.

Linearity in A is what makes unbounded growth possible, and in this sense, unbounded growth is more like an assumption than a result of the model. In what follows, it will become clear that the marginal product of human capital H_Y employed in the manufacturing sector grows in proportion to A. If A were replaced in equation (3) by some concave function of A—that is, if the marginal productivity of human capital in the research sector does not continue to grow in proportion to A—human capital employed in research would shift out of research and into manufacturing as A becomes larger. This will cause the rate of growth to slow down. Whether opportunities in research are actually petering out, or will eventually do so, is an empirical question that this kind of theory cannot resolve. The specification here, in which unbounded growth at a constant rate is feasible, was chosen because there is no evidence from recent history to support the belief that opportunities for research are diminishing. Moreover, linearity in A is convenient analytically, and assumptions about what will happen in the far future for values of A that are very large relative to the current level have very little effect on the question of interest: How do other variables in the model affect the rate of growth of A?

The crucial feature of the specification used here is that knowledge enters into production in two distinct ways. A new design enables the production of a new good that can be used to produce output. A new design also increases the total stock of knowledge and thereby increases the productivity of human capital in the research sector. The owner of a design has property rights over its use in the production of a new producer durable but not over its use in research. If an inventor has a patented design for widgets, no one can make or sell widgets without the agreement of the inventor. On the other hand, other inventors are free to spend time studying the patent application for the widget and learn knowledge that helps in the design of a wodget. The inventor of the widget has no ability to stop the inventor of a wodget from learning from the design of a widget. This means that the benefits from the first productive role for a design are completely excludable, whereas the benefits from the second are completely nonexcludable. In an overall sense, this means that the nonrival design inputs are partially excludable.

In theory and in practice, there is always some ambiguity about what constitutes a design for a new and different good and what constitutes a copy of an existing design. In the model, this ambiguity is artificially resolved by the form of the production function Y. This functional form implies that new goods are never close substitutes for existing goods since all the producer durables enter into production in an additively separable fashion. (See Pascoa [1986] for a discussion of this general property of "no neighboring goods.") Although it greatly simplifies the analysis, this is not a realistic feature of the model. In particular, it rules out the possibility of obsolescence. Nonetheless, the general results here should be robust to more careful modeling of the nature of the interaction between different specialized producer durables. What matters for the results is that the knowledge is a nonrival good that is partially excludable and privately provided.

At the aggregate level, H_A and H_Y are related by the constraint $H_Y + H_A = H$. According to these equations, any person can devote human capital to either the final-output sector or the research sector. Implicitly this formulation neglects the fact that L and H are supplied jointly. To take the equations used here literally, one must imagine that there are some skilled persons who specialize in human capital accumulation and supply no labor.

To fix notation for prices, let spot prices at any point in time be measured in units of current output and let r denote the interest rate on loans denominated in goods. Let P_A denote the price of new designs, and let w_H denote the rental rate per unit of human capital. Because goods can be converted into capital one for one, the spot price for capital is one and its rate of return is r. Because of the assumption that anyone engaged in research can freely take advantage of the entire existing stock of designs in doing research to produce new designs, it follows from equation (3) that P_A and w_H are related by $w_H = P_A \delta A$.

Once a design has been produced, a large number of potential suppliers of the new good bid for the right to do so. Each of these firms takes the price P_A for designs, the price of one for capital goods, and the interest rate as given, but if it begins production it sets prices to maximize profits. Formally, it helps to let the rental price $p(i)$ for the ith durable lie in the range $\mathbb{R}_+ \cup \{\infty\}$ for its durable good. If no firm produces good i, its price can be understood to be $p(i) = \infty$.

Faced with a price list $\{p(i): i \in \mathbb{R}_+\}$ for all the producer durables, including infinite prices for the durables that have not been invented yet, the representative final-output firm chooses a profit-maximizing quantity $x(i)$ for each durable. Because it is a constant-returns-to-scale firm, its input demands are defined only after the scale of operation is pinned down. Let L and H_Y be the total amounts of labor and human capital that are used in the production of final-output goods. (The split of total H between H_Y and H_A remains to be determined.) Given values for H_Y and L, it is possible to derive the aggregate demand for the durables from a maximization problem that is conditional on them:

$$\max_x \int_0^{\infty} [H_Y^{\alpha} L^{\beta} x(i)^{1-\alpha-\beta} - p(i)x(i)]di.$$

Differentiating under the integral sign leads to an inverse demand function

$$p(i) = (1 - \alpha - \beta)H_Y^{\alpha} L^{\beta} x(i)^{-\alpha-\beta}. \tag{4}$$

(There is an important technical issue about what it means to the final-goods producer if prices change on a set of values of i that has measure zero. Because of the symmetry in this model, eq. [4] can readily be derived by a limiting argument. For a general discussion of this issue, see Pascoa [1986, 1990].)

The demand curve in equation (4) is what the producer of each specialized durable takes as given in choosing the profit-maximizing price to set. Faced with given values of H_Y, L, and r, a firm that has already incurred the fixed-cost investment in a design will choose a level of output x to maximize its revenue minus variable cost at every date:

$$\begin{aligned} \pi &= \max_x p(x)x - r\eta x \\ &= \max_x (1 - \alpha - \beta)H_Y^{\alpha} L^{\beta} x^{1-\alpha-\beta} - r\eta x. \end{aligned} \tag{5}$$

The flow of rental income is $p(x)$ times x. The cost is the interest cost on the ηx units of output needed to produce x durables. To keep the analysis simple, it is convenient to assume that the capital is putty-putty, so that the firm can solve this problem at every point in time, converting units of durables back into general capital, and avoid the interest cost if it decides to supply fewer units. At any date, the only sunk cost is the initial expenditure on the design. This assumption is harmless because the demand for durables is stationary in equilibrium, so no disinvestment ever takes place.

The monopoly pricing problem specified in equation (5) is that of a firm with constant marginal cost that faces a constant elasticity demand curve. The resulting monopoly price is a simple markup over marginal cost, where the markup is determined by the elasticity of demand, $\bar{p} = r\eta/(1 - \alpha - \beta)$. The flow of monopoly profit is $\pi = (\alpha + \beta)\bar{p}\bar{x}$, where $\bar{x}$ is the quantity on the demand curve (4) implied by the price $\bar{p}$.

Each producer of specialized durables must rent its output to a large number of final-goods producers that can operate at any scale. By assumption, it is not possible for the producer to monitor the use of its durables. As a result, price discrimination is not feasible. The best the firm can do is charge the simple monopoly price.

The decision to produce a new specialized input depends on a comparison of the discounted stream of net revenue and the cost P_A of the initial investment in a design. Because the market for designs is competitive, the price for designs will be bid up until it is equal to the present value of the net revenue that a monopolist can extract. At every date t, it must therefore be true that

$$\int_t^\infty e^{-\int_t^\tau r(s)ds}\pi(\tau)d\tau = P_A(t). \qquad (6)$$

If P_A is constant (as it will be in the equilibrium described below), this condition can be put in a more intuitive form. Differentiating with respect to time t yields

$$\pi(t) - r(t)\int_t^\infty e^{-\int_t^\tau r(s)ds}\pi(\tau)d\tau = 0.$$

Substituting in the expression for P_A from equation (6) yields

$$\pi(t) = r(t)P_A. \qquad (6')$$

This equation says that at every point in time, the instantaneous excess of revenue over marginal cost must be just sufficient to cover the interest cost on the initial investment in a design. (This formulation of the intertemporal zero profit constraint is taken from Grossman and Helpman [1989*c*].)

The solution of the model for a balanced growth equilibrium given in Section V shows that the technology described above implies a negatively sloped linear relation between the rate of growth of output and the rate of return on investment. To close the model, it remains to specify preferences that imply a parallel relation between the rate of growth of consumption and the marginal rate of intertemporal substitution. This relation is easily derived for Ramsey consumers with discounted, constant elasticity preferences:

$$\int_0^\infty U(C)e^{-\rho t}dt, \quad \text{with } U(C) = \frac{C^{1-\sigma} - 1}{1 - \sigma} \quad \text{for } \sigma \in [0, \infty).$$

The implied intertemporal optimization condition for a consumer faced with a fixed interest rate r is that $\dot{C}/C = (r - \rho)/\sigma$. Preferences enter the solution of the model only through this relation between the consumption growth rate and the interest rate.

The consumers are endowed with fixed quantities of labor L and human capital H that are supplied inelastically. At time 0, consumers own the existing durable-goods-producing firms, and the net revenues of these firms are paid to consumers as dividends. Final-goods firms earn zero profits and own no assets, so they can be ignored in the specification of endowments.

An equilibrium for this model will be paths for prices and quantities such that (i) consumers make savings and consumption decisions taking interest rates as given; (ii) holders of human capital decide whether to work in the research sector or the manufacturing sector taking as given the stock of total knowledge A, the price of designs P_A, and the wage rate in the manufacturing sector w_A; (iii) final-goods producers choose labor, human capital, and a list of differentiated durables taking prices as given; (iv) each firm that owns a design and manufactures a producer durable maximizes profit taking as given the interest rate and the downward-sloping demand curve it faces, and setting prices to maximize profits; (v) firms contemplating entry into the business of producing a durable take prices for designs as given; and (vi) the supply of each good is equal to the demand.

IV. Discussion of the Model

A reasonable intuition for the behavior of this model can be inferred by considering the Solow (1956) model, in which the evolution of A is given exogenously, and the Uzawa (1965) model, in which the evolution of A is determined by the allocation of resources between a research sector and a final-goods sector. For a fixed amount of A, and therefore a fixed set of producer durables, the model is almost identical to the Solow model. Because of the symmetry in the model, all the durable goods that are available are supplied at the same level, henceforth denoted as $\bar{x}$. If they were not, it would be possible to increase profits in the producer durable sector by reducing the output of high-output firms and diverting the capital released in this way to low-output goods. Since A determines the range of durables that can be produced and since η units of capital are required per unit of durable goods, it is possible to solve for $\bar{x}$ from the equation $K = \eta A\bar{x}$. Then output Y can be written as

$$\begin{aligned} Y(H_A, L, x) &= H_Y^{\alpha} L^{\beta} \int_0^{\infty} x(i)^{1-\alpha-\beta}\, di \\ &= H_Y^{\alpha} L^{\beta} A \bar{x}^{1-\alpha-\beta} \\ &= H_Y^{\alpha} L^{\beta} A \left(\frac{K}{\eta A}\right)^{1-\alpha-\beta} \\ &= (H_Y A)^{\alpha} (LA)^{\beta} (K)^{1-\alpha-\beta} \eta^{\alpha+\beta-1}. \end{aligned} \tag{7}$$

The last line of this equation shows that the model behaves just like the neoclassical model with labor and human capital augmenting technological change. In particular, it exhibits the usual diminishing returns to capital accumulation. Given the assumed form of preferences, a fixed level of A will lead to an equilibrium with a steady state in which the level of K is determined by the requirement that the marginal product of capital is equal to the discount rate. If A grew at an exogenously specified exponential rate, the economy would converge to a path on which K grows at the same exponential rate as A, just as it does in the Solow model. Along the transition path, the ratio of K to A would change, which implies that r and $\bar{x}$ would change as well. Along the balanced growth path, r, $\bar{x}$, and the ratio of K to A are all constant.

The nonconvexity evident in the expression for final output as a function of the primary inputs of the model (H, L, K, and A) is supported in a decentralized equilibrium that relies on monopolistic competition. In contrast, the nonconvexity present in equation (3) describing output of designs is supported through competition with external effects that arise from knowledge spillovers. In each case, the nonconvexity arises because the nonrival good A is an input in production. In the final-output sector, A matters indirectly because of its effects on the availability of the new $x(\cdot)$ goods. In the research sector, A enters directly.

Both spillovers and price setting seem essential to capturing the features of knowledge in a model of growth. There is little doubt that much of the value to society of any given innovation or discovery is not captured by the inventor, and any model that missed these spillovers would miss important elements of the growth process. Yet it is still the case that private, profit-maximizing agents make investments in the creation of new knowledge and that they earn a return on these investments by charging a price for the resulting goods that is greater than the marginal cost of producing the goods.

V. Solution of the Model for a Balanced Growth Equilibrium

The strategy for characterizing the model that is followed here is to solve for an equilibrium in which the variables A, K, and Y grow at constant exponential rates. This is generally referred to as a balanced growth equilibrium. The intuition from the Solow model suggests that such an equilibrium will exist if A grows at a constant exponential rate. The intuition from the Uzawa model suggests that it is possible for A to grow at an exponential rate because equation (3) for $\dot{A}$ is linear in A. It will grow at a constant rate if the amount of human capital H_A that is devoted to research stays constant. Verifying that a balanced growth equilibrium exists therefore reduces to the problem of showing that prices and wages are such that H_Y and H_A remain constant as Y, K, C, and A grow.

By focusing only on balanced growth paths, the analysis neglects the transient dynamics that arise when the economy starts from a ratio of K to A that differs from the ratio that is maintained along the balanced growth path. One should be able to study convergence to the balanced growth ratio of K to A using the tools used for studying the Solow and Uzawa models, but this analysis is not attempted here.

The first step in the solution of the model is to derive the relation between the growth rate of output and the rate of return on investment. Coupled with the relation between interest rates and growth rates implied by the preference side of the model, this will determine the rate of growth and interest rate. Given the intuition from the Solow model and the results derived in my earlier model of differentiated inputs (Romer 1987), it follows that along the balanced growth path, the ratio of K to A should be constant, which implies that $\bar{x}$ is constant as well. Because of the accumulation of both K and A, the wage paid for human capital in the final-output sector will grow in proportion to A, but by equation (3), the productivity of human capital in research also grows in proportion to A. Since the productivity of human capital grows at the same rate in both sectors, H_Y and H_A will remain constant if the price P_A for new designs is constant.

Figure 1 illustrates the behavior of the inputs in the model. All the producer durables that have been designed up to time t are used at the level $\bar{x}$ that indicates the height of the rectangle. The width is the measure of the number of designs or durables in use, $A(t)$. The area $A(t)\bar{x}$ is equal to total capital divided by η. Over time, $\bar{x}$ remains constant and A grows at a constant exponential rate.

It remains to check that this description of a balanced growth path is consistent with all the equilibrium conditions. As noted in the dis-

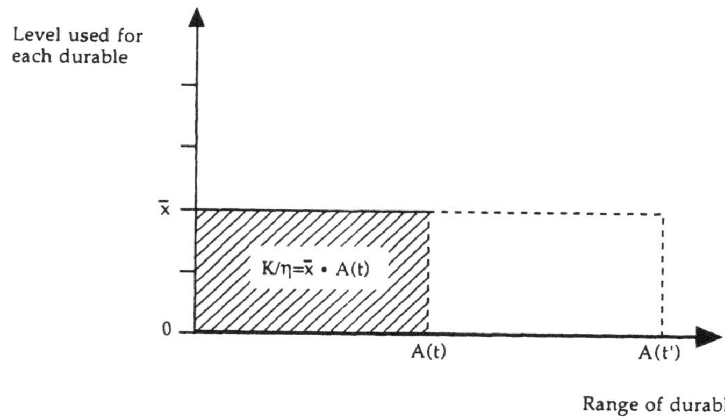

FIG. 1.—Producer durables used in production at time t and $t' > t$

cussion following the monopoly profit problem (5), the flow of profit that can be extracted by the seller of any particular durable input is $\pi = (\alpha + \beta)\bar{p}\bar{x}$. Since the present discounted value of this stream of profit must equal the price P_A of the design, it follows that

$$P_A = \frac{1}{r}\pi = \frac{\alpha + \beta}{r}\bar{p}\bar{x} = \frac{\alpha + \beta}{r}(1 - \alpha - \beta)H_Y^{\alpha}L^{\beta}\bar{x}^{1-\alpha-\beta}. \quad (8)$$

The last equality in this expression follows by using equation (4) to evaluate $\bar{p}$ in terms of $\bar{x}$.

The condition determining the allocation of human capital between the final-output and research sectors says that the wages paid to human capital in each sector must be the same. In the final-output sector, the wage for human capital is its marginal product. Since human capital receives all the income from the research sector, the wage there is $P_A\delta A$. To equalize returns to human capital in both sectors, $H_Y = H - H_A$ must be chosen so that

$$\begin{aligned} w_H = P_A\delta A &= \alpha H_Y^{\alpha-1}L^{\beta}\int_0^{\infty}\bar{x}^{1-\alpha-\beta}\,di \\ &= \alpha H_Y^{\alpha-1}L^{\beta}A\bar{x}^{1-\alpha-\beta}. \end{aligned} \quad (9)$$

Substituting P_A from equation (8) into equation (9) and simplifying yield

$$H_Y = \frac{1}{\delta}\frac{\alpha}{(1 - \alpha - \beta)(\alpha + \beta)}r. \quad (10)$$

For a fixed value of $H_A = H - H_Y$, the implied exponential growth rate for A is δH_A. From the monopoly pricing problem, we know that

$\bar{x}$ is constant if r is. The intuition from the Solow model suggests, and examination of the expression for final output

$$Y = H_Y^{\alpha} L^{\beta} \int_0^{\infty} \bar{x}^{1-\alpha-\beta}\, di = H_Y^{\alpha} L^{\beta} A \bar{x}^{1-\alpha-\beta}$$

from equation (7) shows, that output grows at the same rate as A if L, H_Y, and $\bar{x}$ are fixed. If $\bar{x}$ is fixed, then K must grow at the same rate as A, because total usage of capital is $A\bar{x}\eta$. Let g denote the growth rate of A, Y, and K. Since K/Y is a constant, the ratio

$$\frac{C}{Y} = 1 - \frac{\dot{K}}{Y} = 1 - \frac{\dot{K}}{K}\frac{K}{Y}$$

must also be constant. The common growth rate g for all these variables is therefore

$$g = \frac{\dot{C}}{C} = \frac{\dot{Y}}{Y} = \frac{\dot{K}}{K} = \frac{\dot{A}}{A} = \delta H_A.$$

Together with equation (10), the constraint $H_Y = H - H_A$ implies a relation between the growth rate g and the interest rate r:

$$g = \delta H_A = \delta H - \frac{\alpha}{(1 - \alpha - \beta)(\alpha + \beta)} r, \tag{11}$$

which can be simplified as

$$g = \delta H_A = \delta H - \Lambda r, \tag{11'}$$

where Λ is a constant that depends on the technology parameters α and β,

$$\Lambda = \frac{\alpha}{(1 - \alpha - \beta)(\alpha + \beta)}. \tag{12}$$

Implicitly, the allocation of H between the two sectors is constrained by the requirement that H_A must be nonnegative and that H_Y can be no larger than H. This implies that g is nonnegative. If this constraint is binding, equation (10) will hold as an inequality.

To close the model, it remains to impose the relation between the growth rate g and the interest rate r implied by the preference side of the model, $g = \dot{C}/C = (r - \rho)/\sigma$. Combined with equation (11), this gives an expression for g in terms of the fundamentals of the model,

$$g = \frac{\delta H - \Lambda \rho}{\sigma \Lambda + 1}, \tag{13}$$

where Λ is as defined in equation (12).

The expression for the growth rate suggests a minor technical restriction. For the integral in the consumer's preferences to be finite, the rate of growth of current utility $(1 - \sigma)g$ must be less than the discount rate ρ. Thus, for $\sigma \in [0, 1)$, $(1 - \sigma)\delta H/(\Lambda + 1)$ must be less than the discount rate ρ. If this does not hold, the integral can be infinite and some kind of overtaking criterion must be used to describe the behavior of the consumer.

VI. Interpretation and Welfare Properties of the Solution

Almost all the content of the model is contained in equation (11), which summarizes the effects of the technological side of the model, including the effects of imperfect competition in the market for producer durables. Part of the intuition behind equation (11) is easy to grasp. The opportunity cost of human capital is the wage income that can be earned instantaneously in the manufacturing sector. The return to investing human capital in research is a stream of net revenue that a design generates in the future. If the interest rate is larger, the present discounted value of the stream of net revenue will be lower. Less human capital will be allocated to research, and the rate of growth will be lower.

Other aspects of the intuition behind equation (11) are more subtle. The surprising feature of equation (11) is that neither L nor the parameter η is present. Equation (4) shows that an increase in L increases the demand faced by each monopolistic firm selling a durable good. From the monopoly problem (5), it is clear that a reduction in η reduces the costs of the monopolist and increases output $\bar{x}$. In either case, the stream of net revenue generated by a new design increases. Nonetheless, the amount of human capital devoted to research increases neither when L increases nor when η falls.

In a partial equilibrium analysis, one would always expect any change that increases the return to an activity to increase the allocation of resources to that activity. What a general equilibrium analysis emphasizes is that any intervention that increases the return to one activity can very well increase the return to some other activity that competes with the first activity for resources. An increase in L or a reduction in η (which increases $\bar{x}$) raises the return to human capital employed in manufacturing at the same time that it raises the return to human capital in research. For the functional forms used here, these two effects exactly cancel. The allocation of human capital between research and manufacturing does not change when L and η change.

This kind of exact cancellation is not a robust feature of the model.

If slightly different functional forms were used, the net effect of an increase in L or a decrease in η could be either to increase or to reduce the amount of human capital used in the research sector. For example, in an extension of this model (Romer 1990), I show how an increase in L could reduce research effort and the rate of growth. The correct inference from this model is that the effect of an increase in L or a decrease in η on the rate of growth is ambiguous, something that a priori theorizing cannot resolve.

This ambiguity contrasts with the strong and robust implication that reductions in the interest rate will speed up growth. Equivalently, from equation (12), it follows that any change in the preference parameters that acts to reduce the interest rate (an increase in patience captured by a decrease in the discount rate ρ, or an increase in the intertemporal rate of substitution, as captured by a decrease in σ) will increase research and growth. This implication follows directly from the assumption that the benefits of research come largely in the future and that the costs are incurred immediately.

From a policy point of view, the difference between a reduction in η and a reduction in the equilibrium interest rate is very important. A direct subsidy to investment in physical capital financed by a lump-sum tax is mathematically equivalent to a reduction in η. (An illustration of this easy result is given in Romer [1989].) On the basis of an intuition developed for the one-sector model, economists often identify the marginal product of physical capital with the market interest rate. If they were the same, anything that increases the capital stock and reduces the marginal product of physical capital would have the same growth-enhancing effect as a reduction in the interest rate. The intuition that subsidies to physical capital accumulation will also speed up growth in A was reinforced by the first generation of endogenous growth models. In Arrow's (1962*a*) model of learning by doing and in my first model (Romer 1986), the rate of growth of A was forced by assumption to be the same as the rate of growth of K. As a result, an intervention such as an investment tax credit that increased the accumulation of K necessarily increased the accumulation of A as well.

The model presented here shows that when the decision to invest in physical capital is uncoupled from the decision to invest in research, the effects of a subsidy to physical capital are quite different from the effects of a reduction in the market interest rate. If the fundamental policy problem is that we have too many lawyers and MBAs and not enough engineers, a subsidy to physical capital accumulation is a weak, and possibly counterproductive, policy response.

In the previous paper with differentiated producer durables (Romer 1987), I found that an increase in scale measured by L would increase the rate of growth. Here, an increase in L has no effect.

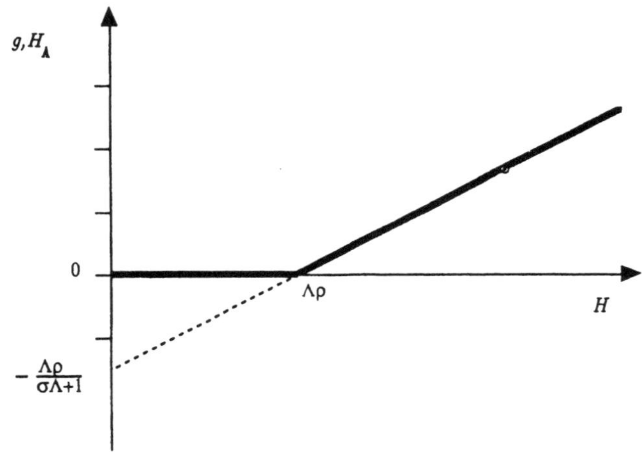

FIG. 2.—Growth rate and amount of human capital in research as a function of total human capital (for $\delta = 1$).

Instead, an increase in scale as measured by total human capital H has the effect of speeding up the rate of growth. This effect is illustrated in figure 2, which plots the rate of growth and the amount of human capital used in research as a function of total human capital. Both of these models have an underlying form of increasing returns in research. As a result, an increase in a scale variable induces an increase in the rate of growth. Human capital is the relevant scale variable in this model because it is the input that is used most intensively in research.

That the research sector in this model exhibits increasing returns is clear from equation (4). If the research technology exhibited constant returns to scale, a doubling of both the human capital and the stock of knowledge would leave the marginal product of human capital in research unchanged. Under the specification used here, a doubling of both leads to an increase in the marginal product of human capital in research. As a result, a permanent increase in the total stock of human capital in the population leads to an increase in the ratio of A to K and a more than proportional increase in the amount of human capital that is devoted to the research sector, as illustrated in figure 2. This implication is of interest from both a historical and a cross-sectional point of view. It is surely the case that the total level of human capital and the fraction of human capital devoted to research are higher now than they were at any time in the past. Moreover, the fraction of human capital devoted to research is apparently highest in the most developed countries of the world.

An extreme possibility illustrated in figure 2 is that if the total level of human capital is too small, stagnation may arise. If H is too low, the nonnegativity constraint on H_A is binding and growth does not take place. In this case, all the feasible growth rates for A are too small relative to the discount rate to justify the sacrifice in current output necessary for growth to take place. This result offers one possible way to explain the wide variation in growth rates observed among countries and the fact that in some countries growth in income per capita has been close to zero. This explanation is reminiscent of the explanation for the absence of growth in prehistoric time that is offered by some historians and anthropologists: civilization, and hence growth, could not begin until human capital could be spared from the production of goods for immediate consumption. This model cannot offer a complete explanation for these observations because it treats the stock of H (and of L) as given, but it does suggest directions for further work.

The effects of a subsidy to capital can be contrasted with a policy designed to encourage research. A subsidy to employment in the research sector that is financed through lump-sum taxes has the same effects on growth as an increase in the productivity parameter δ in equation (3). In the long run, the subsidy will cause an increase in the growth rate, a fall in P_A, and a reduction in $\bar{x}$ and in the ratio of K to A.

There are two reasons to expect that too little human capital is devoted to research. The most obvious reason is that research has positive external effects. An additional design raises the productivity of all future individuals who do research, but because this benefit is nonexcludable, it is not reflected at all in the market price for designs. The second and an equally important reason why too little human capital is devoted to research is that research produces an input that is purchased by a sector that engages in monopoly pricing. The markup of price over marginal cost forces a wedge between the marginal social product of an input used in this sector and its market compensation. Equation (7) shows that a new design (i.e., an additional unit of A) increases output at every date by an amount $H_Y^{\alpha}L^{\beta}\bar{x}^{1-\alpha-\beta}$; from the calculation in (5), the producer of a design captures only a fraction $1 - \alpha - \beta$ of this net benefit to society.

Both of these effects cause human capital to be undercompensated. The marginal product of capital in the manufacturing sector is equal to the wage w_H, but the marginal product in the research sector is higher than the wage because the price of the patent captures only part of the social value of the patent. From equation (9), (10), or (11), it follows that an increase in total human capital would lead to an increase in the amount of human capital employed in the research sector. As a result, in equilibrium, the marginal value of an additional unit of human capital is higher than the market wage. For simplicity, the model here has treated the stock of total human capital as being exogenously fixed, so undercompensating human capital has no welfare effects. In a more realistic model in which human capital is accumulated endogenously, the supply will be too low. As a result, a second-best policy for a government that cannot affect the allocation of human capital between different sectors would be to subsidize the production of human capital.

Within the confines of the model, the social optimum can be achieved by subsidizing the accumulation of A. Demonstrating this result rigorously starting from arbitrary initial conditions forces the analysis to depart from consideration of balanced growth paths. Any intervention designed to move an economy from one balanced growth path to another must consider the transition dynamics along the way, and an explicit analysis of these dynamics is beyond the scope of this paper. It is easy, however, to compare the growth rate along the balanced growth path that emerges from the equilibrium with the one that would emerge from the solution to a social planning problem. Because of the symmetry between the different producer durables in this economy, the optimal level of $x(i)$ is the same for all i between zero and A. This level is related to K and A by the constraint that $K = \eta x A$. Using this to express x in terms of K and A, we can therefore write the social planning problem for this economy as

$$\max \int_0^\infty \frac{C^{1-\sigma} - 1}{1 - \sigma} e^{-\rho t} dt,$$

subject to

$$\dot{K} = \eta^{\alpha+\beta-1} A^{\alpha+\beta} H_Y^\alpha L^\beta K^{1-\alpha-\beta} - C,$$

$$\dot{A} = \delta H_A A,$$

$$H_Y + H_A \leq H.$$

As is shown in the Appendix, the balanced growth solution to the first-order necessary conditions for this problem has a growth rate g^* that is given by

$$g^* = \frac{\delta H - \Theta\rho}{\Theta\sigma + (1 - \Theta)}, \tag{14}$$

where $\Theta = \alpha/(\alpha + \beta)$. The coefficient Λ from equation (12) is equal to this coefficient Θ times the markup from the monopoly sector, $1/(1 - \alpha - \beta)$. This accounts for part of the difference between the equilibrium and the socially optimal rate of growth. However, there is an additional effect because the term in the denominator in equation (14) also replaces the constant 1 from equation (13) with the expression $1 - \Theta$. This change reflects the effect of correcting for the external effects associated with the production of new ideas. Both of these changes in the expression for the rate of growth—the substitution of Θ for Λ and the substitution of $1 - \Theta$ for 1—cause the socially optimal allocation of human capital to research to be higher, and this causes the socially optimal rate of growth to be higher.

VII. Growth, Trade, and Research

The final observations about this model pertain to its implications for growth, trade, and research. These can be seen most simply by comparing the balanced growth equilibrium for two identical closed economies that operate in isolation with the balanced growth equilibrium that would obtain if the economies had always been fully integrated. In isolation, the common growth rate is given by g from equation (13), with H set equal to the amount of human capital in each country. In the second case, the growth rate is found by replacing H by total worldwide human capital, $2H$. Both the fraction H_A/H of total worldwide human capital devoted to research and the rate of growth increase.

This thought experiment suggests why a decision to engage in trade may be important even for a country that has a large population, such as China or India. If access to a large number of workers or consumers were all that mattered, having a large population would be a good substitute for trade with other nations. The model here suggests that what is important for growth is integration not into an economy with a large number of people but rather into one with a large amount of human capital. Many of the details of trade between different economies of this kind remain to be worked out,[2] but since growth seems to be correlated with the degree of integration into worldwide markets but not closely related to population size or density, the results from this model seem promising.

The most direct test of this implication of the model would come from a controlled experiment in which the level of research activity was monitored both before and after a country was opened to trade with the rest of the world. Sokoloff (1988) reports historical data on a natural experiment that comes close to this test. He finds cross-sectional variation that supports the model: counties in the United States in the early nineteenth century that had access to navigable waterways had higher rates of patenting than counties that did not.

[2] Since the first draft of this paper was written, Grossman and Helpman (1989*a*, 1989*b*) have explored detailed models of trade and growth with this kind of underlying technology.

More convincingly, he shows that over time, the introduction of water transportation (because of either the construction of a new canal or the dredging of a river) was followed by a sharp increase in the rate of patenting in counties adjacent to the waterway.

In subsequent work, Sokoloff and Khan (1989) examine the time-series variability in patenting for specific individuals. They find that there was a substantial group of people with broad general knowledge who moved in and out of research in response to aggregate disturbances (primarily business cycles and a trade embargo by the British) just as one would expect if human capital that could be used to do research had alternative uses in manufacturing and commercial trade. They conclude that at least in the nineteenth century, there was a fairly elastic short-run supply of human capital for use in research. Combined with the prior evidence that patenting activity does respond to changes in the size of the market, this offers some assurance that the basic mechanisms described in the model are relevant for historical experience.

VIII. Conclusions

The model presented here is essentially the one-sector neoclassical model with technological change, augmented to give an endogenous explanation of the source of the technological change. The most robust welfare conclusion from the model is that because research projects exchange current costs for a stream of benefits in the future, the rate of technological change is sensitive to the rate of interest. Although all the research is embodied in capital goods, a subsidy to physical capital accumulation may be a very poor substitute for direct subsidies that increase the incentive to undertake research. In the absence of feasible policies that can remove the divergence between the social and private returns to research, a second-best policy would be to subsidize the accumulation of total human capital.

The most interesting positive implication of the model is that an economy with a larger total stock of human capital will experience faster growth. This finding suggests that free international trade can act to speed up growth. It also suggests a way to understand what it is about developed economies in the twentieth century that permitted rates of growth of income per capita that are unprecedented in human history. The model also suggests that low levels of human capital may help explain why growth is not observed in underdeveloped economies that are closed and why a less developed economy with a very large population can still benefit from economic integration with the rest of the world.

Appendix

Calculation of the Balanced Growth Social Optimum

To derive the necessary conditions for the social optimization problem given in the text, construct the current-valued Hamiltonian:

$$\mathcal{H} = \frac{C^{1-\sigma}}{1-\sigma} + \lambda[\eta^{\alpha+\beta-1}A^{\alpha+\beta}(H - H_A)^{\alpha}L^{\beta}K^{1-\alpha-\beta} - C] + \mu\delta H_A A.$$

The necessary conditions follow by maximizing $\mathcal{H}$ with respect to the control variables C and H_A, and from the equations for the evolution of the multipliers λ and μ:

$$\dot{\lambda} = \rho\lambda - \frac{\partial\mathcal{H}}{\partial K}, \quad \dot{\mu} = \rho\mu - \frac{\partial\mathcal{H}}{\partial A}.$$

The first-order condition for maximizing $\mathcal{H}$ with respect to C gives the usual expression relating marginal utility and the multiplier λ:

$$C^{-\sigma} = \lambda. \tag{A1}$$

If the symbol Δ is used to represent the term $\eta^{\alpha+\beta-1}A^{\alpha+\beta}(H - H_A)^{\alpha}L^{\beta}K^{1-\alpha-\beta}$ from the Hamiltonian, the first-order condition for maximizing $\mathcal{H}$ with respect to H_A can be written as

$$\Delta = (H - H_A)\frac{\delta\mu}{\alpha\lambda}A. \tag{A2}$$

Then with equation (A2), the evolution equation for μ can be simplified to yield

$$\frac{\dot{\mu}}{\mu} = \rho - \delta\left(\frac{\alpha+\beta}{\alpha}H - \frac{\beta}{\alpha}H_A\right). \tag{A3}$$

For a balanced growth equilibrium, it must be the case that $\dot{\mu}/\mu = \dot{\lambda}/\lambda$ and that $\dot{C}/C = \dot{A}/A$. With equation (A1), these can be combined to yield $-\sigma(\dot{A}/A) = \dot{\mu}/\mu$. Then combining the evolution equation for A, $\dot{A}/A = \delta H_A$, with equation (A3) gives an equation in H, H_A, and the basic parameters of the model:

$$-\sigma\delta H_A = \rho - \delta\left(\frac{\alpha+\beta}{\alpha}H - \frac{\beta}{\alpha}H_A\right). \tag{A4}$$

Equation (14) in the text can be derived from this equation by solving for H_A and using the fact that the growth rate g is given by $g = \delta H_A$.

References

Abramovitz, Moses. "Resource and Output Trends in the United States since 1870." *A.E.R. Papers and Proc.* 46 (May 1956): 5–23.

Arrow, Kenneth J. "The Economic Implications of Learning by Doing." *Rev. Econ. Studies* 29 (June 1962): 155–73. (*a*)

———. "Economic Welfare and the Allocation of Resources for Invention." In *The Rate and Direction of Inventive Activity: Economic and Social Factors,* edited by Richard Nelson. Princeton, N.J.: Princeton Univ. Press (for NBER), 1962. (*b*)

Becker, Gary S.; Murphy, Kevin M.; and Tamura, Robert. "Human Capital, Fertility, and Economic Growth." *J.P.E.*, this issue.
Cornes, Richard, and Sandler, Todd. *The Theory of Externalities, Public Goods, and Club Goods.* Cambridge: Cambridge Univ. Press, 1986.
Dasgupta, Partha, and Stiglitz, Joseph E. "Learning-by-Doing, Market Structure and Industrial and Trade Policies." *Oxford Econ. Papers* 40 (June 1988): 246–68.
Dixit, Avinash K., and Stiglitz, Joseph E. "Monopolistic Competition and Optimum Product Diversity." *A.E.R.* 67 (June 1977): 297–308.
Ethier, Wilfred J. "National and International Returns to Scale in the Modern Theory of International Trade." *A.E.R.* 72 (June 1982): 389–405.
Gollop, Frank M., and Jorgenson, Dale W. "U.S. Productivity Growth by Industry." In *New Developments in Productivity Measurement and Analysis,* edited by John W. Kendrick and Beatrice N. Vaccara. Chicago: Univ. Chicago Press (for NBER), 1980.
Griliches, Zvi. "Issues in Assessing the Contribution of Research and Development to Productivity Growth." *Bell J. Econ.* 10 (Spring 1979): 92–116.
Grossman, Gene M., and Helpman, Elhanan. "Comparative Advantage and Long Run Growth." Working Paper no. 2809. Cambridge, Mass.: NBER, January 1989. (*a*).
———. "Growth and Welfare in a Small Open Economy." Working Paper no. 2970. Cambridge, Mass.: NBER, May 1989. (*b*)
———. "Product Development and International Trade." *J.P.E.* 97 (December 1989): 1261–83. (*c*)
Heckman, James J. "A Life-Cycle Model of Earnings, Learning, and Consumption." *J.P.E.* 84, no. 4, pt. 2 (August 1976): S11–S44.
Helpman, Elhanan, and Krugman, Paul R. *Market Structure and Foreign Trade: Increasing Returns, Imperfect Competition, and the International Economy.* Cambridge, Mass.: MIT Press, 1985.
Jones, Larry, and Manuelli, Rodolfo. "A Model of Optimal Economic Growth." Working paper. Stanford, Calif.: Stanford Univ., 1988.
Jorgenson, Dale W.; Gollop, Frank M.; and Fraumeni, Barbara M. *Productivity and U.S. Economic Growth.* Cambridge, Mass.: Harvard Univ. Press, 1987.
Judd, Kenneth L. "On the Performance of Patents." *Econometrica* 53 (May 1985): 567–85.
Kendrick, John W. "Productivity Trends: Capital and Labor." *Rev. Econ. and Statis.* 38 (August 1956): 248–57.
King, Robert G., and Rebelo, Sergio. "Business Cycles with Endogenous Growth." Working paper. Rochester, N.Y.: Univ. Rochester, 1987.
Lucas, Robert E., Jr. "On the Mechanics of Economic Development." *J. Monetary Econ.* 22 (July 1988): 3–42.
Maddison, Angus. *Phases of Capitalist Development.* Oxford: Oxford Univ. Press, 1982.
Nordhaus, William D. *Invention, Growth, and Welfare: A Theoretical Treatment of Technological Change.* Cambridge, Mass.: MIT Press, 1969.
Pascoa, Mario. "Monopolistic Competition and No-Neighboring-Goods." Working Paper no. 86-14. Philadelphia: Univ. Pennsylvania, Center Analytic Res. Econ. and Soc. Sci., 1986.
———. "Noncooperative Equilibrium and Chamberlinian Monopolistic Competition." *J. Econ. Theory* (1990), in press.
Rebelo, Sergio. "Long Run Policy Analysis and Long Run Growth." Working paper. Rochester, N.Y.: Univ. Rochester, 1988.

Romer, Paul M. "Increasing Returns and Long-Run Growth." *J.P.E.* 94 (October 1986): 1002–37.

———. "Growth Based on Increasing Returns Due to Specialization." *A.E.R. Papers and Proc.* 77 (May 1987): 56–62.

———. "Capital Accumulation in the Theory of Long Run Growth." In *Modern Business Cycle Theory,* edited by Robert J. Barro. Cambridge, Mass.: Harvard Univ. Press, 1989.

———. "Capital, Labor, and Productivity." *Brookings Papers Econ. Activity* (1990), in press.

Rosen, Sherwin. "A Theory of Life Earnings." *J.P.E.* 84, no. 4, pt. 2 (August 1976): S45–S67.

Schumpeter, Joseph A. *Capitalism, Socialism, and Democracy.* New York: Harper, 1942.

Shell, Karl. "Toward a Theory of Inventive Activity and Capital Accumulation." *A.E.R. Papers and Proc.* 56 (May 1966): 62–68.

———. "A Model of Inventive Activity and Capital Accumulation." In *Essays on the Theory of Optimal Economic Growth,* edited by Karl Shell. Cambridge, Mass.: MIT Press, 1967.

———. "Inventive Activity, Industrial Organisation and Economic Growth." In *Models of Economic Growth,* edited by James A. Mirrlees and Nicholas H. Stern. New York: Wiley, 1973.

Sokoloff, Kenneth L. "Inventive Activity in Early Industrial America: Evidence from Patent Records, 1790–1846." *J. Econ. Hist.* 48 (December 1988): 813–50.

Sokoloff, Kenneth L., and Khan, Zorina. "The Democratization of Invention during Early Industrialization: Evidence from the United States, 1790–1846." Working paper. Los Angeles: Univ. California, 1989.

Solow, Robert M. "A Contribution to the Theory of Economic Growth." *Q.J.E.* 70 (February 1956): 65–94.

———. "Technical Change and the Aggregate Production Function." *Rev. Econ. and Statis.* 39 (August 1957): 312–20.

Uzawa, Hirofumi. "Optimum Technical Change in an Aggregative Model of Economic Growth." *Internat. Econ. Rev.* 6 (January 1965): 18–31.

Wilson, Robert. "Informational Economies of Scale." *Bell J. Econ.* 6 (Spring 1975): 184–95.

[8]

Journal of Monetary Economics 22 (1988) 3–42. North-Holland

ON THE MECHANICS OF ECONOMIC DEVELOPMENT*

Robert E. LUCAS, Jr.

University of Chicago, Chicago, IL 60637, USA

Received August 1987, final version received February 1988

This paper considers the prospects for constructing a neoclassical theory of growth and international trade that is consistent with some of the main features of economic development. Three models are considered and compared to evidence: a model emphasizing physical capital accumulation and technological change, a model emphasizing human capital accumulation through schooling, and a model emphasizing specialized human capital accumulation through learning-by-doing.

1. Introduction

By the problem of economic development I mean simply the problem of accounting for the observed pattern, across countries and across time, in levels and rates of growth of per capita income. This may seem too narrow a definition, and perhaps it is, but thinking about income patterns will necessarily involve us in thinking about many other aspects of societies too, so I would suggest that we withhold judgment on the scope of this definition until we have a clearer idea of where it leads us.

The main features of levels and rates of growth of national incomes are well enough known to all of us, but I want to begin with a few numbers, so as to set a quantitative tone and to keep us from getting mired in the wrong kind of details. Unless I say otherwise, all figures are from the World Bank's *World Development Report* of 1983.

The diversity across countries in measured per capita income levels is literally too great to be believed. Compared to the 1980 average for what the World Bank calls the 'industrial market economies' (Ireland up through Switzerland) of U.S. $10,000, India's per capita income is $240, Haiti's is $270,

*This paper was originally written for the Marshall Lectures, given at Cambridge University in 1985. I am very grateful to the Cambridge faculty for this honor, and also for the invitation's long lead time, which gave me the opportunity to think through a new topic with the stimulus of so distinguished an audience in prospect. Since then, versions of this lecture have been given as the David Horowitz Lectures in Israel, the W.A. Mackintosh Lecture at Queens University, the Carl Snyder Memorial Lecture at the University of California at Santa Barbara, the Chung-Hua Lecture in Taipei, the Nancy Schwartz Lecture at Northwestern University, and the Lionel McKenzie Lecture at the University of Rochester. I have also based several seminars on various parts of this material.

and so on for the rest of the very poorest countries. This is a difference of a factor of 40 in living standards! These latter figures are too low to sustain life in, say, England or the United States, so they cannot be taken at face value and I will avoid hanging too much on their exact magnitudes. But I do not think anyone will argue that there is not enormous diversity in living standards.[1]

Rates of growth of real per capita GNP are also diverse; even over sustained periods. For 1960–80 we observe, for example: India, 1.4% per year; Egypt, 3.4%; South Korea, 7.0%; Japan, 7.1%; the United States, 2.3%; the industrial economies averaged 3.6%. To obtain from growth rates the number of years it takes for incomes to double, divide these numbers into 69 (the log of 2 times 100). Then Indian incomes will double every 50 years; Korean every 10. An Indian will, on average, be twice as well off as his grandfather; a Korean 32 times. These differences are at least as striking as differences in income levels, and in some respects more trustworthy, since within-country income comparisons are easier to draw than across-country comparisons.

I have not calculated a correlation across countries between income levels and rates of growth, but it would not be far from zero. (The poorest countries tend to have the lowest growth; the wealthiest next; the 'middle-income' countries highest.) The generalizations that strike the eye have to do with variability within these broad groups: the rich countries show little diversity (Japan excepted – else it would not have been classed as a rich country in 1980 at all). Within the poor countries (low and middle income) there is enormous variability.[2]

Within the advanced countries, growth rates tend to be very stable over long periods of time, provided one averages over periods long enough to eliminate business-cycle effects (or corrects for short-term fluctuations in some other way). For poorer countries, however, there are many examples of sudden, large changes in growth rates, both up and down. Some of these changes are no doubt due to political or military disruption: Angola's total GDP growth fell from 4.8 in the 60s to −9.2 in the 70s; Iran's fell from 11.3 to 2.5, comparing the same two periods. I do not think we need to look to economic theory for an account of either of *these* declines. There are also some striking examples

[1] The income estimates reported in Summers and Heston (1984) are more satisfactory than those in the World Development Reports. In 1975 U.S. dollars, these authors estimate 1980 U.S. real GDP per capita at $8000, and for the industrialized economies as a group, $5900. The comparable figures for India and Haiti are $460 and $500, respectively. Income differences of a factor of 16 are certainly smaller, and I think more accurate, than a factor of 40, but I think they are still fairly described as exhibiting 'enormous diversity'.

[2] Baumol (1986) summarizes evidence, mainly from Maddison (1982) indicating apparent convergence during this century to a common path of the income levels of the wealthiest countries. But De Long (1987) shows that this effect is entirely due to 'selection bias': If one examines the countries with the highest income levels at the *beginning* of the century (as opposed to currently, as in Maddison's 'sample') the data show apparent *divergence*!

of sharp increases in growth rates. The four East Asian 'miracles' of South Korea, Taiwan, Hong Kong and Singapore are the most familiar: for the 1960–80 period, per capita income in these economies grew at rates of 7.0, 6.5, 6.8 and 7.5, respectively, compared to much lower rates in the 1950's and earlier.[3,4] Between the 60s and the 70s, Indonesia's GDP growth increased from 3.9 to 7.5; Syria's from 4.6 to 10.0.

I do not see how one can look at figures like these without seeing them as representing *possibilities*. Is there some action a government of India could take that would lead the Indian economy to grow like Indonesia's or Egypt's? If so, *what*, exactly? If not, what is it about the 'nature of India' that makes it so? The consequences for human welfare involved in questions like these are simply staggering: Once one starts to think about them, it is hard to think about anything else.

This is what we need a theory of economic development *for*: to provide some kind of framework for organizing facts like these, for judging which represent opportunities and which necessities. But the term 'theory' is used in so many different ways, even within economics, that if I do not clarify what I mean by it early on, the gap between what I think I am saying and what you think you are hearing will grow too wide for us to have a serious discussion. I prefer to use the term 'theory' in a very narrow sense, to refer to an explicit dynamic system, something that can be put on a computer and *run*. This is what I mean by the 'mechanics' of economic development – the construction of a mechanical, artificial world, populated by the interacting robots that economics typically studies, that is capable of exhibiting behavior the gross features of which resemble those of the actual world that I have just described. My lectures will be occupied with one such construction, and it will take some work: It is easy to set out models of economic growth based on reasonable-looking axioms that predict the cessation of growth in a few decades, or that predict the rapid convergence of the living standards of different economies to a common level, or that otherwise produce logically possible outcomes that bear no resemblance to the outcomes produced by actual economic systems. On the other hand, there is no doubt that there must be mechanics other than the ones I will describe that would fit the facts about as well as mine. This is why I have titled the lectures '*On* the Mechanics ...' rather than simply '*The* Mechanics of Economic Development'. At some point, then, the study of development will need to involve working out the implications of competing theories for data other than those they were constructed to fit, and testing these implications against observation. But this is getting far ahead of the

[3]The World Bank no longer transmits data for Taiwan. The figure 6.5 in the text is from Harberger (1984, table 1, p. 9).

[4]According to Heston and Summers (1984), Taiwan's per-capita GDP growth rate in the 1950s was 3.6. South Korea's was 1.7 from 1953 to 1960.

story I have to tell, which will involve leaving many important questions open even at the purely theoretical level and will touch upon questions of empirical testing hardly at all.

My plan is as follows. I will begin with an application of a now-standard neoclassical model to the study of twentieth century U.S. growth, closely following the work of Robert Solow, Edward Denison and many others. I will then ask, somewhat unfairly, whether this model *as it stands* is an adequate model of economic development, concluding that it is not. Next, I will consider two adaptations of this standard model to include the effects of human capital accumulation. The first retains the one-sector character of the original model and focuses on the interaction of physical and human capital accumulation. The second examines a two-good system that admits specialized human capital of different kinds and offers interesting possibilities for the interaction of trade and development. Finally, I will turn to a discussion of what has been arrived at and of what is yet to be done.

In general, I will be focusing on various aspects of what economists, using the term very broadly, call the 'technology'. I will be abstracting altogether from the economics of demography, taking population growth as a given throughout. This is a serious omission, for which I can only offer the excuse that a serious discussion of demographic issues would be at least as difficult as the issues I will be discussing and I have neither the time nor the knowledge to do both. I hope the interactions between these topics are not such that they cannot usefully be considered separately, at least in a preliminary way.[5]

I will also be abstracting from all monetary matters, treating all exchange as though it involved goods-for-goods. In general, I believe that the importance of financial matters is very badly over-stressed in popular and even much professional discussion and so am not inclined to be apologetic for going to the other extreme. Yet insofar as the development of financial institutions is a limiting factor in development more generally conceived I will be falsifying the picture, and I have no clear idea as to how badly. But one cannot theorize about everything at once. I had better get on with what I do have to say.

2. Neoclassical growth theory: Review

The example, or model, of a successful theory that I will try to build on is the theory of economic growth that Robert Solow and Edward Denison developed and applied to twentieth century U.S. experience. This theory will serve as a basis for further discussion in three ways: as an example of the *form* that I believe useful aggregative theories must take, as an opportunity to

[5] Becker and Barro (1985) is the first attempt known to me to analyze fertility and capital accumulation decisions *simultaneously* within a general equilibrium framework. Tamura (1986) contains further results along this line.

explain exactly what theories of this form can tell us that other kinds of theories cannot, and as a possible theory of economic development. In this third capacity, the theory will be seen to fail badly, but also suggestively. Following up on these suggestions will occupy the remainder of the lectures.

Both Solow and Denison were attempting to account for the main features of U.S. economic growth, not to provide a theory of economic development, and their work was directed at a very different set of observations from the cross-country comparisons I cited in my introduction. The most useful summary is provided in Denison's 1961 monograph, *The Sources of Economic Growth in the United States*. Unless otherwise mentioned, this is the source for the figures I will cite next.

During the 1909–57 period covered in Denison's study, U.S. real output grew at an annual rate of 2.9%, employed manhours at 1.3%, and capital stock at 2.4%. The remarkable feature of these figures, as compared to those cited earlier, is their *stability* over time. Even if one takes as a starting point the trough of the Great Depression (1933) output growth to 1957 averages only 5%. If business-cycle effects are removed in any reasonable way (say, by using peak-to-peak growth rates) U.S. output growth is within half a percentage point of 3% annually for any sizeable subperiod for which we have data.

Solow (1956) was able to account for this stability, and also for some of the relative magnitudes of these growth rates, with a very simple but also easily refineable model.[6] There are many variations of this model in print. I will set out a particularly simple one that is chosen also to serve some later purposes. I will do so without much comment on its assumed structure: There is no point in arguing over a model's assumptions until one is clear on what questions it will be used to answer.

We consider a closed economy with competitive markets, with identical, rational agents and a constant returns technology. At date t there are $N(t)$ persons or, equivalently, manhours devoted to production. The exogenously given rate of growth of $N(t)$ is λ. Real, per-capita consumption is a stream $c(t)$, $t \geq 0$, of units of a single good. Preferences over (per-capita) consumption streams are given by

$$\int_0^{\infty} e^{-\rho t} \frac{1}{1-\sigma} \left[c(t)^{1-\sigma} - 1 \right] N(t) \, dt, \tag{1}$$

[6]Solow's 1956 paper stimulated a vast literature in the 1960s, exploring many variations on the original one-sector structure. See Burmeister and Dobell (1970) for an excellent introduction and survey. By putting a relatively simple version to empirical use, as I shall shortly do, I do not intend a negative comment on this body of research. On the contrary, it is exactly this kind of theoretical experimentation with alternative assumptions that is needed to give one the confidence that working with a particular, simple parameterization may, for the specific purpose at hand, be adequate.

where the discount rate ρ and the coefficient of (relative) risk aversion σ are both positive.[7]

Production per capita of the one good is divided into consumption $c(t)$ and capital accumulation. If we let $K(t)$ denote the total stock of capital, and $\dot{K}(t)$ its rate of change, then total output is $N(t)c(t) + \dot{K}(t)$. [Here $\dot{K}(t)$ is net investment and total output $N(t)c(t) + \dot{K}(t)$ is identified with net national product.] Production is assumed to depend on the levels of capital and labor inputs and on the level $A(t)$ of the 'technology', according to

$$N(t)c(t) + \dot{K}(t) = A(t)K(t)^{\beta}N(t)^{1-\beta}, \tag{2}$$

where $0 < \beta < 1$ and where the exogenously given rate of technical change, $\dot{A}/A$, is $\mu > 0$.

The resource allocation problem faced by this simple economy is simply to choose a time path $c(t)$ for per-capita consumption. Given a path $c(t)$ and an initial capital stock $K(0)$, the technology (2) then implies a time path $K(t)$ for capital. The paths $A(t)$ and $N(t)$ are given exogenously. One way to think about this allocation problem is to think of choosing $c(t)$ at each date, given the values of $K(t)$, $A(t)$ and $N(t)$ that have been attained by that date. Evidently, it will not be optimal to choose $c(t)$ to maximize current-period utility, $N(t)[1/(1-\sigma)][c(t)-1]^{1-\sigma}$, for the choice that achieves this is to set net investment $\dot{K}(t)$ equal to zero (or, if feasible, negative): One needs to set some *value* or *price* on increments to capital. A central construct in the study of *optimal* allocations, allocations that maximize utility (1) subject to the technology (2), is the *current-value Hamiltonian* H defined by

$$H(K,\theta,c,t) = \frac{N}{1-\sigma}[c^{1-\sigma}-1] + \theta[AK^{\beta}N^{1-\beta} - Nc],$$

which is just the sum of current-period utility and [from (2)] the rate of increase of capital, the latter valued at the 'price' $\theta(t)$. An optimal allocation must maximize the expression H at each date t, provided the price $\theta(t)$ is correctly chosen.

The first-order condition for maximizing H with respect to c is

$$c^{-\sigma} = \theta, \tag{3}$$

which is to say that goods must be so allocated at each date as to be equally valuable, on the margin, used either as consumption or as investment. It is

[7]The inverse σ^{-1} of the coefficient of risk aversion is sometimes called the intertemporal elasticity of substitution. Since all the models considered in this paper are deterministic, this latter terminology may be more suitable.

known that the price $\theta(t)$ must satisfy

$$\dot{\theta}(t) = \rho\theta(t) - \frac{\partial}{\partial K} H(K(t), \theta(t), c(t), t)$$

$$= \left[\rho - \beta A(t) N(t)^{1-\beta} K(t)^{\beta-1}\right] \theta(t), \tag{4}$$

at each date t if the solution $c(t)$ to (3) is to yield an optimal *path* $(c(t))_{t=0}^{\infty}$.

Now if (3) is used to express $c(t)$ as a function $\theta(t)$, and this function $\theta^{-1/\sigma}$ is substituted in place of $c(t)$ in (2) and (4), these two equations are a pair of first-order differential equations in $K(t)$ and its 'price' $\theta(t)$. Solving this system, there will be a one-parameter family of paths $(K(t), \theta(t))$, satisfying the given initial condition on $K(0)$. The *unique* member of this family that satisfies the transversality condition:

$$\lim_{t \to \infty} e^{-\rho t} \theta(t) K(t) = 0 \tag{5}$$

is the optimal path. I am hoping that this application of Pontryagin's Maximum Principle, essentially taken from David Cass (1961), is familiar to most of you. I will be applying these same ideas repeatedly in what follows.

For this particular model, with convex preferences and technology and with no external effects of any kind, it is also known and not at all surprising that the *optimal* program characterized by (2), (3), (4) and (5) is also the unique *competitive equilibrium* program, provided either that all trading is consummated in advance, Arrow–Debreu style, *or* (and this is the interpretation I favor) that consumers and firms have rational expectations about future prices. In this deterministic context, rational expectations just means perfect foresight. For my purposes, it is this equilibrium interpretation that is most interesting: I intend to use the model as a positive theory of U.S. economic growth.

In order to do this, we will need to work out the predictions of the model in more detail, which involves solving the differential equation system so we can see what the equilibrium time paths look like and compare them to observations like Denison's. Rather than carry this analysis through to completion, I will work out the properties of a *particular* solution to the system and then just indicate briefly how the rest of the answer can be found in Cass's paper.

Let us construct from (2), (3) and (4) the system's *balanced growth path*: the particular solution $(K(t), \theta(t), c(t))$ such that the rates of growth of each of these variables is constant. (I have never been sure exactly what it is that is 'balanced' along such a path, but we need a term for solutions with this constant growth rate property and this is as good as any.) Let κ denote the rate of growth of per-capita consumption, $\dot{c}(t)/c(t)$, on a balanced growth path. Then from (3), we have $\dot{\theta}(t)/\theta(t) = -\sigma\kappa$. Then from (4), we must have

$$\beta A(t)N(t)^{1-\beta}K(t)^{\beta-1} = \rho + \sigma\kappa. \tag{6}$$

That is, along the balanced path, the marginal product of capital must equal the constant value $\rho + \sigma\kappa$. With this Cobb–Douglas technology, the marginal product of capital is proportional to the average product, so that dividing (2) through by $K(t)$ and applying (6) we obtain

$$\frac{N(t)c(t)}{K(t)} + \frac{\dot{K}(t)}{K(t)} = A(t)K(t)^{\beta-1}N(t)^{1-\beta} = \frac{\rho + \sigma\kappa}{\beta}. \tag{7}$$

By definition of a balanced path, $\dot{K}(t)/K(t)$ is constant so (7) implies that $N(t)c(t)/K(t)$ is constant or, differentiating, that

$$\frac{\dot{K}(t)}{K(t)} = \frac{\dot{N}(t)}{N(t)} + \frac{\dot{c}(t)}{c(t)} = \kappa + \lambda. \tag{8}$$

Thus per-capita consumption and per-capita capital grow at the common rate κ. To *solve* for this common rate, differentiate either (6) or (7) to obtain

$$\kappa = \frac{\mu}{1-\beta}. \tag{9}$$

Then (7) may be solved to obtain the constant, balanced consumption–capital ratio $N(t)c(t)/K(t)$ or, which is equivalent and slightly easier to interpret, the constant, balanced net savings rate s defined by

$$s = \frac{\dot{K}(t)}{N(t)c(t) + \dot{K}(t)} = \frac{\beta(\kappa + \lambda)}{\rho + \sigma\kappa}. \tag{10}$$

Hence along a balanced path, the rate of growth of per-capita magnitudes is simply proportional to the given rate of technical change, μ, where the constant of proportionality is the inverse of labor's share, $1 - \beta$. The rate of time preference ρ and the degree of risk aversion σ have no bearing on this long-run growth rate. Low time preference ρ and low risk aversion σ induce a high savings rate s, and high savings is, in turn, associated with relatively high output *levels* on a balanced path. A thrifty society will, in the long run, be wealthier than an impatient one, but it will not grow faster.

In order that the balanced path characterized by (9) and (10) satisfy the transversality condition (5), it is necessary that $\rho + \sigma\kappa > \kappa + \lambda$. [From (10), one sees that this is the same as requiring the savings rate to be less than capital's share.] Under this condition, an economy that begins on the balanced path will find it optimal to stay there. What of economies that begin *off* the balanced path – surely the normal case? Cass showed – and this is exactly why the balanced path is interesting to us – that for *any* initial capital $K(0) > 0$, the optimal capital–consumption path $(K(t), c(t))$ will converge to the balanced path asymptotically. That is, the balanced path will be a good approximation to any actual path 'most' of the time.

Now given the taste and technology parameters (ρ, σ, λ, β and μ) (9) and (10) can be solved for the asymptotic growth rate κ of capital, consumption and real output, and the savings rate s that they imply. Moreover, it would be straightforward to calculate numerically the approach to the balanced path from any initial capital level $K(0)$. This is the exercise that an idealized planner would go through.

Our interest in the model is positive, not normative, so we want to go in the opposite direction and try to infer the underlying preferences and technology from what we can observe. I will outline this, taking the balanced path as the model's prediction for the behavior of the U.S. economy during the entire (1909–57) period covered by Denison's study.[8] From this point of view, Denison's estimates provide a value of 0.013 for λ, and two values, 0.029 and 0.024 for $\kappa + \lambda$, depending on whether we use output or capital growth rates (which the model predicts to be equal). In the tradition of statistical inference, let us average to get $\kappa + \lambda = 0.027$. The theory predicts that $1 - \beta$ should equal labor's share in national income, about 0.75 in the U.S., averaging over the entire 1909–57 period. The savings rate (net investment over NNP) is fairly constant at 0.10. Then (9) implies an estimate of 0.0105 for μ. Eq. (10) implies that the preference parameters ρ and σ satisfy

$$\rho + (0.014)\sigma = 0.0675.$$

(The parameters ρ and σ are not separately identified along a smooth consumption path, so this is as far as we can go with the sample averages I have provided.)

These are the parameter values that give the theoretical model its best fit to the U.S. data. How good a fit *is* it? Either output growth is underpredicted or capital growth overpredicted, as remarked earlier (and in the theory of growth, a half a percentage point is a *large* discrepancy). There are interesting secular changes in manhours per household that the model assumes away, and labor's share is secularly rising (in all growing economies), not constant as assumed. There is, in short, much room for improvement, even in accounting for the secular changes the model was designed to fit, and indeed, a fuller review of

[8] With the parameter values described in this paragraph, the half-life of the approximate linear system associated with this model is about eleven years.

the literature would reveal interesting progress on these and many other fronts.[9] A model as explicit as this one, by the very nakedness of its simplifying assumptions, invites criticism and suggests refinements to itself. This is exactly why we prefer explicitness, or why I think we ought to.

Even granted its limitations, the simple neoclassical model has made basic contributions to our thinking about economic growth. Qualitatively, it emphasizes a distinction between 'growth effects' – changes in parameters that alter growth rates along balanced paths – and 'level effects' – changes that raise or lower balanced growth paths without affecting their slope – that is fundamental in thinking about policy changes. Solow's 1956 conclusion that changes in savings rates are level effects (which transposes in the present context to the conclusion that changes in the discount rate, ρ, are level effects) was startling at the time, and remains widely and very unfortunately neglected today. The influential idea that changes in the tax structure that make savings more attractive can have large, sustained effects on an economy's growth rate sounds so reasonable, and it may even be true, but it is a clear implication of the theory we have that it is not.

Even sophisticated discussions of economic growth can often be confusing as to what are thought to be level effects and what growth effects. Thus Krueger (1983) and Harberger (1984), in their recent, very useful surveys of the growth experiences of poor countries, both identify inefficient barriers to trade as a limitation on growth, and their removal as a key explanation of several rapid growth episodes. The facts Krueger and Harberger summarize are not in dispute, but under the neoclassical model just reviewed one would not expect the removal of inefficient trade barriers to induce sustained increases in growth rates. Removal of trade barriers is, on this theory, a level effect, analogous to the one-time shifting upward in production possibilities, and not a growth effect. Of course, level effects can be drawn out through time through adjustment costs of various kinds, but not so as to produce increases in growth rates that are both large and sustained. Thus the removal of an inefficiency that reduced output by five percent (an enormous effect) spread out over ten years in simply a one-half of one percent annual growth rate stimulus. Inefficiencies are important and their removal certainly desirable, but the familiar ones are level effects, not growth effects. (This is exactly why it is not paradoxical that centrally planned economies, with allocative inefficiencies of legendary proportions, grow about as fast as market economies.) The empirical connections between trade policies and economic growth that

[9]In particular, there is much evidence that capital stock growth, as measured by Denison, understates true capital growth due to the failure to correct price deflators for quality improvements. See, for example, Griliches and Jorgenson (1967) or Gordon (1971) These errors may well account for all of the 0 005 discrepancy noted in the text (or more!).

Boxall (1986) develops a modification of the Solow–Cass model in which labor supply is variable, and which has the potential (at least) to account for long-run changes in manhours.

Krueger and Harberger document are of evident importance, but they seem to me to pose a real paradox to the neoclassical theory we have, not a confirmation of it.

The main contributions of the neoclassical framework, far more important than its contributions to the clarity of purely qualitative discussions, stem from its ability to *quantify* the effects of various influences on growth. Denison's monograph lists dozens of policy changes, some fanciful and many others seriously proposed at the time he wrote, associating with each of them rough upper bounds on their likely effects on U.S. growth.[10] In the main, the theory adds little to what common sense would tell us about the *direction* of each effect – it is easy enough to guess which changes stimulate production, hence savings, and hence (at least for a time) economic growth. Yet most such changes, quantified, have *trivial* effects: The growth rate of an entire economy is not an easy thing to move around.

Economic growth, being a summary measure of all of the activities of an entire society, necessarily depends, in some way, on everything that goes on in a society. Societies differ in many easily observed ways, and it is easy to identify various economic and cultural peculiarities and imagine that they are keys to growth performance. For this, as Jacobs (1984) rightly observes, we do not need economic theory: 'Perceptive tourists will do as well.' The role of theory is not to catalogue the obvious, but to help us to sort out effects that are crucial, quantitatively, from those that can be set aside. Solow and Denison's work shows how this can be done in studying the growth of the U.S. economy, and of other advanced economies as well. I take success at this level to be a worthy objective for the theory of economic development.

3. Neoclassical growth theory: Assessment

It seems to be universally agreed that the model I have just reviewed is not a theory of economic development. Indeed, I suppose this is why we think of 'growth' and 'development' as distinct fields, with growth theory defined as those aspects of economic growth we have some understanding of, and development defined as those we don't. I do not disagree with this judgment, but a more specific idea of exactly where the model falls short will be useful in thinking about alternatives.

If we were to attempt to use the Solow–Denison framework to account for the diversity in income levels and rates of growth we observe in the world today, we would begin, theoretically, by imagining a world consisting of many

[10] Denison (1961, ch. 24). My favorite example is number 4 in this 'menu of choices available to increase the growth rate': '0.03 points [i.e., 0.03 of one percentage point] maximum potential ... Eliminate all crime and rehabilitate all criminals.' This example and many others in this chapter are pointed rebukes to those in the 1960s who tried to advance their favorite (and often worthy) causes by claiming ties to economic growth.

economies of the sort we have just described, assuming something about the way they interact, working out the dynamics of this new model, and comparing them to observations. This is actually much easier than it sounds (there isn't much to the theory of international trade when everyone produces the same, single good!), so let us think it through.

The key assumptions involve factor mobility: Are people and capital free to move? It is easiest to start with the assumption of *no* mobility, since then we can treat each country as an isolated system, just like the one we have just worked out. In this case, the model predicts that countries with the same preferences and technology will converge to identical levels of income and asymptotic rates of growth. Since this prediction does not accord at all well with what we observe, if we want to fit the theory to observed cross-country variations, we will need to postulate appropriate variations in the parameters (ρ, σ, λ, β and μ) and/or assume that countries differ according to their initial technology levels, $A(0)$. Or we can obtain additional theoretical flexibility by treating countries as differently situated relative to their steady-state paths. Let me review these possibilities briefly.

Population growth, λ, and income shares going to labor, $1-\beta$, do of course differ across countries, but neither varies in such a way as to provide an account of income differentials. Countries with rapid population growth are not systematically poorer than countries with slow-growing populations, as the theory predicts, either cross-sectionally today or historically. There are, certainly, interesting empirical connections between economic variables (narrowly defined) and birth and death rates, but I am fully persuaded by the work of Becker (1981) and others that these connections are best understood as arising from the way decisions to maintain life and to initiate it *respond* to economic conditions. Similarly, poor countries have lower labor shares than wealthy countries, indicating to me that elasticities of substitution in production are below unity (contrary to the Cobb–Douglas assumption I am using in these examples), but the prediction (9) that poorer countries should therefore grow more rapidly is not confirmed by experience.

The parameters ρ and σ are, as observed earlier, not separately identified, but if their joint values differed over countries in such a way as to account for income differences, poor countries would have systematically much higher (risk-corrected) interest rates than rich countries. Even if this were true, I would be inclined to seek other explanations. Looking ahead, we would like also to be able to account for sudden large changes in growth rates of individual countries. Do we want a theory that focuses attention on spontaneous shifts in people's discount rates or degree of risk aversion? Such theories are hard to refute, but I will leave it to others to work this side of the street.

Consideration of off-steady-state behavior would open up some new possibilities, possibly bringing the theory into better conformity with observation, but I do not view this route as at all promising. Off steady states, (9) need not hold and capital and output growth rates need not be either equal or constant, but it still follows from the technology (2) that output growth (g_{yt}, say) and capital growth (g_{kt}, say), both per capita, obey

$$g_{yt} = \beta g_{kt} + \mu.$$

But g_{yt} and g_{kt} can both be measured, and it is well established that for no value of β that is close to observed capital shares is it the case that $g_{yt} - \beta g_{kt}$ is even approximately uniform across countries. Here 'Denison's Law' works against us: the insensitivity of growth rates to variations in the model's underlying parameters, as reviewed earlier, makes it hard to use the theory to account for large variations across countries or across time. To conclude that even large changes in 'thriftiness' would not induce large changes in U.S. growth rates is really the same as concluding that differences in Japanese and U.S. thriftiness cannot account for much of the difference in these two economies' growth rates. By assigning so great a role to 'technology' as a source of growth, the theory is obliged to assign correspondingly minor roles to everything else, and so has very little ability to account for the wide diversity in growth rates that we observe.

Consider, then, variations across countries in 'technology' – its level and rate of change. This seems to me to be the one factor isolated by the neoclassical model that has the potential to account for wide differences in income levels and growth rates. This point of departure certainly does accord with everyday usage. We say that Japan is technologically more advanced than China, or that Korea is undergoing unusually rapid technical change, and such statements seem to mean something (and I think they do). But they cannot mean that the 'stock of useful knowledge' [in Kuznets's (1959) terminology] is higher in Japan than in China, or that it is growing more rapidly in Korea than elsewhere. 'Human knowledge' is just human, not Japanese or Chinese or Korean. I think when we talk in this way about differences in 'technology' across countries we are not talking about 'knowledge' in general, but about the knowledge of particular people, or perhaps particular subcultures of people. If so, then while it is not exactly wrong to describe these differences by an exogenous, exponential term like $A(t)$ neither is it useful to do so. We want a formalism that leads us to think about individual decisions to acquire knowledge, and about the consequences of these decisions for productivity. The body of theory that does this is called the theory of 'human capital', and I am going to draw extensively on this theory in the remainder of these lectures. For the moment, however, I simply want to impose the terminological convention that 'technology' – its level and rate of change – will be used to refer to something common to all countries, something 'pure' or 'disembodied', something whose determinants are outside the bounds of our current inquiry.

In the absence of differences in pure technology then, and under the assumption of no factor mobility, the neoclassical model predicts a strong tendency to income equality and equality in growth rates, tendencies we can observe within countries and, perhaps, within the wealthiest countries taken as a group, but which simply cannot be seen in the world at large. When factor mobility is permitted, this prediction is very powerfully reinforced. Factors of production, capital or labor or both, will flow to the highest returns, which is to say where each is relatively scarce. Capital–labor ratios will move rapidly to equality, and with them factor prices. Indeed, these predictions survive differences in preference parameters and population growth rates. In the model as stated, it makes no difference whether labor moves to join capital or the other way around. (Indeed, we know that with a many-good technology, factor price equalization can be achieved without mobility in *either* factor of production.)

The eighteenth and nineteenth century histories of the Americas, Australia and South and East Africa provide illustrations of the strength of these forces for equality, and of the ability of even simple neo-classical models to account for important economic events. If we replace the labor–capital technology of the Solow model with a land–labor technology of the same form, and treat labor as the mobile factor and land as the immobile, we obtain a model that predicts exactly the immigration flows that occurred and for exactly the reason – factor price differentials – that motivated these historical flows. Though this simple deterministic model abstracts from considerations of risk and many other elements that surely played a role in actual migration decisions, this abstraction is evidently not a fatal one.

In the present century, of course, immigration has been largely shut off, so it is not surprising that this land–labor model, with labor mobile, no longer gives an adequate account of actual movements in factors and factor prices. What *is* surprising, it seems to me, is that capital movements do not perform the same functions. Within the United States, for example, we have seen southern labor move north to produce automobiles. We have also seen textile mills move from New England south (to 'move' a factory, one lets it run down and builds its replacement somewhere else: it takes some time, but then, so does moving families) to achieve this same end of combining capital with relatively low wage labor. Economically, it makes no difference which factor is mobile, so long as one is.

Why, then, should the closing down of international labor mobility have slowed down, or even have much affected, the tendencies toward factor price equalization predicted by neoclassical theory, tendencies that have proved to be so powerful historically? If it is profitable to move a textile mill from New England to South Carolina, why is it not more profitable still to move it to Mexico? The fact that we do see *some* capital movement toward low-income countries is not an adequate answer to this question, for the theory predicts that *all* new investment should be so located until such time as return and real wage differentials are erased. Indeed, why did these capital movements not take place during the colonial age, under political and military arrangements that eliminated (or long postponed) the 'political risk' that is so frequently cited as a factor working against capital mobility? I do not have a satisfactory answer to this question, but it seems to me a major – perhaps *the* major – discrepancy between the predictions of neoclassical theory and the patterns of trade we observe. Dealing with this issue is surely a minimal requirement for a theory of economic development.

4. Human capital and growth

To this point, I have reviewed an example of the neoclassical model of growth, compared it to certain facts of U.S. economic history, and indicated why I want to use this theory as a kind of model, or image, of what I think is possible and useful for a theory of economic development. I have also described what seem to me two central reasons why this theory is not, as it stands, a useful theory of economic development: its apparent inability to account for observed diversity across countries and its strong and evidently counterfactual prediction that international trade should induce rapid movement toward equality in capital–labor ratios and factor prices. These observations set the stage for what I would like to do in the rest of the lectures.

Rather than take on both problems at once, I will begin by considering an alternative, or at least a complementary, engine of growth to the 'technological change' that serves this purpose in the Solow model, retaining for the moment the other features of that model (in particular, its closed character). I will do this by adding what Schultz (1963) and Becker (1964) call 'human capital' to the model, doing so in a way that is very close technically to similarly motivated models of Arrow (1962), Uzawa (1965) and Romer (1986).

By an individual's 'human capital' I will mean, for the purposes of this section, simply his general skill level, so that a worker with human capital $h(t)$ is the productive equivalent of two workers with $\frac{1}{2}h(t)$ each, or a half-time worker with $2h(t)$. The theory of human capital focuses on the fact that the way an individual allocates his time over various activities in the current period affects his productivity, or his $h(t)$ level, in future periods. Introducing human capital into the model, then, involves spelling out both the way human capital levels affect current production and the way the current time allocation affects the accumulation of human capital. Depending on one's objectives, there are many ways to formulate both these aspects of the 'technology'. Let us begin with the following, simple assumptions.

Suppose there are N workers in total, with skill levels h ranging from 0 to infinity. Let there be $N(h)$ workers with skill level h, so that $N=\int_0^\infty N(h)\,\mathrm{d}h$. Suppose a worker with skill h devotes the fraction $u(h)$ of his non-leisure time to current production, and the remaining $1-u(h)$ to human capital accumulation. Then the effective workforce in production – the analogue to $N(t)$ in (2) – is the sum $N^e=\int_0^\infty u(h)N(h)h\,\mathrm{d}h$ of the skill-weighted manhours devoted to current production. Thus if output as a function of total capital K

and effective labor N^e is $F(K, N^e)$, the hourly wage of a worker at skill h is $F_N(K, N^e)h$ and his total earnings are $F_N(K, N^e)hu(h)$.

In addition to the effects of an individual's human capital on his own productivity – what I will call the *internal effect* of human capital – I want to consider an *external effect*. Specifically, let the *average* level of skill or human capital, defined by

$$h_a = \frac{\int_0^\infty hN(h)\,dh}{\int_0^\infty N(h)\,dh},$$

also contribute to the productivity of all factors of production (in a way that I will spell out shortly). I call this h_a effect external, because though all benefit from it, no individual human capital accumulation decision can have an appreciable effect on h_a, so no one will take it into account in deciding how to allocate his time.

Now it will simplify the analysis considerably to follow the preceding analysis and treat all workers in the economy as being identical. In this case, if all workers have skill level h and all choose the time allocation u, the effective workforce is just $N^e = uhN$, and the average skill level h_a is just h. Even so, I will continue to use the notation h_a for the latter, to emphasize the distinction between internal and external effects. Then the description (2) of the technology of goods production is replaced by

$$N(t)c(t) + \dot{K}(t) = AK(t)^\beta[u(t)h(t)N(t)]^{1-\beta}h_a(t)^\gamma, \qquad (11)$$

where the term $h_a(t)^\gamma$ is intended to capture the external effects of human capital, and where the technology level A is now assumed to be constant.

To complete the model, the effort $1 - u(t)$ devoted to the accumulation of human capital must be linked to the rate of change in its level, $\dot{h}(t)$. Everything hinges on exactly how this is done. Let us begin by postulating a technology relating the growth of human capital, $\dot{h}(t)$, to the level already attained and the effort devoted to acquiring more, say:

$$\dot{h}(t) = h(t)^\zeta G(1 - u(t)), \qquad (12)$$

where G is increasing, with $G(0) = 0$. Now if we take $\zeta < 1$ in this formulation, so that there is diminishing returns to the accumulation of human capital, it is easy to see that human capital cannot serve as an alternative engine of growth to the technology term $A(t)$. To see this, note that, since $u(t) \geq 0$, (12) implies that

$$\frac{\dot{h}(t)}{h(t)} \leq h(t)^{\zeta-1}G(1),$$

so that $\dot{h}(t)/h(t)$ must eventually tend to zero as $h(t)$ grows no matter how much effort is devoted to accumulating it. This formulation would simply complicate the original Solow model without offering any genuinely new possibilities.

Uzawa (1965) worked out a model very similar to this one [he assumed $\gamma = 0$ and $U(c) = c$] under the assumption that the right-hand side of (12) is *linear* in $u(t)$ ($\zeta = 1$). The striking feature of his solution, and the feature that recommends his formulation to us, is that it exhibits sustained per-capita income growth from endogenous human capital accumulation alone: no external 'engine of growth' is required.

Uzawa's linearity assumption might appear to be a dead-end (for our present purposes) because we seem to see diminishing returns in observed, individual patterns of human capital accumulation: people accumulate it rapidly early in life, then less rapidly, then not at all – as though each additional percentage increment were harder to gain than the preceding one. But an alternative explanation for *this* observation is simply that an individual's lifetime is finite, so that the return to increments falls with time. Rosen (1976) showed that a technology like (12), with $\zeta = 1$, is consistent with the evidence we have on individual earnings. I will adapt the Uzawa–Rosen formulation here, assuming for simplicity that the function G is linear:

$$\dot{h}(t) = h(t)\delta[1 - u(t)]. \tag{13}$$

According to (13), if no effort is devoted to human capital accumulation, $[u(t) = 1]$, then none accumulates. If all effort is devoted to this purpose $[u(t) = 0]$, $h(t)$ grows at its maximal rate δ. In between these extremes, there are *no* diminishing returns to the stock $h(t)$: A given *percentage* increase in $h(t)$ requires the same effort, no matter what level of $h(t)$ has already been attained.

It is a digression I will not pursue, but it would take some work to go from a human capital technology of the form (13), applied to each finite-lived individual (as in Rosen's theory), to this same technology applied to an entire infinitely-lived typical household or family. For example, if each individual acquired human capital as in Rosen's model but if *none* of this capital were passed on to younger generations, the 'household's' stock would (with a fixed demography) stay constant. To obtain (13) for a family, one needs to assume both that each individual's capital follows this equation *and* that the initial level each new member begins with is proportional to (not equal to!) the level already attained by older members of the family. This is simply one instance of a general fact that I will emphasize again and again: that human capital accumulation is a *social* activity, involving *groups* of people in a way that has no counterpart in the accumulation of physical capital.

Aside from these changes in the technology, expressed in (11) and (13) to incorporate human capital and its accumulation, the model to be discussed is identical to the Solow model. The system is closed, population grows at the fixed rate λ, and the typical household has the preferences (1). Let us proceed to the analysis of this new model.[11]

In the presence of the external effect $h_a(t)^\gamma$, it will not be the case that optimal growth paths and competitive equilibrium paths coincide. Hence we cannot construct the equilibrium by studying the same hypothetical planning problem used to study Solow's model. But by following Romer's analysis of a very similar model, we can obtain the optimal and equilibrium paths separately, and compare them. This is what I will now do.

By an *optimal* path, I will mean a choice of $K(t)$, $h(t)$, $H_a(t)$, $c(t)$ and $u(t)$ that maximizes utility (1) subject to (11) and (13), *and* subject to the constraint $h(t) = h_a(t)$ for all t. This is a problem similar in general structure to the one we reviewed in section 2, and I will turn to it in a moment.

By an *equilibrium* path, I mean something more complicated. First, take a path $h_a(t)$, $t \geq 0$, to be given, like the exogenous technology path $A(t)$ in the Solow model. Given $h_a(t)$, consider the problem the private sector, consisting of atomistic households and firms, would solve if each agent *expected* the average level of human capital to follow the path $h_a(t)$. That is, consider the problem of choosing $h(t)$, $k(t)$, $c(t)$ and $u(t)$ so as to maximize (1) subject to (11) and (13), taking $h_a(t)$ as exogenously determined. When the solution path $h(t)$ for this problem coincides with the given path $h_a(t)$ – so that actual and expected behavior are the same – we say that the system is in equilibrium.[12]

The current-value Hamiltonian for the optimal problem, with 'prices' $\theta_1(t)$ and $\theta_2(t)$ used to value increments to physical and human capital respectively, is

$$
\begin{aligned}
&H(K, h, \theta_1, \theta_2, c, u, t) \\
&= \frac{N}{1-\sigma}(c^{1-\sigma} - 1) + \theta_1\left[AK^\beta(uNh)^{1-\beta}h^\gamma - Nc\right] \\
&\quad + \theta_2[\delta h(1-u)].
\end{aligned}
$$

In this model, there are two decision variables – consumption, $c(t)$, and the time devoted to production, $u(t)$ – and these are (in an optimal program)

[11] The model discussed in this section (in contrast to the model of section 2) has not been fully analyzed in the literature. The text gives a self-contained derivation of the main features of balanced paths. The treatment of behavior off balanced paths is largely conjecture, based on parallels with Uzawa (1965) and Romer (1986)

[12] This formulation of equilibrium behavior in the presence of external effects is taken from Arrow (1962) and Romer (1986) Romer actually carries out the study of the fixed-point problem in a space of $h(t)$, $t \geq 0$, paths Here I follow Arrow and confine explicit analysis to balanced paths only.

selected so as to maximize H. The first-order conditions for this problem are thus:

$$c^{-\sigma} = \theta_1, \tag{14}$$

and

$$\theta_1(1-\beta)AK^{\beta}(uNh)^{-\beta}Nh^{1+\gamma} = \theta_2\delta h. \tag{15}$$

On the margin, goods must be equally valuable in their two uses – consumption and capital accumulation [eq. (14)] – and time must be equally valuable in its two uses – production and human capital accumulation [eq. (15)].

The rates of change of the prices θ_1 and θ_2 of the two kinds of capital are given by

$$\dot{\theta}_1 = \rho\theta_1 - \theta_1\beta AK^{\beta-1}(uNh)^{1-\beta}h^{\gamma}, \tag{16}$$

$$\dot{\theta}_2 = \rho\theta_2 - \theta_1(1-\beta+\gamma)AK^{\beta}(uN)^{1-\beta}h^{-\beta+\gamma} - \theta_2\delta(1-u). \tag{17}$$

Then eqs. (11) and (13) and (14)–(17), together with two transversality conditions that I will not state here, implicitly describe the optimal evolution of $K(t)$ and $h(t)$ from any initial mix of these two kinds of capital.

In the *equilibrium*, the private sector 'solves' a control problem of essentially this same form, but with the term $h_a(t)^{\gamma}$ in (11) taken as given. Market clearing then requires that $h_a(t) = h(t)$ for all t, so that (11), (13), (14), (15) and (16) are necessary conditions for equilibrium as well as for optimal paths. But eq. (17) no longer holds: It is precisely in the valuation of human capital that optimal and equilibrium allocations differ. For the private sector, in equilibrium, (17) is replaced by

$$\dot{\theta}_2 = \rho\theta_2 - \theta_1(1-\beta)AK^{\beta}(uN)^{1-\beta}h^{-\beta}h_a^{\gamma} - \theta_2\delta(1-u).$$

Since market clearing implies $(h(t) = h_a(t)$ for all t, this can be written as

$$\dot{\theta}_2 = \rho\theta_2 - \theta_1(a-\beta)AK^{\beta}(uN)^{1-\beta}h^{-\beta+\gamma} - \theta_2\delta(1-u). \tag{18}$$

Note that, if $\gamma = 0$, (17) and (18) are the same. It is the presence of the external effect $\gamma > 0$ that creates a divergence between the 'social' valuation formula (17) and the private valuation (18).

As with the simpler Solow model, the easiest way to characterize both optimal and equilibrium paths is to begin by seeking balanced growth solutions of both systems: solutions on which consumption and both kinds of capital are growing at constant percentage rates, the prices of the two kinds of capital are declining at constant rates, and the time allocation variable $u(t)$ is constant. Let us start by considering features that optimal and equilibrium paths have in common [by setting aside (17) and (18)].

Let κ denote $\dot{c}(t)/c(t)$, as before, so that (14) and (16) again imply the marginal productivity of capital condition:

$$\beta A K(t)^{\beta-1}(u(t)h(t)N(t))^{1-\beta}h(t)^{\gamma} = \rho + \sigma\kappa, \tag{19}$$

which is the analogue to condition (6). As in the earlier model, it is easy to verify that $K(t)$ must grow at the rate $\kappa + \lambda$ and that the savings rate s is constant, on a balanced path, at the value given by (10). For the derivation of these facts concerning physical capital accumulation, it is immaterial whether $h(t)$ is a matter of choice or an exogenous force as was technological change in the earlier model.

Now if we let $\nu = \dot{h}(t)/h(t)$ on a balanced path, it is clear from (13) that

$$\nu = \delta(1-u), \tag{20}$$

and from differentiating (19) that κ, the common growth rate of consumption and per-capita capital is

$$\kappa = \left(\frac{1-\beta+\gamma}{1-\beta}\right)\nu. \tag{21}$$

Thus with $h(t)$ growing at the fixed rate ν, $(1-\beta+\gamma)\nu$ plays the role of the exogenous rate of technological change μ in the earlier model.

Turning to the determinants of the rate of growth ν of human capital, one sees from differentiating both first-order conditions (14) and (15) and substituting for $\dot{\theta}_1/\theta_1$ that

$$\frac{\dot{\theta}_2}{\theta_2} = (\beta-\sigma)\kappa - (\beta-\gamma)\nu + \lambda. \tag{22}$$

At this point, the analyses of the efficient and equilibrium paths diverge. Focusing first on the efficient path, use (17) and (15) to obtain

$$\frac{\dot{\theta}_2}{\theta_2} = \rho - \delta - \frac{\gamma}{1-\beta}\delta u. \tag{23}$$

Now substitute for u from (20), eliminate $\dot{\theta}_2/\theta_2$ between (22) and (23), and solve for ν in terms of κ. Then eliminating κ between this equation and (21) gives the solution for the *efficient* rate of human capital growth, which I will call ν^*:

$$\nu^* = \sigma^{-1}\left[\delta - \frac{1-\beta}{1-\beta+\gamma}(\rho-\lambda)\right]. \tag{24}$$

Along an equilibrium balanced path (18) holds in place of (17) so that in place of (23) we have

$$\frac{\dot{\theta}_2}{\theta_2} = \rho - \delta. \tag{25}$$

Then by the same procedure used to derive the efficient growth rate ν^* from (23), we can obtain from (25) the equilibrium growth rate ν:

$$\nu = [\sigma(1-\beta+\gamma)-\gamma]^{-1}[(1-\beta)(\delta-(\rho-\lambda))]. \tag{26}$$

[For the formulas (24) and (26) to apply, the rates ν and ν^* must not exceed the maximum feasible rate δ. This restriction can be seen to require

$$\sigma \geq 1 - \frac{1-\beta}{1-\beta+\gamma}\frac{\rho-\lambda}{\delta}, \tag{27}$$

so the model cannot apply at levels of risk aversion that are too low (that is, if the intertemporal substitutability of consumption is too high).[13] When (27) holds with equality, $\nu = \nu^* = \delta$; when the inequality is strict, $\nu^* > \nu$, as one would expect.]

Eqs. (24) and (26) give, respectively, the efficient and the competitive equilibrium growth rates of human capital along a balanced path. In either case, this growth increases with the effectiveness δ of investment in human capital and declines with increases in the discount rate ρ. (Here at last is a connection between 'thriftiness' and growth!) In either case, (21) gives the corresponding rate of growth of physical capital, per capita. Notice that the theory predicts sustained growth whether or not the external effect γ is positive. If $\gamma = 0$, $\kappa = \nu$, while if $\gamma > 0$, $\kappa > \nu$, so that the external effect induces more rapid physical than human capital growth.

For the case $\sigma = 1$, the difference between efficient and equilibrium human capital growth rates is, subtracting (26) from (24),

$$\nu^* - \nu = \frac{\gamma}{1-\beta+\gamma}(\rho-\lambda).$$

[13] If utility is too nearly linear (σ is too near zero) and if δ is high enough, consumers will keep postponing consumption forever. [This does not occur in Uzawa's model, even though he assumes $\sigma = 0$, because he introduces diminishing returns to $1 - u(t)$ in his version of (13).]

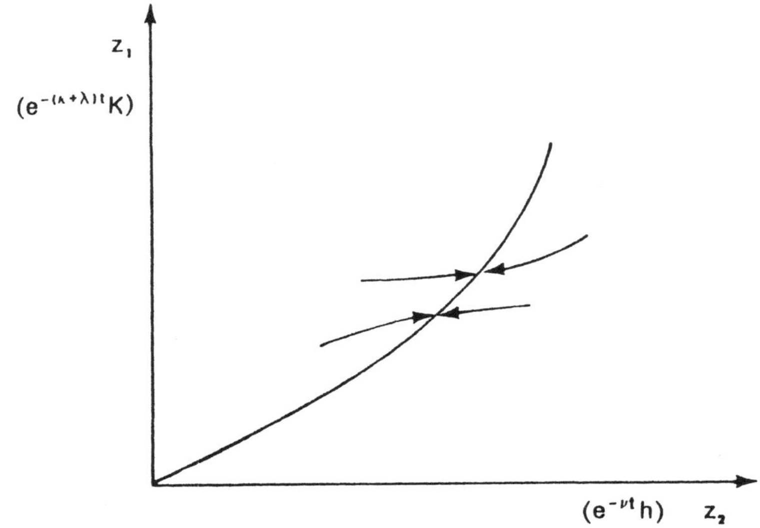

Fig 1

Thus the inefficiency is small when either the external effect is small ($\gamma \simeq 0$) or the discount rate is low ($\rho - \lambda \simeq 0$).

Eqs. (21), (24) and (26) describe the asymptotic rates of change of both kinds of capital, under both efficient and equilibrium regimes. What can be said about the *levels* of these variables? As in the original model, this information is implicit in the marginal productivity condition for capital, eq. (19). In the original model, this condition – or rather its analogue, eq. (6) – determined a unique long-run value of the normalized variable $z(t) = e^{-(\kappa+\lambda)t}K(t)$. In the present, two-capital model, this condition defines a curve linking the *two* normalized variables $z_1(t) = e^{-(\kappa+\lambda)t}K(t)$ and $z_2(t) = e^{-\nu t}h(t)$. Inserting these variables into (19) in place of $K(t)$ and $h(t)$ and applying the formula (21) for κ, we obtain

$$\left(\beta A N_0^{1-\beta} u^{1-\beta}\right) z_1^{\beta-1} z_2^{1-\beta+\gamma} = \rho + \sigma\kappa. \qquad (28)$$

It is a fact that *all* pairs (z_1, z_2) satisfying (28) correspond to balanced paths. Let us ask first what this locus of (normalized) capital combinations looks like, and second what this means for the dynamics of the system.

Fig. 1 shows the curve defined by (28). With no external effect ($\gamma = 0$) it is a straight line through the origin; otherwise ($\gamma > 0$) it is convex. The position of the curve depends on u and κ, which from (20) and (21) can be expressed as functions of ν. Using this fact one can see that increases in ν shift the curve to the right. Thus an efficient economy, on a balanced path, will have a higher level of human capital (z_2) for any given level of physical capital (z_1), since $\nu^* > \nu$.

The dynamics of this system are not as well understood as those of the one-good model, but I would conjecture that for any initial configuration $(K(0), h(0))$ of the two kinds of capital, the solution paths (of either the efficient or the equilibrium system) $(z_1(t), z_2(t))$ will converge to *some* point on the curve in fig. 1, but that this asymptotic position will depend on the initial position. The arrows in fig. 1 illustrate some possible trajectories. Under these dynamics, then, an economy beginning with low levels of human and physical capital will remain *permanently* below an initially better endowed economy.

The curve in fig. 1 is *defined* as the locus of long-run capital pairs (K, h) such that the marginal product of capital has the common value $\rho + \sigma\kappa$ given by the right side of (19). Along this curve, then, returns to capital are constant and also constant over time even though capital stocks of both kinds are growing. In the absence of the external effect γ, it will also be true that the real wage rate for labor of a given skill level (the marginal product of labor) is constant along the curve in fig. 1. This may be verified simply by calculating the marginal product of labor from (11) and making the appropriate substitutions.

In the general case, where $\gamma \geq 0$, the real wage increases as one moves up the curve in fig. 1. Along this curve, we have the elasticity formula

$$\frac{K}{w}\frac{\partial w}{\partial K} = \frac{(1+\beta)\gamma}{1-\beta+\gamma},$$

so that wealthier countries have higher wages than poorer ones for labor of any given skill. (Of course, workers in wealthy countries are typically also more skilled than workers in poor countries.) In all countries, wages at each skill level grow at the rate

$$\omega = \frac{\gamma}{1-\beta}\nu.$$

Then taking skill growth into account as well, wages grow at

$$\omega + \nu = \frac{1-\beta+\gamma}{1-\beta}\nu = \kappa,$$

or at a rate equal to the growth rate in the per-capita stock of physical capital.

The version of the model I propose to fit to, or estimate from, U.S. time series is the equilibrium solution (21), (26) and (10). As in the discussion of Solow's version λ, κ, β and s are estimated, from Denison (1961), at 0.013, 0.014, 0.25 and 0.1, respectively. Denison also provides an estimate of 0.009 for the annual growth rate of human capital over his period, an estimate based mainly on the changing composition of the workforce by levels of education and on observations on the relative earnings of differently schooled workers. I will use this 0.009 figure as an estimate of ν, which amounts to assuming that human capital is accumulated to the point where its private return equals its social (and private) cost. (Since schooling is heavily subsidized in the U.S., this assumption may seem way off, but surely most of the subsidy is directed at early schooling that would be acquired by virtually everyone anyway, and so does not affect the margins relevant for my calculations.) Then the idea is to use (10), (21) and (26) to estimate ρ, σ, γ and δ.

As was the case in the Solow model, ρ and σ cannot separately be identified along steady-state paths, but eq. (10) (which can be derived for this model in exactly the same way as I derived it for the model of section 2) implies $\rho + \sigma\kappa = 0.0675$. Eq. (21) implies $\gamma = 0.417$. Combining eqs. (21) and (26) yields a relationship involving γ, ν, β, δ, λ and $\rho + \sigma\kappa$, but not ρ or σ separately. This relationship yields an estimate for δ of 0.05. The implied fraction of time devoted to goods production is then, from (20), $u = 0.82$.

Given these parameter estimates, the *efficient* rate of human capital growth can be calculated, as a function of σ, from (24). It is: $\nu^* = 0.009 + 0.0146/\sigma$. Table 1 gives some values of this function and the associated values of u^* and $\kappa^* = (1.556)\nu^*$. Under log utility ($\sigma = 1$), then, the U.S. economy 'ought' to devote nearly three times as much effort to human capital accumulations as it does, and 'ought' to enjoy growth in per-capita consumption about two full percentage points higher than it has had in the past.

One could as easily fit this model to U.S. data under the assumption that *all* returns to human capital are internal, or that $\gamma = 0$. In this case ν, ν^* and κ have the common value, from (21), (24) and (26), $\sigma^{-1}[\delta - (\rho - \lambda)]$, and the *ratio* of physical to human capital will converge to a value that is independent of initial conditions (the curve in fig. 1 will be a straight line). Identifying this common growth rate with Denison's 0.014 estimate for κ implies a u value of 0.72, or that 28% of effective workers' time is spent in human capital

Table 1

σ	ν^*	u^*	κ^*
1	0.024	0.52	0.037
2	0.016	0.68	0.025
3	0.014	0.72	0.022

accumulation. Accepting Denison's estimate of a 0.009 growth rate of human capital due to schooling, this would leave 0.005 to be attributed to other forms, say on-the-job training that is distinct from productive activities.

What can be concluded from these exercises? Normatively, it seems to me, very little: The model I have just described has *exactly* the same ability to fit U.S. data as does the Solow model, in which equilibrium and efficient growth rates coincide. Moreover, it is clear that the two models can be merged [by re-introducing exogenous technical change into (11)] to yield a whole class of intermediate models that also fit data in this same rough sense. I am simply generating new possibilities, in the hope of obtaining a theoretical account of cross-country *differences* in income levels and growth rates. Since the model just examined is consistent with the *permanent* maintenance of per-capita income differentials of any size (though not with differences in growth rates) some progress toward this objective has been made. But before returning to empirical issues in more detail, I would like to generate another, quite different, example of a system in which human capital plays a central role.

5. Learning-by-doing and comparative advantage

The model I have just worked through treats the decision to accumulate human capital as equivalent to a decision to withdraw effort from production – to go to school, say. As many economists have observed, on-the-job-training or learning-by-doing appear to be at least as important as schooling in the formation of human capital. It would not be difficult to incorporate such effects into the previous model, but it is easier to think about one thing at a time so I will just set out an example of a system (again, for the moment, closed) in which *all* human capital accumulation is learning-by-doing. Doing this will involve thinking about economies with many consumption goods, which will open up interesting new possibilities for interactions between international trade and economic growth.[14]

Let there be two consumption goods, c_1 and c_2, and no physical capital. For simplicity, let population be constant. The ith good is produced with the Ricardian technology:

$$c_i(t) = h_i(t)u_i(t)N(t), \qquad i = 1, 2, \tag{29}$$

where $h_i(t)$ is human capital specialized to the production of good i and $u_i(t)$ is the fraction of the workforce devoted to producing good i (so $u_i \geq 0$ and $u_1 + u_2 = 1$). Of course, it would not be at all difficult to incorporate physical capital into this model, with (29) replaced by something like (11) for each good i. Later on, I will conjecture the behavior of such a hybrid model, but it will be simpler for now to abstract from capital.

[14] The formulation of learning used in this section is taken from Krugman (1985).

In order to let $h_i(t)$ be interpreted as a result of learning-by-doing, assume that the growth of $h_i(t)$ increases with the effort $u_i(t)$ devoted to producing good i (as opposed to increasing with the effort *withdrawn* from production). A simple way to do this is

$$\dot{h}_i(t) = h_i(t)\delta_i u_i(t). \tag{30}$$

To be specific, assume that $\delta_1 > \delta_2$, so that good 1 is taken to be the 'high-technology' good. For the sake of discussion, assume at one extreme that the effects of $h_i(t)$ in (29) and (30) are entirely external: production and skill accumulation for each good depend on the average skill level in that industry only.

As was the case with (13), the equation for human capital accumulation in the model discussed earlier, (30) seems to violate the diminishing returns we observe in studies of productivity growth for particular products. Learning-by-doing in any particular activity occurs rapidly at first, then more slowly, then not at all. Yet as in the preceding discussion, if we simply incorporate diminishing returns into (30), human capital will lose its status as an engine of growth (and hence its interest for the present discussion). What I want (30) to 'stand for', then, is an environment in which new goods are continually being introduced, with diminishing returns to learning on each of them separately, and with human capital specialized to old goods being 'inherited' in some way by new goods. In other words, one would like to consider the inheritance of human capital within 'families' of goods as well as within families of people.[15]

Under these assumptions of no physical capital accumulation and purely external human capital accumulation, the individual consumer has no intertemporal tradeoffs to decide on, so all we need to know about his preferences is his current-period utility function. I will assume a constant elasticity of substitution form:

$$U(c_1, c_2) = \left[\alpha_1 c_1^{-\rho} + \alpha_2 c_2^{-\rho}\right]^{-1/\rho}, \tag{31}$$

where $\alpha_i \geq 0$, $\alpha_1 + \alpha_2 = 1$, $\rho > -1$, and $\sigma = 1/(1+\rho)$ is the elasticity of substitution between c_1 and c_2. (Please note that the parameters ρ and σ represent completely different aspects of preferences in this section from those they represented in sections 2–4.) With technology and preferences given by (29)–(31), I will first work out the equilibrium under autarchy and then turn to international trade considerations.

Take the first good as numeraire, and let $(1, q)$ be the equilibrium prices in a closed economy. Then q must equal the marginal rate of substitution in

[15] Stokey (1987) formulates a model of learning on an infinite family of produced and potentially producible goods that captures exactly these features.

consumption, or

$$q = \frac{U_2(c_1,c_2)}{U_1(c_1,c_2)} = \frac{\alpha_2}{\alpha_1}\left(\frac{c_2}{c_1}\right)^{-(1+\rho)}.$$

Solving for the consumption ratio,

$$\frac{c_2}{c_1} = \left(\frac{\alpha_2}{\alpha_1}\right)^{\sigma} q^{-\sigma}. \tag{32}$$

Hence both goods will be produced, so that (29) plus profit maximization implies that relative prices are dictated by the human capital endowments: $q = h_1/h_2$. Then (29) and (32) together give the equilibrium workforce allocation as a function of these endowments,

$$\frac{c_2}{c_1} = \frac{u_2 h_2}{u_1 h_1} = \left(\frac{\alpha_2}{\alpha_1}\right)^{\sigma}\left(\frac{h_2}{h_1}\right)^{\sigma},$$

or

$$\frac{1-u_1}{u_1} = \left(\frac{\alpha_2}{\alpha_1}\right)^{\sigma}\left(\frac{h_2}{h_1}\right)^{\sigma-1}. \tag{33}$$

The dynamics of this closed economy are then determined by inserting this information into eq. (30). Solving first for the autarchy price path, $q(t) = h_1(t)/h_2(t)$, we have

$$\frac{1}{q}\frac{\mathrm{d}q}{\mathrm{d}t} = \frac{1}{h_1}\frac{\mathrm{d}h_1}{\mathrm{d}t} - \frac{1}{h_2}\frac{\mathrm{d}h_2}{\mathrm{d}t} = \delta_1 u_1 - \delta_2(1-u_1),$$

or

$$\frac{1}{q}\frac{\mathrm{d}q}{\mathrm{d}t} = (\delta_1+\delta_2)\left[1 + \left(\frac{\alpha_2}{\alpha_1}\right)^{\sigma} q^{1-\sigma}\right]^{-1} - \delta_2. \tag{34}$$

Solving this first-order equation for $q(t) = h_1(t)/h_2(t)$, given the initial endowments $h_1(0)$ and $h_2(0)$, determines the workforce allocation at each date [from (33)] and hence, from (30), the paths of $h_1(t)$ and $h_2(t)$ separately.

It will come as no surprise to trade theorists that the analysis of (34) breaks down into three cases, depending on the elasticity of substitution σ between the two goods. I will argue below, on the basis of trade considerations, that the interesting case for us is when $\sigma > 1$, so that c_1 and c_2 are assumed to be good

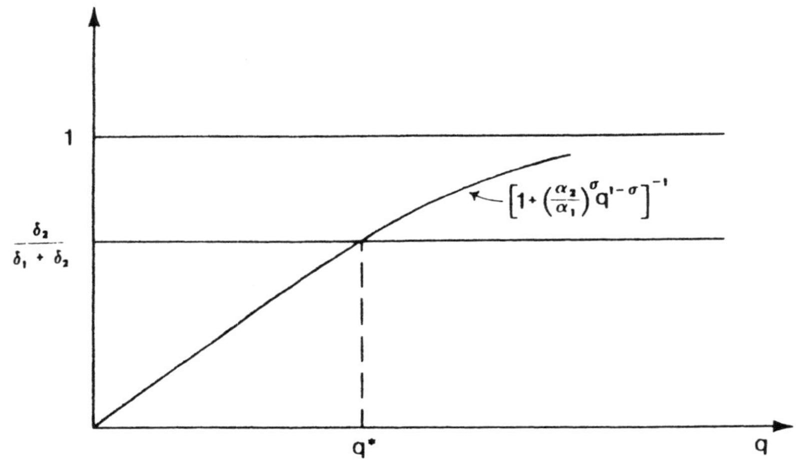

Fig 2

substitutes. But in order to make this case, we need all three possibilities in front of us. Refer to fig. 2.

The figure is drawn for the case $\sigma > 1$, in which case the function $[1 + (\alpha_2/\alpha_1)^{\sigma} q^{1-\sigma}]^{-1}$ has the depicted upward slope. To the left of q^*, $\mathrm{d}q/\mathrm{d}t < 0$, so $q(t)$ tends to 0. To the right, $\mathrm{d}q/\mathrm{d}t > 0$, so $q(t)$ grows without bound. Thus the system in autarchy converges to specialization in one of the two goods [unless $q(0) = q^*$]. The choice of *which* good to specialize in is dictated by initial conditions. If we are initially good at producing c_1 [if $q(0) > q^*$], we produce a lot of it, get relatively better and better at producing more of it, eventually, since c_1 and c_2 are good substitutes, producing vanishingly small amounts of c_2.

If the goods are poor substitutes, $\sigma < 1$, the curve in fig. 2 slopes down and q^* becomes a *stable* stationary point. At this point, the workforce is so allocated as to equate $\delta_1 u_1$ and $\delta_2 u_2$.

In the borderline case of $\sigma = 1$, the curve is flat. The workforce is initially allocated as dictated by the demand weights, $u_i = \alpha_i$, $i = 1, 2$, and this allocation is maintained forever. The autarchy price grows (or shrinks) at the constant rate $(1/q)(\mathrm{d}q/\mathrm{d}t) = \alpha_1\delta_1 - \alpha_2\delta_2$ forever.

As we learn how to produce computers more and more cheaply, then, we can substitute in their favor and consume more calculations and fewer potatoes, or we can use this benefit to release resources from computer production so as to consume more potatoes as well. The choice we take, not surprisingly, depends on whether these two goods are good substitutes or poor ones.

As was the case with the human capital model of the preceding section, it is obvious that the equilibrium paths we have just calculated will *not* be efficient. Since learning effects are assumed to be external, agents do not take them into account. If they did, they would allocate labor toward the 'high δ_i' good, relative to an equilibrium allocation, so as to take advantage of its higher growth potential.

Thus, except for the absence of physical capital, this closed economy model captures very much the same economics as does the preceding one. In both cases, the accumulation of human capital involves a sacrifice of current utility. In the first model, this sacrifice takes the form of a decrease in current consumption. In the second, it takes the form of a less desirable *mix* of current consumption goods than could be obtained with slower human capital growth. In both models the equilibrium growth rate falls short of the efficient rate and yields lower welfare. A subsidy to schooling would improve matters in the first. In the second, in language that is current in the United States, an 'industrial policy' focused on 'picking winners' (that is, subsidizing the production of high $\alpha_i\delta_i$ goods) would be called for. In the model, 'picking winners' is easy. If only it were so in reality!

The introduction of international trade into this second model leads to possibilities that I think are of real interest, though I have only begun to think them through analytically. The simplest kind of world to think about is one with perfectly free trade in the two final goods and with a continuum of small countries, since in that case prices in all countries will equal world prices $(1, p)$, say, and each country will take p as given. Fig. 3 gives a snapshot of this world at a single point in time. The contour lines in this figure are intended to depict a joint distribution of countries by their initial human capital endowments. A country is a point (h_1, h_2), and the distribution indicates the concentration of countries at various endowment levels.

At a given world price p, countries above the indicated line are producers of good 2, since for them $h_1/h_2 < p$ and they maximize the value of their production by specializing in this good. Countries below the line specialize in producing good 1, for the same reason. Then for each p one can calculate world supply of good 1 by summing (or integrating) the h_1 values below this price line, and the world supply of good 2 by summing the h_2 values above the line. Clearly, the supply of good 2 is an increasing function of p and of good 1 a decreasing function, so that the ratio c_2/c_1 of total quantities supplied increases as p increases.

Now world relative demand, with identical homothetic preferences, is just the same decreasing function of p that described each country's demand in the autarchic case: $c_2/c_1 = (\alpha_2/\alpha_1)^{\sigma} p^{-\sigma}$. Hence this static model determines the equilibrium world relative price p uniquely. Let us turn to the dynamics.

Those countries above the price line in fig. 3 are producing only good 2, so their h_1 endowments remain fixed while their h_2 endowments grow at the rate δ_2. Each country below the price line will produce only good 1, so that its h_2 is

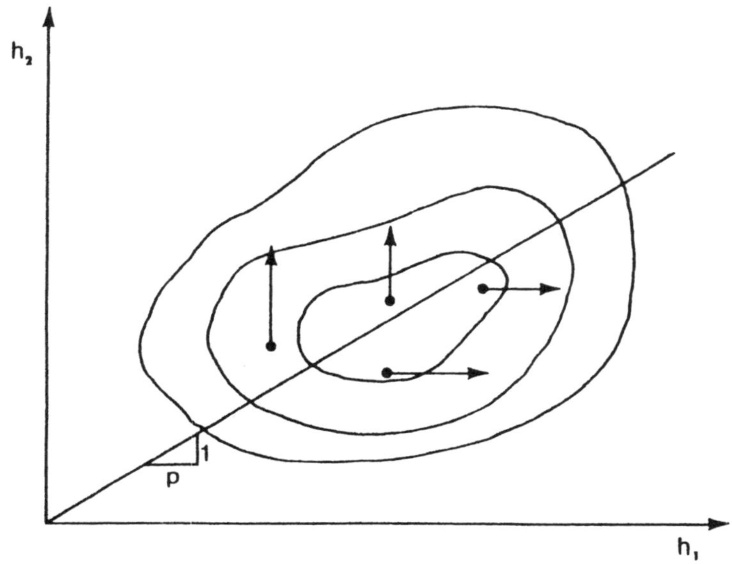

Fig. 3

constant while h_1 grows at the rate δ_1. Thus each country's (h_1,h_2) coordinates are changing as indicated by the arrows in fig. 3, altering the distribution of endowments that determines goods supplies over time. These movements obviously intensify the comparative advantages that led each country to specialize in the first place. On the other hand, as the endowment distribution changes, so does the equilibrium price p. Is it possible that these price movements will induce any country to switch its specialization from one good to the other?

A little reflection suggest that if anyone switches, it will have to be a producer of the high-δ good: good 1. The terms of trade are moving against good 1 (in the absence of switching) since its supply is growing faster. The issue again turns on the degree of substitutability between the two goods. If σ is low, the terms of trade may deteriorate so fast that a marginal good 1 producer may switch to producing good 2: he is getting relatively better at producing good 1, but not fast enough. The inequality that rules this possibility out is

$$\sigma \geq 1 - \frac{\delta_2}{\delta_1}. \tag{35}$$

I have already said that I think $\sigma > 1$ is the interesting case, so I want to accept (35) for the rest of the discussion.

Under (35) – that is, with no producer switching – we can read the dynamics of prices right off the relative demand schedule:

$$\frac{1}{p}\frac{\mathrm{d}p}{\mathrm{d}t} = \frac{\delta_1 - \delta_2}{\sigma}. \tag{36}$$

With relative price movements determined, the growth rates of real output in all countries is also determined. Measured in units of good 1, output of the good 1 producers grows at the rate δ_1. Output of the good 2 producers, also measured in units of good 1, grows at the rate $\delta_2 + (1/p)(\mathrm{d}p/\mathrm{d}t) = \delta_2 + (\delta_1 - \delta_2)/\sigma$. In general, then countries in equilibrium will undergo constant but not equal growth rates of real output.

Which countries will grow fastest? The condition that producers of the high-δ good, good 1, will have faster real growth is just

$$\delta_1 > \delta_2 + \frac{\delta_1 - \delta_2}{\sigma},$$

which is equivalent to the condition: $\sigma > 1$. That is, producing (having a comparative advantage in) high-learning goods will lead to higher-than-average real growth *only* if the two goods are good substitutes. Since it is exactly this possibility that the model is designed to capture, the case $\sigma > 1$ seems to me the only one of potential interest. If the terms-of-trade effects of technological change dominated the direct effects on productivity (which would be the case if $\sigma < 1$), those countries with rapid technological change would enjoy the slowest real income growth. There may be instances of such 'immiserizing growth', but if so they are surely the exceptions, not the rule. (These are the 'trade considerations' I mentioned earlier.)

This simple model shares with the model of section 4 the prediction of constant, endogenously determined real growth rates. In addition, it offers the possibility of different growth rates across countries, though differences that are not systematically related to income levels. In the equilibrium of the model, production patterns are dictated by comparative advantage: Each country produces goods for which its human capital endowment suits it. Given a learning technology like (30), countries accumulate skills by doing what they are already good at doing, intensifying whatever comparative advantage they begin with. This aspect of the theory will tend to lock in place an initial pattern of production, with rates of output growth variable across countries but stable within each country. There is no doubt that we observe forces for stability of this type, but there seem to be offsetting forces in reality that this model does not capture.

One of these has to do with the composition of demand. With homothetic utility the composition of world demand will remain fixed as income grows. In fact, we know that income elasticities for important classes of goods differ significantly from unity (contrary to the assumption of homotheticity). (We know, for example, that demand shifts systematically away from food consumption as income grows.) This force will 'create' comparative advantages in the production of other goods as time passes, altering world production patterns and growth rates as it does.

Another, I would guess more important, force has to do with the continual introduction of new goods and the fall-off of learning rates on old goods. By modeling learning as occuring at fixed rates on a fixed set of goods, I have here abstracted from important sources of change in world trade patterns. Modifying the model to incorporate possibilities of these two types is an entirely practical idea, given current theoretical technology, but the general equilibrium possibilities for such a modified system have not as yet been worked out.[16]

The present model provides a simple context for discussing two popular 'strategies' for economic development: 'import substitution' and 'export promotion'. Consider first a country with $q = h_1/h_2$ currently to the right of q^* in fig. 2, but with (h_1, h_2) lying above the equilibrium world price price line in fig. 3. Under free trade, this country will specialize in the production of good 2 forever. Under autarchy (which is just the extreme version of an import substitution policy) this country will specialize in producing good 1. Eventually its expertise in this protected industry will grow to the point where it will have a comparative advantage in good 1 under free trade, and the maintenance of autarchy will no longer serve any purpose, but this need not be so from the beginning.

I hasten to add that this is only one theoretical possibility among many. Another possibility is an initial q value below q^* in fig. 2. In this case, autarchy will not provide nurture for the infant industry, but will rather permanently cut off the country from consuming the high-learning good. Within the context of this model, then, there is no substance-free way to deduce useful guides for trade and development policies. One needs to know something about the actual technological possibilities for producing different goods in different places in order to arrive at definite conclusions.

I take an 'export promotion' strategy to mean something slightly different: the manipulation through taxes and subsidies of the terms of trade p faced by a country's producers. With this kind of flexibility, one need not simply choose between world price p and autarchy price q, but can rather set any production incentives and hence choose any growth rate between the two extremes in the free trade equilibrium. Obviously, even with this flexibility it does not follow

[16] Again, see Stokey (1987)

that 'growth-increasing' and 'welfare-improving' policies will necessarily coincide, but they certainly might.

My objective in this section has been to offer one example of a theoretical model in which rates of growth differ across countries, and not to offer policy advice. The case for infant industry protection based on external effects that this model formalizes is the classic one, and it does not become either more or less valid, empirically, by being embedded in a slightly new framework. But is it possible, I wonder, to account for the large cross-country differences in growth rates that we observe in a theoretical model that does *not* involve external effects of the sort I have postulated here? I have not seen it done.

6. Cities and growth

My concern to this point has been almost exclusively with the aggregate mechanics of economic development, and I am afraid the discussion in these lectures will not get much beyond these mechanics. But I believe a successful theory of development (or of anything else) has to involve more than aggregative modeling, and I would like both to explain what I mean by this and to indicate where one might look to extend the analysis to a deeper and more productive level.

The engine of growth in the models of sections 4 and 5 is *human capital*. Within the context of these two models, human capital is simply an unobservable magnitude or force, with certain assumed properties, that I have postulated in order to account for some observed features of aggregative behavior. If these features of behavior were *all* of the observable consequences of the idea of human capital, then I think it would make little difference if we simply re-named this force, say, the Protestant ethic or the Spirit of History or just 'factor X'. After all, we can no more directly measure the amount of human capital a society has, or the rate at which it is growing, than we can measure the degree to which a society is imbued with the Protestant ethic.

But this is *not* all we know about human capital. This same force, admittedly unobservable, has also been used to account for a vast number of phenomena involving the way people allocate their time, the way individuals' earnings evolve over their lifetimes, aspects of the formation, maintenance and dissolution of relationships within families, firms and other organizations, and so on. The idea of human capital may have seemed ethereal when it was first introduced – at least, it did to me – but after two decades of research applications of human capital theory we have learned to 'see' it in a wide variety of phenomena, just as meteorology has taught us to 'see' the advent of a warm front in a bank of clouds or 'feel' it in the mugginess of the air.

Indeed, for me the development of the theory of human capital has very much altered the way I think about physical capital. We can, after all, no more directly measure a society's holdings of physical capital than we can its human capital. The fiction of 'counting machines' is helpful in certain abstract contexts but not at all operational or useful in actual economies – even primitive ones. If this was the issue in the famous 'two Cambridges' controversy, then it has long since been resolved in favor of this side of the Atlantic.[17] Physical capital, too, is best viewed as a force, not directly observable, that we postulate in order to account in a unified way for certain things we *can* observe: that goods are produced that yield no immediate benefit to consumers, that the production of these goods enhances labor productivity in future periods, and so on.

The fact that the postulates of both human and physical capital have many observable implications outside the contexts of aggregate models is important in specific, quantitative ways, in addition to simply giving aggregative theorists a sense of having 'microeconomic foundations'. For example, in my application of a human capital model to U.S. aggregative figures, I matched the U.S. observations to the predictions of a competitive model (as opposed to an efficient one) in spite of the fact that education, in the U.S., involves vast government intervention and is obviously not a competitive industry in any descriptive sense. Why not instead identify the observed paths with the model's efficient trajectories? The aggregative data have *no* ability to discriminate between these two hypotheses, so this choice would have yielded as good a 'fit' as the one I made. At this point, I appealed to the observation that most education subsidies are infra-marginal from the individual's point of view. This observation could stand considerable refinement before it could really settle this particular issue, but the point is that aggregate models based on constructs that have implications for data *other* than aggregates – models with 'microeconomic foundations' if you like – permit us to bring evidence to bear on questions of aggregative importance that cannot be resolved with aggregate theory and observations alone. Without the ability to do this, we can do little more than extrapolate past trends into the future, and then be caught by surprise every time one of these trends changes.

The particular aggregate models I have set out utilize the idea of human capital quite centrally, but assign a central role as well to what I have been calling the *external effects* of human capital. This latter force is, it seems to me, on a quite different footing from the idea of human capital generally: The twenty years of research I have referred to earlier is almost exclusively concerned with the *internal* effects of human capital, or with investments in human capital the returns to which accrue to the individual (or his immediate family). If it is this research that permits us to 'see' human capital, then the external effects of this capital must be viewed as remaining largely invisible, or visible at the aggregative level only. For example, in section 4 I arrived at an estimate of $\gamma = 0.4$ for the elasticity of U.S. output with respect to the external effects of human capital on production. Does this seem a plausible number?

[17] That is, the English side

Or, putting the question in a better way: Is $\gamma = 0.4$ consistent with other evidence? But *what* other evidence? I do not know the answer to this question, but it is so central that I want to spend some time thinking about where the answer may be found. In doing so, I will be following very closely the lead of Jane Jacobs, whose remarkable book *The Economy of Cities* (1969) seems to me mainly and convincingly concerned (though she does not use this terminology) with the external effects of human capital.

I have been concerned with modeling the economic growth of *nations*, considered either singly or as linked through trade. In part, this was a response to the form of the observations I cited at the beginning: Most of our data come in the form of national time series, so 'fitting the facts' is taken to mean fitting national summary facts. For considering effects of changes in policies the nation is again the natural unit, for the most important fiscal and commercial policies are national and affect national economies in a uniform way. But from the viewpoint of a *technology* – like (11) – through which the average skill level of a group of people is assumed to affect the productivity of each individual within the group, a *national* economy is a completely arbitrary unit to consider. Surely if Puerto Rico were to become the fifty-first state this would not, by itself, alter the productivity of the people now located in Puerto Rico, even though it would sharply increase the average level of human capital of those politically defined as their fellow citizens. The external effects that the term h_a^{γ} in (11) is intended to capture have to do with the influences people have on the productivity of others, so the *scope* of such effects must have to do with the ways various groups of people interact, which may be affected by political boundaries but are certainly an entirely different matter conceptually.

Once this question of the scope of external effects is raised, it is clear that it cannot have a single correct answer. Many such effects can be internalized within small groups of people – firms or families. By dealing with an infinitely-lived family as a typical agent, I have assumed that such effects are dealt with at the non-market level and so create no gap between private and social returns. At the other extreme, basic discoveries that immediately become common property – the development of a new mathematical result say – are human capital in the sense that they arise from resources allocated to such discoveries that could instead have been used to produce current consumption, but to most countries as well as to most individual agents they appear 'exogenous' and would be better modelled as $A(t)$ in section 2 than as $h_a(t)$ in section 4.

If it were easy to classify most external productivity effects as either global in scope or as so localized as to be internalizable at the level of the family or the firm, then I think a model that incorporated internal human capital effects only plus other effects treated as exogenous technical change would be adequate. Such a model would fit time series from advanced countries about as well as any I have advanced, being an intermediate model to those I discussed in sections 2 and 4, which were in turn not distinguishable on such data alone.

Such a model would, I think, have difficulty reconciling observed pressures for immigration with the absence of equivalent capital flows, but perhaps this anomaly could be accounted for in some other way.

But we *know* from ordinary experience that there are group interactions that are central to individual productivity and that involve groups larger than the immediate family and smaller than the human race as a whole. Most of what we know we learn from other people. We pay tuition to a few of these teachers, either directly or indirectly by accepting lower pay so we can hand around them, but most of it we get for free, and often in ways that are mutual – without a distinction between student and teacher. Certainly in our own profession, the benefits of colleagues from whom we hope to learn are tangible enough to lead us to spend a considerable fraction of our time fighting over who they shall be, and another fraction travelling to talk with those we wish we could have as colleagues but cannot. We know this kind of external effect is common to all the arts and sciences – the 'creative professions'. All of intellectual history is the history of such effects.

But, as Jacobs has rightly emphasized and illustrated with hundreds of concrete examples, much of economic life is 'creative' in much the same way as is 'art' and 'science'. New York City's garment district, financial district, diamond district, advertising district and many more are as much intellectual centers as is Columbia or New York University. The specific ideas exchanged in these centers differ, of course, from those exchanged in academic circles, but the process is much the same. To an outsider, it even *looks* the same: A collection of people doing pretty much the same thing, each emphasizing his own originality and uniqueness.

Considerations such as these may convince one of the existence of external human capital, and even that it is an important element in the growth of knowledge. But they do not easily lend themselves to quantification. Here again I find Jacobs's work highly suggestive. Her emphasis on the role of cities in economic growth stems from the observation that a city, economically, is like the nucleus of an atom: If we postulate only the usual list of economic forces, cities should fly apart. The theory of production contains nothing to hold a city together. A city is simply a collection of factors of production – capital, people and land – and land is always far cheaper outside cities than inside. Why don't capital and people move outside, combining themselves with cheaper land and thereby increasing profits? Of course, people like to live near shopping and shops need to be located near their customers, but circular considerations of this kind explain only shopping centers, not cities. Cities are centered on wholesale trade and primary producers, and a theory that accounts for their existence has to explain why these producers are apparently choosing high rather than low cost modes of operation.

It seems to me that the 'force' we need to postulate account for the central role of cities in economic life is of exactly the same character as the 'external human capital' I have postulated as a force to account for certain features of aggregative development. If so, then land rents should provide an indirect measure of this force, in much the same way that schooling-induced earnings differentials provide a measure of the productive effects of internal human capital. It would require a much more detailed theory of the external effects of human capital than anything I have provided to make use of the information in urban land rents (just as one needs a more detailed theory of human capital than that in section 4 to utilize the information in earnings data), but the general logic is the same in the two cases. What can people be paying Manhattan or downtown Chicago rents *for*, if not for being near other people?

7. Conclusions

My aim, as I said at the beginning of these lectures, has been to try to find what I called 'mechanics' suitable for the study of economic development: that is, a system of differential equations the solution to which imitates some of the main features of the economic behavior we observe in the world economy. This enterprise has been taken about as far as I am able to take it, at present, so I will stop and try to sum up what the main features of these mechanics are and the sense in which they conform to what we observe.

The model that I think is central was developed in section 4. It is a system with a given rate of population growth but which is acted on by no other outside or exogenous forces. There are two kinds of capital, or state variables, in the system: physical capital that is accumulated and utilized in production under a familiar neoclassical technology, and human capital that enhances the productivity or both labor and physical capital, and that is accumulated according to a 'law' having the crucial property that a constant level of effort produces a constant growth rate of the stock, independent of the level already attained.

The dynamics of this system, viewed as a single, closed economy, are as follows. Asymptotically, the marginal product of physical capital tends to a constant, given essentially by the rate of time preference. This fact, which with one kind of capital defines the long-run stock of that capital, in the two-capital model of section 4 defines a curve in the 'physical capital–human capital plane'. The system will converge to this curve from any initial configuration of capital stocks, but the particular point to which it converges will depend on initial conditions. Economies that are initially poor will remain poor, relatively, though their long-run rate of income growth will be the same as that of initially (and permanently) wealthier economies. A world consisting of such economies, then, each operating autarchically, would exhibit uniform rates of growth across countries and would maintain a perfectly stable distribution of income and wealth over time.

If trade in capital goods is introduced into this model world economy, with labor assumed immobile, there will be no tendency to trade, which is to say no systematic tendency for borrowing and lending relationships to emerge between rich and poor countries. Put another way, the long-run relationship between the two kinds of capital that holds in each country implies the same marginal productivity of physical capital, no matter what the level of capital that has been accumulated. The picture I have given for a world of closed economies thus carries over without change to a world with free trade in capital goods.

If labor mobility is introduced, everything hinges on whether the effects of human capital are internal – affecting the productivity of its 'owner' only – or whether they have external benefits that spill over from one person to another. In the latter case, and only in the latter case, the wage rate of labor at any given skill level will increase with the wealth of the country in which he is employed. Then if labor can move, it will move, flowing in general from poor countries to wealthy ones.

The model I have described fits the evidence of the last century for the U.S. economy as well as the now standard neoclassical model of Solow and Denison, which is to say, remarkably well. This is of course no accident, for the mechanics I have been developing have been modeled as closely as possible on theirs. It also fits, about as well, what seem to me the main features of the world economy: very wide diversity in income levels across countries, sustained growth in per-capita incomes at all income levels (though not, of course, in each country at each income level), and the absence of any marked tendency for growth rates to differ systematically at different levels of income. The model is also consistent with the enormous pressures for immigration that we observe in the world, even with its extreme assumptions that assign no importance to differences in endowments of natural resources and that permit perfectly free trade in capital and consumption goods. As long as people at each skill level are more productive in high human capital environments, such pressures are predicted to exist and nothing but the movement of people can relieve them.

Though the model of section 4 seems capable of accounting for *average* rates of growth, it contains no forces to account for diversity over countries or over time within a country (except for arbitrary shifts in tastes or technology). Section 5 develops a two-commodity elaboration of this model that offers more possibilities. In this set-up, human capital accumulation is taken to be specific to the production of particular goods, and is acquired on-the-job or through learning-by-doing. If different goods are taken to have different potentials for human capital growth, then the same considerations of comparative advantage that determine which goods get produced where will also dictate each country's rate of human capital growth. The model thus admits the possibility of wide and sustained differences in growth rates across countries, differences that one would not expect to be systematically linked to each country's initial capital *levels*.

With a fixed set of goods, which was the only case I considered, this account of cross-country differences does not leave room for within-country changes in growth rates. The comparative advantages that dictate a country's initial production mix will simply be intensified over time by human capital accumulation. But I conjecture that a more satisfactory treatment of product-specific learning would involve modeling the continuous introduction of new goods, with learning potentials on any particular good declining with the amount produced. There is no doubt that we observe this kind of effect occuring in reality on particular product lines. If it could be captured in a tractable aggregative model, this would introduce a factor continuously shaking up an existing pattern of comparative advantages, and offer some interesting possibilities for shifts over time in a country's growth rate, within the same general equilibrium framework used in section 5.

If such an analysis of trade-related shifts in growth rates should turn out to possible, this would be interesting, because the dramatic recent development success stories, the 'growth miracles' of Korea, Taiwan, Hong Kong and Singapore (not to mention the ongoing miracle of Japan) have all been associated with increases in exports, and more suggestively still, with exports of goods not formerly produced in these countries. There is surely no strain in thinking that a model stressing the effects of learning-by-doing is likely to shed light on these events.

A successful theory of economic development clearly needs, in the first place, mechanics that are consistent with sustained growth and with sustained diversity in income levels. This was the objective of section 4. But there is no one pattern of growth to which all economies conform, so a useful theory needs also to capture some forces for change in these patterns, and a mechanics that permits these forces to operate. This is a harder task, certainly not carried out in the analysis I have worked through, but I think the analysis of section 5 is a promising beginning.

Acknowledgements

The fact that a fairly well known economist is willing to speak so broadly on a topic of such enormous importance, about which he obviously knows very little, has proved a great stimulus to discussion whenever these lectures have been given. I have received many more interesting reactions than I will ever be able to follow up on, or even to acknowledge. But I would like to thank Nancy Stokey for her criticism of preliminary drafts, Arnold Harberger, Jane Jacobs, Akiva Offenbacher, Theodore Schultz and Robert Solow for their comments, Richard Manning for his very able assistance, and Edward Prescott and Sherwin Rosen for stimulating discussions of all aspects of economic development over many years before and after these lectures were first given.

Finally, I would like to thank Robert King and Charles Plosser for encouraging the publication of this awkwardly-sized (too long for an article, too short for a book) paper. In response to their suggestions, I have retained the lecture style in this version, making for the most part only minimal changes. (Section 5 is the only exception: I found a much better framework than the one used in the original, and so have replaced much of the original text.) Their hope, and mine, is that without being definitive on any aspect of the problem we may be productively provocative on many.

References

Arrow, Kenneth J., 1962, The economic implications of learning by doing, Review of Economic Studies 29, 155–173

Baumol, William J., 1986, Productivity growth, convergence, and welfare: What the long-run data show, American Economic Review 76, 1072–1085.

Becker, Gary S., 1964, Human capital (Columbia University Press for the National Bureau of Economic Research, New York).

Becker, Gary S., 1981, A treatise on the family (Harvard University Press, Cambridge, MA).

Becker, Gary S. and Robert J. Barro, 1985, A reformulation of the economic theory of fertility, Unpublished working paper (University of Chicago, Chicago, IL).

Boxall, Peter J., 1986, Labor and population in a growth model, Unpublished doctoral dissertation (University of Chicago, Chicago, IL).

Burmeister, Edwin and A. Rodney Dobell, 1970, Mathematical theories of economic growth (Macmillan, New York).

DeLong, Bradford, 1987, Have productivity levels converged?, Unpublished working paper (MIT, Cambridge, MA).

Denison, Edward F., 1961, The sources of economic growth in the United States (Committee for Economic Development, New York).

Gordon, Robert J., 1971, Measurement bias in price indexes for capital goods, Review of Income and Wealth, Income and wealth series 17.

Griliches, Zvi and Dale W. Jorgenson, 1967, The explanation of productivity change, Review of Economic Studies 34, 249–282.

Harberger, Arnold C., ed., 1984, World economic growth (ICS Press, San Francisco, CA).

Jacobs, Jane, 1969, The economy of cities (Random House, New York).

Jacobs, Jane, 1984, Cities and the wealth of nations (Random House, New York).

Krueger, Anne O., 1983, The developing countries' role in the world economy, Lecture given at the University of Chicago, Chicago, IL.

Krugman, Paul, 1985, The narrow moving band, the Dutch disease and the competitive consequences of Mrs. Thatcher: Notes on trade in the presence of dynamic scale economies, Unpublished working paper (MIT, Cambridge, MA).

Kuznets, Simon, 1959, Six lectures on economic growth (The Free Press, Glencoe).

Maddison, Angus, 1982, Phases of capitalist development (Oxford University Press, New York).

Romer, Paul M., 1986, Increasing returns and long-run growth, Journal of Political Economy 94, 1002–1037.

Rosen, Sherwin, 1976, A theory of life earnings, Journal of Political Economy 84, 545–567.

Schultz, Theodore W., 1963, The economic value of education (Columbia University Press, New York).

Stokey, Nancy L., 1987, Learning-by-doing and the introduction of new goods, Unpublished working paper (Northwestern University, Evanston, IL).

Summers, Robert and Alan Heston, 1984, Improved international comparisons of real product and its composition: 1950–1980, Review of Income and Wealth, Income and wealth series 30.

Tamura, Robert, 1986, On the existence of multiple steady states in one sector growth models with intergenerational altruism, Unpublished working paper (University of Chicago, Chicago, IL).

Uzawa, Hirofumi, 1965, Optimum technical change in an aggregative model of economic growth, International Economic Review 6, 18–31.

Part IV
Knowledge Spillover in Cities

[9]

GEOGRAPHIC LOCALIZATION OF KNOWLEDGE SPILLOVERS AS EVIDENCED BY PATENT CITATIONS*

ADAM B. JAFFE
MANUEL TRAJTENBERG
REBECCA HENDERSON

We compare the geographic location of patent citations with that of the cited patents, as evidence of the extent to which knowledge spillovers are geographically localized. We find that citations to domestic patents are more likely to be domestic, and more likely to come from the same state and SMSA as the cited patents, compared with a "control frequency" reflecting the pre-existing concentration of related research activity. These effects are particularly significant at the local (SMSA) level. Localization fades over time, but only very slowly. There is no evidence that more "basic" inventions diffuse more rapidly than others.

The last decade has seen the development of a significant body of empirical research on R&D spillovers.[1] Generally speaking, this research has shown that the productivity of firms or industries is related to their R&D spending, and also to the R&D spending of other firms or other industries. In parallel, economic growth theorists have focused new attention on the role of knowledge capital in aggregate economic growth, with a prominent modeling role for knowledge spillovers (e.g., Romer [1986, 1990] and Grossman and Helpman [1991]).

We know very little, however, about where spillovers go. Is there any advantage to nearby firms, or even firms in the same country, or do spillovers waft into the ether, available for anyone around the globe to grab? The presumption that U. S. international competitiveness is affected by what goes on at federal laboratories and U. S. universities, and the belief that universities and other research centers can stimulate regional economic growth[2] are predicated on the existence of a geographic component to the

*We gratefully acknowledge support from the Ameritech Foundation, via the Ameritech Fellows program of the Center for Regional Economic Issues at Case Western Reserve University, and from the National Science Foundation through grant SES91-10516. We thank Neil Bania, Ricardo Caballero, Michael Fogarty, Zvi Griliches, Frank Lichtenberg, Francis Narin, seminar participants at NBER and Case Western Reserve University, and two anonymous referees for helpful comments. Any errors are the responsibility of the authors.

1. E.g., Jaffe [1986] and Bernstein and Nadiri [1988, 1989]. For a recent survey and evaluation of this literature, see Griliches [1991].

2. See, e.g., Minnesota Department of Trade and Economic Development [1988]; Dorfman [1988]; Feller [1989]; and Smilor, Kozmetsky, and Gibson [1988].

The Quarterly Journal of Economics, August 1993

spillover mechanism. The existing spillover literature, however, is virtually silent on this point.[3]

In the growth literature it is typically assumed that knowledge spills over to other agents within the country, but not to other countries.[4] This implicit assumption begs the question of whether and to what extent knowledge externalities are localized. As emphasized recently by Krugman [1991], acknowledging the importance of spillovers and increasing returns requires renewed attention by economists to issues of economic geography. Krugman revives and explores the explanations given by Marshall [1920] as to why industries are concentrated in cities. Marshall identified three factors favoring geographic concentration of industries: (1) the pooling of demands for specialized labor; (2) the development of specialized intermediate goods industries; and (3) knowledge spillovers among the firms in an industry. Krugman believes that economists should focus on the first two of these, partially because he perceives that "[k]nowledge flows, by contrast, are invisible; they leave no paper trail by which they may be measured and tracked, and there is nothing to prevent the theorist from assuming anything about them that she likes" [Krugman, p. 53].

But knowledge flows do sometimes leave a paper trail, in the form of citations in patents. Because patents contain detailed geographic information about their inventors, we can examine where these trails actually lead. Subject to caveats discussed below relative to the relationship between citations and spillovers, this allows us to use citation patterns to test the extent of spillover localization. We examine citations received by patents assigned to universities, and also the citations received by a sample of domestic corporate patents. If knowledge spillovers are localized within countries, then citations of patents generated within the United States should come disproportionately from within the United States. To the extent that regional localization of spillovers is

3. Jaffe [1989] provides evidence that corporate patenting at the state level depends on university research spending, after controlling for corporate R&D. Mansfield [1991] surveyed industrial R&D employees about university research from which they benefited. He found that they most often identified major research universities, but that there was some tendency to cite local universities even if they were not the best in their field.

4. The existence of this implicit assumption was noted by Glaeser, Kallal, Scheinkman, and Shleifer [1991]: "After all, intellectual breakthroughs must cross hallways and streets more easily than oceans and continents." Grossman and Helpman [1991] consider international knowledge spillovers explicitly.

important, citations should come disproportionately from the same state or metropolitan area as the originating patent.[5]

The most difficult problem confronted by the effort to test for spillover-localization is the difficulty of separating spillovers from correlations that may be due to a pre-existing pattern of geographic concentration of technologically related activities. That is, if a large fraction of citations to Stanford patents comes from the Silicon valley, we would like to attribute this to localization of spillovers. A slightly different interpretation is that a lot of Stanford patents relate to semiconductors, and a disproportionate fraction of the people interested in semiconductors happen to be in the Silicon valley, suggesting that we would observe localization of citations even if proximity offers no advantage in receiving spillovers. Of course, the ability to receive spillovers is probably one reason for this pre-existing concentration of activity. If it were the *only* possible reason, then, under the null hypothesis of no spillover localization we should still see no localization of citations. As discussed above, however, there are other sources of agglomeration effects that could explain the geographic concentration of technologically related activities without resort to localization of knowledge spillovers. For this reason, we construct "control" samples of patents that are not citations but have the same temporal and technological distribution as the citations. We then calculate the geographic matching frequencies between the citations and originating patents, and between the controls and originating patents. Our test of localization is whether the citation matching frequency is significantly greater than the control matching frequency. Since the "control" matching frequency is, itself, likely to be partly the result of spillover-localization, we believe this to be a conservative test for the existence of localization.

The first section of the paper describes patents and patent citations, explains the construction of the control samples, and

5. Glaeser, Kallal, Scheinkman, and Shleifer [1991] characterize the "Marshall-Arrow-Romer" models as focusing on knowledge spillovers within the firms in a given industry. They examine the growth rate of industries in cities as a function of the concentration of industrial activity across cities, within-city industrial diversity, and within-city competition. They find that within-city diversity is positively associated with growth of industries in that city, while concentration of an industry within a city does not foster its growth. They interpret this contrast to mean that spillovers across industries are more important than spillovers within industries. As is discussed below, there is evidence from the R&D spillover literature to suggest that across-industry knowledge spillovers are, indeed, important. In this study, we do not consider the *industrial* identity of either generators or receivers of spillovers, though we do have some information on their *technological* similarity.

considers more carefully how citations might be used to infer spillovers. The second section presents the results of the tests of geographic localization. The following section examines whether the probability of geographic localization of any given citation can be explained by attributes of the originating or citing patents, or of relationships between them. A concluding section follows.

I. Experimental Design

A. Patents and Patent Citations[6]

A patent is a property right in the commercial use of a device.[7] For a patent to be granted, the invention must be nontrivial, meaning that it would not appear obvious to a skilled practitioner of the relevant technology, and it must be useful, meaning that it has potential commercial value. If a patent is granted,[8] a public document is created containing extensive information about the inventor, her employer, and the technological antecedents of the invention, all of which can be accessed in computerized form. Among this information are "references" or "citations." It is the patent examiner who determines what citations a patent must include. The citations serve the legal function of delimiting the scope of the property right conveyed by the patent. The granting of the patent is a legal statement that the idea embodied in the patent represents a novel and useful contribution over and above the previous state of knowledge, as represented by the citations. Thus, in principle, a citation of Patent X by Patent Y means that X represents a piece of previously existing knowledge upon which Y builds.

The examiner has several means of identifying potential citations. The applicant has a legal duty to disclose any knowledge of the prior art that she may have. In addition, the examiner is supposed to be an expert in the technological area and be able to

6. All of the data we use relate to patents granted by the U. S. patent office. About 40 percent of U. S. patents are currently granted to foreigners. Other countries, of course, also grant patents, leading to some ambiguity in the meaning of phrases like "U. S. patent" and "foreign patent." We shall use the phrase "U. S. patent" to mean a patent granted by the U. S. patent office, regardless of the residence of the inventor. We shall use the phrase "domestic patent" to refer to a patent granted (by the U. S. patent office) to an inventor residing in the United States. We shall use the phrase "foreign patent" to refer to a patent granted by the U. S. patent office to non-U. S. residents.

7. Ideas are not patentable; nor are algorithms or computer programs, though a chip with a particular program coded into it might be. The definition of a device was recently broadened to include genetically engineered organisms.

8. There is no public record of unsuccessful patent applications.

identify relevant prior art that the applicant misses or conceals. The framework for the search of the prior art is the patent classification system. Every patent is assigned to a nine-digit patent class (of which there are about 100,000) as well as an unlimited number of additional or "cross-referenced" classes. An examiner will typically begin the search of prior art using her knowledge of the relevant classes. For the purpose of identifying distinct technical areas, we utilize aggregations of subclasses to a three-digit level; at this level there are currently about 400 technical classes.[9]

For this study, we begin with two cohorts of "originating" patents, one consisting of 1975 patent applications and the other of 1980 applications. In each cohort we include all patents granted to U. S. universities and two samples of U. S. corporate patents[10] chosen to match the university patents by grant date and technological distribution. These sets of originating patents were chosen because we conjectured that the extent of geographic localization might differ depending on the nature of the originating institution. As discussed below, such differences turn out to be minor. The 1975 originating cohort contains about 950 patents that had received a total of about 4750 citations by the end of 1989. The 1980 originating cohort contains about 1450 that had received about 5200 citations by the same time.

B. Construction of "Control" Samples

The main idea of this paper is to compare the geographic location of the citations with the originating patent that they cite. But to make such a comparison meaningful, we have to consider how often we would expect them to match under some "null" hypothesis. That is, we need to compare the probability of a patent matching the originating patent by geographic area, *conditional* on its citing the originating patent, with the probability of a match *not conditioned on the existence of a citation link.* This noncitation-conditioned probability gives us a baseline or reference value against which to compare the proportions of citations that match.

9. Examples of three-digit patent classes are "Batteries, Thermoelectric and Photoelectric"; "Distillation: Apparatus"; "Robots"; seventeen distinct classes of "Organic Compounds"; and the ever-popular "Whips and Whip Apparatus."

10. The "top corporate" sample consists of patents granted to the 200 top-R&D-performing firms in the United States, as reported in S.E.C. 10-k forms and compiled by Compustat. The "other corporate" sample contains patents assigned to U. S. corporations that are not universities and not in the "top corporate" sample.

We call this baseline or reference probability the "control frequency."

Two considerations drove our choice for constructing the control frequency. First, the fraction of U. S. patents granted to foreigners has been climbing steadily during the period under study here. We do not want to conclude that citations are initially localized, but that this localization fades over time, simply because of this aggregate trend. Second, countries (and cities and states) differ in their areas of technological focus. Although such technological specialization is probably due, in part, to geographic localization of spillovers, we want to be conservative and test whether spillovers are localized *relative to what would be expected given the existing distribution of technological activity.*

To derive a control frequency that would be immune to contamination from either aggregate movements over time or localization based on the pre-existing concentration of technological activity, we went back to the patent data base and found a "control patent" to correspond to each of the citing patents. For each citing patent, we identified all patents in the *same patent class* with the *same application year* (excluding any other patents that cited the same originating patent). We then chose from that set a control patent whose grant date was as close as possible to that of the citing patent. This process yielded, for each set of citing patents, a corresponding control sample of equal size, whose distribution across time and technological areas is essentially identical to that of the citation data set. Each control patent is paired with a particular citing patent, allowing us to compare the geographic location of the control patent with that of the originating patent cited by its counterpart in the citing dataset. The frequency with which these control patents match geographically with the originating patent is an estimate of the frequency with which a randomly drawn patent that is not a citation, but has the same technological and temporal profile as the citation, matches geographically.

To put it slightly differently, when we calculate the frequency with which the citations match the geographic location of the cited patents, we are estimating the probability of geographic match for two patents, *conditional on there being a citation link and also conditional on the technological nature and timing of the citation.* When we calculate the frequency with which the "control" patents match geographically with the cited patents, we are estimating the probability of geographic match for two patents, *conditional only on the technological nature and timing of the citation.* If the citation match frequency is significantly higher, then that implies that citations are localized even after controlling for timing and technology.

C. Issues Relating to the Use of Citations to Infer Spillovers

With the construction of the control samples, we believe that we have designed a very clean test of the extent to which patent citations are geographically localized. Before going on, we must address the validity of drawing inferences about knowledge spillovers from patent citations. For discussion purposes, we can classify the links that might exist between two inventions into one of three groups: spillovers accompanied by citations, citations that occur where there was no spillover, and spillovers that occur without generating a citation. Our experiment uses the first set, but clearly the other two are non-empty. The key question is whether and to what extent we expect that either of the latter two groups would be systematically more or less localized than the group we examine. Though there are a number of considerations, all difficult to quantify, we believe that on balance it is reasonable to draw inferences about spillovers from citations.

As a general consideration, it is important to keep in mind that any analogy between patent citations and academic article citations cannot be taken too far. Academics may cite a friend (or neighbor) just to be nice, since the price of doing so is infinitesimal, or even negative if a longer list of references is perceived as making the research look more thorough. An inventor who did the same in a patent application is, in effect, leaving money lying on the table: if those citations are included in the final patent the inventor has reduced the scope of her monopoly. Further, the patent examiner should not include such citations in the patent even if the inventor did put them there. Thus, it does not seem that "gratuitous" citations are a serious concern.

A deeper problem is created by "real" citations that are not spillovers. For example, suppose that a firm gets a patent on an invention and then contracts with another firm to make some part of it, or a machine necessary to make it, or any other aspect of the downstream development. It is possible that such a contractor might later get a patent on a related technology. To the extent that the flow of rents between these parties is governed by a complete contract, there could conceivably be no externality running from the original inventor to the contractor. If we now add to this hypothetical contract the assumption that such contracted development is relatively likely to be localized, we have the potential for the observed localization of citations to be greater than the true localization of knowledge spillovers.[11]

Although such "internalized spillovers" surely exist, it is likely that most citations that are not spillovers are of a different sort: citations (added by the examiner) to previous patents of which the citing inventor was unaware. Clearly, no spillover occurs in this case. Further, it seems likely that citations of this sort should not be any more geographically localized than the control patents. If many citations are in this category, it introduces "noise" into the citations as a measure of spillovers, and biases the results *away* from finding significant localization. Our a priori belief is that this category is much larger than the previous one, suggesting that spillovers are, on balance, probably more localized than citations, but readers with different beliefs should interpret our results accordingly.

Finally, there are an enormous number of spillovers with no citations, since only a small fraction of research output is ever patented. In particular, much of the results of very basic research cannot be patented. It is plausible that basic research generates the largest spillovers,[12] and also that basic research is communicated via mechanisms that are less likely to be localized, such as international journals. For this reason, it is probably appropriate to view our results as related to applied research, and to exercise care in extrapolating to the localization of spillovers from extremely basic research.

D. Geographic Assignment of Patents

The preceding discussion has presumed that the "location" of a patent is an unambiguous construct. The patent data contain the country of residence of each inventor, and the city and state of

11. As discussed below, we focus on tests of localization that exclude citing patents that are owned by the same organization as the originating patent, precisely because such "self-citations" do not represent an externality. Citations by other organizations that have an economic relationship with the original inventor could be viewed as similar to self-citations. There is, however, a significant difference: we expect that, in general, the contract between the two parties will be quite incomplete, making it more likely than not that the citing organization could capture some rents from the original invention and hence benefit from at least a partial spillover.

12. This question is analyzed in detail in Trajtenberg, Henderson, and Jaffe [1992].

residence for U. S. inventors.[13] Use of this information is complicated by the fact that patents can have multiple inventors who can live in different places. The following procedure was followed:

1. For U. S. *inventors,* city/state combinations were placed in counties using a commercially available city directory; each U. S. inventor was then assigned to an SMSA[14] based on state and county. For this purpose an additional "phantom" SMSA was created in each state, encompassing all counties in the state outside of defined SMSAs. Approximately 98 percent of inventors were successfully assigned to SMSAs.

2. Assignments of each *patent* to a country, a state, and an SMSA were then made based on pluralities of inventors. So, for example, a patent with one inventor living in Bethesda, MD, one in Alexandria, VA, and one in rural Virginia would be assigned VA for its state and Washington, DC, for its SMSA. Ties were assigned arbitrarily, except that ties between true SMSAs and phantom SMSAs were resolved for the true one and ties between United States and foreign were resolved in favor of foreign.[15]

II. Results on Extent of Localization

As a prelude to the geographic analysis, Table I and Figure I present some descriptive data about the citations and their relationship to the originating patents. Table I shows that about 80–90 percent of the 1975 patents and 70–80 percent of the 1980 patents had received at least one citation by the end of 1989, with the higher proportion in each case applying to the university patents. Mean citations received (including zeros) were four–six for 1975 and three–four for 1980, again with the higher numbers corresponding to the university patents.[16] The average lag between the originating application year and the application year of the citing

13. Published data on the geographic distribution of U. S. patents (including those cited in the popular press) are based on the location of the organization or individual to which the patent is "assigned" by the inventor (usually her employer). For analysis of localization, the location of the assignee is not a desirable datum. There is an ambiguity in such data relating to the way employees of multinational corporations make their assignments. An employee of "Honda" in the United States could assign her patent to "Honda U.S.A., Inc." or she could assign it to the parent company in Japan. In the former case, the patent office would call it a domestic patent, and in the latter case a foreign patent; similarly for IBM Switzerland.

14. These assignments were made based on the 1981 SMSA definitions. In areas where Consolidated Metropolitan Statistical Areas were defined in 1981, these were used; elsewhere Metropolitan Statistical Areas were used. Hence we use the generic term "SMSA."

15. At the country level, 98 percent of patents were assigned unanimously. At the state level, 90 percent were assigned unanimously; an additional 4 percent had more than half of inventors in a single state. At the SMSA level, 86 percent were assigned unanimously, and an additional 6 percent had a clear majority. Overall, 4.5 percent of patents were assigned to "phantom" SMSAs.

16. Our companion paper [Trajtenberg, Henderson, and Jaffe, 1992] explores in detail the use of citation intensity and related measures for measuring the basicness and appropriability of inventions.

TABLE I
DESCRIPTIVE STATISTICS

Originating dataset	Percent receiving citations	Total no. of citations	Mean citations received	Average citation lag[a,b]	Percent self-citations[b]	Percent same patent class[b,c]
1975						
University	88.6	1933	6.12	6.53	5.6	54.3
Top corporate	84.2	1476	4.70	7.17	18.6	55.7
Other corporate	82.3	1341	4.22	7.82	9.1	57.5
1980						
University	79.9	2093	4.34	4.36	8.9	56.3
Top corporate	79.9	1701	3.54	4.41	24.6	58.3
Other corporate	74.1	1424	2.95	4.46	12.6	57.2

a. Application year of citing patent minus application year of originating patent.
b. For those patents receiving any citations.
c. Comparison is at the three-digit level (see text).

patent is 6.5 to 8 years for the 1975 cohort, and a little over 4 years for the 1980 cohort.

The inference that a citation indicates a possible knowledge spillover is much less clear in the case where the citing patent is owned by the same organization as the originating patent. For this reason, we distinguish what we call "self-citations." A self-citation is defined as a citing patent assigned by its inventors to the same party as the originating patent, which is, by construction, either a university or a domestic corporation. Not surprisingly, the self-citation rate differs for the different sources of originating patents, with universities having the lowest and top corporations the highest rates.[17] Finally, Table I shows that 55 to 60 percent of citations have a primary patent class that is the same as the primary patent class of the originating patent, indicating that the originating and citing patents are *technologically* close to one another.

Figure I provides additional detail on the distribution of lags between originating and citing patents, again defined as the difference in application years. The figure shows that citations are few in the early years,[18] and reach a plateau after about three years.

17. The apparent increase in self-citation rates between 1975 and 1980 is probably spurious; self-citations tend to come earlier than other citations. See Trajtenberg, Henderson, and Jaffe [1992] for more on this issue.

18. Patents are typically granted one to three years after application; thus, a citation lag of zero or one implies that the citing patent may well have been applied for before the originating patent was actually granted. Pending applications are not public, so in this case the citation would almost surely have been identified by the examiner.

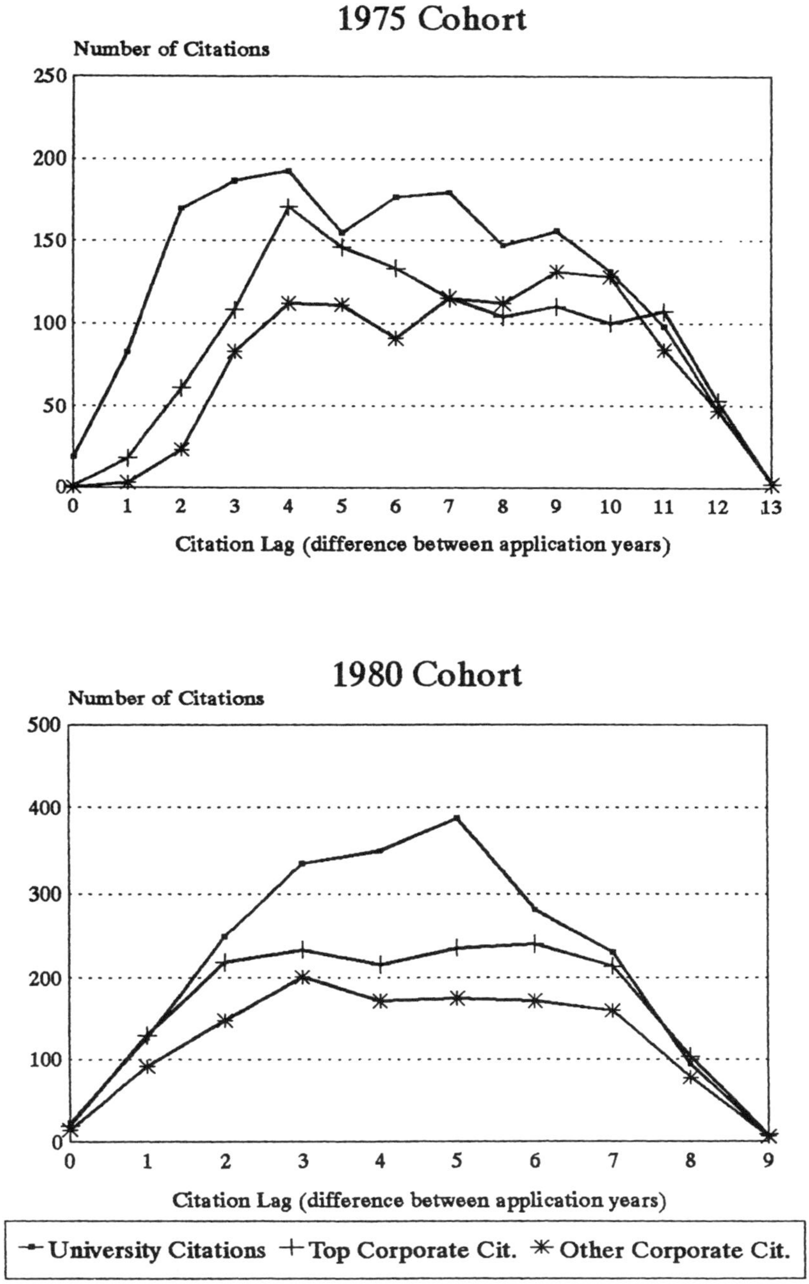
1975 Cohort
Number of Citations
250
200
150
100
50
0
0 1 2 3 4 5 6 7 8 9 10 11 12 13
Citation Lag (difference between application years)
1980 Cohort
Number of Citations
500
400
300
200
100
0
0 1 2 3 4 5 6 7 8 9
Citation Lag (difference between application years)
University Citations
Top Corporate Cit.
Other Corporate Cit.

FIGURE I

TABLE II
SMSA DISTRIBUTIONS FOR SOME DATASETS

Location	1975 University originating	1975 Top corporate originating	Citations to 1975 university	Citations to 1975 top corporate	Controls for citations to 1975 university	All citations to patents from in NY SMSA
Foreign	—	—	31.8	31.4	35.8	31.2
Boston	15.0	3.1	7.5	4.6	5.1	4.0
Los Angeles/ Anaheim	7.0	4.8	9.0	5.7	6.1	3.9
San Francisco/ Oakland	5.1	1.4	3.8	3.7	6.1	3.5
Madison, WI	4.2	—	1.6	—	0.5	0.6
Philadelphia/ Wilmington	4.2	9.3	5.4	8.2	4.5	9.1
Rural Iowa	3.8	—	1.6	0.6	0.2	—
San Jose	3.5	2.8	4.0	3.4	—	1.9
New York/ NJ/CT	3.2	13.5	9.7	11.7	13.7	28.5
Salt Lake City	3.2	—	2.1	—	0.5	0.4
Detroit/ Ann Arbor	2.6	2.4	2.6	1.7	1.7	1.2
Minneapolis/ St. Paul	1.3	5.2	2.8	2.9	1.9	2.1
Chicago	1.9	4.2	3.9	5.7	5.6	4.2
Albany	0.6	3.1	1.9	2.1	1.3	0.8

All figures are percentages. SMSA percentages for citations and controls are relative to *domestic* total.

It is not possible to tell for sure from these data when (if ever) that plateau tails off; the apparent tail-off in both panels of the figure is due at least in part to the 1989 observational cutoff.[19] For 1975 the higher citation rate for university patents is particularly pronounced in the early years; this pattern is not apparent in the 1980 cohort.

Before getting to our formal test of localization, an examination of Table II is useful to get a sense for the extent of geographic concentration in these data. It shows the fraction of patents coming from abroad and from a selection of major U. S. SMSAs for

19. The dropoff in both panels corresponds approximately to application year 1987 (1975 + 12 and 1980 + 7). Typically, a significant fraction of applications have not been granted within two years, so when we looked in 1989 this fraction of 1987 applications were not yet granted.

several of the datasets. Not surprisingly, a measurable fraction of university patents comes from Madison, WI; this is not true for corporate patents. A measurable (though smaller) fraction of the *citations* of university patents comes from Madison, and this fraction is larger than that for the controls. Indeed, the controls for the university citations look generally "more like" corporate patents than do the citations, suggesting that localization may be present. Other qualitative evidence of localization is apparent in the table, including the high percentage of NY SMSA citations that come from the NY SMSA.

The basic test of localization is presented in Table III. For each geographic area and each originating dataset, it presents the proportion of citations that geographically matched the originating patent. These proportions are shown both with and without self-citations. The matching proportions for the control samples are then shown, as well as a t-statistic testing the equality of the control proportions and the citation proportions (excluding self-citations).[20]

We focus first on the 1975 results on the left of the table. Starting with the country match, we find that citations *including self-citations* are domestic about 6 or 7 percent more often than the controls. Excluding self-citations eliminates this difference for the top corporate citations and cuts it roughly in half for the others. The remaining difference between the citations excluding self-cites and the controls is only marginally significant statistically.

Looking at the 1975 results for states, we find that citations of university patents come from the same state about 10 percent of the time; this rises to 15 percent for other corporate and 19 percent for top corporate. Excluding self-citations, however, makes a big difference. The university and top corporate proportions are cut to 6–7 percent, and the other corporate to just over 10. For the university and other corporate cohorts, the matching frequencies excluding self-citations are significantly greater than the matching control proportions.

20. Let p_c be the probability that a citation comes from the same geographic unit as the originating patent; let p_o be the corresponding probability for a randomly drawn patent in the same patent class (control). We test H_o:$p_c = p_o$ versus H_a:$p_c > p_o$ using the test statistic:

$$t = \frac{\hat{p}_c - \hat{p}_o}{\sqrt{[\hat{p}_c(1 - \hat{p}_c) + \hat{p}_o(1 - \hat{p}_o)]/n}},$$

where $\hat{p}_c$ and $\hat{p}_o$ are the sample proportion estimates of p_c and p_o. This statistic tests for the difference between two independently drawn binomial proportions; it is distributed as t.

TABLE III
GEOGRAPHIC MATCHING FRACTIONS

	1975 Originating cohort			1980 Originating cohort		
	University	Top corporate	Other corporate	University	Top corporate	Other corporate
Number of citations	1759	1235	1050	2046	1614	1210
		Matching by country				
Overall citation matching percentage	68.3	68.7	71.7	71.4	74.6	73.0
Citations excluding self-cites	66.5	62.9	69.5	69.3	68.9	70.4
Controls	62.8	63.1	66.3	58.5	60.0	59.6
t-statistic	2.28	−0.1	1.61	7.24	5.31	5.59
		Matching by state				
Overall citation matching percentage	10.4	18.9	15.4	16.3	27.3	18.4
Citations excluding self-cites	6.0	6.8	10.7	10.5	13.6	11.3
Controls	2.9	6.8	6.4	4.1	7.0	5.2
t-statistic	4.55	0.09	3.50	7.90	6.28	5.51
		Matching by SMSA				
Overall citation matching percentage	8.6	16.9	13.3	12.6	21.9	14.3
Citations excluding self-cites	4.3	4.5	8.7	6.9	8.8	7.0
Controls	1.0	1.3	1.2	1.1	3.6	2.3
t-statistic	6.43	4.80	8.24	9.57	6.28	5.52

Number of citations is less than in Table I because of missing geographic data for some patents. The t-statistic tests equality of the citation proportion excluding self-cites and the control proportion. See text for details.

At the SMSA level, 9 to 17 percent of total citations are localized. This again drops significantly when self-citations are excluded, but 4.3 percent of university citations, 4.5 percent of top corporate citations, and 8.7 percent of other corporate citations are localized excluding self-cites. This compares with control matching proportions of about 1 percent, and these differences are highly significant.

The results for citations of 1980 patents (right side of Table III) are even stronger and more significant. For every dataset, for every geographic level, the citations are quantitatively and statistically significantly more localized than the controls. The general increase in the proportion of U. S. patents taken by foreigners is reflected in a decline of 3 to 6 percent in the control percentages matching by country. The citation matching percentages actually rise, however, particularly for top corporate citations. It is impossible to tell from this comparison whether this represents a real change, or whether it is the result of the 1980 citations having shorter average citation lags. Since this gets to the issue of explaining which citations are localized, we postpone discussion until the next section.

Before moving on, the results on the extent of localization can be summarized as follows. For citations observed by 1989 of 1980 patents, there is a clear pattern of localization at the country, state, and SMSA levels. Citations are five to ten times as likely to come from the same SMSA as control patents; two to six times as likely excluding self-citations. They are three to four times as likely to come from the same state as the originating patent; roughly twice as likely excluding self-cites. Whereas about 60 percent of control patents are domestic, 70 to 75 percent of citations and 69 to 70 percent of citations excluding self-cites are domestic. Once self-cites are excluded, universities and firms have about the same domestic citation fraction; at the state and SMSA level there is weak evidence that university citations are less localized. For citations of 1975 patents, the same pattern, but weaker, emerges for citations of university and other corporate patents. For top corporate there is no evidence of localization at the state or country levels, though the SMSA fraction is significantly localized. Thus, we find significant evidence that citations are even more localized than one would expect based on the pre-existing concentration of technological activity, particularly in the early years after the originating patent.

III. Factors Affecting the Probability of Localization

The contrast between the 1975 and 1980 results suggests that localization of early citations is more likely than localization of later ones. This accords with intuition, since whatever advantages are created by geographic proximity for learning about the work of others should fade as the work is used and disseminated. Another hypothesis that is implicit in the previous discussion is that citations that represent research that is technologically similar to the originating research are more likely to be localized, because the individuals pursuing these related research lines may be localized. In addition, attributes of the originating invention or the institution that produced it may affect the probability that its spillovers are localized.

To explore these issues, we pooled the citations (excluding self-cites) to university and corporate patents for each cohort, and ran a probit estimation with geographic match/no match between the originating and citing patents as the dependent variable. As independent variables we included the log of the citation lag (set to zero for lags of zero), dummy variables for top corporate and other corporate originating patents, interactions of the lag and these dummies, and a dummy variable equal to unity if the citation has the same primary class as the originating patent. We also included a dummy variable that is unity if the *control* patent corresponding to this citation matches geographically with the originating patent, to control for the general increase over time in the fraction of U. S. patents granted to foreigners.

We also included two variables relating to the originating patent suggested by our work on basicness and appropriability of inventions [Trajtenberg, Henderson, and Jaffe, 1992]. The first, "generality" is one minus the Herfindahl index across patent classes of the citations received.[21] It attempts to capture the extent to which the technological "children" of an originating patent are diverse in terms of their own *technological* location. Thus, an originating patent with generality approaching one has citations that are very widely dispersed across patent classes; generality of zero corresponds to all citations in a single class. We argue elsewhere that generality is one aspect of the "basicness" of an invention. One might hypothesize that basic research results are less likely to be localized, because their spread is more likely to be through communication mechanisms (e.g., journals) that are not localized. The other variable characterizing the originating invention is the fraction of the originating patent's citations that were

21. Let C_{ik} be the number of citations received by patent i from subsequent patents whose primary patent class is k, and let C_i be the total number of citations received by patent i. The measure of generality is then

$$G_i = 1 - \sum_k \left(\frac{C_{ik}}{C_i}\right)^2.$$

TABLE IV
GEOGRAPHIC PROBIT RESULTS

	Country match		State match		SMSA match	
	1975	1980	1975	1980	1975	1980
Dummy for control Sample match	0.139 (0.045)	0.085 (0.041)	0.396 (0.124)	0.300 (0.102)	*	0.283 (0.172)
Log of citation lag	−0.078 (0.049)	0.094 (0.056)	−0.264 (0.073)	0.198 (0.079)	−0.123 (0.057)	0.037 (0.086)
Dummy for top corporate	−0.114 (0.168)	−0.010 (0.127)	−0.383 (0.249)	0.013 (0.177)	−0.234 (0.288)	−0.208 (0.200)
Dummy for other corporate	0.069 (0.209)	0.053 (0.134)	−0.214 (0.277)	−0.007 (0.189)	0.325 (0.291)	−0.042 (0.207)
Log-lag *top corp. dummy	0.046 (0.091)	−0.016 (0.086)	0.226 (0.138)	0.007 (0.115)	0.102 (0.156)	0.156 (0.131)
Log-lag *other corp. dummy	0.008 (0.108)	−0.026 (0.091)	0.307 (0.147)	0.036 (0.124)	0.037 (0.155)	0.039 (0.138)
Dummy for matching patent class	−0.085 (0.050)	0.069 (0.045)	−0.013 (0.073)	0.034 (0.058)	−0.057 (0.080)	−0.016 (0.068)
Generality of origin patent	0.092 (0.091)	0.177 (0.088)	0.026 (0.136)	−0.140 (0.111)	0.013 (0.150)	−0.298 (0.130)
Origin fraction Self-citations	−0.813 (0.180)	0.162 (0.124)	0.815 (0.246)	0.883 (0.134)	1.174 (0.237)	0.828 (0.154)
# of observations	3581	4217	3573	4215	3566	3972
# of matches	2363	2925	256	490	197	298
Log likelihood	−2269	−2559	−894	−1459	−736	−1022

*The number of observations for which the control patent matched at the SMSA level was so small that this parameter could not be estimated.

Standard errors are in parentheses. All equations also included five technological field dummies.

self-cites. We take a high proportion of self-cites as evidence of relatively successful efforts by the original inventor to appropriate the invention. We expect that the nonself-citations to such a patent are more likely to be confined to suppliers, customers, or other firms that the inventing firm has a relationship with, and may therefore tend to be localized.

Finally, the extent of localization depends fundamentally on the mechanisms by which information flows, and these mechanisms may be different in different technical fields. For this reason, we also included dummy variables for broad technological fields.[22]

The results are presented in Table IV. Because of the presence

22. (1) Drugs and Medical Technology; (2) Chemicals and Chemical Processes Excluding Drugs; (3) Electronics, Optics, and Nuclear Technologies; (4) Mechanical Arts; and (5) All Other.

of the interaction terms between the lag and the corporate dummies, the coefficient on the lag itself corresponds to the fading of localization of citations of university patents. There is evidence in the 1975 results of such fading. This effect is statistically significant at the state and SMSA levels; its quantitative significance is discussed further below. For the citations of corporate patents, the interaction terms measure the difference between their fading rates and those of university citations. These terms are generally not statistically significant. In only one case (other corporate, 1975) could we reject the hypothesis of equality of fading rates at traditional confidence levels. There is, however, weak evidence that the corporate citations do not fade as rapidly as those of university patents, at least at the state and SMSA levels. The coefficients on the corporate dummies themselves capture differences in the predicted probability of localization for citations with lags of zero or one year. These are all insignificant, and there is no clear pattern.

The matching-patent-class and generality variables do not work well. The effects are generally insignificant, and show no consistent pattern. The effect of the self-citation fraction, however, is strong and puzzling. At the state and local level, there is a very significant effect in the predicted direction: citations of patents with a high self-citation fraction are more likely to be localized. This is *not* just saying that self-citations are localized, since they are excluded; it is the other citations that are more localized. At the country level, at least in 1975, this effect is reversed and is significant. Taking all results together, it suggests that for patents with a lot of self-citations, the nonself-citations are more likely to be foreign, but those that are domestic are more likely to be in the same state and SMSA as the originating patent.

The 1980 results are disappointing. The coefficient on the time lag term switches sign, though it is generally insignificant. One possibility is that these citations span too short a time period to capture the lag effect well. To test this possibility, we reran the estimation in Table IV on the 1975 citations, excluding all that were granted after 1984. This analysis tells us what we would have believed about citations of 1975 patents if we had looked for them only as long as we have looked for the citations of the 1980 patents. The results (not reported here) looked "more like" the 1980 results than the original 1975 results did. In particular, the coefficient on the lag term was insignificant, and was positive at the SMSA level.

TABLE V
PREDICTED LOCALIZATION PERCENTAGES OVER TIME (BASED ON 1975 PROBIT RESULTS FOR CITATIONS OF UNIVERSITY PATENTS)

	Predicted percentage for:		
Lag	Same country	Same state	Same SMSA
0 or 1 year	67.1	9.7	4.8
5 years	65.5	6.5	4.0
10 years	64.6	5.3	3.7
25 years	63.5	4.0	3.3

Thus, it may be that the "perverse" results for the 1980 sample would go away if we had later citations to include.

A probit coefficient does not have an economically meaningful magnitude, because of the need to standardize the variance of the underlying error distribution. However, we can calculate what the coefficients imply about changes in the predicted probabilities. This is done in Table V, using the 1975 lag coefficient.[23] Table V was constructed by calculating the predicted localization probability using the results of Table IV, evaluating the citation lag at different values, and evaluating the other independent variables at the mean of the data. It shows that the estimates correspond to a reduction in the localization fraction after, for example, ten years, from 67.1 percent to 64.6 percent at the country level, 9.7 percent to 5.3 percent at the state level, and 4.8 percent to 3.7 percent at the SMSA level.

IV. DISCUSSION AND CONCLUSION

Despite the invisibility of knowledge spillovers, they do leave a paper trail in the form of citations. We find evidence that these trails, at least, are geographically localized. The results, particularly for the 1980 cohort, suggest that these effects are quite large and quite significant statistically. Because of our interest in true externalities, we have focused on citations excluding self-cites. For some purposes, however, this is probably overly conservative. From the point of view of the Regional Development Administra-

23. As discussed above, this is the point estimate of the lag coefficient for citations of university patents. The point estimates are different for the corporate originating patents, but we have not performed separate calculations for each dataset.

tor, it may not matter whether the subsequent development that flows from an invention is performed by the inventing firm, as long as it is performed in her state or city. Our results are also conservative because we attribute none of the localization present in the control samples to spillovers, despite the likelihood that spillovers are, indeed, one of the major reasons for the pre-existing concentration of research activity.

We also find evidence that geographic localization fades over time. The 1980 citations, which have shorter average citation lags, are systematically more localized than the 1975 citations. By using a probit analysis, we produced estimates of the rate of fading. These estimates seem to suggest a rate of fading that is both smaller than one would expect, and smaller than would be necessary to explain the difference between the 1975 and 1980 overall matching fractions. One possibility is that the difficulty of measuring the rate of fading is due to the "contamination" of citations by the patent examiner. As noted in footnote 18, it is particularly likely that citations with very short lags were added by the examiner. If we believe that such citations are less likely to represent spillovers and less likely to be localized, then this would tend to bias toward zero our measure of the effect of time on localization.[24]

We find less evidence of the effect of technological area on the localization process. Citations in the same class are no more likely to be localized. Overall, there is not really any evidence in these data that the probability of coming from a given geographic location conditional on patent class is different from the unconditional probability. This may be due to the arbitrary use of the "primary" patent class, to the exclusion of the "cross-referenced" classes. There is no legal difference in significance between the primary and cross-referenced classes, and in many cases the examiners do not place any significance on which class is designated primary. In future work we hope to explore whether using the full range of information contained in the cross-referenced classes provides a better technological characterization of the patents.

In this context it is worth noting that part of what is going on is probably that knowledge spillovers are not confined to closely related regions of technology space. Approximately 40 percent of

24. Attempts to test for this possibility by including a dummy variable for age 0 or 1, as well as other explorations of possible nonlinearity in the match-lag relationship, were inconclusive.

citations do not come from the same primary patent class; even at the level of the five broad "technological fields" listed in footnote 21, 12 to 25 of percent of citations are across fields. This is consistent with Jaffe [1986], which found that a significant fraction of the total "flow" of spillovers affecting firms' own research productivity comes from firms outside of the receiving firm's immediate technological neighborhood.

We find surprisingly little evidence of differences in localization between the citations of university and corporate patents. The largest difference is that corporate patents are more often self-cited, and self-cites are more often localized. The probit results do not allow rejection of the hypothesis that the initial localization rates for nonself-citations are indistinguishable for the different groups. They do provide some weak evidence that this initial localization is more likely to fade for the university patents, at least at the state and local levels.

In order to provide a true foundation for public policy and economic theorizing, we would ultimately like to be able to say more about the mechanisms of knowledge transfer, and about something resembling social rates of return at different levels of geographic aggregation. The limitations of patent and citation data make it difficult to go much farther with such questions within this research approach. Ex post, the vast majority of patents are seen to generate negligible private (and probably social) returns. In future work we plan to identify a small number of patents that are extremely highly cited. It is likely that such patents are both technologically and economically important [Trajtenberg, 1990]. Case studies of such patents and their citations could prove highly informative about both the mechanisms of knowledge transfer, and the extent to which citations do indeed correspond to externalities in an economic sense.

HARVARD UNIVERSITY
TEL-AVIV UNIVERSITY
MASSACHUSETTS INSTITUTE OF TECHNOLOGY

REFERENCES

Bernstein, J., and M. Nadiri, "Interindustry R&D Spillovers, Rates of Return, and Production in High-Tech Industries," *American Economic Review Papers and Proceedings*, LXXVIII (May 1988), 429–34.

Bernstein, J., and M. Nadiri, "Research and Development and Intra-industry Spillovers: An Empirical Application of Dynamic Duality," *Review of Economic Studies*, LVI (April 1989), 249–67.

Dorfman, N., "Route 128: The Development of a Regional High-Technology

Economy," D. Lampe, ed., *The Massachusetts Miracle: High Technology and Economic Revitalization* (Cambridge, MA: MIT Press, 1988).
Feller, Irwin, "R&D Theories and State Advanced Technology Programs," Paper prepared for the American Association for the Advancement of Science Annual Meeting, January 1989.
Glaeser, Edward, H. D. Kallal, J. A. Scheinkman, and A. Shleifer, "Growth in Cities," NBER Working Paper No. 3787, July 1991.
Griliches, Z., "The Search for R&D Spillovers," NBER Working Paper No. 3768, July 1991.
Grossman, G., and E. Helpman, *Innovation and Growth in the Global Economy* (Cambridge, MA: M.I.T. Press, 1991).
Jaffe, Adam, "Technological Opportunity and Spillovers of R&D: Evidence from Firms' Patents, Profits and Market Value," *American Economic Review,* LXXVI (1986), 984–1001.
——, "Real Effects of Academic Research," *American Economic Review,* LXXIX (1989), 957–70.
Krugman, P., *Geography and Trade* (Cambridge: M.I.T. Press, 1991).
Mansfield, E., "Sources and Characteristics of Academic Research Underlying Industrial Innovations," mimeo, University of Pennsylvania, 1991.
Marshall, A., *Principles of Economics* (London: Macmillan, 1920).
Minnesota Department of Trade and Economic Development, Office of Science and Technology, *State Technology Programs in the United States,* 1988.
Romer, P., "Increasing Returns and Long-Run Growth," *Journal of Political Economy,* XCIV (1986), 1002–37.
——, "Endogenous Technological Change," *Journal of Political Economy,* XCVIII (1990), S71–S102.
Smilor, R., G. Kozmetsky, and D. Gibson, *Creating the Technopolis: Linking Technology, Commercialization and Economic Development* (Cambridge, MA: Ballinger Publishing Co., 1988).
Trajtenberg, M., "A Penny for Your Quotes: Patent Citations and the Value of Innovations," *Rand Journal of Economics,* XXI (1990), 172–87.
Trajtenberg, M., R. Henderson, and A. Jaffe: "Ivory Tower Versus Corporate Lab: An Empirical Study of Basicness and Appropriability," NBER Working Paper No. 4146, 1992.

[10]

R&D Spillovers and the Geography of Innovation and Production

By David B. Audretsch and Maryann P. Feldman*

More than most other economic activities, innovation and technological change depend upon new economic knowledge. Thus, Paul Romer (1986), Paul Krugman (1991a, b), and Gene Grossman and Elhanan Helpman (1991), among others, have focused on the role that spillovers of economic knowledge across agents and firms play in generating increasing returns and ultimately economic growth. In fact, several recent studies have identified the existence of spatially-mediated knowledge spillovers. An important finding of Adam B. Jaffe (1989), Zoltan Acs et al. (1992, 1994), and Feldman (1994a, b) is that investment in R&D by private corporations and universities "spills over" for third-party firms to exploit. If the ability to receive knowledge spillovers is influenced by distance from the knowledge source, then geographic concentration should be observed, especially in industries where knowledge spillovers are likely to play a more important role. The purpose of this paper is to examine the extent to which industrial activity clusters spatially and to link this geographic concentration to the existence of knowledge externalities. Of course, as Jaffe et al. (1993) point out, one obvious explanation why innovative activity in some industries tends to cluster geographically more than in other industries is that the location of production is more concentrated spatially. Thus, in explaining why the propensity for innovative activity to cluster geographically varies across industries, we need first to explain, and then to control for, the geographic concentration of the location of production.

As Alfred Marshall (1920) and, later Krugman (1991b) argue, there may be geographic boundaries to information flows or knowledge spillovers, particularly tacit knowledge, among the firms in an industry. Although the cost of transmitting information may be invariant to distance, presumably the cost of transmitting knowledge rises with distance. That is, proximity and location matter. While there is considerable evidence supporting the existence of knowledge spillovers, neither Jaffe (1989), Jaffe et al. (1993), nor Acs et al. (1992, 1994), and Feldman (1994a) actually examine the propensity for innovative activity to cluster spatially. But implicit in the knowledge production function model is the assumption that innovative activity should concentrate geographically in those industries where the direct knowledge-generating inputs are the greatest and where knowledge spillovers are the most prevalent. No one, to date, has examined the underlying propensity for industrial activity to cluster spatially. While one of the central themes in the industrial organization literature is to explain the degree of concentration of economic activity within an industry (F. M. Scherer and David Ross, 1990), the focus has typically been on the extent of dispersion across different enterprises and establishments within a single spatial unit—the country. The emerging importance of location as a unit of observation argues for examining both production and innovation within a geographic context. We empirically

* Audretsch: Wissenschaftszentrum Berlin für Sozialforschung and the Centre for Economic Policy Research, Reichpietschufer 50, D-10785 Berlin, Germany; Feldman: Institute for Policy Studies, Johns Hopkins University, Baltimore, MD 21218. This article was written while Maryann Feldman was visiting at the Heinz School of Public Policy and Management, Carnegie Mellon University. We thank Richard Baldwin, Paul Krugman, James Markusen, and participants at the CEPR Conference on the "Location of Economic Activity: New Theories and New Evidence," 17–20 December, 1993, Vigo, Spain, for their useful comments. We also thank Jim Adams, Zvi Griliches, Bronwyn Hall, Frank Lichtenberg, Richard Nelson and Mike Scherer and the participants in the discussion at the 1995 AEA Meetings. We would also like to thank the anonymous referees for useful comments and suggestions. Gail Cohen Shaivitz provided invaluable research assistance.

test for the importance of geographic location to different types of industries by linking the geographic concentration in manufacturing industries to industry specific characteristics, most notably the relative importance of knowledge spillovers.

In the following section of this paper, we examine the spatial distribution of innovative activity as well as the geographic concentration of production. An empirical model is specified in Section II, and the results are presented in Section III. In the final section, we provide a summary and conclusion. The empirical evidence suggests that, even after controlling for the degree of geographic concentration in production, innovative activity tends to cluster more in industries where knowledge spillovers play a decisive role. Although such industries also tend to exhibit a greater geographic concentration of production, the results suggest that the propensity for innovative activity to cluster is more attributable to the role of knowledge spillovers and not merely the geographic concentration of production.

I. The Spatial Distribution of Innovation and Production

To measure the spatial distribution of innovative activity we rely on the most recent and most ambitious major data base that provides a direct measure of innovative activity. The United States Small Business Administration (the Small Business Administration's Innovation Data Base or the SBIDB) compiled a data base of 8,074 commercial innovations introduced in the United States in 1982. A private firm, The Futures Group, compiled the data and performed quality control analyses for the United States Small Business Administration. A data base consisting of innovations by four-digit standard industrial classification (SIC) industries was formed from the new product announcement sections in over 100 technology, engineering and trade journals that span every industry.[1] These data were used by Acs and Audretsch (1988, 1990) to analyze the relationships between firm size and technological change, and market structure and technological change, and by Acs et al. (1992, 1994), Feldman (1994a, b), and Feldman and Richard Florida (1994) to examine the geography of innovation.

We adopt the state as the spatial unit of observation. While this is at best a crude proxy of the relevant economic market,[2] it does have one obvious appeal other than that it conforms to a number of data sources—the most relevant unit of policy-making is at the level of the state. Still, states are certainly not an entirely satisfactory unit of observation for the analysis of spatial phenomena. The analyses of spatial processes are handicapped by a lack of data for what might be considered to be the ideal observation. Certainly considerable progress would be made if data sources identifying innovation activity at the city or county level were made available.

Using the citation data base described above, an innovation is attributed to the state in which the establishment responsible for the development of that innovation is located. Some innovations are, in fact, developed by subsidiaries or divisions of companies with headquarters in other states. Since headquarters may announce new product innovations, the data base discriminates between the location of the innovating establishment and the location of the larger, innovating entity (Edwards and Gordon, 1984). For our purposes, the state identifier of the establishment is used to investigate the spatial distribution of innovation. Of the total number of innovations recorded in the data base, 4,200 were manufacturing innovations with information specifying location.[3]

Figure 1 shows the distribution of innovations by states. California is the state in which

[1] A detailed description of the U.S. Small Business Administration's Innovation Citation Data Base can be found in chapter two of Acs and Audretsch (1990), as well as in Acs and Audretsch (1988).

[2] As Krugman (1991b p. 57) emphasizes, "States aren't really the right geographical units," because of the lack of concordance between economic markets and political units.

[3] The SBIDB contains a total of 4,476 innovations in manufacturing industries. Of these, there are 276 innovations which are not used because they were developed by establishments outside the United States or did not have complete location information.

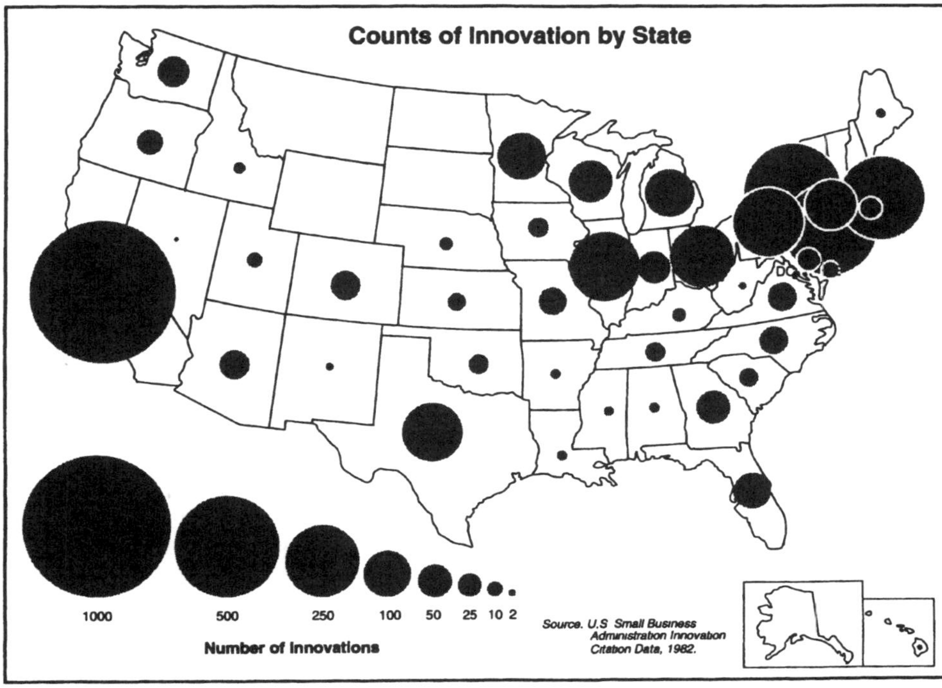

FIGURE 1. NUMBER OF INNOVATIONS BY STATE

the greatest number of innovations were registered, followed by New York, New Jersey, and Massachusetts. A particularly striking feature shown in Figure 1 is that the bulk of innovative activity in the United States occurs on the coasts, and especially in California and in New England. By contrast, no innovative activity is registered in certain Midwestern states such as North Dakota, South Dakota, Montana, and Wyoming.[4] States in the traditional manufacturing belt such as Ohio, Illinois, Michigan, and Pennsylvania are not at all particularly innovative. Thus, while the location of manufacturing activity may explain the spatial distribution of innovative activity to some degree, it is certainly not the only factor.

This presentation of the aggregate geographic distribution of innovative activity in the United States obscures the propensity for innovative activity to cluster spatially within specific industries. Thus, the distribution of innovative activity for the seven most innovative four-digit standard industrial classification (SIC) industries is shown in Table 1. A striking result is that the spatial concentration of innovative activity in particular industries is considerably greater than for all of manufacturing. For example, in the computer industry, 342 of the 821 innovations recorded, or 41.7 percent, are in California. And an additional 10 percent are recorded in Massachusetts.

[4] Of course, simply comparing the absolute amount of innovative activity across states ignores the fact that the manufacturing base of some states is larger than others. Presumably one of the most important determinants of innovative activity is the location of manufacturing activity. Additional information on the geographic distribution of the innovation data can be found in Feldman (1994b) and Feldman and Florida (1994).

TABLE 1—GEOGRAPHIC DISTRIBUTION OF INNOVATIVE ACTIVITY FOR MOST INNOVATIVE INDUSTRIES

SIC[a]	Industry[b]	State	Number of innovations	State share of industry innovations	Industry share of state innovations
3573	Computers	California	342	41.7	35.1
	(n = 821)	Massachusetts	78	9.5	21.7
		New York	58	7.1	12.7
		Texas	39	4.8	23.1
		New Jersey	38	4.6	8.9
		Illinois	28	3.4	12.1
3823	Process control instruments	California	80	17.2	8.2
	(n = 464)	Massachusetts	61	13.1	16.9
		New York	45	9.7	9.9
		Pennsylvania	40	8.6	16.5
		Illinois	32	6.9	13.9
3662	Radio and TV communications	California	105	31.0	10.8
	equipment	New York	40	11.8	8.8
	(n = 339)	Massachusetts	32	9.4	8.9
3674	Semiconductors	California	84	48.8	8.6
	(n = 172)	Massachusetts	17	9.9	4.7
		Texas	13	7.6	7.7
3842	Surgical appliances	New Jersey	43	28.3	10.1
	(n = 152)	California	17	11.2	1.7
		Pennsylvania	10	7.9	4.1
2834	Pharmaceuticals	New Jersey	50	39.4	11.7
	(n = 127)	New York	18	14.2	3.9
		Pennsylvania	10	7.9	4.1
		Michigan	8	6.3	7.1
3825	Measuring instruments for	California	37	32.2	3.8
	electricity	Massachusetts	22	19.1	16.9
	(n = 115)	New York	13	11.3	2.9

[a] The SIC is the standard industrial classification used in the U.S. Small Business Administration's Innovation Citation Data Base.
[b] The total number of innovations recorded in the four-digit industry is listed in parentheses.

Thus, these two states alone account for over one half of all the innovations in the computer industry. At the same time, the last column indicates that innovations in the computer industry accounted for slightly more than one third of all the innovations in California and a little more than one fifth of all innovations in Massachusetts. Similarly, nearly 40 percent of the 127 innovations in the drug industry (pharmaceuticals) were recorded in New Jersey, while an additional 14 percent were made in New York. Thus, over one-half of pharmaceutical innovations were in the New Jersey-New York area. At the same time, pharmaceutical innovations account for over one tenth of all innovations registered in New Jersey.

To measure the extent to which manufacturing in specific industries is concentrated geographically and the extent to which innovative activity tends to cluster spatially, we follow Krugman's (1991b) example and calculate Gini coefficients for the geographic concentration of innovative activity and for the location of manufacturing.[5] Table 2 provides

[5] The locational Gini coefficients for production are based on industry value-added. We calculate the amount of value-added in an industry and state divided by the national value-added for the industry. This ratio is normalized by the state share of total manufacturing value-added in order to account for the overall distribution of manufacturing activity. An industry which is not geographically concentrated more than is reflected by the overall distribution of manufacturing value-added would have a coefficient of 0. The closer the industry coefficient is to 1, the more geographically concentrated the industry would be. Cases in which state or industry data have been suppressed are omitted from the analysis. The Gini coefficients for innovation are based on the count of innovation in a state and industry and are calculated in a similar way. Further details are available from the authors upon request.

TABLE 2—GEOGRAPHIC CONCENTRATION OF PRODUCTION AND INNOVATIVE ACTIVITY FOR MANUFACTURING SECTORS (MEAN GINI COEFFICIENTS)[a]

Manufacturing sector	Value added	Employment	Innovations
Food and beverages	0.6973	0.5584	0.2567
	(0.1685)	(0.1828)	(0.2226)
Tobacco	0.6589	0.4137	0.3319
	(0.2559)	(0.1444)	(0.2043)
Textiles	0.7040	0.5670	0.1659
	(0.1149)	(0.1430)	(0.2347)
Apparel	0.6179	0.5160	0.0583
	(0.1589)	(0.1687)	(0.1469)
Lumber	0.6309	0.5605	0.1180
	(0.1007)	(0.1208)	(0.1235)
Furniture	0.5815	0.4632	0.4204
	(0.1373)	(0.1366)	(0.2347)
Paper	0.6036	0.5580	0.2363
	(0.1525)	(0.1568)	(0.3253)
Printing	0.5977	0.5325	0.1762
	(0.1491)	(0.1485)	(0.2220)
Chemicals	0.7003	0.5987	0.3881
	(0.1612)	(0.1790)	(0.1945)
Petroleum	0.6786	0.4766	0.2598
	(0.1512)	(0.1493)	(0.3674)
Rubber and plastics	0.5771	0.4569	0.3932
	(0.3089)	(0.2434)	(0.1952)
Leather	0.7186	0.5552	0.0646
	(0.1150)	(0.1300)	(0.1119)

[a] Standard deviations are given in parentheses.

the weighted mean Gini coefficients for value-added, employment, and innovative activity within each broad two-digit SIC manufacturing sector. Those sectors exhibiting the greatest geographic concentration of manufacturing activity include primary metals, transportation equipment, textiles, food and beverages, leather and chemicals. By contrast, those manufacturing sectors exhibiting the highest propensity for innovative activity to cluster spatially include transportation equipment, instruments, and electronics. That the propensity for innovative activity to spatially cluster cannot be simply explained by the geographic concentration of the location of manufacturing activity is evident from Table 2. This points to the importance of controlling for the geographic concentration of production in explaining the propensity for innovative activity to spatially cluster.

II. The Model

Why should innovations tend to cluster spatially more in some industries than in other industries? One obvious answer is simply that the location of production is more geographically concentrated in some industries than in others. This raises the issue of endogeny. Jaffe et al. (1993) identify two related critical issues which must be considered in trying to identify why the propensity for innovative activity to cluster spatially varies across industries. First, the extent to which the location of production is geographically concentrated must be controlled for, so that the relevant question becomes: *even after accounting for the geographic concentration of the production location, why does the propensity for innovative activity to cluster vary across industries?* And second, in trying to account for the degree to which the location of production is geographically concentrated, an important factor is the role which knowledge spillovers play in the industry. It is only after the geographic concentration of production has been controlled for, that the degree to which innovative activity clusters spatially can be addressed. Thus, to explain the propensity for innovative activity to cluster spatially we begin with the extent to which production is geographically concentrated.

While it is not possible to directly measure the extent to which knowledge externalities exist, as Kenneth J. Arrow (1962) and Krugman (1991a) point out, it is possible to identify industries in which new economic knowledge plays a relatively more important role. This is done on the basis of the industry R&D intensity, or R&D-sales ratio. The crucial assumption we make here is Arrow's (1962) argument that knowledge spillovers are more important in, and reflected at least to some degree by, highly R&D-intensive industries. By contrast, such knowledge externalities, while perhaps still present, play a less important role where the creation of new economic knowledge, as reflected by R&D intensity, is negligible. Thus, the location of production would be expected to be more concentrated in those industries where knowledge spillovers are prevalent, that is in industries which are R&D intensive.

Similarly, skilled workers endowed with a high level of human capital are a mechanism by which economic knowledge is transmitted. The greater the extent to which the industry

TABLE 3—DESCRIPTION OF VARIABLES

Variable	Description and source	Mean	Standard deviation
Gini of production	Gini coefficient of four-digit SIC industry value-added across states, weighted by national value-added for the industry in 1982 (U.S. Department of Commerce, Bureau of the Census, 1982 Economic Census).	0.56	0.13
Gini of innovation	Gini coefficient of four-digit SIC industry count of innovations across states, weighted by national innovation count for the industry in 1982 (Edwards and Gordon, 1984).	0.30	0.23
Natural resources	Share of total industry inputs purchased from mining and agriculture in 1976 (Input-output data as provided by U.S. International Trade Commission databank).	0.09	0.16
Scale	Mean size of the largest establishments accounting for one half of the industry value-of-shipments divided by industry value of shipments in 1982 (U.S. Department of Commerce, Bureau of the Census, 1982).	2.13	3.97
Transportation costs	Radius of the mean distance shipped in 1967 (Commodity Transport Survey of the United States Census of Transportation for 1967, taken from Weiss [1991]).	7.9549	4.0574
Industry R&D/sales	Industry expenditures on research and development divided by sales in 1977 (Line of Business Survey, U.S. Federal Trade Commission, 1977).	1.66	1.69
Skilled labor	Share of industry employment accounted for by professional and kindred workers, managers and administrators, plus craftspeople and kindred workers in 1970 (U.S. Department of Commerce, Bureau of the Census, 1972).	0.35	0.09
University research	Expenditures on university research for departments relevant to industry (Yale Survey of Industrial Managers in Levin et al. [1987] and National Science Foundation's (NSF) Survey of Science Resources Survey)	17.5946	13.851

work force is composed of skilled workers, the more important knowledge spillovers are likely to be. Thus, industries which rely on a higher component of skilled workers should tend to exhibit a greater tendency towards spatial concentration of industrial location.

Of course, while knowledge externalities may be important in influencing the degree to which the location of production is spatially concentrated, they are certainly not the only factors. Krugman (1991a) points out that the extent to which the location of production is geographically concentrated will be shaped by transportation costs. Transportation costs are inversely related to the mean distance shipped, so that a higher value of transportation costs should be associated with a lower geographic concentration of production. Similarly, industries which are highly dependent upon natural resource inputs are also going to tend to be geographically concentrated—presumably close to the source of those inputs. Augustus Loesch (1954) and Victor R. Fuchs (1962) argue that firms in industries with a high dependency on natural resource inputs will tend to locate in close proximity to those resources. Therefore, a higher content of natural resource inputs in an industry should result in a greater geographic concentration of the location of production. In addition, Robert C. Shelburne and Robert W. Bednarzik (1993) argue that industries which are more capital-intensive will tend to be geographically concentrated, since production will be concentrated among fewer enterprises. That is, as capital intensity and the importance of scale economies rise, fewer large establishments will be able to exist at a level of output in excess of the minimum efficient scale (MES) level of output.

The main hypothesis of this paper suggests that innovative activity will tend to cluster in industries where new economic knowledge plays an especially important role. In estimating the main influences on the geographic

TABLE 4—CORRELATION MATRIX

	Gini of production	Gini of innovation	Natural resources	Scale	Transportation costs	Industry R&D/sales	Skilled labor	University research
Gini of production	1.0	—	—	—	—	—	—	—
Gini of innovation	0.0090	1.0	—	—	—	—	—	—
Natural resources	0.1292	−0.1130	1.0	—	—	—	—	—
Scale	−0.2370	−0.1158	0.1009	1.0	—	—	—	—
Transportation costs	0.2076	0.1013	−0.1400	0.3010	1.0	—	—	—
Industry R&D/sales	0.3241	0.2254	−0.2148	0.2420	0.3225	1.0	—	—
Skilled labor	0.0563	0.2540	−0.2848	0.0997	0.0702	0.3961	1.0	—
University research	0.8730	0.5100	0.1199	−0.1077	0.0819	0.1946	0.0239	1.0

concentration of innovation we consider three sources of economic knowledge—industry R&D, skilled labor, and the size of the pool of basic science for a specific industry. Conceptually, there are great differences in the scope and commercial applicability of university research undertaken in different fields. Academic research will not necessarily result in useful knowledge for every industry; however, scientific knowledge from certain academic departments is expected to be more important for certain industries than for others. To capture the relevant pool of knowledge, academic departments are assigned to industries using the survey of industrial R&D managers by Richard C. Levin et al. (1987).[6] For example, basic scientific research in medicine, biology, chemistry and chemical engineering is found to be relevant for product innovation in drugs (SIC 2834).

III. The Results

Descriptive statistics for the variables used to estimate the model are provided in Table 3, and Table 4 provides the correlation matrix. There are 163 four-digit SIC industries for which comparable data for the different measures could be compiled.

Table 5 presents the regression results using ordinary-least-squares (OLS) estimates. Table 6 presents the results estimating the system of equations using three-stage least squares (3SLS).[7] The statistical results are generally quite consistent between the OLS and 3SLS methods of estimation.

In equation (1) of Table 5, the positive and statistically significant coefficient of the measure of natural resource utilization suggests that the degree to which inputs in an industry are composed of natural resources clearly tends to shape the geographic concentration of production. Resource dependent industries tend to be more geographically concentrated. The negative and statistically significant coefficient on the scale measure suggests that industries tend to be less, and not more, geographically concentrated when scale economies play a more important role. This result emerges even after controlling for the size of the industry. One explanation for this result

[6] To measure the relevance of a discipline to an industry a survey of industrial R&D managers was used. The question was asked, "How relevant were the basic sciences to technical progress in this line of business over the past 10–15 years?" The survey uses a Likert scale of 1 to 7 to assess relevance. We consider relevant science to be those academic departments that are rated with a relevance greater than a value of 5 on the scale.

[7] The system of equations was also estimated using two-stage least-squares estimation. The differences in the standard errors indicate the presence of cross-equation correlation. Thus, we estimate the model with 3SLS. The instruments used include all of the exogenous variables appearing on the right-hand side of the equations in the model.

TABLE 5—OLS REGRESSION RESULTS ESTIMATING GINI COEFFICIENTS ACROSS STATES[a]

	Gini of production			Gini of innovation		
	(1)	(2)	(3)[b]	(4)	(5)	(6)[b]
Gini of innovation	—	0.768 (0.143)	−0.125 (−1.741)	—	—	—
Natural resources	0.326 (4.950)	0.330 (5.261)	0.384 (5.058)	—	—	−0.108 (−1.228)
Scale	−0.137 (−4.162)	−0.160 (−4.173)	−0.244 (−0.695)	—	—	−0.007 (1.986)
Transportation costs	1.223 (4.439)	1.419 (4.838)	1.741 (5.631)	—	—	0.006 (1.674)
Industry R&D/sales	0.455 (7.791)	0.436 (7.170)	0.608 (2.860)	0.469 (2.137)	0.565 (2.405)	0.543 (2.341)
Skilled labor	1.094 (15.044)	1.058 (12.483)	1.318 (15.031)	0.466 (4.910)	0.657 (4.581)	0.645 (4.686)
University research	—	—	0.034 (2.147)	0.108 (7.920)	0.116 (8.093)	0.118 (8.139)
Gini of production	—	—	—	—	−0.119 (−1.587)	−0.146 (−1.741)
Sample size	163	163	163	163	163	163
R^2	0.951	0.952	0.970	0.827	0.908	0.921
Standard error	0.15034	0.15079	0.18601	0.21443	0.21469	0.18487

[a] t values are given in parentheses.

[b] Columns (3) and (6) provide the unrestricted regression results. When the regression estimated for the Gini of production in column (2) is compared with the estimate of the unrestricted regression in column (3), the F test statistic (2,156) of 0.087 is computed for the overidentifying restrictions. Similarly, when the regression estimated for the Gini of innovation in column (5) is compared to the estimate of the unrestricted regression in column (6), the F test statistic (3,156) equals 2.662 for the overidentifying restrictions.

may be that this measure limits the size of the market to the United States, but when many manufacturing industries are global in scale, this measure will be understated.

There is little ambiguity concerning the two measures that are the primary focus of this paper—the relative importance of industry R&D in an industry and the extent to which the labor force is composed of skilled workers. The coefficient of industry R&D is positive and clearly statistically significant, supporting the hypothesis that industries where new economic knowledge tends to play a more important role will have a higher propensity to cluster together. Similarly, industries where skilled labor is relatively important also tend to exhibit a greater degree of geographic concentration of production.

An alternative specification which includes the Gini coefficient of innovation is presented in equation (2) of Table 5. To the extent that geographic proximity between R&D labs and production facilities is important to gain the benefits of R&D, we might expect that industries with closely clustered innovations may also have closely clustered factories. However, the coefficient of this variable cannot be considered to be statistically different from 0.

Equations (4) and (5) of Table 5 present the OLS regression results estimating the Gini coefficients of innovative activity across states. Equation (4) indicates that, when the extent to which production activity is geographically concentrated is not controlled for, the coefficients of all three types of knowledge-generating measures included—industry R&D, skilled labor, and university research—are positive and statistically significant. However, industries in which new economic knowledge plays a more important role also tend to exhibit a greater degree of spatial concentration. That is, without first controlling for the extent to which the location of production is geographically concentrated, it is not at all clear whether

TABLE 6—3SLS REGRESSION RESULTS ESTIMATING GINI COEFFICIENTS ACROSS STATES[a]

	Gini of production	Gini of innovation	Gini of production	Gini of innovation
Gini of innovation	—	—	0.224	—
			(0.416)	
Natural resources	0.331	—	0.347	—
	(5.145)		(5.261)	
Scale	−0.160	—	−0.166	—
	(−4.333)		(−3.589)	
Transportation costs	1.432	—	1.506	—
	(5.052)		(3.974)	
Industry R&D/sales	0.440	0.572	0.460	0.557
	(7.290)	(2.421)	(6.723)	(2.341)
Skilled labor	1.075	0.687	1.193	0.683
	(14.846)	(3.707)	(5.599)	(3.736)
University research	—	0.119	—	0.140
		(7.887)		(3.480)
Gini of production	—	−0.135	—	−0.158
		(−1.247)		(−1.391)
Sample size	163	163	163	163
Standard error	0.15523	0.21733	0.15571	0.21767

[a] t values are given in parentheses.

the greater propensity for innovative activity to cluster in industries where knowledge spillovers are more prevalent is attributable to the fact that knowledge externalities are more conducive to innovative activity or simply that the firms are already located within a relatively tight geographic area.

Thus, in equation (5) of Table 5, the Gini measure of the value-added across states is included. The coefficient of this variable is statistically insignificant, but the coefficients of the other explanatory variables remain virtually unchanged. Which is to say that, even after controlling for the extent to which the location of production is geographically concentrated, the three knowledge-generating variables are still found to have a significant impact on the propensity for innovative activity to cluster spatially.

Table 6 presents the 3SLS method of estimation. In the first set of equations the Gini coefficient of innovative activity across states is endogenous in both equations. In the second set of equations a fully specified simultaneous version of the model is presented in which the Gini coefficients of both innovative activity and production are endogenously included. Most importantly, the propensity for innovative activity to spatially cluster is found to be the result of new economic knowledge and not merely the existing geographic concentration of production.

The positive coefficients of industry R&D, skilled labor, and university research, even after controlling for the degree of concentration of production, using both the OLS and 3SLS methods of estimation, are certainly consistent with the following hypothesis. The propensity for innovative activity to cluster will tend to be higher in industries where new economic knowledge plays a more important role. Presumably, it is in such industries where new economic knowledge which generates innovative activity is transmitted tacitly through what has been described as knowledge spillovers. Therefore, innovative activity is more likely to occur within close geographic proximity to the source of that knowledge, be it a university research laboratory, the research and development department of a corporation, or exposure to the knowledge embodied in a skilled worker.

Industries where new economic knowledge plays a more important role also tend to exhibit a greater geographic concentration of production. However, based on the statistical results reported above it appears that the propensity for innovative activity to cluster spatially is more attributable to the influence of knowledge spillovers and not merely the geographic concentration of production.

IV. Conclusions

This paper examines the geography of innovation and production. In particular, by examining the concentration of economic phenomena, we re-focus the lens from the usual product dimension to a geographic or spatial dimension. A key assumption we make in examining the link between knowledge spillovers in an industry and innovative activity clustering spatially is that knowledge externalities are more prevalent in industries where new economic knowledge plays a greater role. New economic knowledge is captured by industry R&D, university R&D, and skilled labor.

One obvious complication in testing for this link is that innovative activity will be more geographically concentrated in industries where production is also geographically concentrated, simply because the bulk of firms are located within close proximity. Even more problematic, though, is the hypothesis that new economic knowledge will tend to shape the spatial distribution of production as well as that of innovation. Indeed, we find that a key determinant of the extent to which the location of production is geographically concentrated is the relative importance of new economic knowledge in the industry. Even after controlling for the concentration of production we find evidence that industries in which knowledge spillovers are more prevalent—that is where industry R&D, university research and skilled labor are the most important—have a greater propensity for innovative activity to cluster than industries where knowledge externalities are less important.

REFERENCES

Acs, Zoltan J. and Audretsch, David B. "Innovation in Large and Small Firms: An Empirical Analysis." *American Economic Review*, September 1988, *78*(4), pp. 678–90.

———. *Innovation and small firms*. Cambridge, MA: MIT Press, 1990.

Acs, Zoltan J.; Audretsch, David B. and Feldman, Maryann P. "Real Effects of Academic Research: Comment." *American Economic Review*, March 1992, *82*(1), pp. 363–67.

———. "R&D Spillovers and Recipient Firm Size." *Review of Economics and Statistics*, May 1994, *76*(2), pp. 336–40.

Arrow, Kenneth J. "Economic Welfare and the Allocation of Resources for Invention," in Richard R. Nelson, ed., *The rate and direction of inventive activity*. Princeton, NJ: Princeton University Press, 1962, pp. 609–26.

Edwards, Keith L. and Gordon, Theodore J. "Characterization of Innovations Introduced on the U.S. Market in 1982." The Futures Group, prepared for the U.S. Small Business Administration under Contract No. SBA-6050-0A-82, March 1984.

Feldman, Maryann P. "Knowledge Complementarity and Innovation." *Small Business Economics*, October 1994a, *6*(5), pp. 363–72.

———. *The geography of innovation*. Boston: Kluwer Academic Publishers, 1994b.

Feldman, Maryann P. and Florida, Richard. "The Geographic Sources of Innovation: Technological Infrastructure and Product Innovation in the United States." *Annals of the Association of American Geographers*, May 1994, *84*(2), pp. 210–29.

Fuchs, Victor R. *Change in the location of manufacturing in the United States since 1929*. New Haven: Yale University Press, 1962.

Grossman, Gene and Helpman, Elhanan. *Innovation and growth in the global economy*. Cambridge, MA: MIT Press, 1991.

Jaffe, Adam B. "Real Effects of Academic Research." *American Economic Review*, December 1989, *79*(5), pp. 957–70.

Jaffe, Adam B.; Trajtenberg, Manuel and Henderson, Rebecca. "Geographic Localization of Knowledge Spillovers as Evidenced by Patent Citations." *Quarterly Journal of Economics*, August 1993, *63*(3), pp. 577–98.

Krugman, Paul. "Increasing Returns and Economic Geography." *Journal of Political Economy*, June 1991a, *99*(3), pp. 483–99.

———. *Geography and trade*. Cambridge, MA: MIT Press, 1991b.

Levin, Richard C.; Klevorick, Alvin K.; Nelson, Richard R. and Winter, Sidney G. "Appropriating the Returns from Industrial Research and Development." *Brookings Papers on Economic Activity,* 1987 (3), pp. 783–820.

Loesch, Augustus. *The economics of location.* New Haven: Yale University Press, 1954.

Marshall, Alfred. *Principles of economics,* 8th ed. London: Macmillan, 1920.

Romer, Paul. "Increasing Returns and Long-Run Growth." *Journal of Political Economy,* November 1986, *94*(5), pp. 1002–37.

Scherer, F. M. and Ross, David. *Industrial market structure and economic performance,* 3rd ed. Boston: Houghton Mifflin, 1990.

Shelburne, Robert C. and Bednarzik, Robert W. "Geographic Concentration of Trade-Sensitive Employment." *Monthly Labor Review,* June 1993, *116*(6), pp. 3–13.

Weiss, Leonard W. "The Geographic Size of Markets in Manufacturing," in David B. Audretsch and Hideki Yamawaki, eds., *Structure, conduct, and performance.* New York: New York University Press, 1991, pp. 64–91.

[11]

JOURNAL OF URBAN ECONOMICS **42**, 422–448 (1997)
ARTICLE NO. UE972032

Local Geographic Spillovers between University Research and High Technology Innovations

Luc Anselin

Regional Research Institute and Department of Economics, West Virginia University, P.O. Box 6825, Morgantown, West Virginia 26506-6825

Attila Varga

Regional Research Institute, West Virginia University, P.O. Box 6825, Morgantown, West Virginia 26506-6825, and Department of Economics, Janus Pannonius University, Pecs, Hungary

and

Zoltan Acs

Office of Advocacy, U.S. Small Business Administration, Washington, DC 20416, and Merrick School of Business, University of Baltimore, Baltimore, Maryland 21201-5779

Received May 20, 1996; revised October 1996

This paper re-examines the empirical evidence on the degree of spatial spillover between university research and high technology innovations. The familiar Griliches–Jaffe knowledge production function is estimated at both the state and the metropolitan statistical areas (MSA) level and extended with more precise measures of spatial spillover. Alternatives based on the gravity potential and covering indices are formulated for Jaffe's "geographical coincidence index" and found to provide strong evidence of local spillovers at the state level. At the MSA level, a distinction is made between research and development activities and university research in the MSA and in the surrounding counties. Evidence is found of local spatial externalities between university research and high technology innovative activity, both directly and indirectly via private research and development. © 1997 Academic Press

INTRODUCTION

The accumulation of knowledge and its spillover into productive capacity through technological change is a central theme in the new theory of endogenous economic growth (e.g., Romer [66–68], Grossman and Helpman [38, 39]). An interesting aspect of this perspective has been the

0094-1190/97 $25.00

renewed attention to the geographic scope of the spillovers between knowledge creation and production, or, the extent of Marshallian spatial externalities (e.g., as exemplified in the new economic geography of David and Rosenbloom [21], Krugman [46], Glaeser *et al.* [31], and others). In urban and regional economics, a large number of empirical studies have focused on the creation of knowledge, on research and development activities (R & D), and on technological innovation as determinants of various aspects of local and regional economies. Examples are analyses of the location of R & D facilities and high technology production, and the effect of universities and technology parks on characteristics of the labor market and regional growth (e.g., Malecki [51–53], Markusen, Hall, and Glasmeier [57], Andersson, Anderstig, and Harsman [6], Goldstein and Luger [33], Luger and Goldstein [49], Glasmeier [32], Florax [28]).

An important aspect of studies of technological innovation at the regional scale is the role of spatial interaction and spatial structure, as expressed in the form of organizational networks of innovators, regional innovation complexes and regional knowledge infrastructure (e.g., Stohr [71], Von Hippel [74], Storper and Walker [72], DeBresson and Amesse [23], Feldman [26], Saxenian [69]). Universities play a central role in this process, not only as producers of basic research, but also by creating human capital in the form of higher skilled labor. Both of these aspects have received considerable attention in the literature, from a theoretical as well as from an empirical perspective. The importance of basic (university) research in the stimulation of technological innovation (and higher productivity) is derived from the public good nature of the research, and the resulting positive externalities to the private sector in the form of knowledge spillovers. The initial conceptualization of this process was provided by Arrow [12] and Nelson [59], and further refined by Griliches [35], Nelson [60], Von Hippel [74], Cohen and Levinthal [20], among others (for recent reviews, see, e.g., Dosi [24], Acs and Audretsch [1], Griliches [36, 37], Mansfield [55], Florax [28], Feldman [26]). The role of universities in the development of human capital that is more likely to create technological innovations is highlighted in the work of Bartel and Lichtenberg [16] and Lucas [48], among others. Both the knowledge spillovers and the human capital development constitute important locational attractors for private sector R & D and for high technology production (Malecki [53]).

While conceptually clear, the role of universities as "engines of growth" or "regional boosters" (Florax [28]), is not uniformly confirmed by empirical studies. A significant and positive effect of the presence of universities is found on the location of high technology production, new start ups and R & D facilities in, among others, Malecki [52], Nelson [61], Lund [50], Rees and Stafford [65], Harding [40], Bania, Calkins and Dalenberg [14],

Parker and Zilberman [62], and Sivitanidou and Sivitanides [70], but others have either qualified these findings or did not discover such a relationship. For example, Link and Rees [47], Rees [64], and Malecki and Bradbury [54] point out that the locational attraction may show little relationship to the research activities of universities, but instead may be determined by their teaching and training roles. Others, such as Howells [42], Markusen, Hall, and Glasmeier [57], Glasmeier [32], Florax and Folmer [29], Bania, Eberts, and Fogarty [15], and Beeson and Montgomery [17], find a weak, inconsistent (across sectors), or no relationship at all between the university research activities and high technology (or R & D) location, or other regional variables such as manufacturing investment, local labor market conditions, and high technology start ups. A number of explanations have been offered for this lack of uniform evidence, such as the difficulty of quantifying the inputs and outputs in a "knowledge production function" (Griliches [36]) and various sectoral and spatial aggregation problems (e.g., as argued in Markusen, Hall, and Glasmeier [57, p. 156]).

The goal of our paper is to shed further light on this issue by focusing on the spatial extent of the "local" geographic effects that university research (at both private and public universities) may have on the innovative capacity in a region, both directly and indirectly through its interaction with private sector R & D efforts. Our conceptual framework is based on the knowledge production function of Griliches [35] and is implemented at an aggregate level, as in other recent studies of this phenomenon by Jaffe [44], Acs, Audretsch, and Feldman [2–4], Audretsch and Feldman [13], and Feldman and Florida [27] (see Feldman [26] for a recent review). We extend the empirical evidence in two important respects:

(1) We broaden the cross-sectional basis for empirical analysis by utilizing data for 43 states (compared to 29 in the above mentioned studies) and for 125 metropolitan statistical areas (MSA). This is the first time MSA-level data are used in this type of analysis, which avoids many problems associated with the inappropriate spatial scale of a state as the areal unit of analysis. MSA-level results are obtained by using R & D laboratory employment as a proxy for R & D activity, based on a specially compiled data set.

(2) We focus on more precise measures of local geographic spillovers. At the state level, we introduce several alternatives to Jaffe's [44] "geographic coincidence index" that are more tightly integrated with the existing body of spatial interaction theory, and are able to significantly improve on his results. At the MSA scale, we formalize the spatial extent of the geographic spillovers by means of so-called spatial lag variables that capture the research activities in concentric rings around the MSA as well as in the MSA itself.

In the remainder of the paper, we first introduce the formal model of the knowledge production function and briefly review current empirical evidence on geographic knowledge spillovers of universities. We next elaborate on the data set used in the study. Subsequently, we present the results of our analysis at the state level and at the MSA level, focusing in particular on our alternative definitions of spatial interaction. We conclude with a summary and evaluation of our findings.

THE KNOWLEDGE PRODUCTION FUNCTION

Model

The conceptual framework for analyzing geographic spillovers of university research on regional innovative capacity is based on the knowledge production function of Griliches [35] (see also Jaffe [43, 44]). In essence, this is a two-factor Cobb–Douglas production function that relates an output measure for "knowledge" to two input measures: research and development performed by industry, and research performed by universities. Formally, this is expressed as

$$\log(K) = \beta_{K1} \log(R) + \beta_{K2} \log(U) + \varepsilon_K, \tag{1}$$

where K is a proxy for knowledge output (either patents or innovation counts), R is industry R & D, and U is university research, with ε_K as a stochastic error term. The analysis is typically carried out for aggregate cross-sectional units (e.g., states), possibly for several points in time and/or disaggregated by sector.

Following Jaffe [44], the potential interaction between university and industry research is captured by extending the model with two additional equations that allow for simultaneity between these two variables

$$\log(R) = \beta_{R1} \log(U) + \beta_{R2} Z_1 + \varepsilon_R \tag{2}$$

and

$$\log(U) = \beta_{U1} \log(R) + \beta_{U2} Z_2 + \varepsilon_U \tag{3}$$

where U and R are as before, Z_1 and Z_2 are sets of exogenous local characteristics, and ε_U and ε_R are stochastic error terms.

Jaffe [44] introduced an additional term in Eq. (1) to compensate for the inappropriateness of using states as the units of observation. This acknowledges that states are too large to accurately capture the local spatial interaction between universities and firms. The additional term consists of the product of the logarithms of a "geographic coincidence index" (C) and university research. This term therefore adjusts the magnitude of university research in the state by a measure of the extent to which private and university research coincide in space (we return to a more detailed specification of this term below). The extended "geographic" knowledge production function then becomes

$$\log(K) = \beta_{K1} \log(R) + \beta_{K2} \log(U) + \beta_{K3}[\log(U) \cdot \log(C)] + \varepsilon_K \quad (4)$$

in the same notation as before.

The framework expressed in Eqs. (2)–(4) has become the basis of several empirical investigations, which we will briefly review in the next section. Before proceeding however, note that the geographic interaction term in Eq. (4) is an artifact of using states as the unit of analysis and is not needed when the spatial units of observation correspond more closely to the spatial scale of interaction between firms and universities. For example, this may be the case when the analysis is carried out at the MSA or county scale.

Previous Empirical Evidence

Almost all empirical investigations of geographic knowledge spillovers of universities in the U.S. have been aggregate in nature and based on the Griliches–Jaffe knowledge production function framework applied at the state level. A few notable exceptions are studies that use micro-data derived from surveys or patent information, e.g., Mansfield [56], Jaffe, Trajtenberg, and Henderson [45], and Almeida and Kogut [5]. These studies all provide strong evidence of the importance of spatial interaction at the *local* level. However, this does not find uniform confirmation in the aggregate studies. As noted above, this may be somewhat due to the fact that the unit of analysis—the state—only partially captures this interaction. We argue in this paper that it is also due to the formal specification of local spatial interaction in the form of a geographic coincidence index.

The aggregate empirical analyses in the literature are all based on data for 29 U.S. states. Knowledge output is proxied by patent counts in Jaffe [44] (also replicated in Acs, Audretsch, and Feldman [2]), while Acs, Audretsch and Feldman [2, 3] and Feldman and Florida [27] use a more direct measure of innovative activity, based on a 1982 data set of innovation counts compiled by the U.S. Small Business Administration (see below for further details).

The empirical studies in the literature vary somewhat in terms of research design, but they all find a strong and positive relationship between innovative activity and both industry R & D and university research at the state level. However, the situation is different in terms of the significance of a *local* geographic spillover effect. Overall, the evidence is non-existent, weak or mixed, only pertaining to a few individual sectors (e.g., in Jaffe [44] for only two out of four sectors considered).

We suggest that, in part, this lack in the evidence for local geographic spillovers (which contradicts the strong findings in micro-level studies) is due to the specification of the geographic effects in the model. All studies use the idea initially suggested in Jaffe [44, p. 959], where the product of the logarithm of state-level university expenditures with the logarithm of a geographic coincidence index is used. The latter is obtained as the uncentered correlation coefficient (i.e., without subtracting the mean) between university research and professional employees in R & D labs for the SMSAs in the state (i.e., R & D and university research outside SMSA are not included in the computation of the index). The resulting index is rescaled such that its mean is zero. While the coincidence index may make intuitive sense, it is unrelated to the existing large body of research on spatial interaction and measures of spatial accessibility (for a recent review, see, e.g., Frost and Spence [30], Talen and Anselin [73]). We will therefore consider the extent to which alternative indices provide a clarification of this issue.

DATA AND VARIABLE DEFINITIONS

We extend previous empirical evidence by using a detailed data set on innovation counts and R & D employment at both the state and MSA scale of geographic aggregation. Following Acs, Audretsch, and Feldman [2–4] and Feldman and Florida [27], we use a count of innovations as the dependent variable in the geographic knowledge production function [K in Eq. (4)], constructed from the U.S. Small Business Administration Innovation Database. This data set is a compilation of innovations that were introduced to the U.S. market in the year 1982, based on an extensive review of new product announcements in trade and technical publications. In the original data set, 4200 innovations are identified by the address of the innovating establishment, which allows for aggregation at different spatial scales, such as the state, MSA or county level (for further details on the data set and a discussion of their limitations, see Edwards and Gordon [25], Acs and Audretsch [1, Chapter 2], and Feldman [26]). In our study, we use aggregations at the state, county, and MSA level.

At the sectoral scale, the innovation data were aggregated to the two-digit SIC code level. Our interest focused on innovations in the "high technology" sector as an aggregate. Clearly, this is not an unambiguous determination and a number of different classifications have been suggested in the literature (e.g., Premus [63], Herzog, Schlottmann, and Johnson [41], Malecki [52], Glasmeier [32]). In general, the objective is to identify sectors dominated by the importance of non-routine functions, in contrast to standardized mass production. A number of criteria to identify such sectors have been suggested in the literature, such as the percentage of scientists and engineers employed, and the number of innovations per employee. At the two-digit SIC level, it is virtually impossible to designate sectors as "pure" high technology, although the SIC codes 28 and SIC 35–38 contain most of the commonly identified high technology sectors at the more appropriate 3 and 4 digit level (see, e.g., Herzog, Schlottman, and Johnson [41]). We consequently proceeded to use the aggregate of these five SIC codes as "high technology." To the extent that the sectoral mix in these sectors shows systematic variation over space in terms of its "pure" high tech content, our results on the relationship between innovation and research could be affected. However, we are confident that we will be able to detect such systematic variations by means of careful specification tests for spatial effects (Anselin [7], Anselin and Bera [10]).

Earlier studies of the aggregate knowledge production function were limited to data for 29 U.S. states. This was not due to the lack of data on innovations or patents (for which actual addresses are available), but to data limitations for the explanatory variables in the model, in particular for the variable on private R & D expenditures [R in Eqs. (2) and (4)]. In Jaffe [44], and also in Acs, Audretsch, and Feldman [2–4] and Feldman and Florida [27], this was computed from information on total industry R & D by state, compiled by the U.S. Bureau of the Census on the basis of survey data from the National Science Foundation (NSF). However, this data was only consistently reported for 29 states (for details, see, e.g., Jaffe [44, pp. 968–969]). Instead, we constructed a proxy for industrial R & D activity on the basis of data on professional employment in high technology research laboratories in the Bowker directories (Jaques Cattell Press [19]). While imperfect, this approach allowed us to construct a private R & D variable for 43 U.S. states and for 125 MSAs (see also Bania, Calkins, and Dalenberg [14, pp. 218–219], for a similar approach). As it turns out, our proxy variable is remarkably similar to the R & D expenditure variable used in Jaffe [44], yielding a correlation of 0.91 for the 29 states common to both studies (the data on innovations, private R & D employment, and university research expenditures for 43 U.S. states in 1982 are listed in Appendix A). Clearly, the use of lab employment as a proxy for expenditures assumes a constancy of the labor intensity and capital/labor ratio of R & D across the units of observation. Again, to the extent that this is not the case, it will tend to yield misspecifications in the form of heteroskedastic and/or spatially autocorrelated error terms, which will merit special attention in our analysis.

Our data for university research expenditures [*U* in Eqs. (3) and (4)] follow the common approach in the literature and are compiled from the NSF Survey of Scientific and Engineering Expenditures at Universities and Colleges for the year 1982 (National Science Foundation [58]). In addition, this data set also provides the source for exogenous variables used in the estimation of Eq. (3) at the MSA level: a proxy for the overall academic quality of high technology departments at universities, *RANK*; and a proxy for size, the total enrollment at universities, *ENRL*. We also consider data for total educational expenditures, *EDEXP*, from the County and City Data Book. We match the sectoral aggregation of the two-digit SIC industries to university departments using the same approach as in Feldman [26, p. 58].

In addition, we also included a number of variables compiled from County Business Pattern data for 1982: high technology employment, *HTEMP*; a location quotient for high technology employment, *LQ*; employment in business services (SIC 73), *BUS*; and the percent large firms (i.e., firms with employment exceeding 500), *LARGE*. Following general practice in the literature, the first three variables are included to capture agglomeration economies (see also Feldman and Florida [27]), the last one to assess the effect of firm scale (see also Acs, Audretsch, and Feldman [3]). A proxy for the coincidence of R & D and headquarter locations is a dummy variable for the presence of at least 10 headquarters of Fortune 500 companies in an MSA, *FORTUNE*, compiled from the May 2, 1982 listing in *Fortune Magazine*.

Our final data set only included those MSAs for which there were innovations in the high technology sector as well as both private industry R & D and university research expenditures (see Appendix B for a listing of the MSAs, innovation counts and research data). This results in 125 MSAs being considered that meet the three criteria, an almost complete match of university MSAs (125 out of 130), and a better than 70% portion of both the innovative MSAs (169) and the R & D MSAs (159). This selection leaves out a number of important non-MSA university locations, such as Ithaca, NY, and State College, PA, but it covers more than 80% of the relevant interaction (92% of all university research is in the 130 university MSAs). This limitation also excludes from consideration the joint determination of location and magnitude of high technology innovation and research. On the other hand, it avoids the problem of "zeros," and is motivated by a focus on the strength of interaction between the two forms of research and the generation of innovations where these are present. In other words, we address the strength and spatial extent of local geographic spillovers for metropolitan areas where an innovative complex is already in place. We leave the explanation of the more complex issue of the establishment and absence of these links for future research.

LOCAL GEOGRAPHIC SPILLOVERS AT THE STATE LEVEL

Alternative State-Level Geographic Coincidence Indicators

In addition to the familiar Jaffe indicator of geographic coincidence, we considered four alternative measures of local spatial interaction between university and industry research for states. One is constructed similar to Jaffe's approach, in that the log of a coincidence index is multiplied with the log of university research expenditures at the state level. However, rather than the uncentered correlation coefficient between university research and total professional employees in R & D laboratories for the MSAs in the state, we used the proportion of counties where industry research (proxied by R & D lab employment) and university research are co-located (out of all counties with university research). Formally, for state s,

$$CUR_s = \Sigma_i \delta R_i \cdot \delta U_i / \Sigma_i \delta U_i, \qquad (5)$$

where the summation is over counties, and δR_i and δU_i are Kronecker deltas that take on a value of 1 when respectively $R_i > 0$ and $U_i > 0$, and a value of zero otherwise. In contrast to the Jaffe index, CUR_s also includes information on counties that do not belong to an MSA in the state.

The other three measures are more closely related to traditional accessibility indices from spatial interaction theory (see, e.g., Weibull [75]). The first is a gravity or potential measure with distance decay parameter 2 (see also Sitivanidou and Sivitanides [70] for a slightly different approach) for each industry R & D county in a state with respect to all university counties:

$$GRAV_i = \Sigma_j U_j / (d_{ij})^2, \qquad (6)$$

where i is an index for each industry R & D county in the state, j an index for each university county, U is as before, and d_{ij} is the distance between the two counties. When an industry and university county coincide, no distance decay is applied to the U_i variable to avoid the familiar self-potential problem (e.g., Frost and Spence [30]). To obtain an index for each state, an average is taken over all counties in the state for which $GRAV_i > 0$.

The other two indices are based on the notion of "covering," i.e., a measure of how much university research is carried out in counties within a given distance band of a industry R & D county (note that the university counties do not need to be within the same state):

$$COV_i = \Sigma_j \delta_{ij} \cdot U_j \tag{7}$$

where δ_{ij} takes on a value of 1 when d_{ij} is less than or equal to a critical covering distance, respectively, 50 and 75 miles in our application. Again, to obtain an index for a state, an average is taken over all industry R & D counties in the state.

The computation of the indices was carried out by means of the access functions in the SpaceStat software (Anselin [9]) from data on boundary files for U.S. counties.

Empirical Results

In Table 1, we present the results of the estimation of a cross-sectional regression of the geographic knowledge production function (4) for 43 U.S. states in 1982. All variables are in logarithms. In addition to a replication of the standard Jaffe approach (but using innovation counts as the dependent variable, as in Acs, Audretsch, and Feldman [2, p. 365]), reported in the first column of the table, we also estimated the model with the four new indices of local geographic interaction (reported in columns 3–6).

All regressions yield highly significant and positive coefficients for both private R & D and university research (at $p < 0.01$), confirming the consensus result in the literature. The new gravity and covering indices are positive and significant as well (at $p < 0.05$), whereas the co-location index (5) is only marginally significant (at $p < 0.10$). Jaffe's geographic coincidence index does not turn out to be significant, confirming earlier findings of Acs, Audretsch, and Feldman [2, 3]. For all models, diagnostic tests were carried out for heteroskedasticity, using the White [76] test. In addition, specification tests for spatial autocorrelation were carried out, using the Lagrange multiplier tests for spatial error and spatial lag dependence implemented in the SpaceStat software (for technical details, see Anselin [7–9], Anselin and Hudak [11], Anselin and Bera [10]). The tests for spatial autocorrelation were computed for three different spatial weights matrices which reflect different prior notions on the spatial structure of dependence. The weights matrices have non-zero elements for each observation pair (an observation, row , and its potential neighbor, column) that is assumed to interact. The magnitude of the non-zero elements is either a function of contiguity (the presence of a common border), or some function of the distance between the two observations (for further details, see Anselin [7], Anselin and Bera [10]). The three weights used for the 43 states are based on simple contiguity (common border between states), on distance-based contiguity with a 500 mile cut-off (states are neighbors if there geographic centers are within 500 miles of each other); and on inverse distance squared (the magnitude of the interaction is proxied by the squared inverse distance between the geographic centers of states). Only the results for the most significant diagnostic are reported in Table 1. No evidence of heteroskedasticity was found, but the Jaffe model showed significant spatial error autocorrelation. Maximum likelihood estimation of the Jaffe model with spatial error autocorrelation (reported in column 2) did not alter the substantive interpretation of the model: both private R & D and university research remained strongly significant, but the

TABLE 1
Regression Results for Log(Innovations) at the State Level
(N = 43, 1982)

Model	Jaffe (OLS)	Jaffe (ML-SER)	CUR (OLS)	GRAV (OLS)	COV(50) (OLS)	COV(75) (OLS)
Constant	−7.079	−7.084	−7.117	−10.181	−7.205	−8.386
	(0.100)	(0.093)	(1.032)	(1.574)	(1.001)	(1.103)
Log(R)	0.508	0.549	0.396	0.350	0.451	0.375
	(0.100)	(0.098)	(0.110)	(0.110)	(0.095)	(0.104)
Log(U)	0.574	0.544	0.676	0.633	0.402	0.549
	(0.142)	(0.129)	(0.144)	(0.133)	(0.153)	(0.133)
Log(local spillover)	0.0266	0.0164	0.0463	0.301	0.268	0.255
	(0.030)	(0.027)	(0.026)	(0.121)	(0.109)	(0.100)
λ		0.507				
		(0.218)				
R^2-adj	0.820		0.831	0.841	0.841	0.843
Log-Lik	−41.99	−39.91	−40.62	−39.28	−39.31	−39.08
White	6.29		9.09	5.42	6.91	5.38
LM-Err	5.40		1.01	0.84	0.60	0.52
	(IDIS2)		(D500)	(IDIS2)	(D500)	(D500)
LM-Lag	0.91	0.48	0.56	1.90	0.02	0.48
	(CONT)	(IDIS2)	(CONT)	(D500)	(D500)	(D500)

Notes: Estimated standard errors are in parentheses; critical value for the White statistic with 9 degrees of freedom is 16.92 (p = 0.05); critical value for LM-Err and LM-Lag statistics is 3.84 (p = 0.05); spatial weights matrices are row-standardized: CONT is first order contiguity; D500 is distance-based contiguity for 500 miles; and IDIS2 is inverse distance squared; only the highest values for a spatial diagnostic are reported.

geographic coincidence index was not (the spatial autoregressive coefficient was significant and positive; no evidence was found of any remaining higher order spatial dependence).

Interestingly, relative to the results of the Jaffe model, the effect of the introduction of the gravity and covering interaction indices seems to be to lower the coefficient of private R & D, whereas the coefficient of university research remains virtually unchanged. This may suggest that the original results overestimated the pure effect of private R & D on innovation, in the sense that some of it may be due to local geographic interaction with university research. The new indices seem to capture this interaction, while the original Jaffe index does not. Also, the introduction of the new indices removes any misspecification in the form of spatial autocorrelation, which is present in the original Jaffe formulation.

In sum, these results suggest that the spatial range of interaction between private R & D and university research reaches beyond the county (or MSA) where the R & D is carried out. Even the crude averages at the state level provide consistent evidence of this. Parenthetically, our results also demonstrate the usefulness of high technology lab employment as a proxy for private R & D activity, since no significantly different relationship was found compared to earlier results in the literature that used a purer measure of R & D expenditures.

LOCAL GEOGRAPHIC SPILLOVERS AT THE MSA LEVEL

Spatially Lagged Variables

The use of R & D lab employment as a proxy for private R & D activity allows us to carry out an analysis of the geographic knowledge production function at a smaller spatial scale than the state level. We constructed a data base for 125 MSAs in the U.S. for which some innovative and research activity was present (see Appendix B for a listing). Given the indication of a wider range of spatial interaction than purely within-county between university and private R & D from the results at the state level, we constructed two new variables that we refer to as *spatial lags*. These variables are designed to capture the effect of, respectively, university research and private R & D in counties surrounding the MSA, within a given distance band from the geographic center of the MSA. Specifically, for any MSA i, the spatial lags $U50_i$ and $R50_i$ are the sums of, respectively, the university research and private R & D in those counties surrounding the MSA (and not part of the MSA) whose geographic centers are within 50 miles of the geographic center of the core MSA county. Similar measures were computed for a 75 mile range as well (*U75*, *R75*). These measures correspond to our notion of covering used in the state level analysis.

Since the analysis is carried out at the scale at which we assume that the spatial interaction takes place (MSA and possibly its surrounding counties), there is no need to create an artificial index of geographic coincidence, as is the case at the state level. In fact, by explicitly including both the research magnitude at the MSA as well as for surrounding counties, we are able to get a much more precise insight into the spatial extent of local geographic spillovers.

Estimation Issues

Our model consists of three equations, the knowledge production function for K (1), an industry research equation, R (2), and a university research equation, U (3). The knowledge production function contains both R and U as explanatory variables and will be extended with the two spatial lag variables, *U50* (or *U75*) and *R50* (or *R75*). Both R and U equations contain the other as explanatory variable, as well as the spatial lags. The form of this system of equations raises a number of issues with respect to estimation and identification.

First, while the system is recursive between K and R and U, the exogeneity of the latter two in the knowledge production function should not be taken on faith. In fact, misspecifications (e.g., errors in variables) could easily lead to endogeneity and must be checked. We address this by means of the Durbin–Wu–Hausman test for exogeneity (e.g., Davidson and MacKinnon [22, pp. 237–242]). We also take this approach to test the extent to which R and U are endogenous to each other in Eqs. (2) and (3).

Second, even in a purely recursive system, ordinary least squares estimation of the knowledge production function would only be legitimate in the absence of interequation correlation, i.e., correlation between the error terms of the equations (e.g., Greene [34], p. 600). We check this by means of a Lagrange multiplier (LM) test on the diagonality of the error covariance matrix for the least squares residuals (Breusch and Pagan [18]).

Third, the spatially lagged variables *R50* (or *R75*) in the R equation, and *U50* (or *U75*) in the U equation could be endogenous as well. In spatial econometric models, such variables are often referred to as spatially lagged dependent variables and it is well known that their presence causes OLS to be biased and inconsistent (Anselin [7]). In our model however, the locations to which these lags pertain (counties surrounding the MSA, but not part of the MSA) are not part of the observation set (the MSAs), unless they belong to the MSA counties of a neighboring MSA.

There is some degree of overlap for 71 out of the 125 MSAs and we assess the extent to which this may cause endogeneity by means of a Durbin–Wu–Hausman (DWH) test.

Finally, the use of a cross-sectional sample potentially leads to spatial autocorrelation in the regression equations. We assess this by means of a Lagrange multiplier test for spatial error dependence using three spatial weights based on distance: the same 50 and 75 mile cut-offs as used in the construction of the lag variables, and a squared inverse distance weights matrix. These tests are only valid when the explanatory variables in the regression are exogenous and should be interpreted with caution when this is not the case. They are used here to assess the extent to which remaining unspecified spatial spillover may be present, even after the inclusion of the spatial lags (provided that the latter are exogenous, which turns out to be the case in our study).

Empirical Results

We first focus on the knowledge production function in Eq. (1), i.e., without a geographic coincidence index. The estimation results for a cross-sectional regression on 125 MSAs in 1982 are reported in Table 2. Three models are presented: a bare-bones Jaffe model, a spatial model which also includes the spatial lags for university and private R & D (only the most significant of, respectively, *U50* and *U75* and *R50* and *R75* are reported), and an extended spatial model, which includes local economic characteristics as explanatory variables as well.

The standard Jaffe model (in column 1 of Table 2) confirms the strong significance (at $p < 0.01$) of both private R & D and university research on the level of innovative activity in an MSA. However, compared to the analysis at the state level, there is a clear dominance of the coefficient of private R & D over university research, indicating an elasticity that is almost five times higher (compared to roughly equal elasticities at the state level in Table 1). There is no evidence of heteroskedasticity, but there is a strong indication of misspecification in the form of a spatial lag (at $p < 0.05$). When the spatial variables *U50* and *R75* are added to the Jaffe specification (instead of a coincidence index at the state level), the model fit improves slightly, with a positive and significant effect for *U50*, but an insignificant effect for the research spatial lag. Neither heteroskedasticity, nor spatial dependence remain. In the full model (in column 3), the spatial lag for university research, with a 50 mile range *(U50)* remains positive and significant ($p < 0.05$). However, there is no evidence that the effect of private R & D on MSA innovative activity spills over from outside the MSA (the coefficient of *R75* is negative, but not significant). The addition

TABLE 2
OLS Regression Results for Log(Innovations) at the MSA Level
(N-125, 1982)

Model	Jaffe	Spatial	Extended spatial
Constant	−1.045	−1.134	−1.407
	(0.146)	(0.172)	(0.212)
Log(R)	0.540	0.504	0.277
	(0.054)	(0.055)	(0.057)
Log($R75$)		0.001	−0.027
		(0.041)	(0.037)
Log(U)	0.112	0.132	0.093
	(0.036)	(0.036)	(0.034)
Log($U50$)		0.037	0.032
		(0.018)	(0.015)
Log(LQ)			0.652
			(0.163)
Log(BUS)			0.332
			(0.057)
Log($LARGE$)			−0.337
			(0.094)
$RANK$			0.202
			(0.101)
R^2-adj	0.599	0.611	0.725
White	1.18	9.02	37.85
LM-Err	2.69	2.18	0.10
	(D75)	(D75)	(D50)
LM-Lag	5.62	1.48	2.08
	(D50)	(D75)	(D75)

Notes: Estimated standard errors are in parentheses; critical values for the White statistic with, respectively 5, 20 and 35 degrees of freedom are 11.07, 31.41, and 49.52 ($p = 0.05$); critical values for LM-Err and LM-Lag statistics are 3.84 ($p = 0.05$) and 2.71 ($p = 0.10$); spatial weights matrices are row-standardized: D50 is distance-based contiguity for 50 miles; D75 is distance-based contiguity for 75 miles; and IDIS2 is inverse distance squared; only the highest values for a spatial diagnostic are reported.

of the local economic variables causes the elasticity with respect to private R & D to drop substantially, while the one for university research stays roughly the same. All three local economic variables are highly significant (with $p < 0.001$) and have the expected sign: positive for specialization in high technology (LQ) and the importance of business services (BUS), and negative for the presence of large firms ($LARGE$). Note that the negative sign for the presence of large companies confirms earlier evidence in Acs, Audretsch, and Feldman [3, p. 338] that smaller firms tend to be more innovative. In other words, *ceteris paribus*, MSAs dominated by the presence of larger firms tend to show less innovative activity. A fourth variable is added to correct for potential unmeasured quality effects that may cause inter-equation correlation. Following Jaffe [44], we use the rank of the universities as a proxy for quality. This variable turns out to be positive and significant ($p < 0.05$).

For the full specification, we further tested the exogeneity of each of the four variables *R*, *U*, *R75*, and *U50*, using the DWH test for a two stage least squares estimation with Log(*HTEMP*), Log(*ENRL*), Log(*EDEXP*), and *FORTUNE* as instruments. We failed to reject the null hypothesis for each (none achieved a *p* value less than 0.20). In other words, there was no evidence *against* exogeneity of these variables. In addition, the LM tests on diagonality of the error covariance between *K* and *R* and *K* and *U* were, respectively, 1.17 ($p = 0.28$) and 1.73 ($p = 0.19$), which did not provide evidence against the null hypothesis of no inter-equation error correlation.

The OLS estimation results in column 3 of Table 2 can thus be reliably interpreted to indicate the strong influence of university research on innovative activity in an MSA, not only of university research in the MSA itself, but also in the surrounding counties. By contrast, the effect of private R & D seems to be contained within the MSA itself. Of course, private and university R & D are not independent, and we turn to their interaction/ simultaneity in Tables 3 and 4.

Following Jaffe [44], we estimated two additional models that explicitly incorporate the potential simultaneity between private R & D and university research. The results in Table 3 show the estimation of the private R & D equation (2) in which, in addition to university research, the spatial lags of both university research and private R & D are included, as well as the log of the high technology employment (*HTEMP*), the *FORTUNE* dummy, and the *RANK* measure as exogenous variables (the latter to control for potential quality effects). Given the strong indication of heteroskedasticity (the White test in the OLS regression of the model yielded 42.54, which is significant at $p < 0.02$), both standard and robust results are reported. University enrollment (*ENRL*) and education expenditures (*EDEX*) were used as instruments in the 2SLS estimation reported in columns 3 and 4 of Table 3, with *U* as the endogenous variable.

The estimates and their significance vary considerably between the standard and robust results on the one hand and the OLS and 2SLS approaches on the other hand, indicating the strong effect of both heteroskedasticity and simultaneity bias. A Durbin–Wu–Hausman (DWH) test on the exogeneity of university research is clearly rejected at $p < 0.05$

TABLE 3
Regression Results for Log(Private R & D) at the MSA Level
($N = 125$, 1982)

Model	OLS	OLS-Robust	2SLS	2SLS-Robust
Constant	−0.952	−0.952	−1.036	−1.049
	(0.408)	(0.622)	(0.418)	(0.457)
Log(*R50*)	0.055	0.055	0.070	0.071
	(0.034)	(0.036)	(0.035)	(0.035)
Log(*U*)	0.155	0.155	0.259	0.257
	(0.047)	(0.065)	(0.067)	(0.078)
Log(*U75*)	0.044	0.044	0.040	0.040
	(0.041)	(0.040)	(0.042)	(0.038)
Log(*HTEMP*)	0.637	0.637	0.567	0.571
	(0.097)	(0.173)	(0.104)	(0.136)
FORTUNE	0.414	0.414	0.419	0.416
	(0.188)	(0.171)	(0.192)	(0.150)
RANK	0.122	0.122	−0.004	−0.007
	(0.163)	(0.154)	(0.176)	(0.154)
R^2	0.585	0.585	0.652	0.651

Notes: Estimated standard errors are in parentheses; Log(*U*) considered as endogenous; instruments in 2SLS estimation are Log(*ENRL*) and Log(*EDEX*). DWH test on exogeneity of Log(*U*), $F(1,117) = 5.27$, $p = 0.023$.

(the statistic was 5.27). The OLS results are therefore only reported for comparison purposes and should not be the basis of any substantive interpretation.

Focusing on the results in column 4 of Table 3, we find a strong positive and significant effect of university research on private R & D ($p < 0.001$), but only within the MSA itself (the coefficient of *U75* is not significant). On the other hand, there is an indication of spatial spill-over in private R & D, supported by a significant (at $p < 0.05$) and positive coefficient of *R50*. Both specialization in high technology and the presence of Fortune 500 companies also contribute significantly to the magnitude of private R & D. Note that the latter has a positive sign, in contrast to the effect of larger firms on innovations. A DWH test on the exogeneity of the spatial lag variable *R50* cannot be rejected [$F(1,117) = 0.078$, $p = 0.78$], hence the results in the last column of Table 3 yield a valid interpretation.

The effect of private R & D on university research is quite different, as borne out by the results for Eq. (3) reported in Table 4, in the same format as in Table 3. There is no evidence of a significant influence of private

TABLE 4
Regression Results for Log(University Research) at the MSA Level
(N = 125, 1982)

Model	OLS	OLS-Robust	2SLS	2SLS-Robust
Constant	−2.355	−2.355	−2.502	−2.244
	(1.715)	(1.846)	(1.749)	(1.702)
Log(*U*75)	0.096	0.096	0.102	0.107
	(0.057)	(0.075)	(0.058)	(0.062)
Log(*R*)	0.052	0.052	−0.025	0.051
	(0.109)	(0.152)	(0.202)	(0.160)
Log(*R*50)	−0.074	−0.074	−0.066	−0.066
	(0.046)	(0.044)	(0.049)	(0.039)
Log(*ENRL*)	1.777	1.777	1.846	1.760
	(0.179)	(0.230)	(0.234)	(0.227)
Log(*EDEXP*)	−1.187	−1.187	−1.176	−1.242
	(0.928)	(1.001)	(0.930)	(0.899)
RANK	0.510	0.510	0.539	0.469
	(0.208)	(0.159)	(0.218)	(0.157)
R^2	0.650	0.650	0.650	0.629

Notes: Estimated standard errors are in parentheses; Log(*R*) considered as endogenous; instruments in 2SLS estimation are Log(*HTEMP*) and *FORTUNE*. DWH test on exogeneity of Log(*R*), $F(1,117) = 0.210$, $p = 0.648$.

R & D in the MSA itself on university research, nor of spatial spillovers. The only strongly significant coefficients are for internal characteristics of the universities, i.e., their size (enrollment) and quality, both with the expected positive sign. In contrast to the R & D equation, there is no significant evidence of heteroskedasticity (White test yields 34.03, with 38.89 as the critical level for 26 degrees of freedom at $p = 0.05$), nor of spatial autocorrelation. More importantly, there is no indication of simultaneity bias (a DWH test on the exogeneity of R yields 0.21, which is clearly non-significant at $p = 0.65$). The correct interpretation should therefore be based on column 1 and the other results are only reported for completeness sake.

In sum, the estimates in Table 3 and 4 support the earlier finding of Jaffe [44, p. 968] about the direction of causality in the relation between private R & D and university research, but now at a much finer level of spatial detail and with an explicit measure of spatial spillovers. There is no evidence that private R & D in the MSA or its surrounding counties is endogenous to the university research equation. By contrast, there is a strong indication that university research in an MSA is endogenous in the private R & D equation.

CONCLUSIONS

In this paper, we have been able to shed additional light on the issue of *local* geographic spillovers between university research and high technology innovations. Our point of departure was Jaffe's [44, p. 968] often cited finding that "there is only weak evidence that spillovers are facilitated by geographic coincidence of universities and research labs within the state." We approached this issue from an explicit spatial econometric perspective and implemented the classic Griliches–Jaffe knowledge production framework for high technology innovations in 43 U.S. states, as well as in 125 MSAs. The latter became possible by using a specially compiled set of data on R & D laboratory employment. This yielded more precise insight into the range of spatial externalities between innovation and R & D in the MSA and university research, both within the MSA and in surrounding counties.

We also refined Jaffe's geographic coincidence index for analysis of local spillover based on data for states. We suggested four alternative measures, three of which were derived from established principles in spatial interaction theory, i.e., the gravity potential and covering index. We showed that the inclusion of the alternative indices in the geographic knowledge production function provided strong and unequivocal evidence of local spatial externalities. In addition, the evidence suggested that these externalities transcended the boundaries of counties.

Overall, we confirmed the positive and significant relationship between university research and innovative activity, both directly, as well as indirectly through its impact on private sector R & D. We found that the spillovers of university research on innovation extended over a range of 50 miles from the innovating MSA, but not with respect to private R & D. We also confirmed earlier findings on the direction causality between university and private research, the former being endogenous to the latter, but not vice versa.

Our findings are important in that they highlight the relevance of a precise consideration of the spatial range of interaction in the analysis of spatial externalities. However, some cautionary remarks are in order as well. Our analysis is limited by the use of a single cross-section. Unfortunately, there is currently no update of the 1982 U.S. SBA innovation data base for later points in time, precluding a more extensive analysis of the space-time dynamics. Also, we have elected to focus on studying the relations between research and innovations in those locations for which both were observed. This leaves aside the issue of why certain locations have research and innovative activity and others do not, especially when one of the two is present, but the other is not. We leave this aspect of the study for a separate paper.

APPENDIX A

Innovations, R & D Lab Employment and University Research for 43 U.S. States—1982

State	Innovations	R & D	University research
Alabama	4	1868	45293
Arizona	38	4086	62206
Arkansas	4	187	12303
California	851	48552	658487
Colorado	40	4244	87230
Connecticut	117	8799	136435
Delaware	11	6173	6962
Florida	49	2298	105177
Georgia	29	1585	102981
Illinois	185	20574	207669
Indiana	38	6970	92986
Iowa	13	1939	77435
Kansas	11	2029	21849
Kentucky	8	571	21224
Louisiana	4	701	48509
Maine	3	430	5527
Maryland	26	5244	131229
Massachusetts	316	24753	364871
Michigan	84	21816	149432
Minnesota	82	7478	103549
Mississippi	4	142	18220
Missouri	23	4863	100942
Nebraska	4	275	21893
Nevada	1	103	3966
New Hampshire	28	624	19368
New Jersey	352	42907	67887
New Mexico	3	339	57745
New York	347	27709	585312
North Carolina	23	5037	111994
Ohio	139	18886	161911
Oklahoma	14	2630	39175
Oregon	29	310	44788
Pennsylvania	197	22702	267614
Rhode Island	22	1083	22035
South Carolina	11	1131	27019
Tennessee	16	1397	55692
Texas	136	9896	296114
Utah	13	1926	62249
Vermont	3	67	14166
Virginia	27	8023	72208
Washington	41	1385	95530
West Virginia	2	1471	13589
Wisconsin	69	4504	127440

Sources: Compiled from U.S. SBA Innovations Data Base; compiled from R. R. Bowker Company Directories; compiled from NSF Survey of Scientific and Engineering Expenditures at Universities and Colleges.

APPENDIX B

Innovations, R & D Lab Employment and University Research for 125 U.S. MSAs—1982

MSA	Innovations	R & D	University research
Akron	7	2989	4504
Albany-Schenectady-Troy	1	2475	33600
Albuquerque	2	273	26064
Allentown-Bethlehem-Easton PA-NJ	7	946	9455
Anaheim-Santa Ana-Garden Grove	108	2904	31037
Ann Arbor	7	1428	84876
Atlanta	26	661	71321
Austin	12	1003	45079
Baltimore	12	2499	88697
Bellingham	1	78	255
Benton Harbor	1	388	19
Binghamton NY-PA	2	152	2988
Birmingham	1	646	35162
Bloomington-Normal	2	16	356
Boston	282	22391	325474
Bridgeport	67	3221	14
Bryan-College Station	2	37	40178
Buffalo	24	4404	25591
Burlington	3	41	13952
Charlotte-Gastonia	6	191	627
Chicago	164	15416	145814
Cincinnati OH-KY-IN	13	1247	26709
Cleveland	54	5195	42792
Colorado Springs	6	726	52
Columbia	1	135	23448
Columbus	20	2733	52252
Dallas-Fort Worth	77	1900	46974
Davenport-Rock Island-Moline Iowa-IL	5	4535	3
Dayton	11	2538	22608
Daytona Beach	1	97	31
Denver-Boulder	26	3386	65379
Detroit	51	11599	21511
El Paso	7	1160	834
Fort Collins	6	83	21717
Fort Lauderdale-Hollywood	9	285	99
Fresno	1	48	276
Gainesville	1	158	45324
Galveston-Texas	2	99	11882
Grand Rapids	4	359	21
Greensboro-Winston-Salem-High Point	5	1010	9159
Greenville-Spartanburg	10	506	9085
Hamilton-Middletown	4	39	709
Hartford	27	2087	52297
Houston	29	2545	120942
Huntsville	3	847	1843
Janesville-Beloit	2	207	49

APPENDIX B
(*Continued*)

MSA	Innovations	R & D	University research
Jersey City	11	709	4352
Johnson City-Kingsport-Bristol Tenn.-VA	2	737	736
Kalamazoo-Portage	5	950	146
Kansas City	12	1325	956
Knoxville	1	356	32526
Lafayette-West Lafayette	1	121	45048
Lancaster	4	671	69
Lansing-East	4	140	35951
Lincoln	2	127	13236
Lorain-Elyria	2	204	75
Los Angeles-Long Beach	161	20226	199028
Louisville	7	414	4992
Madison	4	741	107915
Melbourne-Titusville-Cocoa	11	81	779
Memphis TN.-AK.-MS.	3	122	799
Miami	4	177	32677
Milwaukee	34	2130	18832
Minneapolis-St. Paul	80	7181	103235
Nashville-Davidson	5	113	20604
Nassau-Suffolk	120	2534	31881
New Bedford	6	95	569
New Brunswick-Perth Amboy-Sayreville	30	2338	20408
New Haven-West Haven	19	1708	83974
New London-Norwich	1	1659	156
New Orleans	1	320	14394
New York NY-NJ	222	11136	285776
Essex County	143	26780	17500
Newburgh-Middletown	3	324	206
Newport News-Hampton	2	310	1676
Norfolk-Virginia Beach-Portsmouth	1	136	2819
Northeast	2	25	70
Oklahoma City	1	204	19703
Orlando	5	514	4704
Paterson-Clifton-Passaic	25	3098	226
Peoria	1	27	146
Philadelphia PA-NJ	139	16967	134499
Phoenix	29	3809	8037
Pittsburgh	39	4357	68160
Pittsfield	2	14	354
Portland	1	286	463
Portland OR-WA	22	285	20750
Portsmouth-Dover-Rochester	5	166	5468
Providence-Warwick-Pawtucket	15	480	22036
Provo-Orem	3	281	3419
Raleigh-Durham	8	2317	101136
Reading	1	423	123
Reno	1	26	3226
Riverside-San Bernardino-Ontario	13	193	9494

APPENDIX B
(Continued)

MSA	Innovations	R & D	University research
Rochester	32	4554	60999
Sacramento	7	311	32634
St. Louis	13	3312	69825
Salem	1	44	20
Salt Lake City	10	1585	40785
San Antonio	3	1674	19539
San Diego	59	4823	79857
San Francisco-Oakland	75	7856	154824
San Jose	374	9727	127897
Santa Barbara-Santa Maria-Lompoc	9	1945	10269
Santa Cruz	2	30	7168
Seattle-Everett	34	798	85514
South Bend	5	166	11192
Spokane	3	11	230
Springfield	3	131	19
Springfield-Chicopee-Holyoke	3	449	29455
Stockton	2	10	648
Syracuse	9	1020	31844
Tacoma	2	269	76
Tampa-St. Petersburg	12	668	4790
Toledo OH-MI	6	1129	6705
Trenton	29	2836	24680
Tucson	9	209	53599
Tulsa	12	89	1317
Waco	3	37	363
Washington, DC	21	8274	38311
Waterloo-Cedar Falls	1	14	5
Wichita	5	1490	909
Wilmington DE-NJ-MD	11	6706	6962
Worcester	17	861	4628
Youngstown-Warren	1	167	75

Sources: Compiled from U.S. SBA Innovations Data Base; compiled from R. R. Bowker Company Directories; compiled from NSF Survey of Scientific and Engineering Expenditures at Universities and Colleges.

REFERENCES

1. Z. Acs and D. Audretsch, "Innovation and Small Firms," MIT Press, Cambridge, MA (1990).
2. Z. Acs, D. Audretsch, and M. Feldman, Real effects of academic research: Comment, *American Economic Review*, **81**, 363–367 (1991).
3. Z. Acs, D. Audretsch, and M. Feldman, R & D spillovers and recipient firm size, *The Review of Economics and Statistics*, **76**, 336–340 (1994).
4. Z. Acs, D. Audretsch, and M. Feldman, R & D spillovers and innovative activity, *Managerial and Decision Economics*, **15**, 131–138 (1994).

5. P. Almeida and B. Kogut, "The geographic localization of ideas and the mobility of patent holders," Working Paper, Department of Management, The Warton School, University of Pennsylvania (1995).
6. A. E. Andersson, C. Anderstig, and B. Harsman, Knowledge and communications infrastructure and regional economic growth, *Regional Science and Urban Economics*, **20**, 359–376 (1990).
7. L. Anselin, "Spatial Econometrics, Methods and Models," Kluwer Academic, Boston (1988).
8. L. Anselin, "SpaceStat, a Program for the Analysis of Spatial Data," National Center for Geographic Information and Analysis, University of California, Santa Barbara, CA (1992).
9. L. Anselin, "SpaceStat Version 1.80 User's Guide," Regional Research Institute, West Virginia University, Morgantown, WV (1995).
10. L. Anselin and A. Bera, Spatial dependence in linear regression models with an introduction to spatial econometrics, *in* "Handbook of Economic Statistics" (D. Giles and A. Ullah, Eds.), Marcel Dekker, New York, (1997).
11. L. Anselin and S. Hudak, Spatial econometrics in practice: A review of software options, *Regional Science and Urban Economics*, **22**, 509–536 (1992).
12. K. Arrow, Economic welfare and the allocation of resources for invention, *in* "The Rate and Direction of Inventive Activity" (R. Nelson, ed.), Princeton University Press, Princeton, NJ (1962).
13. D. B. Audretsch and M. P. Feldman, Knowledge spillovers and the geography of innovation and production, *American Economic Review*, **86**, 630–640 (1996).
14. N. Bania, L. N. Calkins, and D. R. Dalenberg, The effects of regional science and technology policy on the geographic distribution of industrial R & D labs, *Journal of Regional Science*, **32**, 209–228 (1992).
15. N. Bania, R. Eberts, and M. S. Fogarty, Universities and the startup of new companies: Can we generalize from Route 128 and Silicon Valley? *The Review of Economics and Statistics*, **75**, 761–766 (1993).
16. A. Bartel and F. Lichtenberg, The comparative advantage of educated workers in implementing new technology, *The Review of Economics and Statistics*, **69**, 1–11 (1987).
17. P. Beeson and E. Montgomery, The effects of colleges and universities on local labor markets, *The Review of Economics and Statistics*, **75**, 753–761 (1993).
18. T. Breusch and A. Pagan, The LM test and its applications to model specification in econometrics, *Review of Economic Studies*, **47**, 239–254 (1980).
19. Jaques Cattell Press, Industrial Research Laboratories of the United States, 17th Edition, R. R. Bowker Company, New York and London (1982).
20. W. M. Cohen and D. A. Levinthal, Innovation and Learning, *The Economic Journal*, **99**, 569–596 (1989).
21. P. David and J. Rosenbloom, Marshallian factor market externalities and the dynamics of industrial localization, *Journal of Urban Economics*, **28**, 349–370 (1990).
22. R. Davidson and J. G. MacKinnon, "Estimation and Inference in Econometrics," Oxford University Press, New York (1993).
23. C. DeBresson and F. Amesse, Networks of innovators: A review and introduction to the issues, *Research Policy*, **20**, 363–380 (1992).
24. G. Dosi, Sources, procedures and microeconomic effects of innovation, *Journal of Economic Literature*, **26**, 1120–1171 (1988).
25. K. Edwards and T. J. Gordon, "Characterization of Innovations Introduced on the U.S. Market in 1982, Report Prepared for the U.S. Small Business Administration," The Futures Group, Washington, DC (1984).

26. M. Feldman, "The Geography of Innovation," Kluwer Academic, Boston (1994).
27. M. Feldman and R. Florida, The geographic sources of innovation: Technological infrastructure and product innovation in the United States, *Annals of the Association of American Geographers*, **84**, 210–229 (1994).
28. R. Florax, "The University: A Regional Booster? Economic Impacts of Academic Knowledge Infrastructure," Avebury, Aldershot (1992).
29. R. Florax and H. Folmer, Knowledge impacts of universities on industries: An aggregate simultaneous investment model, *Journal of Regional Science*, **32**, 437–466 (1992).
30. M. E. Frost and N. A. Spence, The rediscovery of accessibility and economic potential: The critical issue of self-potential, *Environment and Planning A*, **27**, 1833–1848 (1995).
31. E. L. Glaeser, H. D. Kallal, J. A. Scheinkman and A. Schleifer, Growth in cities, *Journal of Political Economy*, **100**, 1126–1152 (1992).
32. A. Glasmeier, "The High-Tech Potential. Economic Development in Rural America," Center for Urban Policy Research, New Brunswick, NJ (1991).
33. H. Goldstein and M. I. Luger, Science/technology parks and regional development theory, *Economic Development Quarterly*, **4**, 64–78 (1990).
34. W. H. Greene, "Econometric Analysis," Macmillan, New York (1993).
35. Z. Griliches, Issues in assessing the contribution of R & D to productivity growth, *Bell Journal of Economics*, **10**, 92–116 (1979).
36. Z. Griliches, Patent statistics as economic indicators: A survey, *Journal of Economic Literature*, **28**, 1661–1707 (1990).
37. Z. Griliches, The Search for R & D Spillovers, *Scandinavian Journal of Economics*, **94**, 29–47 (1992).
38. G. Grossman and E. Helpman, "Innovation and Growth in the Global Economy," MIT Press, Cambridge, MA (1991).
39. G. Grossman and E. Helpman, Endogenous innovation in the theory of growth, *Journal of Economic Perspectives*, **8**, 23–44 (1994).
40. C. F. Harding, Location choices for research labs: A case study approach, *Economic Development Quarterly*, **3**, 223–234 (1989).
41. H. W. Herzog, A. M. Schlottman and D. J. Johnson, High-technology jobs and worker mobility, *Journal of Regional Science*, **26**, 445–459 (1986).
42. J. R. Howells, The location of research and development: Some observations and evidence from Britain, *Regional Studies*, **18**, 13–29 (1984).
43. A. Jaffe, Technological opportunity and spillovers of R & D: Evidence from firms' patents, profits and market value, *American Economic Review*, **76**, 984–1001 (1986).
44. A. Jaffe, Real effects of academic research, *American Economic Review*, **79**, 957–970 (1989).
45. A. Jaffe, M. Trajtenberg and R. Henderson, Geographic localization of knowledge spillovers as evidenced by patent citations, *Quarterly Journal of Economics*, **108**, 577–598 (1993).
46. P. Krugman, Increasing returns and economic geography, *Journal of Political Economy*, **99**, 483–499 (1991).
47. A. Link and J. Rees, Firm size, university based research and returns to R & D, *Small Business Economics*, **2**, 25–32 (1990).
48. R. Lucas, On the mechanics of economic development, *Journal of Monetary Theory*, **22**, 3–39 (1988).
49. M. I. Luger and H. A. Goldstein, "Technology in the Garden: Research Parks in Regional Economic Development," University of North Carolina Press, Chapel Hill, NC (1991).

50. L. Lund, "Locating Corporate R & D Facilities," Conference Board Report No. 892, The Conference Board, New York (1986).
51. E. Malecki, Federal R & D spending in the United States of America: Some impacts on metropolitan economies, *Regional Studies*, **16**, 19–35 (1981).
52. E. Malecki, Research and development and the geography of high-technology complexes, *in* "Technology, Regions and Policy" (J. Rees, ed.), Rowman and Littlefield, Totowa, NJ (1986).
53. E. Malecki, "Technology and Economic Development," Longman Scientific and Technical, Essex (1991).
54. E. Malecki, and S. L. Bradbury, R & D facilities and professional labour: labour force dynamics in high technology, *Regional Studies*, **26**, 123–136 (1992).
55. E. Mansfield, Academic research and industrial innovation, *Research Policy*, **20**, 1–12 (1991).
56. E. Mansfield, Academic research underlying industrial innovations: Sources, characteristics and financing, *The Review of Economics and Statistics*, **77**, 55–65 (1995).
57. A. Markusen, P. Hall, and A. Glasmeier, "High Tech America: The What, How, Where and Why of Sunrise Industries," Allen and Unwin, Boston (1986).
58. National Science Foundation, "Academic Science and Engineering R & D Expenditures, Fiscal Year 1982," Data obtained from GASPAR data files (1982).
59. R. Nelson, The simple economics of basic scientific research, *Journal of Political Economy*, **67**, 297–306 (1959).
60. R. Nelson, The role of knowledge in R & D efficiency, *Quarterly Journal of Economics*, **97**, 453–470 (1982).
61. R. Nelson, Institutions supporting technical advance in industry, *American Economic Review*, **76**, 186–189 (1986).
62. D. D. Parker and D. Zilberman, University technology transfers: impacts on local and U.S. economies, *Contemporary Policy Issues*, **11**, 87–99 (1993).
63. R. Premus, "Location of high technology firms and regional economic development," U.S. Government Printing Office, Washington, DC (1982).
64. J. Rees, State technology programs and industry experience in the United States, *Review of Urban and Regional Development Studies*, **3**, 39–59 (1991).
65. J. Rees and H. Stafford, Theories of regional growth and industrial location: Their relevance for understanding high-tech complexes, *in* "Technology, Regions and Policy" (J. Rees, ed.), Rowman and Littlefield, Totowa, NJ (1986).
66. P. M. Romer, Increasing returns and long-run growth, *Journal of Political Economy*, **94**, 1002–1037 (1986).
67. P. M. Romer, Endogenous technological change, *Journal of Political Economy*, **98**, S72–102 (1990).
68. P. M. Romer, The origins of endogenous growth, *Journal of Economic Perspectives*, **8**, 3–22 (1994).
69. A. Saxenian, "Regional Advantage: Culture and Competition in Silicon Valley and Route 128," Harvard University Press, Cambridge, MA (1994).
70. R. Sivitanidou and P. Sivitanides, The intrametropolitan distribution of R & D activities: Theory and empirical evidence, *Journal of Regional Science*, **35**, 391–415 (1995).
71. W. Stohr, Regional innovation complexes, *Papers of the Regional Science Association*, **59**, 29–44 (1986).
72. M. Storper and R. Walker, "The Capitalist Imperative: Territory, Technology and Industrial Growth," Basil Blackwell, Oxford (1989).

73. E. Talen and L. Anselin, Assessing spatial equity: An evaluation of measures of accessibility to playgrounds, *Environment and Planning A*, (1997).
74. E. Von Hippel, "The Sources of Innovation," Oxford University Press, New York (1988).
75. J. Weibul, An axiomatic approach to the measurement of accessibility, *Regional Science and Urban Economics*, **6**, 357–379 (1976).
76. H. White, A heteroskedastic-consistent covariance matrix estimator and a direct test for heteroskedasticity, *Econometrica*, **48**, 817–838 (1980).

Part V
Human Capital and the Growth of Cities

[12]

EXPLORATIONS IN ECONOMIC HISTORY **33**, 384–413 (1996)
ARTICLE NO. 0021

The Talk of the Town: Human Capital, Information, and the Growth of English Cities, 1861 to 1961

CURTIS J. SIMON

Clemson University

AND

CLARK NARDINELLI

U.S. Food and Drug Administration

The growth of cities virtually always accompanies modern economic growth. Many observers attribute the relationship to the rise of urban factories or improvements in transportation. We believe that information-based human capital, particularly as embodied in business professionals, provides a better explanation for the growth of cities. Cities grew because concentrated human capital raised productivity. In a study of English cities from 1861 to 1961 we found that cities whose work forces contained high proportions of business professionals grew more rapidly. The talk of the bourgeoisie, not the smoke of the factory, was the defining characteristic of the modern city economy.

Cities and modern economic growth go hand in hand. As Lampard (1955, p. 81) put it, with the coming of the industrial revolution "many towns and villages in Europe and North America ceased to be mere regional markets for craftsmen and cultivators; they became vibrant centers for almost all the manufacturing, servicing, and distributive functions developed in an expanding economy." Kuznets (1966) called urbanization the most conspicuous change in living conditions brought about by modern economic growth. In his classic book on 19th-century cities, Weber (1899, p. 1) wrote that it was a "common observation" that "the most remarkable social phenomenon of the present century is the concentration of population in cities."

The obvious historical correlation between urbanization and modern economic growth underlies most theories of why modern cities grew. Popular explanations for the long-run growth of cities include economies of scale in industry, agglomeration effects, transportation improvements, and external economies of one sort or another.

We believe that the standard stories of city growth leave out an important

0014-4983/96 $18.00

factor—knowledge-based human capital. By knowledge-based human capital, we do not mean broad attributes, such as literacy or average skill—although such attributes doubtless contribute. The human capital that concerns us embodies those skills that have to do with the production and spread of information. We believe that information-oriented professionals, such as brokers, accountants, and lawyers, played a particularly important role in the growth of modern cities. The spread of information, rather than physical production or consumption, is the activity most concentrated in modern cities. People in cities talk; the talk leads to the creation of knowledge.[1] Cities where the "talk is good," meaning that it carries useful information, grow more rapidly than cities where the talk is mostly noise.

The idea that modern cities are largely devoted to information suggests a connection between modern economic growth and urbanization. Kuznets (1966, p. 9) identified "the extended application of science to problems of economic production" as the "epochal innovation" of the modern age. In the modern world, cities spread the useful knowledge so important to production. As the role of knowledge increased, cities necessarily played an ever larger role in the economy and therefore grew ever larger. As Hohenberg and Lees put it:

> [U]rban development is a dynamic process whose driving force is the ability to put information to work. After 1850 the large cities became the nurseries as well as the chief beneficiaries of an explosion in knowledge-centered economic growth. (Hohenberg and Lees, 1985, p. 205)

Because cities differed in their ability to "put information to work," it is not surprising that their rates of growth differed as well.

We found that during the era of modern growth after 1850, those English cities possessing high proportions of business professionals grew more rapidly than the rest. We do not argue that only human capital and information as embodied in business professionals mattered; we mean only that after all the geographical, commercial, political, and industrial factors are taken into account, knowledge-based human capital exerted a systematic influence on the long-run growth of cities. Location and other geographical attributes, such as harbors, rivers, and climate, also played a major role in that growth. Before we turn to the regression evidence, we will first show how the information story fits in with other explanations for the long-run growth of cities.

EXPLANATIONS OF THE GROWTH OF CITIES

The most influential theories of city growth are built on the notions of central places, transportation improvements, manufacturing base, internal economies of scale, and external economies. We shall consider each explanation in turn.

[1] The creation of knowledge cannot be separated from its spread. In a recent article, McCloskey (1994) argued that talk was the main economic activity of towns. The discourse of the town should not, however, be thought of as a simple exchange of information. As city folk talk to each other they create new knowledge.

Central Places

Many urban histories stress the importance of a central place in the geography of economic activity. In their history of European urbanization, Hohenberg and Lees (1985, pp. 47–73) employed a modified version of central place theory based on Clark and Slack's (1976) classification of English towns into market towns, county towns, provincial capitals, and national capitals. According to Hohenberg and Lees, central place theory works best when combined with the notion of cities as elements of trade networks.

Central place theory's strongest selling point is simplicity. Cities have an assigned role and size based on their relationship to the central place. In one version of the model, economic activity on a flat featureless plain generates specialization, which in turn leads to markets for the exchange of surplus output. The growth of cities depends on the growth of their hinterlands and of the urban system itself. Cities surrounded by productive hinterlands and dense trading networks grow more rapidly than cities not so situated.

The historical development of most cities does not reflect the geographical determinism of the simple models. Indeed, Hohenberg and Lees argued that the central place model serves better as a starting point than as a full model of city growth.[2] Central place theory explains the existence of city systems far better than it explains the growth of particular cities.[3] Something more must be added to explain the historical growth of cities.

Transportation

Improved transportation during the industrial revolution made it possible for English cities to grow ever larger. Ports, railroad centers, and major crossroads all grew because of the falling cost of transporting goods and materials to and from transportation centers. Canals and railroads enabled cities to grow in places not reachable by natural waterways. Transportation improvements, then, increased the size of well-situated existing cities and made growth possible in formerly isolated areas, such as the coal fields of Lancashire and Yorkshire.

The notion that transportation led to urbanization is plausible but incomplete. Transportation improvements explain overall urban growth much better than they explain why Bristol grew more rapidly than Liverpool after 1861. When we look at the success of individual ports, a host of factors come into play, including the port's hinterland, its entrepreneurs, and the business and other services provided there. Major port cities served as major information centers as well as transportation hubs. Insurance, banking, and related services originated in ports because of the ready availability of information. Information must have played as large a part in the growth of port cities as did the loading and unloading of physical goods.

[2] See, for example, de Vries (1984). De Vries used a measure of a city's *potential* for interacting with other cities and areas.

[3] For an example of an application of the city-system approach to 19th-century England and Wales, see Robson (1973).

The connection between transportation and other economic activities is more explicit in the case of man-made transportation, such as canals and railroads. Here, transportation improvements followed economic opportunities. As Trumball put it:

> Canals could be constructed to exploit potential opportunities, linking places at will by deliberate, rational economic calculation. This greatly changed and indeed reversed the relationship between water transport and the economy: bulk water transport could be brought to favourable manufacturing sites instead of a forced movement in the opposite direction. (Trumball, 1987, pp. 539–540)

The canal system followed lines dictated by the location of coal and other resources used by the rapidly expanding manufacturing sector. The railroads also followed the existing distribution of economic activity. The role of transportation, then, cannot be considered in isolation from those forces that created the "favorable manufacturing sites."

Manufacturing Base

In the manufacturing base story, one or more basic products create all economic activity in a city or region. All other economic activity rests on basic manufacturing. According to the manufacturing base advocates, during the industrial revolution cities with prosperous manufacturing sectors grew faster, for two reasons. First, as ancillary occupations arose around the manufacturing base, the city grew. Second, as exports of the manufactured good increased, the expanding base pulled other economic activity along with it—a classic case of export-led growth. The manufacturing base model implies that cities that started with proportionally large and successful manufacturing sectors should have grown faster, other things the same.

Export-led growth from manufacturing explains the growth of many 19th-century English cities, particularly in Lancashire and the West Riding. The rapid growth of textile exports transformed the villages and towns of the northern counties into an urban belt surrounding the coal fields. The textile-led urban growth lasted longer and affected more cities than most examples of export-led city growth. Total textile output increased by a factor of 15 between 1800 and 1872 (Deane and Cole, 1969, p. 213); the consumption of raw cotton increased by a factor of 34 over the 19th century (Mitchell and Deane, 1971, p. 179). Although the rate of expansion of the industry slowed during the second half of the 19th century, the movement of factories into cities and the transformation of factory villages into towns made the southeast Lancashire conurbation the fastest growing urban area in Britain.

The growth of the cities where raw cotton and wool became cloth may be history's best example of export-led city growth. Cotton workers alone made up the third largest occupation in the 1861 census, ranking behind only agricultural laborers and domestic servants. Although one can find other examples of export-led city growth (steel and Sheffield, shipbuilding and Sunderland), none were as long-lasting and widespread as that of the textile cities. For that reason, instances of textile-led growth should be regarded as exceptional rather than as typical examples of the phenomenon. Most successful cities had far more diversified industrial structures.

Export-led growth tends to dissipate because after the initial burst associated with the growth of ancillary occupations, the city should grow only as fast as the demand for its export, and demand eventually levels off or declines. The typical S-shaped curve that describes the historical growth of a particular industry, then, should describe city growth as well. In many cases it does; the experiences of many Lancashire cities after 1911 once again supply good examples. Many cities, however, continued to grow long after the original export good had stagnated and declined. The manufacturing base story cannot explain why some cities grew in the very long run.[4]

Internal Economies of Scale

It is possible that the introduction of factories that could take advantage of economies of scale led to the birth of modern industrial cities. Large factories employed many workers, who in turn swelled the population of the cities. Economies of scale imply that cities grew because larger and larger factories settled in them.

We do not accept the internal economies of scale explanation of city growth. One obvious objection is that cities preceded large-scale factory production by several thousand years. That objection is weakened by the fact that industrial cities grew more rapidly and achieved sizes much larger than even the greatest preindustrial cities. As Weber (1899) demonstrated and other scholars, such as Bairoch (1988), restated on a grander scale, the 19th century witnessed an explosion of urbanization to accompany the industrial revolutions sweeping across the western world.

A bigger problem with tying cities to economies of scale is that during the 19th century, when cities grew most rapidly, the scale economies associated with modern industry were modest. In the British cotton industry, for example, the largest firms at mid-century employed about 1500 workers—hardly enough to explain the size of Manchester. Moreover, the largest firms were often in the countryside. Gatrell (1977) found modest scale economies in Lancashire cotton factories in the first half of the 19th century. In Manchester itself the average cotton textile firm had but 260 employees in 1841. The largest on-site employer in 1841 was in Preston, not Manchester, and employed 1422 people (Gatrell, 1977, p. 118). Indeed, Gatrell's data showed that the largest factories got smaller as Manchester itself grew larger. Lloyd-Jones and Le Roux (1980) revised some of Gatrell's findings but supported the basic conclusion on the size of the largest

[4] A city can grow in the long run because of a succession of exports. In such a case, the manufacturing base argument begs the important question of why a particular city proves to be a good location for the succession of industries.

firms. Although average plant size increased in most industries after 1841, small and medium-sized factories continued to be the rule in English cities throughout the second half of the 19th century (Clapham, 1939, v. 2, pp. 114–133). Furthermore, as plant size increased in many industries during the 20th century, the rate of urban growth slowed down. In 1961, no single plant accounted for more than a negligible fraction of the work force in any major city.

The growth of larger and larger establishments, then, contributed little to the growth of English cities after 1861. Large factories scattered over the countryside would lead to some concentration of population, but could not have led to the degree of concentration eventually found in the cities of the 19th and 20th centuries. Internal economies of scale can explain large factories, not large cities. To explain the size of cities, we need to explain why factories, workshops, and other employers located in cities.

External Economies

Although internal economies of scale will not necessarily create great cities, *external* economies might well do the trick. The external economies associated with various industries clearly explain the geographical concentration of various forms of production. The concentration of steel in Sheffield and cotton textiles in Manchester are obvious examples. The growth of intermediate producer services has long been recognized as a form of external economy associated with the geographical concentration of industry (Stigler, 1968, pp. 129–141). An increase in steel production in a region increases the number of specialized producers of auxiliary services for the steel industry, leading to a further increase in steel output. Allen's (1929) study described how workshops producing similar goods concentrated in the various parts of Birmingham and the Black Country. Allen pointed out that the concentration of industry occurred because of the bunching together of workshops, not the establishment of large factories. Allen also emphasized the great increase in productivity brought about by the external economies associated with the concentration of trades.

External economies from manufacturing fit the casual evidence better than the alternative theories of city growth. Even so, the story is at best incomplete, mainly because it encounters the same objection as the export-base story—the fortunes of cities did not mirror the fortunes of their main manufacturing industries. One version of the external economies story, however, will take us considerably further toward understanding why cities grew. That version emphasizes the external economies associated with knowledge. Marshall described the importance of knowledge in the following terms:

> For External economies are constantly growing in importance relatively to Internal in all matters of Trade-knowledge: newspapers, and trade and technical publications of all kinds are perpetually scouting for him and bringing him much of the knowledge he wants—knowledge which a little while ago would have been beyond the reach of anyone who could not afford to have well-paid agents in distant places. (Marshall, 1920, p. 237)

In the above passage it is not clear whether Marshall meant industry-specific or general knowledge. Although both types of knowledge are productive, the external effects of general knowledge provide a better explanation for the long-term growth of a city. We believe that knowledge generates the main agglomeration economies of modern cities.

HUMAN CAPITAL, INFORMATION, AND GROWTH

Growth theorists use human capital and information to explain the economic growth of nations.[5] Productivity growth occurs in those places most capable of innovation. Innovation in turn occurs in those places most capable of using and creating information. At any time, more of the economy is engaged in adopting the latest methods than in creating new methods. The tinkering and slight adjustments necessary to the economic success of an innovation can best be carried out where people can get together and exchange useful information, mainly by talking.

Some recent theories of growth emphasized the external effects of knowledge or human capital. Romer (1986, 1994) based his theory of economic growth on increasing returns from knowledge. Knowledge has the attributes of a public good in that additional use does not exhaust the stock. Knowledge, however, is costly to acquire; although the great ideas and inventions travel across geographical boundaries, the small ideas that make the great ideas useful are often highly local. As two contributors to contemporary growth theory put it:

> While it may appear to the casual observer that knowledge always flows rapidly and costlessly around the globe, the reality is sometimes different. The concentration of high-technology industries in particular locations such as the Silicon Valley and Route 128 suggests that some benefit exists from physical proximity to other researchers. Perhaps this is because new ideas are spread by skilled personnel whose geographic mobility is somewhat restricted, or because firms that are geographically close are exposed more often to the products of their nearby rivals.
>
> The existence of local or national technological externalities introduces an important role for history in the determination of dynamic comparative advantage. Such spillovers can generate a self-perpetuating process whereby an initial lead, however generated, is sustained indefinitely into the future, regardless of a country's relative factor endowments. (Grossman and Helpman, 1994, pp. 39–40)

Some of the knowledge generated in Manchester might have as much of an effect on Birmingham as on Manchester itself. For much knowledge and information, however, the benefits will be largely confined to the city of origin. The small bits of information necessary to turn an idea into a working machine could not be assembled in most places. There were far more cities where people understood the

[5] Growth theories incorporate human capital in a number of ways. It is not our purpose to support any one proposed means of transmission. Several of the new theories would serve equally well—as would several "old" growth theories.

principle of the steam engine than there were cities where people knew how to make one.

In Lucas's (1988, 1990) model of growth, the amount of human capital in a society affects the productivity of each worker, apart from the worker's individual human capital. The acquisition of human capital by individual workers generates external effects, or spillovers of knowledge. The spillovers increase with the concentration of human capital. It is plausible that such external economies are strongest within a city, because the costs of acquiring and exchanging information—particularly small bits of information—are lower in cities. If human capital generates external economies, economies whose industries use human capital more intensively should grow faster, all else the same.

Lucas, citing the diamond and advertising districts of New York, suggested that the very existence of cities indicated that they were "mainly and convincingly concerned with the external effects of human capital" (Lucas, 1988, p. 37). Cities reduce the costs of collecting and spreading information. The ability to meet with other people at low cost is important because much economic activity requires face-to-face contact. As McCloskey put it, much of what white collar workers (the "bourgeoisie") do is talk; the "bourgeoisie works with its mouth, and it depends on word of mouth" (1994, p. 186). The creativity of the market economy—the increasing returns so important in modern growth theory—in large part arises from what happens when people with information get together and talk. The talk is necessary to turn information into productive knowledge. The so-called communications revolution may someday eliminate the need for face-to-face talk, although it has certainly not yet done so. In the England of 1861 or 1911, no close substitute for talk existed.

Loans provide a good example of the economic value of talk. Lending and borrowing ought to be a purely numerical transaction, requiring only the verification of earnings, collateral, and other tangible financial factors. Yet face-to-face meetings accompany virtually all significant loans. The major exception, credit card loans, proves the rule; the high interest typically charged for such loans reflects their anonymity. McCloskey might argue that face-to-face meetings enable borrowers to persuade lenders that they are worthy of trust.

In cities, firms produce goods and services using city-specific knowledge in addition to general knowledge and the more traditional factors of production.[6] Much of the knowledge has the attributes of a public good. As new knowledge is created, productivity in the city rises. The productivity-enhancing effects of new knowledge will be greater, and the productivity of all factors of production will be correspondingly higher, the more people are involved in the production and dissemination of information. Cities with relatively large initial concentrations of human capital should grow more rapidly, because the concentration increases productivity, which induces migration into the city. Furthermore, concentrations

[6] The benefits of human capital could also arise because nonrivalrous factors of production, such as methods and ideas, can only be exploited by those with the required human capital.

of people with high levels of human capital produce new knowledge at a faster rate, which further raises productivity in the city, leading to still more in-migration.

If there were no costs associated with agglomeration, increasing returns to information would eventually lead to all production being carried out in one or a few large cities. The costs of congestion, however, limit the growth of the city. Indeed, the bad spillover effects of population concentration grow in much the same way as the good spillover effects, increasing as population increases.[7] The Newcastle newspaper proprietor Joseph Cowen aptly described the trade-off between the two effects in 19th-century England:

> When people declaim in doleful numbers against the noise and dirt of the busy centers of population, they should remember the liberty we enjoy as a consequence of the mental activity and enterprise which have been generated by the contact of mind with mind brought together in great towns. (Quoted in Briggs, 1965, p. 65)

The "noise and dirt" eventually slow down the growth generated by the contact of "mind with mind."[8]

Human capital and information did not accumulate in certain cities as a result of deliberate policy. The city's initial economic conditions, not conscious policies, largely determined the long-run outcome. Cities built around industries that employed (directly or indirectly) human capital and information grew more rapidly than cities built around low human capital industries.

The human capital and knowledge spillover stories of overall economic growth, then, can be turned into similar stories of city growth. It is important to remember that the definition of growth differs in the two cases. At the level of the nation, the growth of real income per person indicates economic success. Within a nation, however, migration tends to make real wages and incomes equal across cities—at least in the long run. Over periods of 50 years or more, one should look at quantities (that is, population) rather than prices (real income in this case). The measure of a city's success, or its economic growth, must be its rate of population growth, not its level of income per person.[9]

The history of cities lends much support to the notion that information explains

[7] In Appendix 1, we sketch an algebraic model of city growth based on the productivity of information and the congestion costs of population growth. For a historical description of the growth of the information-based modern city, see Hohenberg and Lees (1985), pp. 199–214.

[8] Williamson (1990) provided estimates of the cost of noise and dirt. He estimated wage premiums of 8 to 30%. The upper range was much smaller than the increased productivity of workers in the cities. With the productivity effect stronger than the crowding effect, cities grew rapidly despite the absolute level of urban disamenities. Williamson's results are consistent with the notion that English cities generated information-based (or other) positive external effects. Otherwise, most production would have taken place away from cities.

[9] We do not suggest that real income per person will be identical across cities. Even in a steady state long-run equilibrium, measured real incomes will differ with the demographic and other characteristics of cities.

a large part of city growth.[10] Smith (1984) explained Amsterdam's rise to prominence in the 17th century as the result of its role as an information exchange. The leading cities of later centuries followed the pattern set by Amsterdam. We now take it for granted that the largest city in a region or nation will also be the headquarters of information-intensive industries such as publishing, fashion, and finance. By contrast, the main export of a region may or may not be concentrated in its largest city. Manchester's share of cotton textile employment and output fell steadily throughout the 19th century. By 1901, the combined population of Blackburn, Bolton, Burnley, Oldham, and Preston was about the same as that of Manchester, but each city employed more textile workers than Manchester itself.

The well-documented correlation between inventiveness and urbanization provides other evidence that cities grew because they dominated the exchange of information. Several studies of American patents, for example, found higher frequency of patents in urban than in rural areas.[11] Sokoloff (1988) took the analysis a step further and showed the frequency of patents to be greater in major than in minor urban areas in early industrial America. The greater amounts of information in cities apparently generated higher levels of patents per person.

The creation of new knowledge and the dissemination of existing information are closely related. Cities that are not part of the flow of existing information are unlikely to be prime locations for the creation of useful knowledge. Without innovation—new knowledge—the information flowing within and between cities would soon grow stale and its effects on city productivity growth would probably disappear. It is therefore impossible to say whether the flow of information or the creation of new knowledge contributes more to overall urban growth. It is also difficult to say which plays a larger part in the growth of individual cities. Although the creation of new knowledge may cause faster growth within a city, new knowledge is not necessarily created in the fastest growing cities. We suspect, but cannot demonstrate, that the growth of a particular city depends more on the flow of information through and within the city than on the amount of new knowledge actually created within its borders.

The history of English cities since the industrial revolution provides further evidence that information-spreading talk gave modern cities their productive advantage. We now look in some detail at the growth of English cities in the century after 1861.

THE DATA: ENGLISH CITIES IN 1861 AND 1911

The 1861 Base Year

We gathered the data on occupations in English and Welsh cities from the published censuses. For 1861, we used the 79 cities and towns for which the

[10] Hohenberg and Lees described various ways that cities had advantages in "information-intensive production" (1985, p. 200).

[11] Higgs (1971) showed that the frequency of patents in American states increased substantially when the proportion of the population living in cities increased. The greater availability of information accounted for at least part of the greater inventiveness of city dwellers.

TABLE 1
Selected Descriptive Statistics on English and Welsh Cities

	Mean	Standard deviation	Minimum	Maximum
Growth rates (percent per year)				
1861–1911:79 Cities	1.23	0.78	−0.074	3.42
1861–1961:79 Cities	0.86	0.39	−0.35	2.20
1861–1961:96 Cities	0.97	0.66	−0.35	4.01
1911–1961:126 Cities	0.46	0.62	−1.22	2.19
1911–1961:92 Cities	0.38	0.64	−1.22	2.19
Explanatory variables: 1861 base year (fraction of labor force, except rail lines)				
Business professionals	0.0067	0.003	0.002	0.018
Professionals	0.0145	0.006	0.0048	0.0299
Water transportation	0.016	0.024	0.000075	0.108
Railroads	0.005	0.003	0.00086	0.0152
Telegraphs	0.0002	0.0002	0	0.00094
Crafts and construction	0.061	0.018	0.024	0.157
Textiles	0.059	0.097	0.004	0.38
Mining and manufacturing	0.042	0.04	0.012	0.259
Railroad lines	2.582	1.429	0	7
1911 base year, all cities (fraction of labor force, except rail lines)				
Business professionals	0.021	0.013	0.008	0.095
Professionals	0.025	0.011	0.012	0.066
Communications	0.005	0.005	0.001	0.028
Railroads	0.029	0.019	0.006	0.105
Roads	0.029	0.0085	0.01	0.057
Water transportation	0.011	0.019	0	0.093
Mines	0.032	0.089	0	0.659
Metals	0.089	0.082	0.011	0.326
Vehicles	0.026	0.037	0.002	0.238
Pottery	0.008	0.027	0	0.218
Crafts	0.007	0.007	0.001	0.05
Construction	0.063	0.017	0.034	0.11
Chemicals	0.01	0.01	0.001	0.057
Textiles	0.06	0.129	0	0.56
Railroad lines	2.128	1.438	0	7

published census contained separate data on occupations.[12] We used later census data on those same cities to compute the population growth rates for the subsequent 50 and 100 years.[13] We recognize the problems inherent in using the census definition of the city. An ideal measure would be to define the economi-

[12] Appendix 2 lists the 79 cities for which the *Census of England and Wales 1861* gave detailed information on occupations of the people. The sample of cities looks forward rather than backward. The census included those cities that were large or important in 1861; we then studied how they grew subsequently. The minor towns of 1861 that became large later were excluded.

[13] Table 1 contains descriptive statistics on city growth rates and occupations.

cally meaningful "city," which in some cases would correspond to the census definition of the city, in others to a metropolitan area, and in still other cases to a portion of the official city. The impossibility of drawing a satisfactory line between city and not-the-city led us to settle for the census boundaries. For the period we looked at, 1861 to 1961, official city size probably made a decent proxy for the true city. The suburban push, the rise of conurbations, and direct state interference with urban patterns all accelerated after 1961.

Boundaries changed for many cities over the century. Extensions of city boundaries posed a problem.[14] If a city attracted people, the small extensions of boundary formed part of the population growth to be explained and should not be excluded. In other cases, however, annexations may have represented more of a political than an economic change. In our statistical work, we allowed city boundaries to change over time. With two exceptions the changes were small. The two exceptions, Birmingham and Coventry, both grew much larger than our hypothesis predicted. In leaving the population figures alone for those two cities, we chose the method biased against our story.

For the most part, we used occupations to characterize the economic structure of cities. The occupational census data from 1861, as Lee (1994) pointed out, classified many workers according to the industries that employed them rather than according to their true occupations.[15] Other practices, such as including the professional's assistants or clerks, led to over-counting in some professions. We can only trust that the broad trends we tried to identify will show through despite the flawed classifications.[16]

The story we tell about city growth involves people exchanging information; we therefore needed to identify occupations that embodied information-intensive human capital. As Mitch (1992, 1993) demonstrated, relatively few 19th-century occupations required literacy. Those few occupations, however, may have been responsible for a disproportionate amount of the creation and spread of information in the cities. We tried two measures of information-intensive professionals for 1861: (1) business professionals; and (2) the general census category of professionals, with religious employment excluded. The business professional category, which accounted for about two-thirds of 1% of the labor force, included bankers, brokers, insurance agents, accountants, auctioneers, commercial travelers, and miscellaneous other workers. The general professional category, which accounted for about 2.5% of the labor force, included lawyers, doctors, dentists, druggists,

[14] The following cities annexed significant new areas in the first half of the 20th century: Birmingham, Coventry, Middlesbrough, Plymouth, Southampton, Stockport, Swansea, Oxford, Sheffield, Sunderland, Wolverhampton, Huddersfield, and Leicester.

[15] The census characterized occupations according to (1) skill, talent or intelligence used; (2) tools, instruments, machinery, and structures used; (3) materials worked; (4) processes of manufacturing; and (5) products created. The first criterion interests us most. According to the *Census of England and Wales 1861:* General Report, p. 30, in cases where the criteria differed, "materials worked" took precedence.

[16] For fuller discussions of the problems of occupational classifications in the 19th-century censuses, see Lee (1979) and Wrigley (1972).

actors, artists, teachers, musicians, writers, engineers, and scientists.[17] Neither occupational group—general professionals or bankers and brokers—precisely corresponded to the number of workers who created and spread useful information in the city. We can only hope that our categories were highly correlated with the true information sector. One indication that our occupational measures were not too far off as a measure of human capital comes from later censuses, for which the data on education can be compared to data on occupational and social classes. In 1951, the correlation for 157 cities between high-status occupations (business and general professionals as well as other occupations requiring high levels of human capital) and the proportion of the work force leaving school before age 15 was -0.862 (Moser and Scott, 1961).

Telegraph employment gave us a plausible proxy for the spread of information. Although telegraph employees themselves were not the high human capital workers suggested by our hypothesis, the use of telegraphs might have been highly correlated with information spillovers and human capital in the city. Telegraph employment was closely related to information movement itself, whereas professional employment or literacy represent only the potential flow of information.[18]

We also collected data on employment in manufacturing and transportation. We divided manufacturing into five categories: mining and manufacturing, crafts and construction, textiles, dress, and processing. Mining and manufacturing employments included mining, iron and steel, metals, glass, pottery, and various miscellaneous manufacturers. Crafts and construction included instruments, precision tools, watches, clocks, shipbuilding, the building trades, and a host of minor occupations. Textiles included all cloth but excluded dress and other clothing, which we included as a separate category. Processing included working with food and drinks, animal substances, and vegetables. We divided transportation into three categories: railroad, roads, and water transportation. Roads included providers of all overland transportation except railroads. Water transportation mainly included employments associated with ports, canals, and docks.

The importance of railroads in the 19th century led us to construct an alternative to employment as a measure of a city's railroad connections. We chose the number of rail lines radiating from the city in order to indicate a city's place in the transportation, and by implication, the urban hierarchy. We counted the lines, using Ferneyhough (1975) and White (1969).[19] Our method was wholly visual; we matched lines to cities on maps. Because the maps referred to different years, we did not construct it specifically for the year 1861 but for the mid-19th century.[20] We also found it impossible, partly because of the increasing complex-

[17] Mitch (1992, pp. 11–42, 213– 214), for the most part, included both of our categories as well as clergy in his measure of occupations that required literacy.

[18] The actual number of messages carried would be preferable as an index of information flow through a city, but we found no data on messages.

[19] We also consulted Kellett (1969), but it provided little that was useful for our purpose.

[20] We used maps for different years, although all were from the middle years of the century.

ity of the system, to duplicate the method for the early 20th century rail system. The railroad system as established by the late 19th century served as the railroad variable in the regressions we report below.

We intended for manufacturing occupations to capture the effects of the manufacturing base and economies of scale. Transportation stories of city growth usually emphasize ports or railroads. Because ports and railway centers may also have played major roles in the exchange and creation of information, employment in those categories may reflect more than transportation. With the data available, it was difficult to capture other basic services. Although in a city with good sewage and internal transportation congestion costs would rise less steeply and—all else the same—population would grow more rapidly, we could not separate employments that corresponded to infrastructure from similar occupations. As a proxy for these services (or infrastructure), we tried employment in waterworks.

As a way to pick up the effects of location, we included regional variables. The hinterlands of a city helped to determine its rate of growth; the regional variables should pick up some of the hinterland effect. The regions as defined by the 1861 census included London, South Eastern, South Midlands, Eastern, South Western, West Midlands, North Midlands, North Western, Yorkshire, Northern, and Wales (with Monmouthshire).

The 1911 Base Year

For the 1911 base year, we collected data on 126 English and Welsh cities.[21] As before, we relied on the published census. Although the changing occupational categories in the data made it impossible for us to replicate the categories used in 1861, the categories were nonetheless close enough to make comparisons possible. The larger number of cities in the 1911 base year reflected the increasing absolute size of the urban population. Many of the classification problems that plagued the 1861 occupational census had been eliminated by 1911, which gave us more confidence in the results.

Several differences between the occupational data in the two base years must be noted. The business professional category remained more or less the same, though the count was no doubt more accurate. The professional category in 1911, however, included clergy.[22] The category also continued to be a catch-all that included many occupations that could not be described as professional. The census itself warned about the use of "professionals":

> Little value for statistical purposes can, however, be attached to the Order as a whole, having regard to the dissimilarity of the professional occupations which it comprises and to the fact that assistants and subordinates who would not ordinarily be regarded as professional

[21] We took the data on occupations from Table 15 of the *Census of England and Wales 1911:* Occupations and Industries. Table 1 contains the descriptive statistics for the 1911 base year data.

[22] The eight professional categories were clerical, legal, medical, teaching, literary–scientific–political, engineers and surveyors, art–music–drama, and exhibitions.

persons are included. (*Census of England and Wales 1911:* Occupations and Industries, Part 1, p. xviii)

Differences across cities in this imperfect measure may nevertheless have led to some differences in growth.

We expanded the industrial category to include new industries, such as chemicals and motor vehicles. The industrial occupations in general became more subdivided. The industrial occupations we included were census categories IX (mines and quarries), X (metals, machines, implements, and conveyances—we split conveyances off into a separate category called "vehicles"), XI (precious metals, watches, instruments, and games—we call the category "crafts"), XII (building and construction), XIV (brick, cement, pottery, and glass), XV (chemicals), XVIII (textiles). The transportation categories, railroad, road, and water transportation, were the same as those of 1861. The railroad lines measure was the same as that used in the earlier period, although we included more cities. We replaced telegraph employment with "communications" employment, which included telegraphs, telephones, and other involved in the conveyance of messages. Because it excluded some government communications workers and included workers engaged in carrying people and goods, the communications category for 1911 was inferior to the telegraph category of 1861 as an indicator of the flow of information into and out of a city.

We adopted the census's changed definition of regions. The 10 regions now included Northern, East and West Ridings, North Western, North Midland, Midland, Eastern, Southern, South Western, Wales, and London and South Eastern. The regional classification for 1911 more closely reflected economic ties than geographic proximity.

WHAT THE GROWTH REGRESSIONS SHOW

Our main city regressions covered the periods 1861–1911, 1861–1961, and 1911–1961. We report one regression for 1861–1911, three for 1861–1961, and two for 1911–1961.[23] We also ran but do not report some regressions covering the years 1891–1961 and 1901–1961. For a further check on our results, we compared the regressions with similar regressions for American cities.

1861 to 1911

The basic regressions for English cities took the form

Annual Rate of Population Growth = Constant + β_1 1861 City Size + β_2 Railroad Lines + β_3 Business Professional Share + $\Sigma_i \eta_i$ Other Employment Shares + $\Sigma_j \gamma_j$ Regional Dummies + error.

[23] We ran dozens of other regressions that contained subsets or combinations of the explanatory variables. For the main variables, the results differed little from one specification to another. We therefore decided to report only those regressions that contained all of the main occupations and regions.

The other employment shares included general professionals, textiles, dress, mining and manufacturing, crafts and construction, processing, waterworks, water transportation, telegraphs, and roads. Regressions with railroad employment rather than railroad lines produced similar results, so we do not report them. We ran but do not report regressions that included financial employment and lawyers as separate categories. None of the unreported variables explained any city growth, either separately or in combination with other variables. The beginning-of-period size of the city captured the effects of location and earlier political events that could have caused the city to be large by 1861.[24] The size and employment variables were all transformed by dividing through by their standard deviations. The reported regression coefficients show the effect on the growth rate of a one-standard deviation change in the independent variable.[25]

Table 2, column 2 contains the regression results for the 79 cities for 1861 to 1911. The strength of the textile variable reflected the role of the textile industry in shaping the urban map of Lancashire and the West Riding. Although the heroic era of the textile industry had passed, it continued to attract population and other occupations. Mining and other manufacturing, lumped together in the regressions, proved weaker, perhaps because the decline of an export product was as common as the rise. One of the more surprising results was the strong effect of crafts and construction employment. Although the share of crafts workers might be a proxy for something else, we suspect that the workers themselves may have mattered. Instrument makers, for example, possibly generated important external economies. A high proportion of the work force in building trades may have indicated a rapid expansion of private infrastructure.

We found that employment on ports and docks also explained much of city growth after 1861, although other types of transportation—road and railroad—employment did not. We believe that the information hypothesis may explain the difference between railroads and ports. Ports and docks—particularly ports—served as more important information centers in a British economy that dealt increasingly in world markets. The coefficient on the employment share in telegraphs, which reflected the actual flow of one type of information, proved large.

As our theory led us to expect, cities with relatively large populations of business professionals in 1861 grew relatively rapidly over the next 50 years. We

24 The effect of beginning-of-period size is ambiguous. On the one hand, cities in particularly good locations or with particularly desirable amenities might be expected to grow more rapidly than smaller cities. On the other hand, cities with large populations might suffer from higher congestion costs and housing prices, which would tend to make them grow more slowly than smaller cities. Rather than indicating the unimportance of initial city size, the weak performance of the variable in many of our regressions may reflect these offsetting effects.

25 The t statistics were all calculated using heteroscedasticity consistent standard errors. The standard errors were transformed as

$$\frac{\hat{u}_t^2}{1 - \hat{H}_t},$$

where $\hat{H}_t = X_t(X'X)^{-1}X_t'$. See Davidson and MacKinnon (1993), p. 554.

ran many regressions with varying degrees of success; the regression coefficients on business professionals varied with the specification, but the results differed little from those reported. As Table 2 shows, a one standard deviation increase in the share of business professionals in 1861 increased the annual growth rate over the next 50 years by around 0.38 percentage points. With an average growth rate of 1.23%, business professionals clearly mattered to the success of the city. Cites with proportionally large professional classes, such as Newport and Gloucester, grew rapidly; cities with proportionally small professional classes, such as Stafford, Chatham, and Rochester, did not.

1861 to 1961

We used two sets of cities and explanatory variables for the 100-year regressions. One set of regressions used the same 79 cities and the employment shares from 1861. Those regressions are reported in Table 2, columns 3 and 4. Table 3 contains regression results for 96 cities with employment shares from 1911. The regressions with 1861 employment shares took the same form as those for the 1861–1911 period. The coefficient on business professionals implied that a one standard deviation increase in employment increased the annual growth rate by around 0.14 to 0.23 percentage points, less than for 1861–1911 but still a good-sized fraction of the average growth rate of 0.86%. The effect was larger in the regression that excluded railroad lines.

The regressions using employment shares from 1911 differed from those using the beginning year shares. The greater availability of data for 1911 enabled us to expand the sample to 96 cities. The categories, as we explained above, differed from those of earlier years, which may explain some of the differences between the results of Tables 2 and 3. The biggest qualitative changes were in the effects of textiles and communications. Textile employment appeared to have no effect on the growth rates of the cities in Table 3. Communications employment, for reasons we can only guess at, was negatively associated with growth.

The employment of business professionals—the bourgeoisie—appeared stronger in the regressions based on 1911 data. A one standard deviation increase in their employment share was associated with approximately a 0.64 percentage point rise in the growth rate. With the average growth rate at 0.97%, the effect of business professionals was significant by any criterion. If we regard the midpoint as representing a kind of average value, we can see that the cities where accountants, brokers, bankers, and others could congregate and share information enjoyed a decided advantage over other cities. Blackpool and Bristol were examples of successful cities with relatively large business professional classes.

1911 to 1961

We ran regressions on city growth over the years 1911–1961 based on the occupational data from the 1911 census and the earlier railway measure extended

TABLE 2
The Growth of English Cities, 1861 Base Year

Independent variables	1861–1911	1861–1961	1861–1961
Constant	1.753	0.837	1.260
	[4.7]	[3.2]	[7.6]
1861 city size	−0.756	−0.484	−0.481
	[−1.6]	[−2.1]	[−2.0]
Railroad lines	0.048	0.116	
	[0.6]	[2.2]	
Employment shares			
Business professionals	0.383	0.143	0.228
	[2.5]	[1.7]	[3.2]
Professionals	−0.161	0.006	0.006
	[−1.0]	[0.1]	[0.1]
Telegraphs	0.170	0.080	0.134
	[2.8]	[1.2]	[2.3]
Roads	−0.014	−0.002	−0.062
	[−0.1]	[0.0]	[−0.9]
Water transportation	0.296	0.209	0.172
	[1.9]	[2.4]	[2.1]
Mining and manufacturing	0.147	0.056	0.072
	[1.1]	[0.7]	[0.8]
Crafts and construction	0.273	0.142	0.195
	[3.0]	[2.6]	[3.9]
Textiles	0.450	0.258	0.303
	[2.8]	[2.9]	[3.0]
Dress	0.209	0.094	0.119
	[2.0]	[1.7]	[1.9]
Processing	−0.115	−0.69	−0.090
	[−0.8]	[−1.3]	[−1.8]
Waterworks	0.070	0.061	0.074
	[0.9]	[1.3]	[1.5]

to more cities. We used two different groups of cities: 126 cities for which we had 1911 and 1961 data, and a subgroup of 92 cities with populations over 50,000 in 1911. The rationale behind the second group was to take the leading cities of 1911 and ask which ones grew more rapidly over the next half-century. Although the cities were different, the method used to select the last group corresponded more closely to that used to select the 1861 group.

As can be seen in Table 4, the results differed for the two groups of cities. Before we look at the differences, we note the similarities. For all regressions, the negative coefficients on textile employment indicated that the leading textile cities (in relative terms) declined steadily in the twentieth century—as is well known. Most industrial employments appeared to have negative or negligible effects on growth. Construction, which is forward-looking, was the exception. Canal, dock, and port employment continued to exert a positive influence on growth. In the regressions with regional variables, all but Wales were always positive, with the

TABLE 2—*Continued*

Independent variables	1861–1911	1861–1961	1861–1961
Regions			
London	4.570	3.170	3.44
	[1.3]	[1.7]	[1.8]
Northern	−1.237	−0.658	−0.751
	[−3.2]	[−2.8]	[−2.9]
Yorkshire	−0.774	−0.461	−0.682
	[−1.7]	[−1.6]	[−2.4]
North Western	−1.248	−0.870	−0.872
	[−3.4]	[−4.0]	[−3.4]
North Midland	−0.264	−0.267	−0.321
	[−0.6]	[−1.1]	[−1.2]
West Midland	−0.917	−0.075	−0.282
	[−2.1]	[−0.3]	[−1.1]
Eastern	−0.662	−0.301	−0.497
	[−2.1]	[−1.3]	[−2.2]
South Western	−0.868	−0.230	−0.459
	[−2.3]	[−1.0]	[−2.1]
South Midland	0.148	0.349	0.239
	[0.3]	[1.3]	[1.0]
Wales	−0.816	−0.403	−0.633
	[−1.6]	[−1.2]	[−1.7]
Regression statistics			
R^2	0.50	0.54	0.52
Observations	79	79	79

Note. The dependent variable is the annual rate of population growth. The numbers in brackets are *t* statistics calculated using heteroscedasticity-consistent standard errors. The independent variables are normalized using their standard deviations.

Midland regional variable easily the strongest.[26] Because London and the South East was the missing region, the regional results indicate a movement of population away from the capital city and its surrounding areas. This result contrasted with that for 1861 to 1911, when London and the South East (the missing region) enjoyed a clear advantage over the rest of England and Wales.

Both sets of regressions in Table 4 indicated that business professionals exerted a strong positive effect on the growth of cities. The coefficients from the 126-city regression imply that a one standard deviation increase in business professionals in the city's work force increased the rate of growth by 0.23 percentage points, which is large when compared to the mean city growth rate of 0.46%. For the 92-city group, a one standard deviation jump in the employment share increased the growth rate by about the same percentage. In regressions not reported in the table, we varied the combinations of control variables with relatively little effect

[26] The Midland counties of 1911 were Herefordshire, Shropshire, Staffordshire, Warwickshire, and Worcestershire. The strong positive coefficient on the Midland reflected the rapid growth of Birmingham, Coventry, and some nearby cities.

TABLE 3
The Growth of English Cities, 1861 to 1961:
1911 Base Year

Independent variables	
Constant	1.023
	[5.94]
1861 city size	1.800×10^{-2}
	[0.2]
Railroad lines	9.000×10^{-3}
	[0.1]
Employment shares	
Business professionals	0.640
	[9.4]
Professionals	−0.153
	[−1.4]
Communications	−0.220
	[−1.7]
Roads	−0.058
	[−0.6]
Water transportation	0.216
	[2.3]
Mines	−0.068
	[−0.8]
Metals	0.171
	[1.8]
Vehicles	0.090
	[1.2]
Pottery	0.107
	[2.7]
Crafts	0.051
	[0.8]
Construction	0.343
	[2.9]
Chemicals	−0.020
	[−0.3]
Textiles	0.063
	[0.6]

on the coefficients on business professionals. Although we do not report them, the coefficients of interest did not change when railroad employment replaced railroad lines. The conclusion that business professionals greatly contributed to city growth in the half-century after 1911, then, is robust.

From our standpoint, the changed effect of the general professional category stands out as an important difference between the 1911 to 1961 regressions and those for earlier years. The regressions imply that after 1911, professionals—teachers, physicians, engineers, lawyers—started to have a positive effect on the growth of cities. In (unreported) regressions where the cities with the strongest business professional effects were excluded, the general professional class coeffi-

TABLE 3—*Continued*

Independent variables	
Regions	
Northern	−0.149
	[−0.5]
East & West Riding	0.181
	[0.6]
North Western	0.077
	[0.2]
North Midland	0.122
	[0.3]
Midland	0.037
	[0.1]
Eastern	0.028
	[0.1]
Southern	0.328
	[1.6]
South Western	−0.254
	[−1.4]
Wales	−0.395
	[−1.1]
Regression statistics	
R^2	0.57
Observations	96

Note. The dependent variable is the annual rate of population growth. The numbers in brackets are *t* statistics calculated using heteroscedasticity-consistent standard errors. The independent variables are normalized using their standard deviations.

cients became even stronger.[27] Because the result occurred only in the regressions with some cities excluded, it would be hasty to conclude that the change in regressions indicated a change in the economy.

Although we do not report them here, we also ran regression similar to those in Table 4 for 62 cities over the period 1901 to 1961 and for 94 cities over the period 1891 to 1961.[28] The results of those regressions were quite similar to those in Table 4 for the 92 "large" cities. These results further increase our belief in the growing relative importance of the information occupations—the talkers. The strong effect of *both* professional categories indicates the growing importance of human capital and information in the British economy during the first half of the 20th century.

[27] In the regressions that excluded the cities that most strongly supported the role of business professionals, the coefficients on general professionals increased and gained statistical significance.

[28] For the 1901–1961 regressions, we used data on occupations from the 1901 census. For the 1891–1961 regressions we used 1911 data on occupations.

TABLE 4
The Growth of English Cities, 1911 to 1961: 1911 Base Year

Independent variables	All cities	"Large" cities
Constant	0.113	−0.026
	[0.8]	[−0.1]
1911 city size	0.063	0.108
	[0.6]	[1.5]
Railroad lines	0.088	0.049
	[2.0]	[1.0]
Employment shares		
Business professionals	0.233	0.200
	[2.9]	[1.8]
Professionals	0.045	0.21
	[0.4]	[1.3]
Communications	−0.318	−0.316
	[−3.4]	[−2.3]
Roads	−0.130	−0.158
	[−2.2]	[−2.1]
Water transportation	0.255	0.254
	[3.4]	[2.7]
Mines	−0.080	−0.089
	[−1.6]	[−1.2]
Metals	0.002	0.027
	[0.0]	[0.4]
Vehicles	0.054	0.101
	[0.6]	[1.1]
Pottery	0.013	0.021
	[0.2]	[0.4]
Crafts	−0.021	−0.017
	[−0.3]	[−0.2]
Construction	0.236	0.266
	[2.6]	[2.0]
Chemicals	−0.086	−0.021
	[−1.7]	[−0.3]
Textiles	−0.146	−0.075
	[−1.9]	[−0.6]

Comparison With American Results

English cities with greater initial proportions of the work force in high human capital, information-based occupations grew faster, other things the same. The relationship should hold for cities in other, similar advanced industrial countries. To see if the relationship held for American cities between 1900 and 1960, we collected data on professional employment from the censuses on occupations in American cities in 1900. Professional employment included lawyers, journalists, bankers, insurance workers, architects, electricians, engineers, and scientists—about 2.5% of the 1900 labor force.

As in the English city regressions, we used occupations to control for other factors that affected growth. Other explanatory variables included railroad employ-

TABLE 4—*Continued*

Independent variables	All cities	"Large" cities
Regions		
Northern	−0.060	0.082
	[−0.3]	[0.2]
East & West Riding	0.315	0.601
	[1.6]	[2.1]
North Western	0.221	0.501
	[1.0]	[1.8]
North Midland	−0.032	0.404
	[−0.2]	[1.1]
Midland	0.640	0.996
	[3.2]	[2.5]
Eastern	0.348	0.308
	[1.7]	[0.8]
Southern	0.246	0.449
	[0.9]	[1.3]
South Western	0.213	0.422
	[1.3]	[1.4]
Wales	−0.360	0.120
	[−1.5]	[0.4]
Regression statistics		
R^2	0.54	0.58
Observations	126	92

Note. The dependent variable is the annual rate of population growth. The numbers in brackets are *t* statistics calculated using heteroscedasticity-consistent standard errors. The independent variables are normalized by their standard deviations.

ment, manufacturing employment, and employment by telegraph and telephone companies. We also included the size of the city in 1900 as an explanatory variable. Finally, because the United States center of population moved steadily westward and southward, we included regions as variables.

The regressions for the United States produced positive coefficients on professional employment share, indicating that cities with high human capital and information indeed grew faster. The professional variable used in the 1900 base-year regressions for American included occupations from both of the professional categories in the 1911 English regressions. As with the English results, the 20th century appeared to be the information century. Regression results from different economies and different starting years support the notion that, all else the same, the more information-intensive a city, the more rapid its population growth.

CONCLUSION

As Hayek put it, "the economic problem of society is mainly one of rapid adaptation to changes in the particular circumstances of time and place" (1945, p.

524). Hayek believed that although only the "man on the spot" can know the particular circumstances of time and place, he cannot decide on the basis of his limited knowledge. "There still remains the problem of communicating to him such further information as he needs to fit his decisions into the whole pattern of changes of the economic system" (Hayek, 1945, p. 525). Hayek stressed the role played by prices in the spread of useful information; we stress the role of the talk of the bourgeoisie.

Data from the occupational censuses of cities to some degree explained the growth of English cities over periods of 50 to 100 years. We found that, in addition to such conventional sources of growth as ports, textiles, and other manufacturing, human capital played an important role in the long run. The providers of information did not include a large proportion of city population, but their presence may have made the difference between fast growth and stagnation. An early and perhaps chance investment in information services and the professionals who provide them gave a city a continuing advantage over its rivals well into the 20th century. Professional employment mattered because cities are where people talk. Much of the talk no doubt contributed little to productivity. The more business professionals in the city, however, the more likely that the talk of the town led to the innovations that increased productivity.

We cannot predict whether the relationship between information and the growth of cities will last. Although the growing importance of information in production implies that the importance of those who spread it will steadily increase, it does not necessarily follow that cities will play as large a role. Many people believe that in the age of electronic information the face-to-face transfer of information—what made cities useful in the first place—will become obsolete. Indeed, much of the recent technology aims to duplicate the ordinary face-to-face talk of the coffee houses and central business districts of the past. In our opinion, however, the modern economy has a long way to go before electronic communication can replace old-fashioned talk. For some purposes, face-to-face talk may be irreplaceable.

The knowledge produced by face-to-face meetings of people with high levels of human capital, rather than raw materials or large-scale manufacturing, then, may well be the driving force behind the typical city's long-run population growth. Our results also shed some indirect light on the role of the entrepreneur in the growth of the city. The people in our human capital category spread useful information. Although they may not themselves have been the source of the innovations that generated a city's growth, they may have been necessary auxiliaries for a much smaller group—the innovators or entrepreneurs—who were the underlying source of growth and change. We may inadvertently have identified conditions for entrepreneurs to flourish.[29]

[29] Chinitz (1961) suggested that entrepreneurship played a role in the success of cities. The factors he identified as congenital to entrepreneurship—a diverse industry mix and competition—may well be correlated with professional or information-based employment.

APPENDIX 1: A MODEL OF CITY GROWTH

We illustrate our notions about city growth with a brief model. In Simon and Nardinelli (1994), we developed the model in more detail. We assume that each city produces an export good (with a Cobb–Douglass production function) using labor, land, and information,

$$M = L_u^{\alpha} T^{\beta} I^{\gamma},\ \alpha + \beta + \gamma = 1, \tag{1}$$

where L_u is the amount of unskilled labor used to produce the good M, T is the fixed amount of land, and I is the quantity of information services, all at time t (we omit time subscripts). The amount of information in the city, I, is a function of the number of skilled workers with high human capital, L_s, and the average human capital per skilled worker, h:

$$I = hL_s. \tag{2}$$

The term skilled worker is perhaps misleading. The distinguishing feature of such workers is that they embody information. We assume that much of the information is city-specific.

In equilibrium the wage of skilled workers must equal the wages of unskilled workers multiplied by one plus the cost of acquiring human capital, c.

$$W_s = W_u(1 + c) \tag{3}$$

The profit-maximizing condition for employment in manufactures implies that marginal revenue product is equal to the wage. The ratio of unskilled to skilled wages should therefore be equal to the ratio of marginal products, or

$$W_u/W_s = (\alpha/\gamma)(L_s/L_u). \tag{4}$$

If (3) and (4) are combined to solve for L_s we get

$$L_s = AL_u, \tag{5}$$

where $A = (\alpha/\gamma)(1 + c)^{-1}$. We can express total city population, $L = L_u + L_s$, as

$$L = L_u + AL_u. \tag{6}$$

The movement of wages over time [from (1) and (2)] is

$$W_u^* = (\alpha - 1)L_u^* + \gamma h^* + \gamma L_s^*, \tag{7}$$

where * indicates the rate of change of the variable. In terms of total labor (or population), we can write

$$W_u^* = (\alpha + \gamma - 1)L^* + \gamma h^* + Z^*, \tag{8}$$

where

$$Z^* = (\alpha - 1)[(1 + A)^*]^{-1} + \gamma[A/(1 + A)]^*.$$

The wages of unskilled workers, then, decrease with an increase in the number of workers, but increase with an increase in the amount of human capital per skilled worker. The size of the effects depends on the degree of diminishing marginal returns to labor, the share of information services in output, and the degree of substitutability between information services. If *c*, the cost of acquiring human capital, is constant, the wages of skilled workers will grow at the same rate as those of unskilled workers ($Z^* = 0$). If the cost of acquiring human capital falls as the level rises, then unskilled wages will grow more rapidly than skilled wages.

People in the model maximize utility. We assume a Cobb–Douglass utility function with two goods: an internationally traded consumption good, *G*, and housing, *H*. We assume that there is a fixed amount of residential land in the city. As population grows, then, housing per person falls. In the simple model housing per person can be thought of as a proxy for congestion, pollution, and other disamenities associated with larger size. The utility of a city's resident is

$$U = G^{\theta 1} H^{\theta 2}, \tag{9}$$

where θ_1 and θ_2 are the exponents on goods and housing in a Cobb–Douglass utility function. As the city grows, city-specific knowledge grows and wages rise. Because the level of city-specific knowledge will be higher in some cities, as the model now stands the city with a head start will grow faster indefinitely. We introduce a check on city growth by having residents bid up the price of land, causing housing consumption and utility to fall, other things the same.[30] The rate of growth of utility in the city is then given by

$$U^* = \theta_1 W_u^* - \theta_2 L^*. \tag{10}$$

Long-run equilibrium requires that utility be equal across cities at each time, implying that utility grows at the same rate across cities once equilibrium is achieved. If we combine Eq. (8) with (10) the growth of utility over time can be expressed as

$$U^* = [\theta_1(\alpha + \gamma - 1) - \theta_2]L^* + \theta_1\gamma h^* + \theta_1 Z^*. \tag{11}$$

If cities were identical in their initial conditions, they would grow at the same rate. Differences in growth rates, however, will arise if initial conditions differ. We assume that initially, some cities have a higher concentration of human capital, which in the model translates into a larger number of people with skill and a greater amount of information. The initial concentration of human capital might reflect the demand for factors by the city's export industry. Whatever the cause, over a given historical time period, the rate of growth of cities with higher concentrations of human capital should be higher as people migrate to the higher information cities. The in-migration continues until the decline in housing per person chokes it off.

In the messy real world of cities, cities grow as a result of chance, exogenous

[30] The bidding up of housing prices stands in for all of the costs of rising population, including congestion, pollution, crime, and so on.

changes in demand for their products, deliberate government policies, events in the hinterland, and a host of other factors. Our goal is not to provide an explanation for all of city growth. Rather, our work aims to show that the human capital variable, though sometimes overwhelmed by other factors, plays a systematic and continuing role in generating city growth.

APPENDIX 2: CITIES IN THE 1861 SAMPLE

London	Truro	Manchester & Salford
Canterbury	Bridgwater	Oldham
Chatham & Rochester	Bath	Preston
Dover	Bristol	Bradford
Maidstone	Gloucester	Halifax
Brighton	Hereford	Huddersfield
Chichester	Shrewsbury	Leeds
Portsmouth	Stafford	Sheffield
Southampton	Wolverhampton	Wakefield
Winchester	Dudley	Hull
Reading	Worcester	York
Oxford	Birmingham	Durham
Northampton	Coventry	Gateshead
Bedford	Leicester	South Shields
Cambridge	Boston	Sunderland
Wisbech	Lincoln	Newcastle-upon-Tyne
Colchester	Newark	Tynemouth
Bury St. Edmunds	Nottingham	Carlisle
Ipswich	Derby	Whitehaven
Kings Lynn	Chester	Kendal
Norwich	Macclesfield	Newport
Yarmouth	Stockport	Cardiff
Salisbury	Blackburn	Methyr Tydfil
Dorchester	Bolton	Swansea
Poole	Lancaster	Pembroke
Exeter	Liverpool	Carnarvon
Plymouth & Devonport		

APPENDIX 3: CITIES IN THE 1911 SAMPLE

London	Bristol	Stoke on Trent
Liverpool	West Ham	Salford
Manchester	Bradford	Portsmouth
Birmingham	Hull	Leicester
Sheffield	Newcastle-upon-Tyne	Plymouth & Devonport
Leeds	Nottingham	Cardiff

APPENDIX 3—*Continued*

Bolton	Halifax	Rotherham
Croydon	St. Helens	Ealing
Willesden	Wolverhampton	Hastings
Rhondaa	Walsall	Tynemouth
Sunderland	Northampton	Bury
Oldham	Rochdale	Blackpool
Tottenham	Wigan	Acton
East Ham	Hornsey	Lincoln
Blackburn	Newport	Enfield
Brighton	York	Yarmouth
Birkenhead	Methyr Tydfil	Darlington
Leyton	Bournemouth	Wimbledon
Walthamstow Derby	Wallasey	Dewsbury
Norwich	Ilford	Oxford
Southampton	Reading	Eastbourne
Preston	Grimsby	Gillingham
Gateshead	Ipswich	Stockton on Teas
Swansea	Warrington	Southport
Stockport	Smethwick Bootle	Wakefield
South Shields	West Bromwich	Dudley
Huddersfield	Edmonton	Swindon
Coventry	West Hartlepool	Bath
Burnley	Barrow in Furness	Barnsley
Middlesbrough	Southend on Sea	Gloucester

The following cities are included only in the set of 126 cities.

Exeter	Macclesfield	Bridgwater
Worcester	Rochester	Bury St. Edmunds
Carlisle	Shrewsbury	Boston
Dover	Canterbury	Newark
Colchester	Stafford	Pembroke
Chatham	Winchester	Kendal
Lancaster	Hereford	Chichester
Cambridge	Salisbury	Truro
Bedford	Kings Lynn	Wisbech
Chester	Whitehaven	Dorchester
Poole	Durham	Carnarvon
Maidstone		

REFERENCES

Allen, G. C. (1929; 1965), *The Industrial Development of Birmingham and the Black Country, 1860–1927*. New York: Kelley.

Bairoch, P. (1988), *Cities and Economic Development.* Translated by Christopher Brauder. Chicago: Univ. of Chicago Press.

Briggs, A. (1965), *Victorian Cities.* New York: Harper & Row.

Chinitz, B. (1961), "Contrasts in Agglomeration: New York and Pittsburgh." *American Economic Review* **51** (May), 279–289.

Clapham, J. H. (1939), *An Economic History of Modern Britain.* Cambridge, UK: Cambridge Univ. Press.

Clark, P., and Slack, P. (1976), *English Towns in Transition 1500–1700.* Oxford: Oxford Univ. Press.

Davidson, R., and MacKinnon, J. G. (1993), *Estimation and Inference in Econometrics.* New York: Oxford Univ. Press.

Deane, P., and Cole, W. A. (1969), *British Economic Growth 1688–1959.* Cambridge, UK: Cambridge Univ. Press.

De Vries, J. (1984), *European Urbanization, 1500–1800.* Cambridge, MA: Harvard Univ. Press.

Ferneyhough, F. (1975), *The History of Railways in Britain.* Reading: Osprey.

Floud, R., and McCloskey, D. N. (Eds.) (1994), *The Economic History of Britain Since 1700.* Cambridge, UK: Cambridge Univ. Press. Second ed.

Gatrell, V. A. C. (1977), "Labour, Power, and the Size of Firms in Lancashire Cotton in the Second Quarter of the Nineteenth Century." *Economic History Review* **30** (Feb.), 95–139.

Great Britain, *Census of England and Wales, 1861, 1901, 1911, 1961.*

Grossman, G. M., and Helpman, E. (1994), "Endogenous Innovation in the Theory of Growth." *Journal of Economic Perspectives* **8** (Winter), 23–44.

Hayek, F. A. (1945), "The Use of Knowledge in Society." *American Economic Review* **35** (Sept.), 519–530.

Higgs, R. (1971), "American Inventiveness, 1870–1920." *Journal of Political Economy* **79** (June), 661–667.

Hohenberg, P., and Lees, L. (1985), *The Making Of Urban Europe, 1000–1950.* Cambridge, MA: Harvard Univ. Press.

Kellet, J. R. (1969), *The Impact of Railways on Victorian Cities.* London: Routledge and Kegan Paul.

Kuznets, S. (1966), *Modern Economic Growth.* New Haven/London: Yale Univ. Press.

Lampard, E. (1955), "The History of Cities in the Economically Advanced Areas." *Economic Development and Cultural Change* **3** (Jan.), 81–136.

Lee, C. H. (1979), *British Regional Employment Statistics 1841–1971.* Cambridge, UK: Cambridge Univ. Press.

Lee, C. (1994), "The Service Industries." In R. Floud and D. N. McCloskey (Eds.), *The Economic History of Britain since 1700.* Cambridge, UK: Cambridge Univ. Press. Second ed., Vol. 2, pp. 117–144.

Lloyd-Jones, R., and Le Roux, A. A. (1980), "The Size of Firms in the Cotton Industry: Manchester, 1815–41." *Economic History Review* **33** (February), 72–82.

Lucas, R. E., Jr. (1988), "On the Mechanics of Economic Development." *Journal of Monetary Economics* **22** (July), 3–42.

Lucas, R. E., Jr. (1990), "Why Doesn't Capital Flow from Rich to Poor Countries?" *American Economic Review* **80** (May), 92–96.

Marshall, A. (1920; reprinted 1990), *Principles of Economics.* Philadelphia: Porcupine Press. Eighth ed.

McCloskey, D. (1994), "Bourgeois Virtue." *American Scholar* (Spring), 177–191.

Mitch, D. M. (1992), *The Rise of Popular Literacy in Victorian England.* Philadelphia: Univ. of Pennsylvania Press.

Mitch, D. (1993), "The Role of Human Capital in the First Industrial Revolution." In J. Mokyr (Ed.), *The British Industrial Revolution: An Economic Perspective.* Boulder: Westview. Pp. 267–307.

Mitchell, B. R., and Deane P. (1971), *Abstract of British Historical Statistics.* Cambridge, UK: Cambridge Univ. Press.

Mokyr, J. (Ed.) (1993), *The British Industrial Revolution: An Economic Perspective.* Boulder: Westview.

Moser, C. A., and Scott W. (1961), *British Towns.* Edinburgh/London: Oliver and Boyd.

Robson, B. T. (1973), *Urban Growth: An Approach.* London: Methuen.

Romer, P. M. (1986), "Increasing Returns and Long-Run Growth." *Journal of Political Economy* **94** (Oct.), 1002–1037.

Romer, P. M. (1994), "The Origins of Endogenous Growth." *Journal of Economic Perspectives* **8** (Winter), 2–22.

Simon, C. J., and Nardinelli C. (1994), "Information and the Rise of Cities: Evidence from American Urban History." Working paper.

Smith, W. D. (1984), "The Function of Commercial Centers in the Modernization of European Capitalism: Amsterdam as an Information Exchange in the Seventeenth Century." *Journal of Economic History* **44** (Dec.), 985–1005.

Sokoloff, K. L. (1988), "Inventive Activity in Early Industrial America: Evidence from Patent Records, 1790–1846." *Journal of Economic History* **48** (Dec.), 813–850.

Stigler, G. J. (1968), *The Organization of Industry.* Homewood, IL: Irwin.

Turnbull, G. (1987), "Canals, Coal and Regional Growth During the Industrial Revolution." *Economic History Review* **40** (Nov.), 537–560.

Weber, A. F. (1899; reprinted 1963), *The Growth of Cities in the Nineteenth Century.* Ithaca: Cornell Univ. Press.

White, H. P. (1963), *A Regional History of the Railways of Great Britain.* London: Phoenix House. Vols. 1–8.

Williamson, J. G. (1990), *Coping with City Growth during the British Industrial Revolution.* Cambridge, UK: Cambridge Univ. Press.

Wrigley, E. A. (Ed.) (1972), *Nineteenth-Century Society.* Cambridge, UK: Cambridge Univ. Press.

[13]

ELSEVIER Journal of Monetary Economics 36 (1995) 117–143

JOURNAL OF Monetary ECONOMICS

Economic growth in a cross-section of cities

Edward L. Glaeser*,[a,b,c], José A. Scheinkman[d], Andrei Shleifer[a]

[a] *Harvard University, Cambridge, MA 02138, USA*
[b] *Hoover Institution, Stanford, CA 94305, USA*
[c] *National Bureau of Economic Research, Cambridge, MA 02138, USA*
[d] *University of Chicago, Chicago, IL 60637, USA*

(Received September 1993; final version received May 1995)

Abstract

We examine the relationship between urban characteristics in 1960 and urban growth between 1960 and 1990. Income and population growth move together, and both types of growth are (1) positively related to initial schooling, (2) negatively related to initial unemployment, and (3) negatively related to the initial share of employment in manufacturing. Racial composition and segregation are uncorrelated with urban growth across all cities, but in cities with large nonwhite communities segregation is positively correlated with population growth. Government expenditures (except for sanitation) are uncorrelated with growth; government debt is positively correlated with later growth.

Key words: Growth; Cities; Regions; Education

JEL classification: O40; R11

1. Introduction

Over the last 30 years, the growth experiences of United States cities have varied widely. The population of some cities grew significantly while other cities

---

* Corresponding author.

The title is partially borrowed from Barro (1991). Participants in the Harvard Urban Workshop, the Harvard Growth Workshop, the NBER Summer Institute, and one anonymous referee provided helpful comments. David Mare and Melissa McSherry provided excellent research assistance, and the NSF provided research suppport.

SSDI 0304393295012064

virtually disappeared. Some dispersion of growth experiences can be explained by geographic factors, such as the movement of population west and south. But what are the economic forces that explain city growth over the last 30 years in a cross-section of the United States cities?

In this paper, we empirically investigate this question. We examine how the growth experiences of 203 large U.S. cities relate to their location, initial population, initial income, past growth, output composition, unemployment, inequality, racial composition, segregation, size and nature of government, and the education of their labor force. The primary purpose of our analysis is descriptive: we want to understand which cities grew between 1960 and 1990.

As a description, this analysis continues an extensive regional growth literature, including the studies by Borts (1960), Kain and Neidercorn (1962), Mills (most recently, 1992), and others. Our focus on human capital as a determinant of city growth is particularly closely related to Chinitz (1962), who emphasized the connection between urban success and the transmission of enterpreneurial skills. Similar arguments have been made by Jacobs (1969) and Marshall (1890).

In addition, our analysis aims to contribute to the recent studies of economic growth. Starting with Baumol (1986), Delong (1988), and Barro (1991), economists have compared income growth experiences of different countries as a function of their characteristics. These studies typically find weak evidence of convergence of incomes between countries, and stronger evidence that education, physical investment, political stability, and openness to trade contribute to growth. Other studies, including Barro and Sala-i-Martin (1991) and Blanchard and Katz (1992), look at growth experiences of U.S. states. Barro and Sala-i-Martin find much more income convergence between states than between countries. Blanchard and Katz show that employment growth in different states is very persistent, while unemployment is not.[1]

Looking at cities complements looking at countries and states in three ways. First, unlike countries, cities are completely open economies; there is tremendous movement of capital, labor, and ideas between cities. Cities are more specialized (and less arbitrary) economic units than states, and hence it may make more sense to study the movement of resources and convergence between cities than between states. National boundaries that bar factor mobility and national policies that encourage industrial diversification eliminate the gains from factor mobility. These forces complicate work on cross-national studies. Cities allow us to look at economic growth without these concerns.

Second, many of the cross-sectional – particularly the cross-national – studies of growth argue that ideas are important for growth. Various versions of this

---

[1] More precisely, they show that unemployment rate deviations from long-run state average unemployment levels are not persistent.

theory focus on the external benefits of physical capital or disembodied knowledge (Romer, 1986), human capital (Lucas, 1988), particular industries such as manufacturing or high-technology industries (Porter, 1991), and various other spillovers. Glaeser, Kallal, Scheinkman, and Shleifer (1992) found evidence that cross-industry intellectual externalities were particularly important for urban growth. This paper complements the earlier one. We now focus on whole cities, rather than individual industries, and on the sources of city-wide externalities, such as those from human capital.

Third, recent studies of economic growth across countries have focused on political and social, as well as economic determinants of growth. For example, several studies starting with Barro (1991) have shown that political instability is bad for growth, while Alesina and Rodrick (1994) and others have argued that inequality is bad for growth. DeLong and Shleifer (1993) have shown that limited government, as opposed to absolutist government, strengthened the growth of medieval cities. This paper will use the political and social characteristics of cities to provide further evidence on the importance of political and social factors for growth.

Section 2 presents a formal framework for our empirical work. Section 3 describes our data. Section 4 then presents our results on the relationship between economic characteristics of cities and their growth. Section 5 focuses on social and political characteristics of cities. Section 6 concludes.

2. A framework

This section provides a formal setting for our empirical work.[2] Cities will be treated as separate economies that share common pools of labor and capital. Differences in urban growth experiences cannot then come from savings rates or exogenous labor endowments. Because of our assumption of mobile labor and capital, cities differ only in the 'level of productivity' and the 'quality of life'. Total output in a city is given by

$$A_{i,t} f(L_{i,t}) = A_{i,t} L_{i,t}^{\sigma}, \tag{2.1}$$

where $A_{i,t}$ represents the level of productivity in city i at time t,[3] $L_{i,t}$ denotes population of city i at time t, $f(.)$ is a common across cities Cobb–Douglas production function. The coefficient of this production function, σ, is a nation-wide production parameter.

[2] This framework is an extension of the framework used in Glaeser et al. (1992).

[3] We interpret $A_{i,t}$ broadly, to allow for the possibility that social, technological, and political forces all determine the overall productivity of a city.

The labor income of a potential migrant will be the marginal product of labor:[4]

$$W_{i,t} = \sigma A_{i,t} L_{i,t}^{\sigma-1}. \tag{2.2}$$

Total utility equals wages multiplied by a quality of life index. We assume that this index is declining in the size of the city, or using a simple functional form:

$$\textit{Quality of life} = Q_{i,t} L_{i,t}^{-\delta}, \tag{2.3}$$

where $\delta > 0$. Quality of life is meant to capture a wide range of factors including crime, housing prices, and traffic congestion. Total utility of a potential migrant to city i is

$$\textit{Utility} = \sigma A_{i,t} Q_{i,t} L_{i,t}^{\sigma-\delta-1}. \tag{2.4}$$

We initially assume free migration across cities. This assumption ensures constant utilities across space at a point in time, so each individual's utility level in each city must equal the reservation utility level at time t, which we denote $\underline{U}_t$. Thus, for each city:

$$\log\left(\frac{\underline{U}_{t+1}}{\underline{U}_t}\right) = \log\left(\frac{A_{i,t+1}}{A_{i,t}}\right) + \log\left(\frac{Q_{i,t+1}}{Q_{i,t}}\right) + (\sigma - \delta - 1)\log\left(\frac{L_{i,t+1}}{L_{i,t}}\right). \tag{2.5}$$

We also assume that

$$\log\left(\frac{A_{i,t+1}}{A_{i,t}}\right) = X'_{i,t}\beta + \varepsilon_{i,t+1}, \tag{2.6a}$$

$$\log\left(\frac{Q_{i,t+1}}{Q_{i,t}}\right) = X'_{i,t}\theta + \zeta_{i,t+1}, \tag{2.6b}$$

where $X_{i,t}$ is a vector of city characteristics at time t which determines the growth in both the quality of life in the city and the growth of city level productivity. Combining (2.5), (2.6a), and (2.6b) yields

$$\log\left(\frac{L_{i,t+1}}{L_{i,t}}\right) = \frac{1}{1+\delta-\sigma} X'_{i,t}(\beta + \theta) + \chi_{i,t+1}, \tag{2.7}$$

$$\log\left(\frac{W_{i,t+1}}{W_{i,t}}\right) = \frac{1}{1+\delta-\sigma} X'_{i,t}(\delta\beta + \sigma\theta - \theta) + \bar{\omega}_{i,t+1}, \tag{2.8}$$

[4] Assuming wages are average output makes no qualitative difference for the calculations.

E L. Glaeser et al / Journal of Monetary Economics 36 (1995) 117–143 121

where χ_{it} and $\bar{\omega}_{it}$ are error terms uncorrelated with urban characteristics.[5] The net result is that employment growth regressions can be interpreted as showing how city level variables (the X's) determine the sum of quality of life and productivity growth. The wage growth regressions can be thought of as showing a weighted average of the productivity growth and $\sigma - 1$ times quality of life growth.

One difficulty in interpreting wage growth regressions is that they may be reflecting population composition changes as well as compensation changes. While our model presumes homogeneous labor (like most growth models), heterogeneity of labor is a principal feature of urban growth. We will handle this problem by (1) discussing how our results might be interpreted in a heterogeneous labor model and (2) examining movements of population subgroups to allow different urban characteristics to attract different people differently.

Migration vs. convergence

A negative correlation between initial wages and wage growth (convergence) might occur because (1) technology improves more slowly in advanced cities (real convergence) or (2) because the in-migration of labor to high-wage regions causes the wages in those regions to decline. For the second explanation to make sense, migration of labor must respond slowly to shocks in local labor demand. To examine this question, this subsection presents a model with migration costs and delayed labor supply responses to local shocks.

We assume that the quality of life for potential migrants declines not only in the level of population but also in the growth rate of population. This decline might occur because the costs of migration are rising in the number of in-migrants. A negative connection between the quality of life and growth might also occur because it takes time to build certain public goods, or basic infrastructure, or housing. The residents of quickly growing cities may suffer in terms of quality of life until those cities are built up. Quality of life is now given by

$$\textit{Quality of life} = Q_{i,t} L_{i,t}^{-\delta_1} \left(\frac{L_{i,t}}{L_{i,t-1}} \right)^{-\delta_2}. \tag{2.3'}$$

In this case the correctly specified wage growth equation is[6]

$$\log\left(\frac{W_{i,t+1}}{W_{i,t}}\right) = \frac{X'_{i,t}(\delta_1\beta + \delta_2\beta + \sigma\theta - \theta)}{1 + \delta_1 + \delta_2 - \sigma} - \frac{\delta_2(1-\sigma)}{1 + \delta_1 + \delta_2 - \sigma} \log\left(\frac{L_{i,t}}{L_{i,t-1}}\right) + \bar{\omega}_{i,t+1}. \tag{2.8'}$$

[5] Formally, $\chi_{i,t+1} = (-\log(\underline{U}_{t+1}/\underline{U}_t) + \varepsilon_{i,t+1} + \zeta_{i,t+1})/(1 + \delta - \sigma)$ and $\bar{\omega}_{i,t+1} = ((1-\sigma) \times \log(\underline{U}_{t+1}/\underline{U}_t) + \delta\varepsilon_{i,t+1} + (\sigma - 1)\zeta_{i,t+1})/(1 + \delta - \sigma)$.

[6] The definition of $\bar{\omega}_{i,t+1}$ in (2.8′) has been changed appropriately.

If lagged growth rates of population are omitted from the regressions and $\delta_2 \neq 0$, then the coefficients on initial characteristics will be biased (and in particular, the coefficient on initial wages should be biased downwards since current wages are positively correlated with lagged growth). Including lagged population growth rates into a wage growth regression is a test of the existence of this bias. As long as the coefficient on the lagged population growth rate is zero, we can accept that $\delta_2 = 0$. If the coefficient is nonzero, then we must concern ourselves with the possibility that convergence comes from in-migration slowly meeting labor demand and wages being driven down by in-migration.

Controlling for lagged growth rates may have substantial costs. If the first version of our model is correct, and our measures of the X variables are imperfect, and the true X variables are correlated with lagged growth (perhaps because lagged growth came about because of lagged X variables), then controlling for lagged growth rates will decrease the signal to noise ratio in the X variables and spuriously lower the coefficients on these variables.

3. Data description

The analysis in this paper is based on a sample of 203 U.S. cities between 1960 and 1990, although we also use some information from 1950. The data were hand-collected from County and City Data Books (1950, 1960, 1970), the 1990 Census (earlier censi provided the basis for the Data Books – the data are absolutely comparable), and from Taeuber and Taeuber (1965). Sample selection is primarily determined by our reliance on Taeuber and Taeuber for racial characteristics of cities: they looked at all United States cities which in 1960 had over 1,000 occupied housing units with a nonwhite head. The resulting sample of 203 cities includes all but one of the largest 100 cities in the United States, but oversamples Southern cities in the next 100. We have verified that none of the principal results of this paper change if we restrict attention to the largest 100 cities.

Table 1 presents the means and correlations of the variables used in this study (Appendix I provides variable definitions). The average city population in our sample of 203 cities is 269,000 in 1960, growing to an average of 288,000 in 1990, which amounts to only 8.5% growth over the whole period. East St. Louis shrunk the most in this sample (70% over 30 years) whereas Las Vegas grew the most (139% over 30 years). In this sample, 40% of the cities are in the South, 27% are in the Central Region, 19% are in the Northeast, and the remaining 14% are in the West. In 1960, these cities have an average of 25.5% of their activities in manufacturing, an unemployment rate of 5.4%, black population equal to 18% of total, and the average median years of schooling of the population above 25 years old in these cities is 10.8 years.

Table 1
Means and standard deviations, city variables

	Mean	Std. dev.	Minimum	Maximum
Changes				
Growth in log of population 1960–1990	0.085	0.396	– 0.691	1.389
Growth in log of population 1950–1960	0.037	0.054	– 0.065	0.309
1960 Starting variables				
Log of population	11.864	0.931	10.168	15.867
Per capita income ($1000)	1.993	0.4088	1.150	3.872
Migrants per capita	17.154	8.588	4.2	42.6
Geographical				
South	0.397	0.490	0	1
Central	0.270	0.445	0	1
Northeast	0.191	0.394	0	1
Race				
Segregation index	86.19	7.46	60.4	98.1
Weighted segregation index	1.589	1.090	0.145	5.054
% nonwhite	18.09	11.91	2.1	57.5
Labor force				
Unemployment rate	5.4	1.7	1.9	10.5
Manufacturing share of employment	0.255	0.118	0.043	0.589
Education				
Median years of schooling	10.833	1.116	8.4	13.7
% of pop with 16 + yrs of school	8.9	4.533	2.5	38.7
% of pop with 12–15 yrs of school	33.781	6.463	17.4	49.5
% of pop with less than 5 yrs of school	8.333	3.948	1.9	21.4
Income inequality				
% of pop earning under $3000	19.437	7.411	5.0	41.6
% of pop earning over $10,000	15.98	6.383	5.7	43.8
Government				
Per capita revenue	103.639	64.423	11.775	581.353
Property tax share of revenue	0.445	0.175	0.082	0.876
Intergov funds share of revenue	0.161	0.109	0.0002	0.516
Per capita expenditure	110.002	66.107	13.931	634.169
Police share of expenditure	0.121	0.043	0.039	0.240
Highway share of expenditure	0.133	0.070	0.014	0.428
Sanitation share of expenditure	0.143	0.083	0.026	0.534
Per capita debt	205.326	164.684	8.146	1679.698

However, as Table 1 illustrates, there is a great deal of variation in all of these variables.

We use the segregation index from Taeuber and Taeuber (1965). This index is defined as the percentage of nonwhites who would have to move so that the

124 *E.L. Glaeser et al. / Journal of Monetary Economics 36 (1995) 117–143*

Table 2

Raw correlations, city variables

	GR6090	*GR5060*	*LNPOP*	*PCINC*	*STH*	*CENT*	*NE*	*SEG*
Changes								
Growth in log of population 1960–1990	1	0.52	−0.08	−0.02	0.26	−0.28	−0.32	0.08
Growth in log of population 1950–1960	0.52	1	−0.02	0.00	0.23	−0.15	−0.35	0.23
1960 starting variables								
Log of population	−0.08	−0.02	1	0.11	−0.13	0.04	0.01	−0.02
Per capita income (000)	−0.02	0.00	0.11	1	−0.47	0.07	0.22	−0.40
Geographical								
South	0.26	0.23	−0.13	−0.47	1	−0.49	−0.39	0.54
Central	−0.28	−0.15	0.04	0.07	−0.49	1	−0.30	0.12
Northeast	−0.32	−0.35	0.01	0.22	−0.39	−0.30	1	−0.42
Race								
Segregation index	0.08	0.23	−0.02	−0.40	0.54	0.12	−0.42	1
% nonwhite	−0.04	−0.17	−0.02	−0.38	0.51	−0.21	−0.16	0.34
Labor force								
Unemployment rate	−0.25	−0.28	0.01	−0.33	−0.19	0.01	0.11	−0.13
Manufacturing share of employment	−0.45	−0.35	−0.02	−0.01	−0.48	0.40	0.32	−0.19
Education								
Median years of schooling	0.39	0.45	0.01	0.58	−0.08	0.00	−0.27	−0.07
Government								
Per capita revenue	−0.07	−0.09	0.26	0.20	−0.14	−0.17	0.34	−0.26
Per capita expenditure	−0.02	−0.05	0.22	0.14	−0.08	−0.20	0.28	−0.21
Per capita debt	0.22	0.23	0.23	−0.08	0.21	−0.19	−0.05	0.17

Table 2 (continued)

	%NONW	*UNEM*	*MERG*	*MEDYRS*	*REV*	*EXP*	*DEBT*
Changes							
Growth in log of population 1960–1990	– 0.04	– 0.25	– 0.45	0.39	– 0.07	0.22	– 0.02
Growth in log of population 1950–1960	– 0.17	– 0.28	– 0.35	0.45	– 0.09	0.23	– 0.05
1960 starting variables							
Log of population	– 0.02	0.01	– 0.02	0.01	0.26	0.23	0.22
Per capita income (000)	– 0.38	– 0.33	– 0.01	0.58	0.20	– 0.08	0.14
Geographical							
South	0.51	– 0.19	– 0.48	– 0.08	– 0.14	0.21	– 0.08
Central	– 0.21	0.01	0.40	0.00	– 0.17	– 0.18	– 0.20
Northeast	– 0.16	0.11	0.33	– 0.27	0.34	– 0.05	0.28
Race							
Segregation index	0.34	– 0.13	– 0.19	– 0.07	– 0.26	0.17	– 0.21
% nonwhite	1	0.13	– 0.19	– 0.38	0.05	0.12	0.10
Labor force							
Unemployment rate	0.13	1	0.29	– 0.48	– 0.02	– 0.11	0.01
Manufacturing share of employment	– 0.19	0.29	1	– 0.50	0.06	– 0.20	0.03
Education							
Median years of schooling	– 0.38	– 0.48	– 0.50	1	– 0.11	– 0.02	– 0.14
Government							
Per capita revenue	0.05	0.02	0.06	– 0.11	1	0.51	0.97
Per capita expenditure	0.12	– 0.11	– 0.20	– 0.02	0.51	1	0.58
Per capita debt	0.10	0.01	0.03	– 0.14	0.97	0.58	1

white/nonwhite distribution is uniform over blocks of a city.[7] A value of 0 would correspond to a perfectly integrated city, a value of 100 would correspond to a city where no whites live on blocks with any nonwhites.

The correlation matrix in Table 2 points to some obvious and some less obvious results. The omitted dummy in this sample is for Western cities. The correlation matrix shows that Southern cities in the sample tend to have relatively smaller populations, somewhat lower median education levels, dramatically lower per capita incomes, lower unemployment, lower commitment to manufacturing, and much higher segregation than average. In contrast, Northeastern cities are large, have surprisingly low median levels of education, higher per capita income, higher unemployment, a much higher fraction of output in manufacturing, and lower segregation. Educated cities tend to have a higher per capita income, lower unemployment, and lower share of employment in manufacturing. In general, the correlations are what one might expect.

A more casual look at the data is provided in Appendix II which describes the data in more detail for four cities: Phoenix, Arizona; Greensboro, North Carolina; Yonkers, New York; and East Chicago, Illinois. Phoenix's population has almost doubled in the past 30 years, and it is now one of the 10 largest cities in America. Phoenix is now, and was in 1960, a well educated city[8] whose success has not come in traditional manufacturing (which was always a small part of the economy) but rather in services and high-tech areas (electronic machinery is its largest industry). Government has acted to increase Phoenix's population in two ways: (1) a growing state government provided more employment for workers in Phoenix[9] and (2) municipal (and state) authorities have kept a strong commitment to laissez faire.[10]

Durham is also an urban success (albeit on a smaller scale), growing 55.7% over the sample period. Durham benefited from its access to a center of human capital (Duke University),[11] low wages, and the usual regional advantages that made southern cities successful in this period.

If in 1960 Yonkers, N.Y., existed on a featureless plain, it would have contained three of the ingredients which have typically led to disaster among post-war American cities: an orientation towards manufacturing, a high unemployment rate, and a low level of education. However, Yonkers held its ground,

[7] The index assumes only nonwhites moving.

[8] The median years of education were 11.8.

[9] Ades and Glaeser (1995) argue that capital cities grow because of tax dollars raised in the hinterland that are spent on services and bureaucracy in the capital.

[10] Local government expenditures per capita are only 60% of the national average.

[11] While Durham's median years of schooling were low (9.9 years), it had a high (12.3%) percent of college graduates.

losing only 0.3% of its population since 1960. The explanation is simple – the flight to the suburbs from New York saved Yonkers from precipitous decline.[12]

Suburbanization could not save East Chicago. Despite its proximity to a major city, which has been rapidly suburbanizing (both North and West), East Chicago's population fell 60% since 1960 because (1) an extraordinarily high share of its labor force is in manufacturing (58.9%, more than double that of Yonkers) and (2) a low level of its population are high-school graduates (27.3%, or about one-half of Yonkers).

4. Economic determinants of city growth

In this paper, the primary measure of city growth is the growth of population.[13] Population growth might not be appropriate for a country because population is relatively immobile and differences in population growth mainly reflect differences in fertility. Across cities, population growth captures the extent to which cities are becoming increasingly attractive habitats and labor markets. Income growth is a natural measure of productivity growth across countries because labor is immobile. When labor is mobile as it is across U.S. cities – and across U.S. states – the situation is radically different. Within the U.S. economy, migration responds strongly to growth opportunities (Blanchard and Katz, 1992). As we saw in the model, income growth will capture some portion of productivity growth, but it will also capture declines in quality of life. Income growth is therefore a less straightforward measure of urban success.

One concern about our population measure is the movement of population to city suburbs, even when people continue to work in the city. For example, when people avoid pollution, bad schools or racial tension in a city, they move to the suburbs. Because of this suburbanization, the city's population falls even if its employment doesn't drop. To address this concern, we examine not only the growth of city populations, but also the growth of the populations of standard metropolitan statistical areas (SMSAs) as a function of city characteristics.

Table 3 presents the basic results on city and SMSA population growth as a function of initial conditions. The first column presents the regression of city population growth on only the initial population, initial income per capita, and geographical dummies. The regression confirms the well-known fact that (the omitted) Western cities grew the fastest, Southern cities grew the second fastest, Central cities grew third fastest, and Northeastern cities grew the slowest during

[12] This example suggests that cities' outcomes are often a function of neighboring areas attributes. We have dealt with this problem by running regressions both for cities and for SMSAs.

[13] Using employment growth provides qualitatively equivalent results in almost all cases.

Table 3
City growth and initial growth conditions, dependent variable: growth in log of variable (1960–1990)

Variable	(1) City population	(2) City population	(3) City population	(4) SMSA[a] population	(5) City income	(6) City income
Intercept	0.831	0.295	0.620	0.361	15.578	15.562
Log (population 1960)	−0.045	−0.016	−0.034	0.007	−0.013	−0.012
	(0.025)	(0.024)	(0.024)	(0.026)	(0.01)	(0.01)
Per capita income 1960 ($1000)	0.068	−0.052	0.010	0.102	−0.012	−0.010
	(0.066)	(0.063)	(0.064)	(0.116)	(0.026)	(0.027)
Migrants per capita 1960		0.020				
		(0.003)				
Growth in log of population 1950–1960			0.251			0.023
			(0.045)			(0.019)
Geographical dummies						
South	−0.209	−0.214	−0.166	−0.107	0.18	0.177
	(0.081)	(0.074)	(0.076)	(0.082)	(0.031)	(0.032)
Central	−0.529	−0.374	−0.387	−0.430	0.004	0.006
	(0.079)	(0.076)	(0.076)	(0.076)	(0.03)	(0.03)
Northeast	−0.619	−0.364	−0.405	−0.374	−0.031	−0.021
	(0.082)	(0.085)	(0.083)	(0.088)	(0.031)	(0.034)
N	203	203	197	133	201	201
Adj. R^2	0.296	0.411	0.373	0.268	0.368	0.364

Numbers in parentheses are standard errors.

[a]SMSA regression excludes Las Vegas SMSA.

this sample period. These regional dummies cannot be easily rendered insignificant with our economic variables, suggesting that weather and other regional characteristics have played an important role in migration and hence the growth of cities.

This regression also shows that population of larger cities grew slower, a finding which is not robust. More interestingly, there is no evidence that the population of richer cities grew slower, which is not consistent with the basic convergence idea that capital (and with it, labor) should move to regions where wages are lower. In later regressions, we use regional dummies, initial population, and initial income per capita as permanent controls.

The middle two regressions in Table 3 begin to address the issue of persistence of growth rates. They show that cities that attracted migrants in the past[14] and cities that grew faster between 1950 and 1960, also continued to grow faster after 1960. These results are highly statistically significant and consistent with the similar findings at the state level of Blanchard and Katz (1992). We return to persistence of growth rates later when we split the sample period. The fourth regression in Table 3 repeats the basic specification for SMSAs.[15]

Again, there is no evidence of convergence. One interesting finding is that Northeastern SMSAs did relatively better than Northeastern cities. We believe that this finding occurs because the movement of population to the suburbs is more important in the Northeast than in other regions. SMSAs also show less population convergence than cities. Indeed, limits on available land should make congestion more important in the densely populated inner city than in the more open SMSA.

Regression (5) in Table 3 presents the results for per capita income growth. The only significant coefficient is the South dummy, which confirms the well-known result that Southern incomes grew faster. In the West, population grew the fastest; in the South, income did. As before, there is no significant evidence of convergence.

Regression (6) repeats (5) controlling for past population growth. This procedure follows Eq. (2.8′) and is meant to estimate δ_2. The insignificant correlation between lagged population growth and future income growth fails to reject the hypothesis that $\delta_2 = 0$, or that the second effect is statistically insignificant. These results suggest that our findings on income growth are actually reflecting productivity movements, not a gradual return to nationwide average wages from slow migration.[16]

---

[14] By definition these cities have had high levels of gross population growth.

[15] We excluded Las Vegas from all SMSA regressions. We found that Las Vegas was an extreme outlier that influenced many of our regressions in perverse ways. Moreover, we believe that since we have not included gambling-related controls which would explain Las Vegas' outlier status, it is appropriate to exclude Las Vegas from the regressions.

[16] None of our subsequent income growth results change if we control for lagged employment growth; we do not report those findings.

Table 4
City growth and manufacturing, dependent variable: growth in log of variable (1960–1990)

Variable	(1) City total population	(2) City manufacturing employment	(3) City non-manufacturing employment	(4) SMSA[a] total population	(5) City income
Intercept	1.167	− 8.398	5.622	0.560	15.703
Log (population 1960)	− 0.050 (0.025)	0.776 (0.076)	− 0.417 (0.041)	0.035 (0.025)	− 0.015 (0.01)
Per capita income 1960 ($1000)	0.012 (0.067)	− 0.062 (0.208)	− 0.019 (0.095)	0.007 (0.112)	− 0.033 (0.027)
Manufacturing share 1960	− 0.799 (0.253)	− 2.482 (0.783)	− 0.670 (0.355)	− 0.977 (0.230)	− 0.283 (0.096)
Geographical dummies					
South	− 0.244 (0.080)	0.144 (0.246)	− 0.233 (0.114)	− 0.107 (0.077)	0.167 (0.031)
Central	− 0.433 (0.083)	− 0.416 (0.258)	− 0.341 (0.119)	− 0.293 (0.078)	0.039 (0.032)
Northeast	− 0.516 (0.087)	− 0.729 (0.269)	− 0.389 (0.124)	− 0.227 (0.089)	0.003 (0.033)
N	203	189	189	133	201
Adj. R^2	0.326	0.456	0.406	0.354	0.392

Numbers in parentheses are standard errors.

[a]SMSA regression excludes Las Vegas SMSA.

Table 4 examines the relationship between city population growth and the share of initial employment in the manufacturing sector. The first column shows that the population of cities significantly involved in manufacturing grew slower than that of cities less involved in manufacturing. This result holds true even controlling for regions and other initial conditions identified in Table 3. The next two columns show that a high initial exposure to manufacturing had an adverse effect on the growth of both manufacturing and nonmanufacturing employment in a city, and not just on the growth of the city's population. Moreover, manufacturing cities' populations relatively declined not just because people escaped to the suburbs: the population of whole SMSAs grew slower as well. Finally, manufacturing cities relatively declined not only in terms of population, but also in terms of per capita income. A one standard deviation increase in manufacturing's share of employment is associated with a 3.4% decline in income between 1960 and 1990.

These results suggest that cities followed the fortunes of the industries that they were exposed to initially. Nonmanufacturing activities did not move into cities where manufacturing declined; rather, the population of these cities (relatively) declined through emigration, and incomes fell as well. These results seem particularly supportive of a vintage capital model, where cities that invested in older types of capital do not replace that capital as it becomes obsolete because (1) existing capital represents a sunk investment and (2) the pre-existing capital 'crowds out' newer capital.[17] As the capital becomes more out-of-date, the marginal product of labor, the wage rate, and eventually population fall.

Table 5 focuses on initial unemployment and subsequent population growth. High initial unemployment reduces subsequent population growth of both the city and its SMSA. The middle columns show that a high initial unemployment rate adversely affects the growth rate of both employed and unemployed populations, and not just the total urban population. There are two natural interpretations of this effect: (1) cities with high unemployment rates declined as workers responded to business cycle shocks through emigration, or (2) unemployment is proxying for omitted human capital variables, and cities with high unemployment lacked the skilled labor forces necessary for success in our period.[18] The last regression in Table 5 also shows that initial unemployment reduces subsequent income growth; a one standard deviation drop in the unemployment rate creates a 3.75% fall in income.

Table 6 presents results on the initial education levels and subsequent population growth. The first column of Table 6 shows that initial median years of schooling exert a positive and significant influence on the subsequent population growth, using our standard controls as well as the initial unemployment rate and manufacturing share.[19] Once we control for education, we also get the income convergence result – namely that cities with higher initial per capita income grow less. This result is similar to Barro's (1991) finding of conditional convergence.[20] A closer inspection of what it is about schooling that matters

[17] This crowding out may occur because scarce resources (such as land) are used with the older capital.

[18] We believe that the strong persistence in unemployment rates over time (more than 80% between 1960 and 1990) makes the interpretation that unemployment is proxying for omitted human capital variables more plausible.

[19] Miracky (1994) also finds a correlation between schooling levels and urban growth between 1977 and 1990.

[20] This result is not changed when we control for past population growth rates, so we do not believe that it is simply the outcome of wages falling as migrants catch up with demand.

Table 5
City growth and unemployment, dependent variable: growth in log of variable (1960–1990)

Variable	(1) City population	(2) City unemployed	(3) City employed	(4) SMSA[a] population	(5) City income
Intercept	1.777	4.544	3.787	0.963	15.944
Log (population 1960)	– 0.050	– 0.172	– 0.166	0.030	– 0.015
	(0.024)	(– 0.048)	(0.045)	(0.025)	(0.009)
Per capita income 1960	– 0.116	– 0.354	– 0.278	– 0.050	– 0.084
($1000)	(0.074)	(0.149)	(0.141)	(0.111)	(0.0297)
Unemployment rate	– 0.057	– 0.164	– 0.086	– 0.047	– 0.022
1960	(0.016)	(0.031)	(0.030)	(0.017)	(0.006)
Manufacturing share	– 0.631	– 0.623	– 1.381	– 0.888	– 0.225
1960	(0.250)	(0.490)	(0.463)	(0.227)	(0.095)
Geographical dummies					
South	– 0.370	– 0.121	– 0.243	– 0.172	0.116
	(0.085)	(0.168)	(0.159)	(0.079)	(0.033)
Central	– 0.521	– 0.067	– 0.269	– 0.356	0.006
	(0.084)	(0.165)	(0.156)	(0.079)	(0.032)
Northeast	– 0.570	– 0.462	– 0.321	– 0.266	– 0.018
	(0.086)	(0.169)	(0.160)	(0.160)	(0.032)
N	203	201	201	133	201
Adj. R^2	0.364	0.268	0.203	0.387	0.426

Numbers in parentheses are standard errors.

[a]SMSA regression excludes Las Vegas SMSA.

shows that the percentage of the population with 12 to 15 years (high-school graduates or some college) is more important than the percentage of the population with over 16 years (college graduates). This result suggests the importance of a well-educated labor force, not just of the top of the education distribution. Interestingly, median years of schooling are less significant if we look at the SMSA population, suggesting that schooling may have contributed to suburbanization and not just city growth.

One interesting question in the theory of economic growth is whether the average or the total quantity of human capital speeds up economic growth. One can build human capital spillover models in the spirit of Lucas (1988) in which growth is proportional to either total or average education. As a rough test of

Table 6
City growth and education, dependent variable: growth in log of variable (1960–1990)

Variable	(1) City population	(2) City population	(3) City population	(4) SMSA[a] population	(5) City income
Intercept	0.819	– 1.108	1.104	0.422	15.664
Log (population 1960)	– 0.042	– 0.043	– 0.040	0.038	– 0.012
	(0.024)	(0.025)	(0.024)	(0.025)	(0.009)
Per capita income 1960 ($1000)	– 0.212	– 0.223	– 0.235	– 0.177	– 0.155
	(0.085)	(0.087)	(0.093)	(0.136)	(0.034)
Unemployment rate 1960	– 0.044	– 0.044	– 0.042	– 0.044	– 0.018
	(0.017)	(0.017)	(0.017)	(0.017)	(0.0065)
Manufacturing share 1960	– 0.353	– 0.322	– 0.300	– 0.686	– 0.144
	(0.276)	(0.281)	(0.284)	(0.259)	(0.105)
Median years of schooling 1960	0.080	0.075		0.059	0.024
	(0.035)	(0.035)		(0.038)	(0.013)
Median schooling 1960 * Log (population 1960)		– 0.015			
		(0.023)			
Percent of population with 12–15 years of schooling			0.014		
			(0.006)		
Percent of population with 16 + years of schooling			0.007		
			(0.008)		
Geographical dummies					
South	– 0.309	– 0.320	– 0.278	– 0.152	0.133
	(0.089)	(0.090)	(0.092)	(0.079)	(0.034)
Central	– 0.489	– 0.503	– 0.461	– 0.353	0.016
	(0.085)	(0.088)	(0.086)	(0.079)	(0.032)
Northeast	– 0.476	– 0.490	– 0.424	– 0.243	0.01
	(0.094)	(0.097)	(0.101)	(0.089)	(0.036)
N	203	203	203	133	201
Adj. R^2	0.377	0.375	0.379	0.395	0.432

Numbers in parentheses are standard errors.

[a]SMSA regression excludes Las Vegas SMSA.

these theories, Table 6 presents a regression in which both the average education and a proxy for the total stock of education are included. Our proxy for the total stock of education is (median years of schooling in the city minus average median years of schooling across cities) times (log of population minus average log of population). The results suggest that average education matters and total education does not.[21]

Finally, the last regression in Table 6 examines the effect of education on subsequent per capita income growth. The basic result, similar to that for countries, is that cities with higher median years of schooling show faster subsequent per capita income growth. Here a one standard deviation rise in median years of schooling raises income 2.78% over the period.

Overall, the results on income growth closely resemble the results on population growth. Unemployment lowers growth. Manufacturing lowers growth. Initial education raises growth. The fact that income and population results are similar suggests that either (1) movements in wages and population are generally determined by productivity changes, and not quality of life changes [i.e., in Eqs. (2.7) and (2.8) the β terms are much larger than the θ terms in absolute value], or (2) the same factors that cause increases in productivity cause decreases in quality of life [i.e., the β terms have the reverse sign as the θ terms]. We find the second explanation less plausible, since it requires one to believe that unemployment makes cities a more pleasant place to live. Income results may also reflect changes in labor force composition. Since cities with growing populations have also had increases in the percent college-educated, it is possible that the correlation between income growth and population growth occurs because population growth is accompanied by an upgrading in the human capital quality of the population.[22]

Tables 3 through 6 have presented some results on the growth of cities over 1960–1990. Since we have some data for 1950, we can actually look at two twenty-year periods, 1950–1970 and 1970–1990, as is done in Table 7. Initial unemployment rates and manufacturing shares negatively affect employment growth in both subperiods, although the effect of the initial manufacturing share is insignificant in the second subperiod. High unemployment appears to be worse for growth after 1970. The effect of median years of schooling on subsequent growth has increased after 1970 (although this result is not statistically significant). This result is consistent with studies that show the increased importance of education after 1970 in other contexts (for example, Murphy and Welch, 1991).

---

[21] Population growth and income growth are both correlated with growth in percent college educated; we do not report these results. Cities become more appealing in this way as they grow. However, we did not find the growth in percent college educated (or any other meaningful measure of human capital stocks) to be significantly related (in either direction) to our measures of initial conditions.

[22] We cannot with our data look at income within subgroups of the population, so it is difficult for us to distinguish whether rises in income represent higher income for all the residents of the city or whether they represent a shift in the human capital composition of the labor force. A preliminary examination of other data has suggested that compositional shifts explain a significant, but not overwhelming, portion of income changes.

Table 7
City growth and education for two subperiods, dependent variable: 20-year growth in log of population

Variable	(1) City population (1950–1970)	(2) City population (1950–1970)	(3) City population (1970–1990)	(4) SMSA[a] population (1970–1990)
Intercept	1.314	1.187	0.372	0.206
Log (initial population)	– 0.073	– 0.073	– 0.019	– 0.021
	(0.023)	(0.023)	(0.014)	(0.014)
Median income 1950	– 0.034	0.038		
	(0.065)	(0.034)		
Per capita income 1970 ($1000)			– 0.096	– 0.101
			(0.033)	(0.033)
Initial median years of schooling	0.032	0.038	0.064	0.081
	(0.034)	(0.024)	(0.023)	(0.024)
Initial unemployment rate	– 1.32	– 1.797	– 3.512	– 0.044
	(1.477)	(1.444)	(1.182)	(0.012)
Initial manufacturing share	– 0.536	– 0.493	– 0.197	– 0.141
	(0.262)	(0.261)	(0.172)	(0.173)
% nonwhite	– 0.386		– 0.268	
	(0.265)		(0.106)	
Geographical dummies				
South	– 0.068	– 0.128	– 0.249	– 0.29
	(0.111)	(0.103)	(0.06)	(0.058)
Central	– 0.219	– 0.24	– 0.441	– 0.462
	(0.102)	(0.102)	(0.054)	(0.054)
Northeast	– 0.401	– 0.413	– 0.38	– 0.406
	(0.096)	(0.096)	(0.061)	(0.061)
N	197	197	202	202
Adj. R^2	0.351	0.347	0.496	0.482

Numbers in parentheses are standard errors.
[a]SMSA regression excludes Las Vegas SMSA.

In summary, the results for cities are surprisingly similar to the results for countries in that cities with 'good' characteristics, such as low manufacturing exposure, high education, and low unemployment, grow faster. There is little evidence of that bigger cities have a lower population growth or that richer cities have a lower income growth (although income convergence is stronger once we control for city characteristics).

5. Population growth and social characteristics of cities

In this section, we examine the effect of several social and political characteristics on the growth of U.S. cities, 1960–1990. The social measure we use to proxy for social tension in a city are income (and education), inequality, racial composition, and segregation of the population. We also look at the effect of composition of government spending on subsequent growth.

Table 8 presents the results on city population growth and inequality, controlling for the initial conditions that have been shown to matter so far. Controlling for mean income, the percent of the population that has income above $10,000 in 1960 raises city growth – although this result disappears if we look at the SMSA. In addition, the percent of the population that has income below $3,000 in 1960 has no effect on city growth, but substantially raises SMSA growth. This result suggests that, in this period, social tensions resulting from substantial poverty in a city manifested themselves through suburbanization. At the same time, the second column presents the surprising result that a high percentage of uneducated people is associated with higher city growth, which might mean that an abundance of inexpensive, low human capital labor attracted capital, or, alternatively, that we are not fully accounting for regional variation.

Regression (5) shows the connection between schooling inequality and income growth. Urban income growth is positively related to the share of college-educated workers. Median years of schooling becomes insignificant when college-educated workers share is controlled for.

Table 9 presents the results on city growth and racial composition. Without any but our standard Table 3 controls, percent nonwhite weakly slows down city growth, although this result is not particularly robust, and disappears when more controls such as initial unemployment, manufacturing share, and median years of schooling are added. Obviously, race in this sample is correlated with the initial economic characteristics of a city.

The segregation index has no effect on subsequent growth. However, weighted segregation, defined as the segregation index times the percentage of the population that is black (to differentiate between segregated cities with a large number of blacks and those with a few) has a positive influence on future growth. Another way of seeing this result is by looking only at cities with over 10% nonwhites. For these cities, segregation positively affects growth, perhaps because segregation lessened racial discord or because whites demanded segregation. This result is inconsistent with Benabou (1993), who argues that under certain conditions segregation is costly to growth because of diminished knowledge spillovers.

Table 10 examines the relationship between city growth and the levels and composition of government receipts and expenditures in 1960. It does not appear that either 1960 per capita revenue or 1960 per capita expenditure

Table 8
City growth and inequality, dependent variable: growth in log of variable (1960–1990)

Variable	(1) City population	(2) City population	(3) City population	(4) SMSA[a] population	(5) City income
Intercept	1.007	– 0.083	0.854	– 1.458	15.871
Log (population 1960)	– 0.037	– 0.042	– 0.045	0.024	– 0.01
	(0.025)	(0.024)	(0.024)	(0.026)	(0.009)
Per capita income 1960 ($1000)	– 0.430	– 0.162	– 0.314	0.437	– 0.147
	(0.158)	(0.093)	(0.122)	(0.219)	(0.037)
Unemployment rate 1960	– 0.043	– 0.046	– 0.044	– 0.043	– 0.015
	(0.017)	(0.017)	(0.0169)	(0.016)	(0.007)
Manufacturing share 1960	– 0.398	– 0.359	– 0.521	– 0.304	– 0.107
	(0.324)	(0.280)	(0.312)	(0.263)	(0.106)
Median years of schooling 1960	0.072	0.143	0.064	0.090	0.004
	(0.039)	(0.047)	(0.038)	(0.037)	(0.018)
% population with income greater than $10,000	0.016			– 0.007	
	(0.010)			(0.012)	
% population with income less than $3,000	0.001			0.026	
	(0.007)			(0.007)	
Percent of population with 5 – years of schooling		0.023			– 0.004
		(0.010)			(0.004)
Percent of population with 16 + years of schooling		– 0.007			0.007
		(0.009)			(0.003)
Median income 1960			0.007		
			(0.006)		
Geographical dummies					
South	– 0.311	– 0.350	– 0.278	– 0.216	0.131
	(0.094)	(0.091)	(0.093)	(0.077)	(0.036)
Central	– 0.471	– 0.447	– 0.479	– 0.351	0.004
	(0.085)	(0.086)	(0.085)	(0.075)	(0.033)
Northeast	– 0.453	– 0.462	– 0.460	– 0.214	0.002
	(0.095)	(0.094)	(0.095)	(0.085)	(0.036)
N	203	203	203	133	201
Adj. R^2	0.379	0.386	0.378	0.460	0.439

Numbers in parentheses are standard errors.

[a]SMSA regression excludes Las Vegas SMSA.

Table 9
City growth and race, dependent variable: growth in log of variable (1960–1990)

Variable	(1) City population	(2) City population	(3) City population	(4) SMSA[a] population	(5) City income
Intercept	0.908	0.938	1.481	0.355	15.627
Log (population 1960)	− 0.043 (0.025)	− 0.042 (0.024)	− 0.043 (0.024)	0.033 (0.027)	− 0.012 (0.009)
Per capita income 1960 ($1000)	0.043 (0.067)	− 0.202 (0.086)	− 0.159 (0.088)	− 0.146 (0.146)	− 0.117 (0.035)
% nonwhite 1960	− 0.005 (0.002)	− 0.002 (0.002)	− 0.048 (0.025)	0.002 (0.003)	0.001 (0.001)
Unemployment rate 1960		− 0.043 (0.017)	− 0.039 (0.017)	− 0.043 (0.017)	− 0.019 (0.007)
Manufacturing share 1960		− 0.402 (0.285)	− 0.429 (0.284)	− 0.653 (0.265)	− 0.128 (0.108)
Median years of schooling 1960		0.070 (0.038)	0.060 (0.038)	0.061 (0.038)	0.027 (0.014)
Segregation index 1960			− 0.006 (0.006)		
Segregation 1960 * % nonwhite 1960			0.529 (0.284)		
Geographical dummies					
South	− 0.150 (0.085)	− 0.289 (0.093)	− 0.296 (0.103)	− 0.162 (0.081)	0.127 (0.036)
Central	− 0.517 (0.078)	− 0.483 (0.085)	− 0.482 (0.092)	− 0.358 (0.079)	0.014 (0.033)
Northeast	− 0.602 (0.082)	− 0.479 (0.094)	− 0.478 (0.094)	− 0.242 (0.089)	0.01 (0.036)
N	203	203	203	133	201
Adj. R^2	0.308	0.375	0.381	0.392	0.431

Numbers in parentheses are standard errors.
[a]SMSA regression excludes Las Vegas SMSA.

significantly influence subsequent population growth. Expenditure levels in 1960 are positively related to later income growth, but this result is not robust. However, 1960 debt levels (holding revenue constant) are positively associated

Table 10
City growth and government expenditure, dependent variable: growth in log of variable (1960–1990)

Variable	(1) City population	(2) City population	(3) City population	(4) City population	(5) City income
Intercept	0.803	0.734	0.771	0.812	15.66
Log (population 1960)	– 0.047 (0.025)	– 0.056 (0.025)	– 0.048 (0.025)	0.042 (0.025)	– 0.018 (0.01)
Per capita income 1960 ($1000)	– 0.225 (0.087)	– 0.183 (0.091)	– 0.229 (0.086)	– 0.216 (0.087)	– 0.125 (0.034)
Median years of schooling 1960	0.085 (0.036)	0.081 (0.036)	0.088 (0.036)	0.077 (0.038)	0.029 (0.014)
Unemployment rate 1960	– 0.044 (0.017)	– 0.035 (0.017)	– 0.043 (0.017)	– 0.044 (0.017)	– 0.018 (0.006)
Manufacturing share 1960	– 0.340 (0.277)	– 0.366 (0.283)	– 0.344 (0.277)	– 0.413 (0.277)	– 0.13 (0.104)
Per capita revenue 1960	0.290 (0.392)	– 0.492 (0.501)			
Property revenue share 1960		0.137 (0.163)			
Intergovernmental revenue share 1960		0.202 (0.235)			
Per capita debt 1960		0.462 (0.175)			
Per capita expenditure 1960			0.408 (0.370)	0.445 (0.466)	0.297 (0.156)
Police share of expenditure 1960				– 0.504 (0.678)	
Highway share of expenditure 1960				– 0.003 (0.388)	
Sanitation share of expenditure 1960				0.557 (0.294)	
Geographical dummies					
South	– 0.305 (0.089)	– 0.325 (0.089)	– 0.303 (0.089)	– 0.329 (0.095)	0.13 (0.034)
Central	– 0.482 (0.085)	– 0.486 (0.085)	– 0.475 (0.085)	– 0.492 (0.090)	0.018 (0.032)
Northeast	– 0.483 (0.095)	– 0.485 (0.104)	– 0.481 (0.094)	– 0.481 (0.097)	– 0.008 (0.037)
N	203	203	203	201	201
Adj. R^2	0.375	0.389	0.378	0.378	0.44

Numbers in parentheses are standard errors.

with subsequent growth.[23] In addition, a higher fraction of expenditures which go to sanitation (perhaps a useful type of government spending) is associated with faster subsequent growth. Interestingly, a higher fraction of expenditures which go to highway construction – a plausible measure of infrastructure spending – is not associated with higher subsequent growth. In sum, there are no striking results on the effect of government revenue or spending on growth, which is in line with much of the evidence on growth of countries.

6. Conclusion

This paper has presented a variety of partial correlations between growth of city populations and these cities' initial characteristics. A key variable for cities, as for countries, is initial education level of the population. Several other economic variables, such as initial unemployment and exposure to manufacturing, proved important as well. We were relatively less successful in identifying important social and political variables; we did find that our weighted segregation measure was positively correlated with income and population growth. In many respects, however, the story of growth of cities is similar to that of the growth of countries. Since cities' differences are not created by different savings rates or different labor force endowments, our results suggest that higher education levels influence later growth not through savings but through influencing the growth of technology.

Many of the most interesting explanations for the connection between growth and initial human capital levels across countries have focused on productive externalities generated by schooling. Since these externalities should be particularly prevalent in cities, finding a connection between growth and initial schooling across urban centers supports the view of schooling as a generator (through spillovers) of growth. The robust relationship between schooling and growth for SMSAs, city employment, and city income growth provides more evidence supporting the positive role of education in economic expansion.

Appendix 1: Description of variables

Population (City): This variable comes from the 1950–1970 county and city data books and is the number of persons living within the political unit of the city. It does not count commuters. This variable is ultimately derived from the census.

[23] This correlation between debt levels and later growth may be because higher expected growth levels made it cheaper to borrow, or because cities anticipating high levels of population growth invested heavily in infrastructure to serve that growth.

Population (SMSA): This variable again comes from the 1950–1970 county and city data books and the 1990 census. Since SMSA definitions have changed over the relevant sample period, we have used the county definitions from 1990 and reconstructed SMSAs in 1960 using the 1990 definitions. The population counts themselves are from the census.

Regions: The four regional classifications are standard. We used the regional definitions given by Taeuber and Taeuber (1965).

Per Capita Income: This variable was created from two variables included in the city and county data books. Aggregate income is described as the amount received by all income recipients 14 years and older. This variable was then divided by the population measure.

Manufacturing's Share: This variable (again from the data books, which will be our sources for all the remaining variables unless noted) represents the share of workers who are employed who worked in manufacturing industries, which means specifically industries with SIC codes 20–39.

Unemployment Rate: This variable refers to the share of the labor force that is currently unemployed. The usual problems about workers who have dropped out of the labor force apply.

Median Years of Schooling: This variable gives the median years of schooling for all persons 25 years and older.

16 +, 12–16, 12 Years of Schooling Share of Population: These variables again refer to the schooling of persons 25 years and over.

Percentage Nonwhite: This refers to the share of the population not classified as whites, and it includes Hispanics.

Segregation Index: This index comes from Taeuber and Taeuber (1965) and is described in the text of our paper and in their book. This index measures the percentage of the nonwhite population that would have to move to achieve perfect integration. The range of the index is between 0 (perfect integration) and 100 (complete segregation – all of the nonwhite population would have to move).

3000 –, 10000 + Share of Population: These variables refer to the shares of the population that earn less than 3,000 or more than 10,000 dollars per year among the working population

Government Expenditures, Revenues, and Debt: These variables are from the county and city data books and are formed by dividing total government expenditures, revenues, and debt by population figures described above.

142 E L. Glaeser et al. / Journal of Monetary Economics 36 (1995) 117–143

Appendix II: Description of four cities

Durham, North Carolina	*1960*	*1990*
Population	78,302	136,611
% high-school graduates	38.2%	78.5%
Per capita income	$1,1647	$10,257
% manufacturing	25.4%	N/A
% unemployment	4.9%	4.5%
% nonwhite	36.3%	48.3%
Per capita government expenditures	$101.2	$468.0

East Chicago, Illinois	*1960*	*1990*
Population	57,669	33,892
% high-school graduates	27.3%	57.7%
Per capita income	$1,821	$7,905
% manufacturing	58.9%	31.5%
% unemployment	4.5%	14.3%
% nonwhite	24.0%	62.0%
Per capita government expenditures	$143.9	$1135.0

Phoenix, Arizona	*1960*	*1990*
Population	439,170	983,403
% high-school graduates	48.7%	64.3%
Per capita income	$2,013	$8,807
% manufacturing	16.6%	25.0%
% unemployment	4.7%	6.9%
% nonwhite	5.8%	18.3%
Per capita government expenditures	$63.52	$111.90

Yonkers, New York	*1960*	*1990*
Population	190,634	188,082
% high-school graduates	47.9%	73.6%
Per capita income	$2,691	$13,112
% manufacturing	25.7%	7.2%
% unemployment	3.7%	4.9%
% nonwhite	4.2%	15.8%
Per capita government expenditures	$187.2	$1322.0

References

Ades, A. and E. Glaeser, 1995, Trade and circuses: Explaining urban giants, Quarterly Journal of Economics CX, 195–228.

Alesina, A. and D. Rodrik, 1994, Distributive politics and economic growth, Quarterly Journal of Economics CIX, 465–490.

Barro, R., 1991, Economic growth in a cross-section of countries, Quarterly Journal of Economics CVI, 407–444.

Barro, R. and X. Sala-i-Martin, 1992, Convergence, Journal of Political Economy 100, 223–251.

Baumol, W.J., 1986, Productivity growth, convergence and welfare: What does the long run data show?, American Economic Review 76, 1072–1085.

Benabou, R., 1993, Working of a city: Location, education and production, Mimeo.

Blanchard, O. and L. Katz, 1992, Regional evolutions, Brookings Papers on Economic Activity 1, 1–76.

Borts, G., 1960, The equalization of returns and regional economic growth, American Economic Review 50, 319–347.

Chinitz, B., 1962, Contrasts in agglomeration: New York and Pittsburgh, American Economic Review, Papers and Proceedings 52, 279–289.

DeLong, J.B., 1988, Productivity growth, convergence and welfare: Comment, American Economic Review 78, 1138–1154.

DeLong, J.B. and A. Shleifer, 1993, Princes and merchants: European city growth before the industrial revolution, Journal of Law and Economics 36, 671–702.

Glaeser, E., H. Kallal, J. Scheinkman, and A. Shleifer, 1992, Growth in cities, Journal of Political Economy 100, 1126–1152.

Jacobs, J., 1969, The economy of cities (Vintage, New York, NY).

Kain, J. and D. Neidercorn, 1963, An econometric model of metropolitan development, Regional Science Association Papers 11, 000–000.

Lucas, R.E., 1988, On the mechanics economic development, Journal of Monetary Economics 12, 3–42.

Marshall, A., 1890 Principles of economics (Macmillan, London).

Mills, E.S., 1992, Forecasting the growth of metropolitan areas, Mimeo.

Miracky, W., 1994, Cities and the product cycle, Mimeo.

Murphy, K. and F. Welch, 1992, The structure of wages, Quarterly Journal of Economics CVII, 285–326.

Porter, M., 1990, The comparative advantage of nations (Free Press, New York, NY).

Romer, P., 1986, Increasing returns and long-run growth, Journal of Political Economy 94, 1002–1037.

Taeuber, A. and K. Taeuber, 1965, Negroes in cities (McGraw-Hill, New York, NY).

[14]

JOURNAL OF URBAN ECONOMICS 34, 380–400 (1993)

Productivity Gains from Geographic Concentration of Human Capital: Evidence from the Cities

JAMES E. RAUCH*

Department of Economics, University of California at San Diego, La Jolla, CA 92093; and National Bureau of Economic Research, Cambridge, MA 02138

Received February 3, 1992; revised July 20, 1992

Based on recent theoretical developments the argument is made that the average level of human capital is a local public good. Cities with higher average levels of human capital should therefore have higher wages and higher land rents. After conditioning on the characteristics of individual workers and dwellings, this prediction is supported by data for Standard Metropolitan Statistical Areas (SMSAs) in the United States, where the SMSA average levels of formal education and work experience are used as proxies for the average level of human capital. The alternative explanations of omitted SMSA variables and self-selection are evaluated. An estimate of the effect of an additional year of average education on total factor productivity is computed. © 1993 Academic Press, Inc.

1. INTRODUCTION

It is commonly believed that individuals do not capture all of the benefits from their ownership of human capital. This belief is often cited to justify government subsidies to formal education. As Schultz [26] states in his survey of the role of education in economic development, "education is widely viewed as a public good (with positive externalities), which increases the efficiency of economic and political institutions while hastening the pace of scientific advance." Recently human capital externalities have come to bear considerable explanatory weight in formal theoretical modeling. Lucas [17] allows the average national level of human capital per worker to act as an externality that Hicks-neutrally shifts the aggregate production function. Azariadis and Drazen [1] follow Lucas in assuming that the average level of human capital is a "social input" to aggregate production in their model.

The microeconomic foundation of this external effect of human capital is the sharing of knowledge and skills between workers that occurs through

*I am grateful for the research assistance of Brad Kamp, Kathy Kiel, Francis Lim, and Jaesun Noh, for discussions with George Borjas and Glenn Sueyoshi, and for the support of NSF Grant SES90-01761. The comments of participants in the April 1991 NBER Growth Conference, Ed Glaeser, and two anonymous referees have also been helpful. I am responsible for any errors.

0094-1190/93 $5.00

both formal and informal interaction. The "diffusion and growth of knowledge" that takes place as a result of that interaction is modeled in a paper by Jovanovic and Rob [14]. In their model individuals augment their knowledge through pairwise meetings at which they exchange ideas. In each time period each individual seeking to augment his knowledge meets an agent chosen randomly from a distribution of agents/ideas. Though a proof requires additional assumptions, intuitively it seems clear that the higher the average level of human capital (knowledge) of the agents, the more "luck" the agents will have with their meetings and the more rapid will be the diffusion and growth of knowledge. If this knowledge concerns technological improvements, we have a microeconomic foundation not only for external effects of human capital on total factor productivity, but also for making those external effects dependent on the average level of human capital.

Given the existence of human capital externalities, economically identical workers will tend to earn higher wages in human capital rich, rather than in human capital poor, countries. This result is consistent with the large realized net migration from the latter to the former countries and unsatisfied demand for further immigration. The problem with inferring that human capital externalities *cause* these wage differentials is that a high average level of human capital is associated with a high level of economic development.[1] A high level of economic development is in turn associated with other factors that tend to cause high wages such as a large and technologically current stock of physical capital per capita. For this reason it would obviously be very difficult to econometrically identify the effects of human capital externalities with any confidence using cross-country data, even if we were to overcome the problems of cross-country data comparability. By looking at different regions within one country, however, we can identify these effects, since if the country has a well-developed electronic communications system the cost of capital and the level of "disembodied" technological knowledge will presumably be the same within its borders. Indeed, Lucas argues that metropolitan areas are the most appropriate units to examine when looking for the productivity-enhancing effects of human capital abundance. He cites the work of Jane Jacobs, *The Economy of Cities* [13], which provides numerous concrete examples of "creative" economic life in cities where external economies generated by interaction among educated and/or experienced individuals are important. It certainly seems reasonable to think that random meetings, as opposed to costly, prearranged ones, would take place within a

[1]Balassa [2] computes a rank correlation coefficient of 0.754 between per capita GNP and the Harbison–Myers index of human resource development for 36 countries evenly divided between developed and less developed.

limited spatial area rather than uniformly distributed over an entire country.[2]

By generating higher wages for a given set of individual characteristics, productivity benefits from geographic concentration of human capital set up strong pressures for migration. Since within a country people are free to migrate (there are usually no immigration or emigration barriers), why are wages not driven to equality for all economically identical agents? The answer given by the simple local public goods model below is that migration to high wage areas leads to higher residential and commercial rents there, which offset the higher wages and allow for a spatial equilibrium where utility levels and production costs are equalized across metropolitan areas. Cities with higher average levels of human capital should therefore have higher wages and higher land rents. After conditioning on the characteristics of individual workers and dwellings, I test this prediction using data for Standard Metropolitan Statistical Areas (SMSAs) in the United States, where the SMSA average levels of formal education and work experience are used as proxies for the average level of human capital. If the data support this prediction, my model also allows me to estimate the *magnitude* of the productivity benefits realized from geographic concentration of human capital.

The rest of this paper is organized as follows. A formal model that treats the average level of human capital as a local public good is presented in Section 2. In Section 3 I estimate the model of Section 2 in reduced form. Section 4 deals with the possibility of omitted variable bias in the results of Section 3 by controlling for additional SMSA local public goods likely to be correlated with the SMSA average level of human capital and also evaluates the argument that the results of Section 3 are due to selection bias. In the concluding section I compute an estimate of the effect of an additional year of average education on total factor productivity and compare my estimate to one that can be obtained from the growth model calibration exercise of Lucas [17].

2. THE FORMAL MODEL

The model of this section is a straightforward adaptation of the local public goods model of Roback [23]. To minimize repetition I therefore am very brief in my description.

[2] One concrete example of how these externalities work was provided to me by a Silicon Valley engineer. He explained that firms locating in the San Jose metropolitan area were able to benefit from what he called "cross-pollenization of ideas" because of the constant movement of both engineers and lower-level technical workers between the firms located there. By dramatically reducing transportation costs, geographic concentration facilitates this movement, which is analogous to the random "meetings" modeled by Jovanovic and Rob.

Assume that households and firms are freely mobile between a fixed number of SMSAs in the sense that their cost of changing locations is zero, although once they choose to locate in a particular SMSA they cannot work (employ workers) in another SMSA. For simplicity intracity commuting is not considered. Each SMSA offers a fixed amount of land L_j and a given bundle of site characteristics s_j, where the subscript j indexes the SMSAs. Households gain utility through use of a composite commodity, local residential land, and local site characteristics. Site characteristics are local public goods (climate is a good example), while use of the composite commodity and land are purchased out of labor earnings, which comprise all of income. (Here I have adopted the assumption of Henderson [11, Chap. 2] that rentiers (owners of land and capital) do not live in SMSAs.) The differences in the level of human capital across households are measured by the amount of efficiency units of labor h with which they are endowed.[3] Households supply their efficiency units inelastically in exchange for the wage per efficiency unit w_j. As in Heckscher–Ohlin trade models, preferences are assumed to be identical and homothetic across households. The assumption of homotheticity ensures that spatial equilibrium can be attained by equation across SMSAs of utility *per efficiency unit*. Thus in spatial equilibrium we have

$$V_j = v(r_j; s_j)w_j = u^0, \tag{1}$$

where V is the indirect utility enjoyed by the owner of one efficiency unit of labor, r is the rent on one unit of land, and u^0 is the common nationwide utility level for owners of one efficiency unit. Firms combine capital, local labor, and local land to produce the composite commodity. The return to capital is fixed by an international capital market. Prices, wages, and rents are normalized on the price of the composite good, so that the price of the composite good is set to unity. Production technology is constant returns to scale in labor, land, and capital, with site characteristics entering production as Hicks-neutral shift parameters. Spatial equilibrium then requires

$$c(r_j, w_j; s_j) = 1, \tag{2}$$

where c is the unit cost function. The cost of capital has been suppressed since it is equal across all SMSAs.

Combining the conditions for spatial equilibrium among consumers and firms leads to Fig. 1. The upward-sloping or consumer equilibrium curve

[3] This is the Lucas [17] formulation: in his model (p. 17) a worker with human capital h is the productive equivalent of two workers with $h/2$ each.

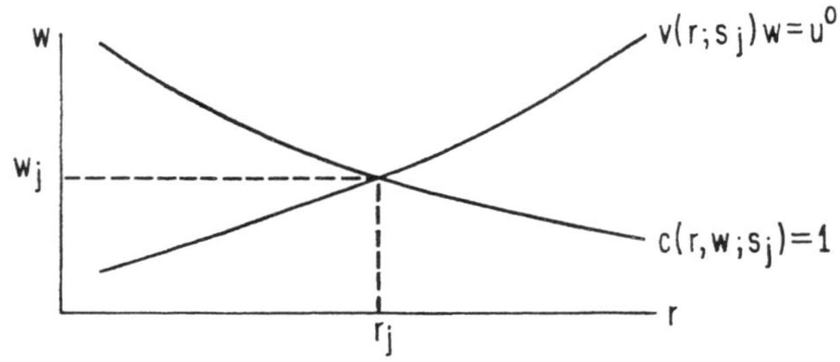

FIG. 1. Determination of the wage and rental rates in city j.

represents the w–r combinations satisfying $v(r; s_j)w = u^0$ for a given bundle of site characteristics s_j, while the downward-sloping or firm equilibrium curve represents w–r combinations satisfying $c(r, w; s_j) = 1$ for a given s_j. With r_j and w_j determined by Eqs. (1) and (2), I can find derived demands for land and labor by firms and for land by households. Given the local average level of human capital $\bar{h}_j$, I can then use the land and labor market clearing conditions to determine the equilibrium values of SMSA production of the composite good (X_j) and SMSA population (N_j), noting that $N_j\bar{h}_j$ gives the measure of all efficiency units of labor available in the SMSA:

$$X_j\, \partial c/\partial r - N_j\bar{h}_j(\partial V/\partial r)/(\partial V/\partial w) = L_j \tag{3}$$

$$X_j\, \partial c/\partial w = N_j\bar{h}_j. \tag{4}$$

It is important to note that, because there is no spatial sorting of consumers by efficiency unit endowment, $\bar{h}_j$ itself is left undetermined by my model. In particular, it is not influenced by w_j or r_j. In the empirical work of Sections 3 and 4 it is simplest to think of $\bar{h}_j$ as historically predetermined.

Consider a site characteristic that increases productivity but has no amenity value. A higher value of that characteristic will shift out the firm equilibrium curve uniformly while leaving the consumer equilibrium curve in place. According to the discussion in the Introduction,[4] the local

[4]A constant productivity differential due to differences in $\bar{h}$ is the result generated by models such as those of Lucas [17] and Azariadis and Drazen [1]. However, the microfoundations of these models suggest that a higher $\bar{h}$ leads to more rapid diffusion and growth of knowledge, implying an increasing productivity differential. One could reconcile the two specifications by supposing that after a certain fixed interval of time knowledge created within an SMSA becomes disembodied (reproduced on blueprints or taught in business schools) and thus available to the nation (or world) at large. In this case overall productivity will increase at the same rate across all SMSAs, but a fixed productivity differential will exist between SMSAs with different average levels of human capital.

average level of human capital can be thought of as just such a site characteristic.[5] As can be seen using Fig. 1, the result is that an SMSA with a higher $\bar{h}$ will have a higher wage per efficiency unit of labor and a higher rent per unit of land, holding other site characteristics constant. This result is robust to a common generalization of the model given by Eqs. (1) and (2) (see, for example, Roback [23] and Beeson and Eberts [3]), which is to allow for local production of housing services that households consume rather than consuming land services directly. Insofar as a higher $\bar{h}$ tends to enhance the productivity of housing service production, the effect of a higher $\bar{h}$ on the wage rate is dampened because the impact on households of the higher rent on land is dampened, causing them to require a smaller wage differential to be compensated for the higher land rental. It can be shown that even if the effect of $\bar{h}$ on housing service productivity is as great (as measured by the absolute value of the cost function elasticity) as its effect on composite commodity productivity, the wage differential will still be positive provided that the share of rent in the cost of housing services is greater than in the cost of composite commodity production.

3. EVIDENCE

The system consisting of Eqs. (1) and (2) that is depicted in Fig. 1 can be estimated in reduced form using hedonic wage and rent equations. In other words, I can estimate the wage of an individual as a function of both his individual characteristics, such as health and marital status, and the characteristics of the SMSA in which he lives, such as the average level of human capital. By controlling for the wage earner's individual characteristics I obtain in effect an equation for w_j in the above model. A similar equation can be estimated for housing expenditure, which serves as a good proxy for r_j when I control for the characteristics of the dwelling. Like the site characteristics used in the empirical local public goods literature, the average level of human capital in an SMSA is exogenous from the point of view of an individual consumer or firm making its location decision. Since $\bar{h}_j$ is not influenced by w_j or r_j, I treat it in the same way that exogenous variables like climate are treated in the empirical local public goods literature (see Blomquist *et al.* [4] for a recent example) and enter it directly into the reduced form regressions. On the basis of the discussion in Section 2 we expect this variable to have a positive coefficient in both the wage and the rent equations. If these coefficients are significant their size will allow me to determine the *importance* of the productivity benefits

[5]The possibility that some amenity value may be associated with $\bar{h}$ will be considered in the empirical discussion of Section 4.1.

realized from geographic concentration of human capital in a manner made precise below.

I need an empirical translation of the theoretical concept of homogeneous human capital. In the empirical labor economics literature, the human capital that an individual accumulates over her lifetime is typically decomposed into two measurable components: education and experience, measured by years of schooling completed and age minus years of schooling minus six, respectively. Suppose now that I measure the average level of human capital in an SMSA by average education and average experience. Since this decomposition could presumably be influenced by government educational policy, it would be interesting to know if these two variables differ in their external effects (if any) on total factor productivity. I therefore investigate this issue below.

The wage and rent equations are estimated using the 1 in 1000 B Public Use Microdata Sample of the 1980 Census of Population. The B Sample is collected on an SMSA basis (as opposed to the A Sample which is collected on a county basis) and is therefore ideal for our purposes. It contains information on all of the usual individual attributes of workers and dwellings known to influence wages and housing expenditures, respectively. I have included in my housing sample all housing units on 10 or fewer acres for which value of the unit or contract rent is reported. In the wage sample are included all individuals aged 16 and over who reported their earnings, hours and weeks, had nonzero wage and salary earnings, and had positive total earnings.[6] The SMSA averages for years of schooling and years of experience are computed from the wage sample. The logarithm of monthly housing expenditures is the dependent variable in the housing equation. For renters, monthly housing expenditure is defined as gross rent including utilities. For owners, reported house value is converted to monthly imputed rent using a 7.85% discount rate taken from a user cost study by Peiser and Smith [21]. Monthly expenditures for utilities are added to obtain gross imputed rent for owners. The dependent variable in the wage equation is the logarithm of average hourly earnings. Average hourly earnings are obtained by dividing annual earnings by the product of weeks worked during the year and usual hours worked per week.[7]

[6]In order to ensure more accurate reporting of home values, Beeson and Eberts [3] limit their sample to individuals who changed addresses between 1975 and 1980. However, for most individuals their largest asset is their home (Pearl and Frankel [20]), so we might expect home owners to be fairly accurate when asked to estimate the value of their units. Cross-sectional studies have in fact found this to be true on average (Kain and Quigley [15]; Robins and West [24]).

[7]Inclusion of non-traded services other than housing in the model of Section 2 might suggest adjusting wages for differences across SMSAs in non-housing cost of living. However,

The wage and rent hedonic equations were estimated using observations on 69,910 individuals and 44,758 households, respectively, residing in 237 SMSAs. I use the same set of variables used by Blomquist *et al.* [4] to control for the characteristics of individual workers and dwellings. The Census control variables listed in Tables 1 and 2 are self-explanatory. The omitted broad occupational category in the wage equation is service. The only non-Census control variable is the percentage of the individual's industry that is unionized, which is taken from a study by Kokkelenberg and Sockell [16].

The standard procedure in the empirical local public goods literature (see, e.g., the work of Roback [23] and Blomquist *et al.* [4] cited above) is to estimate the wage and rent hedonic regressions using ordinary least squares (OLS). Their underlying statistical model is therefore

$$y_{ij} = \alpha + x_{ij}\beta + z_j\gamma + \varepsilon_{ij}, \tag{5}$$

where i indexes the individual workers or dwellings, j continues to index cities, y is the dependent variable (in our case either the log wage or the log rent), x is a vector of observed individual characteristics, z is a vector of observed city characteristics, and ε is an error term that captures the effects of unobserved individual characteristics and satisfies all the properties of the classical regression model. The OLS model implicitly assumes that all relevant city characteristics are observed. If, as seems reasonable to believe, this is not the case, then a more appropriate model is the random effects model

$$y_{ij} = \alpha + x_{ij}\beta + z_j\gamma + \mu_j + \varepsilon_{ij}, \tag{6}$$

where μ is an error term that captures the effects of unobserved city characteristics and also satisfies all the properties of the classical regression model. Equation (6) can be viewed as the outcome of a model where city effects on wages and rents are explained by observed and unobserved city characteristics:

$$\begin{aligned} y_{ij} &= x_{ij}\beta + \delta_j + \varepsilon_{ij} \\ \delta_j &= \alpha + z_j\gamma + \mu_j, \end{aligned} \tag{7}$$

where δ is the city effect on log wages (log rents).

---

separate cost of living indices are reported by the Bureau of Labor Statistics for only 28 metropolitan areas covering 33 SMSAs. Beeson and Eberts [3, p. 451] report that these non-housing cost of living indices vary by only plus or minus 4% of the national average.

Since the error term $\nu_{ij} = \mu_j + \varepsilon_{ij}$ in Eq. (6) is nonspherical,[8] estimation of this model by OLS yields inefficient coefficient estimates and biased and inconsistent estimates of their standard errors.[9] In this paper I estimate Eq. (6) using generalized least squares (GLS). The estimation procedure that follows naturally from the underlying model (7) requires several steps:

(1) Estimate the least-squares dummy variable equation $y_{ij} - \bar{y}_j = (x_{ij} - \bar{x}_j)\beta + \varepsilon_{ij} - \bar{\varepsilon}_j$. The coefficient estimates b serve as useful points of reference and are also employed in the next step. The residuals from the estimated equation are used to compute a consistent estimate of σ_ε^2, the variance of ε_{ij}.

(2) Form $a_{ij} = y_{ij} - x_{ij}b$. $\bar{a}_j$ is then an estimate of $\delta_j + \bar{\varepsilon}_j$. Substitute for δ_j from (7), and estimate the equation $\bar{a}_j = \alpha + z_j\gamma + \mu_j + \bar{\varepsilon}_j$. The estimates of γ are less efficient than the GLS estimates but should be similar. The residuals from the estimated equation are used to compute a consistent estimate of $\sigma_*^2 = \sigma_\mu^2 + [(1/m)\Sigma_j(\sigma_\varepsilon^2/n_j)]$, where m (= 237) is the number of cities and n is the number of individual observations per city.

(3) Use the estimates of σ_ε^2 and σ_*^2 from steps 1 and 2 to compute a consistent estimate of σ_μ^2 and make the appropriate GLS transformation of the data (see, e.g., Greene [9, p. 487]).

In column (1) of Tables 1 and 2, respectively, I report the least-squares dummy variable estimates of the hedonic wage and rent equations from step 1. Except for anomalous negative coefficients giving the effects of condominium status and city sewer connection on log rents, the signs and magnitudes of the coefficients on the characteristics of individual workers and dwellings are as expected and are in line with those obtained by other studies. Column (2) gives the GLS estimates of these equations including the SMSA average levels of education and experience. The coefficients on the individual variables remain virtually unchanged from column (1), and the coefficients on the SMSA variables have the positive signs predicted by the theory of Section 2. All of the latter coefficients are significant at the 1% level except for the coefficient on the SMSA average level of experience in the wage equation, which is insignificant. It appears

[8] Specifically, the variance–covariance matrix for observations within any city j generated by the error term ν_{ij} has off-diagonal terms equal to σ_μ^2, the variance of μ_j, rather than zero.

[9] This problem has only recently been recognized in the empirical literature on local public goods (see, e.g., Gyourko and Tracy [10]). My experience with OLS estimates of this model is in accord with the findings of Moulton [18], who reviewed several studies where equations were estimated using both individual and group level data: the null hypothesis of spherical errors is rejected and the OLS standard errors on the coefficients for the group level variables tend to be much smaller than those estimated using the random effects model, leading to much higher t ratios for these coefficients in the OLS case.

TABLE 1
Wage Equations

Variable (Mean):	(1)	(2)	(3)	(4)	(5)	(6)
Intercept		−0.237	−0.106	−0.138	−0.059	0.024
		(0 207)	(0.200)	(0.199)	(0.190)	(0.245)
Education	0.048	0.048	0.048	0.048	0.048	0.048
(12.86)	(0.001)	(0.001)	(0.001)	(0.001)	(0.001)	(0.001)
Experience	0.035	0.035	0.035	0.035	0.035	0.035
(17.36)	(0.001)	(0.001)	(0.001)	(0.0009)	(0.0009)	(0.0009)
Experience squared	−0.00051	−0.00051	−0.00051	−0.00051	−0.00051	−0.00051
(519.41)	(0.00002)	(0.00002)	(0.00002)	(0.00002)	(0.00002)	(0.00002)
Health (limitation = 1)	−0.115	−0.116	−0.116	−0.116	−0.116	−0.116
(0.049)	(0.012)	(0.012)	(0.012)	(0.012)	(0.012)	(0.012)
Marital status (married = 1)	0.185	0.185	0.185	0.185	0.185	0.185
(0.59)	(0.009)	(0.009)	(0.009)	(0.009)	(0.009)	(0.009)
Enrollment (in school = 1)	−0.096	−0.097	−0.097	−0.097	−0.097	−0.097
(0.14)	(0.008)	(0.008)	(0.008)	(0.008)	(0.008)	(0.008)
Unionization rate	0.0052	0.0052	0.0052	0.0052	0.0052	0.0052
(22.64)	(0.0002)	(0.0002)	(0.0002)	(0.0002)	(0.0002)	(0.0002)
Race (non-white = 1)	−0.110	−0.109	−0.107	−0.107	−0.107	−0.107
(0.15)	(0.010)	(0.010)	(0.010)	(0.010)	(0.010)	(0.010)
Sex (female = 1)	−0.080	−0.080	−0.080	−0.080	−0.080	−0.080
(0.45)	(0.011)	(0.011)	(0.011)	(0.011)	(0.011)	(0.011)
Sex × experience	−0.012	−0.012	−0.012	−0.012	−0.012	−0.012
(7.54)	(0.001)	(0.001)	(0.001)	(0.001)	(0.001)	(0.001)
Sex × (experience squared)	0.00021	0.00021	0.00021	0.00021	0.00021	0.00021
(224.09)	(0.00002)	(0.00002)	(0.00002)	(0.00002)	(0.00002)	(0.00002)
Sex × (marital status)	−0.182	−0.182	−0.182	−0.181	−0.182	−0.182
(0.24)	(0.012)	(0.012)	(0.012)	(0.012)	(0.012)	(0.012)
Sex × race	0.113	0.114	0.114	0.114	0.114	0.114
(0.074)	(0.014)	(0.014)	(0.014)	(0.014)	(0.014)	(0.014)
Sex × (children under 18)	−0.024	−0.025	−0.025	−0.025	−0.025	−0.025
(1.11)	(0.003)	(0.003)	(0.003)	(0.003)	(0.003)	(0.003)
Professional or managerial	0.339	0.339	0.340	0.340	0.340	0.340
(0.22)	(0.010)	(0.010)	(0.010)	(0.010)	(0.010)	(0.010)
Technical or sales	0.174	0.174	0.174	0.175	0.175	0.175
(0.32)	(0.008)	(0.008)	(0.008)	(0.008)	(0.008)	(0.008)
Farming	−0.027	−0.027	−0.030	−0.030	−0.030	−0.029
(0.013)	(0.023)	(0.023)	(0.023)	(0.023)	(0.023)	(0.023)
Craft	0.204	0.203	0.203	0.203	0.204	0.204
(0.12)	(0.011)	(0.011)	(0.011)	(0.011)	(0.011)	(0.011)
Operator or laborer	0.119	0.118	0.118	0.118	0.119	0.119
(0.18)	(0.010)	(0.010)	(0.010)	(0.010)	(0.010)	(0.010)
Education (SMSA average)		0.051[a]	0.034[a]	0.039[a]	0.033[a]	0.028[c]
(12.86)		(0.013)	(0.013)	(0.013)	(0.012)	(0.016)
Experience (SMSA average)		0.0046	0.0071[b]	0.0060[c]	0.0028	0.0017
(17.36)		(0.0036)	(0.0036)	(0.0036)	(0.0035)	(0.0041)
West			0.105[a]	0.102[a]	0.090[a]	0.091[a]
(0.19)			(0.017)	(0.017)	(0.017)	(0.017)
North Central			0.071[a]	0.073[a]	0.078[a]	0.077[a]
(0.28)			(0.015)	(0.014)	(0.014)	(0.014)
Northeast			0.035[b]	0.035[b]	0.032[b]	0.031[c]
(0.22)			(0.017)	(0.017)	(0.016)	(0.016)
Culture per capita				−0.000056[b]	−0.000060[b]	−0.000059[b]
(310.40)				(0.000025)	(0.000025)	(0.000025)
Climate					0.000070	0.000064
(591.68)					(0.000060)	(0.000061)
Coast					0.055[a]	0.050[a]
(0.36)					(0.013)	(0.016)
Population						0.000008
(2027.91)						(0.000014)
Dependent variable Log (hourly wage) (1.70)						

Note. Standard errors are shown in parentheses. Significance levels are given only for the variables below the broken line: [a]Significantly different from zero at the 1% level. [b]Significantly different from zero at the 5% level. [c]Significantly different from zero at the 10% level.

TABLE 2

Rent Equations

Variable (Mean)	(1)	(2)	(3)	(4)	(5)	(6)
Intercept		2.059	2.933	2.916	3.105	2.845
		(0.369)	(0.299)	(0.299)	(0 262)	(0.321)
Dwelling rented						
(renter = 1)	−0.105	−0.104	−0.104	−0.104	−0.104	−0 104
(0.41)	(0.022)	(0.022)	(0.022)	(0.022)	(0.022)	(0.022)
Units at address	−0.0088	−0.0088	−0.0087	−0.0087	−0.0087	−0.0087
(2.71)	(0.0029)	(0.0029)	(0.0029)	(0.0029)	(0.0029)	(0.0029)
× renter	0.0029	0.0030	0.0028	0.0028	0.0029	0.0029
	(0.0030)	(0.0030)	(0.0030)	(0.0030)	(0.0030)	(0.0030)
Age of structure	−0.0052	−0.0052	−0.0052	−0.0052	−0.0052	−0.0052
(23.39)	(0.0002)	(0.0002)	(0.0002)	(0.0002)	(0.0002)	(0.0002)
× renter	0.0017	0.0017	0.0017	0.0017	0.0017	0.0017
	(0.0003)	(0.0003)	(0.0003)	(0.0003)	(0.0003)	(0.0003)
Number of floors	0.016	0.016	0.016	0.016	0.016	0.016
(2.52)	(0.003)	(0.003)	(0.003)	(0.003)	(0.003)	(0 003)
× renter	−0.030	−0.030	−0.030	−0.030	−0.030	−0.030
	(0.003)	(0.003)	(0.003)	(0.003)	(0.003)	(0.003)
Number of rooms	0.097	0.098	0.097	0.098	0.098	0.098
(5.29)	(0.002)	(0.002)	(0.002)	(0.002)	(0.002)	(0.002)
× renter	−0.011	−0.011	−0.011	−0.011	−0.011	−0.011
	(0.004)	(0.004)	(0.004)	(0.004)	(0.004)	(0.004)
Number of bedrooms	0.021	0.021	0.021	0.021	0.021	0.021
(2.46)	(0.004)	(0.004)	(0.004)	(0.004)	(0.004)	(0.004)
× renter	−0.017	−0.017	−0.017	−0.017	−0.017	−0.017
	(0.007)	(0.007)	(0.007)	(0.007)	(0.007)	(0.007)
Number of bathrooms	0.158	0.159	0.159	0.159	0.159	0.159
(1.66)	(0.003)	(0.003)	(0.003)	(0.003)	(0.003)	(0.003)
× renter	−0.014	−0.014	−0.014	−0.014	−0.014	−0.014
	(0.006)	(0.006)	(0.006)	(0.006)	(0.006)	(0.006)
Condominium (= 1)	−0.098	−0.095	−0.096	−0.096	−0.096	−0.096
(0.03)	(0.016)	(0.016)	(0.016)	(0.016)	(0.016)	(0.016)
× renter	0.158	0.158	0.159	0.159	0.159	0.159
	(0.025)	(0.025)	(0.025)	(0.025)	(0.025)	(0.025)
Central air cond. (= 1)	0.165	0.162	0.163	0.163	0.163	0.163
(0.32)	(0.006)	(0.006)	(0.006)	(0.006)	(0.006)	(0.006)
× renter	0.125	0.125	0.125	0.125	0.125	0.125
	(0.009)	(0.009)	(0.009)	(0.009)	(0.009)	(0.009)
City sewer						
connection (= 1)	−0.013	−0.013	−0.013	−0.013	−0.013	−0.013
(0.86)	(0.007)	(0.007)	(0.007)	(0.007)	(0.007)	(0.007)
× renter	−0.037	−0.036	−0.037	−0.037	−0.037	−0.037
	(0.014)	(0.014)	(0.014)	(0.014)	(0.014)	(0.014)
Lot size > 1 acre (= 1)	0.156	0.155	0.155	0.155	0.156	0.156
(0.075)	(0.009)	(0.009)	(0.009)	(0.009)	(0.009)	(0.009)
× renter	−0.183	−0.183	−0.183	−0.183	−0.184	−0.184
	(0.017)	(0.017)	(0.017)	(0.017)	(0.017)	(0.017)
Education (SMSA						
average)		0.199[a]	0.128[a]	0.131[a]	0.112[a]	0.128[a]
(12.87)		(0.023)	(0.019)	(0.019)	(0.017)	(0.021)
Experience (SMSA						
average)		0.027[a]	0.021[a]	0.020[a]	0.013[a]	0.017[a]
(17.48)		(0.006)	(0.005)	(0.005)	(0.005)	(0.005)
West			0.308[a]	0.306[a]	0.276[a]	0.274[a]
(0.19)			(0.028)	(0.028)	(0.025)	(0.025)
North Central			0.139[a]	0 140[a]	0.157[a]	0.159[a]
(0.27)			(0.023)	(0.023)	(0.020)	(0.020)
Northeast			0.246[a]	0.246[a]	0.232[a]	0.235[a]
(0.24)			(0.027)	(0.027)	(0.024)	(0.024)
Culture per capita				−0.000034	−0.000033	−0.000035
(309.87)				(0.000038)	(0.000034)	(0.000034)
Climate					0.000250[a]	0.000277[a]
(595.11)					(0.000087)	(0.000090)
Coast					0.144[a]	0.161[a]
(0.38)					(0.019)	(0.022)
Population						−0.000034
(2242.61)						(0.000024)
Dependent variable Log (monthly housing expenditure) (5.82)						

Note. Standard errors are shown in parentheses. Significance levels are given only for the variables below the broken line: [a]Significantly different from zero at the 1% level. [b]Significantly different from zero at the 5% level. [c]Significantly different from zero at the 10% level.

that the average level of human capital is a productive local public good as described in Section 2. It is also important to note that the average level of education has a much greater productive external effect than the average level of experience. While at this level of empirical analysis I cannot investigate directly the cause of this difference, it appears to be consistent with the microeconomic foundation for external effects of human capital described in the introduction to this paper. There it was argued that the average level of human capital influences productivity indirectly through its effect on sharing of ideas for technological improvements. It stands to reason that the probability that a meeting between agents/ideas in an SMSA will be "productive" is increased more by a year of SMSA average education than by a year of SMSA average experience, since a major part of formal education is concerned with communication skills, i.e., reading, writing, and (to a lesser extent) oral presentation.

4. ALTERNATIVE EXPLANATIONS

4.1. Omitted Variables

It could be argued that the large coefficients on the SMSA average level of education shown in column (2) of Tables 1 and 2 are the result of the association of average education with other exogenous variables that increase wages and/or rents at the SMSA level. In particular, one would expect the average level of education to be higher where there is a high concentration of universities and university-administered federally funded research and development centers. This in turn means that average education should be associated with a high level of federal R & D funding per capita, an exogenous variable that could possibly be expected to have the same local effects on productivity that I have argued are generated by the local average level of human capital. I have therefore aggregated to the SMSA-level 1980 data supplied by the National Science Foundation [19] on federal R & D funding at the level of the individual university and university-administered federally funded R & D center.[10] When this variable was added to the wage and rent equations the estimated coefficient was negative and insignificant. A possible reason for this result is that this variable is very skewed because a few SMSAs (especially those with federally funded R & D centers) receive very large amounts of federal R & D dollars per capita relative to all others.[11] I attempted to correct this problem of skewness as follows. I created a dummy variable that equals zero if an SMSA received no federal R & D dollars and one if it received

[10]The wage sample correlation coefficient between SMSA average education and federal R & D expenditure per capita is 0.22.

[11]For the wage sample the measure of skewness for this variable exceeds 16. When all observations with zero dollars are deleted the measure of skewness still exceeds 15.

positive federal R & D dollars. This dummy variable and its product with the logarithm of federal R & D dollars per capita were then included in the wage and rent equations. Again, the estimated coefficients on the new variables were insignificant. I also experimented with a per capita, weighted sum of major universities present in each SMSA taken from *Places Rated Almanac* [7], without obtaining better results.

It is widely believed that workers in the U.S. South tend to be less educated than workers elsewhere in the country. Indeed, in our wage sample the average level of education for the U.S. Census region South is 12.7 years of schooling, compared to 13.2, 12.8, and 12.9 years for the West, North Central, and Northeast regions, respectively. If there are historical or other reasons independent of average education levels that cause Southern productivity to be lower, failing to control for the region in which workers reside and homes are located could bias upwards the coefficient on SMSA average education in the wage and rent equations. As seen in column (3) of Tables 1 and 2, inclusion of regional dummies does in fact reduce the estimated coefficient on average education substantially in both equations. As expected, residence outside the South has a positive effect on both wages and rents.

Average education may also be correlated with omitted variables that have amenity value. Since amenities shift down the consumer equilibrium curve in Fig. 1, lowering wages and raising rents, such a correlation would bias downwards (upwards) the estimated coefficient on average education in the wage (rent) equation. The obvious candidate for an SMSA-level amenity that is correlated with average education is cultural facilities. *Places Rated Almanac* [7] has compiled a "culture per capita" index based on SMSA possessions of major universities, symphony orchestras, opera companies, dance companies, theaters, public television, fine arts radio, museums, and public libraries (see pp. 293–294 and 320 for exact definition). I have subtracted the major universities component since I argued above that this could have a productivity effect (although I found no evidence for it). This corrected culture per capita index is then included as an explanatory variable in column (4) of Tables 1 and 2. We see that in the wage equation the effects are as expected: the coefficient on culture per capita is negative and significant at the 5% level, while the coefficient on average education rises from 0.034 to 0.039.[12] The rent equation results are not as expected, however, since the coefficient on the corrected culture per capita index is negative, though insignificant.

I turn next to two site characteristics that regularly appear in the local public goods literature and have both productivity and amenity value:

[12]The wage sample correlation coefficient between average education and the corrected culture per capita index is 0.35.

climatic mildness and coastal location. The cost-reducing and utility-enhancing effects of climatic mildness need no explanation. Coastal location is a recreational amenity, and I have argued elsewhere (Rauch [22]) that wages and rents should be higher in port cities due to their privileged access to the gains from international trade. *Places Rated Almanac* [7] has created a rating for SMSA climatic mildness (see p. 4 for exact definition) based on average seasonal temperature variation and the average annual number of heating- and cooling-degree days, very hot and very cold months, freezing days, zero days, and 90-degree days (adjusted for relative humidity). I let a dummy variable for coastal location equal one if an SMSA touches an ocean or any of the Great Lakes and zero otherwise.[13] The climate rating and the coast dummy are entered as explanatory variables in column (5) of Tables 1 and 2. The coefficient on climate is positive and significant at the 1% level in the rent equation and insignificant in the wage equation, indicating that its productivity and amenity effects are of comparable importance, while the coefficient on coast is positive and significant at the 1% level in both the wage and rent equations, indicating that its productivity effects dominate any amenity effects. The coefficient on the SMSA average level of education falls slightly in the wage equation while the coefficient on the SMSA average level of experience becomes completely insignificant.[14] In the rent equation the coefficients on both SMSA average level of human capital variables fall.

In the model of Section 2, SMSA population N_j does not enter the list of relevant SMSA site characteristics s_j. I am thus able to solve the model sequentially by first using Eqs. (1) and (2) to solve for r_j and w_j as functions of s_j and next using Eq. (4) to substitute out for SMSA output X_j in Eq. (3), which then determines N_j as a function of s_j and SMSA land L_j. However, much empirical research in urban economics starts from the premise that city population could be a consumption disamenity (e.g., Roback [23]) or that it could increase productivity (e.g., Henderson [12] and the references cited therein). In this case I cannot omit N_j from the reduced form equations for w_j and r_j, but if I include it these equations suffer from simultaneity bias. Fortunately, Eq. (3) immediately suggests

[13] Ports on the Great Lakes trade directly with Europe via the St. Lawrence Seaway.

[14] Since we saw in our discussion of regional effects that average education is highest in the West and next to lowest in the North Central region, we might expect a positive correlation in our wage sample between SMSA climate rating and SMSA average education, and indeed the correlation coefficient is 0.27. On the other hand, the wage sample correlation coefficient between coastal location and average education is only 0.02. If entered separately, the climate and coast variables both reduce the coefficient on average education in the wage equation to 0.035.

using SMSA land area as an instrument for SMSA population.[15] Total land area, as opposed to the quality of a given plot of land, can reasonably be supposed to have neither productivity value nor amenity value, and can therefore be excluded from Eqs. (1) and (2). It follows that in this case the wage and rent reduced form equations are just identified, and we can estimate them by applying the method of instrumental variables to the data after it has been subjected to the GLS transformation described in section 3.[16]

SMSA population and land area in 1980 are available from the *State and Metropolitan Area Data Book, 1982*, published by the Bureau of the Census. Is land area a good instrument for population? I address this question by running a cross-SMSA regression of population on all the SMSA-level variables in column (5) of Tables 1 and 2 plus land area. In this regression population is expressed in thousands and land area in square miles. The estimated equation, giving t ratios in parentheses, is

$$\begin{aligned} \text{POP} = {} & -6879.8 + \underset{(3.17)}{397.5} \times \text{AVGED} + \underset{(2.11)}{73.9} \times \text{AVGEXP} \\ & - \underset{(2.14)}{430.7} \times \text{W} + \underset{(0.86)}{130.8} \times \text{NC} \\ & + \underset{(1.04)}{185.6} \times \text{NE} + \underset{(0.05)}{0.01} \times \text{CULTPC} \\ & + \underset{(2.09)}{1.38}\ \text{CLIMATE} + \underset{(3.45)}{500.1} \times \text{COAST} \\ & + \underset{(5.40)}{0.15} \times \text{LAND}. \end{aligned}$$

The partial correlation coefficient between SMSA population and land area implied by the t ratio for the latter is 0.34. We can conclude that land area is a useful but far from perfect instrument for population.

With this information in hand, we look at the estimates reported in column (6) of Tables 1 and 2. We see that the coefficients on population in both the wage and rent equations are insignificant, although the signs of the coefficients support the view that population is a consumption dis-amenity. In view of the quality of the instrument we have for population one should hesitate to draw the conclusion that population is not a relevant site characteristic. The coefficients on the SMSA-level variables that were positively associated with population in the cross-SMSA regres-

[15]The model implicitly assumes that land area is given exogenously by arbitrary political boundaries. To the extent that outlying political jurisdictions are added to an SMSA on the basis of population, land area may to some extent be endogenous for the same reasons as population.

[16]This is the "IV-GLS analog" estimation recommended by Bowden and Turkington [6, Chap. 3].

sion all fall slightly in the wage equation and rise slightly in the rent equation. The standard errors for the coefficients on the variables for which that association was significant rise noticeably in both equations. The only important result of these changes is that the coefficient on SMSA average education in the wage equation is now only significant at the 10% level. However, given that the results in column (6) indicate that we have added an irrelevant variable to both the wage and rent equations, we must conclude that the preferred model remains that in column (5).

4.2. *Self-selection*

It is possible that the positive effects of SMSA average education in the wage equations of Table 1 reflect its association with higher unobserved ability of workers rather than higher productivity of a given worker. This could happen because higher average education in an SMSA is positively correlated with the return to unobserved ability or skill, causing higher quality workers to migrate to SMSAs with higher average education levels.[17] A formal model of this spatial selection or sorting process is set out by Borjas *et al.* [5]. They apply the Roy [25] model of occupational choice to selection between regions of residence. Their model can be expressed using the notation of Eq. (7) if we impose more structure on the error term ε_{ij}. Specifically, it is supposed that the log wage of individual i in region (SMSA) j is given by

$$\log w_{ij} = x_{ij}\beta + \delta_j + \varepsilon_{ij} = x_{ij}\beta + \delta_j + \eta_j\psi_i, \tag{7'}$$

where ψ indexes an individual's unobserved ability or skills and is a continuous random variable with mean zero and a range defined over the real number line. The parameter η_j can then be interpreted as the "rate of return" to ability in SMSA j. Other things equal, it is clear that an individual with a high (low) realization of ψ will want to live and work in an SMSA with a high (low) η. In fact, Borjas *et al.* are able to show that

$$E(\psi | \text{choose } j) > E(\psi | \text{choose } k) \qquad \text{if and only if } \eta_j > \eta_k.$$

[17]It could also be that the return to observed human capital, in particular years of schooling, is positively correlated with SMSA average education. Including the cross-product between individual education and SMSA average education in the wage equation resulted in positive but insignificant coefficients on all three terms involving education: individual education, SMSA average education, and the cross-product. It seems that including the cross-product introduces a multicollinearity problem. However, from the separate SMSA wage regressions that were run to compute the "standardized" measure of wage dispersion discussed at the end of this subsection I have separate estimates of the education coefficient for each SMSA. The correlation coefficient between these and SMSA average education is 0.08, which is not significant at even the 10% level.

We can complete the formalization of the selection argument by supposing that η is positively correlated with the average level of formal schooling in an SMSA, in which case this average will have a positive coefficient in the log wage equation that reflects selection bias.

I investigate the selection argument by examining both what it implies and what it does *not* imply. Taking the second approach first, one can note a key difference between the model just outlined and the model of Section 2: the latter is a general equilibrium model that determines SMSA rents as well as wages, while the former is a partial equilibrium model that determines SMSA wage levels only. More specifically, in the model of Section 2 rents and wages adjust so as to make all individuals indifferent concerning in which SMSA they live and work. If the model of Section 2 is correct in that SMSA average education is a site characteristic that increases productivity but has no amenity value, then ceteris paribus an additional year of SMSA average education should raise rents so as to just offset the benefit to consumers from its positive effect on wages. In short, changes in SMSA average education should shift the firm equilibrium curve in Fig. 1 along an unchanged consumer equilibrium curve. On the other hand, the Roy model of spatial selection makes *no* predictions concerning inter-SMSA rent differentials. Thus the apparent effect of SMSA average education on wages, which according to this model is due to selection bias, should have no particular relationship to any effect of SMSA average education on rents. A finding of the relationship predicted by the model of Section 2 would therefore be a remarkable coincidence.

The consumer equilibrium curve is defined by Eq. (1) in Section 2. By logarithmically differentiating this equation and using Roy's identity, we can show that movement along the consumer equilibrium curve implies

$$-\phi_r\left(dr/d\bar{h}\right)/r + \left(dw/d\bar{h}\right)/w \equiv \Delta = 0, \tag{8}$$

where ϕ_r is the share of consumers' (labor) income spent on rent. Note that $(dr/dx)/r$ and $(dw/dx)/w$ are given by the coefficients on independent variable x in the rent and wage equations, since $\log(r)$ and $\log(w)$ are the dependent variables. We should be able to accept the hypothesis that $\Delta = 0$ when we substitute into Eq. (8) the coefficients on SMSA average education reported in column (5) of Tables 1 and 2 and the sample value $\phi_r = 0.30$. We have $-(0.30)(0.112) + (0.033) = -0.0006$ with a standard error of 0.013, ignoring any error in the estimate of ϕ_r. Thus the relationship between the wage and rent effects of SMSA average education is as predicted by the model of Section 2, which I interpret as evidence favoring the productivity interpretation of the wage effect over the selection bias interpretation.

One can also attempt to directly address the implication of the selection model that SMSA average education is positively correlated with the SMSA return to ability η. Following Borjas *et al.*, we note from (7′) that η is proportional to the extent of earnings inequality, so that one can use wage dispersion within each region to measure returns to ability.[18] They compute two measures of wage dispersion, whose construction I imitate exactly. The first is the standard deviation of the log hourly wage, which they call the "unstandardized" dispersion in wages. The alternative "standardized" measure of dispersion is the root mean-square error from SMSA-specific log wage regressions. The unstandardized measure of wage dispersion ranges from 0.54 to 1.11, with a mean of 0.77 and a standard deviation of 0.10, while the standardized measure ranges from 0.28 to 1.04 with a mean of 0.65 and a standard deviation of 0.11. The two measures are highly correlated across SMSAs, with a correlation coefficient of 0.88 that is similar to the correlation coefficient of 0.90 that Borjas *et al.* found across states. I also computed a third measure of wage dispersion that is more directly based on the model (7′), which is the standard deviation by SMSA of the residuals from the least-squares dummy variable (LSDV) equation in column (1) of Table 1.[19] The descriptive statistics for this LSDV measure are very similar to those for the standardized measure. If SMSA average education is positively correlated with returns to ability, as the selection argument requires, then it should be positively correlated with one or more of these measures of wage dispersion. However, the correlation coefficients of SMSA average education with the unstandardized, standardized, and LSDV measures of wage dispersion are, respectively, 0.03, 0.01, and −0.01, none of which is significantly different from zero at even the 10% level.

5. CONCLUSIONS

I began this paper by noting the widespread belief in the existence of positive externalities from formal education. The evidence presented in Sections 3 and 4 supports the existence of these externalities. I am now in the position to answer the question of how *large* the effect of an additional year of average education is on total factor productivity. Using the coefficients on the average level of human capital variables reported in Tables 1 and 2, I can compute an estimate of the percentage cost reduction (total

[18] If the spatial sorting generated by the model (7′) is sufficiently strong, the differences between wage dispersions across regions ex ante will be greatly attenuated ex post. However the descriptive statistics reported show substantial variation across SMSAs in the measures of wage dispersion that I use.

[19] The model (7′) suggests that the error term ε_{ij} might be groupwise heteroskedastic. No significant changes in the wage and rent equations reported in Tables 1 and 2 occur if we adjust the GLS transformation to allow for this.

factor productivity increase) that would result from an additional year of average education, taking the conservative view that this implies an offsetting one year reduction in average experience. By logarithmically differentiating Eq. (2), we obtain

$$[(\partial c/\partial r)/c](dr/d\bar{h}) + [(\partial c/\partial w)/c](dw/d\bar{h}) = -(\partial c/\partial \bar{h})/c, \quad \text{or}$$

$$\theta_{\mathrm{r}}(dr/d\bar{h})/r + \theta_{\mathrm{w}}(dw/d\bar{h})/w = -(\partial c/\partial \bar{h})/c, \tag{9}$$

where θ_{r} and θ_{w} are the land and labor shares in the value of output, respectively. I follow Beeson and Eberts [3, p. 448], who use national accounts data to compute the figures 0.064 and 0.73 for θ_{r} and θ_{w}. Substituting into Eq. (9) these figures and the coefficients on average education and average experience reported in column (5) of Tables 1 and 2, we have (0.064)(0.112) + (0.73)(0.033) = 0.031 for the effect of the SMSA average level of education and (0.064)(0.013) + (0.73)(0.003) = 0.003 for the effect of the SMSA average level of experience, yielding a net effect of an additional year of average education equal to 0.028. The standard error of the net effect is 0.008, ignoring any errors in the estimates of θ_{r} and θ_{w}. My best estimate on the basis of the evidence presented in this paper is therefore that each additional year of SMSA average education can be expected to raise total factor productivity by 2.8% with a standard error of estimate of 0.8%.

Is this estimate reasonable? The only other attempt to quantify the productivity effect of the average level of human capital of which I am aware is in Lucas [17]. He builds a theoretical model of long-run growth where the "engine of growth" is accumulation of human capital, and the average level of human capital raised to a power enters the aggregate production function. Lucas "estimates" that exponent using U.S. data under the assumption that the United States was on the balanced growth path described by his model during the period 1909–1957. (He notes that one could fit this data equally well using a model in which there is no human capital externality.) The result is 0.417, which is basically driven by the difference between the rates of growth of physical and human capital per capita estimated by Denison [8]. This exponent implies that an additional year of average education increases U.S. total factor productivity by 3.2%.[20] My estimate of 2.8% therefore seems very reasonable in

[20]There is no experience component of human capital in Lucas's model or in the estimate of the rate of growth of human capital per capita that he uses, so the average level of human capital identically equals the average level of education. The Cobb–Douglas formulation implies a constant *elasticity* of total factor productivity with respect to average education, which I use in combination with the mean of 12.86 years reported in Table 1 to obtain the effect of a one-year increase in average education.

light of the only other comparable figure available. While one would obviously not want to make too much of the closeness of the two numbers, it is nevertheless encouraging to see this degree of agreement between cross-sectional and time series findings.

A divergence between the private and social returns to schooling follows from the existence of positive externalities from formal education. The coefficients in column (5) of Table 1 imply that the social return exceeds the private return by a factor of (0.048 + 0.033)/0.048, or roughly 1.7. While this seems high, it is again similar to the findings of Lucas [17]. He uses a labor share of 0.75, the average over the period 1909–1957, rather than the figure 0.73 that I use above. His Cobb–Douglas production function then implies that the social marginal product of human capital exceeds its private marginal product by a factor of (0.75 + 0.417)/0.75, or roughly 1.6.

In Section 3 I argue that the difference found between the effects of the SMSA average levels of education and experience supports the hypothesis that my results reflect productivity benefits from geographic concentration of human capital caused by sharing of ideas. Yet I would hesitate to claim that this study has enabled us to "see" sharing of ideas at work: this may require empirical analysis at a still further disaggregated level than the city. Rather it is hoped that this study will serve both as a springboard for more detailed research and as a stimulus for attempts at replication, perhaps using data from countries other than the United States.

REFERENCES

1. C. Azariadis and A. Drazen, Threshold externalities in economic development, *Quarterly Journal of Economics*, **105**, 501–526 (1990).
2. B. Balassa, The changing pattern of comparative advantage in manufactured goods, *Review of Economics and Statistics*, **61**, 259–266 (1979).
3. P. E. Beeson and R. W. Eberts, Identifying productivity and amenity effects in interurban wage differentials, *Review of Economics and Statistics*, **71**, 443–452 (1989).
4. G. C. Blomquist, M. C. Berger, and J. P. Hoehn, New estimates of quality of life in urban areas, *American Economic Review*, **78**, 89–107 (1988).
5. G. J. Borjas, S. G. Bronars, and S. J. Trejo, Self-selection and internal migration in the United States, *Journal of Urban Economics*, **32**, 159–185 (1992).
6. R. J. Bowden and D. A. Turkington, "Instrumental Variables," Cambridge Univ. Press, Cambridge (1984).
7. R. Boyer and D. Savageau, "Places Rated Almanac," Rand McNally, Chicago (1981).
8. E. F. Denison, "The Sources of Economic Growth in the United States," Committee for Economic Development, New York (1961).
9. W. H. Greene, "Econometric Analysis," Macmillan Co., New York (1990).
10. J. Gyourko and J. Tracy, The structure of local public finance and the quality of life, *Journal of Political Economy*, **99**, 774–806 (1991).
11. J. V. Henderson, "Economic Theory and the Cities," Academic Press, Orlando, FL (1985).

12. J. V. Henderson, Efficiency of resource usage and city size, *Journal of Urban Economics*, **19**, 47–70 (1986).
13. J. Jacobs, "The Economy of Cities," Random House, New York (1969).
14. B. Jovanovic and R. Rob, The growth and diffusion of knowledge, *Review of Economic Studies*, **56**, 569–582 (1989).
15. J. F. Kain and J. M. Quigley, Note on owner's estimate of housing value, *Journal of the American Statistical Association*, **67**, 803–806 (1972).
16. E. C. Kokkelenberg and D. R. Sockell, Union membership in the United States: 1973–1981, *Industrial and Labor Relations Review*, **38**, 497–543 (1985).
17. R. E. Lucas, On the mechanics of economic development, *Journal of Monetary Economics*, **22**, 3–42 (1988).
18. B. R. Moulton, Random group effects and the precision of regression estimates, *Journal of Econometrics*, **32**, 385–397 (1986).
19. National Science Foundation, "Academic Science/Engineering: R & D Funds. Detailed Statistical Tables," NSF 84-308, **132**, pp. 31–42, 132.
20. R. B. Pearl and M. Frankel, Composition of the personal wealth of American households at the start of the eighties, *in* "The Collection and Analysis of Economic and Consumer Behavior Data" (S. Sudman and M. A. Spaeth, Eds.), Bureau of Economic and Business Research Laboratory, Univ. of Illinois (1984).
21. R. B. Peiser and L. B. Smith, Homeownership returns, tenure choice and inflation, *American Real Estate and Urban Economics Journal*, **13**, 343–360 (1985).
22. J. E. Rauch, Comparative advantage, geographic advantage, and the volume of trade, *Economic Journal*, **101**, 1230–1244 (1991).
23. J. Roback, Wages, rents, and the quality of life, *Journal of Political Economy*, **90**, 1257–1278 (1982).
24. P. K. Robins and R. W. West, Measurement error in the estimation of home value, *Journal of the American Statistical Association*, **72**, 290–294 (1977).
25. A. D. Roy, Some thoughts on the distribution of earnings, *Oxford Economic Papers*, **3**, 135–146 (1951).
26. T. P. Schultz, Education investment and returns, *in* "Handbook of Development Economics Volume I" (H. Chenery and T. N. Srinivasan, Eds.), North-Holland, Amsterdam (1988).

[15]

Productivity and the Density of Economic Activity

By ANTONIO CICCONE AND ROBERT E. HALL *

To explain the large differences in labor productivity across U.S. states we estimate two models—one based on local geographical externalities and the other on the diversity of local intermediate services—where spatial density results in aggregate increasing returns. Both models lead to a relation between county employment density and productivity at the state level. Using data on gross state output we find that a doubling of employment density increases average labor productivity by around 6 percent. More than half of the variance of output per worker across states can be explained by differences in the density of economic activity. (*JEL* R10)

Differences in average labor productivity across U.S. states are large: in 1988, output per worker in the most productive state was two thirds larger than in the least productive state, and output per worker in the top ten productive states was one third larger than in the ten states which ranked at the bottom. Our purpose in this paper is to look at data on the spatial dimension of externalities and increasing returns to examine how they relate to these differences in average labor productivity. To do so we develop two different models—one based on local geographical externalities and the other on the variety of local intermediate services—where the spatial density of economic activity is the source of aggregate increasing returns.

By density we mean simply the intensity of labor, human, and physical capital relative to physical space. Density is high when there is a large amount of labor and capital per square foot. Density affects productivity in several ways. If technologies have constant returns themselves, but the transportation of products from one stage of production to the next involves costs that rise with distance, then the technology for the production of all goods within a particular geographical area will have increasing returns—the ratio of output to input will rise with density. If there are externalities associated with the physical proximity of production, then density will contribute to productivity for this reason as well. A third source of density effects is the higher degree of beneficial specialization possible in areas of dense activity. Although the idea that denser economic activity had advantages from agglomeration was implicit in a large earlier literature, there does not appear to be any earlier work in which density was an explicit element of the theory, nor has there been empirical work based on measures of density.

The finest level of geographical detail in the United States for which reliable data on value added have been assembled appears to be the state level. Thus the observations on output are for the 50 states and the District of Columbia. But the average density of activity for a state is a meaningless concept. Most of the area of the United States supports essentially no economic activity at all. To get a meaningful measure of density, as well as a sensible specification for the geographical extent of the spillovers, we use much more detailed data by county. This work views the unit of production to be the labor, capital, and land present in a county. Estimation involves dealing with the aggregation from the county to the state level. In effect, we create an index of inputs for each

* Ciccone: Department of Economics, Universitat Pompeu Fabra, Balmes 132, 08008 Barcelona, Spain, and University of California, Berkeley; and Hall: Hoover Institution, Stanford University, Stanford CA, 94305-6072, and National Bureau of Economic Research. This research was supported in part by the National Science Foundation and is part of the Economic Fluctuations Program of the National Bureau of Economic Research. We would like to thank Steve Davis for generously providing data, Chris Wilkins for outstanding research assistance, and the editor and two anonymous referees for comments.

state, adjusted for density at the county level. The index depends on the extent of increasing returns. The estimate of increasing returns is the one that generates a cross-sectional pattern of the input index that most closely matches the pattern of value added across states.

To summarize our results briefly: capital accounts for some of the differences in productivity across U.S. states but leaves most of the variation unexplained. Estimation of our model of locally increasing returns reveals that accounting for the density of economic activity at the county level is crucial for explaining the variation of productivity at the state level. According to our estimates, a doubling of employment density in a county results in a 6-percent increase of average labor productivity. This degree of locally increasing returns can explain more than half of the variation of output per worker across states.

The paper starts by discussing the related literature. We develop the two models and the corresponding method of aggregating the county employment data to state levels. We then describe how differences in density and productivity can persist in equilibrium. Following a description of the county employment data and state value-added data, we discuss identification and estimation. The main results are summarized in Section VI. Section VII extends the models to account for differences in the availability of public capital, externalities from density at the state level, and externalities from output at the county level.

I. Related Literature

The economics of agglomeration began with Alfred Marshall (1920), who emphasized technological spillovers from one firm to another one nearby. J. Vernon Henderson (1974) formalized Marshall's ideas and demonstrated—building on work by Edwin S. Mills (1967)—that, in an equilibrium, disamenities from agglomeration on the side of households may offset the productivity advantages on the side of firms. A second branch of the literature on agglomeration hypothesizes economies of scale internal to firms. Mills was an early contributor. An essential task with internal increasing returns is to offer a coherent theory of the firm and its market. Mills assumed that all goods are produced by monopolists. More recent papers use a monopolistically competitive market structure to study agglomeration with internal increasing returns to scale. H. M. Abdel-Rahman (1988), Masahisa Fujita (1988, 1989) and Francisco L. Rivera-Batiz (1988) employ the well known formalization of monopolistic competition of A. Michael Spence (1976) and Avinash Dixit and Joseph E. Stiglitz (1977) to demonstrate that nontransportable intermediate inputs produced with increasing returns imply agglomeration. In related models, Paul R. Krugman (1991) demonstrates that agglomeration will result even when transportation costs are small, if most workers are mobile, and Ciccone (1992) argues that patterns of agglomeration are reinforced by endogenous technology adoption. The essence of these models is that when local markets are more active, a larger number of producers of the differentiated intermediate inputs break even. The production of final goods is more productive when a greater variety of intermediate inputs is available.

Empirical studies of agglomeration have focused on city and industry size as determinants of productivity and on technological spillovers as a source of agglomeration economies. Leo A. Sveikauskas (1975), David Segal (1976) and Ronald L. Moomaw (1981, 1985) estimated the effect of city population on productivity. Henderson (1986) found that the productivity of firms increases with the size of the industry as measured by industry employment. But all of these studies are seriously flawed by their reliance on unsatisfactory measures of output from the Census of Manufactures. In addition—following past theoretical and empirical work—they focus on the return to city size. We are not aware of any studies that have examined spatial density directly. We believe that density rather than size is a more accurate determinant. Close calls, such as whether San Francisco and Oakland are the same or different cities, have an important effect in empirical work based on city size, but none at all in our approach based on density. Finally, there appears to be little empirical work investigating the role of geographically localized externalities and increasing returns for explaining the differences in labor productivity across U.S. states. The most closely related work is by Gerald A. Carlino and Richard Voith (1992) who find that total factor productivity across U.S. states increases with urbanization.

An important branch of the empirical literature on agglomeration has studied geographic wage differentials. Wages are higher in cities and other dense areas. No study of wage differentials, though (so far as we know), has considered spatial density directly. We do not attempt to summarize this literature. Edward L. Glaeser and David C. Mare (1994) is an important recent contribution which looks at the wages in many cities. U.S. data on the noncash component wages and salaries are not available by state or other geographic breakdown, as far as we know; differences in cash wages may therefore reflect differences in state taxes and other policies that would affect the split between cash and noncash compensation. Still, exploitation of the rich geographic detail available for cash compensation may be a promising area for future research.

II. Models

A. *Increasing Returns from Externalities*

The ideas of this paper are easiest to understand in models without capital, with land and labor are the factors of production. We begin with a model based on externalities developed to show how density affects productivity and how to aggregate across productive units. The model says nothing new about the sources of agglomeration effects. Let $f(n, q, a)$ be the production function describing the output produced in an acre of space by employing n workers (all space is considered equivalent). The acre is embedded in a larger area (a county, in our empirical work) with total output q and total acreage a. The last two variables describe the density externality in a very general way. We make the assumptions that the externality depends multiplicatively on a particular measure of density, namely output per acre, the elasticity of output with respect to density is a constant, $(\lambda - 1)/\lambda$, and the elasticity of output with respect to employment is also a constant, α:

$$(1) \qquad f(n, q, a) = n^{\alpha}\left(\frac{q}{a}\right)^{(\lambda-1)/\lambda}.$$

The labor employed in a county, n_c, is distributed equally among all the acres in the county. Thus total output in county c is $q_c = a_c(n_c/a_c)^{\alpha}(q_c/a_c)^{(\lambda-1)/\lambda}$. The county-wide joint technology is described by the production function obtained by solving this equation for output:

$$(2) \qquad \frac{q_c}{a_c} = \left(\frac{n_c}{a_c}\right)^{\gamma}.$$

Here γ is the product of the production elasticity, α, and the elasticity of the externality, λ: $\gamma = \alpha\lambda$; α measures the effect of congestion and λ measures the effect of agglomeration. Only the product, γ, is identified in our data. Our empirical results show that the net effect favors agglomeration.

We turn now to aggregating to the state level. Let C_s be the set of counties covering state s. Output in state s is $Q_s = \Sigma_{c \in C_s} n_c^{\gamma} a_c^{-(\gamma-1)}$ and hence average labor productivity in the state is

$$(3) \qquad \frac{Q_s}{N_s} = \frac{\sum_{c \in C_s} n_c^{\gamma} a_c^{-(\gamma-1)}}{N_s},$$

where N_s is the number of workers in state s, and we define the factor density index, $D_s(\gamma)$:

$$(4) \qquad D_s(\gamma) = \frac{\sum_{c \in C_s} n_c^{\gamma} a_c^{-(\gamma-1)}}{N_s}.$$

Letting d_c be employment per acre in county c, D_s be employment per acre in state s, and D be employment per acre in the United States, we can decompose the density index into three components:

$$(5) \qquad D_s(\gamma) = D^{\gamma-1}\left(\frac{D_s}{D}\right)^{\gamma-1} \frac{\sum_{c \in C_s} n_c \left(\frac{d_c}{D_s}\right)^{\gamma-1}}{N_s}.$$

That is, the state density effect is the product of a national effect, a state effect—which depends on the relation of average state density to national density—and a factor that depends on the inequality of density across counties within the state. The last factor is the sum of county density relative to state density raised to the power $\gamma - 1$, weighted by county employment. If, for example, average density in state s is identical to average density in the nation, $D_s = D$, then, relative to the nation, productivity in the state depends on the distribution of employment within the state only.

Under neoclassical conditions, with γ less than one, the density factors would predict lower productivity in states with higher average density, and even lower productivity in states with some particularly dense, congested areas. But if agglomeration effects outweigh congestion effects, density has the opposite effect. States with higher average density and higher inequality of density will have higher levels of productivity.

B. *Increasing Returns from a Greater Variety of Intermediate Products in Denser Areas*

The second model hypothesizes increasing returns in the production of local intermediate goods, as in Abdel-Rahman (1988), Fujita (1988, 1989), and Rivera-Batiz (1988). Our development of this model gives density an explicit role. Let the production function for making the final good on an acre of land be

$$f(m, i) = [m^{\beta} i^{(1-\beta)}]^{\alpha}; \tag{6}$$

here m is the amount of labor used directly in making the final good, i is the amount of a composite service input which cannot be transported outside the acre, α describes decreasing returns to the two variable inputs on the acre (congestion effect), and β is a distribution parameter (agglomeration effect). The service composite, i, is produced from individual differentiated services, $x(t)$, indexed by type t, according to the constant elasticity of substitution production function,

$$i = \left(\int_0^z x(t)^{1/\mu} \, dt \right)^{\mu}. \tag{7}$$

Here z describes the variety of intermediate products produced—types 0 through z are available. The parameter $\mu > 1$ controls the substitutability of the intermediate products. The higher μ is, the less one product substitutes for others and the more monopoly power the producer of that product has. Under the standard Spence-Dixit-Stiglitz assumptions of Bertrand competition that, μ is the markup of price as a ratio to marginal cost that the producer will set in order to maximize profit.

We further assume that it takes $x + v$ units of labor to produce x. With labor paid w, the intermediate product maker will charge a price of μw and make a profit of $(\mu - 1)wx - wv$. With free entry to the intermediate product business, this profit will be pushed down to zero—the fixed cost will just offset the operating profit from market power. The level of output at the zero-profit point is

$$x = \frac{v}{\mu - 1}. \tag{8}$$

Putting this common value for all the service inputs into the production function for the service composite, equation (7), we have

$$i = z^{\mu} x. \tag{9}$$

Production of i uses zx units of intermediate inputs, so the productivity of the i-making process is $z^{\mu-1}$. Because $\mu > 1$, productivity rises with the available variety of intermediate goods. Denser acres have greater variety, because more intermediate services producers can break even. The result is a positive relation between density and productivity.

The Cobb-Douglas specification of the final output technology implies that the share of final output paid to labor employed directly is $\alpha\beta$; hence, $wm = \alpha\beta f(m, i)$. The share paid to land is $(1 - \alpha)$. In a free entry equilibrium all output not paid to land accrues to labor, either directly or indirectly through the intermediate service business. Therefore, $wn = \alpha f(m, i)$, where n, as before, is total labor employed in the acre. Combined, these relationships imply that the equilibrium allocation of labor to direct employment in final goods production is governed by the share parameter,

$$m = \beta n. \tag{10}$$

The remaining share $(1 - \beta)n$ of the labor makes intermediate services. Because we know the total amount of labor devoted to intermediate services and the amount of each one produced, we can solve for the variety of those services:

$$(11) \qquad z = (1 - \beta)\frac{\mu - 1}{\mu}\frac{n}{v}.$$

Intermediate product variety, as measured by z, is proportional to density, as measured by the number of workers, n, working on the acre.

Now we can insert the equilibrium value of z into equation (9) to determine i, and then put m and i into the production function for final goods to get the consolidated production function,

$$(12) \qquad \phi n^{\gamma}.$$

Here ϕ is a complicated function of the other constants and the elasticity of the production function is

$$(13) \qquad \gamma = \alpha[1 + (1 - \beta)(\mu - 1)].$$

Again, the parameter α describes congestion effects—lower productivity resulting from crowding more workers onto the same acre. To the extent that the differentiated intermediate goods are important ($\beta < 1$) and they are not good substitutes for each other ($\mu > 1$), there is a countervailing effect favoring higher density, because it makes possible a greater variety of the intermediate products. With a high enough μ and a low enough β, the production function could have increasing returns, where the favorable effect of density outweighs the congestion effect.

In this equilibrium the market provision of intermediate inputs is inefficient due to distortions from monopoly pricing. We have worked out the alternative where the quantity and variety of intermediate services is optimal, either because of government intervention or vertical integration. The resulting elasticity of output with respect to total labor is the same as for the monopolistic competition case.

If we normalize the measurement of the quantities to make $\phi = 1$ and assume, as before, that labor is distributed uniformly across the acres of a county, we have the county production function,

$$(14) \qquad \frac{q_c}{a_c} = \left(\frac{n_c}{a_c}\right)^{\gamma}.$$

Aggregation to the state level proceeds exactly as before. There are no observational distinctions between the externalities model and the intermediate product variety model. Both provide a theoretical foundation for the same estimation procedure in state data.

C. *Capital and Total Factor Productivity*

Now let the production function describing output produced in an acre of space in county c by employing n_c workers and k_c machines be

$$(15) \qquad A_s[(e_c n_c)^{\beta} k_c^{1-\beta}]^{\alpha}\left(\frac{q_c}{a_c}\right)^{(\lambda-1)/\lambda},$$

where A_s is a Hicks-neutral technology multiplier for state s and e_c is a measure of the efficiency of labor at the county level. As before, the elasticity α is less than one by the amount of the share of land in factor payments. The quantities of labor and capital employed in a county, n_c and k_c, are distributed equally among all the acres in the county. Thus total output in county c is

$$(16) \qquad q_c = a_c A_s\left[\left(\frac{e_c n_c}{a_c}\right)^{\beta}\left(\frac{k_c}{a_c}\right)^{1-\beta}\right]^{\alpha} \times \left[\frac{q_c}{a_c}\right]^{(\lambda-1)/\lambda}.$$

Solving for output per acre, we get

$$(17) \qquad \frac{q_c}{a_c} = A_s^{\lambda}\left[\left(\frac{e_c n_c}{a_c}\right)^{\beta}\left(\frac{k_c}{a_c}\right)^{1-\beta}\right]^{\gamma}.$$

Again, γ is the product of the production elasticity, α, which is less than one, and the elasticity from the externality, λ, which is greater than one. If γ exceeds one, agglomeration effects dominate congestion.

To deal with capital, we make the assumption that the rental price of capital, r, is the same everywhere. Then we use the factor demand function to substitute the factor price for the factor quantity. That is,

$$(18) \qquad \frac{k_c}{a_c} = \frac{\alpha(1-\beta)}{r}\frac{q_c}{a_c}.$$

Thus the county technology becomes

$$(19) \qquad \frac{q_c}{a_c} = \phi A_s^{\omega}\left(\frac{e_c n_c}{a_c}\right)^{\theta}$$

where ϕ is a constant that depends on the interest rate, and the elasticities for the technology multiplier for the state, ω, and for labor input for the county, θ, are:

$$(20) \qquad \theta \equiv \frac{\gamma\beta}{1-\gamma(1-\beta)}$$

and

$$(21) \qquad \omega \equiv \frac{\theta}{\alpha\beta}.$$

We assume that labor efficiency depends log-linearly on workers' average years of education h_c, $e_c = h_c^{\eta}$, where η is the elasticity of education. Using this relationship in equation (19) and aggregating to the state level, we obtain

$$(22) \qquad \frac{Q_s}{N_s} = \phi A_s^{\omega} D_s(\theta, \eta)$$

where

$$(23) \qquad D_s(\theta, \eta) = \frac{\sum_{c \in C_s} (n_c h_c^{\eta})^{\theta} a_c^{1-\theta}}{N_s}.$$

Under these alternative assumptions, the index of density has the same functional form as before with the elasticity θ in place of γ and efficiency units of labor instead of raw labor. The underlying value of γ can be calculated from equation (20). The relation between γ and θ for $\beta = .7$ is such that for values close to 1, as found in our empirical work, the overstatement of γ associated with the treatment of capital is small. The extension of the intermediate product variety model is analogous.

Regarding the stochastic specification in equation (22), we assume that state productivity A_s is distributed log-normally around an underlying nationwide level. We also allow for mismeasurement in state productivity, assuming that the measurement error has a log-normal distribution with zero mean. Using this stochastic specification in equation (22) and taking logarithms yields

$$(24) \qquad \log\frac{Q_s}{N_s} = \log\phi + \log D_s(\theta, \eta) + u_s.$$

Here u_s is the sum of the measurement error, and it is ω times the deviation of state productivity from the underlying level in the nation. We assume that the errors of different states are uncorrelated.

III. Equilibrium

How can states or counties be in equilibrium with different densities? This question arises if θ exceeds 1. Under neoclassical assumptions, density should be equal everywhere. The marginal product of labor is lower in a denser area, and there are arbitrage profits or a higher standard of living available by moving a worker from a dense area to a less dense one. On the other hand, with θ greater than one, the worker is more productive when moved to a denser area. Absent other considerations, the only equilibrium is for employment to concentrate in a single county.

The simplest answer, and a realistic one, is that some workers prefer to live in areas that turn out to be less dense. These workers are willing to accept the lower wages in those locations. The preference could be, but need not be, a preference for lower density itself. The preference could also take the form of devotion to a location that is not an agglomeration point. Furthermore, as households value land, its price drives a wedge between the product wage and the consumption wage. In equilibrium, there are no incentives to move for either firms or households. The marginal cost of production is equalized across all counties as the decrease in marginal cost associated with higher density is offset by higher product efficiency wages and higher land prices. Households find that differing product wages are counterbalanced by any of the considerations described above.

IV. Data

The data needed for estimation are available for the year 1988. The data cover the private nonproprietary economy. That is, data on labor input at the county level includes only employees, not the self-employed. The corresponding measure of output at the state level is gross state product (GSP) less Proprietors' Income (U.S. Bureau of Economic Analysis, 1985). We use GSP at sellers' prices; Indirect Business Taxes are excluded from the output measure. Data on employment by county are compiled by the Regional Economic Measurement Division (U.S. Bureau of Economic Analysis, 1991). Data on the area of each county are from U.S. Bureau of the Census (1989).

Data on GSP are conceptually far superior to those used in previous work on spatial differences in productivity. Moomaw (1985), Sveikauskas (1975), and Segal (1976) all measure output as the concept of value added or total value of production used in the Census of Manufactures. This concept omits all services either purchased in the market or obtained from corporate headquarters. It is hard to see how the Census of Manufactures value added could be used for any purpose in production economics, but it is a particularly unusable concept for agglomeration issues. Because there is likely to be less vertical integration in big cities or in dense areas, firms in those places are likely to purchase more services than do their counterparts in less dense areas. Moreover, a plant in a dense area is more likely to be close to its corporate headquarters and therefore more dependent on it for transferred services. For both reasons, studies using Census of Manufactures value added will overstate the productivity advantage of cities or dense areas. The research of Henderson (1986) uses total value of production, also from the Census of Manufacturers. Compared to the value-added data this concept has the added disadvantage of double-counting inputs traded within an industry. Our data are based on a careful allocation of purchased and transferred services by industry at the state level. Gross state output is a much more satisfactory measure of output than is the Census of Manufactures concept of value added. Our theoretical formulation assumes that all land is equivalent. Therefore, we use data on state output and county employment which exclude the agricultural and the mining sector. In some states, natural resources are sufficiently important for local economic activity to make our output measure unrealistic—despite our adjustment for mining output at the state level and mining employment at the county level. On this basis, we excluded all states where the mining contributes more than 15 percent of our output measure. These states are Alaska, Louisiana, West Virginia and Wyoming; on average mining stands at 21 percent of our output measure in these states compared to 1 percent in all other states.

Our data on education comes from two sources. At the state level, we have data from the Annual Demographic File of the Current Population Survey (U.S. Bureau of the Census, 1988) for 1988. Our education measure at the state level weighs the workers' years of education by the number of hours worked in 1988. At the county level, we have data from the 1990 Census of Population (U.S. Bureau of the Census, 1992). Our education measure at the county level is average years of education.

V. Identification and Estimation

We make two alternative identifying assumptions. First we assume that the random element of output per worker is uncorrelated with density and average education levels. This assumption amounts to saying that density and education are measured with little error and do not respond to the random element of productivity. Because it appears that much of the noise in productivity across states comes from measurement error, this assumption is not as strong as it may seem at first. Under this identifying assumption, we estimate the returns-to-scale parameter, θ, and the elasticity of average product with respect to education, η, by nonlinear least squares.

If there are true differences in the determinants of productivity across states, it is not realistic to assume that density is uncorrelated with those differences. States with natural features that make them more productive (for example, climate or geographic features suited to transportation) will attract workers because wages will be higher. Our alternative identifying assumption is that there is an exogenous characteristic of states that can function as an instrumental variable for the density index. The corresponding estimator is nonlinear instrumental variables. All of our candidate instruments rest on the hypothesis that the original sources of agglomeration in the United States have remaining influences only on the preferences of workers about where to live; they are not related to modern differences in productivity not explained by our model. The characteristics we use are:

1. Presence or absence of a railroad in the state in 1860, from John F. Stover (1961).
2. Population of the state in 1850, from U.S. Bureau of the Census (1975).
3. Population density of the state in 1880, from U.S. Bureau of Census (1975).
4. Distance from the eastern seaboard of the U.S., from Rand McNally (1993).

The direct measures of agglomeration—population and population density—are eligible as instruments if the main sources of agglomeration in the 18th and 19th centuries are not related to the residuals in our equation. Thus our hypothesis is that the early patterns of agglomeration in the United States did not reflect factors which significantly contribute to productivity today but have a remaining influence mainly through the legacy of agglomeration. Railroads became an important historical factor in agglomeration in the second half of the 19th century. Our hypothesis is that the development of railroads was not driven by modern productivity differences not accounted for by our model. Finally, we include distance from the eastern seaboard as an exogenous determinant of agglomeration in the 18th and 19th centuries.

The nonlinear instrumental variables estimator is discussed by Takeshi Amemiya (1983). The nonlinear model is

$$y_s = f_s(\beta) + u_s \tag{25}$$

where y is productivity, f describes our nonlinear specification, and β is the vector of parameters. Z is a matrix of values of the instrumental variables. The nonlinear instrumental variables estimator minimizes

$$[y - f(\beta)]'Z(Z'Z)^{-1}Z'[y - f(\beta)]. \tag{26}$$

When the number of instruments is the same as the number of parameters, the estimator is just the solution to the orthogonality condition,

$$[y - f(\beta)]'Z = 0. \tag{27}$$

When the instrument is a dummy variable, the estimator is based on grouping. For example, with our instrument for the presence of a railroad in 1860, the estimator of the density parameter is the value that explains the difference in the average level of productivity in states that did have railroads compared to states that did not.

The estimated covariance matrix of β is

$$\hat{\sigma}^2[F'Z(Z'Z)^{-1}Z'F]^{-1} \tag{28}$$

where $\hat{\sigma}$ is the standard error of the residuals and F is the matrix of derivatives of the model with respect to the parameters.

VI. Results

Table 1 gives the results using the county level education data. The least squares estimate of θ is 1.052 with a standard error of .008.[1] The elasticity of labor efficiency with respect to education, η, is .41 with a standard error of .40. The R^2 of the regression is 55 percent. Steven J. Davis (1992), using data on individuals, estimates the elasticity of earnings with respect to education to lie between .80 and 1.35. Our estimate is one standard error away from Davis's lower estimate. The

[1] Some readers have been surprised at the small size of the standard error. It is much smaller, for example, than the standard error of the OLS regression of state productivity on our density index. The reason is that θ predicts zero slope at $\theta = 1$. Small changes in θ correspond to large changes in the density index, so the parameter is correspondingly well estimated.

TABLE 1—ESTIMATION RESULTS

Instrument	Density elasticity, θ (standard error)	Education elasticity, η (standard error)	R^2
None (NLLS)	1.052 (0.008)	0.410 (0.396)	0.551
Eastern seaboard	1.055 (0.017)	0.460 (0.51)	0.548
Railroad in 1860	1.061 (0.011)	0.330 (0.450)	0.537
Population in 1850	1.060 (0.015)	0.350 (0.510)	0.539
Population density in 1880	1.051 (0.019)	0.530 (0.550)	0.549
All	1.06 (0.01)	0.060 (0.82)	0.536

Notes: The equation estimated is (24). The data are value added for 46 states and Washington DC. For the 46 states we have used data on employment and average years of education at the county level.

instrumental variable estimates for θ is 1.06 making joint use of all instruments. This estimate implies that doubling the employment density in a county increases labor productivity by 6 percent. The estimated value for γ is about 1.04 which implies that doubling employment density in a county results in a 4-percent increase of total factor productivity. Using the education data available at the state level, the estimating equation in (24) simplifies to

$$(29) \qquad \log \frac{Q_s}{N_s} = \log \phi + \eta \log h_s + \log D_s(\theta) + u_s$$

where $D_s(\theta)$ is defined in equation (4). The nonlinear least squares estimate of θ and η in equation (25) are 1.051 and .51, with standard errors of .008 and .45 respectively. Using instrumental variables, the estimates are 1.058 and .36 and the standard errors .011 and .49.

Table 2 shows the factor density index $D_s(\theta)$ evaluated at $\theta = 1.058$, average years of education of workers at the state level, and the private gross state product per worker in all sectors except farming and mining. The states are ranked in declining order of density. The densest area for which reliable output data is available is Washington, DC. Not surprisingly, New York ranks second. It is the extreme concentration of employment in New York City that gives the high value of the density measure. In fact, New York City comprises the densest county in the United States, New York county (with a factor density index of 1.94), and three of the 10 next densest counties (Bronx county, Kings county, and Queens county).

Our estimated density index for New York county implies, for example, that workers in New York county are 22 percent more productive than workers in New York state, the state with the highest average productivity in our sample. The other dense states are the highly urbanized states of the northeast plus Illinois and California. The least dense states are the thinly populated states of northern New England, the south, and the southwest. It is important to note that density is not just a measure of the inequality of distribution of the work force across counties—it is also dependent on the actual density in the counties where employment is significant. The third column of Table 2 shows output (in 1988 dollars) per worker by state. Output per worker in New Jersey ($44,488), the most productive state, is two-thirds higher than in South Dakota ($26,196), the least productive state. Average output per worker in the ten most productive states ($38,782) is one-quarter higher than in the ten least productive states ($31,578). The positive partial correlation of density and productivity is immediately apparent from Table 2

TABLE 2—DENSITY, EDUCATION, AND PRODUCTIVITY FOR $\theta = 1.058$

State	Density index	Years of education	Productivity (1988 $)
District of Columbia	1.67	14.0	43,164
New York	1.59	13.3	41,921
New Jersey	1.48	13.4	44,488
Massachusetts	1.47	13.4	37,296
Illinois	1.46	13.2	39,150
Maryland	1.45	13.6	34,439
Rhode Island	1.43	12.7	30,055
Connecticut	1.42	13.5	41,927
California	1.42	12.9	40,723
Pennsylvania	1.40	13.2	34,661
Top 10 average	1.48	13.3	38,782
Ohio	1.40	13.1	36,553
Virginia	1.40	13.2	35,986
Delaware	1.40	13.3	35,223
Michigan	1.39	13.2	39,001
Missouri	1.38	13.1	34,520
Hawaii	1.38	13.3	34,485
Minnesota	1.37	13.2	35,494
Florida	1.36	13.1	30,808
Georgia	1.36	12.7	35,407
Texas	1.36	12.6	36,798
Colorado	1.35	13.6	33,342
Indiana	1.34	12.9	34,721
Wisconsin	1.34	13.2	33,495
Tennessee	1.33	12.6	33,169
North Carolina	1.32	12.8	32,677
Kentucky	1.32	12.7	34,406
Utah	1.31	13.6	32,160
Washington	1.31	13.6	32,661
Nebraska	1.30	13.2	30,323
New Hampshire	1.30	13.4	33,668
Oklahoma	1.29	13.0	33,567
Oregon	1.29	13.3	32,713
South Carolina	1.28	12.8	29,623
Kansas	1.27	13.4	36,223
Alabama	1.27	12.4	32,980
Arizona	1.25	13.2	33,579
Iowa	1.25	13.1	32,318
Average	1.33	13.1	34,071
Maine	1.24	13.1	33,097
Vermont	1.23	13.4	33,733
Arkansas	1.23	12.5	32,150
Mississippi	1.21	12.8	32,707
New Mexico	1.21	12.6	31,249
Nevada	1.20	12.9	36,234
Idaho	1.17	12.7	29,861
South Dakota	1.15	13.0	26,196
North Dakota	1.12	13.3	30,248
Montana	1.10	13.3	30,302
Bottom 10 average	1.19	13.0	31,578

Note: Education is at the state level. The density index as defined in equation (4), uses raw employment at the county level, evaluated at $\theta = 1.058$.

and Figure 1. There are a number of outliers that call for further investigation: most conspicuous is Rhode Island, which is just as dense as its neighbors but has a productivity level of the least dense state. Nevada, on the other hand, has a much higher productivity

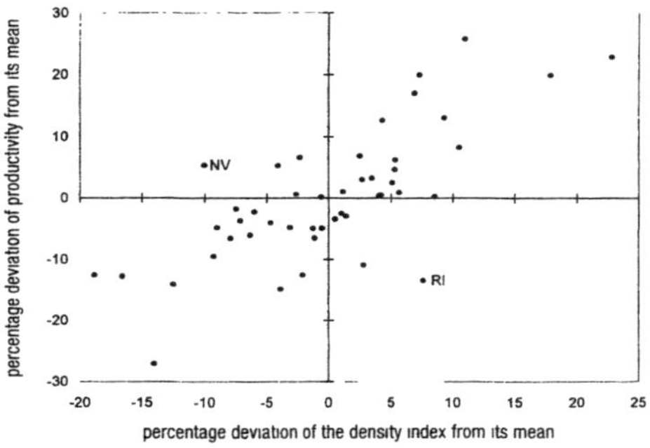

FIGURE 1. DENSITY AND PRODUCTIVITY BY STATE FOR $\theta = 1.058$

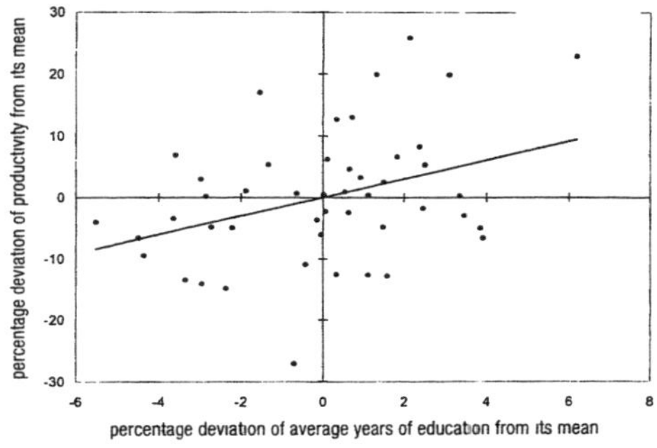

FIGURE 2. EDUCATION AND PRODUCTIVITY BY STATE

level than its low densities would predict. Among the counties with the smallest density indices are Garfield county (Montana), Kimball county (Nebraska), Newton county (Arkansas) and La Paz county (Arizona). We estimate that workers in the 15 counties with lowest density produce on average less than half the output of a worker in New York City.

Figure 2 plots average years of education against productivity at the state level. We have estimated equation (29) without density effects ($\theta = 1$). Using the education data at the state level, we find the elasticity of output per worker with respect to average education in equation (29) is 1.5 with a standard error of .5. The R^2 of the regression is .09. Using the education data at the county level to estimate equation (24) under the same hypothesis, $\theta = 1$, results in an elasticity of labor efficiency with respect to education of 1.2 with a standard error of .6 and a R^2 of .1. Education is a significant determinant of productivity in both cases and the estimate using data at the county level is within the range estimated by Davis (1992).

Figure 3 summarizes the results of our instrumental variables regression of equation (25). It plots estimated output per efficiency unit of labor at the state level against estimated density for $\theta = 1.058$. It is seen that Rhode Island remains an outlier despite the fact that its workers are significantly less educated than workers in neighboring states.

Table 3 decomposes the estimated density index $D_s(1.058)$ as defined in equation (4) into a state effect and a distribution effect along the lines described in equation (5). The column headed *state effect* gives the part of the state effect arising from the average density of the entire state. For example, if the density of employment in Massachusetts fell to the national level, while the distribution of employment over the counties remained unchanged, then this would result in a 15-percent drop in average product. The *distribution effect* measures the part of the state productivity effect attributable to an unequal distribution of employment over counties. For example, productivity in New York would fall by 19 percent if employment were to be allocated uniformly across the area of the state. Colorado, Nebraska, Missouri, Texas and Utah are examples of states with great inequality across counties but low overall density, because their major metropolitan areas have relatively high levels of employment per acre.

VII. Extensions

A. *Public Capital*

Some of the differences in average labor productivity across U.S. states may be due to differences in the amount of public capital that is available. To consider this possibility, we extend the basic model to account for public capital. We assume that the services from the public capital available in state s, g_s, enter county level production with constant elasticity, δ. Then, the county level production function in equation (1) becomes

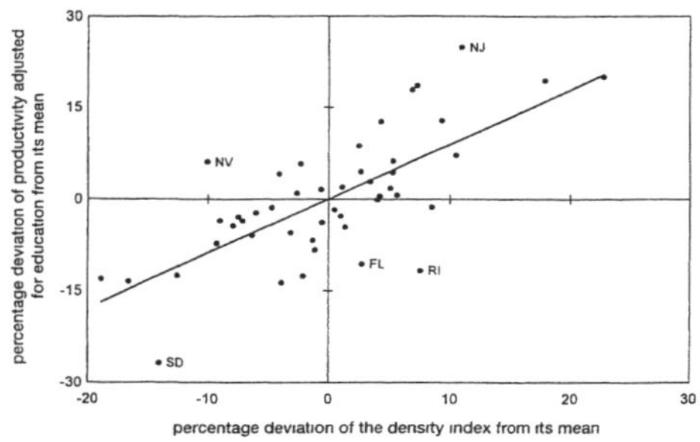

FIGURE 3. DENSITY AND PRODUCTIVITY ADJUSTED FOR EDUCATION, $\theta = 1.058$

$$(30)\quad f(n, q, a) = n^{\alpha}\left(\frac{q_c}{a_c}\right)^{(\lambda-1)/\lambda} g_s^{\delta}.$$

Solving for state level productivity we get that

$$(31)\quad \frac{Q_s}{N_s} = g_s^{\lambda\delta} D_s(\gamma),$$

where $D_s(\gamma)$ is defined in equation (4). Extending the model for exogenous differences in total factor productivity and differences in human and physical capital along the lines of Section II, we get the following estimating equation:

$$(32)\quad \log\frac{Q_s}{N_s} = \psi + \omega\delta \log g_s + \log D_s(\theta, \eta) + u_s$$

where $D(\theta, \eta)$ and ω are defined in equations (23) and (21) above, and ψ is a constant. Table 4 shows the results of estimating equation (32) using the value of public capital in streets and highways in 1988 at the state level from Douglas Holtz-Eakin (1993). The density elasticity falls somewhat, but public capital does not affect productivity at the state level significantly. We have also used other, more inclusive measures of public capital with the same results. Our negative results on public capital are in line with Teresa Garcia-Mila et al. (1994).

B. *State versus Distribution Effects*

As a further exploration of the role played by the state and distribution effects in Table 3 for agglomeration in the United States, we consider an extension of the basic model which allows for influences from density at the state level as well as at the county. The extended model provides a specification test for our earlier model. If there were important state effects not captured by our earlier model it would suggest that we need to consider externalities that cross county boundaries. In the extended model, the production function in equation (1) for output produced in an acre of space becomes

$$(33)\quad f(n, q, a) = n^{\alpha}\left(\frac{q_c}{a_c}\right)^{(\lambda-1)/\lambda}\left(\frac{Q_s}{A_s}\right)^{\kappa},$$

where κ denotes the elasticity with respect to average output per acre in the state, Q_s/A_s. Solving for state productivity along the lines of equations (2) and (3) yields

$$(34)\quad \frac{Q_s}{N_s} = D_s^{\lambda\kappa/(1-\lambda\kappa)} D_s(\gamma)^{1/(1-\lambda\kappa)}.$$

State productivity is determined by average employment density in the state, D_s, and the density index, $D_s(\gamma)$, as defined in equation (4). For $\kappa = 0$ the model reduces to the one discussed at the beginning of Section II.

Extending the model along the lines in Section II results in the following estimating equation:

$$(35)\quad \log\frac{Q_s}{N_s} = \delta + \xi \log D_s + (\xi + 1)\log D_s(\theta, \eta) + u_s,$$

where δ is a common constant across states and ξ is the elasticity of productivity with respect to state density,

$$(36)\quad \xi \equiv \frac{\kappa\omega}{1 - \kappa\omega}.$$

Table 5 shows the results of estimating equation (35). With nonlinear least squares, we obtain an estimate for ξ of 0.005 with a standard

TABLE 3—PRODUCTIVITY, STATE AND DISTRIBUTION EFFECTS FOR $\theta = 1.058$

State	State effect (percent)	Distribution effect (percent)	Productivity (1988 $)
District of Columbia	36	0	43,164
New York	9	19	41,921
New Jersey	16	4	44,488
Massachusetts	15	5	37,296
Illinois	6	13	39,150
Maryland	11	7	34,439
Rhode Island	16	1	30,055
Connecticut	14	2	41,927
California	5	11	40,723
Pennsylvania	7	8	34,661
Top 10 average	8	12	38,782
Ohio	7	7	36,553
Virginia	4	11	35,986
Delaware	10	4	35,223
Michigan	4	10	39,001
Missouri	−1	14	34,520
Hawaii	4	9	34,485
Minnesota	−2	15	35,494
Florida	6	6	30,808
Georgia	2	10	35,407
Texas	−2	14	36,798
Colorado	−6	17	33,342
Indiana	4	6	34,721
Wisconsin	1	9	33,495
Tennessee	2	7	33,169
North Carolina	3	5	32,677
Kentucky	−1	9	34,406
Utah	−9	18	32,160
Washington	−1	9	32,661
Nebraska	−8	16	30,323
New Hampshire	3	3	36,688
Oklahoma	−5	12	34,567
Oregon	−6	12	32,713
South Carolina	1	3	29,623
Kansas	−6	11	36,223
Alabama	−1	5	32,980
Arizona	−6	9	33,579
Iowa	−3	6	32,318
Average	−1	11	34,071
Maine	−4	6	33,097
Vermont	−2	2	33,733
Arkansas	−5	5	32,150
Mississippi	−4	4	32,707
New Mexico	−12	13	31,249
Nevada	−11	10	36,234
Idaho	−12	9	29,861
South Dakota	−13	9	26,196
North Dakota	−13	6	30,248
Montana	−16	7	30,302
Bottom 10 average	−10	9	31,578

Note: The density index as defined in equation (4), uses raw employment at the county level, evaluated at $\theta = 1.058$ decomposed into a state effect and a distribution effect as described in equation (5).

TABLE 4—ESTIMATION RESULTS WITH PUBLIC CAPITAL

Instrument	Density elasticity, θ (standard error)	Education elasticity, η (standard error)	Public capital elasticity, $\omega\delta$ (standard error)
None (NLLS)	1.046 (0.010)	0.570 (0.430)	0.021 (0.015)
All	1.056 (0.012)	0.480 (1.040)	0.017 (0.023)

Notes: The equation estimated is (32). The data are value added for 46 states and Washington, DC. For the 46 states we have used data on employment and education at the county level. The data for public capital at the state level is the capital in streets and highways for 1988 from Holtz-Eakin (1993).

error of .015 and an estimate for θ of 1.047 with a standard error of .018. With nonlinear instrumental variables, the estimates for ξ and θ are $-.019$ and 1.084 respectively with standard errors of .023 and .028. Both estimates suggest rather precisely that there is no state effect at all once the county effects are considered. Although we believe that a specification with some cross-county effects would be an improvement over our model, the absence of incremental state density effects indicates that our current specification captures most density effects. This issue is investigated further in Christopher J. Wilkins (1994) who considers different specifications for the geographical extent of externalities.

C. *Size versus Density Effects*

Finally, our framework allows us to consider size versus density effects at the county level. To do so we consider a further extension of the basic model, where the production function (1) accounts for both an externality from output density in the county and output in the county. We assume that the elasticity of firm level output with respect to county output also enters as a constant, ν:

$$(37) \quad f(n, q, a) = n^{\alpha}\left(\frac{q_c}{a_c}\right)^{(\lambda-1)/\lambda} q_c^{\nu}.$$

Solving for state level productivity, we obtain

$$(38) \quad \frac{Q_s}{N_s} = \frac{\sum_{c \in C_s} (n_c^{\gamma} a_c^{-(\gamma-1)})^{1/(1-\nu\lambda)}}{N_s}.$$

When $\nu = 0$, the framework reduces to the basic model with no size externalities at the county level. When $\gamma = 1$, the framework displays no density effects but only size effects. Extending the model for differences in physical and human capital and exogenous total factor productivity, we obtain

$$(39) \quad \log \frac{Q_s}{N_s} = \delta + \log D_s(\theta, \eta, \sigma) + u_s,$$

where $D_s(\theta, \eta, \sigma)$ is defined analogous to equation (38) with efficiency units of labor, $n_c h_c^{\eta}$, in place of raw labor, n_c,

$$(40) \quad D_s(\theta, \eta, \sigma) = \frac{\sum_{c \in C_s} ((n_c h_c^{\eta})^{\theta} a_c^{-(\theta-1)})^{\sigma}}{N_s},$$

where

$$(41) \quad \sigma = \frac{1}{1 - \nu\omega}.$$

Table 6 shows the results of estimating equation (40). The nonlinear least squares estimate of θ is 1.035 with a standard error of 0.013. The county size parameter σ is estimated to be 1.029 with a standard error of 0.019. The R^2 of the regression is 0.58. The nonlinear instrumental variables estimate of the density parameter is 1.046 with a standard error of 0.023 and the county size parameter is 1.026 with a standard error of 0.039. These estimates suggest that density externalities are more important than size externalities at the

TABLE 5—ESTIMATION RESULTS WITH STATE DENSITY EFFECTS

Instrument	County density elasticity, θ (standard error)	Education elasticity, η (standard error)	State density elasticity, ξ (standard error)
None (NLLS)	1.047	0.450	0 005
	(0.018)	(0.420)	(0.015)
All	1.084	0.235	(0.019)
	(0.028)	(0.498)	(0.023)

Notes: The equation estimated is (35). The data are value added for 46 states and Washington, DC. For the 46 states we have used data on employment and education at the county level for the density index $D_s(\theta, \eta)$. Average density D_s is total employment in the state over its area.

TABLE 6—ESTIMATION RESULTS WITH SIZE EFFECTS

Instrument	County density elasticity, θ (standard error)	Education elasticity, η (standard error)	County size elasticity, σ (standard error)
None (NLLS)	1.035	0.259	1.029
	(0.013)	(0.398)	(0.019)
All	1.046	0.140	1.026
	(0.023)	(0.82)	(0.039)

Notes: The equation estimated is (39). The data used is value added for 46 states and Washington, DC. For the 46 states we have used data on employment and education at the county level.

county level. This issue is investigated further in Wilkins (1994) for size externalities at the level of metropolitan statistical areas and standard metropolitan statistical areas.

VIII. Concluding Remarks

Increasing returns to density play a crucial role for explaining the large differences in average labor productivity across U.S. states. We estimate that doubling employment density in a county increases average labor productivity by 6 percent. This degree of locally increasing returns can explain more than half of the variation in labor productivity across U.S. states. Our instrumental variables estimates rest on the hypothesis that the patterns of agglomeration in the 18th and the middle of the 19th century did not reflect factors which significantly contribute to productivity today but have a remaining influence mainly through the legacy of agglomeration. Our estimates control for labor quality at the county level and for differences in the available public capital at the state level. We also compare increasing returns to density with increasing returns to size at the county level and find that increasing returns to density describes the data better than increasing returns to size.

Our work can be extended in several directions. Because our data are available by industry it would be possible to distinguish between economies from localization—which arise from the collocation of firms in the same industry—and economies from urbanization—where favorable effects arise from the general diversity and scale of urban areas. Our approach can also be used to examine dynamic externalities. Work by Edward L. Glaeser et al. (1992) and Henderson (1994) quantifies dynamic externalities by looking at the behavior of industry employment. But small productive externalities may have large employment effects. This makes it difficult to infer the significance of spatial externalities for productivity from data on the spatial concentration of industry employment. Because our data are available since 1962 our approach could be extended to examine the effect of dynamic spillovers on productivity. Our empirical work also suggests that rising density over time may be an important factor in growth. Large U.S. cities are denser now than in earlier centuries, and a much larger fraction of the population is employed in cities or other dense areas. Our estimates could be applied to historical data on the distribution of employment by county to measure the part of total growth that can be associated with rising density.

REFERENCES

Abdel-Rahman, H. M. "Product Differentiation, Monopolistic Competition and City Size." *Regional Science and Urban Economics*, February 1988, *18*(1), pp. 69–86.

Amemiya, Takeshi. "Nonlinear Regression Models," in Zvi Griliches and Michael D. Intrilligator, eds., *Handbook of econometrics*. New York: North-Holland, 1983, pp. 333–89.

Carlino, Gerald A. and Voith, Richard. "Accounting for Differences in Aggregate State Productivity." *Regional Science and Urban Economics*, November 1992, *22*(4), pp. 597–617.

Ciccone, Antonio. "Cities and the Economy: Aggregate Production, Urban Productivity and the Distribution of Economic Activity." Mimeo, Stanford University, 1992.

Davis, Steven J. "Cross-Country Patterns of Change in Relative Wages." National Bureau of Economic Research (Cambridge, MA) Working Paper No. 4085, 1992.

Dixit, Avinash and Stiglitz, Joseph E. "Monopolistic Competition and Optimum Product Diversity." *American Economic Review*, June 1977, *67*(3), pp. 297–308.

Fujita, Masahisa. "A Monopolistic Competition Model of Spatial Agglomeration: Differentiated Product Approach." *Regional Science and Urban Economics*, February 1988, *18*(1), pp. 87–124.

———. *Urban economic theory: Land use and city size*. Cambridge: Cambridge University Press, 1989.

Garcia-Mila, Teresa; McGuire, Therese J. and Porter, Richard. "The Effect of Public Capital in State-Level Production Functions Reconsidered." *Review of Economics and Statistics*, 1995, *77* (forthcoming).

Glaeser, Edward L.; Kallal, Hedi D.; Scheinkman, Jose A. and Shleifer, Andrei. "Growth in Cities." *Journal of Political Economy*, December 1992, *100*(6), pp. 1126–52.

Glaeser, Edward L. and Mare, David C. "Cities and Skills." Working Paper in Economics E-94-11, Hoover Institution, Stanford University, 1994.

Henderson, J. Vernon. "The Sizes and Types of Cities." *American Economic Review*, September 1974, *64*(4), pp. 640–56.

———. "Efficiency of Resource Usage and City Size." *Journal of Urban Economics*, January 1986, *19*(1), pp. 47–70.

———. "Externalities and Industrial Development." Manuscript, Brown University, 1994.

Holtz-Eakin, Douglas. "State-Specific Estimates of State and Local Government Capital." *Journal of Regional Science and Urban Economics*, April 1993, *23*(2), pp. 185–209.

Krugman, Paul R. *Geography and trade*. Cambridge, MA: MIT Press, 1991.

Marshall, Alfred. *Principles of economics*, 8th Ed. London: MacMillan, 1920.

Mills, Edwin S. "An Aggregative Model of Resource Allocation in a Metropolitan Area." *American Economic Review* May 1967 (Papers and Proceedings), *57*(2), pp. 197–210.

Moomaw, Ronald L. "Productivity and City Size: A Critique of the Evidence." *Quarterly Journal of Economics*, November 1981, *96*(4), pp. 675–88.

———. "Firm Location and City Size: Reduced Productivity Advantages as a Factor in the Decline of Manufacturing in Urban Areas." *Journal of Urban Economics*, January 1985, *17*(1), pp. 73–89.

Rand McNally and Company. *The new international atlas*. Chicago: Rand McNally, 1993.

Rivera-Batiz, Francisco L. "Increasing Returns, Monopolistic Competition, and Agglomeration Economies in Consumption and Production." *Regional Science and Urban*

Economics, February 1988, *18*(1), pp. 125–53.

Segal, David. "Are there Returns to Scale in City Size?" *Review of Economics and Statistics*, August 1976, *58*(3), pp. 339–50.

Spence, A. Michael. "Product Selection, Fixed Costs and Monopolistic Competition." *Review of Economic Studies*, June 1976, *43*(2), pp. 217–35.

Stover, John F. *American railroads*. Chicago: University of Chicago Press, 1961.

Sveikauskas, Leo A. "The Productivity of Cities." *Quarterly Journal of Economics*, August 1975, *89*(3), pp. 393–413.

U.S. Bureau of Economic Analysis. "Experimental Estimates of Gross State Product by Industry," Staff Paper 42. Washington, DC: U.S. Department of Commerce, 1985.

______. Regional Economic Measurement Division. *REIS: regional economic information system*. Washington, DC: Economics and Statistical Administration, 1991 (CD-ROM computer file).

U.S. Bureau of the Census. "Historical Statistics of the U.S., Colonial Times to 1970." Washington, DC: U.S. Department of Commerce, 1975.

______. *Current population survey: Annual demographic file, 1988*. Ann Arbor, MI: Inter-University Consortium for Political and Social Research, 1988 (computer file).

______. Data User Services Division. *County and city data book*. Washington, DC: Department of Commerce 1989 (computer file).

______. "Educational Attainment in the U.S.: March 1989 and March 1988," Current Populations Reports, Series P-20, No. 451. Washington, DC: Government Printing Office, 1991.

______. Data User Services Division. *Census of population and housing (1990): Summary tape file 3A*. Washington, DC, 1992 (computer file).

Wilkins, Christopher J. "Estimating the Size of Geographical Externalities." Manuscript, Stanford University, 1994.

Part VI
Diversity vs. Specialization in Cities

[16]

Growth in Cities

Edward L. Glaeser

Harvard University

Hedi D. Kallal

New York University

José A. Scheinkman

University of Chicago

Andrei Shleifer

Harvard University

Recent theories of economic growth, including those of Romer, Porter, and Jacobs, have stressed the role of technological spillovers in generating growth. Because such knowledge spillovers are particularly effective in cities, where communication between people is more extensive, data on the growth of industries in different cities allow us to test some of these theories. Using a new data set on the growth of large industries in 170 U.S. cities between 1956 and 1987, we find that local competition and urban variety, but not regional specialization, encourage employment growth in industries. The evidence suggests that important knowledge spillovers might occur between rather than within industries, consistent with the theories of Jacobs.

We are grateful to the National Science, Sloan, and Bradley foundations for financial support; to Robert Barro, Gary Becker, Carol Heim, Vernon Henderson, Lawrence Katz, Robert Lucas, Sherwin Rosen, Julio Rotemberg, T. W. Schultz, and George Tolley for helpful comments; and to Christopher Kim for research assistance.

[*Journal of Political Economy*, 1992, vol 100, no 6]

I. Introduction

Some historians have argued that most innovations are made in cities (Jacobs 1969; Bairoch 1988). The cramming of individuals, occupations, and industries into close quarters provides an environment in which ideas flow quickly from person to person. Jacobs (1969, 1984) argues that these interactions between people in cities help them get ideas and innovate. In fact, without an opportunity to learn from others and thus improve one's own productivity, there would be little reason for people to pay high rents just to work in a city. Easy flow of ideas might explain how cities survive despite the high rents.

Such a dynamic view of cities fits nicely with the recent work on economic growth, which views externalities (and particularly externalities associated with knowledge spillovers) as the "engine of growth" (Romer 1986; Lucas 1988). If geographical proximity facilitates transmission of ideas, then we should expect knowledge spillovers to be particularly important in cities. After all, intellectual breakthroughs must cross hallways and streets more easily than oceans and continents. This paper uses a new data set on American cities and industries to test the new growth theories. Because these theories are most compelling in the context of city growth, our focus on cities gives them the benefit of the doubt.

We focus on three theories. All these theories deal with technological externalities, whereby innovations and improvements occurring in one firm increase the productivity of the other firms without full compensation. The theoretical foundations of such knowledge spillovers have been modeled by many authors, starting with Loury (1979) and Dasgupta and Stiglitz (1980). Griliches (1979) surveys the empirical literature on the role of knowledge spillovers.

The Marshall-Arrow-Romer (MAR) externality concerns knowledge spillovers between firms in an industry. Arrow (1962) presents an early formalization; the paper by Romer (1986) is a recent and influential statement. Applied to cities by Marshall (1890), this view says that the concentration of an industry in a city helps knowledge spillovers between firms and, therefore, the growth of that industry and of that city. A good example would be computer chips in Silicon Valley (Arthur 1989). Through spying, imitation, and rapid interfirm movement of highly skilled labor, ideas are quickly disseminated among neighboring firms. The MAR theory also predicts, as Schumpeter (1942) does, that local monopoly is better for growth than local competition, because local monopoly restricts the flow of ideas to others and so allows externalities to be internalized by the innovator. When externalities are internalized, innovation and growth speed up.

Porter (1990), like MAR, argues that knowledge spillovers in specialized, geographically concentrated industries stimulate growth. He insists, however, that local competition, as opposed to local monopoly, fosters the pursuit and rapid adoption of innovation. He gives examples of Italian ceramics and gold jewelry industries, in which hundreds of firms are located together and fiercely compete to innovate since the alternative to innovation is demise. Porter's externalities are maximized in cities with geographically specialized, competitive industries.

Jacobs (1969), unlike MAR and Porter, believes that the most important knowledge transfers come from outside the core industry. As a result, variety and diversity of geographically proximate industries rather than geographical specialization promote innovation and growth. One example is the brassiere industry, which grew out of dressmakers' innovations rather than the lingerie industry. Jacobs also favors local competition because, like Porter, she believes that it speeds up the adoption of technology.

These theories of *dynamic externalities* are extremely appealing because they try to explain *simultaneously* how cities form and why they grow. MAR's and Porter's theories, in particular, predict that industries should specialize geographically to absorb the knowledge spilling over between firms. In addition, they predict that regionally specialized industries should grow faster because neighboring firms can learn from each other much better than geographically isolated firms. In contrast, Jacobs's theory predicts that industries located in areas that are highly industrially diversified should grow faster. Despite their differences, all these theories have implications for *growth rates* of industries in different cities. In this respect, they are different from the more standard location and urbanization externality theories that address the formation and specialization of cities (Henderson 1986) but not city *growth*.

We examine the predictions of the various theories of knowledge spillovers and growth using a new data set on geographic concentration and competition of industries in 170 of the largest U.S. cities. We focus on the largest industries because one of the strongest implications of growth models (such as MAR) is that externalities in these models are sources of permanent income growth. If such externalities are permanent and important, we should see them in the largest industries. If, alternatively, externalities are important only early in an industry's life cycle and disappear as an industry matures, our empirical work would not pick them up, but they are not sources of permanent growth either.

We ask which industries in which cities have grown fastest between 1956 and 1987 and why. All three theories of dynamic externalities focus on knowledge spillovers but differ in where they believe the source of externalities is and what makes the capture of these externalities most effective. The theories are not always mutually exclusive but rather offer different views of what is most important. By testing empirically in which cities industries grow faster, as a function of geographic specialization and competition, we can learn which, if any, externalities are important for growth.

Our results can be briefly summarized. Our findings are based on a cross section of city-industries (e.g., New York apparel and textiles, Philadelphia apparel and textiles, Philadelphia electrical equipment), where knowledge spillovers should be easier to find than by looking at whole cities. In a cross section of city-industries, we find that, as measured by employment, industries grow slower in cities in which they are more heavily overrepresented. For example, the primary-metals industry grew rapidly in Savannah, Georgia, where it was not heavily represented in 1956, and declined in Fresno, California, where it was heavily overrepresented. These results do not favor the local within-industry externality theory of MAR and Porter, according to which industries should grow faster precisely in places in which they are overrepresented.

We also find that industries grow faster in cities in which firms in those industries are smaller than the national average size of firms in that industry. If we take the view that spreading the same employment over more firms increases local competition between these firms and therefore the spread of knowledge, this result supports Porter's and Jacobs's view that local competition promotes growth. One could also take the view that smaller firms grow faster, which, however, is not strictly compatible with the MAR model or with other evidence. Finally, city-industries grow faster when the rest of the city is less specialized. This result supports Jacobs's view that city diversity promotes growth as knowledge spills over industries. The evidence is thus negative on MAR, mixed on Porter, and consistent with Jacobs.

If MAR externalities are not important, why are so many cities specialized in a few industries? There are many other externalities that explain regional specialization and city formation but that do not specifically focus on knowledge spillovers and growth. For example, Marshall (1890) has argued that firms in the same industry often locate next to each other to share various inputs, including specialized labor. Many other "localization" externalities are discussed by Lichtenberg (1960), Henderson (1986, 1988), Arthur (1989), and Rotemberg and Saloner (1990), among others. Henderson (1986) in particular presents empirical evidence indicating that output per labor-hour is higher in firms that have other firms from the same industry located nearby. Static localization externalities can thus easily account for city specialization, but not for growth.

Finally, some work also explains why firms might want to locate in places in which local demand is high, what Henderson (1986) calls "urbanization" externalities. These models tend to predict that firms in *different* industries should locate next to each other, which suggests that they cannot be the complete story of cities. Lichtenberg (1960), Murphy, Shleifer, and Vishny (1989), and Krugman (1991*a*, 1991*b*) all present models of such externalities. Like localization externalities, urbanization externalities explain patterns of industry location rather than of growth. Wheat (1986) finds strong evidence that manufacturing employment grows faster in regions with more rapid population growth. We present some less aggregate evidence also pointing to the importance of urbanization externalities.

Section II of the paper presents the predictions of different views of externalities and city growth. Section III describes the data. Section IV presents results for the growth of city-industries. Section V deals with localization and urbanization externalities. Section VI presents conclusions.

II. Theories of Dynamic Externalities

The models of city growth we consider stress the role of dynamic externalities, and more specifically knowledge spillovers, for city growth. According to these models, cities grow because people in cities interact with other people, either in their own or in other sectors, and learn from them. Because they pick up this knowledge without paying for it, these knowledge spillovers are externalities. The frequency of interaction with other people is ensured by their proximity in a city. Because this proximity makes externalities particularly large in a city, all the models predict that cities grow faster than rural areas in which externalities are less important because people interact less.

The theories of city growth that we present differ along two dimensions. First, they differ in whether knowledge spillovers come from within the industry or from other industries. Second, they differ in their predictions of how local competition affects the impact of these knowledge spillovers on growth.

The MAR theory of spillovers focuses on spillovers within industry. Knowledge accumulated by one firm tends to help other firms' technologies, without appropriate compensation. In Silicon Valley, microchip manufacturers learn from each other because people talk and gossip, products can be reverse engineered, and employees move between firms. In New York, fashion designers move between firms and take their knowledge with them. The same was true of the Bangladeshi shirt industry in the 1980s, where hundreds of firms were founded by people who were initially employed by *one* joint venture with a Korean firm. Physical proximity facilitates this free information transmission. In this case, industries that are regionally specialized and benefit most from the within-industry transmission of knowledge should grow faster. Cities that have such industries should grow faster as well.

In MAR models of externalities, innovators realize that some of their ideas will be imitated or improved on by their neighbors without compensation. This lack of property rights to ideas causes innovators to slow down their investment in externality-generating activities, such as research and development. If innovators had a monopoly on their ideas, or at least if they had fewer neighbors who imitated them immediately, the pace of innovation and growth would rise. The MAR models tend to imply that whereas local competition is bad for growth, local concentration is good for growth because innovators internalize the externalities (see Romer 1990).

The effect of local competition is the primary difference between MAR's and Porter's models. In Porter's model, local competition accelerates imitation and improvement of the innovator's ideas. Although such competition reduces the returns to the innovator, it also increases pressure to innovate: firms that do not advance technologically are bankrupted by their innovating competitors. Porter believes that the second effect is by far the more important. Ruthless competition between local competitors leads to rapid adoption of the innovations of others and to improvement on them, and so generates industry growth. In contrast, local monopolies lead a quieter life as their managers consume perquisites rather than risk innovation. Porter gives striking examples of Italian ceramics and gold jewelry industries, the German printmaking industry, and many others that grew through rampant imitation of new technologies and improvement on them. All these industries are highly geographically concentrated, presumably to facilitate the flow of ideas and imitation.

Because MAR and Porter agree that the most important technological externalities occur within industry, they also agree that regional specialization is good for growth both of the specialized industries and of the cities they are in. However, MAR would argue that local monopoly is good because it allows internalization of externalities. In contrast, Porter would argue that local competition is good because it fosters imitation and innovation. In our empirical work, we shall look at the effects of both specialization and local competition on the growth of industries in cities.

The third theory that stresses knowledge spillovers is that of Jacobs (1969). Jacobs's idea is that the crucial externality in cities is cross-fertilization of ideas across different lines of work. New York grain and cotton merchants saw the need for national and international financial transactions, and so the financial services industry was born. A San Francisco food processor invented equipment leasing when he had trouble finding financing for his own capital; the industry was not invented by the bankers. In a more systematic account, Rosenberg (1963) discusses the spread of machine tools across industries and describes how an idea is transmitted from one industry to another. Scherer (1982) presents systematic evidence indicating that around 70 percent of inventions in a given industry are used outside that industry. Much evidence thus suggests that knowledge spills over across industries. Because cities bring together people from different walks of life, they foster transmission of ideas.

In Jacobs's theory, industrial variety rather than specialization is conducive to growth, because in diversified cities there is more interchange of different ideas. She contrasts Manchester, a specialized textile city that eventually declined, with broadly diversified Birmingham, which eventually flourished. Bairoch (1988) supports Jacobs by arguing that "the diversity of urban activities quite naturally encourages attempts to apply or adopt in one sector (or in one specific problem area) technological solutions adopted in another sector" (p. 336).

In the debate between local monopoly and competition, Jacobs comes squarely on the side of competition. She writes that "monopolies gratuitously harm cities and suppress what their economies are capable of achieving . . . . Extortionate prices, harmful though they most certainly are, are the least of disadvantages of monopolies, for monopolies forestall alternate methods, products and services" (1984, p. 227). Like Porter, Jacobs favors local competition because it stimulates innovation.

The three theories can be summarized using a simple economic model that will guide the empirical work. Suppose that a firm in some industry in a given location has a production function of output given by $A_t f(l_t)$, where A_t represents the overall level of technology at time t measured nominally (so changes in A represent changes in technology and changes in price), and l_t is the labor input at time t. The basic production function $f(l_t)$ abstracts from capital inputs. Allowing for only one input means that we may not capture labor-saving technological innovations and that we shall not capture innovations that result only in further accumulation of physical capital. We unfortunately do not have a measure of total productivity that would allow us to measure different types of technological progress. Each firm in this industry takes technology, prices, and wages, w_t, as given and maximizes

$$A_t f(l_t) - w_t l_t, \tag{1}$$

and so it sets the labor input to equate the marginal product of labor to its wage:

$$A_t f'(l_t) = w_t. \tag{2}$$

We can rewrite (2) in terms of growth rates as

$$\log\left(\frac{A_{t+1}}{A_t}\right) = \log\left(\frac{w_{t+1}}{w_t}\right) - \log\left[\frac{f'(l_{t+1})}{f'(l_t)}\right]. \tag{3}$$

The level of technology A_t in a city-industry is assumed to have both national components and local components:

$$A = A_{\text{local}} A_{\text{national}}. \tag{4}$$

The growth rate will then be the sum of the growth of national technology in this industry and the growth of local technology:

$$\log\left(\frac{A_{t+1}}{A_t}\right) = \log\left(\frac{A_{\text{local},t+1}}{A_{\text{local},t}}\right) + \log\left(\frac{A_{\text{national},t+1}}{A_{\text{national},t}}\right). \tag{5}$$

The growth of the national technology is assumed to capture the changes in the price of the product as well as shifts in nationwide technology in the industry, and the local technology is assumed to grow at a rate exogenous to the firm but depending on the various technological externalities present in this industry in the city:

$$\log\left(\frac{A_{\text{local},t+1}}{A_{\text{local},t}}\right) = g(\text{specialization, local monopoly, diversity, initial conditions}) + e_{t+1}. \tag{6}$$

In equation (6), *specialization* is a measure of concentration of that industry in that city, which MAR and Porter believe raises the rate of technological progress; *local monopoly* is a measure of appropriability of innovation, which raises technological progress according to MAR and reduces it according to Porter; and *diversity* measures the variety of activities that the city pursues, which according to Jacobs speeds up technological progress. In Section IV below, we discuss further the inclusion of certain initial conditions. In the analysis below, we try to measure specialization, local monopoly, and city diversity empirically.

If we set $f(l) = l^{1-\alpha}$, $0 < \alpha < 1$, we can combine (3), (5), and (6) to obtain

$$\alpha \log\left(\frac{l_{t+1}}{l_t}\right) = -\log\left(\frac{w_{t+1}}{w_t}\right) + \log\left(\frac{A_{\text{national},t+1}}{A_{\text{national},t}}\right) + g(\text{specialization, competition, diversity, initial conditions}) + e_{t+1}. \tag{7}$$

Growth in nationwide industry employment is assumed to capture changes in nationwide technology and prices. Workers are assumed to participate in a nationwide labor market so that wage growth will just be a constant across city-industries. Equation (7) then allows us to associate the growth of employment in an industry in a city with measures of technological externalities given by the theories.[1]

We should note that this specification of the three models is restrictive in an important respect; namely, it assumes that knowledge spillovers are constant over time and therefore affect both mature and young industries. One could argue, alternatively, that industries have a life cycle, and externalities are important only at the beginning, when new products are introduced. In this case, the function g is different across industries and in particular depends on their point in the product cycle. This view of temporary externalities is broadly inconsistent with the theories like Romer's (1990), which use externalities to explain permanent growth. Since the main purpose of this paper is to shed light on these theories, we stick with the specification such as (7). Our empirical work includes many old as well as young industries, and hence our results can in no way reject any model of industry life cycle and temporary externalities.

III. The Data

Construction of the Data Set

Our data set was constructed from the 1956 and 1987 editions of *County Business Patterns* (CBP), produced by the Bureau of the Census. The year 1956 was chosen because it was the first year with comprehensive data; 1987 was the last year available. The 1956 data were assembled by hand from hard copy; 1987 data are available on computer tapes.

The data set contains the information on employment, payroll, and number of establishments[2] by two-digit industry for every county in the United States. We obtained wages by dividing payroll by employment. Since we focus on cities rather than counties, we aggregate data across counties into metropolitan area units as described below. When

[1] We are assuming a national labor market for simplicity. If workers participate in an industrywide labor market, then the growth in industrywide employment may pick up changes in these wages. If workers participate in local labor markets (which seems unrealistic given large amounts of migration and the long time periods involved in our data), then it would be wage changes, as much as employment changes, that reflect changes in local technologies. Our wage change regressions in table 4 below are based on this possibility.

[2] We use the 1956 census definition of an establishment, which is an actual firm rather than a plant. In later years, the census redefined the establishment to be a plant. For our purposes, a firm is what is appropriate, so the 1956 definition is good.

counties are aggregated, the wage number is total payroll in a city divided by total employment. Since we run cross-section regressions, we keep all variables nominal.

Cities were constructed from a list of the top 170 standard metropolitan areas (SMAs) in 1956 in the United States contained in CBP. In some cases, an SMA contains several counties; in others (only in New England), several SMAs split a single county. The problem is to decide which counties should be included in a given city since it would be impractical to include all counties in an SMA. We included in each city the largest counties that cover the SMA until their combined payroll added up to at least 80 percent of the total payroll of the SMA *in both 1956 and 1987.*[3] This procedure makes sure that if substantial growth of employment occurred in counties in an SMA that were small or nonexistent in 1956, these counties are included in the city. The multicounty unit arrived at using this procedure is the city we focus on.[4]

For each city constructed through aggregating counties, we use data on the six largest two-digit industries, where size is measured by 1956 payroll. We use only six industries because we are interested in regionally specialized industries; also, hand collection limits how many industries we can take. This choice of industries creates a bias against including small, young, and dynamic industries that have not yet made it into the top six. However, as we mentioned above, the theories we are testing do not just apply to industries in the early years of the product cycle.

In some cases in which an industry in a county has only a few establishments, for confidentiality reasons, CBP does not reveal exact information on employment in that county-industry. Instead, it typically presents the range in which the employment in a given industry in a given county lies, such as 0–20 or 5,000–10,000. In a few cases in which the employment in a given county-industry is below 50, CBP presents no employment number at all. To construct our sample, we had to address this problem of missing variables, which is particularly

[3] Including all the counties that are part of the SMA would be extremely time consuming because to find out which industries in an SMA are the largest—a procedure we use to construct the data—we would have to add up by hand employment in all the potentially largest industries over all the counties. Adding extra counties also significantly worsens missing data problems discussed below. To simplify the first problem and to avoid the second, we have restricted the subset of counties included in the city.

[4] This procedure for constructing a city might introduce errors for larger cities, which cover numerous counties. To test whether this problem is responsible for our results, we repeated the analysis for the smallest 75 percent of the cities in the sample. The results were very similar to the ones reported below. For this reason, we use the whole sample in the results reported in this paper.

severe when a city contains several counties only one of which has missing data.

We addressed this problem as follows. If exact data were missing for some county-industry in 1956, we simply omitted that industry from the sample and replaced it by the next largest industry in that city for which complete data were available for all counties. The missing data problem was not significant in 1956, however, since we are selecting the largest city-industries as of 1956.

If exact data were missing for some county-industry in 1987, we estimated the employment in that county-industry at the midpoint of the range provided by CBP. For example, if it reported the employment in a county-industry to be between zero and 20, we used 10; if the number was between 5,000 and 10,000, we used 7,500. In a multicounty city, we then added these estimates to precise employment numbers for the counties for which they were available. In the few county-industries with employment under 50 for which CBP did not even provide a range, we used 25 as the employment number. This procedure enabled us to compute employment for all but four of the 170 × 6 = 1,020 city-industries. The reason we had to drop four city-industries is that "ordnance and accessories," an industry that occurred four times in our sample in 1956, was discontinued as a qualified two-digit industry by 1987. Of the 1,016 city-industries in this sample, employment in 833 was provided exactly, and employment in the other 183 was estimated as described above.[5]

Although CBP presents ranges of employment by county-industry, it does not provide any information on payroll in the cases in which exact employment numbers are omitted. As a result, we cannot estimate wages for these observations. Consequently, the wage regressions we present below are estimated on 833 city-industries for which we have exact data on employment and payroll.

Description of the Data

Since we are using a new data set, it may be helpful to present a simple description of the data. This is done in table 1. Panel A of the table describes the five smallest and the five largest cities in our sample as of 1956, their employment in 1956 and 1987, and the six largest industries in each of them. Note first that the largest city—New York—has employment of over 4 million, and the smallest—Laredo, Texas—has under 7,500 employees. Clearly our procedure

[5] In addition, we adjusted for many three- and four-digit industries, which were reassigned among two-digit industries. The primary changes occurred with reassignments among industries 50 and 51.

TABLE 1

DESCRIPTION OF THE DATA

A. CITIES IN 1956

CITY	EMPLOYMENT 1956	EMPLOYMENT 1987	SIX LARGEST INDUSTRIES
			Five Largest Cities in 1956
New York	4,065,062	5,449,561	Apparel, business services, printing, special trade contractors, durable wholesale trade, nondurable wholesale trade
Chicago	1,919,757	2,778,180	Metal products, food and kindred products, electric equipment, nonelectric machinery, printing, durable wholesale trade
Los Angeles	1,710,325	3,546,393	Electric equipment, fabricated metal products, nonelectric machinery, special trade contractors, transportation equipment, durable wholesale trade
Philadelphia	1,085,524	1,287,820	Apparel, electric equipment, fabricated metal products, food and kindred products, nonelectric machinery, durable wholesale trade
Detroit	1,063,284	1,567,641	Fabricated metal products, nonelectric machinery, primary metal industries, special trade contractors, transportation equipment, durable wholesale trade
			Five Smallest Cities in 1956
Laredo, Tex.	7,458	25,397	Apparel, apparel stores, auto dealers, general merchandise, transport services, durable wholesale trade
San Angelo, Calif.	12,188	29,720	Auto dealers, communications, general contractors, general merchandise, special trade contractors, nondurable wholesale trade
Ogden, Utah	13,958	40,715	Auto dealers, communications, general contractors, general merchandise, special trade contractors, durable wholesale trade
Fort Smith, Ark.	19,089	55,057	Auto dealers, food and kindred products, food stores, special trade contractors, durable wholesale trade, nondurable wholesale trade
Sioux Falls, S.D.	19,096	59,398	Auto dealers, food and kindred products, insurance, trucking, durable wholesale trade, nondurable wholesale trade

TABLE 1 (*Continued*)

B. 10 Largest City-Industries

City	Industry	Employment
New York	Apparel	366,928
New York	Durable wholesale trade	241,754
Detroit	Transportation equipment	233,761
Los Angeles	Transportation equipment	228,619
New York	Nondurable wholesale trade	157,833
New York	Printing	151,905
New York	Business services	143,043
Chicago	Electric machinery	125,425
Chicago	Nonelectric machinery	121,847
New York	Special trade contractors	114,267

C. Most Common City-Industries

Industry	Number of Appearances in Sample
Durable wholesale trade	146
Food and kindred products	78
Nonelectric machinery	76
Special trade contractors	70
Transportation equipment	60
Nondurable wholesale trade	59
Automotive dealers and service stations	55
Fabricated metal products	48
Primary metal industries	46
Electric machinery	38

D. Industry Growth

City-Industry	Growth	Nondiversity	Competition	Concentration
	Five Fastest-Growing City-Industries*			
Albuquerque Business services	3.325	.217	1.500	1.090
San Jose, Calif. Electric machinery	2.765	.290	.835	2.582
San Jose, Calif. Durable wholesale trade	2.407	.310	1.008	.883
San Jose, Calif. Transportation equipment	2.403	.311	.930	.876
Atlantic City, N.J. Hotels	2.345	.373	.418	11.221
	Five Slowest-Growing City-Industries			
Scranton, Pa. Anthracite coal mining	−5.417	.387	.931	113.139
Manchester, N.H. Leather products	−5.161	.331	.272	19.559
Wilkes-Barre–Hazleton, Pa. Tobacco products	−5.078	.466	.279	21.193
Hamilton-Middletown, Ohio Primary metal industries	−4.813	.513	.326	4.271
Gadsden, Ala. Textile mills	−4.714	.406	.185	4.876

* Growth is log(employment in 1987/employment in 1956) Nondiversity is city's other top five industries' share of 1956 total city employment. Competition is establishments per employee relative to establishments per employee in the U.S. industry. Concentration is the city-industry's share of city employment relative to U.S. industry's share of U.S. employment in 1956

of looking at SMAs gets us down to fairly small places. The panel also shows a great variety of top industries across cities, although wholesale durables and nondurables are big in many of them.

Panel B describes the 10 largest city-industries in our sample. New York City apparel is the largest city-industry in the United States in 1956 with over 350,000 employees. Transportation equipment in Detroit, which is of course autos, is the third largest. New York City appears six and Chicago two times on this list.

Panel C describes the most common city-industries in the sample. Wholesale trade in durables is the most common: it appears in 146 cities. A few other service categories appear as well, but the predominant type of most common sector is manufacturing. In particular, nonelectrical machinery, primary metals, fabricated metals, transportation equipment, and electric equipment all appear quite often. The typical stories of externalities apply to many of these industries. In most of our analysis, we have pooled manufacturing and services, although we discuss below what happens when wholesale trade is removed from the sample, as well as how services are different from manufacturing.

Panel D lists the five fastest-growing and five fastest-declining city-industries in terms of employment. The panel gives three impressions. First, rapidly declining city-industries were more regionally concentrated than the rapidly growing ones. Second, industries grew faster in diversified cities than in specialized ones. Third, fast-growing city-industries were more competitive, as measured by establishments per employee, than shrinking city-industries. All these three impressions turn out to be our general empirical findings.

The panel also shows that the fastest-growing city-industries tended to be in the South, West, and Southwest, whereas the slowest-growing city-industries were often in the East and the Midwest. This finding points to some basic economic forces at work, such as capital moving to low-wage areas. A cynic might say that temperature determined city growth or only that we are observing the decline of U.S. manufacturing. These objections are not valid since we control for location in the South and we compare how fast the same industry grows in different cities.

More important, the decline in certain industries (notably steel, but possibly autos as well) may be related to the theories discussed in this paper. Both steel and auto production were regionally concentrated (autos in Michigan and steel in Pennsylvania). Both industries had only moderate levels of competition. In both industries, innovation was arguably lacking, particularly in areas in which these industries were concentrated.

The steel industry, according to Reutter (1988), has missed opportunities that were exploited by their competitors in less sterile environments. Big steel lost market share not only to foreign competitors but to American minimills located in nontraditional areas such as Roanoke, Virginia, or Florida. Use of concrete in construction also hurt steel badly. The first major use of concrete took place in a Hempstead, Long Island, shopping complex, far away from the traditional steel mills. Both concrete and shopping malls were major postwar innovations coming not from the established building material centers (such as Pennsylvania) but from smaller, more diversified areas. The steel industry may have declined not just because of foreign competition or some exogenous decline in manufacturing but, in part, because of forces stressed by Porter and Jacobs. Our statistical work suggests that this story of steel is a rule rather than an exception.

IV. Results on the Growth of Industries across Cities

If externalities are important for growth, then the clearest way to find these effects is by looking at the growth of the same sectors in different cities and checking in which cities these sectors grow faster. The unit of observation is then an industry in a city, and we look at the growth rates of these industries as a function of our measures of knowledge spillovers. The sample includes 1,016 observations on the top six two-digit 1956 industries in 170 cities.[6] Table 2 describes the variables. The mean of employment growth is zero, indicating that in an average city-industry in our sample employment did not grow. The standard deviation of this number, one, indicates the enormous dispersion of growth records. This dispersion may reflect the decline of some mining and manufacturing industries and the growth of services.

Equation (7) suggests that employment growth in an industry in a city may depend on the specialization of that industry in that city, local competition in the city-industry, and city diversity. Our measure of specialization of an industry in a city is the fraction of the city's employment that this industry represents in that city, relative to the share of the whole industry in national employment:

$$\text{specialization} = \frac{\text{industry employment in city/total employment in city}}{\text{industry employment in U.S./total employment in U.S.}}. \qquad (8)$$

[6] We have also performed the analysis using cities as a unit of observation. These regressions have not produced any statistically significant coefficients on our measures of externalities, although these measures tend to be much cruder for cities than they are for city-industries.

TABLE 2

VARIABLE MEANS AND STANDARD DEVIATIONS

Variable	Mean	Standard Deviation	Number of Observations
Log(employment in 1987/employment in 1956) in the city-industry	.00236	1.004	1,016
Log(U.S. employment in 1987/U.S. employment in 1956) in the industry outside the city	.308	.459	1,016
Wage in the city-industry in 1956 in thousands of dollars per quarter	1.063	.244	1,016
Employment in the city-industry in 1956 (in millions)	.0097	.0228	1,016
City-industry's share of city employment relative to U.S. industry's share of U.S. employment in 1956	3.367	9.019	1,016
Establishments per employee in the city-industry relative to establishments per employee in the U.S. industry	.752	.416	1,016
City's other top five industries' share of 1956 total city employment	.351	.100	1,016
Log(wage in 1987/wage in 1956) in the city-industry	1.649	.208	833
Log(U.S. wage in 1987/U.S. wage in 1956) in the industry outside the city	1.645	.144	833
Wage in the city in 1987	4.600	.663	833
Log(employment in 1987/employment in 1956) in the city	.980	.424	170
Wage in the city in 1956	.864	.114	170
Employment in the city in 1956	.118	.298	170
Employment growth in the four biggest industries	−.0312	.648	170

This variable measures how specialized a city is in an industry relative to what one would expect if employment in that industry was scattered randomly across the United States. The variable corrects for situations in which a city-industry is large only because the city is large. Because we are looking at the largest industries and because of regional specialization, the mean of this variable is 3.37. In our cities, top industries are overrepresented relative to what one would expect if they were randomly scattered over the United States. It is interesting to note that the maximum of this variable is 182.35 for anthracite mining in the Wilkes-Barre and Hazleton (Pennsylvania) SMA. The prediction of both MAR and Porter is that high specialization of an industry in a city should speed up growth of that industry in that city.

Our measure of local competition of an industry in a city is the number of firms per worker in this industry in this city relative to the number of firms per worker in this industry in the United States:

$$\text{competition} = \frac{\text{firms in city-industry/workers in city-industry}}{\text{firms in U.S. industry/workers in U.S. industry}}. \quad (9)$$

A value greater than one means that this industry has more firms relative to its size in this city than it does in the United States. One interpretation of the value greater than one is that the industry in the city is locally more competitive than it is elsewhere in the United States. Alternatively, a value of the competition variable greater than one can mean that firms in that industry in that city are just smaller than they are on average in the United States. It is very hard to distinguish smaller firms from more competitive firms using our data. Unfortunately, we do not have data on the output of individual firms and so cannot construct concentration ratios. Since we are looking at industries with large employment in their respective cities, we expect the mean of this competition variable to be below one; in fact it is .75. In a liberal interpretation of Porter, a higher value of this measure of competition should be associated with faster growth.

Finally, to address Jacobs's theory, we need a measure of a variety of industries in the city outside the industry in question. The measure we use is the fraction of the city's employment the largest five industries other than the industry in question account for in 1956. The mean of this ratio is .35: cities are not well diversified. The lower this ratio, the more diverse the city is and the faster the industry in question should grow according to Jacobs.

Table 3 presents our results for employment growth across city-industries, with 1,016 observations. We include as controls in the regressions the 1956 log of wage and the log of employment in the city-industry, a dummy variable indicating a southern city, and the national employment growth in that industry. Some analysts have argued that firms move to low-wage areas (or workers move to high-wage areas). Including the 1956 wage controls for either of these effects, even though this control variable is not strictly consistent with the assumption of a national labor market. High observed initial employment reduces employment growth because of either measurement error or more serious economic factors. The MAR view is somewhat incompatible with the presence of real (as opposed to measurement-induced) mean reversion, but since we are not correcting for potential measurement problems, we do not use such mean reversion as evidence against the MAR externalities.

We also include national employment change in the industry to correct for demand shifts. Various studies, including Terkla and Doeringer (1991) and Blanchard and Katz (1992), show that changes

TABLE 3

CITY-INDUSTRY EMPLOYMENT GROWTH BETWEEN 1956 AND 1987

	Log(Employment in 1987/Employment in 1956) in the City-Industry			
Dependent Variable	(1)	(2)	(3)	(4)
Constant	−.423	−.923	−.181	−.513
	(.129)	(.129)	(.159)	(.149)
Log(U.S. employment in 1987/ U.S. employment in 1956) in the industry outside the city	1.140	1.209	1.237	1.148
	(.059)	(.052)	(.055)	(.056)
Wage in the city-industry in 1956	.0137	.0226	.0379	.027
	(.109)	(.104)	(.109)	(.104)
Employment in the city-industry in 1956 (in millions)	−2.898	−3.280	−3.91	−4.080
	(1.099)	(1.055)	(1.131)	(1.073)
Dummy variable indicating presence in the South	.426	.416	.370	.378
	(.057)	(.054)	(.058)	(.055)
City-industry's share of city employment relative to industry's share of U.S. employment in 1956	−.0128	...	...	−.00799
	(.003)			(.003)
Establishments per employee in the city-industry relative to establishments per employee in the U.S. industry in 1956	...	.587	...	.561
		(.057)		(.057)
City's other top five industries' share of total city employment in 1956	...	...	−.894	−.913
			(.259)	(.245)
Adjusted R^2	.392	.439	.387	.450
Number of observations	1,016	1,016	1,016	1,016

NOTE.—Standard errors of parameter estimates are in parentheses beneath these estimates.

in demand for a region's output are the principal determinant of employment growth in that region. Since we are looking at city-industries rather than whole regions, we must correct for changes in national industry demand, which we measure by national industry employment. We thus are looking at whether steel has grown faster or slower than average in particular cities, controlling for how steel employment changed in the nation as a whole. This correction is particularly important for traditional manufacturing industries, many of which have declined in the postwar United States.

The control variables tend to have the expected signs. High initial employment in an industry in a city leads to slower growth of that industry's employment. Employment in an industry in a city grows faster when employment in that industry in the whole country grows faster. It is interesting to note that the coefficient on national industry employment growth is above one. Factors shifting employment in national industries seem to be more influential in urban than in rural areas. Southern cities also grew significantly more than cities outside the South (see also Wheat 1986). Initial wages in a city-industry are uncorrelated with subsequent employment growth.

The results on externalities reveal several interesting findings. Equation 1 in table 3 shows that industries that are more heavily concentrated in the city than they are in the United States as a whole grow slower. The effect is statistically significant but qualitatively small. As we raise the measure of specialization by one standard deviation (9.02), cumulative growth of employment over 30 years slows by 12 percent total, which is about one-ninth of a standard deviation. This result is the opposite of the prediction of the MAR model. Not only do we fail to find positive evidence in favor of MAR, but the data point in the opposite direction: geographic specialization reduces growth.

In equation 2 in table 3, the coefficient on the competition variable is positive and very significant. More firms per worker in a city-industry relative to the national average leads to higher growth of that city-industry, consistent with Porter's and Jacobs's hypothesis. Going from as many to twice as many firms per worker as the national average (2.5 standard deviations) raises growth of employment in the city-industry by 59 percent over 30 years, which is almost two-thirds of a standard deviation. Of course, another interpretation of this finding is that smaller firms grow faster. However, recent evidence (Davis and Haltiwanger 1992) indicates that in fact smaller firms do not grow faster once one takes account of the fact that they have a higher probability of death than larger firms. We should also mention that the "small firms grow faster" model is inconsistent with the MAR view that monopolies that internalize externalities are good for growth. So even though the positive evidence in favor of competition is somewhat ambiguous, the negative evidence on MAR is more clear-cut.

Equation 3 in table 3 shows that industries in cities in which other large industries are relatively small grow faster. As we reduce the share of city employment taken up by the five largest industries other than the one in question by .1 (a standard deviation), cumulative employment growth in the city-industry over 30 years falls by 9 percent (one-tenth of a standard deviation). This result suggests that not having dominant industries as neighbors, or alternatively having a greater variety of neighbors, helps own growth. This finding is consistent with the importance of knowledge spillovers stressed by Jacobs from outside the industry.

Equation 4 in table 3 uses all measures of externalities simultaneously. The results remain statistically significant. They confirm our finding that industry overrepresentation hurts its growth. The fact that the coefficient has the wrong sign relative to what MAR predicts and is statistically significant is evidence against the importance of permanent within-industry knowledge spillovers for growth. Competition within the city-industry continues to exert a positive influence on growth of its employment, and the coefficient hardly changes from equation 2 in table 3. The result that concentration of other industries in the city hurts the growth of an industry's employment continues to be strong. The overall results are not favorable to MAR, mixed on Porter, and favorable to Jacobs.

It is also possible that our results support a neoclassical model rather than any kind of externality. Many of our findings are consistent with the observation that industries move to regions in which they are not present. Thus employment moves south, as Wheat (1986) has also found, and grows slower where it is high to begin with. The result that employment in an industry grows higher in a diversified city might reflect the fact that such cities are less crowded and hence cheaper to locate in. Competition for space and labor might thus explain many of our findings. At the same time, the very existence of cities despite the high rents is hard to explain without externalities. And if externalities do matter, our evidence points against intraindustry spillovers and in favor of knowledge spillovers across sectors à la Jacobs.

We have checked the robustness of these results in a number of ways. First, our results might be driven by the mining industries, which exhibit extraordinary regional specialization and have declined sharply in postwar years in part because prices fell and in part because mineral stocks were depleted. We have run the regressions in table 3 without the mining industries, and the results are similar in terms of sign patterns and statistical significance.

Second, one could argue that knowledge spillovers are more important in manufacturing than in services because technological progress is more rapid in manufacturing. Without subscribing to this objection, we tested it empirically by splitting the industries into manufacturing and nonmanufacturing. Our results hold qualitatively for both subsamples. All coefficients on the externality variables remain significant in the nonmanufacturing regression, and all but the coefficient on the urban variety variable are significant for manufacturing. If anything, the results appear to be stronger for nonmanufacturing.

Third, we divided manufacturing into ubiquitous industries (fabricated metals, nonferrous metals, nonelectrical machinery, etc.) and more specialized industries (electrical equipment, transport equipment, primary metals, pulp and paper, textiles and apparel, leather products, etc.). Specialized industries presumably produce primarily for export, so technological change can spur their growth, unfettered by the limits of local demand. Ubiquitous industries are presumably aimed toward local consumption, and their growth will be largely limited by local demand.[7] If this reasoning is correct, the effect of externalities should be more pronounced in more specialized manufacturing industries.

We have run our regressions using the two manufacturing subsamples separately. The negative effect of our specialization measure on industry growth is stronger and more statistically significant for ubiquitous manufacturing, but the effect remains negative, though insignificant, for more specialized manufacturing as well. There is certainly no evidence that the coefficient is positive, as MAR would suggest.

Although we have measured industry growth using employment growth, a better measure would be productivity growth. Since we do not observe output, it is hard for us to measure productivity. However, in the world in which some productivity gains accrue to labor, one rough measure of productivity growth might be city-industry wage growth. While this measure is not compatible with the model of national labor markets that we presented, it is compatible with models of locally more segmented markets. Even so, measuring productivity growth through wage growth is very imperfect. First, productivity increases might accrue only partly to labor, especially as migration occurs; in the long run, identical workers must be indifferent between cities. Second, declining industries might fire their less able and experienced workers first, creating an artificial rise in wages. Third, certain technological innovations (e.g., the assembly line) might make it easier to hire less expensive workers. Fourth, rent sharing might also be a factor, especially in those industries (steel, coal, or autos) with heavy union involvement. A further problem with the wage data is that we have no estimates of wages for counties without precisely reported employment numbers. The restriction of the sample to 833 observations could induce a sample selection bias. Given these objections, our results on wage growth should be interpreted as at best secondary to employment growth results.

Table 4 presents the findings in the same format as table 3. High initial wages in a city-industry reduce wage growth, but high initial employment in a city-industry helped wage growth, although as we saw before it hurt employment growth. This result might reflect a selection effect: high employment leads to employment cuts, which for reasons of seniority affect least well paid workers most, leading to an increase in the average wage of those who remain employed.

Equation 1 in table 4 shows that city-industry specialization has no

[7] This division was suggested by Vernon Henderson.

TABLE 4

City-Industry Wage Growth between 1956 and 1987

Variable	Log(Wage in 1987/Wage in 1956) in the City-Industry			
	(1)	(2)	(3)	(4)
Constant	.332 (.065)	.379 (.064)	.398 (.069)	.440 (.068)
Log(U.S. wage in 1987/U.S. wage in 1956) in the industry outside the city	.961 (.043)	.975 (.042)	.959 (.043)	.973 (.042)
Wage in the city-industry in 1956	−.270 (.027)	−.270 (.027)	−.266 (.027)	−.267 (.027)
Employment in the city-industry in 1956 (in millions)	1.025 (.270)	1.111 (.266)	.849 (.276)	.938 (.271)
Dummy variable indicating presence in the South	.0175 (.013)	.0161 (.013)	.0094 (.013)	.0085 (.013)
City-industry's share of city employment relative to industry's share of U.S. employment in 1956	.00053 (.0007)	...	...	−.00023 (.0007)
Establishments per employee in the city-industry relative to establishments per employee in the U.S. industry in 1956	...	−.0850 (.014)	...	−.0845 (.014)
City's other top five industries' share of 1956 city employment	...	...	−.172 (.060)	−.161 (.059)
Adjusted R^2	.3832	.4099	.3889	.4139
Number of observations	833	833	833	833

Note.—Standard errors of parameter estimates are in parentheses beneath these estimates.

effect on wage growth, which does not support the MAR view, although the coefficient no longer has the wrong sign. When other measures of externalities are added in equation 4 in table 4, the coefficient is insignificant and has the wrong sign. City-industry competition reduces wage growth (the coefficient is significant), which is inconsistent with the view that competition contributes to productivity growth that accrues to the workers. We do not think that the latter position can be ascribed to Porter. Finally, diversity in a city helps wage growth of the industry, consistent with Jacobs's view that productivity growth is helped by diversity.

None of the evidence we have presented supports the importance of within-industry knowledge spillovers for growth. If such spillovers are particularly pronounced at geographical proximity, the evidence is detrimental to the theories of MAR and Porter that focus on these spillovers. We end this discussion with a word of caution, however. We are looking at large, mature cities that are not growing very fast and are in many cases declining, making ours a very special sample. Within-industry knowledge spillovers may not matter for such mature industries, though they may be much more important at the early stages of an industry. For example, these spillovers might be very important when a new industry is born and organizes itself in one location, but unimportant as this industry matures and geographical proximity becomes less important for the transmission of knowledge. Our data, unfortunately, cannot address this industry life cycle model. At the least, however, we are rejecting the strong version of the MAR theory, which predicts that within-industry knowledge spillovers lead to permanent self-sustaining growth in cities.

V. Static External Economies: Localization and Urbanization

Localization

The evidence we have presented suggests that diversity, and not specialization, contributes to growth. This result raises an important problem: If geographical specialization does not contribute to growth, why is it so prevalent? In this section, we address this problem. We also look for the evidence of urbanization externalities: those that make different industries locate next to each other to form a city.

There are several reasons for regional specialization that are not dynamic externalities that contribute to growth. Most obviously, natural resource or transport advantages often favor a particular location, and those apply equally to all firms in the industry. For example, the oil industry at the turn of the century was located in Ohio, near the discovered oil. Bairoch (1988) reports that during the Industrial Revolution many new cities located near the supplies of energy. One could also argue that the auto industry located in the Midwest in part to economize on transport costs of inputs.

But in addition to these natural reasons for specialization, there are several static externalities that contribute to specialization but not to growth. Perhaps most important is the idea of saving on moving inputs, suggested by Marshall (1890). A whole industry might locate near the place of common suppliers both to reduce the cost of getting supplies and to have a closer flow of information to suppliers. In addition, many firms producing specialized products that are subject to wildly fluctuating firm demand but more stable industry demand would locate together. By doing so, they enable specialized labor to move easily between firms without moving between cities, as in the previously mentioned case of the New York fashion industry. More generally, when firms share any input that is not costlessly mobile, it pays them to locate together near that input and so save on moving the input (see Lichtenberg 1960; Henderson 1986, 1988). More recently, Rotemberg and Saloner (1990) argued that firms locate together to commit to compete for labor and not pay monopsony wages. This effective commitment enables firms to attract labor in the first place. There are clearly many reasons for regional specialization other than knowledge spillovers.

Our findings are consistent with the importance of localization externalities as long as the location of firms next to each other to take advantage of these externalities is finished when our sample begins. In this case, there is no reason that regionally specialized industries should grow faster. If, in contrast, we observed young industries, in which entry of firms to take advantage of localization externalities was still taking place, we would still expect employment in regionally specialized industries to grow faster as entry takes place. Our results would then reject the importance of localization externalities, just as they reject the MAR-Porter models. Since we are focusing on the largest city-industries, however, the assumption that they are mature seems reasonable. Our findings do not reject the localization externalities playing a role in determining regional specialization.

Urbanization

Although cities are usually specialized in a few lines of work, they also typically pursue many other activities outside the main lines. Many of these activities are entirely unrelated to each other. This suggests another type of externality operating in a city. Firms locate in a city because local demand is high there, and so they can sell some of their output without incurring transport costs. This is obviously most important for high-fixed-cost industries. Lichtenberg (1960) argues that this externality explains why the insurance industry once located in New York City. Murphy et al. (1989) discuss such pecuniary externalities; Krugman (1991*a*, 1991*b*) models city formation based on local demand. Henderson (1986) refers to these effects as "urbanization" externalities and presents empirical evidence suggesting that they are not important for productivity.

These models imply that when an industry grows, it raises local payrolls and therefore local demand and so helps the growth of other possibly unrelated industries in that city, which adjust to higher demand. As a result, growth rates of different industries in a city are positively correlated. This argument is most compelling for local services, which probably grow when city exports grow.

The argument against urbanization externalities is crowding. When an industry in a city grows, it raises wages and rents and so makes it more expensive for other industries to expand in that city. Con-

TABLE 5

EMPLOYMENT GROWTH OF SMALLER INDUSTRIES BETWEEN 1956 AND 1987

Dependent Variable	Log(Employment in 1987/Employment in 1956) in the City Outside the Four Biggest Industries
Constant	1.410
	(.191)
1956 employment outside the four biggest industries (in millions)	−1.96
	(.083)
1956 wage outside the four biggest industries	−.455
	(.226)
Employment growth in the four biggest industries	.458
	(.033)
R^2	.5910
Number of observations	170

NOTE.—Standard errors of parameter estimates are in parentheses beneath these estimates.

versely, when an industry in a city shrinks, it frees up land and labor and so makes growth of other industries more attractive. Urbanization externalities and crowding have the opposite implications for the data.

Our data enable us to test these predictions, as shown in table 5. The dependent variable in the regression is employment growth in the city outside the four largest industries, and the key independent variable is employment growth in these four largest industries. The evidence indicates very strongly that small industries grow when large industries do. A 1 percent increase in the four-industry employment growth leads to a 0.5 percent increase in employment growth outside these industries. We replicated this result for several combinations of dependent and independent variables. The evidence consistently points in favor of aggregate demand spillovers and against crowding. Of course, another possible interpretation of this finding is that there are some city effects that attract all industries to some cities, but recall that we at least control for the 1956 wage level in our sample. Overall, the results support the role of urbanization externalities in city growth, consistent with the theoretical work of Murphy et al. (1989) and Krugman (1991*a*, 1991*b*) as well as with much of the empirical work in urban economics.

VI. Conclusion

The results presented in this paper allow some tentative conclusions. We have shown that at the city-industry level, specialization hurts, competition helps, and city diversity helps employment growth. Our best interpretation of this evidence is that interindustry knowledge spillovers are less important for growth than spillovers across industries, particularly in the case of fairly mature cities. The Jacobs-Rosenberg-Bairoch model, in which knowledge transmission takes the form of adoption of an innovation by additional sectors, seems to be the most consistent with the evidence.

An important objection to these results that we have mentioned already is that we are looking at a particular period in U.S. history in which traditional manufacturing industries have fared poorly because of import competition and at particular very mature cities. Our results may then not be applicable for more dynamic time periods or places. On this theory, MAR externalities matter the most when industries grow. We cannot address this objection with our data.

The evidence suggests that cross-fertilization of ideas across industries speeds up growth. The growth of cities is one manifestation of this phenomenon, but there may be others. The results would imply, for example, that open societies, with substantial labor mobility across industries, will exhibit a greater spread of ideas and growth. Similarly, the cross-fertilization perspective argues in favor of such labor flows as immigration and migration across areas. If Jane Jacobs is right, the research on growth should change its focus from looking inside industries to looking at the spread of ideas across sectors. As a final point, however, we recall that our evidence on externalities is indirect, and many of our findings can be explained by a neoclassical model in which industries grow where labor is cheap and demand is high.

References

Arrow, Kenneth J. "The Economic Implications of Learning by Doing." *Rev. Econ. Studies* 29 (June 1962): 155–73.

Arthur, W. Brian. "Silicon Valley Locational Clusters: When Do Increasing Returns Imply Monopoly?" Working paper. Sante Fe, N.M.: Santa Fe Inst., 1989.

Bairoch, Paul. *Cities and Economic Development: From the Dawn of History to the Present.* Chicago: Univ. Chicago Press, 1988.

Blanchard, Olivier J., and Katz, Lawrence F. "Regional Evolutions." *Brookings Papers Econ. Activity,* no. 1 (1992).

Dasgupta, Partha, and Stiglitz, Joseph E. "Uncertainty, Industrial Structure, and the Speed of R&D." *Bell J. Econ.* 11 (Spring 1980): 1–28.

Davis, Steve J., and Haltiwanger, John. "Gross Job Creation, Gross Job Destruction, and Employment Reallocation." *Q.J.E.* 107 (August 1992).

Griliches, Zvi. "Issues in Assessing the Contribution of Research and Development to Productivity Growth." *Bell J. Econ.* 10 (Spring 1979): 92–116.

Henderson, J. Vernon. "Efficiency of Resource Usage and City Size." *J. Urban Econ.* 19 (January 1986): 47–70.

———. *Urban Development: Theory, Fact, and Illusion.* New York: Oxford Univ. Press, 1988.

Jacobs, Jane. *The Economy of Cities*. New York: Vintage, 1969.

———. *Cities and the Wealth of Nations: Principles of Economic Life*. New York: Vintage, 1984.

Krugman, Paul. "Cities in Space: Three Simple Models." Manuscript. Cambridge: Massachusetts Inst. Tech., 1991. (*a*)

———. "Increasing Returns and Economic Geography." *J.P.E.* 99 (June 1991): 483–99. (*b*)

Lichtenberg, Robert M. *One-Tenth of a Nation: National Forces in the Economic Growth of the New York Region*. Cambridge, Mass.: Harvard Univ. Press, 1960.

Loury, Glenn C. "Market Structure and Innovation." *Q.J.E.* 93 (August 1979): 395–410.

Lucas, Robert E., Jr. "On the Mechanics of Economic Development." *J. Monetary Econ.* 22 (July 1988): 3–42.

Marshall, Alfred. *Principles of Economics*. London: Macmillan, 1890.

Murphy, Kevin M.; Shleifer, Andrei; and Vishny, Robert W. "Industrialization and the Big Push." *J.P.E.* 97 (October 1989): 1003–26.

Porter, Michael E. *The Competitive Advantage of Nations*. New York: Free Press, 1990.

Reutter, Mark. *Sparrows Point: Making Steel: The Rise and Ruin of American Industrial Might*. New York: Simon & Schuster, 1988.

Romer, Paul M. "Increasing Returns and Long-Run Growth." *J.P.E.* 94 (October 1986): 1002–37.

———. "Endogenous Technological Change." *J.P.E.* 98, no. 5, pt. 2 (October 1990): S71–S101.

Rosenberg, Nathan. "Technological Change in the Machine Tool Industry, 1840–1910." *J. Econ. Hist.* 23 (December 1963): 414–43.

Rotemberg, Julio, and Saloner, Garth. "Competition and Human Capital Accumulation: A Theory of Interregional Specialization and Trade." Manuscript. Cambridge: Massachusetts Inst. Tech., 1990.

Scherer, Frederic M. "Inter-Industry Technology Flows in the United States." *Res. Policy* 11 (August 1982): 227–45.

Schumpeter, Joseph A. *Capitalism, Socialism, and Democracy*. New York: Harper, 1942.

Terkla, D. G., and Doeringer, P. B. "Explaining Variations in Employment Growth: Structural and Cyclical Change among States and Local Areas." *J. Urban Econ.* 29 (May 1991): 329–48.

Wheat, Leonard F. "The Determinants of 1963–77 Regional Manufacturing Growth: Why the South and West Grow." *J. Regional Sci.* 26 (November 1986): 635–59.

[17]

Industrial Development in Cities

Vernon Henderson

Brown University

Ari Kuncoro

University of Indonesia

Matt Turner

University of Toronto

This paper uses data for eight manufacturing industries in 1970 and 1987 to test for and characterize dynamic production externalities in cities. We find evidence of both MAR externalities, which are associated with past own industry employment concentration, and Jacobs externalities, which are associated with past diversity of local total employment. More specifically, for mature capital goods industries, there is evidence of MAR externalities but none of Jacobs externalities. For new high-tech industries, there is evidence of Jacobs and MAR externalities. These findings are consistent with notions of urban specialization and product cycles: new industries prosper in large, diverse metropolitan areas, but with maturity, production decentralizes to smaller, more specialized cities. For mature industries, there is also a high degree of persistence in individual employment patterns across cities, fostered by both MAR externalities and persistence in regional comparative advantage.

The literature on endogenous growth models argues that dynamic information externalities are the driving force for technological inno-

Support of the National Science Foundation and also the Woods Hole Oceanographic Institute (Turner) for this work is gratefully acknowledged. We acknowledge helpful comments of a referee and the editor. We also benefited from comments of participants in seminars at British Columbia, Harvard, Kentucky, and the National Bureau of Economic Research.

[*Journal of Political Economy*, 1995, vol 103, no 5]

vation and hence economic growth (Romer 1986). Since these externalities arise from both intended and unintended communications among economic agents over time, their effects should be more readily observed in places in which communications are focused. As noted by Lucas (1988), this intuition suggests that cities provide a natural laboratory to study the nature and extent of these externalities. This paper is an empirical investigation of the nature of dynamic externalities and their implications for urban development, for a selection of both mature and new high-tech industries.

The study of externalities in urban environments has a long history (e.g., Hoover 1937; Chinitz 1961), but empirical work has focused on static externalities, such as immediate information spillovers about current market conditions. There are two types of static externalities: localization economies in which a firm benefits from local firms in just the same industry, and urbanization economies in which a firm benefits from overall local urban scale and diversity. Both types suggest benefits of being in bigger cities, but as we know, city sizes are limited by overall local congestion and commuting costs (Mills 1967). Given this, with localization economies, cities will tend to specialize to enhance own industry agglomeration benefits relative to commuting and congestion costs. Their sizes will depend on their products and the associated degree of localization economies (Henderson 1974). On the other hand, if urbanization economies dominate for an industry, firms will seek more diversified larger cities. Thus in a snapshot of cities, we see smaller and medium-size textile, apparel, transport equipment, primary metals, food processing, pulp and paper, and so forth type cities in which localization economies dominate; but industries such as high-fashion apparel, upper-end publishing, and many business services subject to urbanization economies agglomerate in very large metropolitan areas (Henderson 1988).

Dynamic externalities deal with the role of prior information accumulations in the local area on current productivity and hence employment. Such accumulations are fostered by a history of interactions and cultivated long-term relationships, which lead to a buildup of knowledge ("local trade secrets"), available to firms just in a local area. As in their static counterparts, there are two types of dynamic externalities. With terminology used in Glaeser et al. (1992), dynamic externalities may be Marshall-Arrow-Romer (MAR) (localization) economies, which derive from a buildup of knowledge associated with ongoing communications among local firms in the same industry, or Jacobs (1969) (urbanization) economies, which derive from a buildup of knowledge or ideas associated with historical diversity. As shown in a snapshot of cities, dynamic externalities can have implications similar to those of static externalities. For example, in the steady state, an industry with a higher degree of MAR externalities is likely to locate in a larger specialized city type than an industry with lower MAR externalities. However, dynamic externalities have broader implications concerning industrial development over time. They help provide an explanation for the location and growth patterns of both mature and newer industries, which we observe in a series of snapshots of an economy.

For mature industries we observe a very high degree of persistence in employment patterns across cities over time. There is very slow mean reversion, or "convergence" in the raw data on individual industry employments. This persistence occurs despite both high plant and employment turnover rates for individual manufacturing industries (see, respectively, Dunne, Roberts, and Samuelson 1989*a*, 1989*b*; Davis and Haltiwanger 1991) and despite strong evidence that plants relocate as local wages and demand conditions change (see Herzog and Schlottmann 1991). If employment concentrations across cities were determined solely by a random draw of just current economic conditions, we would expect strong reversion to the mean over time in individual industry employment levels across cities. We shall show that part of the glue that holds employment concentrations of an industry in specific cities over time is MAR externalities. Cities with historical concentrations of an industry and related local knowledge accumulations will offer a more productive environment for establishments in that industry than those without them. They will be able to better compete for and, over time, retain plants and employment in that industry.

For newer high-tech industries, persistence is not so relevant, since industrial location patterns are new. Instead we observe growth in the number of cities having an industry and in employment levels within those cities. The natural question concerns what historical environments have an advantage in the ongoing race to attract newer industries, recognizing that for any new industry only a few cities will ultimately be winners. Part of the answer will lie with dynamic externalities. In particular, we shall show that new high-tech industries are more likely to take root in cities with a history of industrial diversity, suggesting that Jacobs externalities are important for these industries. Taken together, our results for mature and new industries are consistent with urban product cycle notions. New products are developed in large diverse metro areas with a diversified skill base, but matured product lines subject to MAR effects eventually decentralize to smaller, more specialized metro areas, with lower wage and land costs.

In previous studies of dynamic externalities, Jaffe, Trajtenberg, and Henderson (1993), in studying patent citations, find that such externalities are localized and diffuse slowly over space. In studying employment growth for 170 standard metropolitan statistical areas

(SMSAs) between 1956 and 1987 for the six largest industries in each city in 1956, Glaeser et al. (1992) conclude that, for all economic activity lumped together, dynamic externalities are only Jacobs in nature. For total manufacturing, Miracky (1992) finds some evidence of both MAR and Jacobs externalities for the period 1977–87.

Our data describe employment growth patterns in eight specific manufacturing industries in 224 metropolitan areas between 1970 and 1987. We examine the five key traditional capital goods industries that cities tend to specialize in and whose products are widely traded across cities: primary metals, machinery, electrical machinery, transport equipment, and instruments. We also examine three "new" high-tech industries—computers, electronic components, and medical equipment—to study key aspects of their recent development in cities.

Our methodology and basic results differ from those of previous studies. As already implied, we find intuitively appealing results—only MAR externalities for mature industries and both MAR and Jacobs externalities for new high-tech industries—that are consistent with notions of persistence and urban product cycles. Relative to previous work, we find that it is critical (1) to distinguish between industries and to include newer industries in the sample since externalities vary by industry and stage of product development, (2) to include for each industry the entire available sample of cities, and (3) to incorporate other traditional considerations besides externalities that affect local industry growth, such as local labor market and regional product demand conditions.

In Section I of the paper, we present and estimate the model for five traditional heavy-manufacturing industries, to explore the nature of dynamic externalities for those industries. In Section II, we turn to the three newer high-tech industries. Section III presents conclusions.

I. Traditional Industries

This section examines 1987 employment patterns in five two-digit manufacturing industries: machinery, electrical machinery, primary metals, transportation, and instruments. Information on these industries for 1970 and 1987 is presented in table B1 in Appendix B. On the basis of uncensored 1970 data, it appears that these industries are found in all cities. Except for instruments, these industries have stagnated or declined nationally since 1970.

To test for dynamic externalities, we model 1987 city employment in each industry as a function of historical (1970) and current conditions in cities. For an industry in location i at time t, the equilibrium employment level is such that the local wage rate equals the value of marginal product (VMP), or $W_{it} = A_{it}(\cdot)f'(N_{it}; \ldots)P_{it}(\cdot)$. Industry output is $A_{it}(\cdot)f(N_{it}; \ldots)$, and N_{it} is employment in the industry in location i in time t. The term $A_{it}(\cdot)$ represents the state of technology for that industry in location i in time t; W_{it} is the nominal wage rate, which varies enormously across localities for workers with similar skills; and P_{it} is the price of output, given by the inverse demand function $P_{it}(\cdot) = P(N_{it}, MC_{it})$. From the literature,[1] we know that for a city, $P_{it}(\cdot)$ is downward sloping in local industry output (represented by N_{it}); and its other arguments, MC_{it}, include regional characteristics, access to major urban market centers, and local metro area demand for capital good products.

For the arguments of $A_{it}(\cdot)$, the traditional production externality literature focuses on static externalities that arise from current local own industry scale, or employment, N_{it}. Dynamic externalities deal with aspects of the historical urban environment: in particular, own industry employment in some base period, N_{i0}, concentration of that employment, ρ_{i0}, and diversity of the environment, D_{i0}. Cicone and Hall (1993) as well as Glaeser et al. (1992) argue that concentration, ρ_{i0} (vs. levels, N_{i0}), may better represent the potential for MAR externalities since concentration facilitates spillover or "network" information flows among relevant firms and the development of location-specific knowledge, relative to a location with diffuse economic activity. A diversity measure represents the potential for Jacobs externalities, where, for example, product development may prosper in an environment in which there is a history and tradition of economic and social interaction among diverse economic sectors. Combining static and dynamic externalities, in summary, for A_{it} we have $A_{it} = A(N_{it}, N_{i0}, \rho_{i0}, D_{i0}, \ldots)$. Substituting the expressions for $P_{it}(\cdot)$ and $A_{it}(\cdot)$ into the VMP equals the wage equation, and inverting, we get a reduced-form equation

$$N_{it} = N(W_{it}, MC_{it}, N_{i0}, \rho_{i0}, D_{i0}, \ldots). \tag{1}$$

Note that this equation can also be transformed into a growth rate equation, where $\log(N_{it}/N_{i0}) = \tilde{N}(\cdot)$. In equation (1), since $\partial[A(\cdot)f'(\cdot)P(\cdot)]/\partial N_{it} < 0$ for equilibrium in labor markets, a well-behaved equilibrium requires $\text{sign}[\partial N_{it}/\partial N_{i0}] = \text{sign}[\partial A_{it}(\cdot)/\partial N_{i0}]$. This comment applies to any historical argument of $A_{it}(\cdot)$. For example, if MAR effects matter, then an increase in ρ_{i0} increases worker productivity and thus increases local industry employment for a given wage.

[1] See Herzog and Schlottman (1991). As a specific example, for machinery and electrical machinery, Henderson (1993) finds that, on average, sales to Mexico decline by, respectively, 2.6 percent and 1.1 percent for each 55-mile distance increment of a U.S. plant site from the Mexican border.

Glaeser et al. (1992) arrive at a formulation such as (1) by forming the ratio for wage equals VMP equations in years t and 0 and asserting that the growth in $A(\cdot)$, $\log(A_{it}/A_{i0})$, is a function only of ρ_{i0}, D_{i0}, and N_{i0}. The resulting reduced-form equation corresponds to equation (1), except W_{i0} and MC_{i0} are added. Our results on dynamic externalities in equation (1) with and without the W_{i0} and MC_{i0} arguments are almost identical.

A. *Data and Estimation*

The data pertain to 1970 and 1987, and sources and definitions are discussed in Appendix A. We focus on the 224 SMSAs in 1970 that can be matched to 1987 Metropolitan Statistical Areas (MSAs) or Primary Metropolitan Statistical Areas (PMSAs). These proposed estimations raise a variety of econometric issues. First, equation (1) may contain fixed effects that are correlated with historical variables. Second, our sample of 224 cities from 1970 excludes 81 new cities since 1970, which raises selectivity issues. Later we note efforts to test and account for these problems.

A critical issue for any implementation of equation (1) is that our 1987 *Census of Manufactures* data on employment levels are censored. A city reports zero employment for an industry if employment is less than 250 in that industry in that city in 1987. We therefore use a Tobit formulation in estimation of (1), presuming that all our industries have in fact some minimal employment in any city (see the 1970 numbers in table B1), but about 30 percent of MSAs for the typical industry in 1987 are censored (table B1).

In the rest of this section, we present our basic results on equation (1) for the five two-digit capital goods industries. The presentation is divided into two parts. First we focus on the results concerning dynamic externalities and the effect of history, controlling for current wages, W_{it}, and market conditions, MC_{it}. The estimated coefficients for W_{it} and MC_{it} in equation (1) are presented in table B3 and are briefly discussed in subsection *C*. In that subsection we also deal with other estimation issues.

B. *Basic Results*

In table 1 we present results for equation (1) pertaining just to the effect of history on current employment. These results measure the nature and extent of dynamic externalities. In table 1, history is described by three measures: two relating to own industry employment in 1970 and one relating to industrial diversity of the city in 1970. As discussed earlier, for own industry activity, concentration is thought to represent MAR externalities more directly since it facili-

TABLE 1

DETERMINANTS OF LOG(1987 OWN INDUSTRY EMPLOYMENT): HISTORICAL CONDITIONS

	Machinery	Electrical Machinery	Primary Metals	Transportation	Instruments
Log(1970 own industry employment)	.547*	.592*	.647*	.532*	.365*
	(.070)	(.075)	(.065)	(.102)	(.087)
Concentration 1970: ratio of own industry employment	8.416*	6.672*	4.002**	9.879*	14.926*
to total civilian employment	(2.213)	(2.615)	(2.161)	(3.911)	(7.112)
Index 1970: lack of diversity	.246	−.578	−.250	−.564	−1.904
	(.548)	(.881)	(1.05)	(1.207)	(1.275)

NOTE.—Standard errors are in parentheses.
* Significant at the 5 percent level.
** Significant at the 10 percent level.

tates spillover or "network" information flows among relevant firms. The own industry level measure represents remaining persistence in employment levels.

In table 1, for the own industry level measure, the coefficients on log(1970 own industry employment) are about .5. To get a perspective on these coefficients, it is helpful to interpret them in a Barro and Sala-i-Martin (1991) growth framework, where $\ln(N_{87}/N_{70}) = \alpha - \delta \ln N_{70}$. In that context, a coefficient of .5 in equation (1) states that $1 - \delta = .5$. The implied δ of .5 suggests an annual rate of convergence, β, of about 4 percent a year ($\delta = 1 - \exp[\beta T]$ for $T = 17$ years). This is relatively quick "mean reversion." This quick reversion occurs only in estimation of equation (1) because we have controlled for other historical and current market conditions in tables 1 and B3. Table B2 shows that if we do not control for these conditions, there is much greater overall persistence in individual industry employment patterns. Convergence rates in the basic data are about 1 percent for individual industries and are near zero for manufacturing as a whole. We shall comment on this further below.

The degree of past concentration, the basic MAR measure, strongly affects current employment. Past concentration in table 1 is measured by the ratio of own industry employment to total local civilian employment in 1970. The coefficients for this variable are positive, large, and significant, indicating that historical concentration of the own industry creates an environment conducive to attracting current producers. Consider machinery. When 1970 machinery employment is held fixed, a one-standard-deviation (.030) increase in the share of machinery in overall local employment in 1970 increases 1987 machinery employment by 25 percent. If we compare a city with 20,000 workers in 1970 who constitute 2 percent of the workforce with a city with 10,000 machinery workers who constitute 8 percent of the workforce in 1970, the second city would have a larger workforce in machinery in 1987, ceteris paribus. For our five industries, table 2 calculates the percentage change in 1987 employment of a one-standard-deviation (see App. A) increase in the degree of own industry concentration in 1970. The MAR effects are largest for transport equipment and machinery. Both tend to be found in larger types of specialized cities (Henderson 1988).

As an alternative measure of concentration, we used the ratio of own industry employment to total local effective urban land area using land area data from a 1972 Department of Transportation study of urban land use, intracity highways, and roads (*1974 Transportation Report*). This measure attempts to capture the spatial (density) connotation of MAR externalities. This variable is always positive in

TABLE 2

MAR EFFECTS: PERCENTAGE IMPACT ON CURRENT EMPLOYMENT OF A ONE-STANDARD-DEVIATION INCREASE IN 1970 OWN INDUSTRY CONCENTRATION

Machinery	Electrical Machinery	Primary Metals	Transportation Equipment	Instruments
25%	22%	16%	31%	19%

sign and has reasonable statistical significance;[2] but including it with the employment concentration measure adds no explanatory power, with the employment measure dominating. As another alternative, from a technical point of view, including both a level and an employment concentration measure in equation (1) would be similar to just having level measures for 1970 own industry employment and for 1970 total civilian employment.[3] Doing so produces the expected negative (and generally significant) coefficient for 1970 total civilian employment.

Besides own industry employment and concentration, historical diversity of the urban industrial base might positively affect current employment in an industry, suggesting the presence of Jacobs-type dynamic externalities. Accordingly, we constructed separate diversity measures for all other historical economic activity, all other manufacturing activity, and relevant wholesale, business, and professional service activities. In table 1, we report the results for one of these measures: diversity of all other manufacturing in 1970 based on a Hirschman-Herfindahl index for about 50 three-digit manufacturing industries. The Hirschman-Herfindahl index for city i for two-digit industry k, HHI_{ik}, is

$$\text{HHI}_{ik} = \sum_{j \notin k} s_{ij}^2, \tag{2}$$

[2] The coefficients and standard errors for our five industries in table 1 are, respectively, .0034 (.0020), .0100 (.0031), .0093 (.0041), .0015 (.0017), and .0030 (.0036) for a typical mean and standard deviation of the variable of 19.7 and 28.2 (for electrical machinery).

[3] Our results on past concentration and MAR effects differ from those in Glaeser et al. We consistently get a positive impact using either our specification of the estimating equation or theirs (i.e., whether we control for current or 1970 economic conditions or both). They get a negative impact. However, the samples (and time periods) differ. Glaeser et al. include an industry only if it is one of the six largest industries in a city. So, by definition in their sample, all industries have high degrees of concentration in 1956, whereas we cover the whole spectrum. Second, nontraded sectors such as wholesaling, retailing, and traditional services are heavily represented in their sample, with manufacturing industries accounting for about one-third of their observations. In a recent paper, Miracky (1992) finds that past concentration has a significant impact in manufacturing, but not for all industries lumped together.

where s_{ij} is the share in city i of three-digit manufacturing industry j in local all other manufacturing employment. An increase in HHI_{ik} reflects less diversity in the environment. If employment in 50 other three-digit industries is evenly distributed, the index has a value of .02; if it is concentrated all in one other three-digit industry, it has a value of one. As is apparent in table 1, diversity of all other manufacturing activity has no consistent or significant impact on 1987 employment. Further, as detailed in Henderson, Kuncoro, and Turner (1992), we experimented extensively with the other measures of historical diversity noted above, as well as specific historical levels and shares of interconnected activities for each industry based on input-output tables and locational associations. *None* of these measures had a consistent or significant impact on 1987 employment, whether entered separately, in pairs, all together, and so forth. Indications of Jacobs effects of any kind are absent. The strongest possibility in table 1 is for instruments, our only growing industry. This could suggest that Jacobs externalities might occur for rapidly developing industries, a notion we explore in Section II.

C. *Other Results and Issues*

1. Output and Labor Market Controls in Equation (1)

Table B3 presents results on output and labor market conditions. Output market conditions are represented by regional dummies, an access measure, and a measure of local demand for the industry. Regional dummies represent the effect of regional demand conditions, with 1970 employment controlled for (in table 1). Typically, from 1970, relative to the West (the constant term) and South, industries have declined in the Northeast and Midwest, as manufacturing has shifted with regional population and demand shifts. Access is distance from an MSA to the nearest of the 30 Rand McNally national business centers. Distance to markets matters, repressing the demand for a locality's output. Finally, we measure local demand for the products of our capital goods industries by employment in all other manufacturing. Major portions of the output of our industries are inputs for other manufacturing industries. For current all other manufacturing employment, a 1 percent increase in employment increases own industry employment by 0.3–1.0 percent. A static externality interpretation for this variable is also possible since other manufacturing could enhance productivity in any one industry by enhancing static information flows.

The second block of variables in table B3 represents current labor market conditions.[4] Higher current wages generally reduce demand substantially. A 1 percent rise in wages reduces employment by about 1 percent on average. Correspondingly, improvements in labor force quality as measured by increases in educational attainment of the adult population generally increase the demand by an industry to locate in a city. This variable can also have a static externality interpretation, where a more educated local population enhances an industry's productivity. Finally, we experimented with other measures of current economic conditions, such as state fuel and electrical prices, diversity of the local industrial environment (see eq. [2] above), median residential rents, coastal location, and contiguity of PMSAs in a consolidated metropolitan area. All were insignificant and inconsistent in their own impact and had no consistent or strong impacts on reported variables in tables 1 and B3.

2. Persistence

What variables are responsible for the drop in the coefficient of log(1970 own industry employment) from around .8 to .9 for the basic data in table B2 to around .5 in table 1? Experimentation revealed that log(1987 all other manufacturing employment) in equation (1) (see table B3) is responsible for much of the drop in log(1970 own industry employment), with the historical concentration measure responsible for the rest. Omitting log(1987 all other manufacturing), a key derived-demand variable, would lead to an overstatement of inherent persistence of own industry employment.[5] In table B2, there is a very high degree of persistence in overall other manufacturing employment in cities. Our estimates in table 1 suggest that this persistence translates into a persistent derived-demand base for individual industries, explaining a lot of the persistence in the raw data for these individual industries.

Why is all other manufacturing persistent? Some preliminary work suggests that regional comparative advantage forms part of the explanation. First, the 1987 ratio of local total manufacturing to total civil-

[4] Data availability at the time of sample construction required us to use 1982 manufacturing wages and 1980 educational measures as current labor market conditions.

[5] There could be an issue of whether inclusion of log(1987 all other manufacturing employment) in eq. (1) presents problems because log(1970 own industry employment) and (1970 own industry employment)/(1970 total local employment) are also included. These variables themselves are not inordinately correlated. However, 1987 and 1970 all other manufacturing employment are very highly correlated. Thus we could be in trouble if 1970 total civilian employment and 1970 total manufacturing employment move together or form a constant ratio. In fact, they do not. For example, the 1970 ratio of manufacturing to total local employment has a one-standard-deviation interval of .13–.35 about the mean.

ian employment can be partially explained (R^2 = .31 for an ordinary least squares [OLS] regression) by exogenous regional characteristics such as regional dummies, coastline dummy, state iron ore resources, SMSA rail dummy, SMSA heating degree days, log(1970 state electricity price), and state agricultural workers. Second, for machinery, for example, if we replace log(1987 all other manufacturing) in equation (1) with its exogenous regional determinants just listed above, the coefficient of log(1970 own industry employment) is .58, which is comparable to that in table 1 and much lower than the raw persistence coefficient of .86 in table B2.

3. Endogeneity and Selectivity

As noted earlier, some of the right-hand-side variables in equation (1) listed in tables 1 and B3 are potentially endogenous. Preliminary specification tests (Smith and Blundell 1986) provided modest evidence of a problem, particularly for wages. In Henderson et al. (1992), we present two-stage estimates of the coefficients (of course, there is an issue of finding instruments not influenced by any city-industry fixed effect).[6] The two-stage estimates tend to differ in three ways. With one exception, the wage elasticity for employment in table B3 becomes much more negative. For log(1970 own industry employment), there is generally a modest drop in the coefficient, indicating reduced persistence. Finally, with one exception, the coefficient of past concentration rises in two-stage work, sometimes quite significantly. In estimation, there is also an issue of sample selection bias because, since 1970, 81 new MSAs or PMSAs have arisen that are excluded from our sample for lack of information about 1970. We conducted some two-stage Heckman-type selection tests on OLS employment regressions and found no evidence of selectivity bias.

[6] Instruments include region dummies, 1987 access to a national business center, state industrial fuel price, percentage urban in 1970, percentage 18 years or older in 1970, percentage 65 years or older in 1970, percentage of adults with 4 years of high school or some college in 1970, percentage of adults with college or more in 1970, percentage of female-headed families in 1970, percentage housed in single-family homes in 1970, percentage housed in pre-1950 housing in 1970, mean January temperature, annual rainfall, cooling degree days, 1970 distance and driving time to a national business center, railway dummy, 1970 cost of electric power by state, state farm population in 1970, percentage of manufacturing workers with no high school in 1970, percentage of manufacturing workers with college in 1970, multiple-name SMSA in 1970, state land area, regional iron resources in 1970, regional oil reserves in 1970, percentage with high school or more in 1980, percentage with college or more in 1980, and ratio of black to total population in 1980.

II. New High-Tech Industries

For the traditional manufacturing industries in Section I, there is strong evidence of MAR externalities but little evidence of Jacobs externalities. However, the analysis of dynamic externalities may differ when one looks at new or rapidly growing industries. At a two- or three-digit level it is not possible to identify entirely new industries. However, we identify three three-digit industries—electronic components, medical equipment, and computers—as having undergone sufficient transformation and growth so as to be treated as something akin to "new." Most of their marketed products for 1987 did not exist in 1970 or were not in commercial production. Employment in these industries more than doubled from 1970 to 1987.[7]

Examining these industries raises some econometric and conceptual issues. While major two-digit industries have positive employment in virtually all cities, our three high-tech industries do not. Given this, a discrete-continuous choice analysis is more appropriate than a Tobit formulation. The discrete event is whether a city has an industry or not in 1987; and, conditional on that, the continuous event is the employment level in equation (1).[8] For the discrete event, on the basis of the wage equals VMP equation from Section I, for cities with the industry (I_{it} = 1 vs. 0) we are estimating,

$$\text{prob}(I_{it} = 1) = \text{prob}[W_{it} < A_{it}(\cdot)f'(\tilde{N}_{it}, \ldots)P_{it}(\cdot)], \tag{3}$$

where $\tilde{N}_{it}$ is some minimal employment level (e.g., $\tilde{N}_{it} = 1$) or Inada conditions do not apply.

A. *Where High-Tech Industries Are Found*

This subsection focuses on the discrete choice analysis of where new high-tech industries chose to locate in 1987, on the basis of 1970 labor market and industrial environment conditions. What aspects of local historical industrial environments helped determine the urban winners in the race to attract and retain significant employment in high-tech industries?

Because of censoring of 1987 data, for the discrete event, we can

[7] In theory, we could subtract these three-digit components from the two-digit industries in estimation in the previous section. In practice, we could not because of data censoring (nonreporting below 250 employees) and use of employment intervals for observations censored for confidentiality.

[8] In Henderson et al. (1992), we report joint maximum likelihood estimation of the determinants of 1987 employment levels. There, MAR effects for high-tech industry employment levels are weaker than for the mature industries in table 1 since the exact concentration of the predecessor industry in many cities is less important for these industries in determining specific employment levels today.

examine only whether or not employment is greater than or equal to 250 in a city. We define this level as "significant employment." In table 3 we note that less than half of the 224 MSAs have significant employment in any of these industries, even in 1987. These industries are highly concentrated in cities such as Phoenix, Austin, Boulder, Anaheim, San Jose, Tucson, Minneapolis, Boston, and Poughkeepsie (International Business Machines) for computers and Phoenix, Dallas, Anaheim, Binghamton, Los Angeles, San Jose, Chicago, and Boston for electronic components. Some of these cities—such as Austin, Tucson, and Binghamton—had next to zero employment in these industries in 1970. Table 3 also indicates the relevant data on exit and entry in terms of significant employment. For electronic components, there is considerable mobility, both in and out. Given sufficient exits, we can look also at conditional probabilities, asking whether the way a variable affects the probability of having significant 1987 employment varies by whether the city had significant employment or not in 1970.

Probit results for wage and market condition variables for equation (3) are reported in table B4; they have expected signs. In table 4, we focus on the effects of the historical industrial environment, such as measures of predecessor own industry activity (MAR effects) and diversity (Jacobs effects). For own industry activity, since many cities in 1970 have zero activity, we focus on two measures, a dummy, I_{i0}, for significant 1970 employment (more than 249 workers) and the concentration measure for own industry 1970 employment, ρ_{i0}. For all industries in columns 1, 4, and 7 of table 4, having employment in the predecessor industry in 1970 raises the probability of having significant 1987 employment. Similarly from columns 3, 6, and 9, the greater the 1970 foothold or concentration of an industry in a city, the greater the 1987 probability of having the industry. The MAR effects definitely affect whether the city has the industry or not. To quantify effects, for a representative city (average sample values for all continuous variables) in a specific region, we calculate the effect

TABLE 3

LOCATION OF HIGH-TECH INDUSTRIES: NUMBER OF CITIES

	1970 Employment ≥ 250	1987 Employment ≥ 250	Entry: 1970 Employment < 250, 1987 Employment ≥ 250	Exit: 1970 Employment ≥ 250, 1987 Employment < 250
Computers	48	70	27	5
Electronic components	90	109	34	15
Medical instruments	64	89	39	7

TABLE 4

DETERMINANTS OF HIGH-TECH LOCATIONS: HISTORICAL INDUSTRIAL ENVIRONMENT

	ELECTRONIC COMPONENTS			MEDICAL EQUIPMENT			COMPUTERS		
	(1)	(2)	(3)	(4)	(5)	(6)	(7)	(8)	(9)
I_{i0} dummy: significant employment own industry	1.305*	.412		1.450*	−.313		1.454*	1.445*	
	(.251)	(.551)		(.308)	(.732)		(.368)	(.704)	
Concentration past (ρ_{i0})			13.037			64.731			319.362*
			(8.990)			(40.123)			(146.219)
Diversity:									
Overall (D_{i0})	−4.116*	−6.125*	−6.636*	−1.860	−3.372**	−2.755	−3.223	−3.233	−3.735
	(1.754)	(2.374)	(2.150)	(1.555)	(1.893)	(1.701)	(2.143)	(2.247)	(2.295)
Conditional on significant past employment ($D_{i0} \times I_{i0}$)		7.960**	8.915*		17.976*	9.040*		.094	1.356
		(4.522)	(2.458)		(7.447)	(3.798)		(6.607)	(4.980)
Labor force in higher education or miscellaneous professionals (%)	NA	NA	NA	44.573	51.321**	61.597*	74.544*	74.547*	89.300*
				(28.986)	(30.062)	(31.286)	(35.363)	(35.364)	(36.427)

* Significant at the 5 percent level.
** Significant at the 10 percent level.

of a one-standard-deviation increase in that variable on the probability of having the industry in 1987. To quantify the effects of a dummy variable, we look at its direct impact on the representative city. For example, if our city is in the Northeast, significant employment in electronic components in 1970 (vs. not) raises the probability of having significant employment in 1987 from .30 to .80.

What about the effect of diversity? If 1970 diversity affects 1987 probabilities of attracting the industry, that suggests that Jacobs externalities are important in attracting high-tech industries. The coefficients on 1970 lack of diversity in all manufacturing (based on eq. [2] for 50 three-digit industries) are all fairly large and negative, in columns 1, 4, and 7. Strictly speaking, only one is statistically significant; however, the others are significant at not unreasonable levels. For electronic components, for a representative city in the Northeast without significant 1970 employment, a one-standard-deviation *decrease* in lack of diversity increases the probability of 1987 employment from .30 to .44. But that is not the whole story.

For the diversity measure, we also looked at conditional probabilities, where we allow the slope coefficient to vary with whether the city had significant 1970 employment or not.[9] For high-tech industries in columns 2, 5, and 8, cities without the industry in 1970 have large negative (and reasonably significant) coefficients for lack of diversity. For cities with the industry in 1970, there is generally an offsetting positive effect. For example, for electronic components, for cities in the Northeast without the industry in 1970, a one-standard-deviation decrease in 1970 lack of diversity increases the city's probability of having the industry in 1987 from .44 to .64. Yet, for cities with the industry in 1970, the net coefficient is actually nonnegative, with the impact on probabilities of a change in diversity being negligible. Given the lower number of exits for computers and medical instruments, it is hard to have confidence in the precise conditional results, but the same pattern appears. For all industries, results in columns 3, 6, and 9 are similar to those in columns 2, 5, and 8.[10]

These results suggest a particular twist to Jacobs externalities. They are important for a city's ability to initially attract new industries: diversity provides a fertile breeding ground for childless cities. But they are unimportant for retaining these industries: prior concentration of the own predecessor industry is what matters.

[9] For electronic components, no other variables for differential slope coefficients were close to being significant.

[10] For electronic components we also experimented with two-stage estimation and found no strong effects. For example, if we treat the dummy variable for significant 1970 employment as endogenous in col. 1, the diversity coefficient remains unchanged.

Finally we note that in table 4 for computers and medical equipment, cities with higher concentrations of workers in higher education and miscellaneous professionals also had higher probabilities of having the industry in 1987. This group of people represents the importance to computer and instrument development of interaction with the local basic and applied research sectors of the economy, representing dynamic spillover benefits for our industries. So, for example, for a city in the Northeast without significant 1970 employment, a one-standard-deviation increase in the proportion of higher education and miscellaneous professionals raises the probability of having computers in 1987 from .21 to .31. A similar force may be at work in table B4, where the percentage of adults with a college degree takes on heightened importance, compared to mature industries in table B3. For example, for a Northeast city with no 1970 computers, a one-standard-deviation increase in the proportion of the college-educated population raises the probability of having computers from .21 to .43.

III. Conclusions

Dynamic externalities come from local accumulations of knowledge enhanced by long-term relationships and histories of interactions, creating a stock of "local trade secrets" that benefit local firms. Cities provide a natural laboratory to study the nature and extent of these externalities. In this paper we examined employment growth rates for five key traditional capital goods industries. For these industries we observe a very high degree of persistence in employment patterns. While persistent regional demand and comparative advantage explain some of this, MAR dynamic externalities are critical also. In particular, employment growth in traditional manufacturing industries is higher in cities with high past employment concentrations in the own industry. A history of industrial diversity did not have a significant effect on any of these traditional industries, except for instruments, suggesting that Jacobs externalities are not so important for mature industries.

For newer high-tech industries, we do not focus on persistence since the industry is new. We observe very rapid growth, both in the number of cities hosting a particular industry and in employment levels within those cities. It is natural to ask which historical environments have an advantage in the ongoing race to attract newer industries. Unlike the traditional manufacturing industries, high levels of past industrial diversity increase the probability that a city will attract a high-tech industry. This suggests that Jacobs externalities play an important role in the development of the high-tech sector. But MAR externalities were also found to play a role in the development of the high-tech sector. In particular, while Jacobs externalities are important in attracting new industries, MAR externalities, rather than Jacobs externalities, are important for retaining the industry.

Appendix A

Data Sources and Definitions

1987 Variables

Employment data are taken from the 1987 *Census of Manufactures* supplemented by 1987 *County Business Patterns* (CBP) data. For some cities, 1987 *Census of Manufactures* data are censored for disclosure reasons and give employment figures in intervals. Because the upper open-ended interval starts at just 2,500, these employment data are supplemented with 1987 data from CBP (with appropriate adjustments so 1987 CBP definitions of Standard Industrial Classification 36 and 38 correspond to *Census of Manufactures* definitions). The CBP employment data are less likely to be censored and to have a much more extensive set of intervals for censored data, with the upper open-ended interval starting only at 100,000 employees.

Socioeconomic data are taken from the 1986 *State and Metropolitan Area Data Book* (Bureau of the Census). Access data are based on Rand McNally's 30 national business centers in 1990. National business centers are market centers defined by volume of transactions.

The variables used and their mean and standard deviation are defined as follows.

Manufacturing wage ($000s): average income of production workers in 1982 (mean 17.4; s.d. 3.6).
Percentage of adults with at least high school, for the 1980 population over 25 (mean 0.68; s.d. 0.08).
Percentage of adults with at least college, for the 1980 population over 25 (mean 0.16; s.d. 0.05).
Distance to national business center; straight-line distance (mean 1.71; s.d. 3.29).
Total manufacturing employment (000s) for 1987 (mean 58.4; s.d. 100.8).
Diversity index, for 1987 (mean 0.17; s.d. 0.24 [excludes machinery]).
Regional Northeast, Midwest, and South dummy variables have means of .18, .29, and .35, respectively.

1970 Variables

Employment data are taken from 1970 *Census of Population* Sixth Count Table 1270 on employment by SMSA. Socioeconomic data are taken from the 1972 *City and County Data Book* and the 1972 *Census of Manufactures.*

The variables used and their mean and standard deviation are defined as follows.

Manufacturing wage: average hourly wage of production and nonproduction workers for 1972 (mean 4.6; s.d. 3.6).
Concentration of machinery for 1970 (mean .026; s.d. 030).
Concentration of electrical machinery for 1970 (mean .024; s.d. .033).
Concentration of primary metals for 1970 (mean .023; s.d. .041).
Concentration of transportation for 1970 (mean .022; s.d. .031).
Concentration of instruments for 1970 (mean .0041; s.d. .013).
Concentration of electronic components for 1970 *County Business Patterns* (mean .011; s.d. .026 for 207 metro areas).
Concentration of computers for 1970 (mean .002; s.d. .006 for 207 metro areas).
Concentration of medical equipment for 1970 (mean .0025; s.d. .0099).
Total manufacturing employment (000s) for 1970 (mean 58.8; s.d. 110.8).
Percentage of labor force in higher education and miscellaneous professionals for 1970 (mean .019; s.d. .004).
Total employment (000s) for 1970 (mean 219.1; s.d. 368.3).
Percentage of adults with at least high school for the 1970 population over 25 (mean .55; s.d. .08).
Percentage of adults with at least college for the 1970 population over 25 (mean .11; s.d. .04).
Diversity index for 1970 for 50 manufacturing industries (mean .12; s.d. .08 [excludes electronic components]).

Appendix B

TABLE B1

INDUSTRY EMPLOYMENT PATTERNS ACROSS 224 METROPOLITAN AREAS

	Machinery	Electrical Machinery	Primary Metals	Transportation	Instruments
1970 number of cities with zero employment	0	0 (33)	1	0	5 (64)
1970 number of cities with over 249 employees	196	176 (158)	152	168	102 (108)
1987 number of cities with over 249 employees	206	175	146	166	133
1970 mean employment	6,667	6,504	4,211	7,833	1,344
1970 mean employment in cities with over 249 employees	7,598	8,251	6,133	10,405	2,869
1987 mean employment in cities with over 249 employees	6,948	6,801	3,706	9,180	6,081

SOURCE.—Except for electrical machinery and instruments, all 1970 numbers are taken from the *Census of Population* and all 1987 numbers are taken from the *Census of Manufactures* For electrical machinery and instruments, the numbers in parentheses are taken from *County Business Patterns*.

NOTE.—Different censuses record different levels of specific types of economic activity. The *Census of Population* tends to be broader-based because it incorporates home production, the *County Business Patterns* and *Census of Manufactures* deal with the formal sector. (How would the activities of Steven Jobs and Stephen Wozniak have been classified in 1976?) Differences in the industrial classification schemes and errors in self-reporting also create differences in the numbers.

TABLE B2

(NON-)CONVERGENCE ACROSS METROPOLITAN AREAS

$\log(Y_{87}/Y_{70}) = -\delta \log Y_{70} + \ldots$

	δ	$1 - \delta$	β (%)*	Observations
Total manufacturing employment	.058	.942	.35	224
Per capita manufacturing employment	.061	.939	.37	224
Machinery employment	.136	.864	.85	207
Electrical machinery	.204	.796	1.34	207
Primary metals	.192	.808	1.25	207
Transportation equipment	.127	.873	.79	207
Instruments	.193	.807	1.26	207
Wage rate	.743	.257	7.99	207

NOTE.—Equations for total and per capita manufacturing are estimated as specified by OLS. Standard errors on δ are, respectively, .020 and .054. Equations for individual industries are estimated by Tobit (given that 1987 data are censored) in the form $\log Y_{87} = \alpha + (1 - \delta)\log Y_{70} + \ldots$. Coefficients of $1 - \delta$ have standard errors of, respectively, .029, .035, .045, .042, and .052, so t-statistics are all over 15.0. Included in the estimating equations are regional dummies.

* β is the annual rate of convergence (percentage) and is the solution to $\delta = 1 - e^{-\beta T}$ for $T = 17$ years In contrast to overall manufacturing employment persistence, for manufacturing wages there is strong convergence, so $\beta = 7.99$ percent.

TABLE B3

DETERMINANTS OF LOG (1987 OWN INDUSTRY EMPLOYMENT): MARKET CONDITIONS (N = 207)

	Machinery	Electrical Machinery	Primary Metals	Transportation	Instruments
	Regional Demand				
Intercept	−2.171*	.594	−2.706*	−2.607	1.004
	(.798)	(1.153)	(1.301)	(1.639)	(1.593)
Region:					
Northeast	−.422*	−.678*	.083	−.508**	−.279
	(.152)	(.212)	(.224)	(.284)	(.301)
Midwest	−.356*	−.497*	.530*	−.048	−.299
	(.137)	(.182)	(.193)	(.247)	(.259)
South	.046	−.085	.442**	.288	−.359
	(.145)	(.208)	(.232)	(.283)	(.304)
Distance to national business center	−.045*	−.047*	−.025	−.020	−.130**
	(.014)	(.022)	(.032)	(.023)	(.071)
Log (all other manufacturing employment)	.357*	.223*	.310*	.986*	.641*
	(.074)	(.086)	(.078)	(.058)	(.123)
	Local Labor Markets				
Log (manufacturing wage)	−.939*	−1.534*	−.772**	−.626	−2.345*
	(.241)	(.346)	(.396)	(.500)	(.539)
Adults with at least high school (%)	2.015*	−.107	.851	.044	4.095*
	(.990)	(1.387)	(1.602)	(1.94)	(2.069)
Adults with college (%)	2.678*	5.551*	−2.159	2.350	2.196
	(1.223)	(1.721)	(2.028)	(2.465)	(2.486)
Adjusted R^2 (OLS)	.870	.804	.763	.728	.726

NOTE.—Standard errors are in parentheses. The sample size is reduced from 224 to 207 because of missing observations on wages.
* Significant at the 5 percent level.
** Significant at the 10 percent level.

TABLE B4

OTHER DETERMINANTS OF HIGH-TECH LOCATIONS (N = 224)

Market Conditions Historically	Electronic Components	Medical Instruments	Computers
Constant	−5.354*	−6.856*	−13.444*
	(1.886)	(1.931)	(3.040)
Region:			
Northeast	−.771	−.749	.367
	(.483)	(.511)	(.499)
Midwest	−1.062*	−.689**	−.668
	(.375)	(.411)	(.436)
South	−.847*	−.537	.453
	(.430)	(.452)	(.515)
Distance to national business center	.0011	−.0004	.0006
	(.0006)	(.0007)	(.0007)
Adults with at least high school (%)	−2.916	−5.963*	8.104*
	(2.552)	(2.649)	(3.496)
Adults with college degree (%)	15.169*	16.897*	6.874
	(4.198)	(4.695)	(5.268)
Log(manufacturing wage)	−.280	−.500	−.479
	(.436)	(.522)	(.432)
Log(all other manufacturing employment)	.603*	.829*	.648*
	(.134)	(.154)	(.174)
Predicted correctly (%)	84	83	86

NOTE.—Results correspond to the equation for cols. 1, 4, and 7 in table 4.
* Significant at the 5 percent level.
** Significant at the 10 percent level.

References

Barro, Robert J., and Sala-i-Martin, Xavier. "Convergence across States and Regions." *Brookings Papers Econ. Activity*, no. 1 (1991), pp. 107–58.

Chinitz, Benjamin. "Contrasts in Agglomeration: New York and Pittsburgh." *A.E.R. Papers and Proc.* 51 (May 1961): 279–89.

Cicone, Antonio, and Hall, Robert E. "Productivity and the Density of Economic Activity." Working Paper no. 4313. Cambridge, Mass.: NBER, April 1993.

Davis, Steven J., and Haltiwanger, John C. "Gross Job Creation, Gross Job Destruction, and Employment Reallocation." Working Paper no. 91-5. Chicago: Fed. Reserve Bank, 1991.

Dunne, Timothy; Roberts, Mark J.; and Samuelson, Larry. "The Growth and Failure of U.S. Manufacturing Plants." *Q.J.E.* 104 (November 1989): 671–98. (*a*)

———. "Plant Turnover and Gross Employment Flows in the U.S. Manufacturing Sector." *J. Labor Econ.* 7 (January 1989): 48–71. (*b*)

Glaeser, Edward L.; Kallal, Hedi; Scheinkman, José A.; and Shleifer, Andrei. "Growth in Cities." *J.P.E.* 100 (December 1992): 1126–52.

Henderson, J. Vernon. "The Sizes and Types of Cities." *A.E.R.* 64 (September 1974): 640–56.

———. *Urban Development: Theory, Fact, and Illusion*. New York: Oxford Univ. Press, 1988.

———. "Some Favorable Impacts of a U.S.-Mexico Free Trade Agreement on U.S. Urban Employment." In *The Mexico-U.S. Free Trade Agreement,* edited by Peter M. Garber. Cambridge, Mass.: MIT Press, 1993.

Henderson, J. Vernon; Kuncoro, Ari; and Turner, Matthew. "Industrial Development in Cities." Working Paper no. 4178. Cambridge, Mass.: NBER, October 1992.

Herzog, Henry W., Jr., and Schlottmann, Alan M., eds. *Industry Location and Public Policy.* Knoxville: Univ. Tennessee Press, 1991.

Hoover, Edgar M. *Location Theory and the Shoe and Leather Industries.* Cambridge, Mass.: Harvard Univ. Press, 1937.

Jacobs, Jane. *Economy of Cities.* New York: Vintage, 1969.

Jaffe, Adam B.; Trajtenberg, Manuel; and Henderson, Rebecca. "Geographic Localization of Knowledge Spillovers as Evidenced by Patent Citations." *Q.J.E.* 108 (August 1993): 577–98.

Lucas, Robert E., Jr. "On the Mechanics of Economic Development." *J. Monetary Econ.* 22 (July 1988): 3–42.

Mills, Edwin S. "An Aggregative Model of Resource Allocation in a Metropolitan Area." *A.E.R. Papers and Proc.* 57 (May 1967): 197–210.

Miracky, W. F. "Technological Spillovers, the Product Cycle and Regional Growth." Manuscript. Cambridge: Massachusetts Inst. Tech., 1992.

Romer, Paul M. "Increasing Returns and Long-Run Growth." *J.P.E.* 94 (October 1986): 1002–37.

Smith, Richard J., and Blundell, Richard W. "An Exogeneity Test for a Simultaneous Equation Tobit Model with an Application to Labor Supply." *Econometrica* 54 (May 1986): 679–85.

[18]

ELSEVIER

European Economic Review 43 (1999) 409–429

EUROPEAN ECONOMIC REVIEW

Innovation in cities: Science-based diversity, specialization and localized competition

Maryann P. Feldman[a], David B. Audretsch[b,c,*]

[a] *Institute for Policy Studies, Johns Hopkins University, 3400 N. Charles St., Wyman Park 5th Floor, Baltimore, MD 21218-2696, USA*

[b] *Institute for Development Strategies, Indiana University, Bloomington, IN 47405-2100, USA*

[c] *Centre for Economic Policy Research, London, UK*

Received 1 September 1997; accepted 19 May 1998

Abstract

Whether diversity or specialization of economic activity better promotes technological change and subsequent economic growth has been the subject of a heated debate in the economics literature. The purpose of this paper is to consider the effect of the composition of economic activity on innovation. We test whether the specialization of economic activity within a narrow concentrated set of economic activities is more conducive to knowledge spillovers or if diversity, by bringing together complementary activities, better promotes innovation. The evidence provides considerable support for the diversity thesis but little support for the specialization thesis. © 1999 Elsevier Science B.V. All rights reserved.

JEL classification: O3, O1

Keywords: Innovation; Geography; Economic development; Agglomeration; Spillovers

* Correspondence address: School of Public and Environmental Affairs (SPEA), Institute for Developmental Strategies, Indiana University, Bloomington, IN 47405-2100, USA. Tel.: 1 812 855 6766; fax: 1 812 855 0184; e-mail: daudretsch@indiana.edu.

PII: S0014-2921(98)00047-6

1. Introduction

In proposing a new theory of economic geography, Paul Krugman (1991b, p. 5) asks, 'What is the most striking feature of the geography of economic activity? The short answer is surely concentration ... production is remarkably concentrated in space.' Perhaps in response to Krugman's concern, a literature has recently emerged which focuses on the implications of the concentration of economic activity for economic growth. Models posited by Romer (1986, 1990), Lucas (1993), and Krugman (1991a,b) link increasing returns to scale yielded by externalities within a geographically bounded region to higher rates of growth. The results of Jaffe (1989), Jaffe et al. (1993), Feldman (1994) and Audretsch and Feldman (1996) suggest that R&D and other knowledge spillovers not only generate externalities, but the evidence also suggests that such knowledge spillovers tend to be geographically bounded within the region where the new economic knowledge was created. New economic knowledge may spill over, but the geographic extent of such knowledge spillovers is bounded. Lucas (1993) emphasizes the most natural context in which to understand the mechanics of economic growth is in metropolitan areas where the compact nature of the geographic unit facilitates communication. Indeed, Lucas (1993) asserts that the only compelling reason for the existence of cities would be the presence of increasing returns to agglomerations of resources which make these locations more productive.

None of these studies, however, ask the question, 'Does the specific type of economic activity undertaken within any particular geographic region matter?' This question is important because a recent debate has arisen focusing precisely on the composition of economic activity within an agglomeration and how such externalities will be shaped by that composition of economic activity. One view, which Glaeser et al. (1992) attribute to the *Marshall–Arrow–Romer* externality, suggests that an increased concentration of a particular industry within a specific geographic region facilitates knowledge spillovers across firms.[1] By contrast, Jacobs (1969) argues that it is the exchange of complementary knowledge across diverse firms and economic agents which yields a greater return to new economic knowledge.

There are clear policy implications of this debate in terms of policies directed towards innovation and technological change. If the specialization thesis is correct, then policy should focus on developing a narrow set of economic activities within a geographic region in order to yield greater innovative output. On the other hand, if the diversity thesis is correct, then a geographic region comprised of a diverse set of economic activities will tend to yield greater output

[1] This mirrors an earlier debate summarized by Loesch (1954).

in terms of innovative activity. The key policy concerns would then become how to identify the commonalties and how to foster such diversity.

The purpose of this paper is to penetrate the black box of geographic space by identifying the extent to which the organization of economic activity is either concentrated, or alternatively consists of diverse but complementary economic activities, and how this composition influences innovative output. To consider this question we link the innovative output of product categories within a specific city to the extent to which the economic activity of that city is concentrated in that industry, or conversely, diversified in terms of complementary industries sharing a common science base. We use results on the relevance of academic departments to R&D from Levin et al. (1987) to identify complementary industries across which knowledge spillovers may be realized. We find a tendency for innovative activity in complementary industries sharing a common science-base to cluster together in geographic space. Industries which use the same base of scientific knowledge exhibit a strong tendency to locate together for both the location of production and the location of innovation.

In Section 2, we introduce the main theories alternatively favoring diversity or specialization in generating innovative activity. Issues concerning the measurement of innovative activity, the geographic unit of observation and the concepts of science-based diversity and specialization are examined in Section 3. After the model is presented in the Section 4, the empirical results are provided in Section 5. To explore the validity of our results, we extend our analysis to consider the impact of specialization versus diversity of economic activity within the firm on innovative activity in Section 6. We find considerable evidence rejecting the specialization thesis and in support of the diversity thesis. Based on the findings for both industries within specific cities as well as for individual firms, an organizational structure of economic activities that are diverse, but still complementary, apparently yields a greater innovative output than a specialization of economic activity.

2. Diversity versus specialization

The importance of location to innovation in a world increasingly relying upon E-mail, fax machines, and electronic communications superhighways may seem surprising, and even paradoxical at first glance. The resolution of this paradox lies in the distinction between knowledge and information. While the costs of transmitting information may be invariant to distance, presumably the cost of transmitting knowledge, especially what Von Hipple (1994) refers to as *sticky knowledge*, rises with distance. Von Hipple persuasively demonstrates that highly contextual and uncertain knowledge is best transmitted via face-to-face interaction and through frequent contact. Proximity matters in transmitting knowledge because, as Arrow (1962) pointed out, such sticky knowledge is inherently non-rival in nature and knowledge developed for any particular application can easily spill over and be applied to different use and applications. Indeed, Griliches (1992) has defined knowledge spillovers as 'working on similar things and hence benefiting much from each other's research'.[2]

Despite the general consensus that knowledge spillovers within a given location stimulate technological advance, there is little consensus as to exactly how this occurs. Glaeser et al. (1992) characterize three different models from the literature that would influence the production of innovation in cities. The *Marshall–Arrow–Romer* model formalizes the insight that the concentration of an industry in a city promotes knowledge spillovers between firms and therefore would facilitate innovation in that city-industry observation. This type of concentration is also known as industry localization (Loesch, 1954). An important assumption is that knowledge externalities with respect to firms exist, *but only for firms within the same industry*. Thus, the relevant unit of observation is extended from the firm to the region in the theoretical tradition of the Marshall–Arrow–Romer model and in subsequent empirical studies, but spillovers are limited to occur within the relevant industry. The transmission of knowledge spillovers across industries is assumed to be non-existent or at least trivial.

Restricting knowledge externalities to occur only within the industry may ignore an important source of new economic knowledge–inter-industry spillovers. Jacobs (1969) argues that the most important source of knowledge spillovers are external to the industry in which the firm operates and that cities are the source of innovation because the diversity of these knowledge sources is greatest in cities. Thus, Jacobs develops a theory that emphasizes that the variety of industries within a geographic region promotes knowledge externalities and ultimately innovative activity and economic growth. Of course, there should be some basis for interaction between diverse activities. A common science base facilitates the exchange of existing ideas and generation of new ones across disparate but complementary industries. Thus, in Jacobs' view, diversity rather than specialization is the operative mechanism of economic growth.

A second controversy involves the degree of competition prevalent in the region, or the extent of local monopoly. The Marshall–Arrow–Romer model predicts that local monopoly is superior to local competition because it

[2] Considerable evidence suggests that location and proximity clearly matter in exploiting knowledge spillovers. Not only have Jaffe et al. (1993) found that patent citations tend to occur more frequently within the state in which they were patented but Audretsch and Stephan (1996) and Audretsch and Feldman (1996) found that the propensity for innovative activity to cluster geographically tends to be greater in industries where new economic knowledge plays a more important role.

maximizes the ability of firms to appropriate the economic value accruing from their innovative activity. By contrast, Jacobs (1969) and Porter (1990) argue that competition is more conducive to knowledge externalities than is local monopoly.[3] It should be emphasized that by local competition, Jacobs (1969) does not mean competition within product markets as has traditionally envisioned within the industrial organization literature. Rather, Jacobs is referring to the competition for the new ideas embodied in economic agents. Not only does an increased number of firms provide greater competition for new ideas, but in addition, greater competition across firms facilitates the entry of a new firm specializing in some particular, new, product niche. This is because the necessary complementary inputs and services are likely to be available from small specialist niche firms but not necessarily from large, vertically integrated producers.[4]

3. Measurement issues

Krugman (1991a, p. 53) has argued that economists should abandon any attempts at measuring knowledge spillovers because '... knowledge flows are invisible, they leave no paper trail by which they may be measured and tracked'. But as Jaffe et al. (1993, p. 578) point out, 'knowledge flows do sometimes leave a paper trail' – in particular in the form of patented inventions and new product introductions.

In this paper we rely upon a direct measure of innovative output, rather than on a measure of intermediate output, such as patented inventions.[5] This United States Small Business Administration's Innovation Data Base (SBIDB) is the primary source of data for this paper.[6] The database consists of new product

---

[3] Porter (1990) provides examples of Italian ceramics and gold jewelry industries in which numerous firms are located with a bounded geographic region and compete intensively in terms of product innovation rather than focusing on simple price competition.

[4] A recent series of country studies assembled by Reynolds et al. (1994) show that new-firm start-up rates tend to be greatest in those geographic regions where the average number of firms per employee is the greatest.

[5] Scherer (1983), Mansfield (1984) and Griliches (1990) have all warned that the number of patented inventions is not the equivalent of a direct measure of innovative output. For example, Pakes and Griliches (1984, p. 378) argue that 'patents are a flawed measure of innovative output; particularly since not all new innovations are patented and since patents differ greatly in their economic impact'. In addressing the question, 'Patents as indicators of what?' Griliches (1990, p. 1669) concludes that, 'Ideally, we might hope that patent statistics would provide a measure of the innovative output ... The reality, however, is very far from it. The dream of getting hold of an output measure of inventive activity is one of the strong motivating forces for economic research in this area'.

[6] A detailed description of the SBIDB is contained in Audretsch (1995).

introductions compiled form the new product announcement sections of over 100 technology, engineering and trade journals spanning every industry in manufacturing. From the sections in each trade journal listing new products, a database consisting of the innovations by four-digit standard industrial classification (SIC) industries was formed. These innovation data have been implemented by Audretsch (1995) to analyze the relationship between industry dynamics and technological change, and by Audretsch and Feldman (1996), Feldman (1994) and Feldman and Florida (1994) to examine the spatial distribution of innovation.

There are several important qualifications that should be made concerning the SBIDB. The trade journals report mainly product innovations. Thus, as is the case in the studies by Audretsch (1995) and Audretsch and Feldman (1996), the empirical analyses undertaken in this paper capture product innovation but not process innovation.

Another potential concern might be that the significance and 'quality' of the innovations vary considerably. In fact, each innovation was classified according to one of the following levels of significance: (1) the innovation established an entirely new category of product; (2) the innovation is the first of its type on the market in a product category already in existence; (3) the innovation represents a significant improvement in existing technology; and (4) the innovation is a modest improvement designed to update an existing product. About 87% of the innovations were in this fourth category and most of the remaining innovations were classified in the third category. However, the preliminary nature of such classifications leads us to treat the innovations as being homogeneous. While such an assumption will hopefully be improved upon in the future, it is consistent with the voluminous body of literature treating dollars of R&D, numbers of scientists and engineers, and numbers of patents as being homogeneous.[7]

An important strength of the database is that the innovating *establishment* is identified as well as the innovating *enterprise*. While this distinction is trivial for single-plant manufacturing firms, it becomes important in multi-plant firms. This is because some innovations are made by subsidiaries or divisions of companies with headquarters in other states. Even though the headquarters may announce new product innovations made by the company, the database still identifies the individual establishment responsible for the majority of the work leading to the innovation.

[7] An anonymous referee points out that the implicit assumption that innovations are homogeneous in their impact is found in most studies attempting to measure innovative activity. As Griliches (1990) emphasizes, this limitation has plagued studies using counts of patents, just as it has in those based on R&D expenditures or R&D scientists and engineers.

The innovations from the SBIDB are then classified according to the four-digit SIC industry of the new product and the city where the innovating establishment was located. We adapt either the Consolidated Metropolitan Statistical Area (CMSA) or the Metropolitan Statistical Area (MSA) as the spatial unit of observation. The analysis here is based on 3969 new manufacturing product innovations for which the address of the innovating establishment could be identified.[8]

In 1982 the most innovative city in the United States was New York. Seven hundred and thirty-five, or 18.5%, of the total number of innovations in the country were attributed to firms in the greater New York City area. Four hundred and seventy-seven (12.0%) were attributed to San Francisco and 345 (8.7%) to the Boston area and 333 (8.4%) to the Los Angeles area. In total, 1890, or 45% of the innovations, took place in these four consolidated metropolitan areas. In fact, all but 150 of the innovations included in the database are attributed to metropolitan areas. That is, less than 4% of the innovations occurred outside of metropolitan areas. This contrasts with the 70% of the population which resided in these areas.

Of course, simply comparing the absolute amount of innovative activity across cities ignores the fact that some cities are simply larger than others. Cities vary considerably in terms of measures of city size, and we expect that city scale will have an impact on innovative output. Table 1 presents the number of innovations normalized by the size of the geographic unit. Population provides a crude but useful measure of the size of the geographic unit. Cities in Table 1 are ranked in descending order by innovation rate or the number of innovations per 100,000 population. While New York has the highest count of innovation, it has the third highest innovation rate. The most innovative city in the United States, on a per capita measure of city size, was San Francisco, with an innovation rate of 8.90, followed by Boston, with an innovation rate of 8.69. By contrast, the mean innovation rate for the entire country is 1.75 innovations per 100,000 population. The distribution of innovation rates is considerably skewed. Only 14 cities are more innovative than the national average. Clearly, innovation appears to be a large city phenomenon.

To systematically identify the degree to which specific industries have a common underlying science and technology base, we rely upon a deductive approach that links products on their closeness in technological space. We use the widely acknowledge and established *Yale Survey of R&D managers*. This survey is documented in great detail in Levin et al. (1987) and has been widely used in studies linking various mechanisms of appropriability to R&D activity (Cohen

[8] Feldman (1994) provides a description of the data collection procedure. The results are invariant to using PMSAs.

Table 1
Counts of innovation normalized by population

Consolidated metropolitan statistical area	Innovations (1982)	1980 population (thousands)	Innovations per 100,000 population
San Francisco – Oakland	477	5368	8.886
Boston – Lawrence	345	3972	8.686
New York – Northern New Jersey	735	17539	4.191
Philadelphia – Wilmington	205	5681	3.609
Dallas – Fort Worth	88	2931	3.002
Hartford	30	1014	2.959
Los Angeles – Anaheim	333	11498	2.896
Buffalo – Niagara	35	1243	2.816
Cleveland – Akron	77	2834	2.717
Chicago – Gary	203	7937	2.558
Providence – Pawtucket	25	1083	2.308
Portland – Vancouver	25	1298	1.926
Cincinnati - Hamilton	30	1660	1.807
Seattle – Tacoma	37	2093	1.768
Pittsburgh	42	2423	1.733
Denver – Boulder	28	1618	1.731
Detroit – Ann Arbor	68	4753	1.431
Houston – Galveston	39	3101	1.258
Miami – Fort Lauderdale	13	2644	0.492

Source: 1980 Population is from the Statistical Abstract.

and Levinthal, 1990; Levin et al., 1987). The survey uses a Likert scale of 1 to 7, from least important to most important, to assess the relevance of basic scientific research in biology, chemistry, computer science, physics, mathematics, medicine, geology, mechanical engineering and electrical engineering. We assume any academic discipline with a value greater than 5 to be relevant for a product category. For example, basic scientific research in medicine, chemistry and chemical engineering is found to be relevant for product innovation in drugs (SIC 2834).

We then identify six groups of industries which rely on similar rankings for the importance of the different academic disciplines. These six groups, shown in Table 2, reflect distinct underlying common scientific bases. To facilitate identification of the groupings we assigned a name to each group that reflects not only the underlying science base but also the application to which this knowledge is directed. Thus, what we term as the 'chemical engineering' and 'industrial machinery' groups actually include the same critical academic departments (ranked differently), but applied to different types of industries.

While each industry within the group shares a common scientific base, the geographic space and product space differ across industries. For example, there

Table 2
The common science bases of industry clusters

Cluster	Critical academic departments	Most innovative industries
Agra-business	Chemistry (6.06);	SIC 2013: Sausages
	Agricultural Science (4.65);	SIC 2038: Frozen Specialities
	Computer Science (4.18);	SIC 2087: Flavoring Extracts
	Biology (4.09).	SIC 2092: Packaged Foods
Chemical engineering	Materials Science (5.32);	SIC 3861: Photographic Equipment
	Chemistry (4.80);	SIC 3443: Fabricated Plate Work
	Computer Science (4.50);	SIC 2821: Plastic Materials
	Physics (4.12).	SIC 3559: Special Ind. Machinery
Office machinery	Computer Science (6.75);	SIC 3576: Scales and Balances
	Medical Science (5.75);	SIC 3579: Office Machinery
	Math (5.49);	SIC 3535: Conveyors
	Applied Math (4.64).	SIC 2751: Commerial Printing
Industrial machinery	Materials Science (5.03);	SIC 3551: Food Processing Equipment
	Computer Science (4.76);	SIC 3523: Machinery
	Physics (3.94);	SIC 3546: Hand Tools
	Chemistry (3.88).	SIC 3629: Industrial Apparatus
High-tech computing	Materials Science (5.92);	SIC 3573: Computing Machinery
	Computer Science (5.63);	SIC 3662: Radio/TV Equipment
	Physics (5.45);	SIC 3823: Process Control Instruments
	Math (4.76).	SIC 3674: Semiconductors
Biomedical	Chemistry (5.53);	SIC 3842: Surgical Appliances
	Medical Science (5.47);	SIC 3841: Medical Instruments
	Computer Science (5.32);	SIC 2834: Pharmaceuticals
	Materials Science (5.02).	SIC 3811: Scientific Instruments

are 15 distinct industries included in the biomedical group. On average, each industry contributed 3.22 innovations. Their shared underlying knowledge base consists of chemistry, medical sciences, computer sciences and material sciences. Surgical Appliances (SIC 3842), Surgical and Medical Instruments (SIC 3841), and Pharmaceuticals (SIC 2834) are three of the 15 industries heavily dependent on this common underlying scientific knowledge base. There are 21 industries included in the Agra-Business group, 34 industries included in the Chemical Engineering group, 7 industries in the Office Machinery group and 11 industries included in the Industrial Machinery group. The largest science-based group is what we term High-tech Computing, which includes 80 industries.

The most innovative cities within each science-based industrial cluster are identified in Table 3. This recalls the well-known association between cities and

Table 3
Innovation in science-based industry clusters

Cluster	Prominent cities	Mean industry innovations per 100,000 workers
Agra-business	Atlanta	92.40
	Dallas	41.15
	Chicago	33.03
	St. Louis	91.74
Chemical engineering	Dallas	38.09
	Minneapolis	66.67
	San Francisco	43.89
	Wilmington	85.47
Office machinery	Anaheim-Santa Ana	92.59
	Minneapolis	31.86
	Rochester	72.20
	Stanford	68.40
Industrial machinery	Anaheim-Santa Ana	54.95
	Cincinnati	66.01
	Cleveland	141.51
	Passaic, NJ	90.90
High-tech computing	Boston	73.89
	Houston	62.08
	San Jose	44.88
	Minneapolis	181.74
Biomedical	Boston	38.71
	Cleveland	68.76
	Dallas	35.22
	New York	188.07

industries. For example, Atlanta is a prominent center for innovative activity stemming from the common science base of agra-business. While the national innovation rate was 20.34 innovations per 100,000 manufacturing workers, agra-business in Atlanta was almost five times as innovative.

A Chi-Squared test on the independence of the location of city and science-based industrial activity reveals that neither the distribution of employment nor the distribution of innovative activity is random. Industries which rely on a common science base exhibit a tendency to cluster together geographically with regard to the location of employment and the location of innovation. We conclude that the distribution of innovation within science-based clusters and cities appears to reflect the existence of science-related expertise.

4. Modeling framework

To test the hypothesis that the degree of specialization shapes the innovative output of an industry, we estimate a model where the dependent variable is the number of innovations attributed to a specific four-digit SIC industry in a particular city. To reflect the extent to which economic activity within a city is specialized, we include as an explanatory variable a measure of industry specialization which was used by Glaeser et al. (1992) and is defined as the 1982 share of total employment in the city accounted for by industry employment in the city, divided by the share of United States employment accounted by that particular industry. This variable reflects the degree to which a city is specialized in a particular industry relative to the degree of economic activity in that industry that would occur if employment in the industry were randomly distributed across the United States. A higher value of this measure indicates a greater degree of specialization of the industry in that particular city. Thus, a positive coefficient would indicate that increased specialization within a city is conducive to greater innovative output and would support the *Marshall–Arrow–Romer* thesis. A negative coefficient would indicate that greater specialization within a city impedes innovative output and would support Jacobs' theory that diversity of economic activity is more conducive to innovation than is specialization.

To identify the impact of an increased presence of economic activity in complementary industries, the presence of science-based related industries is included. This measure is constructed analogously to the index of industry specialization, and is defined as the share of total city employment accounted for by employment in the city in industries sharing the science base, divided by the share of total United States employment accounted for by employment in that same science base. This variable measures the presence of complementary industries relative to what the presence would be if those related industries were distributed across the United States. A positive coefficient of the presence of science-based related industries would indicate that a greater presence of complementary industries is conducive to greater innovative output and supports for the diversity thesis. By contrast, a negative coefficient suggests that a greater presence of related industries sharing the same science base impedes innovation and argues against Jacobs' diversity thesis.

The usual concept of product market competition in the industrial organization literature is typically measured in terms of the size-distribution of firms. By contrast, Jacobs' concept of *localized competition* emphasizes the extent of competition for the ideas embodied in individuals. The greater the degree of competition among firms, the greater the extent of specialization among those firms and the easier it will be for individuals to pursue and implement new ideas. Thus, the metric relevant to reflect the degree of localized competition is not the

Table 4
Variable definitions and descriptive statistics

Variable name	Definition	Mean	Standard deviation
Specialization	$\frac{\text{Industry employment in city/total employment in city}}{\text{Industry employment in US/total employment in US}}$	0.96	1.47
Science base diversity	$\frac{\text{Employment in cluster in city/total employment in city}}{\text{Employment in cluster in US/total employment in US}}$	0.37	0.51
Competition	$\frac{\text{Firms in city} - \text{industry/workers in city} - \text{industry}}{\text{Firms in US industry/workers in US industry}}$	0.57	2.08

size of the firms in the region relative to their number (because, after all, many if not most manufacturing product markets are national or at least inter-regional in nature) but rather the number of firms relative to the number of workers. In measuring the extent of localized competition we again adopt a measure used by Glaeser et al. (1992), which is defined as the number of firms per worker in the industry in the city relative to the number of firms per worker in the same industry in the United States. A higher value of this index of localized competition suggests that the industry has a greater number of firms per worker relative to its size in the particular city than it does elsewhere in the United States. Thus, if the index of localized competition exceeds one then the city is locally less competitive than in other American cities.

The data for these measures are from County Business Patterns. Table 4 presents the variable definitions and descriptive statistics for the measures of science-based diversity, specialization and competition.

5. Results

Table 5 presents the regression results based on the 5946 city-industry observations for which data could be collected. The Poisson regression estimation method is used because the dependent variable is a limited dependent variable with a highly skewed distribution.[9]

[9] The number of innovations in a city and product category is either zero or some positive integer. The mean of the distribution is 0.26 and the standard deviation is 2.96.

Table 5
Poisson regression estimation results

	Model 1	Model 2	Model 3	Model 4
Industry specialization	−0.209	−0.334	−0.527	−0.142
	(−8.360)	(−14.522)	(−17.684)	(−5.680)
Science-based related	0.168	0.104	0.089	0.069
industries	(3.812)	(2.122)	(2.405)	(2.091)
Localized competition	−0.175	0.576	0.221	0.168
	(−3.365)	(7.481)	(0.269)	(1.976)
City scale		1.044		1.004
		(28.216)		(20.917)
Technological opportunity			0.079	0.034
			(26.333)	(1.700)
n	5946	5946	5946	5946
Log-likelihood	−1296.793	−901.489	−693.046	−652.264

Note: The t-values of the coefficient is listed in parentheses.

Model 1 provides the results for the three measures which reflect the degree of specialization, diversity and localized competition. The negative and statistically significant coefficient of industry specialization suggests that innovative activity tends to be lower in industries located in cities specialized in economic activity in that industry. The positive and statistically significant coefficient for science-based related industries indicates that a strong presence of complementary industries sharing a common science base is particularly conducive to innovative activity. Taken together, these results provide support for the diversity thesis but not for the specialization thesis. The negative coefficient on the measure of localized competition suggests that less and not more localized competition promotes the innovative activity of an industry in a particular city.

One concern regarding the estimation of Model 1 is that larger cities might be expected to generate a greater amount of innovation in any particular industry, *ceteris paribus*, simply because of a greater degree of economic activity. In addition, the extent of localized competition might tend to be greater as the size of the city grows. Thus, when the total employment is included in the estimation of Model 2, the sign of the coefficient for localized competition switches from negative to positive, suggesting that a greater degree of localized competition is conducive to innovative activity. At the same time, the signs of the coefficients of the measures of industry specialization and science-based related industries remain unchanged.

Another concern regarding the estimation of Model 1 is that some industries are more innovative, or in a higher technological opportunity class than others. But even after controlling for the number of innovations recorded for the relevant industry in the estimation of Model 3 and Model 4, the basic results remain the same.[10]

The results in Table 5 generally provide support that it is diversity and not specialization that is more conducive to innovation. It should be emphasized here that even after we control for city scale and technological opportunity, specialization appears to have a negative effect on innovation, while science-based diversity has a positive impact on innovative output. In addition, the evidence suggests that a greater extent of localized competition and not monopoly tends to promote innovative output. Of course, the cross-sectional city level data do not provide any insight into whether these patterns are stable over time.

6. Diversity versus specialization at the firm level

The debate between specialization and diversity of economic activity and the impact on technological change is also relevant at the level of the firm and allows a further test of the effects of these influences on innovation. We expect a similar result in the organization of innovation within the firm – undertaking innovative activity across a range of science-based complementary activities will lead to greater innovative output than concentrating innovative activity within one industry. With this test we seek to ascertain if analogous processes operate at both the firm and the geographic level.

Jewkes et al. (1958) examined the histories of 61 innovations and found that a variety of different approaches within the firm are often pursued. It appears that diversity in terms of the number and the type of approaches used serves to reduce the uncertainty inherent in innovation by providing a greater number of unique ideas and outcomes. In an empirical examination of this question, Cohen and Malerba (1995) find a strong relationship between technological diversity and the rate of technical advance at the industry level. Our consideration of specialization examines the degree to which the firm focuses its innovative activity on one product category.

In order to model the effects of science-based diversity and specialization at the level, we follow the pioneering study of Scherer (1983) to estimate a

[10] The extent to which these relationships are non-linear was examined in the estimation with the inclusion of quadratic terms for industry specialization and localized competition. The results indicate that, in fact, cities which are highly specialized in economic activity may actually generate more innovative output in that industry. This result does not hold when city scale is included. It appears that scale is more important than the rate of change in the degree of specialization. The quadratic of localized competition suggests that there may be increasing returns to innovative output resulting for increased localized competition. Again, this result does not hold when urban scale is considered.

non-linear model which links R&D[11] to innovative output:

$$I_i = \alpha + \beta_1 RD_i + \beta_2 RD_i^2 + \beta_3 Innovative\ Diversity_i + \beta_4 Innovative\ Specialization_i + \varepsilon,$$

where I represents the total number of innovations attributed to firm I in the 1982 SBIDB.[12] RD represents the R&D expenditures of firm I.[13] To measure the extent to which the innovative activity of firm I is diversified across related product categories we calculate the share of innovations in the same common science base product categories divided by the total number of innovations introduced by the firm. Similarly, specialization, or own industry share, is measured as the proportion's of the firm's innovative activity in the primary industry identified with the firm.

R&D expenditures are taken from the *Business Week* 1975 sample of over 700 corporations for which R&D expenditures play an important role. An important feature of this sample is that it included more than 95% of the total company R&D expenditures undertaken in the United States. There is a seven year lag between the 1982 innovations and the 1975 R&D expenditures. This lag may be somewhat long in view of a number of studies in the literature. However, as long as firm R&D expenditures are relatively stable over a short period of time, differences in the assumed lag structure should not greatly impact the results.[14]

Table 6 indicates the most innovative firms in the database along with the corresponding R&D expenditures, sales, R&D/sales ratio and the number of innovations per R&D dollar. As we might expect a positive relationship can be observed between the size of the R&D budget and innovative output, but there is also great variation in the productivity of R&D. Some firms such as Data General appear to obtain a relatively high degree of innovation per R&D dollar expended. Other firms such as General Electric and RCA exhibit a considerably lower R&D yield.

The regression results from estimating the model of firm innovative activity are shown in Table 7. The positive relationship between R&D inputs and

---

[11] R&D is generally observed to be the most decisive source of economic knowledge generating innovative activity.

[12] A log-log specification was estimated as well as the quadratic specification. The results for both specifications are consistent. We only report the results from the quadratic equations to facilitate comparisons with the earlier studies mentioned in the text.

[13] Note that the firm-level models are for the entire country and not for MSAs or states. It is not possible to identify firm-level R&D by location at this point.

[14] As an anonymous referee emphasizes, future research needs to explore distributed lags between R&D and innovative output.

Table 6
Firm innovation

	Number of innovations	R&D expenditure ($ million)	Sales ($ million)	R&D/ sales ratio	Innovation/ R&D[a]
Hewlett Packard Company	55	981.00	89.6	9.1	5.61
Minnesota Mining & Mfg.	40	3127.00	143.4	4.6	1.28
General Electric	36	13399.00	357.1	2.7	0.27
General Signal	29	548.00	21.2	3.9	5.29
National Semiconductor	27	235.00	20.7	8.8	11.49
Xerox	25	4054.00	198.6	4.9	0.62
Texas Instruments	24	1368.00	51.0	3.7	1.75
Pitney Bowes	22	461.00	10.5	2.3	4.77
RCA	21	4790.00	113.6	2.4	0.44
IBM	21	14437.00	946.0	6.6	0.15
Digital Equipment	21	534.00	48.5	9.1	3.93
Gould	20	773.00	23.1	3.0	2.59
Motorola	19	13112.00	98.5	7.5	1.45
Wheelabrator Frye	18	332.00	2.0	0.6	5.42
United Technologies	18	3878.00	323.7	8.3	0.46
Hoover	18	594.00	4.3	0.7	3.03
Honeywell	18	2760.00	164.2	5.9	0.65
Rockwell International	17	4943.00	31.0	0.6	0.34
Johnson & Johnson	17	2225.00	97.9	4.4	0.76
Eastman Kodak	17	4959.00	312.9	6.3	0.34
Data General	17	108.00	11.6	10.8	15.74
Exxon	16	4486.05	187.0	0.4	0.36
Du Pont	16	7222.00	335.7	4.6	0.22
Stanley Works	15	464.00	3.5	0.7	3.23
Sperry Rand	15	3041.00	163.5	5.4	0.49
Pennwalt	15	714.00	15.7	2.2	2.10
North American Philips	14	1410.00	22.5	1.6	0.99
Harris	14	479.00	21.1	4.4	2.92
General Motors	14	3572.05	1113.9	3.1	0.39

[a] Scaled by 100.

innovative output can be observed in Model 1. The negative coefficient of the quadratic term suggests that although innovative output tends to respond positively to increased investments in R&D inputs, the rate of increase in innovative output diminishes as R&D inputs increase.

When the measure of innovation diversity within industries sharing a common science base is included in Model 2, the positive coefficient provides support for the hypothesis that diversification across complementary economic activities is conducive to greater innovative output. When the measure of innovation specialization is included in Model 3, the positive coefficient suggests

Table 7
Regression results estimating firm innovative activity

	1	2	3	4
R&D	0.02804	0.0119	0.0178	0.0081
	(6.051)	(3.320)	(1.624)	(2.481)
R&D^2	−1.6945	−0.4157	−0.8940	−0.0323
	(−2.603)	(−0.878)	(−1.732)	(−0.075)
Innovative diversity	–	3.3081	–	9.2466
		(9.510)		(9.988)
Innovative specialization	–	–	2.8116	−7.4357
			(6.218)	(−6.819)
Number of observations	209	203	203	203
R^2	0.189	0.466	0.350	0.568
F	23.980	57.905	35.677	64.980

Notes: The t-value of the coefficient is listed in parentheses. The coefficients of R&D^2 have been divided by 100,000 for presentation purposes.

that greater specialization in innovation yields greater innovative output. When both specialization and diversity are included together in Model 4, the coefficient of specialization exhibits a negative coefficient suggesting that greater innovation specialization is less conducive to greater innovative output. On the other hand, holding R&D expenditures constant, greater innovative diversity within the common science base results in more innovative output.

The firms can also be grouped according to major two-digit SIC sectors. Results for six specific industrial sectors are listed in Table 8. There is interesting variation in these results across manufacturing sectors. For example, in instruments and telecommunications diminishing returns to R&D inputs can be observed. By contrast, increasing returns to R&D can be observed in the group of firms classified as conglomerates, and in electrical equipment and transportation no significant relationship can be observed between R&D and innovative output.

While the links between R&D inputs and innovative output vary substantially across sectors in Table 8, the relationships between specialization, innovation diversity and innovation remain remarkably constant across sectors. For all six sectors the coefficient of the measure of science-based innovation diversity remains positive and statistically significant, and the coefficient of the measure of innovation specialization also remains negative and statistically significant in all six sectors. Thus, the main finding that diversity in innovation activities within a common science base tends to promote innovative output more than does the specialization of innovation within just one single industry holds across a broad range of industrial sectors.

Table 8
Regression estimating firm innovative activity for specific sectors

Sector	R&D	R&D²	Innovative diversity	Innovative specialization	n	R^2	F
Instruments	0.1796	−46.7294	7.7780	−7.3747	23	0.756	13.918
	(2.863)	(−2.344)	(3.167)	(−2.311)			
Telecommunications	0.0426	−81.0826	12.1458	−10.5068	31	0.681	13.857
	(1.757)	(−1.955)	(3.995)	(−2.888)			
Pharmaceuticals	0.0579	−21.2919	5.0999	−4.1984	24	0.732	12.951
	(1.162)	(−0.496)	(1.840)	(−1.501)			
Electrical equipment	0.0328	−2.6788	11.8368	−9.5924	27	0.694	12.455
	(0.908)	(−0.249)	(3.471)	(−2.310)			
Transportation	0.0025	0.5888	10.1231	−8.7190	20	0.859	22.934
	(0.357)	(0.880)	(4.985)	(−3.960)			
Conglomerates	−0.1074	0.0019	7.5160	−5.5113	25	0.926	62.866
	(−2.671)	(6.970)	(3.941)	(−2.671)			

Notes: The t-value of each coefficient is listed in parentheses. The coefficients of R&D² are multiplied by 100,000 for presentation purposes.

Table 9
Regression results estimating firm innovative activity for specific science bases

	High-tech computing (1)	High-tech computing (2)	Biomedical (3)	Biomedical (4)
R&D	0.0100	0.0059	0.6000	0.0501
	(2.293)	(1.497)	(2.056)	(1.723)
R&D²	−0.2928	−0.1683	−25.7484	−18.6446
	(−0.522)	(0.300)	(−1.222)	(−0.889)
Innovation share within science base	3.7691	9.8348	1.4761	5.2878
	(7.831)	(8.524)	(1.925)	(2.120)
Innovation share within industry	–	−8.2150	–	−4.0457
		(−5.670)		(−1.601)
Number of observations	134	134	32	32
R^2	0.454	0.563	0.680	0.708
F	35.983	41.493	35.677	64.980

Notes: The t-value of the coefficient is listed in parentheses. The coefficients of R&D² have been divided by 100,000 for presentation purposes.

The results are confirmed in Table 9, which groups the firms according to two of the largest science bases – high-tech computing and biomedical products. Not only is the knowledge production function found to hold for firms in each of these distinct science bases, but again, diversity in innovation across economic activities with a common science base is found to increase innovation.

7. Conclusions

The nature and utility of knowledge is at the heart of the economics of R&D, innovation and technological change. Whether diversity or specialization of economic activities better promotes technological change has been the subject of a heated debate in the economics literature. This paper has attempted to shed light on that debate by linking the extent of diversity versus specialization of economic activities to innovative output. By focusing on innovative activity for particular industries at specific locations, we find compelling evidence that specialization of economic activity does not promote innovative output. Rather, the results indicate that diversity across complementary economic activities sharing a common science base is more conducive to innovation than is specialization. In addition, the results indicate that the degree of local competition for new ideas within a city is more conducive to innovative activity than is local monopoly.

A second perspective explored in this paper is the effect of diversity and specialization at the firm level. The results indicate that innovative activity tends to be lower when that innovation is specialized within a narrow industry than when it is diversified across a complementary set of industries sharing a common science base. Thus, the results at both the level of the firm as well as for the industry across geographic space present a consistent view of the returns to specialization versus diversity of economic activity. Our results suggest that diversity across complementary industries sharing a common base – a crucial qualification – results in greater returns to R&D.[15]

Increasingly scholars of technological change realize that external sources of knowledge are critical to innovation. Our results suggest that the boundaries of the firm are but one means to organize and harness knowledge. An analogous means of organizing economic activity are spatially defined boundaries. Geographic location may provide another useful set of boundaries within which to organize innovation. Geography may provide a platform upon which knowledge may be effectively organized.

[15] We underscore the descriptive nature of these results since there may be alternative explanations. Specifically, the relationships may be endogenous in a way that we have not considered. Firms which are more innovative may be more profitable and therefore, more likely to be able to engage in diverse activities. Similarly, regions which are successful at innovation in one industry may attract other activities. The cross-sectional nature of the data we use here does not allow us to examine these issues.

Acknowledgements

We are grateful to the comments and suggestions of two anonymous referees and an editor of this journal, Elhanan Helpman. An earlier version of this paper was presented at the Center for Economic Policy Research (CEPR) conference on R&D Spillovers at Lausanne, Switzerland, 27–28 January 1995 and at the 1997 annual meetings of the American Economic Association at San Francisco. We would like to thank Zvi Griliches, Bronwyn Hall, Frank Lichtenberg and F.M. Scherer for their comments and suggestions.

References

Arrow, K.J., 1962. Economic welfare and the allocation of resources for innovation. In: Nelson, R.R. (Ed.), The Rate and Direction of Inventive Activity. Princeton University Press, Princeton, NJ, pp. 609–626.

Audretsch, D.B., 1995. Innovation and Industry Evolution. MIT Press, Cambridge.

Audretsch, D.B., Feldman, M.P., 1996. R&D spillovers and the geography of innovation and production. American Economic Review 86 (3), 630–640.

Audretsch, D.B., Stephan, P.E., 1996. Company-scientist locational links: The case of biotechnology. American Economic Review 86, 641–652.

Cohen, W., Levinthal, D., 1990. Absorptive capacity: A new perspective on learning and innovation. Administrative Science Quarterly 35, 1288–1352.

Cohen, W.M., Malerba, F., 1995. Is the tendency to variation a chief cause of progress? Unpublished manuscript, Carnegie-Mellon University.

Feldman, M.P., 1994. The Geography of Innovation. Kluwer Academic Publishers, Boston.

Feldman, M.P., Florida, R., 1994. The geographic sources of innovation: Technological infrastructure and product innovation in the United States. Annals of the Association of American Geographers 84, 210–229.

Glaeser, E.L., Kallal, H.D., Scheinkman, J.A., Shleifer, A., 1992. Growth of cities. Journal of Political Economy 100, 1126–1152.

Griliches, Z., 1990. Patent statistics as economic indicator: A survey. Journal of Economic Literature 28, 1661–1707.

Griliches, Z., 1992. The search for R&D spill-overs. Scandinavian Journal of Economics 94, 29–47.

Jacobs, J., 1969. The Economy of Cities. Random House, New York.

Jaffe, A.B., 1989. Real effects of academic research. American Economic Review 79, 957–970.

Jaffe, A.B., Trajtenberg, M., Henderson, R., 1993. Geographic localization of knowledge spillovers as evidenced by patent citations. Quarterly Journal of Economics 63, 577–598.

Jewkes, J., Sawers, D., Stillerman, M., 1958. The Sources of Innovation. MacMillan, London.

Krugman, P., 1991a. Increasing returns and economic geography. Journal of Political Economy 99, 483–499.

Krugman, P., 1991b. Geography and Trade. MIT Press, Cambridge.

Levin, R.C., Klevorick, A.K., Nelson R.R., Winter, S.G., 1987. Appropriating the returns from industrial research and development. Brooking Papers on Economic Activity, 783–820.

Loesch, A., 1954. The Economics of Location. Yale University Press, New Haven.

Lucas, R.E. Jr., 1993. Making a miracle. Econometrica 61, 251–272.

Mansfield, E.J., 1984. Comment on using linked patent and R&D data to measure interindustry technology flows. In: Griliches, Z. (Ed.), R&D, Patents and Productivity. University of Chicago Press, Chicago, pp. 415–418.

Pakes, A., Griliches, Z., 1984. Patents and R&D at the firm level: A first look. In: Griliches, Z. (Ed.), R&D, Patents and Productivity. University of Chicago Press, Chicago, pp. 139–161.

Porter, M.P., 1990. The Comparative Advantage of Nations. The Free Press, New York.

Reynolds, P., Storey, D., Westhead, P., 1994. Cross-national comparisons of the variation in new firm formation rates. Regional Studies 28, 443–456.

Romer, P., 1990. Endogenous technological Change. Journal of Political Economy 94 (1), 71–102.

Romer, P., 1986. Increasing returns and long-run growth. Journal of Political Economy 94 (5), 1002–1037.

Scherer, F.M., 1983. The propensity to patent. International Journal of Industrial Organization 1 (1), 107–128.

Von Hipple, E., 1994. Sticky information and the locus of problem solving: Implications for innovation. Management Science 40, 429–439.

Part VII
Entrepreneurship and the Growth of Cities

[19]

The Economic Journal, **113** (*January*), 207–225. © Royal Economic Society 2003 Published by Blackwell Publishing, 9600 Garsington Road, Oxford OX4 2DQ, UK and 350 Main Street, Malden, MA 02148, USA

LOW RETURNS IN R&D DUE TO THE LACK OF ENTREPRENEURIAL SKILLS*

Claudio Michelacci

This paper proposes a model of endogenous growth where innovating requires both researchers, who produce inventions, and entrepreneurs who implement them. As research and entrepreneurship compete in the allocation of aggregate resources, the relation between growth and research effort is hump-shaped When entrepreneurs appropriate too little rents from innovation, too few resources are allocated to entrepreneurship and returns to R&D are low because of this lack of entrepreneurial skills. When so, innovation should be promoted by encouraging entrepreneurship rather than research.

Since the Malthusian (1986 [1798]) and Ricardian (1951 [1817]) prophecies of the eventual coming of a stationary state, the spectre of diminishing returns has hovered over economics. Arguably, such diminishing returns may arise because of the progressive exhaustion of investment opportunities which makes the returns to R&D effort decrease over time.[1] According to Schumpeter (1943, 1946) this is, however, just one of two competing explanations of the 'state of decay' of the 'capitalist society'. A 'more plausible' one is that 'the individual leadership of the entrepreneur tends to lose importance'.

According to this view, entrepreneurial ability is as important as research ability in determining the growth rate of an economy. Indeed, Schumpeter (1947) stresses the distinction between the inventor and the entrepreneur: 'The inventor produces ideas, the entrepreneur "gets things done" [...] an idea or scientific principle is not, by itself, of any importance for economic practice'. Hence, in order to grow, an economy requires both researchers who produce inventions and entrepreneurs who implement them.[2] Scientific knowledge has no economic impact unless some effort is made to spread it.[3]

* I would like to thank Philippe Aghion, Samuel Bentolila, Chris Pissarides and Javier Suarez for very helpful discussions and suggestions at different stages of the project. I am also particularly indebted to Christopher Bliss, another Editor and two anonymous referees for valuable and thoughtful comments on previous versions of the paper. The usual disclaimer applies.

[1] Recent empirical evidence seems to support some form of diminishing returns in research over time. For example, Griliches (1990) and Kortum (1993) note that, in the US, the ratio of the number of patent applications to the scientists and engineers involved in R&D has fallen over time in the post-war period, while Jones (1995*a*, *b*) points out that the economy growth rates have remained constant and even declined despite an increase in the amount of R&D effort. Indeed, Griliches (1979) concludes that 'the exhaustion of inventive and technological opportunities remains a major suspect for the productivity slow-down in the 70's'.

[2] The history of innovation is full of examples of successful inventor-entrepreneur partnerships. For example, Mokyr (1990) notices that even James Watt would have hardly succeeded in the implementation of the steam engine without the help of his partner, Matthew Boulton

[3] Economists are well aware of the importance of allocating resources to both research and entrepreneurial activities They must allocate their time between the studying required to write new papers and teaching students, talking with colleagues, presenting work at the NBER meetings, etc.

In order to model the aggregate trade off between allocating resources to research and entrepreneurship, I consider an endogenous growth model where an innovation requires the matching of an entrepreneur with a successful invention. In the model, individuals face the choice of whether to become researchers, who produce inventions, or entrepreneurs who implement them.[4] Hence an economy that allocates too many individuals to the research sector produces too many inventions which will be wasted because there will not be enough entrepreneurs to implement them. Thus the relationship between productivity growth and research effort is hump-shaped.

The existence of search frictions in the process whereby entrepreneurs and researchers are matched produces a natural incompleteness of contracts. Hence, the share of the rents from innovation appropriated by either researchers or entrepreneurs does not necessarily reflect their contribution to innovation. In general equilibrium, however, such rents determine whether individuals choose to become researchers or entrepreneurs. Consequently, the allocation of skill sustained in equilibrium might depart substantially from the socially efficient one. Interestingly, any point on the growth-research effort technological frontier can be sustained as a decentralised equilibrium of the economy.

One of the key results of the paper is that the innovation process can suffer because of the lack of either research effort or entrepreneurial skills. Economic history may provide examples of either case. Islamic civilisation at the end of the twelfth century may be an example of the former. Indeed Mokyr (1990) argues that it declined because it 'was not capable of adding much new to the existing stock of ideas it retrieved and applied so brilliantly'. Mokyr also discusses notorious examples of societies, such as the Greeks and the Romans, that suffered because of the lack of entrepreneurial skills. Such a deficit may also well explain the economic decline of the late Victorian Britain usually attributed to the inability of British entrepreneurship in promoting the diffusion of new and advanced production methods.[5]

Some relation to the literature. By positing that innovation requires the random matching of entrepreneurs with valuable inventions, my model borrows elements from the standard search model, e.g. Pissarides (1990). This setting is, however, extended to capture two important features of the innovation process. First, I introduce a technological externality generated by firm creation that leads to endogenous growth. Second, I allow the skill allocation to be endogenous by

[4] Schumpeter (1949) was concerned that his model resembled a model of 'exogenous' technological growth: 'a stumbling block' of the theory 'may be expressed by saying that the entrepreneur simply does nothing but take advantage of technological progress, which therefore appears, implicitly or explicitly as something that goes along entirely independently of entrepreneurial activity [.] It is perhaps not difficult to understand that technological progress, so obvious in some societies and so nearly absent in others, is a phenomenon that needs to be explained'. My model stresses the role of the entrepreneurial function in an endogenous growth framework.

[5] The 1998 White Paper by the British Government argues that the UK is still lacking in entrepreneurial skills. The recent experience of 'Silicon Fen' – the cluster of over 1,600 high tech firms that have grown up around Cambridge University – is sometimes perceived as a successful attempt to reverse this secular deficit in entrepreneurship.

permitting agents to decide on whether to become an entrepreneur or a researcher.[6] In this context I generalise earlier results, originally due to Diamond (1982) and Hosios (1990), on the efficiency of economies subject to search frictions. In particular I show that restoring efficiency under the presence of technological externalities requires research to have more bargaining power relative to the Hosios's benchmark.

The paper also relates to those strands of the growth literature that have analysed the relation between growth and R&D effort. It shares with Helpman and Trajtenberg (1994) the premise that inventions have widespread economic consequences only if implemented. The finding that research effort can be excessive arises also in Tirole (1988, p.399), in Aghion and Howitt (1992), and in Aghion and Howitt (1998, ch. 6), while the emphasis on the allocation of talent is shared with Baumol (1990) and Murphy *et al.* (1991). It is the interaction between these strands of the literature that is new here.

The model has also some implications for the recent literature which has analysed the relation between growth and the scale of the economy.[7] Jones (1995*b*), Kortum (1997), Young (1998) and Howitt (1999) propose a theoretical solution to the observed absence of relation between the scale of R&D effort and growth. Their models have in common the assumption that the output cost of an innovation is increasing with the level of development. Due to this form of exhaustion of investment opportunities, the amount of research effort increases over time to keep growth constant. Importantly such decreasing returns to R&D arise as the efficient response of a competitive economy to the increase in its scale.[8]

The paper might propose a complementary reading of the same empirical facts. I show theoretically that the returns to R&D might be decreasing if they are the result of an inefficient equilibrium shift that increases the amount of research effort to the detriment of more socially useful entrepreneurial skills. Furthermore, I show that the data are consistent with the existence of a constant-returns-to-scale matching function that links a measure of innovation (number of patent applications) to a measure of research effort (number of scientists involved in R&D) and entrepreneurship (population of self-employed). This implies that, at the optimal allocation, the returns to R&D never decrease since well-balanced increases in the amount of research and entrepreneurial effort always exhibit a constant marginal effect on the innovation rate.

Section 1 expounds the general set-up while Section 2 solves for the steady state equilibrium. Section 3 deals with efficiency. Section 4 carries out some exercises of comparative statics. Section 5 discusses some empirical implications. The conclusions are in Section 6. The Appendix contains some technical derivations.

[6] See Acemoglu (1996, 1997) for an analysis of workers' decisions to accumulate human capital in a labour market subject to search frictions

[7] See Jones (1999) for a survey.

[8] For example, both in Jones (1995*b*) and Kortum (1997) the economy growth rate is the socially optimal one, while in Young (1998) the total amount of R&D effort is optimal even if its allocation among different dimensions of innovation might not be.

1. The Model

The economy is populated by a continuum of agents of size C. Each agent is infinitely lived, risk-neutral and maximises expected returns in output units discounted at rate $r > 0$. At each point in time agents can choose either to become researchers or entrepreneurs. I indicate with f_t the time-t fraction of researchers in the population and I refer to it as the level of *research effort* in the economy.

A researcher discovers *inventions* according to a Poisson process with intensity λ. When he makes a discovery, he starts searching for entrepreneurs able to implement it. Keeping the invention up-to-date requires a flow cost of χx_t per unit of time, where x_t represents the leading technology in the economy at time t and $\chi \geq 0$. I assume that the opportunity of transforming an invention into an innovation vanishes according to a Poisson process with rate of arrival ν. This implies that at each point in time the stock of scientific knowledge, k_t, measured by the number of inventions suitable for economic exploitation, increases in response to the discovery of new inventions, while it tends to fall as the old ones become obsolete. This allows one to capture some key characteristics of the process of accumulating scientific knowledge; see Adams (1990). First, the stock of scientific knowledge is *fundamental* in that its applicability is not immediate. Secondly, it recognises that the stock of knowledge is made up of *heterogenous* pieces of information since the implementation of an invention requires a costly search for a suitable entrepreneurs. Thirdly, the use of the stock of knowledge is *repetitive*, as an invention can give rise to a 'cluster' of innovations concentrated 'in certain sectors and their surroundings' (Schumpeter, 1939, pp. 100–1). Finally it recognises its *time specificity*, as scientific knowledge becomes obsolete as time goes by.[9]

An *innovation* requires an invention discovered by a researcher and an entrepreneur. Once they match, a new firm run by the entrepreneur is created. A firm created at time t accesses the leading technology of that date x_t. A firm can produce, at each point in time t, a flow of goods equal to Px_t and is shut down according to a Poisson process with rate of arrival δ.[10]

The entrepreneur can be either running a firm or working at home; see Benhabib *et al.* (1991). When the entrepreneur runs a firm, he receives a profit flow π_t. Profits of the entrepreneur are chosen, so as to share with the researcher the gains from running the firm, by using a generalised Nash bargaining solution. The bargaining powers of the entrepreneur and the researcher are β and $1 - \beta$, respectively. When the entrepreneur works at home, he produces a flow of goods equal to hx_t where $h \geq 0$ measures the level of human capital of the entrepreneur who uses it according to the technology available at time t. In equilibrium an entrepreneur running a firm has no incentive to search. Hence only entrepreneurs

[9] The model could be extended by assuming that the rate ν at which inventions become obsolete, depends on the rate of growth of the leading technology of the economy x_t so that technological progress causes the obsolescence of scientific knowledge – an effect reminiscent of Schumpeter's idea of creative destruction The main implications of the model would remain however unchanged.

[10] I model firms in this way for analytical convenience In fact the results of the paper would remain qualitatively unchanged if either a firm's death probability or output were assumed to be a function of its age, as they arguably are in reality.

working at home are 'available' for innovating. I will denote their number at time t by s_t and I will refer to it as the amount of *entrepreneurial slackness* available in the economy.

The rate at which free entrepreneurs find suitable inventions is determined by the homogeneous-of-degree-one matching function $m(k_t, s_t)$; see Pissarides (1990). This function is increasing and concave in each of its arguments. The matching function allows one to represent in a parsimonious fashion two key characteristics of the innovation process: the fact that both entrepreneurial skills and inventions are heterogenous, so that the search for viable partnerships is time consuming, and the fact that entrepreneurial skills are a scarce resource for which different researchers compete. The probability that an invention matches with an entrepreneur is given by $q(\theta_t) = m(k_t, s_t)/k_t$ where $\theta_t = k_t/s_t$ will be referred to as the level of *knowledge intensity* in the economy. By analogous considerations it follows that $p(\theta_t) = \theta_t q(\theta_t)$ is the instantaneous probability that a free entrepreneur finds a valuable invention. I also assume that

$$p(0) = q(\infty) = m(0, s_t) = m(k_t, 0) = 0, \quad \text{and} \quad p(\infty) = q(0) = \infty \tag{A1}$$

so as to avoid dealing with uninteresting corner solutions.

Vertical innovations are the only source of growth in this economy. I follow Caballero and Jaffe (1993) and Aghion and Howitt (1998), among others, in assuming that the rate of growth of the technological parameter x_t is given by the product of the size of the innovation, σ, and the frequency of innovation. With the above assumptions, this frequency coincides with the number of innovations introduced in the economy at time t, $\theta_t q(\theta_t) s_t$. Hence the growth rate of the technological parameter x_t is given by $g_t = \sigma\theta_t q(\theta_t) s_t$.[11]

A sufficient condition to assure finite present values is $\sigma C\delta < r$ since it will imply that in steady state equilibrium $g_t < r$. Whereas obtaining a positive equilibrium level of research requires that $(\lambda/v)(1-\beta)(P-h) > 0$. As in Romer (1990), in order to sustain growth we must assign a strictly positive capacity of rent appropriation to researchers, $\beta < 1$, since future rents are the only compensation for the sunk investment in research necessary to produce an invention.

The endogenous variables of the model are the level of knowledge intensity θ, the level of research effort f and the steady state growth rate g.

2. Equilibrium

In this section, I first write down value functions and characterise individual behaviour. I then analyse the level of research effort and the growth rate that arise in a steady state equilibrium.

[11] A natural interpretation is that there is a continuum of increasingly productive techniques indexed by a real number which corresponds to the log of its productivity parameter. Creation of a new firm means that the entrepreneur has succeeded in discovering a new technique and all entrepreneurs can thereafter search for the next technique in the continuum. So, at any point in time, the rate of technological progress is proportional to the flow of newly created firms.

2.1. *The Value of Research and Entrepreneurship*

The value of doing research at time t, R_t, solves

$$R_t = E_{\tau \geq 0}\{[(L_{t+\tau} + \sup(R_{t+\tau}, H_{t+\tau})]e^{-r\tau}\}, \tag{1}$$

where L_t and H_t are the time-t value of an invention and of being an entrepreneur working at home, respectively. $t + \tau$ is the arrival date of the first invention whose arrival rate is λ, while the second term in the right-hand side captures the option a researcher has of becoming an entrepreneur at a later date.

Since an invention becomes obsolete at rate ν and requires a cost of χx_t to remain up-to-date, it has a value L_t to the researcher which follows the asset-type equation

$$(r + \nu)L_t = -\chi x_t + q(\theta)I_t + \dot{L}_t. \tag{2}$$

I_t measures the time-t value to a researcher of an innovation, $\dot{L}_t$ is the time derivative of L_t while $q(\theta)$ is the probability that an invention matches with an entrepreneur. As a firm produces a flow of goods equal to Px_t and entrepreneurs' profits are equal to π_t, I_t can be obtained by solving

$$(r + \delta)I_t = Px_t - \pi_t + \dot{I}_t, \tag{3}$$

where $\dot{I}_t$ represents the time derivative of I_t while δ is the firm's death probability.

Analogously, the value to an entrepreneur of working at home at time t, H_t, solves the asset equation

$$rH_t = hx_t + \theta q(\theta)(E_t - H_t) + \dot{H}_t, \tag{4}$$

where hx_t, E_t and $\dot{H}_t$ are respectively the real flow of goods produced by an entrepreneur working at home, the entrepreneur's value of running a firm and the time derivative of H_t. $\theta q(\theta)$ is the instantaneous probability that a free entrepreneur finds an invention that he can implement. Analogously E_t solves

$$rE_t = \pi_t + \delta[\sup(R_t, H_t) - E_t] + \dot{E}_t, \tag{5}$$

where the above equation embodies the option entrepreneurs have of becoming researchers.

Profits are the outcome of bilateral bargaining between the researcher with the invention and the entrepreneur. Hence they maximise the weighted product of the researcher's and entrepreneur's net return from the creation of a new firm equal to I_t and $E_t - H_t$, respectively. Hence

$$\pi_t = \arg\max(I_t)^{1-\beta}(E_t - H_t)^{\beta},$$

where H_t is the entrepreneur's outside option which from the point of view of the maximisation is taken as constant. As a result, profits are such that

$$I_t = (1 - \beta)(I_t + E_t - H_t) = (1 - \beta)S_t, \tag{6}$$

where $S_t = I_t + E_t - H_t$ is the private net surplus generated by the creation of a new firm. Of this surplus, researchers and entrepreneurs appropriate fractions $1 - \beta$ and β respectively.

2.2. *Free Entry Condition*

In steady state, each variable is growing at the same rate g as the economy so

$$R_t = Rx_t, \quad I_t = Ix_t, \quad L_t = Lx_t, \quad H_t = Hx_t, \quad E_t = Ex_t. \tag{7}$$

From making use of (7) to rewrite (2), (3), (4) and (5) it follows that

$$L = \frac{1}{r - g + \nu}[-\chi + q(\theta)(1 - \beta)S], \tag{2'}$$

$$I = (1 - \beta)S, \tag{3'}$$

$$H = \frac{1}{r - g}[h + \theta q(\theta)\beta S], \tag{4'}$$

$$E = H + \beta S \tag{5'}$$

where $S = I + E - H$ measures (once divided by x_t) the private net surplus associated with the creation of a new firm and is equal to

$$S = \frac{P - h}{r - g + \delta + \theta q(\theta)\beta}. \tag{8}$$

In equilibrium, agents must be indifferent between becoming a researcher or an entrepreneur, so that the *free entry condition* $R_t = H_t$ must hold at each point in time. But then, from using this condition, the fact that $R_t = Rx_t$ and then substituting into (1) it follows that

$$R = \frac{\lambda L}{r - g}.$$

Together with (2′) and (4′) this implies that the condition $R_t = H_t$ reads as

$$h + \theta q(\theta)\beta S = \frac{\lambda[-\chi + q(\theta)(1 - \beta)S]}{r - g + \nu}. \tag{9}$$

The left-hand side of (9) represents the relative profitability of being an entrepreneur as an increasing function of the amount of knowledge intensity θ: the larger the stock of scientific knowledge available per entrepreneur, the more profitable is entrepreneurship. Analogously, an increase in θ reduces the profitability of being a researcher because the implementation of inventions becomes progressively more difficult. As a result the right-hand side of (9), which measures the relative profitability of being a researcher, is decreasing in θ. Hence, for any given level of the steady state growth rate g, there exists a unique value of knowledge intensity θ such that the right-hand side equals the left-hand side. At such value individuals are indifferent between becoming researchers or entrepreneurs.[12]

[12] One can also check that steady state profits π_t are equal to πx_t, where, from substituting (3), (4), (5) into (6), it follows that $\pi = \beta P + (1 - \beta)h + \theta q(\theta)\beta I$.

2.3. *Technological Frontier*

Since firms are closed at rate δ, entrepreneurial slackness, s_t, evolves according to the differential equation

$$\dot{s}_t = \delta[C(l-f) - s_t] - \theta q(\theta)s_t, \tag{10}$$

so that in steady state it is equal to

$$s_t = s = \frac{\delta C(1-f)}{\delta + \theta q(\theta)}. \tag{11}$$

Analogously, the dynamics of the stock of knowledge, k_t, is governed by

$$\dot{k}_t = \lambda Cf - \nu k_t. \tag{12}$$

so that in steady state $k_t = k = (\lambda/\nu)Cf$.

From the assumption that the frequency of innovation is equal to the number of successful matches between k_t and s_t, it follows that the steady state growth rate g of the technological parameter x_t is equal to

$$g = \sigma\theta q(\theta)s = \sigma Cm\left[\frac{\lambda}{\nu}f, \frac{\delta(1-f)}{\delta + \theta q(\theta)}\right], \tag{13}$$

where $\theta = k/s$ solves the simple non-linear equation

$$\theta = \theta(f) = \frac{\lambda}{\nu}\frac{[\delta + \theta q(\theta)]f}{\delta(1-f)}. \tag{14}$$

Equation (14) merely expresses knowledge intensity as the ratio of the steady state amount of scientific knowledge to the amount of slackness given by (11). Notice that (14) defines an function $\theta(f)$ that associates to each f the corresponding value of θ that solves the equation. Such function is increasing since a larger f associates with a larger k and a lower s. But then, after substituting $\theta(f)$ for θ in (13), one can define a function $g = g(f)$ which relates the level of research effort, f, to the steady state growth rate of the economy. Since the function $g(f)$ determines the frontier of technological possibilities of the economy, I refer to it as the *TE-schedule.* Such a schedule is hump-shaped and it satisfies the property that no growth can be the outcome of either too much or too little research effort, $g(0)-g(1) = 0$ (see Propositions 1 and 2 in the Appendix for a formal proof).[13] As both research and entrepreneurial skills are required to innovate, an over-allocation to either factor harms technological process.

2.4. *Equilibrium*

Equations (9) and (13), given the constraint imposed by (14), completely solve the model in the research effort, f, growth rate, g, space. Equation (9) determines, for

[13] Denote by $0 < \eta < 1$ the elasticity of the matching function with respect to the level of entrepreneurial slackness. Then one can show that $g(f)$ is globally strictly concave if, for example, η is weakly increasing in θ. One can also show that $g(f)$ has a unique maximum under the milder condition $-d\eta/d\theta < \eta^2$. For example, both conditions would obviously be satisfied in the case of a Cobb–Douglas matching function

any given level of the growth rate, the relative amount of resources that the economy will end up devoting to research. Given (14) and the associated function $\theta(f)$, (9) allows to define the function $g = e(f)$ which relates the level of research effort that solves the free-entry condition to the given level of the growth rate. Hereafter I will refer to the function $e(f)$ as the *FE-schedule*. Such function maps the zero one interval onto the whole real line (see Proposition 3 in the Appendix). Furthermore $e(f)$ is increasing. To understand why it is so, notice that research requires a longer time than entrepreneurship to become productive and yield some (positive) revenue. Also notice that in the model an increase in the growth rate, g, affects individual decisions through a capitalisation effect that reduces the effective discount rate of agents. But a change in the discount factor always tends to have a greater effect on the investment with the longer time horizon so, as a result of the increase in g, the value of both research and entrepreneurship increase but the value of research increases relatively more so f must rise to restore the free entry condition.

The steady-state equilibrium is defined by the point at which the *FE*-schedule, $e(f)$, crosses the *TE*-schedule, $g(f)$, like at point A in Figure 1. At such a point no researcher has an incentive to become an entrepreneur (or *vice-versa*) and the economy grows at the constant steady state growth rate determined by the technological frontier.

3. Welfare Analysis

In this section, I compare the equilibrium to the constrained social optimum. I define social welfare $\tilde{W}$ as the net present discounted value of the aggregate income flows produced by entrepreneurs and researchers. The social planner

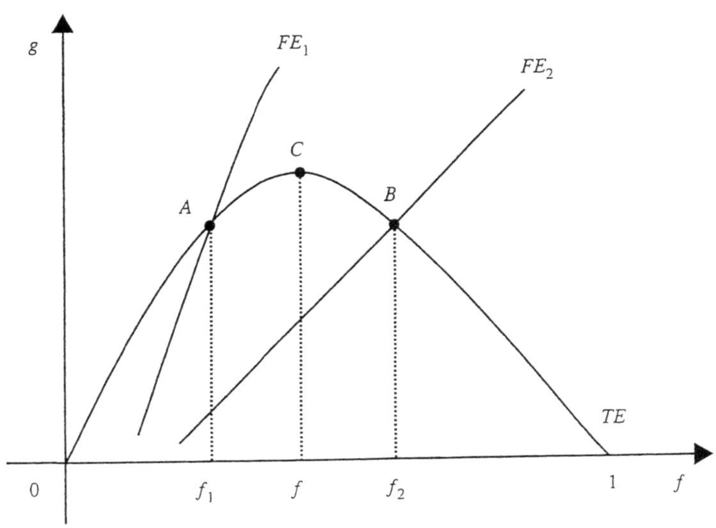

Fig. 1. *Steady State Equilibrium*

takes as given the search frictions involved in the innovation process. At any point in time, the state of the economy is fully summarised by the quantities k_t and n_t that represent the time-t stock of scientific knowledge and total number of firms, respectively. Given the focus on steady state equilibria, I consider time invariant allocations described by a fixed amount of research effort f.

3.1. *The Social Planner Problem*

Once multiplied by x_t,

$$y(k_t, n_t; f) = Pn_t + h[C(1-f) - n_t] - \chi k_t$$

denotes the net income flow produced by the economy at time t. Also notice that the evolution of the number of firms in the economy is governed by

$$\dot{n}_t = m[k_t, C(1-f) - n_t] - \delta n_t \tag{15}$$

while the stock of scientific knowledge grows at a rate $\dot{k}_t$ given by (12). Hence one can implicitly define social welfare as the product of x_t and the function $W(k, n, f)$ which solves

$$rW(k_t, n_t; f) = y(k_t, n_t; f) + W_k\dot{k}_t + W_n\dot{n}_t + g_t W(k_t, n_t; f) \tag{16}$$

where $g_t = \sigma m(k_t, s_t)$ while $W_k = \partial W/\partial k_t$ and $W_n = \partial W/\partial n_t$ measure the social shadow value of an invention and of an additional firm, respectively. Three components account for social welfare in (16): the first is current income, the second is the capital gain associated with changes in the state of the system – the second and third term in (16) – and last the capitalisation effect associated with technological progress, the fourth term in (16).

To characterise the constrained socially optimum allocation, denote by g and n the economy's steady state value of the growth rate and number of firms, respectively. Furthermore let

$$\bar{W} = \frac{Pn + h[C(1-f) - n] - \chi k}{r - g} \tag{17}$$

denote (once divided by x_t) the steady state level of social welfare which is equal to the present discounted value of the level of output net of research costs. Finally let η indicate the elasticity of the matching function with respect to entrepreneurial slackness.[14] Then differentiating (16) with respect to f and after some algebra, the Appendix shows that

$$\frac{dW}{df} = -\frac{C}{r-g}\left\{h + \theta q(\theta)\eta S^E - \frac{\lambda[-\chi + q(\theta)(1-\eta)S^E]}{r - g + \nu}\right\} \tag{18}$$

where

$$S^E = W_n + \sigma\bar{W} = \frac{P - h + \sigma(r - g + \delta)\bar{W}}{r - g + \delta + \theta q(\theta)\eta} \tag{19}$$

[14] Notice that η is in general a function of θ

is the net social surplus associated with the creation of a new firm. Such surplus is equal to the sum of the shadow value of a new firm W_n plus the gains associated with the technological externality generated by innovation, which have value $\sigma\bar{W}$.

3.2. *Results*

I can now state my first result on efficiency: the social optimum always lies on the positively sloped arm of the *TE*-schedule, $g(f)$. To prove this result, the Appendix first shows that the steady state growth rate g is maximised at the 'golden rule' level of research effort f^* such that the level of knowledge intensity $\theta^{*} = \theta(f^*)$ solves

$$\theta^{*} = \frac{\lambda}{\nu}\frac{1-\eta}{\eta}. \tag{20}$$

Then the Appendix shows that the derivative of W with respect to f evaluated at f^* is negative which indeed implies that at the growth maximising allocation the level of research effort in the economy is too large.

The economic intuition of the result is pretty simple. Given the objective of maximising the present discounted value of output, the central planner must equate the marginal benefit of increasing research effort to its marginal cost. Notice that (13) and (15) imply that the value of f that maximises the steady state number of operating firms, n, also maximises the steady state growth rate, g. So the sign of the social marginal benefits of research coincides with that of the derivative of the *TE*-schedule, $g(f)$. But then, since increasing research always involves a strictly positive marginal cost (relative to increasing entrepreneurship), the social optimum must be at a point where the marginal benefits from increasing f are also strictly positive and this is so only on the strictly increasing arm of *TE*-schedule.

It is interesting to compare the level of research effort sustained by the decentralised economy to the social optimal one. This can be done by comparing (9) with (18) and checking when the decentralised economy implements a level of research effort which induces $\mathrm{d}W/\mathrm{d}f = 0$. One can immediately see that the two conditions associate with the same level of research effort only if the private and social net-surplus from the creation of a new firm coincide, $S = S^E$. In the absence of technological externalities, $\sigma = 0$, this occurs when entrepreneurs' bargaining power is such that $\beta = \eta$ so that (8) and (19) are identical. Hosios (1990) first proved that, in an economy with search frictions, a constrained efficient allocation of resources requires bargaining power to reflect each side's contribution to the creation of net surplus as measured by the elasticity of the matching function. Here I have generalised this result to an environment where the allocation of skill is endogenous.

When technological externalities are present, $\sigma > 0$, however, the Appendix shows that entrepreneurs' bargaining power (β must be smaller than η for the decentralised economy to implement the socially efficient allocation. To understand the result, notice that search frictions simultaneously induce a negative externality to agents on the same side of the market (congestion externalities) and a positive one to agents on the other side of the market (thin market externalities). At ($\beta = \eta$ the congestion and thin-market externalities balance exactly. Yet, at this point, technological externalities are still operating and tend to discourage activities, namely research, characterised by a longer time horizon and larger initial investment. Hence restoring efficiency requires $\beta < \eta$. There is only one case where the Hosios's rule $\beta = \eta$ still applies even in the presence of technological externalities. This occurs when the type of investment required to specialise in different activities is the same, that is when the instantaneous return from becoming a researcher or an entrepreneur are equal $h = \chi = 0$, (see Appendix).

4. Comparative Statics

Proposition 4 in the Appendix states that, depending on the parameters of the model, any point on the *TE*-schedule, $g(f)$, can be sustained as an equilibrium of the economy, including those situated on the negatively sloped arm of the *TE*-schedule, like point B in Figure 1. A point like B is associated with a Pareto dominated equilibrium since a reduction in research effort would simultaneously lead to an increase in the output level and in the growth rate (both over the transition path and in the new steady state). These types of equilibria can be sustained for reasons similar to those analysed by Hosios (1990) and Caballero and Hammour (1996). An economy characterised by *ex-ante* competitive relationships but *ex-post* bilateral monopolies has no compelling tendency to coordinate itself towards the social optimum, since there is no reason why the amount of rents appropriated by each party should reflect his contribution to the creation of social surplus.

In equilibria like point B in Figure 1, research effort crowds out more socially useful entrepreneurial skills and the growth process stagnates precisely because of this *lack of entrepreneurial skills*. Indeed, as the amount of research effort rises, the stock of scientific knowledge increases, but it also becomes progressively more difficult to implement inventions both because more researchers are competing for the same resources (congestion externalities) and because an increase in research effort crowds out useful entrepreneurial skills (thin market externalities). When the stock of scientific knowledge is already large while the amount of entrepreneurial skills is low, an increase in research effort reduces the growth rate of the economy, because it misallocates socially useful resources. For example, an increase in research effort can be translated into stagnant or declining growth rates, if the equilibrium of the economy shifts from point A to point B in Figure 1.

Such shifts in the equilibrium can occur, for example, in response to a fall in the bargaining power of entrepreneurs β. A fall in β makes research relatively more profitable for any given level of the growth rate so that the *FE*-schedule shifts to the right while it has no effect on the *TE*-schedule. Such a fall in β might relate to the Schumpeterian (1943, 1946) claim that cultural and sociological shifts tend to destroy the 'protective strata' able to sustain the entrepreneurial function.[15] Since β measures the outcome of a bilateral bargaining problem here specified

[15] Culture, understood as prevailing values and beliefs, is often perceived as an important determinant of the level of entrepreneurship in the economy, see Davidson and Wilklund (1997). In terms of the model, a cultural environment more favourable to entrepreneurship may translate into a larger either β or h. This last case may correspond to societies that reward more (in terms of social status) entrepreneurial 'success' and 'leadership'

parsimoniously, one can also argue that, in the presence of some asymmetric information that favours researchers, β might be affected by the technological content of the innovation: innovations that are more technological advanced might associate with a greater ability of the researcher to appropriate rents and therefore with a lower β.

Particularly interesting are the consequences of a change in the firms' destruction rate δ. A change in δ has generally ambiguous effects on the *FE*-schedule. To see this notice that, a fall in δ requires an increase in the level of knowledge intensity, θ, to restore the free entry condition (9). However, (14) implies that for given f, a lower δ associate with a larger value of θ, since the amount of entrepreneurial slack is reduced. If this last effect is sufficiently strong the *FE*-schedule, $e(f)$, shifts to the left, otherwise f must rise to restore the free entry condition and $e(f)$ shifts to the right.

What is not ambiguous is the impact of a change in δ on the technological frontier of the economy. A fall in δ implies a downward shift of the *TE*-schedule, with the hump of the schedule moving leftward. To prove the downward movement, notice that for given f a fall in δ reduces the steady state level of entrepreneurial slack given by (11) both directly through δ and indirectly through the induced positive effect on θ, see (14). The leftward movement of the hump follows from the fact that entrepreneurial slack falls for any given level of f, so the constraint imposed by the number of free entrepreneurs on the innovation process becomes progressively more binding as f increases. In other words, an economy that liquidates more often (i.e. with a larger δ) has more frequent opportunity to rebuild its productive stock at a higher technological level and grows faster thanks to the intertemporal spill-overs typical of standard endogenous growth models. Thus the model suggests a channel whereby an increase in the pace of reallocation promotes growth by increasing the amount of resources available for innovating.[16]

5. Empirical Implications

I next discuss some empirical implications of the model. The analysis is simply meant to be illustrative of some of the theoretical properties of the model.

If one assumes that the matching function is Cobb–Douglas,

$$m(k_t, s_t) = A(k_t)^{\alpha}(s_t)^{\gamma}, 0 < \alpha, \gamma < 1, \tag{21}$$

then (13) suggests running the following regression

$$\ln(g_t) = \alpha \ln(f_t C_t) + \gamma \ln(s_t) + \alpha \ln\left(\frac{\lambda}{\nu}\right) + \ln \sigma + \ln(A). \tag{22}$$

The model has two important testable implications. First, the parameters α and γ must both be positive numbers strictly between zero and one. Indeed this would imply that both entrepreneurial and research skills affect the innovation rate of

[16] See Mortensen and Pissarides (1998) for a model where the causality between reallocation and growth reverses. In their model an increase in the rate of exogenous technological progress raises the job reallocation rate in the economy

the economy. Second, the matching function (21) exhibits constant returns to scale, that is $\alpha + \gamma = 1$. This would imply that, at the optimal allocation, the returns to R&D can never be decreasing since well-balanced increases in the amount of research and entrepreneurial effort have constant marginal effects on the innovation rate.

To illustrate the methodology, I estimate (22) for the US over the period 1950–90.[17] I follow Griliches (1979, 1990) in taking the total number of patent applications as an index of innovative activity in order to proxy for g_t,[18] and Jones (1995*b*) in considering the ratio of scientists and engineers involved in R&D over the population if working age as a measure of (relative) research effort f_t. As a proxy for the scale of the economy C_t in (22) I take the population if working age. Identifying entrepreneurship is more problematic. According to Schumpeter (1949) the entrepreneurial function can be identified only *a posteriori*.[19] In fact, Evans and Jovanovic (1989) note how, in their sample, 20% of the individuals who switched into self-employment formed incorporated businesses. That is why I follow Evans and Jovanovic (1989), Evans and Leighton (1989) and Blanchflower and Oswald (1998) in using as a proxy of the amount of entrepreneurship, s_t, the population of self-employed.

To control for cyclical disturbances, I also consider some additional specifications with the unemployment rate, the growth rate of *per capita* GDP and an index of the number of vacancies as independent variables.[20] The unemployment rate together with the other cyclical indicators could correct for that part of the population of self employed that does not represent real entrepreneurship but simply people that moved into self-employment from unemployment to find a job.[21]

Table 1 reports the estimates obtained by running the regression (22), allowing for different lags in the relation between research effort, entrepreneurship and innovation. The table shows that, in these data, both implications of the model tends to receive empirical support. In fact, α and γ are both positive and significantly different from zero with α ranging between 0.47 and 0.54 while γ ranges

[17] Data before 1970 are taken from *Historical Statistics of the United States: Colonial Times to 1970*; after 1970 they are from various issues of *The Statistical Abstract of the United States*.

[18] Whether a patent corresponds more to the notion of invention or innovation, is a topic beyond the scope of this paper. The Patents Laws-United State Code, U.S.C. 101, states that 'whoever invents or discovers any new and useful process, machine, manufacture, or composition of matter, or any new and useful improvement thereof, may obtain a patent'. The UK Patents Act 1977 states that the invention to be patented must be 'capable of industrial application'. More generally Cornish (1989) states that 'a patent cannot be granted for a thing or process which however interesting or suggestive it might be to scientists, has no known practical application at the priority date'. The requisite of 'industrial applicability' is not contained in the US law. It seems however, that the requisite of 'usefulness' roughly coincides with that of 'industrial applicability', so that even if a patent still represents something between an invention and an innovation as defined in this paper, it seems to be closer to the latter.

[19] Indeed, he writes: 'our definitions of entrepreneur, entrepreneurial function and so on can only grow out of it *a posteriori*' and again 'when we speak of the entrepreneur we do not mean so much a physical person as we do a function, but even if we look at individuals who at least at some juncture in their lives fill the entrepreneurial function it should be added that these individuals do not form a social class They hail from all the corners of the social universe'

[20] Data on the unemployment rate, growth rate of GDP *per capita* and the number of vacancies are taken from the OECD-CEP data set, see Bagliano *et al* (1991)

[21] Blanchflower and Oswald (1998) document that flows from unemployment into self-employment are large.

Table 1

Results from Running Regression (22) in the Text

Indep variable	Lagged $\ln(f_t C_t)$	Lagged $\ln(s_t)$	Unempl rate	GDP per cap growth rate	Vacancy index	R^2	Wald test (p-value) $\sigma + \gamma = 1$
1lag	0 47 (0.03)	0.24 (0.10)	–	–	–	0.84	0.005
2lags	0.49 (0 03)	0 29 (0.10)	–	–	–	0.86	0.03
3lags	0.50 (0.04)	0.37 (0.10)	–	–	–	0 86	0.24
4lags	0 52 (0 04)	0.48 (0.10)	–	–	–	0 86	0 94
1lag	0.50 (0.04)	0 30 (0 10)	−0.02 (0.008)	–	–	0.86	0 06
2lags	0.53 (0.04)	0 37 (0 10)	−0 02 (0.008)	–	–	0 88	0.36
3lags	0.54 (0 04)	0.45 (0.10)	−0.02 (0.009)	–	–	0.88	0.97
4lags	0.54 (0.05)	0.53 (0.11)	−0.01 (0 009)	–	–	0 87	0 62
1lag	0.51 (0 04)	0.33 (0.10)	−0.02 (0 009)	−0.53 (0.45)	–	0 88	0 15
2lags	0.54 (0.04)	0.39 (0.10)	−0.02 (0.009)	−0.33 (0.42)	–	0 89	0.56
3lags	0.54 (0.04)	0.44 (0.11)	−0 02 (0.009)	0.10 (0.43)	–	0.88	0.92
1lag	0.46 (0.07)	0.32 (0.10)	−0 02 (0 009)	−0.54 (0.45)	0.08 (0.08)	0.88	0 10
2lags	0.47 (0.07)	0 39 (0 10)	−0.02 (0.009)	−0.34 (0.41)	0 10 (0.08)	0.89	0 25
3lags	0.46 (0.07)	0.44 (0.11)	−0.01 (0.01)	0.09 (0.42)	0.10 (0.07)	0.89	0.48

The regression was run using OLS, the dependent variable being the (logged) number of patent applications. Similar results are obtained when allowing for first-order auto-correlation in the residuals and a non-scalar auto-covariance matrix.

from a value of 0.24 to 0.48. This implies that the relation between the number of innovations and research effort $f_t C_t$ turns out to be concave and hump-shaped. Furthermore, Table 1 documents that, if we consider reasonable a lag of over two years before research effort shows up in a patent application, the hypothesis that the matching function (21) exhibits constant returns to scale cannot be rejected even at a 10% level of significance. Moreover, in the specification with the unemployment rate as independent variable, this hypothesis cannot be rejected whatever the number of lags considered.

6. Conclusions

This paper has shown how an increase in the amount of resources devoted to research does not necessarily increase the growth rate of the economy. In a world with rent sharing and inter-temporal spill-overs, an increase in research effort can crowd out more socially useful entrepreneurial skills, reduce the growth rate and ultimately lead to an allocation of skills which is Pareto dominated. Thus, the model might provide some foundations to the Schumpeterian warning that the 'state of decay of the capitalist society' can ultimately be driven by the lack of entrepreneurial skills.

The question on whether the economy's skill allocation is optimal for growth is likely to be particularly relevant in advanced economies where innovation is so *complex* that a single agent fails to possess both the scientific knowledge and the entrepreneurial ability required to innovate. When this is so, individuals must specialise and innovation requires the contribution of different, highly specialised agents. Consequently, the allocation of skills to different tasks changes from being internal to the *individual* to being internal to the *market* and the amount of rents from innovating appropriated by each task may become a crucial determinant of the economy's skill allocation. For instance, the division of rents could affect the choice between allocating resources to accumulating either education or experience as well as any occupational choice, especially among those occupations that are likely to participate to the sharing of the rents from innovation – say, whether to become a manager, a politician, a financier or an entrepreneur.

Investigating whether the actual skill allocation tends to depart from the optimal one as the economy develops and gets more specialised, remains however a question for further research that requires more careful micro-based empirical investigation. Looking at the skill composition either by occupations or along the experience-schooling dimension could be a promising avenue to address this question.

CEMFI, Madrid

Date of receipt of first submission: August 2000
Date of receipt of final typescript: January 2002
Dataset is available for this paper: www.res.org.uk

Appendix

In this Appendix I derive some technical results discussed in the text. Before proceeding, notice that the derivatives of $m(k,s)$ with respect to k and s can be computed as

$$\frac{\mathrm{d}m}{\mathrm{d}k} = (1-\eta)q(\theta) \quad \text{and} \quad \frac{\mathrm{d}m}{\mathrm{d}s} = \eta\theta q(\theta),$$

respectively.

Properties of the function $\theta(f)$

PROPOSITION 1: *Equation (14) defines the level of knowledge intensity θ as an increasing function of f, $\theta = \theta(f)$, with the property that $\theta(0) = 0$ and $\theta(1) = \infty$.*

Proof: By using $\mathrm{d}\theta q(\theta)/\mathrm{d}\theta = (1-\eta)q(\theta)$, total differentiating (14) with respect to θ and f yields

$$\frac{\mathrm{d}\theta}{\mathrm{d}f} = \frac{\delta + \theta q(\theta)}{\delta + \eta\theta q(\theta)} \frac{(\lambda/v)[\delta + \theta q(\theta)] + \delta\theta}{\delta(1-f)} \tag{23}$$

which proves monotonicity. The properties of the function $\theta(f)$ at the boundary follow directly from (14) and (A1) together with the fact that the elasticity of the right hand side of (14) with respect to θ is less than one

Properties of the TE-schedule

PROPOSITION 2: *Given (14), (13) determines the growth rate g, as a function of f, $g = g(f)$ with the property that $g(0) = g(1) = 0$. Steady state growth is maximised at an allocation such that (20) is satisfied*

Proof: Note that (11) together with proposition 1 allow to define entrepreneurial slackness as a decreasing function of f, $s = s(f)$. After substituting in (13), one obtains that

$$g = g(f) = \sigma m\left[\frac{\lambda}{v} Cf, s(f)\right],$$

which explicitly defines the function $g(f)$.

To conclude the proof, notice that from (23), it follows that

$$\frac{\mathrm{d}s(f)}{\mathrm{d}f} = \frac{\mathrm{d}}{\mathrm{d}f}\left[\frac{\delta(1-f)}{\delta+\theta q(\theta)}\right] = -\frac{\delta+(\lambda/v)(1-\eta)q(\theta)}{\delta+\eta\theta q(\theta)},$$

thus the derivative of $g(f)$ reads as

$$\frac{\mathrm{d}g(f)}{\mathrm{d}f} = \frac{\sigma C\delta q(\theta)[(\lambda/v)(1-\eta)-\eta\theta]}{\delta+\eta\theta q(\theta)},$$

which proves (20).

Properties of the FE-schedule

PROPOSITION 3: *Given* (14) *and the associated function* $\theta = \theta(f)$, (9) *defines the growth rate of the economy* g *as an increasing function of* f, $g = e(f)$, *with the properties that* $e(0) = -\infty$ *and* $e(1) = \infty$.

Proof: Multiply the left and right hand side of (9) by $[r - g + \delta + \theta q(\theta)\beta]$. The resulting left hand side of the equation is strictly increasing in θ and equal to $h(r + \delta - g)$ when $\theta = 0$, while the resulting right hand side is strictly decreasing in θ and it approaches infinity when θ goes to zero since $(\lambda/v)(1-\beta)(P-h) > 0$ by assumption. This implies that, for given g, a θ that solves (9) always exists. Furthermore, an increase in g causes a downward (upward) shift in the left (right) hand side so that θ unequivocally increases. As $\theta = \theta(f)$ is increasing, $e(f)$ is also increasing. The properties of the function $e(f)$ at the boundary follow immediately from condition (Al) and the properties of $\theta(f)$ established in Proposition 1.

PROPOSITION 4: *There exist values of the parameters such that any point on the TE-schedule,* $g(f)$, *can be sustained as an equilibrium.*

Proof: Note that, for a given level of g, changes in P, h, χ and β modify the value of θ that solves (9) without affecting the function $\theta(f)$. Hence, one can find a combination of parameters such that arbitrary given values of f and g are on the FE-schedule. Since changes in these parameters do not affect the TE-schedule, any point along $g(f)$ can possibly be sustained as an equilibrium.

Efficiency Results

The derivative of W *with respect to* f. Partially differentiating (16) with respect to k_t and n_t and after using the fact that in steady state $\dot{k}_t = \dot{n}_t = 0$ yields

$$W_k = \frac{-\chi+(1-\eta)q(\theta)(W_n+\sigma\bar{W})}{r-g+v} \tag{24}$$

and

$$W_n = \frac{P-h-\sigma\eta\theta q(\theta)\bar{W}}{r-g+\delta+\eta\theta q(\theta)}. \tag{25}$$

Differentiating (16) with respect to f and evaluating this derivative in steady state yields

$$\frac{\mathrm{d}W}{\mathrm{d}f} = \frac{C}{r-g}[-h-\eta\theta q(\theta)(W_n+\sigma\bar{W})+\lambda W_k]$$

which after using (24) immediately leads to expression (18) in the main text.

Relation between θ' and the social optimum. To prove that the social optimum always lies on the positively sloped arm of the *TE*-schedule, evaluate (18) at $\theta' = [\lambda(l-\eta)]/v\eta$. This immediately yields

$$\frac{\mathrm{d}W}{\mathrm{d}f} = -\frac{C}{r-g}\left[h + \frac{\lambda\chi}{r-g+v} + \frac{(r-g)\eta\theta^* q(\theta^*)S^E}{r-g+v}\right]$$

which is negative.

The efficient sharing rule. I now prove that it exists $\beta < \eta$ such that the decentralised economy implements the optimal allocation. From (9) it follows that

$$h + \frac{\lambda\chi}{r-g+v} = \left[\frac{\lambda(1-\beta)}{r-g+v} - \beta\theta\right]q(\theta)S, \tag{26}$$

which substituted into (18) and after using the definition of S^E in (19) yields

$$\frac{\mathrm{d}W}{\mathrm{d}f} = \frac{C(\beta-\eta)q(\theta)\{\theta(r-g+\delta) - \lambda[r-g+\delta+\theta q(\theta)]/(r-g+v)\}S}{(r-g)[r-g+\delta+\eta\theta q(\theta)]} + \frac{\sigma C q(\theta)(r-g+\delta)[\lambda(1-\eta)/(r-g+v) - \eta\theta]\bar{W}}{(r-g)[r-g+\delta+\eta\theta q(\theta)]}.$$

Evaluating this derivatives at $\beta = \eta$, one can see that the first term is equal to zero while (26) implies that the second term is strictly positive, so that, at this level of β, research effort is too low. Now note that (9) implies that f is negatively related to β, so that β must fall below η for the economy to reach efficiency. There are two exceptions where at $\beta = \eta$, $\mathrm{d}W/\mathrm{d}f = 0$. The former arises when $\sigma = 0$, the latter when $h + \lambda\chi/(r-g+v) = 0$ so that, by (26), $[\lambda(1-\eta)]/(r-g+v) - \eta\theta = 0$.

References

Acemoglu, D. (1996). 'A micro-foundation for social increasing returns in human capital accumulation', *Quarterly Journal of Economics*, vol. 111(3), pp. 779–804.

Acemoglu, D (1997). 'Training and innovation in an imperfect labour market', *Review of Economic Studies*, vol. 64, pp. 445–64.

Adams, J. (1990). 'Fundamental stocks of knowledge and productivity growth', *Journal of Political Economy*, vol. 98, pp. 673–702.

Aghion, P. and Howitt, P. (1992). 'A model of growth through creative destruction', *Econometrica*, vol. 60, pp. 323–51.

Aghion, P. and Howitt, P (1998). *Endogenous Growth Theory*, Cambridge, MA. MIT Press.

Bagliano, F., Brandolini A. and Dalmazzo, A. (1991) 'The OECD-CEP data set', Centre for Economic Performance Working Paper No. 118, June.

Baumol, W (1990) 'Entrepreneurship· productive, unproductive, and destructive', *Journal of Political Economy*, vol. 98(5), pp. 893–921

Benhabib, J., Rogerson, R and Wright, R (1991). 'Homework in macroeconomics: household production and aggregate fluctuations', *Journal of Political Economy*, vol 99(6), pp 1166–87.

Blanchflower, D. G. and Oswald, A. (1998) 'What makes an entrepreneur?', *Journal of Labour Economics*, vol. 16, pp. 26–60.

Caballero, R. and Hammour, M. (1996) 'The fundamental transformation in macroeconomics', *American Economic Review, Papers and Proceedings*, vol 86(2), pp 181–6

Caballero, R and Jaffe, A. (1993). 'How high are the giants shoulders· an empirical assessment of knowledge spillovers and creative destruction in a model of economic growth', in (O Blanchard, and S. Fischer, eds), *NBER Macroeconomics Annual 1993*, Cambridge MA· MIT Press, pp 15–74.

Cornish, W R. (1989). *Intellectual Property· Patents, Copyright, Trade Marks and Allied Rights*, 2nd edn, London. Sweet & Maxwell.

Davidson, P. and Wiklund, J. (1997). 'Values, beliefs and regional variations in new firm formation rates', *Journal of Economic Psychology*, vol 18, pp. 179–99.

Diamond, P (1982). 'Wage determination and efficiency in search equilibrium', *Review of Economic Studies*, vol 49, pp 217–27

Evans, D and Jovanovic, B (1989) 'An estimated model of entrepreneurial choice under liquidity constraints', *Journal of Political Economy*, vol 97(4), pp 808–27.

Evans, D. and Leighton, L (1989). 'Some empirical aspects of entrepreneurship', *American Economic Review*, vol. 79(3) (June), pp 519–35

Griliches, Z (1979) 'Patents recent trends and puzzles', *Brookings Papers on Economic Activity, Microeconomics*, pp 291–330

Griliches, Z (1990). 'Patent statistics as economic indicators a survey', *Journal of Economic Literature*, vol 28 (December), pp 1661–707

Helpman, E and Trajtenberg, M. (1994) 'A time to sow and a time to reap growth based on general purpose technologies', NBER Working Paper Series, No 4854, September

Hosios, A (1990) 'On the efficiency of matching and related models of search and unemployment', *Review of Economic Studies*, vol 57, pp. 279–98.

Howitt, P (1999) 'Steady endogenous growth with population and R&D inputs growing', *Journal of Political Economy*, vol 107(4), pp. 715–30

Jones, C (1995*a*). 'Time series tests of endogenous growth models', *Quarterly Journal of Economics*, vol 110, pp 495–525.

Jones, C. (1995*b*) 'R&D based models of economic growth', *Journal of Political Economy*, vol. 103(4), pp 759–83.

Jones, C (1999). 'Growth: with or without scale effects', *American Economic Review Papers and Proceedings*, vol 89(2), pp 139–44

Kortum, S. (1993) 'Equilibrium R&D and the decline in the patent-R&D ratio', *American Economic Review: Papers and Proceedings*, vol 83(2), pp. 450–7.

Kortum, S (1997) 'A model of research, patenting and productivity growth', *Econometrica*, vol 65(6), pp. 1389–419.

Malthus, T. R. (1986 [1798]). *An Essay on the Principle of Population*, London· W.Pickering.

Mokyr, J. (1990). *The Lever of Riches Technological Creativity and Economic Progress*, Oxford. Oxford University Press.

Mortensen, D T. and Pissarides, C (1998). 'Technological progress, job creation, and job destruction', *Review of Economic Dynamics*, vol 1(4), pp. 733–53.

Murphy, K, Shleifer, A. and Vishny, R (1991). 'The allocation of talent: implications for growth', *Quarterly Journal of Economics*, vol. 56, pp. 503–30.

Pissarides, C. A (1990). *Equilibrium Unemployment Theory*, Oxford: Basil Blackwell

Ricardo, D. (1951 [1817]). *On the Principles of Political Economy and Taxation*, Cambridge· Cambridge University Press

Romer, P (1990). 'Endogenous technological change', *Journal of Political Economy*, vol. 98(5), pp. 71–102.

Schumpeter, J A. (1939). *Business Cycles: a Theoretical, Historical and Statistical Analysis of the Capitalist Process*, New York. McGraw-Hill

Schumpeter, J. A. (1943). 'Capitalism in the postwar world', in (S. E. Harris, ed.), *Postwar Economic Problems*, New York: McGraw-Hill, pp. 113–26.

Schumpeter, J. A. (1946). 'Capitalism', in *Encyclopaedia Britannica*, vol. IV, Encyclopaedia Britannica Inc., pp 801–7.

Schumpeter, J. A. (1947). 'The creative response in economic history', *Journal of Economic History*, vol. 7 (November), pp. 149–59.

Schumpeter, J. A. (1949). 'Economic theory and entrepreneurial history', in (Research Center in Entrepreneurial History, Harvard University) *Change and the Entrepreneur*, Cambridge, MA: Harvard University Press, pp 63–84.

Tirole, J. (1988). *The Theory of Industrial Organization*, Cambridge, MA· MIT Press.

Young. A. (1998). 'Growth without scale effects', *Journal of Political Economy*, vol. 106(1), pp. 41–63.

ELSEVIER

Available online at www.sciencedirect.com
SCIENCE DIRECT

Journal of Urban Economics 56 (2004) 244–278

JOURNAL OF
Urban
Economics

www.elsevier.com/locate/jue

The impact of geographic differences in human capital on service firm formation rates

Zoltan J. Acs [a,*], Catherine Armington [b]

[a] Merrick School of Business, University of Baltimore, Baltimore, MD 21201, USA
[b] 3031 Macomb Street NW, Washington, DC 20008, USA

Received 16 May 2003; revised 9 March 2004

Available online 24 June 2004

Abstract

Although human capital externalities are a key variable in theories of economic growth, there has been little investigation of the mechanism by which these externalities are realized. We examine the relationship between the local levels of human capital and firm formation rates and find that formation rates differ with the share of adults with college degrees, especially for industries that normally require college-educated founders. They also differ strongly with the local concentration of existing establishments in the same sector, especially for industries serving non-local markets, suggesting that an important mechanism is the spillover of relevant knowledge.

JEL classification: R1; L80; J24; M13; O3

Keywords: Firm formation; Knowledge spillovers; Entrepreneurship; Human capital; Regional growth

1. Introduction

This study refines our earlier investigation of the impact of differences in local human capital resources on local differences in new firm formation rates (Armington and Acs [4]). Since the mid-eighties the role of education and human capital externalities has been recognized as a key variable in theories of economic growth. Models posited by Romer [28], Lucas [23] and Krugman [19] link such externalities within a geographically

* Corresponding author.
E-mail address: zacs@ubalt.edu (Z.J. Acs).

0094-1190/$ – see front matter © 2004 Elsevier Inc. All rights reserved.
doi:10.1016/j.jue.2004.03.008

bounded region to higher rates of growth. Lucas [22] emphasizes that the economies of metropolitan areas are a natural context in which to understand the mechanics of economic growth, and an important factor contributing to this growth is the catalytic role of human capital externalities within the cities. While the benefits of human capital to individuals have been extensively studied, economists are now realizing that individuals do not capture all of the benefits from their own human capital. Some benefits spill over to their colleagues and observers—through discussions, example, publications, and even more positive attitudes toward change, risk, and new knowledge.

Several interesting findings provide some groundwork for our study. First, Rauch [25] finds that cities with higher average levels of human capital also have higher wages and land rents. Second, Glaeser et al. [13] find that for a cross section of cities a key economic determinant of growth is level of schooling, just as has previously been found for countries. They suggest that higher education levels influence later growth, not through increased savings, but by promoting higher rates of growth of technology through spillovers. Finally, Simon and Nardinelli [29,30] find historical evidence for both the United States and the United Kingdom that cities with more knowledgeable people grow faster in the long run because (a) knowledge spillovers are geographically limited to the city and (b) knowledge is more productive in the city within which it is acquired.

However, none of these studies asks the question, "What type of activity do agents pursue that leads to faster economic growth?" This question is important because if we wish to explain how growth occurs we need to identify the transmission mechanism from human capital to growth. Jovanovic and Rob [16] develop a model where individual agents augment their knowledge through pairwise meetings at which they exchange ideas. In each time period, each individual seeking to augment his knowledge meets an agent chosen randomly from a distribution of agents. The higher the average level of human capital of the agents the more "luck" the agents will have with their meetings and the more rapid will be the diffusion and growth of knowledge. If this knowledge contributes to technical innovations, new products, processes, or markets, we have a microeconomic foundation not only for the impact of human capital externalities on total factor productivity, but also for making those external effects dependent on both the average level of human capital and the local intensity of individuals with relevant knowledge or examples to share.

This paper extends research reported in Armington and Acs [4] that focused on firm formation in six sectors: distributive, manufacturing, business services, extractive retail trade and local market. The current paper focuses on the rapidly growing service sector and subsectors of service industries that are defined by their educational requirements and primary markets. The authors were fortunate to have limited access, through the Center for Economic Studies of the US Bureau of the Census, to comprehensive US microdata on recent service firm formations, which they grouped into labor market areas for this analysis. These data were crucial for our primary goal—to examine how the sensitivity of firm formation rates to local differences in human capital and other local economic conditions varies as a function of the market segment and location and the entrepreneurial characteristics typical of various industry subsectors. We empirically investigate how the new firm formation rates for various subsectors of service industries are influenced by human capital differences in 394 labor market areas, while controlling for other regional characteristics that are also likely to affect firm formation rates. This analysis contributes to the regional growth literature, with its focus on human capital closely following much earlier work of Jacobs [14] and Marshall [24]. This paper also contributes to the recent cross-sectional literature that argues that new ideas are important for economic growth (Glaeser et al. [12,13]). The focus on new firm formation also contributes to the growing literature on entrepreneurship (Lazear [20], Krueger and Pischke [18], Evans and Jovanovic [10]).

Section 2 presents the data and discusses measurement of the new firm formation rate. Section 3 examines how and why the new firm formation rates in the service sector vary across geographic regions. Section 4 presents the empirical model, and the basic results for the service sector as a whole are in Section 5. Section 6 examines results for nine subsectors of the service sector and our conclusions are briefly discussed in the final section. We find that the extent of human capital already in a region has a significant effect on the new service firm formation rate. The service firm formation rate is even more sensitive to how densely populated (with establishments per thousand people) the local service sector already is. The greater this density is, the more probable are the relevant knowledge spillovers, and the more likely that the resulting new ideas will lead to new firm formations.

2. Measurement of new firm formation rate

2.1. The data

This study uses a new database that the Bureau of the Census has constructed for study of entry, survival, and growth in different types of businesses. The Longitudinal Establishment and Enterprise Microdata (LEEM) file has multiple years of annual data for every US private sector (non-farm) business with employees. The current LEEM file facilitates tracking employment, payroll, and firm affiliation and (employment) size for the more than eleven million establishments that existed at some time during 1989 through 1998. This database was constructed by the Bureau of the Census from its Statistics of US Business (SUSB) files,[1] which were developed from the microdata underlying the aggregate data published in Census' County Business Patterns. These annual data describing establishments were linked together using the SUSB Longitudinal Pointer File, which facilitates tracking establishments over time, even when they change ownership and identification numbers.

The basic unit of the LEEM data is a business establishment (location or plant). An establishment is a single physical location where business is conducted or where services or industrial operations are performed. The microdata describe each establishment for each year of its existence in terms of its employment, annual payroll, location (state, county, and metropolitan area), primary industry, and start year. Additional data for each establishment and year identify the firm (or enterprise) to which the establishment belongs, and the total employment of that firm.

[1] The SUSB data and their Longitudinal Pointer File were constructed by Census under contract to the Office of Advocacy of the US Small Business Administration. For documentation of the SUSB files, see US Small Business Administration [33].

A firm (enterprise or company) is the largest aggregation (across all industries) of business legal entities under common ownership or control. Establishments are owned by legal entities, which are typically corporations, partnerships, or sole proprietorships. Most firms are composed of only a single legal entity that operates a single establishment—their establishment data and firm data are identical, and they are referred to as "single unit" establishments or firms. The single unit businesses are frequently owner-operated. Only 4 percent of firms have more than one establishment, and they and their establishments are both described as multi-location or multi-unit.[2]

New firm formations include both new single-unit firms with less than 500 employees, and the primary locations of new multi-unit firms with less than 500 employees, firm wide. Those new firms that had 500 or more employees in their first year of activity appear to be primarily offshoots of existing companies.[3] Single unit firm formations in year t are identified on the LEEM as non-affiliated establishments with a start-year of t or $t-1$ that had no employment in March of year $t-1$, and had positive employment below 500 in March of year t. This avoids inclusion of either new firms that have not yet actually hired an employee, or firms recovering from temporary inactivity.[4] The start-year is the year that the establishment entered the Census business register. We have also included most of the relatively few multi-unit firms (1500 to 6000 per year) that appeared to start up with less than 500 employees in multiple locations in their first year. We limited multi-unit firm formations to those whose employment in their new primary location constituted at least a third of their total employment in the first year.[5] This rule effectively eliminated the 600 to 1000 new firms each year which were apparently set up to manage existing locations—relatively small new headquarters supervising large numbers of employees in mainly older branch locations which were newly acquired, or perhaps contributed by joint venture partners.

[2] The LEEM data cover all private sector businesses with employees, with the exception of those in agricultural production, railroads, and private households. This is the same universe that is covered in Census' annual County Business Patterns publications, but establishments with positive payroll during a year and no employment in March of that year are not counted for that year for this project.

[3] Annually, there were less than 150 such large apparent births of single-unit firms, with an average of about 1500 employees each. About a third of these larger single unit firms were employee-leasing firms or employment agencies, while the remainder were widely distributed across industries. However, examination of the new firms with 100–499 employees in their first year showed that most seemed to be credible startups, frequently in industries that are associated with large business units, such as hotels and hospitals. Since this study is not concerned with the employment impact of startups, there is no danger of the bulk of the data on smaller startups being swamped by that of a few larger startups that might actually be offshoots of existing businesses. Therefore, the startups with 100 to 499 employees were included, if they qualified otherwise.

[4] About 400,000 new firms generally appear in the business register (with some positive annual payroll) the year before they have any March employment, and we postpone their 'birth' until their first year of reported employment. An average of 90,000 older firms each year with employees in year $t-1$ have no employees in March of year t, but recover some employees the following year.

[5] We tested a similar rule using one-half, and found that the primary difference was in quite small multi-unit firms, where the smaller share was more credible for the first year.

2.2. *The unit of observation*

Within the United States, there are many levels of geographic units that have some economic data associated with them. Common politically defined units include states, counties, cities and towns. But such politically defined units have boundaries that rarely represent the borders of functional economic areas. The US Bureau of the Census has defined census tracts (areas of 3000 to 5000 residents) to facilitate collection of detailed data on where people live, and Metropolitan Statistical Areas (MSAs) for aggregation of politically defined urban units into more functional metropolitan areas. Most of the data collected for these measurement units are based on where people live, rather than where they work or shop.

Data for geographic units based on the location of business establishments (where people work) are needed for measuring the effect of location-specific economic growth, productivity, employment, and other economic factors. These are also collected primarily for various political units—particularly for states and counties. Although local government units (cities and towns) generally collect some economic data, these are rarely comparable across areas, because such data are frequently dependent on local tax laws. The city has the advantage of being a smaller geographic unit, within which there is reasonably integrated economic and social activity, which might be important for spillovers operating in dense areas. However, city boundaries are often quite arbitrary relative to the local patterns of economic activity, and their relatively small size allows them to be substantially influenced by conditions in neighboring political units.[6]

State and county level business data collected by the federal government are generally comparable across all the states, but most states are composed of multiple, diverse economic areas. Therefore analyses of economic data based on states as geographic units usually suffer from aggregation problems due to the diversity of economies with a state. On the other hand, many integrated local economic areas cross both state and county boundaries, and people and businesses flow freely back and forth across these boundaries, so the economic behavior of agents within a given state or county may be significantly affected by unmeasured influences from adjacent areas in other states or counties.

Metropolitan Statistical Areas (MSAs) are multi-county units that are defined to include all of the densely populated areas surrounding the larger cities. These geographic units do a better job of ensuring that people both live and work within their boundaries. However, they are based on residential population densities, without regard for where people work. In addition, they are periodically redefined to keep pace with changing urban population patterns, and they exclude large areas of the country whose local economies are not centered on large cities.

The geographic unit of analysis chosen for this study, *labor market areas* (LMAs), substantially avoids all of the problems associated with the units discussed above. LMAs are aggregations of the 3141 US counties into 394 geographical regions based on the predominant commuting patterns (journey-to-work). Each LMA contains at least one

[6] For example the city of Baltimore is smaller than the County of Baltimore, and many of the people that work in Baltimore city live outside the city limits. In addition, a large proportion of the people in adjacent counties work and shop in other parts of the urban agglomeration of which Baltimore is the center.

central city, along with the surrounding counties that constitute both its labor supply and its local consumer and business market.[7] Many of the 394 LMAs cut across state boundaries, because local economic activity patterns often cross such boundaries. The LMA unit of observation has the advantage of including both the employment location and the residence location of the population and labor force within the same area. Being based on counties, a wide variety of data collected at the county or zip-code level can be aggregated to construct LMA-level data. Finally, the 394 LMAs together cover the whole country, so that their data can be aggregated to US totals, and all areas are represented.[8]

2.3. The sector of inquiry

This paper focuses on the service sector of the US economy. Why do we feel that the service sector is preferable to manufacturing for analysis of new firm formation? First, there has been widespread concern among economists and policy makers alike about the dynamics of the service sector. The slowness of productivity growth in services, together with its rising share in nominal GNP and in employment, has been accused of exerting a major drag on productivity growth of the overall economy and its competitive performance. Second, the service sector has been growing much faster than other sectors, increasing its share of private employment from 28.3% in 1990 to 32.8% in 1998. Third, the broad range of firms in the service sector employ workers with a wide variety of skills, and tend to be more labor-intensive than capital-intensive, so that area differences in human capital may have a stronger impact on the service sector than on more capital-intensive sectors. Fourth, new firm formation rates are much higher in the service sector than in the manufacturing sector (Acs and Armington [2]). Indeed, cities with high concentrations of manufacturing have typically been the slowest growing cities over the past twenty years. Finally, much of the growth in service jobs has been in new firms. While some of these new firms merely replace older establishments that have closed, many others serve new markets, provide new services, or apply innovative techniques to compete with older businesses.

The local economic impact of formation of a new service firm is much broader than the immediate impact we can measure from the number of new jobs they create in the first year. New service firms may be providing the local market with services that were not previously available, or competing with existing providers to drive down prices or improve services. If their services are exportable, the new businesses may be generating income from outside the region, and perhaps contributing to a local specialized cluster that will

[7] These LMAs are defined according to the specification of Tolbert and Sizer [32] for the Department of Agriculture, using the Journey to Work data from the 1990 US Census of Population. They are named according to the largest place within them in 1990. Some LMAs incorporate more than one MSA, while others separate some of the larger MSAs into more than one LMA, depending on the commuter patterns. A few smaller independent (usually rural) Commuting Zones have been appended to adjacent LMAs so that each LMA had a minimum of 100,000 population in 1990. Alaska and Hawaii each are treated as a single integrated LMA. See Reynolds [26] for further discussion of LMAs.

[8] We code the location of each establishment according to its initially specified state and county in the LEEM, since we are focusing on formations. The few businesses that report operating statewide (county = 999), or are missing their county code, have been placed into the largest LMA in each state.

attract yet more businesses and employees. And of course the new firms will buy products and other services from local businesses.

2.4. The firm birth rate

In this paper we study gross new firm formation rates, not the net change in numbers of firms or establishments in an area. To the extent that the ultimate issue is explaining the relative growth of regions, the net change might be of interest, but that is for a much later study. What we are investigating is not the equilibrium result of new firm formation, but the factors accounting for differences in rates of new firm formation, a process that captures local differences in entrepreneurship, restructuring, innovation, industrial evolution and development. The factors contributing to explanations of local differences in firm deaths, plant entry and exit, all of which affect the net numbers of establishments, are far beyond the scope of this paper, and generally not strongly related to local human capital.

Firm birth rates are calculated for each of the 394 LMAs, based on new firm formations during each of three recent time periods—1990 through 1992, 1993 through 1995, and 1996 through 1998.[9] Because the labor market areas vary greatly in size, the absolute numbers of new firm formations must be standardized by some measure of the LMA size before it is meaningful to compare them across areas. When dealing with the whole service sector, firm formation rates are calculated as the number of new firms per thousand members of the labor force in the LMA in the prior year. This labor force approach has a particular theoretical appeal, in that it is based on the theory of entrepreneurial choice proposed by Evans and Jovanovic [10]. A worker starts each new business, and the labor market approach implicitly assumes that the entrepreneur starts the new business in the labor market where he or she lives and previously worked.

When comparing new firm formation rates for different subsectors of the service industry we need to standardize for the differences in sizes of both areas and subsectors. For this purpose we express new firm formation rates in terms of the number of new firms relative to the number of establishments already in existence in that subsector and LMA. This could be termed the ecological approach, because it considers the amount of start-up activity relative to the size of the existing population of businesses.

Two considerations of timing of the firm birth rate data should be noted. While new firms enter the business register underlying the LEEM file on a nearly continuous basis, their employment data are reported only for a pay period in March of each year. Since we require positive employment before recognizing a new firm, if a firm begins hiring after March, we do not count its formation until the following year. Therefore, each specified year's firm formation counts actually represent firms that hired their first employees sometime between April of the prior year and March of the specified year, for an average of nine months lagged reporting (Acs and Armington [1]). Further, Reynolds et al. [27] and others have shown that the time between an individual's decision to create a new firm

---

[9] In fact, birth rates were calculated for each annual period from 1990 through 1998, but these were found to be quite consistent in their rank ordering across LMAs, so the average of the three most recent years was used for most of this analysis. Using period averages serves both to smooth out irregularities and to minimize the possibility of disclosure problems with very small numbers of annual births for the smaller LMAs and subsectors.

and the start of the resulting economic activity averages about two years, and is often longer. With such lags in the initialization and reporting of new firm formations, we do not expect to be able to identify a lag structure between differences in their annual rates and the regional factors associated with these differences, even though we have nine years of annual data on new firm formations.

2.5. *Variations in regional new firm formation rates*

Table 1 shows annual variations in the numbers of new firm formations and the birth rate for service firms in the US. Gross service firm formations were increasing fairly steadily during the 1990s, to just under 200,000 in 1998, which accounts for nearly two fifths of all firm formations. Net service firm births, defined as annual firm formations minus firm deaths during the same year, average only about 25,000 during the 1990s, and vary widely. Net firm births in services accounted for about two thirds of the net firm births in all industries in most of these years (not shown). The rate of new service firm formation per thousand workers in the labor force increased from 1.375 in 1990 to 1.452 in 1997, and it fell slightly in 1992 and again in 1998. Services accounted for 35.1% of all firm formations in 1990, and increased their share to 38.4% of all firm formations in 1998. At the same time, employment in services increased from 28.3% of total to 32.8.

Table 2 looks at some of the regional variation across LMAs in the new firm formation rates, again using the number of new service firm formations per thousand workers in the labor force. Table 2a shows the twenty LMAs that have the highest average birth rates in the period of 1996 through 1998, as well as the twenty lowest. The top twenty LMAs ranked by birth rate had an average annual service firm formation rate of 2.26 per thousand of labor force, while the lower extreme averaged only a third as many new service firm formations, with 0.77 per thousand of labor force. Note that the list of LMAs with the highest birth rates appears to be almost evenly divided between very large LMAs and relatively small LMAs, but all the LMAs in the lowest birth rate group were relatively small.

Table 1
Gross formation and net births of service firms by year, 1990 to 1998, and service firm formation and employment as a share of all industry

Year	Service firm births		Service firm gross formation rates per thou. LF	Service share of all industry	
	Gross formation	Net births		Formation (%)	Employment (%)
1990	170,345	24,521	1.375	35.1	28.3
1991	173,475	24,928	1.378	37.1	29.2
1992	167,266	20,140	1.324	37.0	29.8
1993	174,884	28,546	1.365	36.7	30.6
1994	177,743	27,481	1.376	36.7	30.9
1995	186,050	33,220	1.421	36.9	31.1
1996	192,018	31,812	1.452	37.9	31.4
1997	194,916	19,936	1.452	39.0	32.1
1998	191,911	na	1.445	38.4	32 8

Notes. Gross formation includes all new service firm formations in each year Net births are the excess of firm formations over firm closures each year.

Table 2a
Regional variation in LMA sizes and service firm formation rates, 1996–1998 ranked by annual average service firm formations per thousand 1995 labor force

LMA	Largest place	State	Avg. 1996–1998 formation rate	Service firms '95	1995 Labor force	1995 Population
Top 20 LMAs, average of **2.26** service firm formations per year per 1000 labor force						
287	Laramie	WY	3.276	2250	90,242	157,260
71	West Palm Beach	FL	2.790	12,791	602,263	1,320,841
72	Cape Coral	FL	2.598	4845	251,563	555,042
70	Miami	FL	2.517	36,811	1,794,995	3,559,134
393	Bellingham	WA	2.362	2244	114,745	248,175
69	Sarasota	FL	2.316	5704	280,316	677,113
344	Bozeman	MT	2.280	2255	113,581	214,480
376	Reno	NV	2.260	4421	254,723	489,925
345	Missoula	MT	2.211	2398	126,036	255,454
91	Atlanta	GA	2.188	26,826	1,746,367	3,159,274
352	Grand Junction	CO	2.140	1628	92,686	180,242
289	Denver	CO	2.135	20,972	1,241,321	2,116,579
359	St. George	UT	2.105	1037	82,660	185,658
353	Farmington	NM	2.076	1137	73,850	145,934
354	Flagstaff	AZ	2.065	2173	139,112	288,115
379	Las Vegas	NV	2.060	7083	613,097	1,178,223
75	Daytona Beach	FL	2.025	3614	217,087	517,867
74	Orlando	FL	2.014	11,732	763,432	1,423,362
67	Tampa	FL	2.007	18,150	1,090,154	2,174,602
392	Bend	OR	1.991	1492	95,114	187,506
Bottom 20 LMAs, average of **0.77** service firm formations per year per 1000 labor force						
151	Lorain	OH	0.803	2505	211,001	417,376
139	Kokomo	IN	0.800	1153	95,821	183,584
133	Findlay	OH	0.794	1553	128,032	245,284
225	Appleton	WI	0.793	3497	316,960	518,380
224	Sheboygan	WI	0.785	1147	106,522	190,707
183	Watertown	NY	0.777	1248	105,549	257,062
227	Wausau	WI	0.773	2178	195,815	359,420
187	Sunbury	PA	0.773	1135	89,741	192,916
181	Elmira	NY	0.772	2149	167,177	350,349
128	Greensburg	IN	0.767	658	69,562	130,547
182	Olean	NY	0.767	1378	112,608	241,924
134	Lima	OH	0.764	1679	132,715	261,596
6	North Wilkesboro	NC	0.757	766	74,383	144,671
185	Amsterdam	NY	0.757	652	53,750	111,218
154	Zanesville	OH	0.756	1033	85,927	184,493
237	Galesburg	IL	0.734	878	70,347	147,675
219	Marshalltown	IA	0.708	725	59,299	110,541
178	Oneonta	NY	0.673	996	75,827	160,694
218	Mason City	IA	0.672	1156	81,392	150,274
126	Richmond	IN	0.662	713	55,891	105,835

Table 2b lists the LMAs with the largest and smallest populations in 1995. There is considerable variation in the birth rates of the large LMAs, varying from Miami FL with a birth rate of 2.52 new service firms per thousand of labor force down to Bridgeport CT with

Table 2b
Regional variation in LMA sizes and service firm formation rates, 1996–1998, largest and smallest labor market areas, ranked by 1995 population (formation rates in average annual formations per thousand 1995 labor force)

	Largest place	State	1995 Population	Services		
				Avg. 1996–1998 formation rate	Avg. 1995–1998 empl. change %	Services firms '95
Largest 15 LMAs						
383	Los Angeles	CA	15,273,490	1.61	4.91	109,555
194	New York	NY	10,974,248	1.85	3.75	93,034
243	Chicago IL	IL	7,687,064	1.45	3.91	58,924
113	Arlngtn. Wash. Balt.	VA	5,738,252	1.61	4.82	59,517
196	Newark	NJ	5,488,581	1.79	4.09	50,249
197	Philadelphia	PA	5,424,998	1.35	3.05	41,508
116	Detroit	MI	5,258,367	1.26	3.65	36,185
205	Boston	MA	4,727,659	1.53	3.21	40,779
378	San Francisco	CA	4,335,465	1.88	6.06	39,967
320	Houston	TX	4,007,275	1.61	5.44	28,919
70	Miami	FL	3,559,134	2.52	4.00	36,811
394	Seattle	WA	3,470,732	1.79	5.47	28,764
209	Bridgeport	CT	3,432,869	1.24	3.54	30,209
91	Atlanta	GA	3,159,274	2.19	6.78	26,826
331	Dallas	TX	2,861,201	1.88	7.44	23,953
	Average of 15 largest			1.67	4.68	
Smallest 15 LMAs						
148	Vincennes	IN	112,611	0.99	2.89	782
253	Union City	KY	112,257	0.90	3.91	651
283	North Platte	NE	111,929	1.11	2.24	1002
273	Fairmont	MN	111,436	0.95	1.09	977
185	Amsterdam	NY	111,218	0.76	1.82	652
219	Marshalltown	IA	110,541	0.71	−0.70	725
266	Aberdeen	SD	109,103	1.17	0.64	968
327	Brownwood	TX	107,861	1.20	3.79	751
126	Richmond	IN	105,835	0.66	−0.40	713
291	Salina	KS	105,237	1.14	1.48	920
258	Blytheville	AR	105,214	0.98	8.95	524
245	Fort Leonard Wood	MO	104,561	1.36	2.84	673
101	Thomasville	GA	104,131	1.19	8 29	641
324	Big Spring	TX	103,279	1.07	3.07	657
212	Hutchinson	MN	103,042	0.97	−5.04	742
	Average of 15 smallest			1.00	2.32	

only 1.24. These 15 largest LMAs had an average new firm formation rate of 1.67, with an average corresponding three-year increase in employment of 4.68 percent. At the same time, the smallest 15 LMAs averaged only 1.00 new service firm formations per thousand labor force, with only half the rate of growth in employment. This raises the question, which we will address later, of whether larger places typically have other characteristics which account for their higher service firm formation and higher growth rates, or whether it is the larger size of these economic areas that contribute to their higher average rates of new service firm formation and employment growth.

3. Why do birth rates vary across economic areas?

It is clear from the previous section that the service firm formation rates vary greatly across local economic areas. Recently a growing literature has sought the determinants of such local variation in rates of new firm formation, and has identified a number of factors that contribute to these differences. The agglomeration effects that contribute to new firm formation can come both from demand effects associated with increased local population, income, and business activity, and from supply factors related to the quality of the local labor market and business climate.

Among areas with broadly similar regional demand and business climate characteristics, there are further differences in rates of new firm formation and economic growth that are associated with the specific qualities of their human capital, and the propensity of locally available knowledge to spillover and stimulate innovative activity which culminates in new firm formations. First, highly educated populations provide the human capital embodied in their general and specific skills for implementing new ideas for creating new businesses (Glaeser et al. [12]). Second, they also create an environment rich in local knowledge spillovers, which support another mechanism by which new firm startups are initiated and sustained (Reynolds et al. [27]). Third, to the extent that new firms use skilled labor intensively, they are more likely to be located in cities with concentrations of highly educated labor in order to reduce the costs of hiring a crucial input (Rauch [25]). Thus, regions that are richer in educated people should have more start-up activity. Variation in local new firm formation rates should be positively related to local educational attainment rates. Furthermore, areas, which already have relatively intense development of service businesses, will have higher levels of new service firm formations, resulting in large part from spillovers of relevant specialized knowledge (Littunen [21]). We would expect that areas with relatively high shares of high-school dropouts would have lower rates of new firm formation.

Lazear [20] has contributed insights into one mechanism that contributes to the higher firm formation rates in larger cities, based on the presence of higher levels of individuals with a 'career' life-mode and a college education. Because their dominant value is the advancement of their career, although they are most likely to be working in large hierarchical private or public sector organizations, they will start their own businesses if this becomes the best way in which to benefit from their skills, knowledge and expertise. Individuals can be expected to choose self-employment only if $W(se) > W$. These businesses are often technologically advanced, innovative and with good marketing capabilities. Career mode entrepreneurs are often concentrated in large metropolitan areas and smaller attractive cities.

In fact, the 1990s saw an increase in the incidence of highly educated individuals starting new businesses, especially in the technologically advanced sectors of the economy, like computers, biotechnology, and internet-dependent businesses (Acs et al. [3]). However, there was also an increase in startups of many service businesses using relatively unskilled labor for services such as building cleaning, security, detective, and secretarial services. These may be started by career-oriented individuals who have recognized opportunities or developed new ideas to allow them to compete favorably in these markets, based on their own experiences or on spillovers from others.

New firm startups should be positively associated with higher levels of local human capital (including relevant knowledge spillovers):

$$\text{Firm Birth Rate}_{Lt+2} = \alpha_L + \beta \text{Human Capital}_{Lt} + \gamma [X]_{Lt} + e_{\text{L}} \tag{1}$$

where X is a vector of control variables, the subscript L indexes LMAs, t refers to time and e is stochastic disturbance. The conditioning information set is a vector of exogenous population and business variables specific to each labor market area L.

In a world of perfect information, employed agents confronted with new economic knowledge would not face a choice between developing the innovation as an employee within their existing firm, or taking the idea outside by starting up their own firm with it. However, the asymmetry of such knowledge leads to a host of agency problems spanning incentive structures, monitoring and transaction costs. The existence of such agency costs, and the resistance of bureaucracies to change, provides an incentive for agents with new ideas to form their own new firms. (The potential profit that might accrue to the entrepreneur in excess of his salary (if any) provides a further incentive to take the risks of self-employment.) And further, this same asymmetric nature of information causes the rate of new firm startups to vary from city to city, depending on the underlying knowledge conditions in each (Audretsch [6]).

Similarly, the equilibrium distribution of labor (people) and capital (firms) across cities was constantly changing during this period in which rapid changes in technology and relative demand for outputs caused some industries to become unprofitable and others more profitable (Jorgenson [15]). Across cities the local entrepreneurial response to these changes in supply and demand varied greatly, leading to variations in the firm formation rate, and in the proportion of entrepreneurs.

4. Empirical model

From the above discussion, it should be clear that the major hypotheses concerning the regional variation in firm formation rates deal with differences in levels of human capital and opportunities for spillovers, while controlling for local differences in a set of other regional characteristics which are likely to affect new firm formation rates. To test the basic hypothesis that the new firm formation rates are positively related to the level of human capital in a region, we estimate a regression model where the dependent variable is the average annual new service firm formation rate (dividing gross births by the labor force in thousands) for 1996–1998.[10] This is analogous to the method used by Keeble and Walker [17] and Armington and Acs [4]. The explanatory (independent or exogenous) variables include both the human capital variables discussed below, and the regional control factors discussed later.

[10] Although we have annual firm formation data for 1990 through 1999, we have chosen not to use pooled cross-section time series regressions, because most of the independent variables describing the characteristics of the LMAs change very little over time, and the errors from omitted variables will be nearly identical for each LMA from year to year, so the diagnostic statistics from such an analysis would be very misleading.

4.1. Human capital variables

To measure the level of human capital in each local economy we use two measures of educational attainment in each region, and a measure of the relative intensity of businesses in the same sector. The share of *college graduates* is defined as the number of adults with college degrees in 1990 divided by the total number of adults. This is a proxy measure that covers both technical skills needed in the economy, such as those of engineers and scientists, and skills needed to start and build a business, like finance and marketing and complex reasoning. In 1990, an average of 16 percent of the adult (at least 25 years old) population of the US had a college degree,[11] but this varied from a low of 6 percent to a high of 32 percent across LMAs. Its simple correlation with the new firm formation rates in LMAs is 0.29 and we expect it to be positively related to the birth rate, even after controlling for other important factors (Glaeser et al. [13], Rauch [25], Simon and Nardinelli [30]). Prior US empirical work has presented rather convincing evidence at the individual level that, ceteris paribus, educational attainment levels are positively associated with new business formation (Evans and Leighton [11] and Bates [7]).

The second measure of educational attainment that we use is the *high-school dropout* rate, defined as the percentage of adults (population 25 years or older) without college degrees who do not have high school degrees in 1990.[12] This high-school dropout rate should be a good proxy for the proportion of unskilled and semi-skilled labor in the LMA, and we expect it to be negatively related to the formation rate for most types of service firms. While many personal and business service activities do employ large numbers of high-school dropouts, few of the dropouts have the skills to start and manage a new firm themselves. In fact, the simple correlation between our high-school dropout rate and the new firm formation rate is −0.27. Nationally, 33 percent of non-college adults were high-school dropouts in 1990, and this varied from 17 to 60 percent across LMAs.

Formal education itself does not usually provide either the skills or the inspiration to start a new business. But higher education trains individuals to rationally assess information, and to seek new ideas. Therefore more educated people are more likely to acquire useful local knowledge spillovers from others who are involved in research or in managing some service business. The quantity or probability of potentially useful knowledge spillovers is expected to be a function of the number of similar business establishments, relative to the population of the economic area. *Service-industry intensity* is defined as the number of service establishments in the region divided by the region's population in thousands. The greater the number of establishments relative to the population, the more spillovers should be facilitated due to density of establishments in that industry (Ciccone and Hall [8]).

[11] This number has fallen considerably since then, but more recent data on educational attainment from the 2000 Census of Population had not yet been released at the county level, which is needed to construct the LMA level data.

[12] We used only the adults without college degrees as the base for this calculation, rather than all adults, in order to decrease the negative correlation of the high-school dropout rate with the share of adults with college degrees. College share of adults is correlated −0.70 with high-school dropout share of adults, and this falls to −0.59 when we calculate only the high-school dropout share of non-college adults.

4.2. Regional control variables

The human capital variables whose impact we are analyzing are not the only explanation for differences among LMAs in new firm formation rates. We control for differences in a number of other regional characteristics that are commonly thought to influence the rates at which new firms are formed. Summary statistics are provided in Table 3 for the new firm formation rates, and for all of the regional socio-economic variables that are discussed above and below.

Population growth is the average annual rate of change in the local population in a previous period (calculating the two-year change from the ratio of, for instance, for 1996–1998 firm formations, the 1995 population divided by 1993 population, and taking the square root of that two-year change ratio to calculate the compounded annual change ratio). Population growth may function as both a supply and a demand variable. A growing population increases the supply of potential founders of new businesses, and it increases the demand for consumer services. It captures the extent to which cities are relatively attractive to both migrants and immigrants, for living and for doing business. Population growth in a region stimulates growth in both the quantity and variety of businesses servicing that region's consumers. This growth usually takes place by a combination of expansion of existing businesses and creation of new businesses.

Income growth is the region's average annual rate of increase of personal income per capita in the prior two years, calculated using the same formula as for population. Income growth in excess of population growth captures local growth in labor productivity, and concomitant increases in local average quality of life. There are several mechanisms by which faster growing incomes might contribute to higher rates of new firm formation. Increasing per capita income is likely to increase disposable income, leading to greater

Table 3
Summary statistics on dependent and independent variables for regressions. Observations are 394 labor market areas, covering entire USA

Avg. ann. service firm formations per 1000 labor force	Mean	Std. Dev.	Minimum	Maximum
1996–1998	1.269	0.371	0.662	3.276
1993–1995	1.275	0.352	0.688	3.327
1990-1992	1.233	0.337	0.692	2.785
Independent variables	**Mean**	**Std. Dev.**	**Minimum**	**Maximum**
Human capital in 1990				
College degree/Adult population	0.159	0.050	0.069	0.320
High-school dropouts/Non-college adult popul.	0.329	0.082	0.167	0.598
Intensity of serv. estab./Population (000), 1995	7.620	1.400	3.755	15.548
Regional characteristics in prior period, e.g.				
Population growth ratio, 1993–1995 avg.	1.010	0.010	0.989	1.059
Per capita income growth ratio, 1993–1995 avg.	1.040	0.013	0.969	1.084
Log of population, 1995	12.801	0.940	11.543	16.542
Unemployment rate, 1994–1995 avg.	0.060	0.024	0.020	0.290
Avg. employment per estab., all-industry, 1994	15.097	2 881	8.266	21.237
Intensity of estab./Popul. (000), all-ind., 1994	21.834	3.584	10.774	45 105

demand for a wider range of income-elastic services. In addition, higher income growth rates may enable potential new business founders to raise local capital more easily at lower cost, thereby facilitating new firm formation. Higher growth rates of either population or per capita income during the preceding period are expected to promote higher firm formation rates (Reynolds [26]).

We control for agglomeration effects in each region primarily by including the log of population as a control variable, since we expect proportional differences in population to impact the new firm formation rates (rather than absolute value differences). Agglomeration effects are expected to have a positive impact on the start-up rate. Lucas [23] asserts that the only compelling reason for the existence of cities would be the presence of increasing returns to agglomeration of resources, which make these locations more productive. However, agglomeration effects may be more complex, and have effects that vary across different types of service subsectors.

The *unemployment rate* is calculated for the two-year period prior to our start-up measurement period—for example, for 1996–1998 formations we use the average number of unemployed in 1994 and 1995 divided by the labor force in 1994. Audretsch and Fritsch [5] and Armington and Acs [4] have found conflicting results for this variable—it is not clear whether, or when, the impact of local differences in unemployment rates is negative or positive. The simple correlation between the unemployment rate and the firm birth rate is close to zero and is not statistically significant. Use of the unemployment rate as a measure of local economic distress, or local business health, suggests that higher unemployment should be associated with fewer new firm formations. But a number of studies of new firm formation in the 1980s put heavy emphasis on the possible positive explanatory power of unemployment (Evans and Leighton [11], Storey [31]). Unemployment had then increased significantly in several countries and stayed at very high levels over an extended period. It was suggested that when workers were unemployed they might be more likely to start their own businesses. This activity, in turn, might reduce the unemployment rate, as the resulting new firms employ not only the owners, but also others. This effect of unemployment may dominate in the service industries, with its generally lower capital requirements.

All-industry intensity is the total number of private sector establishments in the region, divided by the region's population. Some prior studies have attempted to assess the potential for positive effects from spillovers using either the population density, or establishment density, measured as the number of units per square mile. But these are more indicative of physical crowding than of communication opportunities for knowledge spillovers. This general business intensity may also be thought of as the ratio of an area's business density (establishments per square mile) to its population density (people per square mile), which is a rough measure of the local pressure on business resources for which any new business must compete. Since we have already taken into consideration the local intensity of establishments in the service sector, we expect that the greater the intensity of all establishments, the higher the local costs of land and labor, and the lower the service firm formation rate will be (Acs et al. [3]). However, if spillovers from other industries are more important than those from similar industries (Glaeser [12]), this all-industry intensity might be positively related to formation rates of new service firms.

Establishment size is a proxy for the broad structure of business in the region. It is measured for 1996 to 1998 formations as the total area employment in 1994 in all industries, divided by the number of all-industry establishments in 1994 in the area. A local business structure with no dominant large firms may offer fewer barriers to entry of new firms and more opportunities for knowledge spillovers. Furthermore, where small firms predominate in a geographical area there is a much broader population of business owners, and more individuals may visualize their own careers as leading to the founding of independent new firms. Thus the average size of area establishments should be negatively related to the new firm formation rates, since larger average size indicates greater dominance by large firms or branch plants (Armington and Acs [4]). Because nearly all young businesses are small, and most large establishments are considerably older than average, differences in the average size of establishments may also be a proxy for the differences in the average age of local business establishments. In fact, over time a high rate of new firm formation will lead to lower average establishment size in an area.

Many of these control variables are correlated with each other, and some may be partially endogenous to others. Although all of the independent variables reflect a time period prior to the firm formations, most of them represent regional differences that are likely to persist over time. The general agglomeration variable, log of area population, is only correlated 0.35 with the formation rate, but it is correlated 0.61 with college share of adults and 0.51 with the average size of area establishments in all industries. Future differences in area rates of growth in population and income certainly result from current differences in area firm formation rates. Spatial differences in unemployment are influenced by local variations in industry mix, demographics (including educational attainment), and other relatively stable factors (such as local unemployment insurance regulations), in addition to the relative health of the local economy and tightness in the labor market, which we use it as an indicator for.[13] In fact, much of the economic geography literature today is concerned with cumulative growth mechanisms in which cause and effect appear to be simultaneous.[14] This study only attempts to sort out a few new details in this complex of interrelationships.

---

[13] The use of deviations from long-term averages of each area's unemployment rate in future work might facilitate isolating the long-term structural causes of local unemployment from shorter term variations, but both contribute to the spatial differences in the relative tightness in the labor market and the health of the local economy. In addition, the long and uncertain lags in the timing of new business formations (between the original formation decision and the registering of employees, which triggers recognition of the startup) preclude the usefulness of time series analysis until much more is understood about both the theory and the facts.

[14] We have also abstained from considering financial variables and regional knowledge factors such as research and development expenditures. The availability of adequate financial resources to fund new firms is an important determinant of new firm formation, which we hope to take into account in subsequent research. Both university-based and industrial research and development activity are probably important contributors to regional new firm start-up rates through spillovers.

5. Empirical results for all services together

Table 4 shows the results of least squares regression on the 1990–1992, 1993–1995 and 1996–1998 average annual firm formation rates for the service sector for 394 labor market areas. We present standardized beta coefficients,[15] so that each parameter indicates the sensitivity of birth rate variation to normalized variation in the corresponding independent variable. The t-ratios shown for each were calculated from the simple estimated standard errors. These were also calculated with a correction for heteroscedasticity, with results that were very similar to the uncorrected standard errors. The estimated coefficients are generally consistent with our expectations, but with several important exceptions. The explanatory and control variables together explain about two-thirds of the regional differences in new service firm formations rates.

Only two of the three human capital variables showed the hypothesized relationships. First, for human capital measured by share of college graduates, the coefficients are positive and statistically significant for all except the 1993–1995 periods, confirming that regions with higher shares of college-educated adults have higher firm formation rates. This positive result on human capital is consistent with previous research (Storey [31]). The 1993–1995 period was one of recovery from the short recession in 1991, which had resulted in a fall in service firm formations in 1992. It appears that the service firm formation rate is less sensitive to the areas' educational attainment levels during such a recovery period.

The positive and statistically significant coefficient for high-school dropouts as a share of the non-college adult population is at first surprising—however it is consistent with our earlier results for the whole economy (Armington and Acs [4]). There we suggested that after controlling for the proportion of adults with college degrees,[16] the additional effect of a greater share of less educated workers is to facilitate the start-up process by providing cheap labor for the new firms. Even the most sophisticated businesses need some relatively uneducated workers to do the manual labor. Thus, the relationship between educational attainment and new firm startups at the regional level may be U-shaped, with both low levels and high levels of education conducive to firm formation and growth. We will examine this issue in greater depth when we analyze subsector data for services formations, distinguishing by educational requirements for founders.

---

[15] These can be calculated from the ordinary coefficients, but it is more illuminating to view them as being estimated from standardized variables. In this case, rather than using the levels, ratios and percents whose means and deviations are shown in Table 3, we would transform each variable by subtracting its mean value (calculated from all 394 LMA values) and then divide this adjusted value by the standard deviation of all 394 values. These transformed values will have a mean of zero and a standard deviation of one, and each value represents the deviation of that particular LMA from the mean. Since the 394 LMAs constitute the universe at a point in time (rather than a sample of areas), it is apparent that the resulting standardized beta coefficients can be interpreted quite simply as measures of the impact of a standard deviation difference in the independent variable on the standardized dependent variable. For example, using standardized variables, if we estimate that $x = 0.1y + 0.5z$, then we can say that each standard deviation in the value of y is associated with 0.1 of a standard deviation of x, and each standard deviation of z is associated with half of a standard deviation of x. Obviously, it follows that x is five times more sensitive to z than to y.

[16] Note that when estimated for 1996–1998 without the high-school dropout rate, the coefficient for college degree falls to 0.10 and when estimated without college, the coefficient on high-school dropout falls to 0.12, while other coefficients remain substantially the same.

Table 4
Regression coefficients for service firm formation rates[a,b] in labor market areas during three consecutive time periods (standardized betas with t-ratios, significant at 0.05 or better, unless starred*)

	1996–1998	1993–1995	1990–1992
Adjusted R^2	0.718	0.658	0.625
Human capital			
College degrees/Adults '90	**0.16**	**0.10***	**0.19**
	3.39	1.79	3.40
High-school dropouts/ Non-college adults '90	**0.16**	**0.21**	**0.14**
	4.21	4.86	3.20
Intensity of service estab./Population	**0.63**	**0.60**	**0.47**
	6.53	5.63	4.26
Regional characteristics[b]			
Population growth	**0.51**	**0.46**	**0.41**
	18.05	14.44	11.46
Per capita income growth	**0.09**	**0.19**	**0.13**
	3.03	5.62	3.77
Population (log)	**0.22**	**0.16**	**0.18**
	4.92	3.25	3.6
Unemployment rate	**0.06***	**−0.09**	**0.17**
	1.64	−2.32	4.08
Avg. size of all estab. (employment)	**−0.34**	**−0.33**	**−0.32**
	−8.06	−8.13	−7.30
Intensity of all estab./Population	**−0.21**	**−0.07***	**−0.03***
	−2.42	−0.71	−0.31
No. of observations	394	394	394

[a] Formation rates are 3-year average formations per 1000 labor force in prior year.
[b] Undated exogenous variables represent prior year, or prior 2 year averages.

Thirdly, the coefficient on intensity of service establishments is positive and statistically significant, suggesting that regions that already a relatively strong supply of service establishments will have higher rates of new service firm formation, as predicted by the theory of regional spillovers (Jovanovic and Rob [16]). Indeed, this factor has the strongest relationship of any of our independent variables. The 0.63 value estimated for 1996–1998 for the standardized coefficient indicates that a locality with a service establishment intensity that is one standard deviation more intense than the mean will be likely to have firm start-up rates that are 0.63 of a standard deviation higher than the mean. When we tried replacing this measure of service establishment intensity with the share of employment in services, the estimates were much weaker, so we conclude that it is important that the local service sector have many business establishments, rather than many employees with service experience.

Furthermore, once we control for the intensity of service establishments, the additional intensity of all establishments is negatively related to service firm formation in 1996–1998, and insignificant in earlier periods. This suggests that startups are facilitated by spillovers from clusters of similar establishments, but that a relatively high intensity of other types of establishments may actually discourage new service firm formation. Business crowding, in general, apparently does not lead to higher rates of service firm startups. These results shed additional light on the debate between diversity and specialization (Glaeser [12]), supporting the view that spillovers have important positive effects within broad industry sectors, but do not play an important positive role across sectors. This finding is consistent with that of Acs et al. [3] who found no spillovers across unrelated industries. We could better distinguish the separate effects of the intensity of related and unrelated industries in the area, and avoid the inflation of the parameter on service intensity by its inclusion in all-industry intensity if we replaced the intensity of all establishments with the intensity of non-service establishments. This substitution was made in exploratory work on 1996–1998 formations, which showed that the parameter on service intensity fell somewhat, but not dramatically, and the estimates for the other parameters in the model remained similar. However, because it was not feasible to use this formulation consistently for the subsequent analysis of subsectors of services, and only limited results could be released in accord with Census confidentiality restrictions, we chose to use the broader all-industry intensity model consistently throughout this research.

While the results for the three time periods shown in Table 4 are broadly similar, there is one additional difference to be noted. The estimated coefficient on the unemployment rate is positive and statistically significant for 1990–1992, when the economy was undergoing a small recession, but it is negative and barely significant during 1993–1995 and insignificant during 1996–1998, suggesting that this positive effect disappears as the economy improves, or as mean unemployment falls. These results are inconsistent with some previous research (Storey [31]) that generally found a negative relationship between unemployment and startups in a cross-sectional analysis. Our results raise the possibility that during recessions more workers turn to entrepreneurship, as the competition for positions as employees is stiffer. Although higher relative unemployment rates were associated with higher relative service formation rates in the subsequent period, there is no evidence that the formerly unemployed workers were the ones starting the new businesses. Moreover, the service firm formation rate actually fell, nationally, during the 1991 recession (as measured in the year ending in March 1992). It may be that the unemployment rate was serving as a proxy for omitted variables in the previous research cited, but those effects were more precisely attributed to the additional variables we have controlled for in this study, robbing the unemployment variable of its apparent effect.

The signs on the other control variables are as expected. Local population growth differences had a very strong positive influence on new service firm formation rates. When local labor force growth was substituted for population growth its estimated parameter was much lower, suggesting that this local growth variable is functioning more as an indicator of growth in demand for services than as an indicator of the supply of either entrepreneurs or labor. Regions that have higher per capita income growth, and those with higher levels of agglomeration (of population) have higher rates of service firm formation. The average size of all local business establishments has a strong negative relationship to service firm formation rates—local dominance by large businesses appears to inhibit the formation of new businesses, while the presence of many smaller businesses may serve both to stimulate competition and to facilitate knowledge spillovers.

6. Subsectors within the service industry

The service sector defined by the Standard Industrial Classification (SIC) system incorporates a huge variety of diverse businesses. Our capacity to disaggregate this sector was severely limited by data disclosure constraints, which allowed it to be divided into no more than 9 subsectors. Our first priority was to better distinguish the relationship of our human capital variables to the start-up rates of various types of service activities, hypothesizing that an important aspect of this linkage is the supply of educationally qualified potential entrepreneurs. Therefore, our primary classification of the 150 4-digit industries was based on the educational requirements expected of the founders of most new firms in each industry code, using three categories for this dimension. The second important industry characteristic to control is its target market, so that we can better account for the effect of local differences in the demand for various types of services. We categorized the market segment served by each of the service industry codes using three categories. Together, these defined 9 service subsectors, within which the service activities are fairly homogeneous with respect to these two dimensions—educational requirement and market segment.

A major factor affecting the supply of new service firms is the availability of individuals with the qualifications generally needed to recognize the opportunities, identify new services, markets, or delivery systems, organize the new firm, and hire the first employees. We therefore expected that the sensitivity of service firm formation rates to the relative supply of adults with various levels of education would differ across service subsectors distinguished by typical educational requirement of their founders. We distinguish activities that are most frequently started by people who do not have college degrees (called 'high school' level for simplicity), from those generally requiring an 'advanced' (graduate, post-graduate, or professional) degree, and assigned the remainder to 'college.' These allocations were based on subjective judgments, using our general knowledge of service industries, supplemented by the detailed descriptions of the 4-digit SIC classes in the 1987 Standard Industrial Classification Manual [9].[17]

An obvious reason for variation across locales in their rates of new service firm startups is variation in local demand for services, so we distinguished three general markets—local consumers, local businesses, and non-local (broader regional, national, or export) markets. Each 4-digit SIC code was assigned to one of these market segment categories, based on close reading of the descriptions of the activities within the definition of the code. It was expected that a substantial portion of the variation in startups of local consumer service firms would be associated with differences in population growth. Similarly, it was

---

[17] We originally hoped to base this classification on the BLS occupational distribution data for each 3-digit industry group, and to use subjective judgment only to distinguish among the 4-digit codes within each 3-digit group. However, we found that many activities requiring academic skills or advanced training for leadership positions reported occupational distributions very heavily weighted toward semi-skilled and unskilled workers. Hospitals and hotels were extreme examples of this contrast between educational requirements for workers and those for the individual responsible for starting the business. Similarly, classification of self-employed workers by SIC was not at all representative of the qualifications of the owners or managers of new employee firms in that SIC. Many self-employed workers serve under contract to large firms, and few need to deal with the management or financial challenges of employee businesses.

expected that locales with more businesses (intensity relative to the population) would be associated with higher rates of local business service firm formations. New service firms serving a broader, non-local market should be considerably less sensitive to these local market differences. Thus, this dimension was expected to improve the control of local variation in demand for new service firms.

The resulting subsector classifications for each 4-digit SIC are listed in Appendix A, where they are ordered by SIC code within each subsector. Data on the number of establishments and employees in each 4-digit SIC in 1995 are included, so that it is easy to pick out the larger industry codes dominating each subsector. This Appendix also shows the net growth rates for numbers of establishments and their employment between 1995 and 1998, as well as the number of new firm formations during 1996 through 1998 per hundred (1995) establishments for each industry code.[18]

Table 5 provides a summary of the diverse firm formation rates and relative sizes (shares of total service employment) for these nine services-subsectors defined according to their market segments and founders' education requirements. Looking first at how the new firm formation rates differ by education requirement, note that they are quite similar for all three categories, ranging only from 8.33 for advanced degrees, up to 9.29 for those types of service industry businesses that are probably founded by individuals with only a high school education. But when we segment the service sector by primary market, we find that the firm formation rate for service businesses that focus on local consumer markets (which account for about 55 percent of employment in services) is only 7.18 new firms per hundred establishments in that market category. At the other extreme, the segment of services that caters to non-local markets was creating new firms at nearly twice that rate—12.66 new firms per hundred existing establishments—but it accounts for only 19 percent of employment in services.

For the nine subsectors defined by the education requirement and the market segment together, the firm formation rate was highest, at 14.78, for businesses in non-local markets with founders normally having advanced degrees. The largest industry groups in this subsector are engineering and management consulting and computer programming services, all of which are subject to rapid innovation and turnover. The subsector requiring the same advanced degree for founders, but serving the local consumer market, had only 5.31 new firms for each hundred existing establishments, and this subsector is dominated by medical offices and religious organizations. Businesses that normally require a college degree for their founder had formation rates that were quite similar across all three of the market segments. Businesses commonly founded by those with no more than a high school degree also showed great variation across market segments, with high formation rates for non-local market (primarily the hotel and motel group), and low ones for the local consumer market (including various repair, cleaning and beauty services and child day care).

[18] Appendix A entries do not sum to the national totals for each subsector because of the infrequent occurrence of establishments that were never classified to the 4-digit level. These were generally assigned to the four digit code that had the most establishments reported within the SIC classification provided, but are not included in the aggregate data in Appendix A.

Z.J. Acs, C. Armington / Journal of Urban Economics 56 (2004) 244–278 265

Table 5
1996–1998 birth rates and relative size of service subsectors defined by market segments and founder's education requirement

Education requirement and market segment	Average annual* firm births per 100 estab. in subsector	Share of services 1995 employment %
All services	8.84	100.0
All education classes		
Local business market	10.66	26.1
Local consumer market	7.18	54.9
Non-local markets	12.66	19.0
High school		
All markets	9.29	30.4
Local business	12.22	9.3
Local consumers	8.42	15.9
Non-local markets	7.86	5.2
College degree		
All markets	9.25	26.1
Local business	8.60	10.2
Local consumers	9.08	13.6
Non-local markets	10.72	2.3
Advanced degree		
All markets	8.33	43.5
Local business	10.31	6.6
Local consumers	5.31	25.4
Non-local markets	14.78	11.5

* The sum of firm births in 1996, 1997, and 1998 is divided by 3, and then divided by the number of establishments in 1995, and then multiplied by 100.

The first subsector regression model reported in Table 6 is a simple pooled regression on average new firm formation rates for 1996 through 1998, where each observation is a subsector in an LMA. Thus, there are 3546 observations, from each of the nine subsectors in each of the 394 LMAs. If we use subscript L to indicate LMA and subscripts E and M to indicate subsectors distinguished by Education and Market, we can specify this model as follows:

$$\begin{aligned}\text{Birth Rate}_{LEM} = f(&\text{Coll}_L, \text{HighSch Drop}_L, \text{Subsector Estab Intensity}_{LEM},\\ &\text{Pop Gro}_L, \text{Income Gro}_L, \text{Pop log}_L, \text{Unempl}_L, \text{Estab Size}_L,\\ &\text{All-Ind Estab Intensity}_L).\end{aligned} \tag{2}$$

Most coefficients fall somewhat, relative to the all-service model results shown in Table 4, suggesting that the independent variables are not equally important to all of the subsectors. The coefficient on the all-industry establishment intensity, which had been somewhat elevated as a result of some collinearity between the all-service-industry intensity and the all-industry intensity, falls substantially in this pooled subsector regression, because there is little multi-collinearity between the individual service sub-sector intensity levels and the all-industry intensity. Using the more detailed subsector

Table 6
Standardized* regression coefficients for service subsector firm formation rates** in labor market areas during 1996–1998, with subsectors defined by education requirement and market segment (betas with t-ratios, significant at 0.05 unless starred*)

		Pooled, with dummy-distinguished exogenous variables					
		Educational requirement			Market segment		
	Pooled 9 subsectors	High school	College degree	Advanced degree	Local business	Local consumer	Non-local
Adj. R^2	0.566			0.660			
Human capital							
College degree/Adults '90	**0.14**	**0.01***	**0.21**	**0.11**			
	7.47	0.41	8.09	4.02			
High-school dropout/ Non-college adults '90	**0.09**	**0.03***	**0.04***	**0.16**			
	5.77	1.26	1.91	6.85			
Intensity of subsector establishments/Popul.	**0.54**				**0.53**	**0.25**	**0.77**
	36.40				18.55	9.25	32.05
Regional characteristics							
Population growth	**0.36**				**0.35**	**0.44**	**0.20**
	32.09				19.43	22.83	11.19
Per capita income growth	**0.07**	**0.06**	**0.08**	**0.06**			
	5.96	3.35	4.44	3.28			
Population (log)	**0.16**	**−0.06**		**0.06**	**0.23**	**0.20**	
	9.18	−2.35		2.18	7.81	7.50	
Unemployment rate	**0.05**	**0.01***	**0.05**	**0.02***			
	3.23	0.47	2.54	1.15			
Avg. size of all estab. (employment)	**−0.26**				**−0.17**	**−0.47**	**−0.14**
	−16.04				−6.87	−19.79	−6.36
Intensity of all industry establishments/Popul.	**−0.06**				**−0.01***	**−0.02***	**−0.07**
	−3.59				−0.22	−0.74	−2.96
No. of obs.	3546			3546			

* All variables are expressed as deviations from mean values for all LMAs within each subsector.
** Subsector firm formation rates are 3-year average formations per 1000 labor force in 1995.

formation rates and subsector intensity rates also reduces the adjusted R-squared, because some of the additional variation in formation rates across subsectors is not as well explained.

Obviously, this simple pooled subsector model estimates only a single coefficient to represent an average of how all subsectors relate to each exogenous variable. But when we discussed the reasons for defining those subsectors, we focused on some expected differences in their coefficients with some of these variables. If we estimated each subsector model separately, we could not easily restrict the coefficients on the locality variables that should be unaffected by subsector differences. Alternatively, we could estimate the model separately for each of the dimensions—education and markets. But that fails to make use of the information we have on how these LMAs differ on both dimensions simultaneously, so the results would be subject to aggregation errors, which could be avoided by making use of both dimensions simultaneously.

In order to allow for variation in the estimated coefficients of variables that should be sensitive to our subsector dimensions, while controlling consistently for other regional characteristics, we expand the independent variables to be subsector-specific for the dimensions we want to test. Naturally, we expected the educational attainment variables to be sensitive to the education requirement dimension. We also wanted to investigate how the intensity of existing establishments in each subsector affected the rate of new firm formation in different market segments. We anticipated that subsectors that differed in education requirement might also differ in their relationship to income growth rates and unemployment rates. Market segment was expected to affect how the start-up rate varied with population growth, the average size of local establishments, and the intensity of all business establishments in the locality. Since little is known about the residual agglomeration effect that is represented by the logarithm of population, we did not try to anticipate whether it would be sensitive to either Market segment or the Educational requirement, and we therefore tested it with both dimensions.

Using the previous notation, this more detailed pooled estimation model has the following form:

$$\begin{aligned}\text{Birth Rate}_{LEM} = f(&\text{Coll}_L * \text{Dum}_E, \text{HighSch Drop}_L * \text{Dum}_E, \text{Subsector Estab}\\ &\text{Intensity}_L * \text{Dum}_M, \text{Pop Gro}_L * \text{Dum}_M, \text{Income Gro}_L * \text{Dum}_E,\\ &\text{Pop log}_L * \text{Dum}_E * \text{Dum}_M, \text{Unempl}_L * \text{Dum}_E, \text{Estab Size}_L *\\ &\text{Dum}_M, \text{All-ind Estab Intensity}_L * \text{Dum}_M). \qquad (3)\end{aligned}$$

Each of the exogenous variables is now in the form of cross products with dummies for the Education requirement or/and the Market segment. The three dummies for each dimension take the usual form of a dummy variable, with a value of zero unless the observation is for the segment specified for that dummy variable. We first standardized all of the nine exogenous variables and the endogenous variable to have a mean of zero and a standard deviation of one, within each of the nine subsectors. Therefore, each represents a relative measure for the LMA, within the subsector. Then we created dummy variables for each of the three values for each of the subsector dimensions—Market and Education. Finally, we multiplied each exogenous variable times the appropriate three dummies to create three specialized exogenous variables for each of the relevant dimensions of the subsectors. Thus the original 9 exogenous variables in the pooled subsector model expand to 30 variables—since log of population has been multiplied by each of 3 education dummies and each of 3 market dummies, and each of the other variables has been multiplied by each of 3 dummies of one type. This allows us to estimate the model across all subsectors simultaneously, while distinguishing among the dimensions we wanted to test for differences in estimated coefficients.

The results of the estimation of this model are shown in the last six columns of Table 6. Looking first at the human capital variables in this estimated model, we see that the relationship between an area's share of adults with college degrees and it service firm formation rate is stronger for the subsectors generally requiring a college education, but tiny and not significant for the formation rate of service businesses requiring only a high-school education for the founder. There is also a significant positive relationship between the share of adults with college degrees and the formation rates of service businesses normally requiring an advanced degree for the founder, since there is substantial similarity between the distribution of college degrees and that of advanced degrees.

The positive and statistically significant coefficient for the relationship of shares of high-school dropouts to formation of new service firms that require advanced degrees might suggest that such businesses are more dependent on having a large pool of unskilled labor. The statistically insignificant coefficients for the impact of the share of high-school dropouts on formation rates in the subsectors of services that require only high-school or college degrees suggests that such businesses are not as sensitive to the supply of unskilled labor. However, having found that higher shares of high-school dropouts are not strongly positively associated with higher formation rates for service businesses requiring less than college education for their founders, but for those requiring advanced degrees, we conclude that the explanation of the positive coefficients on high-school dropout share is a mystery that needs further focused research.

The relative intensity of establishments in the same subsector of services is a significant explanatory variable for all market segments, but the formation of new firms serving non-local markets is particularly sensitive to the prior existence of similar businesses. An area whose subsector intensity of services for non-local markets is one standard deviation above average will tend to have similar new firm formation rates 0.77 of a standard deviation above average. This corroborates the many prior case study analyses that addressed the spillover effects of certain rapidly growing local industry clusters (usually of high-technology firms with non-local markets), and suggests that these spillover effects are particularly important for businesses that are not focusing on local markets. However, there is also a strong clustering effect for local business services, and a smaller, but very significant one for local consumer services.

Most of the estimated coefficients for regional characteristics crossed with education or market dummies were similar to those estimated without such distinctions. However, the differences that appeared are quite illuminating. The population variable was crossed with all six dummies, since we did not have a clear concept of what the additional agglomeration effect was that was being captured by population, only that it was needed to prevent systematic underestimation of formation rates in large labor market areas. The estimated parameters on population for the subsectors of services requiring college degrees of their founders, and for those serving non-local markets were both under 0.02 and insignificant.[19] Perhaps the high coefficient on subsector intensity for non-local services has captured all of the relevant agglomeration effects for that subsector. In contrast, formation rates for services to local markets, both consumers and businesses, are sensitive to the size of the local economic area, even after controlling for both the local population growth rate and the local business intensity.

Distinguishing the impact of population on formation rates by the education requirement for founders, it appears that larger population contributes a bit to the formation rate of service firms requiring advanced degrees, but it slightly reduces the formation rate of

[19] These were later omitted, which had the effect of very slightly strengthening a few of the remaining estimates. Unfortunately, disclosure constraints prevented our showing both sets of results

firms normally started by high-school dropouts. This might be interpreted as additional evidence of the positive effects from greater volumes of knowledge spillovers for highly educated potential entrepreneurs, versus the negative effect of greater competition in larger markets for services provided by less educated entrepreneurs.

The coefficient on unemployment is positive and statistically significant only for service firms normally started by college graduates. This provides some clarification of the conflicting results found in previous studies of the effects of unemployment levels on new firm formation rates. Apparently, after controlling for regional differences in income growth rates, areas with higher unemployment tend to have higher new firm formation of services requiring founders with college degrees, but not those normally founded by high-school dropouts or those with advanced degrees.

Finally, the negative coefficient on average size of local businesses is strongest for formation of new firms serving local consumer markets, suggesting that areas dominated by large businesses are less likely to have a dynamic consumer service subsector. The coefficient on the intensity of all establishments is both tiny and insignificant for formations of firms serving either local businesses or local consumers, contrary to our expectations that a higher intensity of business establishments would lead to higher formation rates for business services. It has a significant negative impact only for formation of new firms serving non-local markets, and that impact is quite small.

These results suggest that the regional differences in new firm formation rates do indeed depend to a large degree on the educational requirements and the market served by the newly formed firms. In particular, the local levels of educational attainment impact primarily the firm formation rates of the types of firms that are normally founded by better educated entrepreneurs, and do not affect start-up rates for those normally founded by individuals with less than a college degree. While formation rates of all service businesses are higher in areas with higher intensities of similar service establishments, new formations of firms serving non-local markets are three times more sensitive to this than those serving local consumer markets, and those serving local business markets are twice as sensitive as those serving local consumers.

7. Conclusions

This paper has used a model of geographic variation in firm birth rates, focusing on their relationship to local human capital and the potential for knowledge spillovers from existing similar businesses. A key variable for the firm birth rate, as for economic growth, both within cities and within countries, is the educational attainment of the labor force. Although the actual knowledge acquired with a college degree seldom suffices as the basis for a successful new business, the analytical methods learned in college facilitate both future acquisition of knowledge and openness to new ideas received as spillovers from other activities in the area. Indeed, after controlling for basic differences in the underlying rates of population growth, the strongest factor accounting for differences in new firm formations was the local intensity of other similar businesses in the area. These results suggest that higher education influences later growth through the increased discovery and implementation of innovative ideas, resulting in more new firm formations.

In addition to the positive impact of higher proportions of adults with college degrees on rates of new firm formation, we also found an additional positive impact of higher proportions of high-school dropouts among the non-college-educated portion of the adult population. However, when we examined this for various subsectors of the service sector, we found it to be strong only for service activities that normally are started and managed by persons with advanced degrees. Therefore that coefficient appears to indicate a stronger tendency to start such businesses in areas that have relatively more unskilled (or cheap) labor resources.

Population growth was the most important of the regional characteristics used to control for other area differences that were likely to affect new firm formation rates. The unemployment rate appears to be unimportant except in years of recession, when higher unemployment rates contribute to higher firm formation rates.

Many of the most interesting explanations for the connection between growth and human capital levels across countries have focused on productive externalities generated by schooling. The potential for these externalities differs greatly across cities in the US, depending on both the levels of education of their work force, and on the strength of the presence of existing businesses in the same industry sector. It appears that an important mechanism by which these externalities contribute to economic growth in cities is through their impact on the level of entrepreneurship. And entrepreneurship provides the catalyst for increasing productivity, as well as increasing diversity and volume of goods and services produced in an area.

This research raises an important issue that should be followed up in future research. Do the economic and human capital characteristics of regions also influence the survival, growth and failure of new firms? In other words, do the same regional characteristics that promote entrepreneurial activity also promote faster growth rates and/or longer or greater survival rates for new businesses?

Acknowledgments

This research was supported by the NSF under Grant # SES-0080316, and was carried out at the Center for Economic Studies (CES), US Bureau of the Census, Washington, DC under the title, "Evaluation of New Service Firm Entries in the SSEL and Analysis of Regional Differences in their Entry Rates," working paper CES 03-06 (2003). All the data shown have been released for public disclosure, in compliance with the confidentiality procedures of the Census Bureau. Research results and conclusions expressed are those of the authors and do not necessarily indicate concurrence by the Bureau of the Census or the Center for Economic Studies. We would like to thank seminar participants at the Uddevalla Symposium 2003 in Uddevalla, Sweden, two anonymous referees and the editor of this journal for valuable comments. All errors and omissions are our responsibility.

Appendix A

Service industry standard industrial classification (4-digit SIC) codes and their subsectors, with 1995 establishment and employment and changes to 1998, and 1996–1998 firm formation rate per 100 establishments in subsector in 1995

Subsectors	Industry SIC	Establishments		Employment		1996–1998 Firm formation rate
		1995	1995–1998 chng. %	1995	1995–1998 chng. %	
Local business market, high-school education						
Linen supply	7213	1194	−4.7	46,950	3.0	4.33
Industrial launderers	7218	1297	4.2	62,821	10.6	2.78
Photocopying and duplicating services	7334	5163	8.0	72,374	17.9	7.72
Secretarial and court reporting	7338	6548	7.5	36,260	25.6	15.50
Disinfecting and pest control services	7342	10,165	2.7	78,782	9.8	7.85
Building cleaning & maintenance services	7349	45,098	6.3	794,517	10.3	16.41
Equipment rental and leasing, n.e.c.	7359	17,891	1.4	167,861	10.5	5.90
Detective & armored car services	7381	11,090	6.1	514,011	13.5	11.69
Security systems services	7382	2980	30.8	57,924	35.6	14.22
Business services, n.e.c.	7389	60,765	15.6	747,252	28.6	16.22
Truck rental and leasing, no drivers	7513	4140	12.5	36,950	34.1	3.40
Refrigeration and air-conditioning services	7623	3557	1.2	26,978	9.7	7.77
Electrical and electronic repair shops	7629	9412	−4.0	64,051	3.9	7.50
Welding repair	7692	5857	0.4	35,066	12.6	8.56
Armature rewinding shops	7694	2335	−6.5	24,459	−3.4	2.56
Local consumer market, high-school education						
Rooming and boarding houses	7021	1523	2.0	10,915	0.0	10.11
Organization hotels & lodging houses	7041	2090	−8.1	14,048	2.9	2.87
Power laundries, family & commercial	7211	1680	3.5	24,998	−4.7	9.62
Garment pressing, & agents for laundries	7212	3061	−0.5	14,198	6.9	8.34
Coin-operated laundries and cleaning	7215	12,473	−0.6	53,307	5.0	9.92
Drycleaning plants, except rug cleaning	7216	20,734	−3.6	165,597	−1.8	6.44
Carpet and upholstery cleaning	7217	7499	6.0	41,019	10.3	15.16
Laundry and garment services, n.e.c.	7219	3069	−3.7	19,159	−5.0	10.01
Photographic studios, portrait	7221	11,628	8.1	71,151	−7.3	7.33
Beauty shops	7231	73,386	−1.3	390,050	4.0	9.21

(*continued on next page*)

Appendix A (*Continued*)

Subsectors	Industry SIC	Establishments		Employment		1996–1998 Firm formation rate
		1995	1995–1998 chng. %	1995	1995–1998 chng. %	
Barber shops	7241	4444	−6.6	15,744	−2.9	6.71
Shoe repair shops & shoeshine parlors	7251	2194	−16.6	6654	−13.4	7.06
Miscellaneous personal services, nec.	7299	16,214	9.9	110,545	11.2	16.83
Medical equipment rental and leasing	7352	2570	6.4	28,256	2.2	7.43
Photofinishing laboratories	7384	6675	−13.4	71,974	−0.7	4.77
Utility trailer and recreational vehicle	7519	661	−2.4	4087	5.1	7.06
Automobile parking	7521	8370	6.2	56,590	13.4	2.22
Top & body repair & paint shops	7532	32,403	1.4	186,647	10.7	7.98
Auto exhaust system repair shops	7533	5203	0.2	26,152	−1.6	4.95
Tire retreading and repair shops	7534	2071	0.5	17,633	−5.4	8.59
Automotive glass replacement shops	7536	4092	20.7	22,346	28.8	8.17
Automotive transmission repair shops	7537	5912	2.0	27,227	9.0	8.56
General automotive repair shops	7538	67,205	5.0	288,119	8.9	10.22
Automotive repair shops, nec.	7539	9615	−4.4	45,270	4.0	5.49
Carwashes	7542	11,290	5.5	113,585	6.8	14.45
Automotive services, nec.	7549	10,818	20.0	83,063	22.9	13.54
Radio and television repair shops	7622	5446	−5.5	31,319	4.9	7.63
Watch, clock, and jewelry repair	7631	1706	1.3	6555	4.1	9.30
Reupholstery and furniture repair	7641	6282	−4.6	24,452	4.1	9.65
Repair services, nec.	7699	34,028	0.6	233,730	6.4	8.70
Motion picture theaters, except drive-in	7832	5610	−4.2	107,422	17.7	2.83
Drive-in motion picture theaters	7833	361	−24.7	3263	−2.1	3.05
Video tape rental	7841	18,707	3.0	154,980	−1.0	6.65
Bowling centers	7933	5608	−7.6	93,357	−7.2	4.10
Sports clubs, managers & promoters	7941	1279	17.6	38,423	19.9	18.11
Racing, including track operators	7948	2437	−1.1	53,692	2.3	12.67
Coin-operated amusement devices	7993	4324	0.3	41,271	29.2	8.33
Amusement parks	7996	771	1.9	72,033	26.1	8.95
Membership sports & recreation clubs	7997	11,751	1.9	261,628	4.5	4.50
Amusement and recreation, nec.	7999	20,840	5.2	357,092	18.3	11.97

(*continued on next page*)

Appendix A (*Continued*)

Subsectors	Industry SIC	Establishments 1995	Establishments 1995–1998 chng. %	Employment 1995	Employment 1995–1998 chng. %	1996–1998 Firm formation rate
Child day care services	8351	49,193	9.2	519,021	13.3	9.66
Residential care	8361	27,495	11.8	537,332	14.3	6.94
Civic, social, and fraternal associations	8641	42,371	−4.1	393,030	2.1	3.00
National market, high-school education						
Hotels and motels	7011	40,179	7.2	1,450,076	4.8	8.22
Sporting and recreational camps	7032	2277	1.7	16,068	12.0	4.98
Recreational vehicle parks & campsites	7033	2778	3.2	16,753	1.0	8.70
Heavy construction equipment rental	7353	3743	17.3	44,202	27.4	7.51
Passenger car rental	7514	3950	−3.9	84,792	21.5	5.64
Local business market, advanced education						
Computer related services, n.e.c.	7379	14,951	76.2	113,906	94.1	39.08
Legal services	8111	151,358	1.9	949,165	3.5	7.68
Accounting, auditing, & bookkeeping serv.	8721	76,299	0.7	553,725	19.2	7.86
Management services	8741	23,077	24.5	429,774	23.8	17.01
Local consumer market, advanced education						
Offices & clinics of medical doctors	8011	183,532	−2.6	1,559,081	9.3	5.10
Offices and clinics of dentists	8021	106,936	2.0	613,709	8.2	4.16
Offices and clinics of doctors of osteop.	8031	7038	−3.7	44,361	2.2	7.88
Offices and clinics of chiropractors	8041	27,009	1.3	88,417	3.6	8.27
Offices and clinics of optometrists	8042	16,151	1.6	74,213	10.1	5.77
Offices and clinics of podiatrists	8043	7574	−1.9	30,081	1.4	5.20
Offices and clinics of health practitioners	8049	22,644	12.7	146,088	28.2	13.54
General medical and surgical hospitals	8062	4382	−0.5	2,919,713	1.9	1.16
Psychiatric hospitals	8063	603	−6.1	90,289	−14.0	2.71
Medical laboratories	8071	7501	8.0	137,977	−1.7	7.43
Specialty outpatient facilities, n.e.c.	8093	10,171	32.6	224,583	18.5	8.28
Elementary and secondary schools	8211	15,158	6.0	550,225	10.6	4.09
Junior colleges and technical institutes	8222	555	−3.8	56,071	0.4	2.88
Religious organizations	8661	148,451	4.3	1,308,329	10.6	4.48

(*continued on next page*)

Appendix A (*Continued*)

Subsectors	Industry SIC	Establishments		Employment		1996–1998 Firm formation rate
		1995	1995–1998 chng. %	1995	1995–1998 chng. %	
National market, advanced education						
Advertising agencies	7311	13,170	3.1	144,935	15.4	10.93
Computer programming services	7371	20,190	47.3	301,811	49.4	25.60
Prepackaged software	7372	5298	41.6	121,341	43.7	19.53
Computer integrated systems design	7373	5310	29.9	101,882	45 5	17.06
News syndicates	7383	485	2.1	8738	15.3	5.02
Botanical and zoological gardens	8031	416	10.6	13,885	19.0	8.65
Specialty hospitals, except psychiatric	8069	621	12.4	203,385	2.6	3.81
Colleges, universities, & prof. schools	8221	2312	7.4	909,798	7.6	3.55
Museums and art galleries	8412	3450	9.6	60,720	11.6	6.44
Professional membership organizations	8621	5778	7.3	58,331	9.7	5.98
Engineering services	8711	38,924	12.8	651,725	17.9	10 60
Architectural services	8712	17,304	7.6	139,428	20.6	9.67
Commercial physical research	8731	5152	17.0	159,564	11.9	14.03
Comm. econ., socio. & educ. research	8732	5114	3.3	117,740	18.5	7.80
Noncommercial research organizations	8733	3216	5.0	82,687	11.4	7.51
Management consulting services	8742	36,378	24.8	356,324	41.4	20.01
Business consulting, n.e.c.	8748	15,458	20.1	123,091	15.4	19.11
Local business market, university education						
Advertising, n.e.c.	7319	2083	17.3	39, 270	25.8	16.07
Commercial photography	7335	3615	−1.5	21, 028	−11.6	8.47
Commercial art and graphic design	7336	11,689	11.0	65, 727	26.4	14.90
Employment agencies	7361	12.314	16.2	382,453	36.0	14.32
Help supply services	7363	17,842	33.9	1,974,710	41.3	10 38
Computer processing & data preparation	7374	7364	14.4	228, 356	18.2	6.81
Computer facilities management	7376	646	31.4	25,674	15.0	12.02
Computer rental and leasing	7377	843	3.4	12,005	16.3	7.55

(*continued on next page*)

Appendix A (*Continued*)

Subsectors	Industry SIC	Establishments		Employment		1996–1998 Firm formation rate
		1995	1995–1998 chng. %	1995	1995–1998 chng. %	
Computer maintenance and repair	7378	4440	1.1	50,987	14.3	12.45
Passenger car Leasing	7515	999	−11.6	11,145	3.5	6.37
Business associations	8611	13,922	0.3	103,424	−3.2	3.75
Labor unions & similar labor organizations	8631	18,159	−6.3	159,167	−1.0	2.08
Surveying services	8713	8503	3.4	53,121	17.0	7.69
Facilities support management services	8744	881	77.2	58,032	10.9	10.18
Local consumer market, university education						
Funeral service and crematories	7261	15,291	1.8	98,423	6.9	2.74
Tax return preparation services	7291	7990	16.8	144,908	−0.6	13.71
Dance studios, schools, and halls	7911	4998	2.9	26,792	11.6	10.21
Theatrical producers and services	7922	5769	10.8	82,003	17.0	13.30
Entertainers & entertainment groups	7929	6143	10.4	69,524	6.6	14.52
Physical fitness facilities	7991	8644	11.2	152,430	19.7	15.93
Public golf courses	7992	3544	16.3	55,261	18.6	8.97
Skilled nursing care facilities	8051	10,546	−12.5	1,115,205	−7.2	0.37
Intermediate care facilities	8052	4452	−12.8	229,765	−9.1	0.26
Nursing and personal care facilities, n.e.c.	8059	2507	−18.9	124,619	−13.6	0.93
Home health care services	8082	11,615	33.3	607,283	16.3	14.74
Kidney dialysis centers	8092	1450	46.0	34,021	25.8	3.66
Health and allied services, n.e.c.	8099	4977	−8.3	99,201	5.5	5.11
Libraries	8231	2078	5.6	20,294	16.2	4.51
Data processing schools	8243	1423	46.2	15,959	58.0	26.68
Business and secretarial schools	8244	651	−12.7	16,251	−8.5	6.45
Vocational schools, n.e.c.	8249	3156	16.5	48,200	20.0	12.53
Schools and educational services, n.e.c.	8299	13,982	15.7	136,885	20.6	14.21
Individual and family services	8322	35,606	19.3	516,329	19.3	8.89
Job training and vocational rehabilitation	8331	7570	3.1	292,757	3.9	3.85
Social services, n.e.c.	8399	16,283	6.9	231,952	1.6	6.16
Political organizations	8651	1539	16.8	7638	24.6	23.93
Membership organizations, n.e.c.	8699	9338	−0.4	77,261	12.5	3.32

(*continued on next page*)

Appendix A (*Continued*)

Subsectors	Industry SIC	Establishments		Employment		1996–1998 Firm formation rate
		1995	1995–1998 chng. %	1995	1995–1998 chng. %	
National market, university education						
Outdoor advertising services	7312	1172	10.3	12,023	7.5	11.77
Radio, TV, publisher representatives	7313	1874	21.0	20,938	40.9	13.73
Adjustment and collection services	7322	5037	−5.5	75,075	14.7	7.41
Credit reporting services	7323	1718	−10.5	29,132	28.8	5.67
Direct mail advertising services	7331	4024	−4.3	83,890	8.0	8.37
Information retrieval services	7375	1082	257.7	25,183	188.9	101.97
Motion picture & video production	7812	7622	7.8	70,992	9.6	14.82
Services allied to motion pictures	7819	2807	25.0	37,321	53.3	15.88
Motion picture & tape distribution	7822	1041	−4.6	18,480	81.4	8.07
Motion picture distribution services	7829	159	−8.8	1095	4.7	6.08
Dental laboratories	8072	7080	−1.7	41,473	2.1	6.13
Testing laboratories	8734	4603	5.0	71,049	11.5	6.75
Public relations services	8743	5037	10.2	38,148	25.6	12.92
Services, n.e.c.	8999	22,167	−2.5	168,746	3.7	7.95

Note. All changes are calculated as the 3-year difference (or the sum of 3 years of firm births), divided by the 1995 base.

References

[1] Z.J. Acs, C. Armington, Longitudinal Establishment and Enterprise Microdata (LEEM) documentation, Center for Economic Studies, US Bureau of the Census, Washington DC, CES 98-9, 1998.
[2] Z.J. Acs, C. Armington, Economic growth and entrepreneurial activity in cities, Regional Studies, in press.
[3] Z.J. Acs, F. FitzRoy, I. Smith, High technology employment and R&D in cities: heterogeneity vs. specialization, Annals of Regional Science 36 (2002) 373–386.
[4] C. Armington, Z.J. Acs, The determinants of regional variation in new firm formation, Regional Studies 36 (2002) 33–45.
[5] D.B. Audretsch, M. Fritsch, The geography of firm births in Germany, Regional Studies 28 (1994) 359–365.
[6] D.B. Audretsch, Innovation, growth and survival, International Journal of Industrial Organizations 13 (1995) 441–458.
[7] T. Bates, Commercial bank financing of white and black owned small business startups, Quarterly Review of Economics and Business 13 (1991) 64–80.
[8] C. Ciccone, R.E. Hall, Productivity and the density of economic activity, American Economic Review 86 (1996) 54–70.
[9] Executive Office of the President, Standard Industrial Classification Manual, Office of Management and Budget, NTIS order No. 87-100012, 1987.
[10] D. Evans, B. Jovanovic, Estimates of a model of entrepreneurial choice under liquidity constraints, Journal of Political Economy 95 (1989) 657–674.
[11] D. Evans, L.S. Leighton, Small business formation by unemployed and employed workers, Small Business Economics 2 (1990) 319–330.
[12] E.L. Glaeser, H. Kallal, J.A. Scheinkman, A. Shleifer, Growth in cities, Journal of Political Economy 100 (1992) 1126–1152.
[13] E.L. Glaeser, J.A. Scheinkman, A. Shleifer, Economic growth in a cross section of cities, Journal of Monetary Economy 36 (1995) 117–143.
[14] J. Jacobs, The Economy of Cities, Random House, New York, 1969.
[15] D. Jorgenson, Information technology and the US economy, American Economic Review 91 (2001) 1–32.
[16] B. Jovanovic, R. Rob, The growth and diffusion of knowledge, Review of Economic Studies 56 (1989) 569–582.
[17] D. Keeble, S. Walker, New firms, small firms and dead firms: spatial patterns and determinants in the United Kingdom, Regional Studies 28 (1994) 411–442.
[18] A.B. Krueger, Jorn-Steffen Pischke, Observations and conjectures on the US employment miracle, National Bureau of Economic Research, 1997.
[19] P. Krugman, Increasing returns and economic geography, Journal of Political Economy 99 (1991) 483–499.
[20] E.P. Lazear, Entrepreneurship, National Bureau of Economic Research, 2002.
[21] H. Littunen, Networks and local environmental characteristics in the survival of new firms, Small Business Economics 15 (2000) 59–71.
[22] R.E. Lucas, On the mechanics of economic development, Journal of Monetary Economics 22 (1988) 3–42.
[23] R.E. Lucas, Making a miracle, Econometrica 61 (1993) 251–272.
[24] A. Marshall, Principles of Economics, 8th Edition, Macmillan, London, 1890.
[25] J. Rauch, Productivity gains from geographic concentration of human capital: evidence from the cities, Journal of Urban Economics 34 (1993) 380–400.
[26] P.D. Reynolds, Autonomous firm dynamics and economic growth in the United States, 1986–1990, Regional Studies 28 (1994) 429–442.
[27] P.D. Reynolds, B. Miller, W.R. Maki, Explaining regional variation in business births and deaths: US 1976–1988, Small Business Economics 7 (1995) 387–407.
[28] P. Romer, Increasing returns and long run economic growth, Journal of Political Economy 94 (1986) 1002–1037.
[29] C. Simon, C. Nardinelli, The talk of the town: human capital, information and the growth of English cities, 1861–1961, Explorations in Economic History 33 (1996) 384–413.
[30] C. Simon, C. Nardinelli, Human capital and the rise of American cities, 1900–1990, Regional Science and Urban Economics 32 (2002) 59–96.

[31] D.J. Storey, The birth of new firms—Does unemployment matter? A review of the evidence, Small Business Economics 3 (1991) 167–178.

[32] C.M. Tolbert, M. Sizer, US commuting zones and labor market areas: a 1990 update, Staff paper No. AGES-9614, Rural Economy Division, Economic Research Service, US Department of Agriculture, Washington, DC, 1996.

[33] US Small Business Administration, Statistics of US Business—Microdata and Tables of SBA/Census Data, Office of Advocacy, Washington, DC, 1999.

[21]

Regional Studies, Vol 38 8, pp 879–891, November 2004

Carfax Publishing
Taylor & Francis Group

Creativity and Entrepreneurship: A Regional Analysis of New Firm Formation

SAM YOUL LEE*, RICHARD FLORIDA* and ZOLTAN J. ACS†‡

**Korea Information Strategy Development Institute, Juam-Dong, Kwachun, Kyunggi-Do 427-710, Korea and H. John Heinz III School of Public Policy and Management, Carnegie Mellon University, 5000 Forbes Avenue, Pittsburgh, PA 15213-3890, USA. Emails: samyoul@andrew.cmu.edu; florida@andrew.cmu.edu*

†*Merrick School of Business, University of Baltimore, 1420 North Charles Street, BC 491, Baltimore, MD 21201, USA. Email: zacs@ubalt.edu*

‡*Max Planck Institute for Research into Economic Systems, Jena D-07745, Germany*

(Received December 2003: in revised form April 2004)

LEE S. Y., FLORIDA R. and ACS Z. J. (2004) Creativity and entrepreneurship: a regional analysis of new firm formation, *Regional Studies* **38**, 879–891. Understanding the factors that promote or mitigate new firm birth is crucial to regional economic development efforts, since a high level of new firm creation significantly contributes to regional economic vitality and is a major signal of a dynamic economy. The literature suggests that various factors such as unemployment, population density/growth, industrial structure, human capital, the availability of financing and entrepreneurial characteristics significantly influence regional variation in new firm birth rates. This study explores whether connections exist among regional social characteristics, human capital and new firm formation. It argues that social diversity and creativity have a positive relationship with new firm formation. Building on the contributions of urbanist Jane Jacobs, Lee, Florida and Gates (2002) showed that social diversity and human capital have positive and significant relationships with regional innovation production measured by per capita patent production. While it is well known that regional human capital stock positively affects new firm formation rates, little attention has been paid to the interaction among social diversity, human capital and entrepreneurship. It is argued that low barriers of entry into the regional labour market (as exhibited in part by the presence of a diverse population) and diverse culture facilitate the influx of a particular kind of human capital that promotes innovation and accelerates information flow, leading to the higher rate of new firm formation. The empirical results support the main hypothesis. By using Longitudinal Establishment and Enterprise Microdata (LEEM), the hypothesis is tested at the Metropolitan Statistical Areas (MSAs) level as well as at the Labor Market Areas (LMAs) level. New firm formation is strongly associated with cultural creativity when controlled for the variables suggested in the literature. Firm formation is positively and significantly associated with the Diversity Index but insignificantly with the Melting Pot Index. The results suggest that one needs to pay attention to the social habitat of a region to boost a regional entrepreneurial dynamics.

Creativity   New firm formation   Entrepreneurship

LEE S. Y., FLORIDA R. et ACS Z. J. (2004) La créativité et l'esprit d'entreprise: une analyse régionale de la création d'entreprise, *Regional Studies* **38**, 879–891. Comprendre les facteurs favorables ou défavorables à la création d'entreprise joue un rôle capital dans le développement économique régional, parce qu'un taux de création élevé contribue de façon très significative à la vitalité économique régionale et constitue un clignotant majeur d'une économie dynamique. La documentation laisse supposer que divers facteurs, tels le chômage, la densité/la croissance de la population, la structure industrielle, le capital humain, la disponibilité du financement, et les caractéristiques de l'esprit d'entreprise influencent sensiblement la variation régionale des taux de création d'entreprise. Cette étude cherche à examiner si, oui ou non, on peut établir une corrélation entre des caractéristiques sociales régionales, le capital humain, et la création d'entreprise. On soutient que la diversité sociale et la créativité sont en corrélation étroite et significative avec la création d'entreprise. Le développement des contributions de l'urbaniste Jane Jacobs, Lee, Florida et Gates (2002) a démontré que la diversité sociale et le capital humain sont en corrélation étroite et significative avec l'innovation régionale, mesurée en termes du nombre de brevets détenus par tête. Alors qu'il est reçu que le stock du capital humain influence de façon positive les taux de création d'entreprise, on prête peu d'attention à l'interaction entre la diversité sociale, le capital humain, et l'esprit d'entreprise. On soutient que les barrières à l'insertion sur le marché du travail régional peu élevées (ce qui laisse supposer jusqu'à un certain point la présence d'une population diverse) et une culture diverse facilitent l'afflux d'un capital humain particulièrement propice à l'innovation et qui accélère le flux d'information, ce qui amène à un taux de création d'entreprise plus élevé. Les résulats empiriques viennent à l'appui de l'hypothèse principal avancé. A partir des données longitudinales et microéconomiques sur les entreprises (Longitudinal Establishment and Enterprise Microdata – LEEM), on cherche à tester l'hypothèse sur le plan métropolitain (Metropolitan Statistical Areas – MSAs) ainsi qu'au niveau des marchés du travail locaux (Labour Market Areas – LMAs). Il s'avère que la création d'entreprise est en corrélation étroite avec la créativité culturelle, une fois contrôlée pour les variables proposées dans

0034-3404 print/1360-0591 online/04/080879-13 ©2004 Regional Studies Association   DOI 10 1080/0034340042000280910
http //www regional-studies-assoc ac uk

la documentation. La création d'entreprise est en corrélation étroite et significative avec l'indice Diversité mais en corrélation faible avec l'indice Melting-pot. Les résultats semblent indiquer qu'il faut faire attention à l'habitat social d'une région pour donner de l'impulsion à la dynamique régionale entrepreneuriale.

Créativité Création d'entreprise Esprit d'entreprise

LEE S. Y., FLORIDA R. und ACS Z. J. (2004) Kreativitat und Unternehmertum: eine Regionalanalyse der Grundung neuer Firmen, *Regional Studies* **38**, 879–891. Verstandnis der Faktoren, welche die Grundung neuer Firmen fördern oder erleichtern, ist unerlaßlich für die Bemuhungen um regionale Wirtschaftentwicklung, da ein höherer Anteil neuer Firmengrundungen wesentlich zur regionalen Wirtschaftsentwicklung beitragt, und ein Hauptanzeichen einer dymanischen Wirtschaft darstellt. Die Literatur deutet darauf hin, daß verschiedene Faktoren, wie Erwerbslosigkeit, Bevölkerungsdichte/wachstum, Industriestruktur, Menschenkapital, Vorhandensein von Finanzierungsmöglichkeiten und unternehmerische Eigenschaften signifkante regionale Unterschiede bei der Rate neuer Firmengrundungen nach sich ziehen. In dieser Studie wird untersucht, ob Verbindungen zwischen regionalen gesellschaftlichen Eigenschaften, Menschenkapital und der Gründung neuer Firmen bestehen. Es wird die These aufgestellt, daß gesellschaftliche Verschiedenartigkeit und Kreativitat ein positives Verhaltnis zur Gründung neuer Firmen aufweisen. Gestutzt auf Beiträge der Urbanistin Jane Jacobs, zeigten Lee, Florida und Gates (2002), daß gemessen an pro-Kopf -patentanmeldung, gesellschaftliche Vielfalt und Menschenkapital positive und signifikante Beziehungen zu regionaler Innovationsproduktion aufweisen. Obschon es wohlbekannt ist, daß Bestande regionalen Menschenkapitals die Rate neuer Firmengrundungen positiv beeinflußt, hat man der Wechselwirkung von gesellschaftlicher Vielfalt, Menschenkapital und Unternehmertum wenig Beachtung geschenkt. Die Autoren stellen die These auf, daß geringe Schranken beim Eintritt in den regionalen Arbeitsmarkt (wie z.T. durch das Vorhandensein einer vielschichtigen Bevölkerung bewiesen wird) und unterschiedliche Kultur den Zustrom einer besonderen Art von Menschenkapital ermöglichen, das Innovation fordert und Informationsstrome beschleunigt, und somit zu einer höheren Rate neuer Firmengründungen führt. Die empirischen Ergebnisse bestatigen die Haupthypothese. Mittels Anwendung der langfristigen Gründungs- und Unternehmensmikrodaten (Longitudinal Establishment and Enterprise Microdata – LEEM) wird die Hypothese auf der Ebene von Statistiken von Großstadtregionen (Metropolitan Statistical Areas – MSAs) wie als auch auf derjenigen der Arbeitsmarktregionen (Labor Market Areas – LMAs) gepruft. Es ergibt sich, daß die Grundung neuer Firmen sich als stark mit kultureller Kreativität verbunden erweist, wenn sie auf in der Literatur vorgeschlagene Variable hin untersucht werden. Firmengrundung steht in positivem und signifkantem Verhältnis zum Diversity Index, bleibt jedoch unbedeutend in Bezug auf den Melting Pot Index. Die Ergebnisse legen nahe, daß dem gesellschaftlichen Lebensraum einer Region Aufmerksamkeit geschenkt werden muß, wenn regionale Unternehmertumsdynamik gefordert werden soll.

Kreativität Gründung neuer Firmen Unternehmertum

LEE S. Y., FLORIDA R. y ACS Z. J. (2004) Creatividad y empresarialidad: un análisis regional de la formación de nuevas empresas, *Regional Studies* **38**, 879–891. Entender los factores que promueven o mitigan el nacimiento de nuevas empresas es crucial para los esfuerzos de desarrollo económico regional, puesto que un nivel alto en la creación de nuevas empresas contribuye de forma significativa a la vitalidad económica regional y es una señal muy importante de una economía dinámica. La literatura sugiere que varios factores como el desempleo, la densidad/crecimiento de la población, la estructura industrial, el capital humano, la disponibilidad de capital financiero, y características de la empresarialidad influyen significativamente en la variación regional de los índices de nacimiento de nuevas empresas. En este estudio exploramos si existen conexiones entre las características sociales regionales, el capital humano, y la formación de nuevas empresas. Argumentamos que la diversidad social y la creatividad tienen un relación positiva con la formación de nuevas empresas. Basándose en las contribuciones de la urbanista Jane Jacobs, Lee, Florida y Gates (2002) mostraron que la diversidad social y el capital humano tienen relaciones positivas y significativas con la producción de innovación regional medida por medio de la producción de patentes per cápita. Mientras que es bien sabido que el stock de capital humano regional afecta positivamente los índices de formación de nuevas empresas, se ha prestado poca atención a la interacción entre la diversidad social, el capital humano y la empresarialidad. Nosotros argumentamos que unas barreras bajas de entrada al mercado laboral regional (como lo demuestra en parte la presencia de una población diversa) y una cultura diversa facilitan el influjo de una cierta clase de capital humano que promueve la innovación y que acelera el flujo de información, llevando a un índice mayor de formación de nuevas empresas. Los resultados empíricos apoyan nuestra hipótesis principal. Utilizando el *Longitudinal Establishment and Enterprise Microdata (LEEM)*, testamos la hipótesis a nivel de las Áreas Metropolitanas para Estadísticas (MSAs) así como a nivel de las Áreas de Mercado Laboral (LMAs). Encontramos que la formación de nuevas empresas está fuertemente asociada con la creatividad cultural cuando ésta está controlada por las variables que se sugieren en la literatura. La formación de empresas está positiva y significativamente asociada con el índice de Diversidad, pero no de forma significativa con el índice Melting Pot. Los resultados sugieren que necesitamos prestar atención al hábitat social de una región para estimular una dinámica de empresarialidad regional.

Creatividad Formación de nuevas empresas Empresarialidad

JEL classifications: O40, R11

INTRODUCTION

Ever since the seminal work of SCHUMPETER (1942), entrepreneurship has been regarded as a major topic in the theory and practice of economic growth and development. Practitioners and politicians are well aware of the importance of entrepreneurship because a significant portion of new employment is created by new firms and often new firms bring 'productive innovation' with them (BAUMOL, 2002). Therefore, it is crucial to understand the factors that promote or mitigate entrepreneurial creativity.

There have been various studies on the determinants of entrepreneurship. Much of the literature on entrepreneurship has investigated the characteristics of successful entrepreneurs. These studies have attempted to explain entrepreneurship by looking into individual characteristics such as personality, educational attainment and/or ethnic origin (for a summary, see STOREY, 1994), the factors associated with new firm formation (REYNOLDS *et al.*, 1994; ARMINGTON and ACS, 2002), the organizational, industrial and geographic factors associated with entrepreneurship (REYNOLDS *et al.*, 1993; SAXENIAN, 1999), and the effect of new firm formation on regional growth and development (STOREY, 1994; KIRCHHOFF *et al.*, 2002).

Others have explored the factors associated with regional variation in new firm formation. These studies have found regional variation in new firm formation to be associated with factors such as population, industrial structure, human capital, university research and development, the availability of financing, and entrepreneurial characteristics (ARMINGTON and ACS, 2002; KIRCHHOFF *et al.*, 2002). STUART and SORENSON (2003, p. 229) look at the effect of social ties on firm founding rate. They argue that new firms are attracted by other firms 'because entrepreneurs find it difficult to leverage the social ties necessary to mobilize essential resources when they reside far from those resources'. However, these researches have paid little attention to the social environment of the place where entrepreneurs live and work.

An influential line of research suggests that cities and regions function as 'incubators' of creativity and innovation and that human capital factors in particular play an important role in spurring regional growth (PARK *et al.*, 1925; JACOBS, 1961; THOMPSON, 1965; LUCAS, 1988). Park *et al.* initially called attention to the role of cities in concentrating and spurring human creativity. Jacobs later explained how cities function as 'open systems' to attract talented people from various backgrounds and stimulate their creative capacities. Lucas formalized the insights of Jacobs to provide a basic theory, arguing that cities function as collectors of human capital, thus generating new ideas and economic growth.

The present paper builds from this line of research, focusing on the underlying social characteristics of a region associated with entrepreneurship or firm formation. Specifically, it explores the effect of factors such as creativity and diversity on new firm formation. While previous studies have examined the effect of human capital on firm formation, they have neglected the factors that might originally effect the concentration of human capital and how such factors affect rates of firm formation. The basic hypothesis is that entrepreneurship is positively associated with regional environments that promote diversity and creativity. It is thus argued that entrepreneurial activity requires not only a productive and supportive business climate along with an educated population, but also a climate where creativity, diversity and innovation are encouraged and valued.

To test this hypothesis, this paper use measures of regional diversity and creativity and it examines the effect of these factors on entrepreneurship while controlling for the effects of well-known factors such as human capital, income change and population. It is argued that regions that are broadly creative and open to diversity possess the broad environment or habitat that promotes innovation and accelerates information flow, leading to the formation of new business. The empirical results support the hypothesis.

THE LITERATURE ON ENTREPRENEURSHIP

Academic approaches to entrepreneurship can be categorized into two major ways. First is to focus on the entrepreneurs and try to explain why a person decides to be an entrepreneur and start a new firm. Second is to explain regional variation in firm formation at an aggregate level by looking at structural variations in geographical areas. The two approaches will be explained below.

Traditionally, studies of entrepreneurship have focused on the individual characteristics of successful entrepreneurs. According to STOREY (1994), these studies focus on the role of factors such as personality, human capital and ethnic origin. Personality studies have found that entrepreneurship is associated with characteristics like entrepreneurial vision, alertness to business opportunities, proactivity and family tradition (BLANCHFLOWER and OSWALD, 1990; CHELL *et al.*, 1991). Human capital studies have found that entrepreneurship is related to educational attainment and work experience (EVANS and LEIGHTON, 1990). Research shows that people with higher educational attainment tend to found new business more often than those with less educational attainment. Other studies have found entrepreneurship to be associated with ethnic origin. LEE (2001) found that Jews and Koreans are more successful entrepreneurs than African-Americans because they enjoyed better access to capital through family or ethnic networks than others. Some studies suggest that immigrants are more likely to be entrepreneurs, arguing that because new immigrants lack networks and contacts in existing businesses and are poor in communication skills and suffer from discrimination, they are more likely to start new firms and be self-employed (YOON, 1997). By using the data from the National Longitudinal Survey of Young Men and Current Population Survey, EVANS and LEIGHTON (1989) found that men with more financial resources and with more confidence in their own ability are more likely to be self-employed.

Another line of research has examined the factors at a regional level, which effect regional variations in new firm formation. Early studies focused on factors such as tax rates, transportation costs and scale economies at the plant level (BARTIK, 1989; KIESCHNICK, 1981). REYNOLDS *et al.* (1994) found that factors such as unemployment, population density, industrial clustering and the availability of financing were important in explaining regional variation in firm birth rates. ARMINGTON and ACS (2002) found that industrial intensity, income growth, population growth and human capital were closely related to new firm formation. KIRCHHOFF *et al.* (2002) found academic research and development expenditure to be significantly associated with rates of new firm formation across regions. A number of studies have suggested that regional rates of entrepreneurship are associated with levels of immigration (REYNOLDS *et al.*, 1995; SAXENIAN 1999; KIRCHHOFF *et al.*, 2002). The entrepreneurship of immigration can be approached in two ways. While most immigrants are less educated and have little skills to success in the USA, some are extremely well educated and equipped with a good set of skills. Although it is hard to find common property between the two groups, one fact in common is that they are risk takers. A study of immigrants in California found that those with a good educational background were involved as founders in 20–25% of new high-technology firm formation in Silicon Valley (SAXENIAN, 1999).

Studies noted the importance of the role of network in entrepreneurship. SAXENIAN (1999) found that extensive networks of Chinese and Indian workers help people start new firms by providing contacts and financial support in Silicon Valley. STUART and SORENSON (2003, p. 229) argue that businesses cluster because geographical proximity enables them to use 'social ties necessary to mobilize essential resources'. Their findings imply that an entrepreneur's social relationship is crucial in using critical business – resources critical to start a firm and set up a new organization.

The present authors are interested in studying entrepreneurship in the context of clustering. The clustering of people and industries has been studied seriously in the literature. Following PARK *et al.*'s (1925) initial attention to the role of cities in concentrating and spurring human creativity, JACOBS (1961) explained how cities function as 'open systems' to attract talented people from various backgrounds and stimulate their creative capacities. Jacobs argued that open and diverse cities attract more talented people, thus spurring creativity and innovation, which are the underlying forces of entrepreneurship. THOMPSON (1965) was among the first to suggest that cities function as 'incubators' of new ideas and innovation. LUCAS (1988) formalizes the insights of Jacobs to provide a basic theory, arguing that cities function as collectors of human capital, thus generating new ideas and economic growth. Following Jacobs, DESROCHERS (2001) argues that economic diversity is a key factor in city and regional growth, as creative people from varied backgrounds come together to generate new and novel combinations of existing technology and knowledge to create innovation and, as a result, new firms. Building on these contributions, LEE *et al.* (2002) show that creativity, diversity and human capital have positive and significant relationships with regional innovation measured as per capita patent production. In addition, FLORIDA (2002) argues that creativity is an important element in regional economic success; and FLORIDA and GATES (2001) find that diversity has a positive association with regional high-technology output and growth.

The present research builds on this line of thinking, arguing that creativity and diversity of a region work together to increase regional capacity to generate entrepreneurial activity. Creativity and diversity are kinds of social infrastructure into which entrepreneurs and policy-makers can tap. Creativity and diversity are quite distinctive since they cannot be easily measured or even defined properly. They are more fundamental than critical resources for entrepreneurship such as tax rate, human capital, venture capital or entrepreneurial zone. It can be regarded as social habitat.

How can diversity promote entrepreneurship? It is argued that more diverse regions tend to have lower entry barriers that make it easier for human capital with various backgrounds to enter the region and stay within it. If one can agree that the central focus of entrepreneurial studies is entrepreneurs themselves, it is natural to think that lower entry barriers can play an important role in attracting creative human capital to come to a region and stay welcomed with a sense of membership. Hence, a more diverse region could enjoy comparative advantage in attracting and retaining creative human capital.

How is creativity related to entrepreneurship? STERNBERG (1999) defines creativity as 'the ability to produce work that is both novel (i.e., original, unexpected) and appropriate (i.e., useful, adaptive concerning task constraints)'. According to STERNBERG and LUBART's (1999) definition, entrepreneurship is a form of creativity and can be labelled as business or entrepreneurial creativity because often new businesses are original and useful. CATTELL and BUTCHER (1968, p. 285) argue that 'creativity is perhaps best acquired by association with creativity'. It is assumed that the presence and concentration of bohemians in an area creates an environment or a milieu that attracts other types of talented or high human capital individuals and promotes creativities of human capital, resulting in business creativity.

DATA AND METHODS

The effect of creativity and diversity on entrepreneurship is examined at a regional level. Two geographic units are used to test the hypothesis. (1) Metropolitan Statistical Areas (MSAs) and Primary Metropolitan Statistical Areas (PMSAs). These cover urban areas of the USA. The present data include information on firm births and deaths for all 320 MSAs/PMSAs and show that 80% of all new firm births occurred within MSAs/PMSAs. However, complete data for all variables are available only for 236 MSAs/PMSAs. In general, the dropped MSAs are smaller in population size than the ones included in the regression. However, when controlled for size, the firm birth per 1 million shows little difference between the two groups. (2) Labor Market Areas (LMAs) are also used. They were defined by the US Department of Agriculture in 1990 (Tolbert and Sizer, 1990) and have been used by Armington and Acs (2002). Since the 3141 counties are aggregated into 394 LMAs based on predominant commuting patterns, the use of the LMA as a unit of observation has an advantage in that it includes residential locations as well as employment locations of populations in the same area. Different from MSAs/PMSAs, LMAs cover the entire USA.

Firm birth

The data on firm formation come from the Longitudinal Establishment and Enterprise Microdata (LEEM) (for a detailed explanation of LEEM, see Armington and Acs, 2001; Acs and Armington 1998; Armington 1998). This file was constructed by the Bureau of the Census, Washington, DC, from its Statistics of US Business (SUSB) files, which were developed from the microdata underlying the aggregate data in Census County Business Patterns. The basic unit of the LEEM data is a business establishment (location or plant). An establishment is a single physical location where business is conducted or where services or industrial operations are performed. The microdata describe each establishment for each year of its existence in terms of its employment, annual payroll, location (state, county and metropolitan area), primary industry and start year. Additional data for each establishment and year identify the firm (or enterprise) to which the establishment belongs, and the total employment of that firm.

A firm (or enterprise or company) is the largest aggregation (across all industries) of business legal entities under common ownership or control. Establishments are owned by legal entities, which are typically corporations, partnerships or sole proprietorships. Most firms are composed of only a single legal entity that operates a single establishment – their establishment data and firm data are identical, and they are referred to as 'single-unit' establishments or firms. The single-unit businesses are frequently owner operated. Only 4% of firms have more than one establishment, and they and their establishments are both described as multi-location or multi-unit. The LEEM data cover all private-sector businesses with employees, with the exception of those in agricultural production, railroads and private households. For MSAs/PMSAs, the present study uses LEEM data for 1997–98. Data from the 1994–96 LEEM are used for LMAs analysis.

Creativity

Creativity is measured by using the Bohemian Index – a measure of the proportion of 'bohemians' and other artistically creative people in a region. It measures the openness of a region to creativity of the sort not directly associated with technological and business-related innovations. This index measures a region's artistic creativity and intellectual dynamism. Regions with higher scores on this measure are expected to be both more attractive to creative people and also to cultivate new ideas and accelerate their flow, which are crucial in forming a new firm. It is a location quotient measure and is based on occupational data from the 1990 Decennial Census 5 Percent Public Use Microdata Sample (PUMS) and includes authors, designers, musicians, composers, actors, directors, painters, sculptors, craft-artists, artist printmakers, photographers, dancers, artists and performers. Florida (2002) shows there is a significantly positive relationship between the Creativity Index and concentrations of high-technology industry.

Diversity

As discussed above, it is assumed that more diverse regions are expected to have advantage in attracting and retaining creative people with unorthodox ideas by lowering the entry barrier and making diverse ideas available. Two measures of diversity are employed.

(1) The Melting Pot Index is a measure of the percentage of the population that is foreign born and is based on data from the 1990 Decennial Census 5 Percent PUMS. Previous studies support the inclusion of the index since they have found a significant and positive effect of immigrants on new firm formation (Reynolds *et al.*, 1995; Saxenian, 1999; Kirchhoff *et al.*, 2002). Since the immigrants usually lack skills, resources and networks, they tend to be more self-employed than non-immigrants. In addition, they bring new ideas and cultures to enrich a region and create new business opportunities.

Table 1. *Summary statistics*

Variable	Observed	Mean	Standard deviation	Minimum	Maximum
Metropolitan Statistical Areas (MSAs)/Primary Metropolitan Statistical Areas (PMSAs)					
Firm birth (all industries)	320	1815 01	3109 84	126 00	27 063 00
Firm birth (manufacturing industries)	320	68 38	144 94	1 00	1694 00
Firm birth (service industries)	320	692 20	1224 94	39 00	9997.00
Firm birth per 1 million people (all industries)	236	2519 33	667 04	1159 17	5049 13
Firm birth per 1 million people (service industries)	236	918 42	294.46	422.76	2398 34
Firm birth per 1 million people (manufacturing industries)	236	91 21	40 77	14 01	271 58
Creativity Index	237	0 92	0 37	0 32	2 90
Diversity Index	236	0 80	1 06	0.00	12.23
Melting Pot Index	252	0 07	0 07	0 00	0 54
Human capital	236	0 21	0 06	0 09	0 42
Population	320	603 772	1 005 119	56 735	8 863 052
Income growth rate	236	0 086	0 052	−0 081	0 300
Patents per 100 000 people	237	201 59	195 00	6 29	1542 04
Population growth rate	236	0 07	0 06	−0 05	0 38
Labor Market Areas (LMAs)					
Firm birth per 1000 people (1995–96)	394	3 741	0 938	2 061	10 177
Establishment size (1994)	394	15 097	2 881	8 266	21 237
Industry intensity (1994)	394	0 022	0 004	0 011	0 045
Income growth	394	1 104	0 033	1 016	1 220
Population growth	394	1 011	0 010	0 980	1 062
Share of proprietors	394	0 206	0 058	0 099	0 448
Unemployment (average 1993–94)	394	0 066	0 025	0 020	0 287
Share of high school dropout	394	0 279	0 080	0 117	0 541
Share of college graduate	394	0 159	0 050	0 069	0 320
Creativity Index	394	0 689	0 284	0 097	1 973
Melting Pot Index	394	0 398	0 538	0 023	4 168

(2) The Diversity (or Gay) Index is used to capture the broader diversity of a region. The index is a measure of the concentration of same-sex male unmarried partners, commonly understood to be gay male couples, in the population and is used to approximate the level of openness or tolerance to newcomers or 'non-conformists' in a region. It is assumed that high concentrations of gay men in a region signal a broader openness towards those who are different, creating lower entry barriers to human capital of various kinds and backgrounds. Based on the 1990 Decennial Census PUMS, the index is constructed as a location quotient of the over- or under-representation of coupled gay men in a region relative to the population. (For more on this measure, see BLACK *et al.*, 2000; and FLORIDA and GATES, 2001).

Human capital

Human capital is measured as the percentage of adults in the population with a bachelor's degree and above. As discussed above, educational attainment has been positively associated with entrepreneurship in the literature. These data come from the 1990 Decennial Census 5 Percent PUMS.

Other variables

Income change is the absolute change between 1990 and 1996. It was expected that higher income change will lead to the formation of more new firms by providing additional financial resources necessary to start a firm. These data come from Bureau of Labor Economics. To control for the size and growth of each region, population and population growth (1990–96) are included. Since a bigger region tends to benefit more from the knowledge-spillover effect, which leads to more innovation and entrepreneurship, it is included in the equation. In addition, the patent variable will be included. It is defined as the number of patents issued per 100 000 people in 1995. Since technology plays an important role in recent venture firm boom, it is expected that there will be a positive relationship between a patent and entrepreneurship.

Variables for LMAs

Establishment size is used to control for the entry barrier in a region. It is assumed that it will be harder to enter the market when the average firm size is bigger. Industry intensity is the total number of private-sector establishments in the region divided by the region's population, which can be interpreted as the industry intensity. The share of high-school dropouts is defined as the percentage of adults without high-school degrees and is a proxy for the proportion of the poorly skilled labour force. The share of proprietors is defined as the number of service establishments in the region divided by the region's population (in thousands) and is expected to capture the knowledge spillovers in the region (for detailed information on these measures, see ARMINGTON and ACS, 2002).

Some studies on organizational birth used Poisson regression or negative binomial regression to study new firm birth since dependent variables do not follow a normal distribution (STUART and SORENSON, 2003). However, the firm birth rate used as a dependent variable in the present study looks close to a normal distribution when it is controlled for the population size, even though it is a slightly skewed. Therefore, bivariate correlation analyses will be used along with multivariate ordinary least squares (OLS) regression models for estimation.

Table 2. Entrepreneurship by state (per 1 million people)

	State	Birth	State	Death	State	Net
1	Colorado	5548	Wyoming	4597	Nevada	1455
2	Wyoming	5349	Colorado	4503	Colorado	1045
3	Nevada	5247	Montana	4495	Delaware	854
4	Montana	5158	District of Columbia	4319	Utah	846
5	Idaho	4769	Florida	4318	Wyoming	752
6	Utah	4690	Vermont	4317	Montana	664
7	Alaska	4648	Alaska	4148	Idaho	645
8	Delaware	4647	Idaho	4123	Georgia	622
9	Florida	4573	Oregon	4053	Washington	574
10	District of Columbia	4435	Washington	3849	Maine	574
11	Washington	4423	Utah	3843	Minnesota	572
12	Oregon	4409	South Dakota	3826	South Carolina	503
13	Vermont	4288	Delaware	3793	Alaska	501
14	Georgia	4260	Nevada	3791	Virginia	486
15	South Dakota	4093	Georgia	3638	New Hampshire	458
16	New Hampshire	4035	Arizona	3590	North Carolina	411
17	Maine	4016	New Hampshire	3577	New Jersey	409
18	Arizona	3988	New Mexico	3576	Texas	399
19	New Jersey	3948	New Jersey	3538	Arizona	398
20	North Carolina	3837	Arkansas	3462	New York	390
21	Texas	3831	Kansas	3458	California	382
22	New Mexico	3821	Maine	3442	Oregon	356
23	California	3729	North Dakota	3437	Massachusetts	335
24	Kansas	3724	Texas	3432	Illinois	305
25	South Carolina	3671	North Carolina	3427	Kentucky	287
26	Minnesota	3614	Oklahoma	3401	South Dakota	267
27	Oklahoma	3607	California	3347	Kansas	266
28	Virginia	3586	Nebraska	3324	Florida	255
29	New York	3584	Hawaii	3323	New Mexico	245
30	Missouri	3531	Missouri	3316	Rhode Island	232
31	Arkansas	3503	Connecticut	3312	Wisconsin	223
32	North Dakota	3498	Maryland	3212	Louisiana	216
33	Nebraska	3466	New York	3194	Mississippi	216
34	Maryland	3367	Tennessee	3174	Missouri	215
35	Tennessee	3366	South Carolina	3167	Oklahoma	205
36	Rhode Island	3344	Rhode Island	3112	Iowa	205
37	Connecticut	3322	Virginia	3100	Tennessee	192
38	Massachusetts	3298	Alabama	3055	Indiana	184
39	Louisiana	3232	Minnesota	3041	Pennsylvania	165
40	Illinois	3218	Louisiana	3016	Maryland	155
41	Alabama	3188	Massachusetts	2963	Michigan	155
42	Iowa	3151	Iowa	2945	Nebraska	142
43	Mississippi	3100	Illinois	2914	Alabama	133
44	Hawaii	3081	Mississippi	2884	District of Columbia	116
45	Indiana	3068	Indiana	2884	Ohio	110
46	Wisconsin	3046	Michigan	2840	North Dakota	61
47	Kentucky	3026	Wisconsin	2823	West Virginia	61
48	Michigan	2995	Kentucky	2739	Arkansas	41
49	Ohio	2782	Ohio	2672	Connecticut	10
50	Pennsylvania	2747	Pennsylvania	2582	Vermont	−29
51	West Virginia	2619	West Virginia	2558	Hawaii	−242

REGIONAL PATTERNS OF FIRM FORMATION

This section examines regional differences in rates of firm formation. ARMINGTON and ACS (2002) explain thoroughly the characteristics of firm birth data at the LMA's level and the variables used in the study. Therefore, the present paper will focus on the findings on MSAs/PMSAs. To control for differences in the size of regions, 'firm birth rate' is defined as firm births per 1 million people. Firm birth rates are calculated for all 320 MSAs/PMSAs using the LEEM between 1997 and 1998. Between 1997 and 1998, 580 803 new firms were created and 524 138 firms ceased to exist.

Table 2 shows variations in new firm formation on a per-capita basis at the state level. It ranges from 5548 (Colorado) to 2619 (West Virginia). The highest rates of firm formation are in Colorado (5548), Wyoming (5349), Nevada (5247), Montana (5138) and Idaho (4769). California (3729) ranks 23rd and Texas (3831) 21st, respectively. Table 3 shows variations at the regional level. Here, firm birth rates range from 6910 (Naples, FL, MSA) to 1322 (Beaver County, PA, PMSA). The top 10 regions are relatively small (under 500 000 population) with the exception of Las Vegas (NV), West Palm Beach and Fort Lauderdale (both FL).

Table 4 shows summary statistics for regions by size. Regions were assigned to three size groups: large regions with populations above 500 000, medium sized regions with between 300 000 and 500 000 people, and small regions with less than 300 000 people. Table 4 suggests that larger regions benefit from their size. The average firm birth rate for large regions is 3076 compared with 2627 for medium-sized regions and 2743 for small regions. These differences are statistically significant at the 95% level. The average net firm birth rate for large regions is 304 compared with 207 for medium-sized regions and 192 for small regions. These differences are statistically significant at the 0.10 level.

Table 3. Entrepreneurship at Metropolitan Statistical Areas (MSAs)/Primary Metropolitan Statistical Areas (PMSAs) (per 1 million people): top 50

Name	Birth
Naples, FL, MSA	6910.0
Wilmington, NC, MSA	5936.0
Boulder–Longmont, CO, PMSA	5857.8
Las Vegas, NV, MSA	5582.9
Boise City, ID, MSA	5496.3
Reno, NV, MSA	5301.0
West Palm Beach–Boca Raton–Delray Beach, FL, MSA	5096.7
Fort Lauderdale–Hollywood–Pompano Beach, FL, PMSA	5041.7
Santa Fe, NM, MSA	4972.5
Bellingham, WA, MSA	4961.7
Atlanta, GA, MSA	4846.0
Denver, CO, PMSA	4835.5
Fort Collins–Loveland, CO, MSA	4749.2
Raleigh–Durham, NC, MSA	4711.3
Austin, TX, MSA	4662.4
Orlando, FL, MSA	4542.6
Sarasota, FL, MSA	4482.0
Billings, MT, MSA	4479.0
Fort Myers–Cape Coral, FL, MSA	4446.3
Portland, ME, NECMA	4421.4
Miami–Hialeah, FL, PMSA	4358.4
Seattle, WA, PMSA	4351.4
San Francisco, CA, PMSA	4279.5
Portland, OR, PMSA	4215.9
Midland, TX, MSA	4183.4
Fayetteville–Springdale, AR, MSA	4126.7
Vancouver, WA, PMSA	4125.1
Charlotte–Gastonia–Rock Hill, NC/SC, MSA	4116.1
Dallas, TX, PMSA	4091.5
Phoenix, AZ, MSA	4053.5
Sioux Falls, SD, MSA	4030.4
Springfield, MO, MSA	3940.3
Fort Pierce, FL, MSA	3907.3
Lafayette, LA, MSA	3892.6
Fort Walton Beach, FL, MSA	3881.0
Colorado Springs, CO, MSA	3879.0
Nashville, TN, MSA	3864.9
Asheville, NC, MSA	3837.0
Middlesex–Somerset–Hunterdon, NJ, PMSA	3830.2
Jacksonville, FL, MSA	3820.3
Medford, OR, MSA	3784.5
Anchorage, AK, MSA	3782.0
Casper, WY, MSA	3740.2
Anaheim–Santa Ana, CA, PMSA	3734.6
Wilmington, DE/NJ/MD, PMSA	3731.5
Salt Lake City–Ogden, UT, MSA	3683.0
Jackson, TN, MSA	3641.9
Santa Rosa–Petaluma, CA, PMSA	3634.5
Columbia, MO, MSA	3612.8
Charlottesville, VA, MSA	3585.2

EXPLAINING REGIONAL DIFFERENCES IN FIRM FORMATION

We now turn to the results of bivariate and multivariate analysis of the factors associated with regional variations in new firm formation. The results of the correlation analysis for MSAs/PMSAs and LMAs are presented in Table 5. The correlations for MSAs/PMSAs indicate that entrepreneurship is most closely associated with the Creativity Index with a correlation coefficient of 0.515. New firm birth per 1 million people is also strongly associated with human capital (0.476). It is moderately related to the Diversity Index (0.332) and the Melting Pot Index (0.169). Entrepreneurship is only moderately associated with patents (0.245) and the size of population (0.181). Firm formation is reasonably associated with income change as the suggested in the literature (0.270). Population growth is highly correlated with new firm birth (0.397). New firm birth in service industries shows similar patterns. However, new firm birth in manufacturing industries shows quite different pictures. Whereas the correlation with the Creativity Index is moderate (0.394), the correlations with the Diversity Index and human capital are quite low (0.156 and 0.160, respectively). In addition, correlation with the Melting Pot Index is quite weak

Table 4. Summary statistics by the size of regions

Variable	Observed	Mean	Standard error	95% Confidence interval	
Regions over 500 000 people					
Firm birth per 1 million people	94	3076.9	87.7	2902.7	3251.1
Firm death per 1 million people	94	2772.7	68.3	2637.1	2908.3
Net firm birth per 1 million people	94	304.3	30.1	244.5	364.0
Regions over 300 000 and less than 500 000 people					
Firm birth per 1 million people	50	2627.4	88.2	2450.1	2804.7
Firm death per 1 million people	50	2420.3	70.8	2277.9	2562.6
Net firm birth per 1 million people	50	207.1	28.3	150.3	264.0
Regions less than 300 000 people					
Firm birth per 1 million people	176	2743.0	69.6	2605.6	2880.4
Firm death per 1 million people	176	2550.3	54.4	2443.0	2657.7
Net firm birth per 1 million people	176	192.7	26.3	140.7	244.6

(0.070). The relatively big size and resource-demanding nature of manufacturing industries may contribute to the difference of manufacturing industries from others.

Entrepreneurship in LMAs shows somewhat different pictures from MSAs/PMSAs. It is strongly related to population growth (0.541) and industry intensity (0.531). Establishment size is negatively related to entrepreneurship (−0.417). The share of college graduates is weakly related with a coefficient of 0.292. The Creativity Index is moderately related (0.300) and the Melting Pot Index is weakly correlated (0.186).

The regression results reported in Table 6 are consistent with the analysis result based on the correlation among variables. To make the comparison easier, Table 6 also reports the β-coefficients of variables. Three OLS regressions for MSAs/PMSAs were run by using three dependent variables: all industries, service industries and manufacturing industries. The first column shows the results for all industries; the second column summarizes the results for manufacturing industries (Codes 20–39); and the last column reports the results for service industries (Standard Industrial Classification Codes 70–89). The all-industries model explains about 47% of the variation in the dependant variable, while the regressions for the service industries and manufacturing industries explain about 56 and 29% of the variation, respectively.

The regression results confirm the main hypothesis. Entrepreneurship is strongly associated with creativity across all three models. The β-coefficient for the Creativity Index is the second largest and is positive and significant at the 1% significance level across all three models. The β-coefficient of the Creativity Index for the all-industries model suggests that 1.000 standard deviation increase in the Creativity Index predicts a 0.262 standard deviation increase in new firm formation per 1 million people. It supports the proposition that that there is a close and positive relationship between entrepreneurship and creativity in a region. The results on diversity measures are mixed. As expected, the Diversity (Gay) Index shows positive signs in all three models, but it is statistically significant only in the service industry. It might be resulted by the Diversity Index's high correlation to the human capital variable (0.692). When the models are estimated without the human capital variable, the coefficient of the Diversity Index becomes bigger and more statistically significant in all industries and service industries. It suggests that the kind of diversity captured by the Diversity Index is quite closely related to the skilled workforce and has a positive impact on the entrepreneurship in service industries. Different from the expectation, the Melting Pot Index (the share of the foreign born) is not significant in all three models. The insignificance of the Melting Pot Index might be explained by the fact that there are two groups of foreign-born people. One group is the well educated and wealthy; the other is the less educated and poor. Since the Melting Pot Index does not differentiate between the two groups, its effect on entrepreneurship might become negated in the measure. The other possibility is that population growth rate takes away the effect of the Melting Pot Index since most of the population increase in the USA comes from immigrants and most of fast-growing cities are also high in terms of the foreign-born people. The fact that the Melting Pot Index becomes statistically significant if the population growth rate is taken out from the regression supports this view.

New firm formation is also closely associated with both income growth rate and human capital as suggested in the literature. Income growth is positive and significant in all industries and service industries. Curiously, the coefficient for human capital in manufacturing industries is negative and significant, but that in service industries it is positive and significant. However, considering that manufacturing industries hire a massive, less-educated workforce and human capital is defined here as the percentage of people who have a bachelor's degree and above, a negative sign becomes less puzzling. In the service industry model, human capital has a positive and significant coefficient.

Table 5. Correlation analysis

	1	2	3	4	5	6	7	8	9	10	11	12	13	14
Metropolitan Statistical Areas (MSAs)/Primary Metropolitan Statistical Areas (PMSAs)														
1 Firm birth (all industries)	1 000													
2 Firm birth (manufacturing industries)	0 938	1 000												
3 Firm birth (service industries)	0 996	0.927	1 000											
4 Firm birth per 1 million people (all industries)	0.317	0.225	0 323	1 000										
5 Firm birth per 1 million people (service industries)	0.384	0 272	0 405	0 938	1.000									
6 Firm birth per 1 million people (manufacturing industries)	0.212	0 294	0.206	0 496	0 403	1 000								
7 Creativity Index	0.480	0.452	0 488	0 515	0 582	0 394	1.000							
8 Diversity Index	0.380	0 317	0 401	0 332	0.414	0 156	0.524	1 000						
9 Melting Pot Index	0.444	0.443	0 441	0.169	0 220	0.071	0 222	0 325	1 000					
10 Human capital	0.340	0 248	0 363	0 476	0 588	0 160	0 692	0 495	0 168	1 000				
11 Population	0 972	0.940	0 965	0.181	0 251	0 158	0 415	0 318	0.417	0.282	1 000			
12 Income growth rate	0 072	0 007	0 080	0.270	0 294	0.175	0.273	0 116	−0 282	0.348	0 010	1.000		
13 Patents per 100 000 people	0 147	0.140	0 163	0 245	0 340	0 307	0 435	0 209	0.085	0.534	0 126	0.276	1 000	
14 Population growth rate	0 038	−0 004	0 045	0.397	0 374	0 177	0.181	0 008	0.151	0.145	−0 063	0 014	0 026	1 000
Labor Market Areas (LMAs)														
1 Firm birth per 1000 people (1995–96)	1 000													
2 Establishment size (1994)	−0.417	1.000												
3 Industry intensity (1994)	0 531	−0.317	1 000											
4 Income growth	0 366	0 195	0 002	1 000										
5 Population growth	0 541	−0.017	0 044	0 699	1 000									
6 Share of proprietors	0.305	−0 635	0 460	−0 190	0 005	1 000								
7 Unemployment (average 1993–94)	−0 019	−0.270	−0 370	−0 172	0 004	−0 205	1.000							
8 Share of high school dropout	−0 194	−0.050	−0 523	0 038	−0 117	−0 193	0.400	1.000						
9 Share of college graduate	0 292	0 221	0 374	0 065	0 229	−0 054	−0 332	−0 701	1 000					
10 Creativity Index	0 300	0.276	0 350	0 113	0 238	−0 111	−0.237	−0 590	0.799	1.000				
11 Melting Pot Index	0 186	0 000	0 024	−0 059	0 234	−0 116	0.351	−0 132	0 398	0.481	1 000			

Table 6. Regression results at Metropolitan Statistical Areas (MSAs)/Primary Metropolitan Statistical Areas (PMSAs)

	Firm birth per 1 million people (1997–98)		
	All industries	Manufacturing industries	Service industries
Creativity Index	476 595 (3 30)***	50 421 (5 09)***	166 095 (2 85)***
Diversity Index	52 158 (1 28)	0 940 (0 34)	32 763 (1 99)**
Melting Pot Index	503 671 (0 85)	−5 545 (0 14)	287 801 (1 20)
Human capital	1651 893 (2 01)**	−236 561 (4.20)***	1161 862 (3.51)***
Population (1990)	−0 000 (0 01)	0 000 (0 67)	0.000 (0 69)
Income growth rate (1990–96)	0 102 (2 54)**	0 004 (1 51)	0 042 (2 59)**
Patents per 100 000 people (1995)	−0 091 (0 43)	0 056 (3 86)***	0 048 (0 56)
Population growth rate (1990–96)	3374 308 (5 81)***	94 757 (2 38)**	1354 182 (5 77)***
Constant	1264.370 (9 53)***	67 006 (7 36)***	293.348 (5 47)***
Observations	236	236	236
R^2	0 41	0 25	0 50
Beta-coefficients			
Creativity Index	0 262***	0 454***	0.207***
Diversity Index	0 083	0 024	0.118**
Melting Pot Index	0 054	−0 010	0 070
Human capital	0 161**	−0 377***	0 256***
Population (1990)	−0 001	0 046	0 039
Income growth rate (1990–96)	0 151**	0 100	0 141**
Patents per 100 000 people (1995)	−0 027	0 269***	0.032
Population growth rate (1990–96)	0 316***	0 145**	0 287***
Constant	1 895***	1 643***	0 996***

Notes Absolute value of the *t*-statistics is in parentheses
Significant at *10%, **5%, ***1%

For service industries, one might expect the sign to be negative because it requires little expertise and thus hires many less-educated people. However, considering that the definition used herein for service industries (Standard Industrial Classification Codes 70–89) also includes industries that require highly skilled labours such as business service, legal service, educational service and heath service, the positive and significant estimate becomes reasonable.

The OLS results show that population growth rate is more important than the size of a region in explaining the regional variation in new firm formation. The coefficients for population are statistically insignificant in all three models. However, population growth rate is positive and significant, which implies that population growth, not size, has a positive relationship with entrepreneurship. Based on the β-coefficients, population growth is the most influential variable. The coefficient for income change is significant and positive in two models except manufacturing industries. It implies that a region's entrepreneurial capacity can benefit from additional financial resource by increased income.

Regression results at LMAs are reported in Table 7. The Creativity Index and Melting Pot Index were added to the variables used by ARMINGTON and ACS (2002). The regression at LMAs shows that establishment size, industry intensity and population growth are strongly related to firm birth. Unemployment rate and income growth have positive and significant effect on entrepreneurship. As expected, the Creativity Index is positive and quite significant, which is consistent with the earlier findings at the MSAs/PMSAs level. However, the Melting Pot Index is negative and statistically significant in Model 3. The inconsistency might be caused by the fact that LMA includes rural areas as well as urban areas, which is not the case with MSAs/PMSAs.

In the model, both the 'share of high-school drop-outs' and the 'share of college graduates' are positive and quite significant across three models, and the effect of the 'share of college graduates' is bigger than that of the 'share of high-school dropouts'. For human capital, ARMINGTON and ACS (2002) found that the ratio of high-school drop-outs has a significant and positive relationship with entrepreneurship in service industries, especially in service firms founded and managed by a small number of better-educated people. The role of less skilled workers in entrepreneurship is worthy of further investigation.

DISCUSSION

The paper has analysed the effect of regional characteristics such as creativity and diversity on new firm formation. It used a new measure of firm formation based on the Longitudinal Establishment and Enterprise Microdata (LEEM) data for 1994–96 at Labor Market Areas (LMAs) and for 1997–98 at Metropolitan Statistical Areas (MSAs) and also introduced some novel measures of creativity (the Bohemian Index) and diversity (the Melting Pot and Gay Indices). Findings confirm the central hypothesis, although with some caveats.

Table 7. Regression results at Labor Market Areas (LMAs)

	Firm birth per 1000 people (1995–96)		
	Model 1	Model 2	Model 3
Establishment size (1994)	−0 118 (7 61)***	−0 106 (6 87)***	−0 118 (7.57)***
Industry intensity (1994)	112 878 (10 25)***	121 447 (11 16)***	113 163 (10 32)***
Income growth	4 671 (3 49)***	4.463 (3 24)***	4 152 (3 05)***
Population growth	31 548 (7 46)***	33 405 (7 64)***	33.356 (7 74)***
Share of proprietors	−0.097 (0.13)	−0 080 (0 11)	−0 025 (0 03)
Unemployment rate (average 1993–94)	2 691 (1 75)*	4 168 (2 43)**	4 135 (2 44)**
Share of high school dropout	2 824 (5 30)***	2 785 (5 12)***	3 022 (5 60)***
Share of college graduates	3 508 (3 35)***	5 815 (6 13)***	3.969 (3 72)***
Creativity Index	0 515 (3 06)***		0 635 (3 57)***
Melting Pot Index		−0 060 (0 87)	−0 146 (2 02)**
Constant	−35 838 (10 80)***	−37 942 (11 03)***	−37 375 (11 02)***
Observations	394	394	394
R^2	0 68	0 67	0 68
Beta-coefficients			
Establishment size (1994)	−0 364***	−0 325***	−0 361***
Industry intensity (1994)	0 431***	0 464***	0 432***
Income growth	0 165***	0 158***	0 147***
Population growth	0 348***	0 369***	0 368***
Share of proprietors	−0 006	−0.005	−0.002
Unemployment rate (average 1993–94)	0 071*	0 111**	0 110**
Share of high school dropout	0 242***	0 238***	0 259***
Share of college graduates	0 189***	0 313***	0 213***
Creativity Index	0 156***		0 192***
Melting Pot Index		−0 035	−0.084**

Notes Absolute value of the *t*-statistics is in parentheses
Significant at *10%, **5%, ***1%.

Overall, new firm formation is associated with creativity. It is also associated with one dimension of diversity, the Diversity Index, but not with other types of diversity associated with the Melting Pot Index. Consistent with the literature, entrepreneurship is also associated with human capital, income change and population growth rate, as suggested in the literature. The findings suggest that the regions that are open and creative and attract human capital enjoy more than dynamic entrepreneurship.

The findings seem to suggest that both scholars and policy-makers should pay more attention to the social context or habitat in which entrepreneurship takes place. Note that the present research is just a start. The authors encourage more research that focuses on the social context of entrepreneurship and firm formation and the way they are affected by factors such as diversity and creativity.

REFERENCES

ACS Z. and ARMINGTON C. (1998) *Longitudinal Establishment and Enterprise Microdata (LEEM) Documentation.* CES 98-9. Center for Economic Studies, US Bureau of the Census, Washington, DC.

ARMINGTON C. (1998) *Statistics of U.S. Businesses – Microdata and Tables of SBA/Census Data on Establishment Size.* Office of Advocacy, US Small Business Administration, Washington, DC.

ARMINGTON C. and ACS Z. (2002) The determinants of regional variation in new firm formation, *Regional Studies* **36**, 33–45.

BARTIK T. (1989) Small business start-ups in the United States: estimates of the effects of characteristics of states, *Southern Economic Journal* **55**, 1004–1018.

BAUMOL W. (2002) *The Free-market Innovation Machine: Analyzing the Growth Miracle of Capitalism.* Princeton University Press, Princeton.

BLACK D., GATES G., SANDERS S. and TAYLOR L. (2000) Demographics of the gay and lesbian population in the United States: evidence from available systematic data sources, *Demography* **37**(2), 139–154.

BLANCHFLOWER D. and OSWALD A. (1990) What Makes an Entrepreneur?, *Journal of Labor Economics* **16**, 26–60.

CATTELL R. B. and BUTCHER H. (1968) *The Prediction of Achievement and Creativity.* Bobbs-Merrill, Indianapolis.

CHELL E., HAWORTH J. M. and BREARLEY S. (1991) *The Entrepreneurial Personality· Concepts, Cases and Categories.* London, Routledge.

DESROCHERS P. (2001) Local diversity, human creativity, and technological innovation, *Growth and Change* **32**, 369–394.

EVANS D. S. and LEIGHTON L. S. (1989) Some empirical aspects of entrepreneurship, *American Economic Review* **79**, 519–535.

EVANS D. S. and LEIGHTON L. S. (1990) Small business formation by unemployed and employed workers, *Small Business Economics* **2**, 319–330.

FLORIDA R. (2002) *The Rise of the Creative Class: And How It's Transforming Work, Leisure, Community and Everyday Life.* Basic Books, New York.

FLORIDA R. and GATES G. (2001) *Technology and Tolerance: The Importance of Diversity to High-technology Growth.* Brookings Institution, Washington, DC.

JACOBS J. (1961) *The Death and Life of Great American Cities.* Random House, New York.

KIESCHNICK M. (1981) *Taxes and Growth: Business Incentives and Economic Development.* Council of State Planning Agencies, Washington, DC.

KIRCHHOFF B., ARMINGTON C., HASAN I. and NEWBERT S. (2002) The influence of R&D expenditures on new firm formation and economic growth (available at: http://www.njit.edu/old/News/Releases/finalreport_10-02-02.pdf).

LEE J. (2001) Entrepreneurship and business development among African Americans, Koreans, and Jews: exploring some structural differences, in CORDERO-GUZMÁN H. R. (Ed.) *Migration, Transnationalization, and Race in a Changing New York*, pp. 258–278. Temple University Press, Philadelphia.

LEE S., FLORIDA R. and GATES G. (2002) *Innovation, Human Capital, and Creativity.* Software Industry Center Working Paper, Carnegie Mellon University, Pittsburgh.

LUCAS R. (1988) On the mechanics of economic development, *Journal of Monetary Economics* **22**, 3–42.

PARK R., BURGESS E. and MCKENZIE R. (1925) *The City.* University of Chicago Press, Chicago.

REYNOLDS P., MILLER B. and MAKI W. (1993) Regional characteristics affecting business volatility in the United States, 1980–4, in KARLSSON C., JOHANNISON B. and STOREY D. (Eds) *Small Business Dynamics*, pp. 78–115. Routledge, New York.

REYNOLDS P., MILLER B. and MAKI W. (1995) Explaining regional variation in business births and deaths. U.S. 1976–88, *Small Business Economics* **7**, 389–407.

REYNOLDS P., STOREY D. and WESTHEAD P. (1994) Cross-national comparison of the variation in new firm formation rates, *Regional Studies* **28**, 443–456.

SAXENIAN A. (1999) *Silicon Valley's New Immigrant Entrepreneurs.* Public Policy Institute of California, San Francisco.

SCHUMPETER J. (1942) *Capitalism, Socialism, and Democracy.* Harper Torchbooks, New York.

STERNBERG J. S. and LUBART T. I. (1999) The concept of creativity: prospects and paradigms, 3, in STERNBERG R. J. (Ed.) *Handbook of Creativity.* Cambridge University Press, New York.

STERNBERG R. J. (1999) *Handbook of Creativity.* Cambridge University Press, New York.

STOREY D. (1994) *Understanding the Small Business Sector.* Routledge, New York.

STUART T. and SORENSON O. (2003) The geography of opportunity: spatial heterogeneity in founding rates and the performance of biotechnology firms, *Research Policy* **32**, 229–253.

THOMPSON W. (1965) *A Preface to Urban Economics.* Johns Hopkins Press, Baltimore.

TOLBERT C. and SIZER M. (1990) *U.S. Communing Zones and Labor Market Areas: A 1990 Update.* Staff Paper No. AGES-9614. Rural Economy Division, Economic Research Service, US Department of Agriculture, Washington, DC.

YOON I. (1997) *On My Own: Korean Businesses and Race Relations in America.* University of Chicago Press, Chicago.

Part VIII
Clusters and the Growth of Cities

[22]

'Old Economy' Inputs for 'New Economy' Outcomes: Cluster Formation in the New Silicon Valleys

TIMOTHY BRESNAHAN,[a] ALFONSO GAMBARDELLA[b] and ANNALEE SAXENIAN[c]

([a]Stanford University, Stanford, CA. Email: tbres@stanford.edu, [b]Sant'Anna School of Advanced Studies, Pisa, Italy. Email: gambardella@sssup.it and [c]University of California, Berkeley. Email: anno@socrates.berkeley.edu)

This paper discusses the results of a two-year research project on the sources of success in regional clusters of entrepreneurship and innovation like Silicon Valley. Our project has studied a number of locations, most of which have shown spectacular rates of growth of information and communcations technology-related activities during the 1990s. Our case studies comprise some emerging regions, notably in Ireland, India, Israel and Taiwan, along with more advanced areas like Northern Virginia in the US, Cambridge, UK, the Scandinavian countries and the Silicon Valley 40 years ago by way of the memory of one of its 'father founders', Gordon Moore. Through visits, interviews and other materials, we uncovered some regularities about the determinants of success of these entrepreneurial-led models of economic growth. We find that the economic factors that give rise to the start of a cluster can be very different from those that keep it going. Agglomeration economies, external effects and 'social increasing returns' of any sort arise almost naturally after a cluster has taken off. But the most difficult and risky part is to get the new clusters started. At that stage, 'old economy' factors like firm-building capabilities, managerial skills, a substantial supply of skilled labor and connection to markets were crucial for the take off of these 'new economy' clusters (including Silicon Valley 40 years ago).

Industrial and Corporate Change Volume 10 Number 4 2001

1. *Introduction*

This paper discusses the sources of success in regional clusters of entrepreneurship and innovation like Silicon Valley. It draws from the results of a two-year research project carried out by a large international and interdisciplinary team. The project has looked at how different forces, including public policy, business strategy, and institutions at regional and national level have combined to encourage the emergence, growth and maintenance of clusters, and how they might achieve the level of positive feedback and ongoing success of Silicon Valley itself. Our viewpoint has been comparative. We have used both in-depth case studies and statistical methods to analyze examples of early, preliminary and partially successful development of regional clusters in information and communications technology (ICT), inside and outside the United States.

The topic of this paper has been difficult to attack until now, both for practical and conceptual reasons. On the practical side, international comparison of Silicon Valley imitators has suffered the difficulties of comparing a roaring success to some bitter failures. Where Silicon Valley is entrepreneurial, decentralized, and only loosely and flexibly connected to broader national institutions, many efforts at imitation have been government sponsored, top down or tightly linked to established firms, perhaps 'national champions'. This wide divergence makes an analytical approach difficult, for one cannot easily investigate success drivers for getting over the positive feedback hump by looking at places that have been a roaring success or at places where so little of the logic is right.[1] As a result, opinion in the policy arena has fractured, with some believing that there is a magic formula for creating a new cluster of innovation and others thinking that the Silicon Valley model is unique and inimitable outside the United States. Much of the same 'all good/all bad' thinking characterizes analyses of Silicon Valley as well. At one stage, the region and its companies could do no wrong, at another, no right.

Against this background, the mission of our project has been to analyze a number of different attempts to gain national economic advantage from regional clusters of development in ICT. We have sought to avoid both the hagiographic 'SV is great' mode and the hypercritical 'there is no new economy' mode. Our main goal has been to assess and possibly identify the sources of long-term economic growth in clusters of industrial activity.

We define a regional cluster simply as a spatial and sectoral concentration of firms; and we measure success by the ability of the cluster as a whole to grow, typically through the expansion of entrepreneurial start-ups. In our project, we have then identified a number of nascent clusters of technology-based innovative activity around the world. They are all distinguished by entrepreneurship and growth, and they all have substantial focus on ICT. In

[1] The study that overcomes this difficulty most successfully (Saxenian, 1994) looks only at one country and still finds itself comparing a region (Silicon Valley) wherein many structures support entrepreneurship to one (Route 128) where few structures do.

short, they are all 'success stories', and in this sense they are all 'young Silicon Valleys'. Indeed, as we shall see below, one nascent cluster in our study is the Silicon Valley of four decades ago, and we asked one of its 'father founders', Gordon Moore, to recount this story. With that one exception, however, we do not yet know if the other centers that we have looked at will achieve the level of commercial and technological success of the mature Silicon Valley. It is not even clear whether these nascent clusters will build sophisticated support and service industries (like venture capital) and experienced entrepreneur-mentors that would make founding a new firm as easy in them as it is in Silicon Valley today.

Our sampling scheme allows us to take up two kinds of questions, but it rules out others. It lets us examine the similarities and differences among emerging clusters, but does not let us contrast these nascent clusters with failed efforts to get clusters going. The reason for this is simple. Unsuccessful cluster attempts, such as Malaysia's Multimedia Super Corridor, have typically failed to provide useful contrasts for the same reason as earlier literature failed. The similarities and differences among the nascent clusters we have selected, however, provide a rich field for investigating other analytically important questions. For instance, what factors are systematic and what are unique feature of the western United States in these emerging regional clusters? Answering this question is critical for breaking out of the 'recipe' mode of analysis that has served policy formation so badly in recent decades.

Specifically, our study examines the forces that create new clusters of entrepreneurship-led growth. More broadly it is an empirical investigation into the microeconomics of growth and trade. Our international and interdisciplinary team has compared nascent regional clusters both inside and outside the United States. We pose two interrelated questions. First, how did these 'imitators' of Silicon Valley initially become centers of ICT-related growth; and second, what accounts for the subsequent success (or lack thereof) of these clusters to ensure that success builds on success in a self-reinforcing fashion. The goal is to understand how factors including national and regional policy, business strategy, and national and regional institutions combine to encourage the growth of regional clusters.

Our team has included the Stanford SIEPR-affiliated economists Tim Bresnahan, Kevin Davis, Michael Horvath and Scott Wallsten; business people/entrepreneurs, notably Ralph Landau and Gordon Moore; and management specialists and economists from other universities or institutions who work on the development of technology clusters, including AnnaLee Saxenian from Berkeley, Suma Athreye from The Open University in the UK, Ashish Arora from Carnegie Mellon University, Erran Carmel from American

University, Catherine de Fontenay from the University of New South Wales, Alfonso Gambardella from the Sant'Anna School of Advanced Studies, Tamah Morant from the University of North Carolina, John Richards from McKinsey & Co., and Salvatore Torrisi from the University of Camerino.

We have looked at a number of different clusters: outside the US in Ireland, Cambridge, UK, Israel, Scandinavia, India and Taiwan, and, within the US, Northern Virginia in the present and Santa Clara County (Silicon Valley) in the 1960s. In spite of their significant differences, India, Ireland and Israel are the prototypical cases of nascent ICT clusters that we have in mind.[2] They have all exhibited a significant acceleration in the production of ICT (especially software and services) during the 1990s. Their ICT growth has been exceptional according to practically all major indicators: annual double-digit growth in the number of new firms, in ICT revenues and employment, in exports; increasing share of ICT of total exports of the country (up to one-third of total exports in the case of Israel). Taiwan has shown a similar pattern in the manufacturing of PCs and related businesses. In the advanced world, Northern Virginia has shown similar features, including double-digit growth figures. Our two special cases are the two European regions. While Cambridge, UK has all the feature of a cluster in the sense we are defining it here, it does not match the patterns, and particularly the growth figures, of the other regions that we have looked at. This is not to say that Cambridge, UK has not been successful. But compared to the other regions (including Silicon Valley), the growth figures have simply not been the same. The Scandinavian story, as we shall see, is interesting because it is the only case in which growth was triggered by a whole different vertical ICT market, wireless hardware.

Our team of researchers has worked with the data on these clusters and we have also visited each of the regions to interview key actors. We have met a number of times over the course of a two-year period in order to review and compare preliminary results and analysis as well as to jointly develop conclusions. In what follows, we discuss the key findings of the entire research.

2. *Agglomeration Economies and External Effects*

Many of the existing theories of clusters of innovative activity focus on external effects and the resulting agglomeration economies.[3] One central

[2] Of course, when we mention India we do not mean the entire country but the few regions that have shown a significant growth of ICT during the 1990s.

[3] Some of the classical references are Krugman (1991), and more recently Porter (1998). Saxenian

feature of clusters of innovative activity is external effects among the technology firms located there. A local external effect is anything that raises the return to particular firms located in a region as a result of the location of other firms in the same region. External effects can be *direct*, as when managers or technologists learn about market or technical developments from colleagues in neighboring firms, when firms in closely related industries serve as one another's customers or suppliers, and so on. External effects can also be *indirect*, as when key inputs are in abundant supply or when the overall level of commercial technology activity is high. These indirect external effects arise from increasing returns to scale in the supply of key inputs such as venture capital, which may locate where entrepreneurship is dense but support the development of new entrepreneurial firms; a thick labor market in technical personnel; or commercially oriented activities in universities or national laboratories, to name just a few. Both direct and indirect external effects generate positive feedback loops that ensure that technology-related firms locate in regions where other technology firms are already located.[4]

These external effects have (at least) two distinct implications. One implication is for economic growth, both within a region and in the broader economy. External effects among innovative firms inventing general-purpose technologies are highly levered mechanisms for increasing the rate of growth of an overall economy. By raising the rate of return to invention, either direct or indirect external effects can, if the field of the invention is important enough, push the commercialization of valuable technologies faster and closer to markets. Similarly, in the case of indirect external effects the economic return to key inputs such as highly skilled labor, the knowledge to be found in universities or the market judgment of venture capitalists can be raised by clustering among firms. A second implication of external effects is that nothing succeeds like success under external effects, and the private return to participation in these clusters of innovation by entrepreneurs, venture capitalists, technologists and those in the key supplier industries can be enormous.[5]

(1994), which we quoted earlier, is another classical citation. Arthur (1990) discusses specifically increasing returns in the context of 'Silicon Valley locational clusters'. There are also, of course, the classical, early references to agglomeration economies that date back to Marshall (1920), Perroux (1950), Myrdal (1957) and Hirschman (1958). Finally, Jaffe *et al.* (1993) and Audretsch and Feldman (1996) provide evidence of the extent of geographically localized knowledge spillovers.

[4] The direct/indirect language follows the standard usage in external effects theories. This mirrors the classical Marshallian distinction, which pointed out three sources of external economies: a thick labor market; specialized input producers (and related increasing returns), which arise because close-by producers entail larger local markets; knowledge spillovers and 'untraded interdependencies'.

[5] This is how thoughtful analysis of agglomeration economies and external effects states what those concepts can explain. See, for example, Fujita *et al.* (1999), who argue against use of 'ad hoc dynamics' to take models too far past the question 'When is a spatial concentration of economic activity sustainable?'

These two points—external economies of scale, and the resulting capture of the rents by producers and the regions in which they are located and headquartered—explain the great interest in agglomeration economies.[6]

But network analysis alone cannot explain how regional clusters emerge. The existence of external effects explains why a region would like to have entrepreneurial-led growth and why the world's consumers would like to have more clusters founded. But examining the positive feedback in an existing successful cluster does not tell you *how* a cluster begins. Moreover, while clusters of entrepreneurial-led growth are valuable social institutions, we know little about why clusters begin where they do or how many clusters will emerge within a given industry. There is a strong element of social increasing returns to scale in the argument, which one would normally expect to lead to a limited number of clusters in any particular industry or technology. Understanding how nascent clusters overcome these limits has been an important part of our study.

The positive feedback logic of the theory and of established and successful clusters makes it difficult for analysts to identify a starting point. Positive feedback, when it is working, appears as a virtuous cycle, and, when it is not working, as a difficult chicken-and-egg problem. For a region that is not yet succeeding, 'nothing succeeds like success' is an empty remark. The related more analytical observation that all of the elements of success (entrepreneurship, venture capital, etc.) feedback to one another positively does not communicate what it is important to get started. Our analysis, by looking at the nascent clusters that exhibit significant signs of growth, lets us address the question of how to start. We return to this question of the factors leading to the start of a cluster in Section 3 below.

The second aspect of external effects, namely the possibility of social increasing returns and the resulting producer rents, also contains the analytical temptation to draw a false dichotomy which positions 'old economy' explanations of growth as alternatives to 'new economy' ones. Here we do not need to linger on a sophisticated distinction between what is old and what is new economy. For the purposes of our discussion the common sense with which these expressions are currently used is sufficient. 'Old economy' is a shorthand for a number of concepts: organizational and firm-building activities, investment in general and industry-specific human capital, larger

[6] For example, in our project, Horvath (2001) deals with the distribution of venture capital in the US, and finds that the venture capital industry is highly concentrated geographically, and that the funds tend to be invested close to where they are collected. This is suggestive of the fact that a venture capital industry tends to arise where there is a large market of potential users. In this respect, it looks very much like a specialized supplier industry in the Marshallian sense that arises locally because of larger local markets (an indirect external effect, using our earlier terminology).

companies and related economies of scale at level of the firms, lengthy periods of investment in capability before their exploitation. 'New economy' means instead entrepreneurship, economies of scale at the level of regions or industries rather than firms, external effects, etc.[7] Our point is that a strong opposition between old and new economy explanations of growth can lead to some analytical errors. Should we seek to explain the success of clusters *only* by external effects? Is this newly discovered feature of supply so important that it trumps all others? Much 'new economy' thinking goes down this false dichotomy path, drawing a distinction between the positive feedback forces that are at work in a cluster—guarantors of instant success in the most fervid 'new economy' thinking—and the supposedly 'ordinary' or 'old economy' work of firm-building, market-building, invention and commercialization.

The studies carried out in our project show that it is an error *either* to focus only on the external effects and conclude that new economy logic supersedes old, *or* to conclude that there is nothing to the positive-feedback external effects story at all. Instead, in the construction and maintenance of successful clusters, the new economy and old economy elements act as complements to one another; neither one can succeed without the other. Many governments have made the analytical error of focusing far too much on the second aspect of external effects, and have viewed clusters of innovative activity as no more than a ticket to producer rents. This has provided the intellectual foundation for largely failed policies that attempt to jump-start growth in clusters by directive policy. In our study, we have taken another look at that kind of policy, but do not dwell on it. Instead, we focus on places—and governments—that have got the mix of policies close to right.

3. *Starting a Cluster*

The first step in understanding the complementary relationship between external effects and ordinary investments is to step back from the already-built, successful Silicon Valley cluster of the present. We need to look, instead, at clusters in the making, whether in Silicon Valley in the 1950s and 1960s, or regions like those that we have looked in our SIEPR project, e.g. Ireland, Israel, India, Northern Virginia and Taiwan in the present. In each of these cases we found that there are some external effects at the nascent stage. Yet

[7] This form of organization distinction, rather than the sectoral one, underlies our use of the old/new economy language.

very important forces for success arise in the ordinary business challenges of building firms and building markets.

Put simply, our argument is that the processes of starting and sustaining a cluster have different economics. Starting a cluster involves, first, building the economic fundamentals for an industry or technology, and, second, finding the spark of entrepreneurship to get it going. Both of these are supported by a number of common elements in the regions that we examined—and it is striking how similar all the nascent clusters are to one another in this regard, and how much more similar to one another they are than to the current established and thriving Silicon Valley cluster. It is particularly significant that the Silicon Valley of 40 years ago is also closer to today's nascent clusters than either is to the Silicon Valley of today.

Many similarities between today's successful clusters and the Silicon Valley of the 1960s arise because founding a new cluster, or the early firms in a new cluster, is a very different entrepreneurial and economic activity than founding a firm in an established cluster. New clusters, including Silicon Valley in the 1960s, offer substantially less support to entrepreneurship in the startup or pioneering phase than does a mature cluster like Silicon Valley today. External effects—namely benefits to particular technology firms that arise from the presence of other firms or of support structures like venture capital—play only a small role in the early phases. Such benefits typically come later in the development of a cluster.

Among others, one similarity across the new clusters that we have examined is that they have taken advantage of a new technological and market opportunity that had not already been exploited: the integrated circuit industry in the Silicon Valley of the 1960s; the internet- and network-security markets exploited in Ireland and Israel today; specific opportunities like software demand following the YK2 or the Euro problem in India; the hardware and equipment opportunities in new kinds of devices like cellphones and PDAs embraced in Scandinavia and Taiwan. The rationale for this is not hard to guess. Markets with substantial producer rents, like ICT today, are characterized by powerful forces that make a direct assault on an existing market position unpromising. In the first place, the advanced technological capabilities and market connections of incumbents may offer a commanding lead. These markets often have barriers to entry created by the deployment of commercialization assets in established markets. At the same time, there can be market forces associated, for example, with *de facto* standard setting, as in much of computing, or political forces that protect existing national champions, as in parts of telephony, which make entry difficult. Broadly speaking, the lesson from our case studies is that to blossom, the new-cluster entrepreneurs have to turn away from established sources of rents to define new ones, and (at least initially) make their relationship with existing technologies and clusters complementary rather than competitive.

In this respect, practically all the technologies developed in the new clusters that we have looked at in our study are complementary to existing ICT technologies, mostly sold by US-based firms or US-linked multinationals. Since ICT is used in large complex systems, it is important for the new entrants to have their inventions and advances linked with other inventions and advances. At the same time, this means that the new clusters can take advantage of the sizable and growing demand for the leading technologies. Our clusters in Ireland, India, Israel and Taiwan come from regions that, for one reason or another, have easier potential interactions with the US market (language, cultural connections, diaspora, etc.). They have then been able to take great advantage of the significant US demand for ICT products, services and components, during the 1990s. Combined with other factors that we shall highlight below (particularly underemployed skilled labor), this has given rise to a powerful mechanism for sparking off growth opportunities in some ICT niches and segments.

At one level, this is no more than the obvious remark that 'demand is important for growth'. In a debate over-emphasizing supply-side factors like agglomeration economies and external effects, however, this remark is often overlooked. That market and demand factors lead to linkages to the US does not reflect a 'US-centric' view. The linkages with the US are merely a powerful example for pointing out how important it can be to be connected to a sizable and growing source of demand. Finding the source of demand that may spark off the growth of the cluster can be critical for its rise, and in many respects it should be one of the policy focuses in this arena.

In brief, to obtain growth one cannot point only to the 'surfing' of agglomeration economies, but also to the underlying 'wave' of technological and market opportunity. For many of the regions that we have examined in our project, the wave was the great expansion of ICT following the commercialization of the internet.

Another similarity across our new clusters is the degree of investment, effort and building needed to set up the background for an innovation cluster's take off. To take up our earlier remarks again, development of the ICT businesses is not the magic exploitation of some 'new economy' rules. Instead, it takes years of firm-building and market-building efforts. The long-term investment in education of a skilled labor force has been critical in a number of regions, notably in Taiwan, Ireland, India and Israel. The supply of highly skilled workers can come from any of a number of sources, but needs to be plentiful: Stanford-trained scientists and engineers to Silicon Valley, military training and Russian émigrés to Israel, alumni of 'beltway bandit' federal contractors in Virginia, and an educated population earning far less than world-standard wages in Ireland and India. While it is critical to invest in the assets that will permit the emergence of a cluster, there is no magic recipe (take one great university, 47 venture capitalists, and . . .) that works. Instead, a number of different routes exist to building the backdrop—technology opportunity, educated labor, flow of entrepreneurial talent, etc.[8]

Sometimes these long-term investments in national or regional capabilities can grow for a long time in what seems like a low-return mode before the take off into cluster growth.

The story of early entrepreneurship and the early integrated circuit industry seen from the perspective of Fairchild and Intel illustrates this point very well. This is the story of the developments that put the Silicon in Silicon Valley, though there was the beginning of a high-tech cluster in Santa Clara County before these times.[9] What was important in determining the success of the early Silicon Valley back in the 1960s? Moore and Davis (2001) bring forward several elements. First, there was a rich technological opportunity in the semiconductor business. Second, there were immediately available markets such as consumer electronics and defense. The creation of general-purpose semiconductors led quickly to sales in those markets, where the value of a miniaturized component was clear. There was also the prospect of whole new uses of semiconductors in information technology industries, a real prospect for long-term growth. Both the technological and the market opportunity were separate from the existing high-tech (tube-based) electronic industry, and provided an advantage to producers located far from existing sellers.

How was this technological and market opportunity exploited? The Moore–Davis story is as far as can be imagined from one in which young entrepreneurs instantly succeed in a supportive environment of external effects. None of the modern institutions of Silicon Valley existed, so none of the incoming benefits of external effects were there. No mentoring from experienced entrepreneurs now working as venture capitalists, no easy access to the required skills or to wise and experienced thoughts about business models, no networks of connections to supply partners and marketing partners—none are part of the early period story. Instead, the story is a one of investment in human capital, firm building and market building processes

[8] In our project, the story of Taiwan is told in Saxenian (2001); Arora *et al.* (2001) discuss India and Ireland; de Fontenay and Carmel (2001) studied Israel.

[9] Hewlett Packard, for example, was established years earlier and it was growing at the time of the founding of the integrated circuit industry in Silicon Valley.

that took a long time and quite serious effort and risk. Allowing for changed circumstances, this same element is present in many of our regions. Building a new capability at the firm level which will lead to local increasing returns and positive feedback does not involve anticipating or exploiting those high-payoff features, it involves investing in the key assets which will permit later collective payoffs.

To summarize, the forces underlying the emergence of a cluster differ from those needed to ensure its continued growth. While increasing returns and external effects can keep a cluster going, the initial spark is more difficult to obtain and more risky to pursue. Our research suggests that these include the importance of being linked to a sizable and growing demand as well as the availability of a proper supply of key factors like, in the case of ICT, skilled labor. Other critical factors are firm- and market-building capabilities. These require significant and systematic efforts by the 'pioneers' of the cluster to promote organizational and technological capabilities of various sorts, create new firms and institutions, etc. Finally, another factor is plain 'luck'. Founders such as Moore recognize that there was considerable uncertainty at the beginning about the potential size of rents and the appropriate firm and industry structure to pursue them. Similarly, the opportunities for many of the newer cluster involved a particular matching of regional supply capability to world demand.

In fact, there is a logical argument for suggesting that luck plays a role in this context. We noted that nascent clusters, and the entrepreneurs operating there, have to bet on new trajectories before they manifest their potential. But this also means that they have to bet on an opportunity before it is clear to everybody else that it is indeed an opportunity. Some degree of risk is therefore unavoidable. At the same time, this means that only some of these opportunities (and most likely few of them) will materialize. Many attempts at creating new clusters and successful new firms in certain industrial or technological trajectories will fail, and they will fail in spite of the fact that the key actors have done all the right things that are to be done in these contexts. In this area it appears that luck and skill are complements; those initiatives that embody a superior business model or technology are more likely to find the 'luck' they need.

This string of similarities across our cases corroborates our earlier remark that the 'new economy'/'old economy' distinction is a false dichotomy. There are two errors. It is just as incorrect to say that clusters of innovative activity can take off as if by magic as it is to deny the huge national and regional advantages that accure from the existence of an established Silicon Valley cluster. The truth is that old economy hard work, both in company building and in regional investments such as education, and new economy external effects are complements—each is more valuable with the other than without it. There is a great deal of truth to the 'external effects' theory that entrepreneurship is easier in a cluster than outside of it. This is the realization of the theoretical idea of 'social increasing returns to scale'. Yet that certainly does not mean that the world can support only one cluster, and the growing pressure on the limited stocks of skilled labor and land in the US and Silicon Valley are powerful forces favoring emergence of new clusters.

4. *No 'Recipes' but Some Deep Regularities*

4.1 Highly Skilled Technical Labor

All our regional stories point to the importance of highly skilled labor as a precondition for the growth of an ICT-based entrepreneurial cluster—Taiwan, Ireland, India and certainly Israel. In some cases, underemployed skilled labor is close to being the 'only' factor (India, but also Ireland or Israel), or at least the one that spurred the rest. Further, some of our other stories reveal that there is a role for universities both as a source of skilled labor and of technologies that are exploited for export and growth—e.g. Cambridge, UK and Silicon Valley in the 1960s.[10]

Taken as a group, however, these stories are not at all encouraging of the simple 'recipe' view of universities and higher education in starting a cluster. Our cases demonstrate that there are a number of different ways to achieve a supply of skilled labor and that it is the ultimate outcome (a highly skilled labor supply), not the particular mechanism (a university) that matters. Large firms can also play a critical role in growing the skill base. While the role of universities like Stanford or Berkeley has been widely emphasized in the Silicon Valley story, one should not neglect the potential training provided by established firms like Hewlett-Packard or Intel. Large firms often nurture technical competencies. For example, individual researchers can use equipment or they can be part of research teams that would hardly be available outside the leading companies. Similarly, many offer managerial training, and possibly even managerial connections. Moreover, this technical and managerial training can encourage spin-offs. This raises a classical problem of 'private' vs. 'social' interest. As pointed out by Moore in his study with Davis, established companies warn against spin-offs. Even though Moore himself notes that one can benefit from spin-off supplier industries (e.g. chip

[10] Athreye (2001) discussed the story of Cambridge, UK in our project.

manufacturing equipment), established companies clearly look unfavorably to the spillovers that they create by training people who would then use these competencies outside the firm.

Our stories pointed out other sources of training, apart from universities and large firms. In Northern Virginia the development of sophisticated technical capabilities in ICT, and particularly in communication technologies, has stemmed from the existing bases of competencies provided by years of contract research for the government and the defense department in the area. Government contracts have for many years nurtured contractor firms, and the related technical skills in such firms. Once defense procurement slowed down, and new internet-related and other communication technology opportunities rose, skilled labor in these fields were available in the area, and this implied the supply of skills that were fruitfully used for launching the cluster. In Israel, the military proved to be the key supply of technical skills, along with some leading technical universities (and particularly the Israel Institute of Technology, or Technion). Finally, the supply of skills can come from outside the region. The clusters in Taiwan and India have drawn heavily on US-educated Chinese and Indian engineers. In Israel as well, immigration (largely from Russia) has been an important source of skill. Similarly, the nascent Silicon Valley attracted engineers from all over the United States (Bresnahan *et al.*, 2001; de Fontenay and Carmel, 2001; Moore and Davis, 2001; Saxenian, 2001).

In sum, the conventional wisdom lumps education, skilled labor and universities in a single idea and often overlooks alternative mechanisms for achieving a skilled labor pool. It is therefore important to stress that a university *per se* is not essential to the emergence of a successful cluster. This qualification is useful for at least two reasons. The first one is that it brings further support to the rejection of a formulaic recipe for the rise of the clusters. Putting a university at the center of the cluster can help, but it is neither a necessary nor a sufficient condition. Second, there can be different sources of skills in different regions, and—given that it is the availability of the skills that matters—regions can look for the most appropriate way (or mix thereof) for acquiring skilled labor, from universities to larger firms and other local as well as distant institutions. The policy implication is also straightforward—forming or attracting skilled labor, rather than a particular means for doing it, is the crucial aim.

4.2 Managerial Labor

The contributions of returning expatriates also suggest that the key human capital investments are managerial as well technical. Most of the nascent clusters we have analyzed are focused on ICT—as is much of the activity in Silicon Valley. They therefore offer difficult management and marketing challenges because these sectors present nontrivial problems of commercialization of scientific and technical opportunity.

Firms in new clusters and sectors develop sources of non-technical (or not narrowly technical) human capital in a variety of ways. Silicon Valley itself, being perhaps more 'first' than the other clusters, has produced an indigenous supply of scientist-managers and engineer-managers by having scientists and engineers learn a second skill set (management), often by experience, in the early stages. Of course, as Silicon Valley matured, it used a number of other models to create dual-knowledge-base managers, including more experiential training (think of Steve Jobs or Larry Ellison,) mentoring by experienced entrepreneurs or venture capitalists of younger managers-to-be, and career moves from established firms in the industry (in and out of Silicon Valley) into new entrepreneurial firms.[11]

Later-growing clusters have not needed such heroic means of creating managerial capabilities as did the early Silicon Valley. ICT markets are more established, and technical people native to a region can achieve training in a number of ways that introduce them both to the management problems of ICT enterprise and to world markets. Multinational enterprises (MNEs) have played an important initial role in this regard—e.g. Taiwan, and possibly Ireland, certainly Israel.[12] Potential engineer-managers can work in MNEs in their home region and gain experience in a second skill set, while also learning about world markets. This mechanism for building managerial human capital can later be deployed in indigenous firms. This is an advantage of late development.

A second source of the same human capital building arises through the return of expatriate engineers and managers working in established clusters elsewhere, especially in Silicon Valley itself. When repatriated, these individuals bring with them experience and knowledge of management in ICT firms, in this case even in ICT startups. And, since their work was overseas (often in the largest market, the US) they return with connections to and knowledge of ICT markets. This mechanism is the long-distance analog of the one that has been going on within Silicon Valley for some time, namely

[11] If anything, this problem was quite clear in the view of the early Silicon Valley, when Gordon Moore and the others, as emphasized in his story, had to hire a fully fledged manager, outside of their technical group, Ed Baldwin, who took up the management for the company and taught them stuff that is nowadays commonly taught in MBA courses (Moore and Davis, 2001).

[12] The more traditional role of MNEs in providing technology transfer and a mechanism for connecting to world markets is not crucial in our cases (e.g. Arora *et al.*, 2001; de Fontenay and Carmel, 2001).

development of human capital within established companies followed by a spinout or a movement of experienced workers to a new firm. The difference is that these spinouts and this labor mobility go to a place where the engineer or manager would like to live—India, Israel or Taiwan—rather than to a site just a few miles away.

4.3 New Firm Formation and Firm Building

We have already noted that firm-formation and firm-building capabilities are important elements of successful clusters. In this respect, the studies in our project show that there is a difference between the growth of firms and the growth in the number of firms in a cluster. Growth in the number of firms may not lead to the sustained creation of substantial economic rents. Similarly, growth in the size of the high-tech sector, as in the early phases in Taiwan and partly the current phase in India, does not necessarily ensure continued growth. The question is whether some firms in the clusters emerge out of the class of small entrepreneurial startups and stake out independent positions in world markets.

To be sure, growth of firms could be a signal rather than a cause of success. A firm that pioneers an important innovation or one that generates a new market will most likely grow. In turn, the growth of local firms can be critical for the success of a cluster. Growing firms create demand for other types of employment and not just technology (e.g. manufacturing, marketing). Moreover, as many technology markets have shown today, most of the rents are downstream. Companies with market power have the resources to make additional investments, with further implications for growth. More generally, larger companies, like Intel or Hewlett-Packard in Silicon Valley, Ericsson and Nokia in Scandinavia, produce backward and forward linkages which systems of smaller firms may not be able to produce, at least to the same extent. These are all important factors for sustaining the growth opportunities.

Our regional studies show that the growth of firms is not a natural outcome of the rise of a cluster. For instance, the new companies could be consulting or service firms established by individuals whose primary job is elsewhere (e.g. university professors). In these cases the effort on firm building would be more limited than those by individuals who would make primary bets on their companies. Also, some entrepreneurs prefer not to have their firms grow. Growth implies new management challenges, possible decentralization of powers and greater tension, risk or responsibilities. It is not uncommon in some locales, as in our case of Cambridge, UK, to find individuals who are not interested in these challenges. If commitments to firm-building are important this could limit the growth of a cluster, particularly compared with the risks and investments in firm building made by the early founders of Silicon Valley and by some of the entrepreneurs that populate the Indian, Israeli or Taiwanese scene today. In sum, we suggest that the growth of companies, and not just the growth in the number of firms, is a signal of the success of the cluster. These are the firms that will become one of the sources of increasing returns for the continuous growth of the cluster, in the form of training for potential spinouts, development of managerial and technical competencies, along with various forms of backward and forward linkages.

The growth (or lack thereof) of local firms may also stem from comparative advantage. Indian engineers and Cambridge boffins may be reacting to the incentives they face, and supplying only what is economically efficient to the world economy (in these cases, the services of highly skilled individuals.) Over time, this would lead to a classical gain from trade, i.e. wages for highly skilled labor far closer to the global level for labor of that type. Israeli high-tech companies, for example, are often bought by American firms and moved to the US as soon as they reach a certain size and stage in their evolution. This may not be bad for the Israeli economy if the companies are paid the right price, notably the present value of the company given its expected opportunities. If the market for companies showed few imperfections in this respect, the gains from trade could produce notable benefits. If they remained in Israel, these companies might be limited by the size of the domestic market. Since development and commercialization (unlike initial research) benefit from exploitation of larger markets, there may be an advantage from selling the companies once they reach the stage when a large market becomes an important asset. This then shifts the problem from a general preoccupation with the acquisition of the best domestic firms, toward the potential imperfections in the international market for companies.

Another regional path is specialization in the creation of new innovative companies, with acquisition and commercial development of those companies or the technologies they create by established firms outside the region. Success on this path requires a fertile field of technological opportunity. If the field matures or is exhausted there may be limits to creating rents from selling new technologies. This too can be seen as a problem of setting the right price for the technology, or for the acquired company. Research suggests that asymmetric information leads to payments for technology licenses that are below the actual present value of the firm and to the acquisition of technology firms at prices below their long-term contribution (Arora *et al.*, 2001). Yet even under these conditions comparative advantage may still be critical. It may be that focusing on the early part of the innovation cycle is the right specialization for Israeli companies. In turn, the advantages of specialization would increase if the imperfection of the global technology markets were reduced, and the companies sold for a price closer to the 'right' price.

4.4 Connection to Markets

We have already noted that our emerging ICT clusters in Israel, India, Ireland and Taiwan all had significant ties with the Unites States, which helped them to exploit the ICT-intensive US growth of the second half of the 1990s. In this respect, it was critical for these regions to position themselves in product spaces that were complementary to the main sources of demand (notably Silicon Valley and the US) rather than directly competing with them (Arora *et al.*, 2001; de Fontenay and Carmel, 2001; Saxenian, 2001).

Our case studies suggest that there can be two patterns of connecting to markets and sources of demand. The first one is that of the emerging countries mentioned above. Here the linkage is given by the relationships with the main market (the US), and the complementarity of the products of the clusters with the existing leading technologies, particularly those produced in the US themselves. The second model is the one epitomized by our case of the Scandinavian countries. The key here is to position the companies in areas not covered by the existing leaders, once again the US firms. The problem is even more complex because, as the Scandinavian stories in our project point out, the issue is not simply to cover products not produced by the leaders. In ICT, the leaders—and particularly the US firms—have been able to occupy entire vertical markets, and in most cases they have occupied them on a global scale. This has left little room for other global players to occupy even parts of these vertical markets. The only opportunity left was to occupy other vertical markets, with potentially global, or at least continental demand, which had not been occupied by the leaders (see Richards, 2001).

Some leading European firms, and particularly the Scandinavian companies Ericsson and Nokia, successfully pursued this strategy. In this respect, the study by John Richards (2001) in our project shows that the success of the Scandinavian model depended in good part on the same factors that we have highlighted in other cases—highly educated workforce, supply of technically and managerially skilled people, and connection to demand (particularly continental demand). The peculiarity of the model, however, is that it focused on a vertical market that was not occupied by the leading US firms, notably the wireless hardware segment of the ICT business. Moreover, as we shall also see later, the opportunities in this area were further raised by one of the most successful institutional European accomplishments in the ICT business during the last decade, i.e. the establishment of the GSM standard for communication throughout the continent. This was a key event for creating a continental demand, which enhanced the growth of this segment and benefited the producers that had occupied this area.

The connections to market and to the leading areas worldwide also open up the question of the potential advantage of being 'out of town' that may accrue to newcomers or to latecomers. The point is that being 'out of town' may imply lower stakes in existing technologies and activities, and this may help new clusters to seek new sources of rents or to tap new product spaces and technologies. Of course, with the advantages of being 'out of town' for novelty come disadvantages for connectedness to world markets. One particular disadvantage for not being connected to the leading clusters, or to the leading centers of technological growth, is that of not being able to take advantage of some of the results of external effects, such as the availability of the 'right' labor. But this also explains the rise of some of the new regions that we have discussed in this paper. As noted, some of them were indeed 'out of town', but they all had local sources, or characteristics—such as excess supply of skilled labor—that made up for the disadvantage they faced.

Thus, on the one hand, countries like Israel, India, Ireland and Taiwan compensated for the disadvantages of being out of town by virtue of focusing on activities that were complementary to those of the United States. They then combined the advantages of being out of town, which enabled them to focus on new areas, with the ability to link with the sources of growth (whether demand or technical linkages and the like). On the other hand, the Scandinavian model implied that the companies had the opportunity (and possibly the fortune) of exploiting a new vertical market, and in this process they also built up the new linkages (technical, with demand, etc.) that were critical for the rise of the new cluster.

In these last several sections, we have emphasized the deep regularities cutting across all our study regions in the domains of human capital of two kinds: market connection and new firm formation. In emphasizing the deep similarities, we do not want to gloss over the considerable differences across these countries in institutions and in market mechanisms. These matter considerably for the ways in which the various deep capability-building goals were accomplished in different places.

5. *Cooperation vs. Competition Among Clusters in the World Economy*

The new clusters that emerged in the late 1990s have had to deal with the dominance of the most important existing technology areas by US firms, largely firms located in the western US. From the perspective of rich places like Northern Virginia or Israel, or poorer places like Taiwan or Ireland, or vastly poorer places like India, linkages to established centers of technology are very important in the growth phase. Maintaining these linkages can be a difficult management problem for companies that have to do something new, but at the same time have to find ways to stay linked to the old. The best solutions appear to be participation in a worldwide production network that is an extension of the Silicon Valley network itself.

We have already noted the importance of finding a new area of technology or product area in which to succeed. This involves cooperation with existing firms and sales of complementary rather than competing products—at least in the short term. This pattern is not, however, a policy recommendation so much as it is an observation of why and how certain more narrowly imitative policies are likely to be ineffective. The real policy implications arise from thinking carefully about the particular sources of advantage for a nascent cluster and why that source might yield short-term complements with the potential to become long-term substitutes.

All the nascent regions we have examined have escaped from the belief that cooperation with existing richer economies is 'colonialist'. Linkages with the US have been critical to all, in one way or another. India and Taiwan are linked to the US via outsourcing of software services and manufacturing respectively. Israel and Taiwan are also linked to the US by a returning group of expatriates who have worked there, and who see the benefits of long-distance collaboration. There is, in these cases, a flow of people and ideas back and forth between the existing cluster and the nascent ones.[13] The diaspora has been particularly valuable to the nascent clusters that we have looked at in emerging markets. Technology transfer (in the narrow sense) is not very important in these cases. Rather what is transferred is primarily organizational models, a valuable piece of understanding for a nascent cluster, and the opportunity to apply that knowledge in a new domain.

This pattern of connection-led growth varies across the areas we study. India began simply by arbitraging the differences in engineering labor costs between that country and the US in outsourcing. There are no guarantees that such an activity will lead to entrepreneurship-led growth, though many valuable assets (such as knowledge of and connections to the US market and

[13] The role of social structures and community ties may also be strong. This is particularly true within bodies of entrepreneur-managers who participate in a brain drain and then a reverse brain drain (Arora *et al.*, 2001; Saxenian, 2001, 2002) or who have common experiences, e.g. in the military (de Fontenay and Carmel, 2001).

up-to-date technical skills) were gained in the process. Taiwan similarly began as a source of low-cost labor for manufacturing PCs, which had become too expensive to make in Silicon Valley. This was strongly complementary to Silicon Valley, and, combined with economic and institutional arrangements inside Taiwan which permitted and encouraged entrepreneurship, led to considerable growth of technical and market capabilities and ultimately an indigenous industry. Entrepreneurs in Ireland, Israel and Northern Virginia have been content to position themselves as niche players, producing in areas where they have an advantage (either individually or regionally) and thinking of themselves as linked to and complementary to Silicon Valley.

This pattern has also meant a significant departure from earlier relationships between the US and other countries' high-tech industries. For example, the pattern looks different from the more competitive, and even confrontational, relationship between the US and Japan in semiconductors in the 1980s. The increasing competitiveness of Japanese semiconductor memory producers directly threatened the market position of the US producers. The case of Cambridge, UK provides another contrast. The Cambridge cluster looks more similar to Silicon Valley in terms of product space than the other emerging countries that we have examined. At the same time, while Cambridge has certainly grown, its growth rates have been steady and 'normal' compared to those of some of the emerging countries, and of Silicon Valley itself. This is why we think of Cambridge as a 'partial' success story when compared to such high-growth clusters.[14] The products of firms in Cambridge were similar to those of Silicon Valley, with its first-mover advantage, and so could only cover spaces in the worldwide market that were not already exploited by the leader. By contrast, companies in the emerging countries engaged in complementary activities. This allowed them to avoid competing directly with Silicon Valley, while taking advantage of the growth rates of the latter through their complementary relationships.[15]

The worldwide economic forces that have enabled these complementary patterns between the US and the emerging countries in ICT are important. First, the ICT industries grew more rapidly and in more varied a way than could have been anticipated. A number of technologies turned out to have significantly larger markets than originally anticipated—the PC and associated complementary hardware and software, and the internet and associated close complements come immediately to mind. Further, many of these new

[14] Another possible explanation of Cambridge's only 'partial' success is the lack of market access. No mechanism like returning expatriates created strong links to major markets and the domestic market was small (Saxenian, 1988).

[15] See Athreye (2001) for details on the Cambridge, UK case.

technologies raised rather than lowered the demand in existing markets—the internet has dramatically increased the demand for PCs and for mainframe computers along with the demand for telecommunications transport and switching services, for example. All this has provided tremendous market and technical opportunities, and put pressure on the existing innovation-supporting resources, notably engineering labor and conveniently located land, in the existing successful clusters in the US.

Immigration and the physical expansion of Silicon Valley have provided partial offsets to these powerful forces, but both face real limits. Not everyone wants to live in the United States, and what had once seemed a vast expanse of cheap land used in fruticulture was discovered to be a valley—with walls. Out of people and out of conveniently located land, Silicon Valley was in a position to cooperate with nascent clusters.

The countries and regions that responded come from the 'also rans' of economic development in the 20th century. Apart from the Scandinavian countries and Northern Virginia, all were relatively poor and peripheral. Why have the less advanced economies responded so well to these opportunities? Even more interestingly, why have many richer countries (e.g. the continental European countries and Japan) been less responsive than these 'also rans'? The first part of the answer was largely given in the previous pages. In particular, our regions offered underemployed skilled labor and connections with the US that helped match their excess supply of skills with the excess demand for (ICT) skills in the US. While our analysis does not extend to countries like Germany, France and Japan, it is clear that they also offered ample, if not greater, supplies of both technical and managerial skills. A strong possibility for explanation is that the human capital in these other countries had substantially higher opportunity costs. A German engineer has lucrative opportunities in existing industries such as automobile, chemicals and electronics. The incentives to set up a risky new venture in a new industry could not easily match the gains from working in such established industries.

To summarize, as new firms or regions emerge in a context where leading firms, regions and countries exist, the former have advantages and disadvantages. The 'also rans' can exploit the opportunities created by the markets and institutions in the ICT sector, if they are well connected to them, without bearing the costs of creating them. However, because the major product spaces have been filled, they can only serve niche markets or focus on complementary activities. But what about the longer (or medium) term future? The initial successes imply that competencies are gradually accumulated. The longer (medium) term implications are not straightforward. Will there eventually be competition with the leaders? In the specific context we are analyzing here (India, Taiwan, etc.) the situation is still fluid, and it is difficult to give definitive answers.

One cannot rule out *a priori* that some leading companies in the emerging countries will eventually compete with companies in the leading countries. The leaders, however, will probably keep moving up the frontier. The situation may turn out to be like the classical one in product markets wherein mature products are taken up by the followers, and the leaders develop more advanced technologies. However, other trajectories are possible. For example, firms in the new countries may specialize according to comparative advantage. The Indian firms, for instance, may further their initial comparative advantage of being able to organize large-scale projects for developing relatively lower end software at low costs. Comparative advantage has certainly played a role in Taiwan, which has developed sophisticated chip-manufacturing capabilities complementary to chip design in the US. The improved functioning of global markets may affect these processes as well. For example, it may be easier today to operate an international market in firms. This would favor the acquisition of companies overseas, and reinforce the existing division of labor, as noted earlier in the case of Israel.

6. *Policy Issues and Conclusions*

The core of good advice to policy is concrete understanding of the forces driving the economy. Two very different bodies of understanding have been brought to bear on high-tech, entrepreneurial-led growth. Should we understand the foundation of new regional clusters of entrepreneurial activity using new economy theories of increasing returns, or should we focus on comparative advantage? Our answer is an element of each. Much of the opportunity for new regions arises because old regions find themselves running up a steeply rising supply curve of land and of highly skilled labor. Even the very substantial migration of highly skilled labor to the existing Silicon Valley cluster has crashed into this classical diminishing-returns phenomenon. As a result, many of the policy implications are simple and classical: invest in education, have open market institutions, tolerate and even encourage multinationals, tolerate and even encourage a brain-drain. For similar reasons, we find no support for the wisdom of protectionist, infant industry, national champion or directive industrial policy programs. Yet that is not the end of the story. The mechanism by which entrepreneurial-led growth takes off and becomes a contributor to regional and even national development is one with a strongly increasing-returns flavor. Once clusters are founded, they do indeed deliver the kinds of opportunities emphasized in the increasing-returns framework. But the big issue is how to start a cluster, and what can policy do (and not do) in this respect.

For one, our case studies clearly show the foolishness of directive public-policy efforts to jump-start clusters or to make top-down or directive efforts to organize them. Clusters of innovative activity do not respond well to being directed, organized or jump-started, entrepreneurship being a quirky thing. For example, the study by Scott Wallsten in our project (Wallsten, 2001) could not empirically find any effect of a public-sponsored program as important as the Small Business Innovation Research (SBIR) program in the US on high-tech employment in US counties (see also Wallsten, 2000). In contrast, accommodative government policies can be an important part of cluster development. Apart from public investments in areas like education, governments played an important supporting (though not leading) role in making entrepreneurship easier in many of our regions, notably in Ireland, Taiwan, Virginia and Israel.

A powerful policy lesson from all of our cases is that none of our clusters engaged in protectionism or 'strategic trade policy'—even while attempting to capture greater shares of rent-producing activities or sectors. In fact, it was the openness of many of these regions to the world economy that has allowed success in the market-connected model. Rather than seek to offer firms a protected domestic market, which in the cases we study would have led to suboptimal scale, these regions have sought to define new niches and maintained a strong export orientation.

One of the most effective policies for regional development that we have encountered in our analyses has been that which encouraged the creation of the GSM standard. The GSM standard was critical for the growth of a continental wireless hardware market in Europe that benefited the Scandinavian firms first, and firms from other European countries later on. Yet the purposes of the policy were not set to encourage regional growth in Scandinavia. The rise of the Scandinavian model, built around the two leading firms Ericsson and Nokia, relied on several of the factors that we have already discussed for the other countries or regions—a substantial supply of skills and a strong policy for education; the role of the larger firms in encouraging the supply of skills; the ability to cover a niche (wireless hardware, and possibly wireless internet software and services) that has eventually become quite important. The formation of a European standard was, however, key for creating a sizable demand. Public policy was not the only factor that gave rise to the standard. It was, however, an important one, as with many standard setting processes (whether technical, legal, or else) that require the coordination of different actors. In sum, rather than interventionist policies, often directed toward the formation of the clusters themselves, and of their firms, the creation of the conditions for the clusters to arise, like a continental demand, can be a much more effective mechanism for promoting the opportunities of growth.

A natural end to our journey into the potentially new micro-determinants of economic growth suggested by the cases of regions like Ireland, India, Israel or Taiwan is to summarize the key findings of our research. Notably, novel factors at the turn of the century are adding relevant new dimensions and relevant new determinants of growth. These include greater opportunities to exploit external effects, as well as economies of scale at the level of regions or industries, rather than mainly at the level of the firms. Yet, the importance of old economic factors, and related determinants of growth, has not vanished. Even within the context of the new models, old economy ingredients, like firm-building capabilities, connection to markets and demand, along with the supply of skills (both technical and managerial), are critical for the new economy forces (external effects, agglomeration economies, etc.) to reach their full potential. In particular, these factors are critical forces for the initial push that is required to start any new entrepreneurial-led growth process. Old economy factors are crucial for new economy outcomes.

This also speaks to public policy. Direct, top-down policies are most likely to fail. Particularly worrisome are policies that would direct at a level of detail such as picking the specific industries or technologies to be sponsored. The right policies have elements of a 'benign neglect', and they allow for a significant decentralization in the choice of the initiatives. They would focus instead on the enabling conditions like the creation of suitable demand and markets (including formation of standards), openness, competition for encouraging the success of skilled people and people with entrepreneurial ambitions, along with policies focused on key supply-side factors and institutions, and on education in the first place.

The stories of the emerging regions that we have studied in our project have also shown that one driver of entrepreneurship in these areas has been the low opportunity costs of local human capital. When human capital has high opportunity costs because of alternative productive uses (employment in large established firms in leading industries like ICT itself, automobile, chemicals, etc.), this is hardly an issue for public policy. The question is far more serious when the opportunity costs for local human capital comes from artificially high wages in relatively unproductive jobs—e.g. excessive levels of employment in public administration, or in intermediaries of various sort, which is typical for instance of countries like Italy or Japan. The right policies for new cluster formation will fight such 'unproductive' opportunity costs, to allow for the full blossoming of true opportunities for their human capital.

Acknowlegements

This paper draws on the results of a two-year research project, 'Silicon Valley and its Imitators', organized by the Stanford Institute for Economic and Policy Research (SIEPR), and carried out by an international and interdisciplinary team. In drafting this paper we benefited from specific, and most helpful comments by Scott Wallsten, along with several fruitful discussions with other team-mates of the project, and particularly Ashish Arora, Suma Athreye, Kevin Davis, Michael Horvath, Gordon Moore, John Richards and Salvatore Torrisi. A special acknowledgment goes to Ralph Landau for his intellectual suggestions and for his encouragement in pursuing this research. All errors remain ours.

References

Arora, A., A. Fosfuri and A. Gambardella (2001), *Markets for Technology: The Economics of Innovation and Corporate Strategy*. MIT Press, Cambridge MA, forthcoming.

Arora, A., A. Gambardella and S. Torrisi (2001), 'In the Footsteps of the Silicon Valley? Indian and Irish Software in the International Division of Labour,' Working Paper, Stanford Institute for Economic Policy Research (SIEPR), Stanford University, Stanford, CA.

Arthur, B. (1990), 'Silicon Valley Locational Clusters: When Do Increasing Returns Imply Monopoly?,' *Mathematical Social Sciences*, 19, 235–251.

Athreye, S. (2001), 'Agglomeration and Growth: A Study of the Cambridge Hi-tech Cluster,' Working Paper, Stanford Institute for Economic Policy Research (SIEPR), Stanford University, Stanford, CA.

Audretsch, D. B. and M. P. Feldman (1996), 'Knowledge Spillovers and the Geography of Innovation and Production,' *American Economic Review*, 86, 630–640.

Bresnahan, T., T. Morant, and S. Wallsten (2001), 'Northern Virginia's High-tech Corridor and the Economics of Regional Development,' Working Paper, Stanford Institute for Economic Policy Research (SIEPR), Stanford University, Stanford, CA.

De Fontenay, C. and E. Carmel (2001), 'Israeli's Silicon Wadi: The Forces Behind Cluster Formation,' Working Paper, Stanford Institute for Economic Policy Research (SIEPR), Stanford University, Stanford, CA.

Fujita, M., P. Krugman and A. J. Venables (1999), *The Spatial Economy: Cities, Regions and International Trade*. MIT Press: Cambridge, MA.

Hirschman, A. O. (1958), 'The Strategy of Economic Development,' *Yale Studies in Economics*, no. 10.

Horvath, M. (2001), 'Imitating Silicon Valley: Regional Comparisons of Innovative Activity Based on Venture Capital Flows,' Working Paper, Stanford Institute for Economic Policy Research (SIEPR), Stanford University, Stanford, CA.

Jaffe, A., M. Trajtenberg, and R. Henderson (1993), 'Geographic Localization of Knowledge Spillovers as Evidenced by Patent Citations,' *Quarterly Journal of Economics*, 63, 577–598.

Krugman, P. (1991), *Geography and Trade*. MIT Press: Cambridge, MA.

Marshall, A. (1920), *Principles of Economics*, 8th edn. Macmillan: London.

Moore, G. and K. Davis (2001), 'Learning the Silicon Valley Way,' Working Paper, Stanford Institute for Economic Policy Research (SIEPR), Stanford University, Stanford, CA.

Myrdal, G. (1957), *Rich Lands and Poor*, World Perspectives Series, vol. 16. Harper Brothers: New York.

Perroux, F. (1950), 'Economic Space: Theory and Applications,' *Quarterly Journal of Economics*, 64, 89–104.

Porter, M. E. (1998), 'Clusters and the New Economics of Competition,' *Harvard Business Review*, Nov.–Dec., 77–90.

Richards, J. (2001), 'Clusters, Competition, and "Global Players" in ICT Markets: The Case of Scandinavia,' Working Paper, Stanford Institute for Economic Policy Research (SIEPR), Stanford University, Stanford, CA.

Saxenian, A. (1988), 'The Cheshire Cat's Grin: Innovation and Regional Development in England,' *Technology Review*, Feb./Mar. 1988.

Saxenian, A. (1994), *Regional Advantage. Culture and Competition in Silicon Valley and Route 128*. Harvard University Press: Cambridge, MA.

Saxenian, A. (2001), 'Taiwan's Hsinchu Region: Imitator and Partner for Silicon Valley,' Working Paper, Stanford Institute for Economic Policy Research (SIEPR), Stanford University, Stanford, CA.

Saxenian, A. (2002) *The New Argonauts: How Entrepreneurs are Linking Technology Markets in a Global Economy*. Harvard University Press: Cambridge, MA, forthcoming.

Wallsten, S. (2000), 'The Effects of Government–Industry R&D Programs on Private R&D: The Case of the Small Business Innovation Research Program,' *Rand Journal of Economics*, 31, 82–100.

Wallsten, S. (2001), 'The Role of Government in Regional Technology Development: The Effects of Public Venture Capital and Science Parks,' Working Paper, Stanford Institute for Economic Policy Research (SIEPR), Stanford University, Stanford, CA.

[23]

INTERNATIONAL REGIONAL SCIENCE REVIEW 25, 1: 8–37 (January 2002)

BIOTECHNOLOGY CLUSTERS AS REGIONAL, SECTORAL INNOVATION SYSTEMS

PHILIP COOKE
Centre for Advanced Studies, University of Wales,
Cardiff, UK, cookepn@cardiff.ac.uk

Today, knowledge economies are a key asset for global competitiveness. Biotechnology is a knowledge-driven sector because it consists of knowledge working on knowledge to create value, decoding in genomics and proteomics being paradigmatic knowledge-based economic activity. Like many other new economy industries such as information and communications technology, new media, and advanced finance, firms cluster in proximity to knowledge sources. In the case of biotechnology, universities are key magnets. But to transfer science from the laboratory bench to the market involves complex, interactive chains of transactions among scientists, entrepreneurs, and various intermediaries. Chief among the latter are investors and lawyers. Proximity to such services and, in biotechnology, research hospitals for clinical trials creates an innovation system. This is best analyzed regionally and locally. This article anatomizes the functioning of regional sectoral innovation systems in Germany, Cambridge, Massachusetts, and Cambridge, U.K.

Innovation is a key competitive weapon in an era of globalization. Firms and enterprise support infrastructures are becoming more knowledge intensive, and policies are being adjusted accordingly. Among the key general findings of the European Union (EU)-Targeted Socio-Economic Research (EU-TSER) project "Regional Innovation Systems: Designing for the Future" (REGIS) (Cooke, Boekholt, and Tödtling 2000) are the following. First, despite globalization and increased foreign ownership, most European businesses are rather strongly regional and national in key business relationships. Significant decision autonomy exists at the regional level, not least because of the predominance of small and medium-sized enterprises (SMEs). Second, all firms, large and small, are confronted by twin competitiveness pressures to raise quality and reduce cost. This impulse drives a great deal of innovation practice. Third, a majority of firms respond initially by organizational innovation, especially quality measures. Fourth, in Europe, many firms rely on the supply chain and their own knowledge sources to innovate products and processes. But there is growing recognition of the importance of universities, research institutes, consultants, and technology-transfer agencies in supplying new knowledge.

Smaller firms show some evidence of recognizing the importance of vertical and horizontal networks for collective learning and innovation. Moreover, at the regional level, particularly where there is a regional governance structure and presence of knowledge centers, finance, and industry clusters, policies are being developed to support clusters by creating economic communities within a multilevel governance structure to develop access to global markets. A new, knowledge-intensive industry in which these characteristics are particularly pronounced is biotechnology. Growing from research laboratories, the industry is characterized by many new start-up firms needing major support from university technology-transfer and licensing agencies, venture capitalists, large firms (as corporate partners), and regional governance bodies, both political and industrial. Centered on the two Cambridges (United States and United Kingdom), successful biotechnology clusters with a full range of systemic interaction mechanisms exist and, while unique in many ways, offer lessons for systemic regional innovation in other sectors and regions. These are being followed by German federal government policies that support regional biotechnology clusters, notably the BioRegio contest. This article reflects on the limitations and capabilities of a strongly public and federal attempt to develop the national innovation system through seeking to build regional innovation systems based on a core technological capability.

In doing this, a useful light is cast on all three cases in respect of the strengths but also limits of regional innovation systems in relation to policies to enhance, support, or build innovative regional clusters. Probably the key limitation on regional initiatives for advanced technology is funding for basic and applied research, since most regional administrations do not have remotely enough of such capital, especially in biotechnology. The second limitation is venture capital and other sources of investment capital for the commercialization stages of biotechnology, although this is less of a limitation in certain cases. Boston and Cambridge (United Kingdom) are interesting instances of world-class science attracting critical mass in venture capital and it may be true also, at present, in Munich. Where the national innovation system cannot function well without regional innovation systems is in respect of the enterprise and innovation support infrastructure, specialized human capital, leading-edge basic and applied research, and the varieties of network relationships that function most effectively in the relatively close proximity of regional clusters.

Recent work by Porter (1998), Audretsch (1998), Krugman (1998), and Best (1999, 2000) confirms the earlier insights of regional scientists like Scott (1993); Saxenian (1994); Storper (1995); Florida (1995); Amin and Thrift (1994); Asheim (1996); Cooke (1995); Braczyk, Cooke, and Heidenreich (1998); and Cooke and Morgan (1998) that clusters offer key competitive advantages over vertical integration in single firms with respect to three key competitiveness variables. These are productivity, which is enhanced by lower transaction costs and untraded interdependencies; innovation, which is dependent on interactive knowledge exchange between a variety of knowledge actors, especially because of the proximity necessary for tacit knowledge exchange; and new business formation, which is massively assisted by the mentoring, role-model provision, learning, communication, and commercialization gains that arise from operating in a cluster setting.

In this article, the first section will argue the case for regional innovation systems, drawing on theoretical and empirical findings from the REGIS project (see Cooke, Boekholt, and Tödtling 2000). This will draw attention to the concepts of interactive innovation, learning, proximity, associational networking, and clustering activities of public and private governance actors, paying respect also to the multilevel governance aspects of innovation systems. The second half of the article consists of some detailed case studies of biotechnology clustering in the United States, the United Kingdom, and Germany, paying attention to the differences and similarities in the processes involved despite their origins in different national innovation systems and the distinctive role of public policy involvement in the three cases. The article concludes with a review of the strengths and limitations of regional innovation systems with respect specifically to the development trajectories of the biotechnology clusters under inspection.

THEORIZING THE NEW REGIONALISM

Speaking in theoretical terms and picking up on points made in the introduction to this article concerning regional advantage, a consensus has formed among writers such as Grabher (1993), Maillat (1995), Sabel (1995), Enright (1996), and Rosenfeld (1997) that accomplished regional economies tend to display certain common features. Among the most important of these are agglomeration economies, institutional learning, associative governance, proximity capital, and interactive innovation (see also Malmberg and Maskell 1997; Johnson 1992; Amin and Thomas 1996; Crevoisier 1997; Edquist 1997a). These are briefly explained below.

AGGLOMERATION ECONOMIES

Since Marshall, the advantages of colocation by firms in single or complementary industries have been well understood. Krugman (1997) itemized these as follows: first, a concentration of producers supports local suppliers of specialized inputs who thus help generate external economies of scale effects; second, agglomerations generate localized skills pools benefiting workers' and firms' flexible labor market opportunities; and third, knowledge spillovers are implied by the existence of agglomeration. In the sphere of regional innovation, these translate into opportunities for lowering transaction costs from uncertainty due to the possibilities for specialist, tacit-knowledge exchange present in the agglomeration (although always subject to efforts to minimize leakage and maximize equivalence from tacit-knowledge exchange with others) (Saxenian 1994; Storper and Scott 1995; Malmberg and Maskell 1997).

Institutional Learning

Institutional learning refers to the institutional setting of norms, routines, "rules of the game," and conventions (after North 1993), whereby it is widely understood that certain practices are acceptable and promote trustful relationships among firms and organizations (which may also help reduce transaction costs). But among the norms of growing importance for firms and enterprise support organizations is the presumption in a globalizing economy, characterized by turbulence and uncertainty, that openness to learning good practice from others is of special importance. In Lundvall and Johnson's (1994) formulation, this is conceived of as the externalized form of the kind of learning more typical of what Argyris and Schon (1978) referred to as the more internalized characteristics of the "learning organization" or firm. It applies equally to organizations that interact with firms, including governance agencies, that must "learn by monitoring" in respect of the performance of the wider economy, their own goals achievement, and that of competitor agencies (Sabel 1995). It goes without saying that such learning is global as well as local.

Associative Governance

Here, reference is made to a networking propensity whereby key regional governance mechanisms, notably the regional administrative bodies, are interactive and inclusive with respect to other bodies of consequence to regional innovation. This may lead to an organizational setting in which, let us say, the regional administration animates or facilitates associativeness among representative bodies inside or outside public governance but does not seek to dominate a process of consensus formation with respect to, say, a readjustment of regional economic strategy. It may involve a government agency letting go of, or at least sharing with legitimate private governance bodies such as chambers of commerce or business associations, a function it may have been responsible for innovating.

Proximity Capital

Proximity capital can be hard or soft, financial or human, and refers to different kinds of infrastructure of relevance to regional innovation. According to Smith (1997), there is a strong association between past investments in a variety of infrastructures and economic performance. Thus, the existence of appropriate communication links such as road, rail, airport, and telecommunication services is crucially important in proximity to industrial agglomerations. As Krugman (1997) put it, quoting U.S. Federal Reserve chairman Alan Greenspan, "the gross domestic product is getting lighter." Hence, "for businesses which depend on personal contact and/or rapid shipment of goods, two locations 500 miles apart but close to major airports with frequent direct flights are effectively closer to each other than two locations on opposite sides of the same large metropolitan area" (Krugman 1997, 44-46). This is material capital, but Crevoisier (1997) referred also to the importance in agglomerations (especially of SMEs) of localized, trustful means of raising venture capital, maybe through local entrepreneurs or "business angels." Intellectual capital from previous investments in universities and research institutes in proximity to complementary firms is yet another form of proximity capital.

Interactive Innovation

As is well known, the concept of interactive innovation is very much associated with the "national systems of innovation" literature (Lundvall 1992; Nelson 1993; Freeman 1994, 1995; Edquist 1997b), but it is of obvious relevance to the regional level, too. Where there is a rich innovation infrastructure—ranging from specialist research institutes to universities, colleges, and technology transfer agencies—and institutional learning is routine, firms have considerable opportunities to access or test knowledge, whether internally or externally generated to the region. Clearly, by no means all innovation interactions can or even should occur locally, but the rise of the "entrepreneurial university" (Smilor, Dietrich, and Gibson 1993) and promotion of the so-called triple helix of interaction between industry, government, and universities as a key feature of the knowledge economy (Etkowitz and Leydesdorff 1997) testify to the practical evolution of interactive innovation processes.

It is in respect to this new regional science approach to thinking about regional economic development that regional studies and innovation studies in the manner of Lundvall, Nelson, Freeman, and Edquist themselves begin to interact. The latter tradition is more overtly evolutionary in its theoretical perspective on economics. But all of what has been described in the new regional science approach is compatible with evolutionary economics. Summarizing Edquist's (1997a) presentation of key concerns of contemporary innovation research in light of the interests of new regional science, especially where innovation is under the microscope, we find that both fields envision innovation and learning processes involving knowledge transfer as a key focus, share an interest in systemic interaction within political economies, and are concerned with questions of path dependence, development trajectories, and the role of institutions and ways they evolve over time (see also Cooke, Uranga, and Etxebarria 1998).

Conditions and Criteria for Regional Innovation Systems

In considering the prospects for regional systems of innovation, Cooke, Uranga, and Extebarria (1998) have explored theoretically the key organizational and institutional dimensions providing for strong and weak regional innovation systems potential. This is a pioneering attempt to specify desirable criteria on which systemic innovation at the regional level may occur. These can be divided into infrastructural and superstructural characteristics.

INFRASTRUCTURAL ISSUES

The first infrastructural issue concerns the degree to which there is regional financial competence. This includes private and public finance. Where there is a regional stock exchange, firms, especially SMEs, may find opportunity in a local capital market. Where regional governments have jurisdiction and competence, a regional credit-based system in which the regional administration can be involved in cofinancing or provision of loan guarantees will be of considerable value, something that is extremely important about the German approach in which the private sector strongly avoids high risk. In the United States and the United Kingdom, private venture capital is the proximate source and main lubricant of commercialization activities. Hence, secured proximity capital can clearly be of great importance, especially as lender-borrower interaction and open communication are seen to be increasingly important features in modern theories of finance. Hence, regional governance for innovation entails the facilitation of interaction between parties, including—where appropriate and available—the competencies of public as well as private resources. Public-private animation of investment can also help build up capability, reputation, trust, and reliability among regional partners.

However, regional public budgets are also important for mobilizing regional innovation potential. We may consider three kinds of budgetary competence for those situations in which at least some kind of regional administration exists. First, regions may have competence to administer decentralized spending. This is where the region is the channel through which central government expenditure flows for certain items. In Europe, much Italian, Spanish, and French regional expenditure is of this kind, although there are exceptions, such as the Italian Special Statute regions and some Spanish regions where taxes are raised and spent regionally. A second category applies to cases in which regions have autonomous spending competence. This occurs when regions determine how to spend a centrally allocated block grant (as in Scotland and Wales in the United Kingdom) or where, as in federal systems, they are able to negotiate their expenditure priorities with their central state and, where appropriate, the EU. The third category is when regions have taxation authority as well as autonomous spending competence since this allows them extra capacity to design special policies to support, for example, regional innovation. The Basque Country in Spain has this competence, as does Scotland. Clearly, the strongest base for the promotion of regional innovation is found when regions have regionalized credit facilities and administrations with autonomous spending and/or taxation authority. Of course, in the United States, states have access to their own sales taxes and powers to vary tax rates on such items as expenditure on research and development (R&D).

A further infrastructural issue concerns the competence regional authorities have for controlling or influencing investments in hard infrastructures such as transport and telecommunications and softer, knowledge infrastructures such as universities, research institutes, science parks, and technology transfer centers. Most regions lack the budgetary capacity for the most strategic of these, but many have competencies to design and construct many of them or, if not, to influence decisions ultimately made elsewhere in respect of them. The range of possibilities is enormous in this respect, so we classify broadly into types of infrastructure over which regions may have more or less managerial or influence capacity. If we think of our three cases, then the federal systems in Germany and the United States have the most influence over infrastructural decisions, including roads and even airport policies. In Germany, basic research funding frequently has a regional (land) component. In the United States, management and funding of public universities is devolved. In the United Kingdom case, regions in England (but not Northern Ireland, Scotland, or Wales) have had only regional development agencies since April 1999, so autonomy is low and dependence on discretionary budgets from the center is still high.

Superstructural Issues

Three broad categories of conditions and criteria can be advanced in respect to superstructural issues. These refer, in general, to mentalities among regional actors or the culture of the region and can be divided into the institutional level, the organizational level for firms, and the organizational level for governance. Together, these help to define the degree of embeddedness of the region, its institutions, and its organizations. Embeddedness is here defined in terms of the extent to which a social community operates in terms of shared norms of cooperation, trustful interaction, and "untraded interdependencies" (Dosi 1988) as distinct from competitive, individualistic, "arm's length exchange," and hierarchical norms. The contention here is that the former set of characteristics is more appropriate to systemic innovation through network or partnership relationships. It is widely thought that American entrepreneurship involves this cultural characteristic, but in biotechnology, as in other cases of high technology, there is cooperation as well as competition, as we shall see. It should also be noted that the work of Saxenian (1994) pointed strongly to the conclusion that a key reason for Silicon Valley's better long-term innovation performance than that of Route 128 Boston was that Silicon Valley was the region with the greater embeddedness. But the resurgence of the latter is linked to Massachusetts's adoption of a cluster policy from which biotechnology and biomedical instruments, for example, have benefited (Porter 1998; Best 2000).

Therefore, if we look first at the institutional level, the atmosphere of a cooperative culture, associative disposition, learning orientation, and quest for consensus would be expected to be stronger in a region displaying characteristics of systemic

TABLE 1. Conditions for Higher and Lower Regional Innovation Systems Potential

Higher Potential	*Lower Potential*
Infrastructural level	
Autonomous taxing and spending	Decentralized spending
Regional private finance	National financial organization
Policy influence on infrastructure	Limited influence on infrastructure
Regional university-industry strategy	Piecemeal innovation projects
Superstructural level	
Institutional dimension	
Cooperative culture	Competitive culture
Interactive learning	Individualistic
Associative consensus	Institutional dissension
Organizational dimension (firms)	
Harmonious labor relations	Antagonistic labor relations
Worker mentoring	Self-acquired skills
Externalization	Internalization
Interactive innovation	Stand alone research and development
Organizational dimension (policy)	
Inclusive	Exclusive
Monitoring	Reacting
Consultative	Authoritative
Networking	Hierarchical

innovation, whereas a competitive culture, individualism, a "not invented here" mentality, and dissension would be typical of nonsystemic, weakly interactive innovation at the regional level. Moving to the organizational level of the firm, those with stronger systemic innovation potential will display trustful labor relations, shopfloor cooperation, and a worker welfare orientation with emphasis on helping workers improve through a mentoring system and an openness to externalizing transactions and knowledge exchange with other firms and organizations with respect to innovation. The weakly systemic firm characteristics would include antagonistic labor relations, workplace division, "sweating," and a "teach yourself" attitude to worker improvement. Internalization of business functions would be strongly pronounced, and innovativeness might be limited to adaptation. Regarding the organization of governance, the embedded region will display inclusivity, monitoring, consultation, delegation, and networking propensities among its policy makers while the disembedded region will have organizations that tend to be exclusive, reactive, authoritarian, and hierarchical. These characteristics are summarized in Table 1.

Clearly, both sets of conditions are ideal types in the sense that it is unlikely that a single region would conform to all of one or the other set of characteristics. However, it could be expected that regions might display tendencies toward one or the other end of the continuum, and in dynamic terms, it might be possible to identify evolutionary tendencies by regions toward one or the other pole, perhaps signifying an element of convergence influenced either by globalization processes or the policy effects of governments or (in Europe) EU programs.

Operationalizing Regional Innovation in the Context of Multilevel Governance

Because, under conditions of globalization and liberal trading, the EU has been conscious of the relative weakness of the European economy in competing with the United States in terms of the commercialization of the fruits of research, a great effort has been made to support and promote the improvement of innovation among firms of all sizes. While the EU's Science & Technology Framework Programme was at first strongly influenced by and mainly directed toward Europe's largest multinationals, the focus was later extended to encompass the interests of SMEs and regional innovation, as the Green Paper on Innovation (European Commission 1995) makes clear. The fact that promoting regional innovation also targets less favored regions and thus helps the Commission to meet its cohesion obligations strengthens this disposition. Moreover, the emergence of innovation promotion as an element of the EU's "structural funds" for implementing regional policies underlines the commitment to regional innovation policy. Experimentation with regional technology plans, regional innovation strategies, and regional information society initiatives also testified to the growing importance of capacity building for innovation at the regional level.

However, as has been stated, the absorptive capacity and organizational competencies in a context of multilevel power relations within different member-states mean that building the capability for regional firms to engage in interactive or even systemic innovation varies considerably. It is well known, for example, that while the wealth disparity within the EU ranges from 1 to 5, that for R&D expenditure ranges from 1 to 11, meaning that there is far less basic innovation activity away from the main metropolitan centers in the larger and more northerly member-states. Moreover, the capability of regional administrations in the Southern member-states in multilevel lobbying and influence to access regional innovation funding can be affected by decision-making structures that remain centralized for some functions even when a wide-ranging program of regional decentralization may have been implemented.

These points are made because multilevel governance (MLG) relationships differ due to member-state constitutional and practical political traditions and conventions. Five key points assist our analysis of the regions studied in the REGIS project. MLG is highly dependent on the presence of strong and established regional governance organizations. MLG for innovation is significantly assisted where the region has a substantial number and diversity of regional and local innovation organizations. Regional and external innovation interaction among firms and other innovation organizations is important for regional innovation potential. The existence of regional scientific, technological, and innovation policies and programs, assisted by the EU and nationally, is important. Finally, the ability to access and use funding for innovation support for regional firms and organizations is crucial for regional innovation promotion.

On this basis, it is clearly necessary to say more about the kinds of organizations that can be found to comprise the organizational innovation support infrastructure in a given region. The two key subsystems in any functioning regional innovation system are (1) the knowledge application and exploitation subsystem and (2) the knowledge generation and diffusion subsystem (Autio 1998). The first is principally, but not only, concerned with firms while the second is more concerned with public organizations like universities, research institutes, technology transfer agencies, and regional and local governance bodies responsible for innovation support practices and policies. However, private investors can be the most important actors outside basic research in highly innovative regions and metropolitan areas. Firms applying and exploiting innovation directly can have vertical and horizontal network linkages; vertical relationships are mainly supplier linkages, whereas horizontal linkages are found typically amongst SMEs who may, on occasion, also be competitors. Many innovation network policies seek to build horizontal linkages, but some also aim to assist the elaboration of vertical supply-chain relationships. Evidence has emerged that venture capitalists do this as a matter of course to watch their investments. Possibly Kleiner, Perkins, Caufield & Byers in Silicon Valley is the most conscious of this through its *keiretsu* building practices (Cooke 2001a, 2001b). In the knowledge generation and diffusion subsystem are technology-mediating organizations, those that mediate with respect to vocational training and workforce skill provision, public research institutes, and educational organizations. Each of these subsystem organizations interacts with the others and with national innovation organizations or the National System of Innovation of their member-state as well as international policy and knowledge-generating organizations such as the EU, on one hand, and non-European universities, research institutes, and firms, on the other hand. Figure 1 is an attempt, based on the work of Autio (1998), to present the structure of a regional innovation system in the abstract.

This model captures the main features and relationships of a functioning regional innovation system operating in an MLG environment. But it only indicates the linkages in a neutral fashion. Empirical research is necessary to capture the variety of degrees of influence and decision-making authority and the presence or absence of weaker and stronger relationships among the diverse possible kinds of application, exploitation, generation, and diffusion elements of specific regions and their degrees of "systemness."

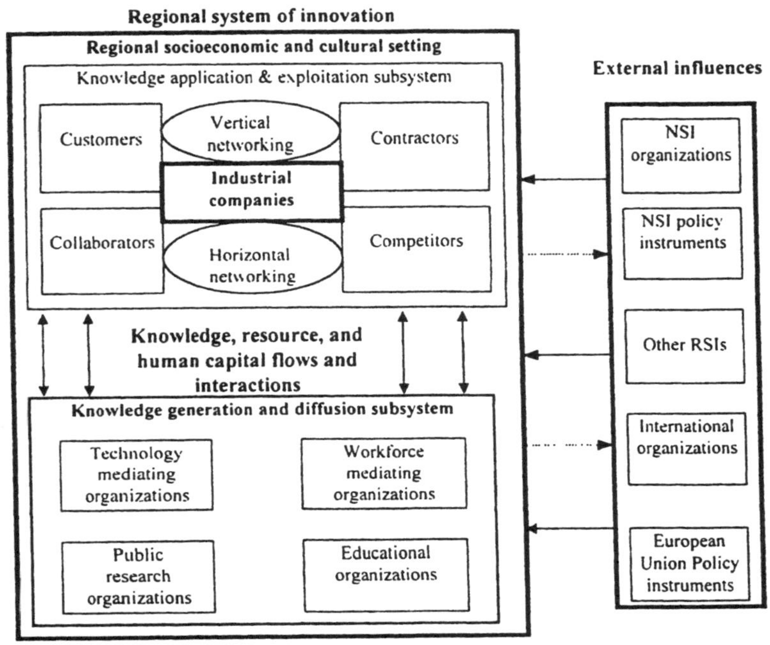

FIGURE 1. Schematic Illustration of the Structuring of Regional Systems of Innovation

EXPLORING BIOTECHNOLOGY CLUSTERING FROM A REGIONAL INNOVATION SYSTEMS VIEWPOINT

We now need to focus in on empirical cases to seek the limitations of regional innovation systems as well as their contribution to sector competitiveness in the context of MLG. Some of the funding limitations at the regional level have already been discussed, particularly with respect to the funding of basic and much applied research in universities and specialist research institutes. Another limitation is the regulatory regime, a matter of national or federal responsibility, although the implementation of certain regulations, in terms of speed, can be a subnational matter. Thus, issues concerning taxation, rules about depreciation of investment, and such issues as the rules governing share options are usually national and affect the general climate for entrepreneurship and rules of competition. For example, the United States is widely understood to have the most benign regulatory regime for market-based entrepreneurship. The United Kingdom has a less benign regime; for example, capital gains tax on the selling of share options by firm founders is set higher than in the United States and is considered a barrier to growth by the biotechnology industry (Department of Trade and Industry 1999). Germany, despite some recent reforms, has a more rigid regulatory regime in relation to depreciation and share options than either the United States or the United Kingdom (Casper and Kettler 2000). Germany also has rules that make it much harder for academics to become entrepreneurs because they are classified as civil servants who may not take a second employment. In Germany, however, the implementation powers of the *länder* mean that, for example, the federal Genetic Engineering Act (which regulates this activity) has been implemented as different speeds—Bavaria being one of the earliest to act—thereby assisting the development of genetic engineering entrepreneurship there as compared with other regions.

However, although it is fairly uncontroversial to state that the United States has the best business climate for the commercialization of scientific research, this does not mean that development of U.S. biotechnology has not been assisted by substantial public funding at both federal and state levels. The role of the National Institutes of Health, with a 1999 research budget of $15.6 billion (increased by $2 billion or 14 percent since 1998), and the National Science Foundation, which supports biological science research with a 1999 budget of $391 million, along with the U.S. Department of Agriculture ($1.6 billion), NASA ($264 million), and the U.S. Department of Energy, which supports the human genome project ($433 million), constitutes a massive public-funding resource from which biotechnology research benefits. Furthermore, the Small Business Innovation Research (SBIR) program, whereby 2.5 percent of the external budget of eleven U.S. federal agencies is available for funding R&D in small firms, is of major importance to new technology companies, including biotechnology. Indeed, one possible weakness of SBIR is that some firms exercise "grantsmanship" and spend much of their time seeking such grants in a dependent, "rent-seeking" manner. So this is not proximity capital, although it certainly arrives in proximity to elite research institutes with some regularity. In basic science funding, multilevel budget governance is well to the fore.

These funds dwarf even the U.S. venture capital industry for biotechnology, which in 1998 was some $1.4 billion. However, this private investment allied to the growth of state-initiated venture funds for biotechnology in California, Massachusetts, Maryland, North Carolina, and Seattle in Washington State means the importance of proximity for private investment is high as well as regionally variable. States, through their economic development initiatives, also operate tax incentives and support programs to assist the sector. California exempts biotechnology firms from the 6 percent state sales tax, North Carolina gives tax exemption for equipment purchases, and the state of Washington gives credits against business taxes for R&D expenditure. Massachusetts is probably the most interventionist, having a 10 percent to 15 percent tax credit on research and a 3 percent investment tax credit on fixed assets, both with lengthy carry-forward periods.

Such is the nature of regional-level support for U.S. biotechnology that a recent mission there by the U.K. Bioindustry Association (BIA) led them to call for a new National Biotechnology Center: "In no case did U.S. (biotechnology) manufacturing plants just spring up. It [*sic*] was kicked into existence by government bodies," said Dr. John Sime (personal communication, August 1999), head of BIA. Maryland and North Carolina were seen as having especially helped to build successful biotechnology clusters. "In both states, biotechnology companies have been encouraged to undertake manufacturing by a supportive regulatory and planning environment, a responsive academic environment and financial incentives," reported a team member (personal communication, August 1999). North Carolina was seen as a model due to the establishment of its North Carolina Biotechnology Center, having been set up with public funds as an independent organization, being financially self-sufficient, and playing a coordinating role between the industry and government, universities, financial institutions, and the media (Cookson and Pilling 1999). It is interesting to note how a key ingredient of North Carolina's regional cluster building of a biotechnology innovation system acts as a recipe for enhancing the United Kingdom's national biotechnology innovation system. This clearly suggests the tenaciousness of a national and centralizing perspective by science in the face of manifest evidence of the importance of the regional in cluster building. U.K. Department of Trade & Industry policy up to and including 2001 was to make disbursements of special cluster, innovation support, and public venture capital funding to England's Regional Development Agencies and U.K. science entrepreneurship funding direct to universities following national contests. Nowadays, the other U.K. countries develop their own distinctive cluster-support policies. So multilevel governance is important for U.K. public innovation funds, acting to offset the large disparities in a U.K. investment system heavily skewed to London and away from the regions.

CAMBRIDGE, MASSACHUSETTS

One of the biggest and most dynamic biotechnology clusters is that in Boston. The science base is exceptionally strong in the Massachusetts Institute of Technology (MIT), Harvard University, Boston University, and Massachusetts General Hospital. Each year some $770 million in basic research funding flows through the system. Leading scientists and academic entrepreneurs, one of whom has been involved with some 350 patent applications, are present. At MIT, in particular, the Technology Licensing Office is a major operation, also involved in assisting at least twenty start-ups per year to be established. Massachusetts has at least 150 venture capitalists, most of them in Boston or Cambridge. The Massachusetts Biotechnology Council is an industry association that organizes common purchasing and other services such as promotion, educational placement, and career development for its 215 member firms. In 1998, there were 132 member firms in the greater Boston area (59 in Cambridge, 73 elsewhere) and 83 outside Route 128, employing some 17,000 people. By 1999, Massachusetts Biotechnology Council's membership had reached 245 biotechnology firms.

Because of proximity and common backgrounds from educational institutions, the level of interfirm and firm-agency interaction is high. In these respects, this industry constitutes an exemplary case of a cluster, although as with high-technology clusters in general, global linkages to other clusters and, particularly, "big pharma" partners or customers are also pronounced. The connection to other centers of biotechnology is testified by the presence in the Massachusetts Biotechnology Council of promotional material from other clusters, including that of the Eastern Region Biotechnology Initiative (ERBI) based in Cambridge (United Kingdom). This association is a major factor in the private governance of the metropolitan regional cluster.

If we look at the biotechnology sector springing mainly from MIT and Harvard in Cambridge, supported by Massachusetts General Hospital and, to a lesser extent, Boston University in Boston, we have to talk of biotechnology nowadays in the greater Boston area, since many start-ups have moved out to Route 128 and even beyond Route 495 to Worcester as the encompassing area. The 1998 geographical breakdown, bearing in mind the 59 firms in Cambridge, was as follows: 132 firms were located east of Route 128 (59 in Cambridge, 16 in Boston, and the remainder between there and Route 128), 58 were located between Route 128 and Route 495 (including 11 in Bedford and 6 in Wilmington), and 25 were located west of Route 495 (including 11 in Worcester). Many of these, especially in the outer locations, were based on science or technology parks, as were many start-ups on the technology park campuses of the key universities. The Massachusetts Biotechnology Park at Worcester has venture capital on site, suggesting that proximity is important for some in meeting demand, despite the presence of large numbers of investment firms in downtown Boston.

The market segment breakdown is that 34 percent of firms are in the therapeutic products sector (meaning they have grown beyond the early stages, typically in platform technologies, including diagnostics), 20 percent are in scientific equipment or supplies, 15 percent are in scientific services, 14 percent are in human diagnostics, 10 percent are in environmental and veterinary services, and 7 percent are in agricultural biotechnology (animal, plant, diagnostic, and transgenics). Perceived industry growth areas are in medical therapeutics (genetically produced protein, vaccines, gene therapy, and human growth hormones), human diagnostics (monoclonal antibodies, biological imaging, DNA probes, biosensors, and polymerase chain reaction), agricultural biotechnology (nutraceuticals, rapid diagnostic testing, and transgenics), and bioinformatics (biological discovery, patient databases, etc.). Seventy-nine firms were founded in the 1980s including Biogen, Genetics Institute, and Genzyme (with more than three hundred employees). A further eighty-eight firms began between 1990 and 1997; the remainder are more recent start-ups or inward investments. Employment grew from 7,682 in 1991 to 16,872 in 1998. As the industry matures, the number of start-ups is decreasing annually. Between 1996 and 1999, seven mergers and acquisitions occurred. Financing of companies in biotechnology is high risk, and analyses show that public investment is strongest at the risky process or product development stage.

Of considerable significance as agents in the regional innovation system, within the knowledge generation and diffusion subsystem, are the following:

- Massachusetts Department of Economic Development: has a key role in business and trade development, improving the business climate (R&D tax credits, investment tax credits), responding to lobbying from industry associations, and providing grants to growth firms and inward investors.
- MIT: is a leading center for biotechnology research and commercialization, campus incubators, and technology parks. MIT Entrepreneurship Center trains scientists in entrepreneurship. MIT Technology Licensing Office identifies technologies suitable for start-ups and introduces technology to potential investors (usually venture capitalists).
- Harvard University: provides Ph.D. programs in biochemistry, biology, biophysics, cell and development biology, genetics, microbiology and molecular genetics, technology, and so on at the Joint Harvard-MIT Division of Health and Technology, the School of Medicine, and the School of Public Health.
- Massachusetts General Hospital and Boston University: conducts research and commercialization at Boston University, Bio Square Technology Park.
- Whitehead Institute of Biomedical Research: is an independent research and teaching institution (affiliated with MIT in teaching) and an international leader in the human genome project, the source of comprehensive, published genome data, which conducts world-leading research in genetics and molecular biology and houses a technology-licensing program and start-up scheme.
- Massachusetts Technology Collaborative: is a state-founded, independent body to foster technology-intensive enterprises and cluster-building strategies.
- Massachusetts Biotechnology Council: is a trade association representing biotechnology firms (162 full and 83 associate members), which provides educational, career, and promotional information to the industry and conducts common-purchasing contracting for biotechnology firm members.

In conclusion, leading exploitation firms such as Genzyme, patenter and inventor of the therapeutic product that controls the genetically caused Gaucher's disease, are closely intertwined with this generation and diffusion system. Moreover, Genzyme, as a founding member of the Partners Healthcare System with Brigham and Women's and Massachusetts General Hospitals on research funded at $400 million by the National Institutes of Health, reinforces the system. Along with Biogen and Genetics Institute, and other internationally known firms such as BASF, Corning and Quintiles, and a host of SMEs and start-ups, this means the greater Boston region is supported by the generation and diffusion organizations and associations already noted and clearly functions as a well-integrated regional innovation system based on a cluster of leading-edge biotechnology businesses. It has a major proximity capital resource in the 150 or so venture capital firms in and around Boston. Lobbying through the Massachusetts Biotechnology Council led the Food and Drug Administration to open an office in the city, testimony to the sector's powers of association.

The Cambridge (United Kingdom) Economy and Biotechnology

As in Boston, the economy around biotechnology is important but by no means overwhelmingly so by comparison with other economic sectors in Cambridgeshire. Thus, Cambridgeshire County Council estimates that in 1998 there were 37,000 high-technology jobs in the area and that these comprised 11 percent of the Cambridgeshire labor market. South Cambridgeshire had about 66 percent of these jobs, while Cambridge city accounted for most of the remainder. The main high-tech activity is R&D, supplying 24 percent of total high-tech employment; electronics has 17 percent; computer services have 13 percent; scientific instrumentation has 8 percent; and biotechnology has 7 percent. Probably the estimate of some 2,600 employees in biotechnology (and chemicals) for the county is a not an unreasonable figure. However, if we inspect the ERBI Biotechnology (1998) *Sourcebook*, the number of core biotechnology firms in Cambridgeshire listed is 36. So the discrepancy between that figure and the estimate of 200 biotechnology firms by Segal, Quince & Wicksteed (SQW) in 1998 needs some qualification. The first qualification can be offered with some confidence: in ERBI's list of Cambridgeshire biotechnology firms there are venture capitalists, research institutes, management consultants, and lawyers. Together these total 96; thus, the cluster support firms and agencies exist in an approximate 2:1 ratio with biotechnology firms. ERBI considers this a significant underestimate and, in its new survey (1999), estimates numbers, in general, about one-third higher. This would take the Cambridgeshire figure to around 50 core biotechnology firms. The second reason for the discrepancy is that the significant number of very small start-ups in incubators and the like are underrepresented in the ERBI figures published thus far. Therefore, we may conclude that Cambridgeshire's core biotechnology industry consists of no less than 50 firms, and the broader cluster (venture capitalists, patent lawyers, etc.) probably consists of not much more than 200 firms, including the core biotechnology firms.

It is quite useful, in trying to categorize the biotechnology sector, to follow the German custom of referring to "red," "green," and "gray" biotechnology. The first is primarily medical and biopharmaceutical, the second is agro-food biotechnology, and the third is environmental. It is clear from both ERBI (1998) data and the SQW estimates that Cambridgeshire specializes in red biotechnology. The two categories of biopharmarceuticals including vaccines and pharmaceuticals largely from chemical synthesis register fourteen and nine Cambridgeshire-based firms, respectively. Examples of the former are Actinova, Amgen, and Hexagen and of the latter, Chiroscience, Napp, and Quadrant. In addition to these two key categories

TABLE 2. Shares of Biotechnology and Services Functions

Distribution	%
Biotechnology firms	
Biopharmaceuticals	41
Instrumentation	20
Agro-food bio	17
Diagnostics	11
Reagents/chemicals	7
Energy	4
Biotechnology services	
Sales and marketing	29
Management consulting	23
Corporate accounting	15
Venture capital	15
Legal and patents	8
Business incubation	10

Source: Eastern Region Biotechnology Initiative (1999).

are direct biotechnology services like clinical trails, diagnostics, and reagent supply. A further eight Cambridgeshire direct-services firms are listed in the ERBI *Sourcebook*, which, it will be recalled, probably underestimates the numbers by about one-third (not counting micro-firms). Cambridgeshire has four green bio firms, but ERBI lists no gray bio firms. It is important to note that Cambridgeshire also hosts twelve biotechnology equipment and instrumentation firms according to ERBI. This is more than two-thirds of such firms in the Eastern region.

The growth in the number of biopharmaceutical firms has been from one to twenty-three over the 1984-97 period, an average of just less than two per year, but the rate has been more like four per year in the past two years of that period. Equipment firms grew from four to twelve in 1984-97 and diagnostics firms from two to eight. Table 2 shows the distribution of technology-based companies in Cambridgeshire and the distribution of support services.

Thus, it is clear that Cambridgeshire has a rather diverse biotechnology processing and development as well as services support structure, even though the industry is relatively young and small. Some of the service infrastructure and perhaps the equipment sector benefits from the earlier development of information technology businesses, many also spinning out from university research in Cambridge. It is notable that 15 percent of biotechnology support services comprise venture capital. For a small city, Cambridge is well supplied with this commodity even though it is less than an hour away from London. This is a striking case of local demand attracting a supply of private investment, something that has been true of Cambridge since the earlier development in the 1980s of its thriving information and communications technology (ICT) industry. As in Boston, basic scientific funding is a largely public affair, although the Wellcome Trust, the world's largest medical charity, has been highly active both independently and in partnership with government in funding bioscience and medical research.

The infrastructure support for biotechnology in and around Cambridge is impressive, much of it deriving from the university and hospital research facilities. The Laboratory of Molecular Biology at Addenbrookes Hospital, funded by the Medical Research Council; Cambridge University's Institute of Biotechnology, Department of Genetics and Centre for Protein Engineering; the Babraham Institute and Sanger Institute, with their emphasis on functional genomics research; and the Babraham and St. John's incubators for biotechnology start-ups and commercialization are all globally recognized facilities, particularly in biopharmaceuticals. However, in the Eastern region are also located important research institutes in the green biotechnology field of agricultural and food biotechnology, such as the Institute for Food Research, John Innes Centre, Institute of Arable Crop Research, and National Institute of Arable Botany. Thus, in research and commercialization terms, Cambridge is well placed in red bio and with respect to basic and applied research but perhaps less so to commercialization and green bio.

Within a 25-mile radius of Cambridgeshire are found many of the big pharma or specialist biopharmaceutical firms with which commercialization development by smaller start-ups and R&D by research institutes must be cofinanced. Firms like Glaxo-SmithKline, Merck, and Aventis in the big pharma category are represented, and in the specialist biopharmaceutical sector, Amgen, Napp, Genzyme Yamaguchi, and Bioglan are represented. Thus, on another of the criteria for successful cluster development—namely, access within reasonable proximity to large customer and funding partner firms—Cambridge is, again, fortuitously positioned.

Finally, with respect to agro-food bio, Aventis, Agrevo, Dupont, Unilever, and Ciba are situated in reasonably close proximity to Cambridge. Hence, the prospects for linkage, although more occluded by public concerns about genetically modified organisms than in the case of health-related biotechnology, are nevertheless propitious in locational terms.

Cambridge has a number of science and technology parks, although the demand for further space is significant. At least eight biopharmaceutical firms are located in Cambridge Science Park. St. John's Innovation Centre, Babraham Bioincubator, Granta Park, the Bioscience Innovation Centre, and Hinxton Science Park are all newly available or planned. Hinxton is home to the U.K. human genome research center and the Sanger Institute, and commercialization will occur in an integrated science park. Most of the newer developments are taking place within short commuting distance of Cambridge itself, on or near main road axes like the M11, A11, A10, and A14. This is evidence of the importance of access for research applications firms to centers of basic research, also reinforcing the point that not everything concerning biotechnology must occur "on the head of a pin" in Cambridge city itself.

The final, important feature of the biotechnology landscape in Cambridge and the surrounding Eastern Region is the presence of both informal and formal networking between firms and research or service organizations and among firms themselves. Of more direct relevance to the biotechnology community are the activities of ERBI. This biotechnology association is the main regional network with formal responsibilities for creating newsletters; organizing network meetings; running an international conference, Web site, sourcebook, and database on the bioscience industry; providing aftercare services for biobusinesses; making intranational and international links (e.g., Boston, Oxford, San Diego); and organizing common purchasing, business-planning seminars, and government and grant-related interactions for firms.

Although substantially smaller than the Boston cluster, Cambridge already has most of the generic features of a sectoral innovation system. The presence of venture capital and other support services, mainly private, were noted. The existence of the regional biotechnology industry association ERBI compares with the Massachusetts Biotechnology Council in Boston, although it is worth noting that its origins lie in a multilevel governance initiative by the U.K. industry ministry (Department of Trade and Industry) and a local public-private enterprise support partnership (CambsTEC Business Link). Clearly, institutional learning from information technology networks established in the region and practices by international biotechnology comparators has played a role in the conception of how to govern the agglomeration and give it more cluster consciousness. Clusters like this are clearly cases of localized sectoral innovation systems possessing global reach. In scientific and commercialization terms, it is Europe's leading biotechnology cluster in a business with expected global turnover of $70 billion in 2000. Because of the sunk costs associated with colocation by venture capitalists, specialist patenting, legal, accountancy and insurance services, the immobility of the key knowledge-driving resource, the university, and the presence of a critical mass of biotechnology firms and entrepreneurs, Cambridgeshire is likely to remain the key biotechnology focus it has become.

The German BioRegio Clusters

We have seen that clustering in biotechnology is perceived as advantageous and successfully practiced in the United States and United Kingdom, something that the federal BioRegio initiative has sought to emulate. In what follows, we shall see a different picture, in which efforts are made by the government to induce learning, stimulate commercialization, and create governance and venture funding to build clusters through the BioRegio contest. The three key BioRegios are Rhineland, comprising Cologne, Düsseldorf, Wuppertal, and Aachen; Rhine-Neckar, including Heidelberg, Mannheim, and Ludwigshafen; and Munich. Jena was given a consolation prize but is least developed as a cluster. The accounts will be provided in that order.

Rhineland BioRegio

Given its history as a heavy industrial region undergoing major restructuring in coal, steel, chemicals, and heavy engineering (particularly in the Ruhr) toward newer growth industries, the land of North Rhine Westphalia has launched numerous technology-orientated initiatives, especially from 1984, when a number of new technological institutes with near-market research functions, technology parks, and innovation networks were set up under the Future Technology Program initiative. Among these was the land's first biotechnology program (*Landesinitiative Bio-und Gentechnik e.V*) to support SMEs. Other sectors receiving support included environmental technology, energy technology, micro-electronics, measurement, information technology, and materials. The biotechnology initiative was superseded in 1994 by the establishment of BioGenTec. This agency is a nonprofit organization with representation from industry, academia, trade unions, and government. It acts as an intermediary body linking biotechnology start-ups, an expert network of 200 members, venture capitalists, and partners from industry. It is seeking to become a commercial company and will sell services to the industry on that basis. A wide range of mainly medical biotechnology areas are prioritized under its program of support, but environmental and agro-food technologies are also supported.

Various networks have been established, including a venture capital network of local but also internationally operational firms and groups (the BioGenTec Atlas [BioGenTec 1998] lists fifteen), a competence and training network, and a management and coaching network. BioGenTec is about to establish biocenters at various locations and organizes an annual international meeting called the BioGenTec Forum. In 2000, an international forum on nanobiotechnology was organized. The research strengths of the land include the Max Planck Institute for Plant Breeding Research at Cologne, which has become the center of green biotechnology, around which larger (e.g., Monsanto, DSV, and Agrevo) and smaller firms are clustered. In 1998, a letter of intent was signed between the governments of North Rhine-Westphalia (NRW) and Saskatchewan, Canada, to improve collaboration in the field of agro-biotechnology. Also in Cologne is the Max Delbrück Laboratory (also part of the Max Planck Society), specializing in plant genetics. The Max Planck Institute for Neurological research, specializing in (photo)receptors, signal transduction, and recombinant proteins, is at Mülheim an der Ruhr. A Helmholtz Institute exists at Aachen (biomedicine and cryobiology), and a Fraunhofer Institute for Environmental Chemistry and Ecotoxicology is located at Schmallenberg. Altogether, the land has some 167 research institutes, many employing relatively small numbers of researchers but with representation across the red, green, and gray biotechnology spectrum.

It is fairly evident that multilevel land and federal programs of support fit together well, as the transition from the Future Technology Program to BioGenTec suggests. Dr. Fritschi, head of the latter organization, reported there is no conflict but rather, on the contrary, they have been complementary sources of funding over the years (Cooke 1999). He noted a spatial distinction in the stage at which particular kinds of public-venture funding occurred such that "interestingly, mostly expanding companies were able to take advantage of federal funding while land funding went directly into start-ups. We also recognize a steady movement of companies into the BioRegio area because of federal and/or land funding" (Dr. E. Fritschi, personal communication, September 1999). BioGenTec has already established regional offices outside the BioRegio area (Münster and Bergkamen, north and east of the Ruhr) to seek to seed new clusters, which with land funding it is hoped will also attract start-ups to get established in these more outlying areas.

It is worth noting that, in terms of the sectoral distribution of companies working in the field of biotechnology, 22 percent are in diagnostics, 12 percent are in pharmarceuticals, 7 percent are in agro-food biotechnology, 18 percent are in environmental protection, 9 percent are in filtration engineering, and 10 percent are in bioanalysis. The last three are primarily engaged in gray or environmental biotechnology activities, making it the largest category. Thus, although it is seen as a "Cinderella" part of biotechnology and one in which it is hard to get university-derived start-ups under way, it remains a strength of this region's biotechnology profile. The origins of this commercial expertise are interesting and reflect well on the restructuring efforts of the NRW government noted earlier. Because of German rules on codetermination, involving management and unions in strategic decisions concerning, for example, company restructuring, management could not simply decide to close down redundant plants but were required to explore alternative trajectories firms might seek to move along. Because of expertise in the mining and steel industries of, for example, filtration and ventilation technologies, it was recognized that adapting these for environmental cleanup, initially in the Ruhr itself but later in former East Germany and into Central and Eastern Europe, would meet a huge potential market need. Moreover, it was known that the EU was set on introducing tougher environmental legislation in 1990, and on advice from NRW politicians, the federal government introduced an equivalent German version in anticipation, thus augmenting market demand for environmental clean-up technologies. Between 1984 and 1994, some six hundred firms turned partly or wholly in this direction, including a number of new start-ups or spin-offs. More than 100,000 jobs were found in this new industry, which itself has been shown to have a clusterlike character (Cooke and Davies 1993; Rehfeld 1995).

Finally, to what extent can clustering be said to be a feature of the Rhineland BioRegio or areas adjacent to it? From responses elicited from the question, it seems that in place are the key conditions of a strong science base, expanding numbers of firms, qualified staff, available physical infrastructure (e.g., the Rechtsrheinisches Technologie Zentrum in Cologne, a 5,000 meter square biotechnology incubator with plans for a C4 quality central laboratory), availability of finance, business support services, a skilled workforce, effective networks, and a supportive policy environment. However, the number of biotechnology start-ups was at a peak of twenty-six in 1990-91 and in decline to a figure of twelve in 1994-95, recovering over 1996-97 after the announcement and including the first operational year of BioRegio (BioGenTec 1998). An International Technology Services Expert Mission report (Department of Trade and Industry 1998) stated that there were eleven start-ups and eight company expansions since 1996 in biotechnology in the Rhineland BioRegio area. This is consistent with the head of BioGenTec's hints of caution about funding only quality projects and disappointment in the gray biotechnology center of Aachen and Jülich, where start-ups had been hard to stimulate. It is also worth noting that QIAGEN, Germany's best firm based in this BioRegio, networks in its "gene alliance" with firms outside it. So, for the present, the Rhineland BioRegio has all the appropriate conditions for stimulating the development of reasonably large numbers of new biotechnology firms, but whether a significant cluster of growing biotechnology firms will appear swiftly must remain doubtful because of the evidence presently available.

Rhine-Neckar-Dreieck

Heidelberg is Germany's oldest university and has one of the best science bases for biotechnology. Two Max Planck Institutes—Cell Biology and Medical Research—are in the region, as is the German (Helmholtz) Cancer Research Center. The European Molecular Biology Laboratory and the European Molecular Biology Organization are there, along with one of Germany's four Gene Centers, the Resource Center of the German Human Genome Project, two further medical genetics institutes, and two plant genetics centers. Three other universities—Mannheim, Ludwigshafen, and Kaiserslautern—and three polytechnics complete the generation and diffusion subsystem. There are a number of Germany's leading big pharma firms nearby, such as BASF/Knoll (Ludwigshafen), Boehringer Mannheim Roche Diagnostics (Mannheim), and Merck (Darmstadt). But the heart of the BioRegio is the Heidelberg-based commercialization organization, the Biotechnology Center Heidelberg (BTH). This is a three-tired organization consisting of a commercial business consultancy, a seed capital fund, and a nonprofit biotechnology liaison and advisory service. Central to BTH's functioning is Heidelberg Innovation GmbH (HI), a commercial consultancy that takes company equity in exchange for drawing up market analyses, business, and financing plans; assisting in capital acquisition; and providing early-phase business support for start-ups. It is a network organization, relaying information, partnering with organizations seeking contact with local biotechnology companies, and linking to research institutes and local authorities.

The key initial financing element of BTH is BioScience Venture. This was established by local big pharma and banks, managed by HI, and acts as a seed fund and lead investor in early start-ups. It also seeks international venture capital to finance second-round developments. Assessments of project viability are made with advice from HI and BioRegio Rhine-Neckar e.V., the third element of BTH. The latter seeks out commercial projects and recommends the most promising for BioRegio public funding support. Business proposals have run at some fifty per year since 1996, but between 1996 and 1998 only nine start-ups had been established, a figure that had risen to seventeen (including biochip and biosoftware firms) by July 1999. The total number of biotechnology SMEs (excluding start-ups) was twenty in July 1998. Most are in the health care sector, with some in plant genetics. The main location for this cluster of some thirty-seven biotechnology firms is the Heidelberg Technology Park for SMEs and the adjoining Biopark on the university's science campus. The Biopark has 10,000 square meters of laboratory and office space plus a further 6,000 on the Production Park nearby, where start-ups move to once they have grown beyond the research phase. A joint venture by local firms and universities has been to establish the Postgraduate BioBusiness Program. This is designed to provide scientists with hands-on experience of business administration through three months' coursework and nine months of practical training in industry (König 1998).

Once more, key ingredients for successful clustering are present, including close proximity for firms on the technology park to both big pharma in Ludwigshafen and Mannheim and leading-edge science in Heidelberg. The land of Baden-Württemberg has a biotechnology initiative but also distributes its funding among the Freiburg BioValley (one of Germany's most dynamic BioRegios), Ulm, and Tübingen-Stuttgart as well as the Rhine-Neckar region. As we have seen elsewhere, BioRegio funding is principally used for start-ups, most of which are currently suffering losses. But through the networklike character of BTH, lead investor capital from BioScience Venture can be tripled by leveraging both federal BioRegio funding and land/corporate venturing funds. Thus, reasonable sums of start-up capital can very easily be raised at low risk to the essentially public lead investor. The land helped fund Heidelberg Technology Park; subsidizes a patenting support initiative, providing grants to universities for making patent applications; and funds a young innovators pre-start-up funding program for university and research institute personnel (Clarke 1998).

Munich

The commercial application of biotechnology in Germany is said by König (1998) to have begun in the 1950s when Boehringer Mannheim moved part of its diagnostics R&D to Munich. More recently, this company invested DM150 million in production facilities for therapeutic Reteplase (cardiac infarction treatment) in a southern part of Munich. But Martinsreid and Grosshadern in the southwestern suburbs mark the center of biotechnology in Bavaria. Hoechst Marion Roussel (merged with Rhone Poulenc Rorer to form Aventis in 2000) opened its Center for Applied Genomic Research there, and the Biotechnology Innovation Center (IZB), funded DM40 million by the Bavarian government, is located nearby with 9,000 square meters of laboratory and office space. The organization responsible for managing the development of biotechnology, BioM, is also located at Martinsried. The area has become a biomedical research campus with eight thousand researchers working in biology, medicine, chemistry, and pharmacy. Unlike Rhine-Neckar's BTH, BioM AG is a one-stop shop with seed financing, administration of BioRegio awards, and enterprise support under one roof. Seed financing is a partnership fund from the Bavarian State government, industry, and banks up to DM300 thousand per company. BioM's investments are tripled by financing from Bayern Kapital, a special Bavarian financing initiative. The latter supplies equity capital as co-investments. The fund has DM80 million for supporting biotechnology activities. Bay BG and BV Bank-Corange-ING Barings Bank have special public/private cofunding pools, and a further eight (from sixteen) Munich venture capitalists in the private-market sector invest in biotechnology. By 1999, sixteen start-ups had been funded to the tune of DM59 million, with a third of this coming from BioRegio sources. BioM is a network organization, reliant on science, finance, and industry expertise for its support committees. It also runs young entrepreneur initiatives, including development of business ideas into business plans and financial engineering plans. Business plan competitions are also run in biotechnology.

The science base in Munich is broad, and seen as Germany's number one high-technology region (it is especially strong also in ICT, see Sternberg and Tamásy 1999), but with special expertise in health-related and agricultural and food biotechnology. There are three Max Planck Institutes of relevance, in Biochemistry, Psychiatry, and the MPI Patent Agency. GSF is the Helmholtz Research Center for Environment and Health, and the German Research Institute for Food Chemistry is a Leibniz Institute. There are three Fraunhofer Institutes, one of Germany's four Gene Centers, two universities, and two polytechnics. The main research-oriented big pharma companies are Roche Diagnostics (formerly Boerhinger Mannheim) and Hoechst Marion Roussel. The work areas of this science community include three-dimensional structural analysis, biosensors, genomics, proteomics, combinatorial chemistry, gene transfer technologies, vaccines, bioinformatics genetic engineering, DNA methods, primary and cell cultures, microorganisms, proteins, enzymes, and gene mapping. The Bavarian commitment to biotechnology (and other new technologies) was realized through its state government decision to privatize parts of its share in power-generation and distribution companies in the 1990s, thereby creating a funding pool to subsidize applied technology developments. The commitment was expressed in permission for biotechnology production facilities being issued with fewer obstacles and delays than in the other German *länder.* Such permissions are *land* and not federal responsibilities, and Bavaria showed its commitment earliest. The Bavarian Ministry of Economics learned the U.S. model of commercialization on the consultancy advice of the Fraunhofer Institute for Systems Innovation, Karlsruhe: venture capital, management support, and start-ups to transfer research results from laboratory to market. As we have seen elsewhere, however, this is mostly sought through public initiative, as with the IZB, which is a combination of incubator and technology park in proximity to the Gene Center and two of the Max Planck Institutes conducting biotechnology research. In common with the other BioRegio winners, the vertical networks from science through (public) funding to start-up are in principle strong; however, as elsewhere, given the almost risk-free funding regime, the number of start-ups is not overwhelming, perhaps because of the quest for quality start-ups in which substantial sums may be individually invested. A further explanation for conservatism is that BioM AG, set up as a corporation, makes investments with its shareholders' (state, industry, and banks) money. Banks hold most shares, looking for high returns and to learn about biotechnology' risks and prospects. Hence, while BioM is the network face of the biotechnology cluster in Munich, its activities are ultimately orchestrated indirectly and directly by the banks, abetted by a fairly risk-averse, mostly publicly funded, venture capital industry and the local pharmaceutical and chemical companies (see Giesecke 1999).

With respect to land and federal relationships on funding, Munich BioRegio once again demonstrates the seamlessness of the fit between programs. This is no surprise since a great deal of "concertation" proceeds between the two levels of government on a constant basis and the last thing either wants is a resort to the Constitutional Court to rule on intergovernmental conflicts. Hence, this is a good instance of the German consensus-oriented mode of policy evolution. Similarly, the consistency with which public, scientific, and industrial partnership characterizes funding or technology-transfer mechanisms is illustrative of the ingrained networking culture that characterizes German governance. As to whether Martinsreid and Munich more widely constitute a cluster, the answer is probably affirmative, although there are conflicting reports as to whether three key firms commercializing biotechnology from Max Planck Institutes are interacting, collaborating companies. Dohse (1999; personal communication, September 1999), suggested that despite their common origin they are not strongly linked. But Clarke (1998) noted that two of the firms, MorphoSys and Micromet, are collaborating on the development of an antibody-based treatment for micrometastatic cancer. MorphoSys was the first firm to receive a BioRegio grant and had previously collaborated successfully with Boeringer-Mannheim on the development of a diagnostic reagent. MorphoSys' business strategy is to focus on the development of horizontal networking. They have no plans to develop therapeutics themselves, aiming to remain a science discovery firm, but rather to let partners carry the risk of drug development. Thus, MorphoSys works with a variety of companies, minimizing its risk profile but potentially benefiting from substantial injections of capital from research funding, milestone payments, and royalties. MediGene is another Munich company that does plan to become a fully integrated biopharmaceutical company. It was a spin-off from a Gene Center in 1994 and has raised DM23 million from the typical German sources: venture capital and state and federal funds. Its expertise is in gene therapy for cancer and cardiovascular diseases. MediGene has alliances with Hoechst on gene therapy vectors and a vaccine for malignant melanoma. Academic-clinical partnerships include the Munich Gene Center, the Munich University Hospital, German Cancer Center at Heidelberg and, in the United States, the National Institutes of Health and Princeton University. Its cofounder Horst Domdey recently

TABLE 3. Stylized Assessment of Different Sectoral Innovation Systems

	United States	*United Kingdom*	*Germany*
Innovation strength	Applications	Discovery	Platform technology
Venture capital	Private	Mostly private	Mostly public
Commercialization	Entrepreneurial	Liberalizing	Highly regulated
Governance	Firm association	Public and SME	Public and large firm
Clustering	Mature	Developing	Immature
Competitiveness	High	Medium	Low

Note: SME = small and medium-sized enterprise.

gave up a chair at Munich University to become head of BioM. Mondogen spun off from the Virus Research Department at the Martinsried Max Planck Institute for Biochemistry. Its founder Peter Hofschneider was director of the department and had cofounded Biogen, one of Boston's oldest biotechnology firms, in the late 1970s. IZB and BioRegion, plus a McKinsey Business Plan competition, led to the founding of Mondogen. Martinsreid is said by Hofschneider to be unlike MIT and Cambridge as a cluster but to have the "seed crystal" of high-tech firms: the main obstacles are the different cultures between German scientists and venture capital speculators.

Prior to the conclusions, we can see the distinctiveness of the three approaches in Table 3. This takes key dimensions of the sectoral innovation system in the form of biotechnology clusters and draws the fairly obvious deduction that private-sector interaction with the science base produces more rapid commercialization but not necessarily invention from which innovation as commercialization subsequently flows.

A conclusion of a companion study (Cooke 2001a, 2001b) to this article is that the United States was early into the commercialization of discoveries made elsewhere in biotechnology, notably the United Kingdom because of its superior, private system of innovation. This compensated at the level of the market for a weaker invention system based on inadequate use of public resources. Germany has weaknesses in both; hence, whether its modest performance—causing major public subsidy to be injected in order to try to catch up—is successful remains to be seen.

CONCLUSIONS

These biotechnology clusters each have exceptionally strong enterprise support infrastructures complementing strong local science bases. Network links among actors are pronounced, with cooperation on finance and services between national and regional, public and private sectors common. In Germany, there are difficulties in getting large numbers of new businesses up and running despite the apparently generous grant aid available. This seems partly explicable by the risk averseness of the lead investors and the conservatism of the banks that are quite closely involved behind the scenes in the management of BioRegio economic development. In all cases, land and federal funding regimes coexist happily, and in some cases, initiatives set up by the lower level of government are easily absorbed into new initiatives, notably BioRegio, emanating from the federal level. Funding is a challenge in the United States and United Kingdom, but firms have more abundant venture capital to access than in Germany, especially private funding. Perhaps one of the most striking features of the government, industry, and science relationship with respect to biotechnology is how interwoven they are into what Etkowitz and Leydesdorff (1997) called "the triple helix," even down to the individual, sometimes small-city level of operations. Communication levels among key actors are thus of a high quality, networks function effectively, and "seed crystals" of well-functioning future clusters have been sown in numerous regions of the United States, United Kingdom, and Germany. The real testing time for these possibly emergent clusters of biotechnology firms in Germany will come when large doses of second-round funding are needed as firms move toward therapeutic-drug production. This will begin occurring seriously around the year 2002. This period is over in the United States, and the industry is maturing, with mergers and acquisitions occurring. In the United Kingdom, probably ten years behind the United States, the first-stage firms have, in some cases, reached take-off and Chrioscience: Celltech placed the United Kingdom's first therapeutic product on the market in 1998.

Clearly, clustering is absolutely central to the growth prospects of biotechnology firms at present. The cases studied here all have in common exceptionally well-developed scientific research bases, associations that manage collective affairs, local venture capital, infrastructure appropriate to biotechnology commercialization, and much national and some regional public funding of diverse aspects of cluster activities. While the United States appears to be the most marketized system, it is clear that behind the scenes major federal funding sustains its leading edge in science. In Germany, the whole cluster system is in general more publicly dependent; without major public funding, even of venture capital, the German industry would not be extensive. Strangely, perhaps, it is the U.K. case of Cambridge that has the least subnational innovation support from the public sector, at least for commercialization, having until a few months ago little by way of regional governance. Venture capital is largely private and relatively abundant. Despite this, it also has the character of a localized regional innovation system based on strong clustering and networking among research and business actors.

Thus, the limitations of regional innovation systems are made particularly clear by this textured analysis of five cases. The funding of basic research is a national innovation system priority and a responsibility that regional innovation systems can only consider at the margin; even then, they need fairly full devolution. The regulatory regime, including laws on laboratory practice (e.g., the Genetic Engineering Act), financial rules, and rates of taxation, are more national than regional but can be implemented differentially or adjusted regionally. But the local-regional level becomes the most important for the evolution of clusters, including the concentration of critical research mass, the formation of networks, development of cluster interactions, and even the commercialization of products. However, with respect to commercialization, links to big pharma, customers, and even venture capital is frequently global as well as national or even regional.

ACKNOWLEDGMENTS

Many people deserve thanks for helping me write this article. The conceptual part of the article grew in consequence of the EU-TSER Regional Innovation Systems: Designing for the Future project. DG12 of the European Commission are thanked, along with our collaborators, particularly Goio Etxebarria and Mikel Gomez Uranga who conceived with me key elements of the regional innovation system concept. Patries Boekholt and Franz Tödtling also contributed in major ways to the published research findings of the REGIS report. For the biotechnology research, I am grateful to the U.K. Minister of Science, Lord Sainsbury, and the U.K. Department of Trade and Industry for appointing me a member of the Biotechnology Clusters Task Force, from which information on the two Cambridges arose following intensive study visits. Dr. Monica Darnbrough, head of the DTI Biotechnology Directorate was also instrumental in commissioning the research on Germany. In conducting that research, I was assisted in major ways by Dirk Dohse, Susanne Giesecke, Gerd Krauss, Thomas Stahlecker, Knut Koschatzky, Olaf Arndt, Edgar Fritschi, and Steffen Reich. All are warmly acknowledged and the usual disclaimer applies.

REFERENCES

Amin, A., and D. Thomas. 1996. The negotiated economy: State and civic institutions in Denmark. *Economy and Society* 25: 255-81.

Amin, A., and N. Thrift. 1994. Living in the global. In *Globalization, institutions and regional development in Europe*, edited by A. Amin and N. Thrift, 1-22. Oxford, UK: Oxford University Press.

Argyris, C., and D. Schon. 1978. *Organisational learning: A theory of action perspective*. Reading, MA: Addison-Wesley.

Asheim, B. T. 1996. Industrial districts as "learning regions": A condition for prosperity? *European Planning Studies* 4: 379-400.

Audretsch, D. 1998. Agglomeration and the location of innovative activity. *Oxford Review of Economic Policy* 14: 18-29.

Autio, E. 1998. Evaluation of RTD in regional systems of innovation. *European Planning Studies* 6: 131-40.

Best, M. 1999. Cluster dynamics in theory and practice, with applications to Singapore/Johor and Penang electronics. Report to the Judge Institute of Management Studies, Cambridge University.

———. 2000. Silicon Valley and the resurgence of Route 128: Systems integration and regional innovation. In *Regions, globalization and the knowledge-based economy*, edited by J. Dunning. Oxford, UK: Oxford University Press.

BioGenTec. 1998. *BioGenTec atlas*. Cologne, Germany: BioGenTec.

Braczyk, H.-J., P. Cooke, and M. Heidenreich, eds. 1998. *Regional innovation systems: The role of governances in a globalized world.* London: UCL Press.

Casper, S., and H. Kettler. 2000. *The road to sustainability in the UK and German biotechnology industries.* London: Office of Health Economics.

Clarke, L. 1998. Setting the pace. *Pharmaceutical Forum* pp. 2-6.

Cooke, P., ed. 1995. *The rise of the rustbelt.* London: UCL Press.

———. 1999. *The German biotechnology sector, the public policy impact and regional clustering: An assessment.* London: Department of Trade & Industry.

———. 2001a. Biotechnology clusters in the UK: Lessons from localisation in the commercialisation of science. *Small Business Economics* 17: 43-59.

———. 2001b. *Knowledge economies: Clusters, learning and cooperative advantage.* London: Routledge.

Cooke, P., P. Boekholt, and F. Tödtling. 2000. *The governance of innovation in Europe: Regional perspectives on global competitiveness.* London: Pinter.

Cooke, P., and S. Davies. 1993. An economic profile of North Rhine-Westphalia. Report to the Welsh Development Agency, Center for Advanced Studies in the Social Sciences, Cardiff University, Cardiff, UK.

Cooke, P., and K. Morgan. 1998. *The associational economy: Firms, regions and innovation.* Oxford, UK: Oxford University Press.

Cooke, P., M. Uranga, and G. Etxebarria. 1998. Regional systems of innovation: An evolutionary perspective. *Environment & Planning, A* 30: 1563-84.

Cookson, C., and D. Pilling. 1999. Biotech trade looks to US to cure confidence crisis. *Financial Times*, 9 September, 11.

Crevoisier, O. 1997. Financing regional endogenous development: The role of proximity capital in the age of globalization. *European Planning Studies* 5: 407-16.

Department of Trade and Industry. 1998. *Our competitive future: Building the knowledge-driven economy.* London: Department of Trade and Industry.

———. 1999. *Biotechnology clusters.* London: Department of Trade and Industry.

Dohse, D. 1999. The BioRegio contest: Results of an empirical investigation. Kiel, Institute of World Economics (mimeo).

Dosi, G. 1988. Sources, procedures and microeconomic effects of innovation. *Journal of Economic Literature.* 26: 1120-71.

Eastern Region Biotechnology Initiative. 1998. *Sourcebook '98.* Cambridge, UK: Eastern Region Biotechnology Initiative.

———. 1999. Background information for Cambridge and Eastern Region biotechnology cluster. Cambridge, UK: Eastern Region Biotechnology Initiative (mimeo).

Edquist, C. 1997a. Introduction: Systems of innovation approaches—their emergence and characteristics. In *Systems of innovation: Technologies, institutions & organizations*, edited by C. Edquist, 1-35. London: Pinter.

———, ed. 1997b. *Systems of innovation: Technologies, institutions and organisations.* London: Pinter.

Enright, M. 1996. Regional clusters and economic development: A research agenda. In *Business networks: Prospects for regional development*, edited by V. Staber, N. Schaefer, and B. Sharma, 190-214. Berlin, Germany: de Gruyter.

Etkowitz, H., and L. Leydesdorff. 1997. *Universities and the global knowledge economy.* London: Pinter.

European Commission. 1995. *The green paper on innovation.* Luxembourg: Commission of the European Communities.

Florida, R. 1995. The industrial transformation of the Great Lakes Region. In *The rise of the rustbelt*, edited by P. Cooke, 162-76. London: UCL Press.

Freeman, C. 1994. The economics of technical change. *Cambridge Journal of Economics* 18: 463-514.

———. 1995. The "National System of Innovation" in historical perspective. *Cambridge Journal of Economics* 19: 5-24.

Giesecke, S. 1999. Determinants of successful S&T policy in a national system of innovation. Vienna, Economics University (mimeo).

Grabher, G. 1993. The weakness of strong ties: The lock-in of regional development in the Ruhr area. In *The embedded firm: On the socio-economics of industrial networks*, edited by G. Grabher, 255-77. London: Routledge.

Johnson, B. 1992. Institutional learning. In *National systems of innovation: Towards a theory of innovation and interactive learning*, edited by B. Lundvall, 23-44. London: Pinter.

König, G. 1998. Nurturing biotech in the regions. *Pharmaceutical Forum* pp. 9-11.

Krugman, P. 1997. Good news from Ireland: A geographical perspective. In *International perspectives on the Irish economy*, edited by A. Gray. Dublin, Ireland: Indecon.

———. 1998. What's new about the new economic geography? *Oxford Review of Economic Policy* 14: 7-17.

Lundvall, B., ed. 1992. *National systems of innovation: Towards a theory of innovation and interactive learning*. London: Pinter.

Lundvall, B., and B. Johnson. 1994. The learning economy. *Journal of Industry Studies* 1: 23-41.

Maillat, D. 1995. Territorial dynamic, innovative milieus and regional policy. *Entrepreneurship and Regional Development* 7: 157-65.

Malmberg, A., and P. Maskell. 1997. Towards an explanation of regional specialization and industry agglomeration. *European Planning Studies* 5: 25-42.

Nelson, R., ed. 1993. *National innovation systems—A comparative analysis*. Oxford, UK: Oxford University Press.

North, D. 1993. Institutions and economic performance. In *Rationality, institutions and economic methodology*, edited by U. Mäki, B. Gustafsson, and C. Knudsen, 242-64. London: Routledge.

Porter, M. 1998. *On competition*. Cambridge, MA: Harvard Business School Press.

Rehfeld, D. 1995. Disintegration and reintegration of production clusters in the Ruhr area. In *The rise of the rustbelt*, edited by P. Cooke, 85-102. London: UCL Press.

Rosenfeld, S. 1997. Bringing business clusters into the mainstream of economic development. *European Planning Studies* 5: 3-23.

Sabel, C. 1995. *Experimental regionalism and the dilemmas of regional economic policy in Europe*. Paris: Organization for Economic Cooperation and Development.

Sabel, C., H. Kern, and G. Herrigel. 1989. *Collaborative manufacturing: New supplier relations in the automobile industry and the redefinition of the industrial corporation*. Cambridge, MA: MIT Press.

Saxenian, A. 1994. *Regional advantage: Culture and competition in Silicon Valley and Route 128*. Cambridge, MA: Harvard University Press.

Scott, A. 1993. *Technopolis: High technology industry and regional development in Southern California*. Oxford: University of California Press.

Segal, Quince & Wicksteed. 1998. *Cambridge phenomenon update*. Cambridge, UK: Segal, Quince & Wicksteed.

Smilor, R., G. Dietrich, and D. Gibson. 1993. The entrepreneurial university: The role of higher education in the United States in technology commercialization and economic development. *International Social Science Journal* 45: 1-11.

Smith, K. 1997. Economic infrastructures and innovation systems. In *Systems of innovation*, edited by C. Edquist. London: Pinter.

Sternberg, R., and C. Tamásy. 1999. Munich as Germany's no.1 high technology region: Empirical evidence, theoretical explanations and the role of small firm/large firm relationships. *Regional Studies* 33: 367-77.

Storper, M. 1995. The resurgence of regional economies, ten years after: The region as a nexus of untraded interdependencies. *European Urban and Regional Studies* 2: 191-221.

Storper, M., and A. Scott. 1995. The wealth of regions: Market forces and policy imperatives in local and global context. *Futures* 27: 505-26.

[24]

ELSEVIER

Research Policy 31 (2002) 849–858

research policy

www.elsevier.com/locate/econbase

When do industries cluster? A proposal on how to assess an industry's propensity to concentrate at a single region or nation

Claus Steinle*, Holger Schiele

Universitat Hannover, Abteilung Unternehmensfuhrung und Organisation, Konigsworther Platz 1, D-30167 Hannover, Germany

Received 21 March 2001; received in revised form 22 May 2001; accepted 13 July 2001

Abstract

Often the literature on clustering implicitly assumes national or regional–sectoral agglomeration to be a general phenomenon applying to all or most businesses. We argue that not all industries are equally affected by the process of clustering. Therefore, the question on how to assess the relevance of clustering for a particular company or industry becomes important.

It is deduced that those industries with a divisible production process and a transportable product or service fulfil the necessary conditions for clustering. A long value-chain including multiple distinct competencies, innovation-intensity characterised by "network-innovations" and the volatility of markets further increase the potential for an industry to cluster.

Besides providing a heuristic for strategic management, this paper alerts regional planners that only in those regions whose economic structure is characterised by industries subject to clustering, the promotion of local network-building is relevant.

Keywords: Conditions for clustering; Innovation; Regional advantages; Globalisation; Strategic management

1. The need for considering the process of clustering

1.1. Centralisation as a consequence of globalisation in industries subject to clustering

In the last few years, a growing interest in the phenomenon of clustering has occurred (as an introduction, see e.g. Roelandt and den Hertog, 1999). It has been claimed that the harmonisation of international markets and reduction of transport costs, somewhat paradoxically, has increased the importance of the proximate environment of a company (Gertler, 1995; Porter, 1998). There is increasing statistical material supporting the assumption that globalisation often means centralisation: instead of spreading evenly around the globe, similar companies tend to cluster in particular nations or even regions (e.g. Patel and Pavitt, 1991; Amendoa et al., 1992; Archibugi and Pianta, 1992; Isaksen, 1997; Patel and Vega, 1999). Globalisation promoted by electronic business could speed up this process even more. In a clustered environment, isolated companies have been found to suffer from a "periphery discount" which can amount to 40% lower returns (Fabiani and Pellegrini, 1998). The meaning of these insights is that companies should be alert, analysing the possibility of any "Silicon valley-type" process of clustering taking place in their industry as well.

Clustering is a multi-dimensional phenomenon. Therefore, an integrative approach, not only covering

* Corresponding author.
E-mail address: holger@schiele-net.de (H Schiele).

0048-7333/02/$ – see front matter © 2002 Elsevier Science B V All rights reserved.
PII: S0048-7333(01)00151-2

that literature with an explicit reference to "clusters" seems to be rewarding. This is particularly true if we want to understand the conditions for clustering. Several converging schools of thought have contributed to our understanding of such processes of concentration. Since the 1980s not only the significance of the "Silicon Valley" (Saxenian, 1991) but also the success of Italian one-product towns known as "industrial districts" has interested many scholars, whose explanation centres around the higher efficiency of these agglomerations due to flexible specialisation (Brusco, 1982; Becattini, 1991). The French milieu-approach has put more emphasis not on cost-reducing efficiency, but on the innovative power of regions (Aydalot and Keeble, 1988; Capello, 1999). Research on isomorphism deepens the understanding of the interaction within value-creating systems, because it offers explanations on how the members of multi-organisational systems adjust their behaviour to each other, which can be observed in well-functioning clusters (DiMaggio and Powell, 1983). Insights deriving from innovation theory have lead to a revival of the theory of national systems of manufactures and innovation (List, 1910; Lundvall, 1992; Freeman, 1995; Furtado, 1997). If there are distinct national systems, then some are better suited for hosting particular industries, which would then cluster in that country. From a different perspective, but in line with the above, scholars dedicated to the analysis of international commodity chains have found some wares to resist a globalisation of production (Hopkins and Wallerstein, 1994; Schoenberger, 1994). Influenced by the filière-approach (Stoffaes, 1980; Morvan, 1991), another model has been Porter's "diamond of competitive advantage" (Porter, 1990, Enright and Tenti, 1990), whose concept tended to dominate the discussion on innovation-induced competitiveness at subcentral level (Cooke et al., 1997).

Porter claims to have developed a new theory of international trade. He explains national specialisation patterns as a result of the coincidence of favourable conditions of demand, factor conditions, related and supporting industries and a sound structure of rivalry between companies in one location. However, his work has been criticised on the grounds that this clustering-approach would not fit for those industries centred on raw materials and offers little insights for strictly domestic industries or those producing non-tradable goods (Yetton et al., 1992; Penttinen, 1994; Rugman and Hodgetts, 1995; Judex, 1996). It becomes clear that not all industries are equally affected by the process of clustering.

Acknowledging the validity of the quoted authors' criticisms, they do not necessarily falsify Porter's theory in particular and clustering approaches in general. However, what is missing—not only in Porter—are conditions for clustering (Yetton et al., 1992; Dunning, 1992). Criteria have to be developed that allow for differentiating between industries where the theories of clustering apply, because they fulfil the conditions for clustering, and those which are only marginally affected by the phenomenon.

1.2. Advantages of companies being located in clusters: reduced costs of co-ordination improving efficiency and innovation

Before proposing an answer to the question of when industries tend to cluster, we have to sketch why companies would derive advantages from being located close to each other. If these advantages of proximity fit with the central success-factors of an industry, clusters offer competitive advantages.

Clustering is the process of development of locally rooted value-creating systems. Several companies at each step of a single value-chain, in addition to related and supporting organisations, agglomerate in one nation or even one single region within a nation, while serving a larger market. Clustering can happen because one value-creating system can impose itself already at the beginning of an industry life cycle or because most companies of an industry, except the members of the strongest cluster, go out of business over time.

Here, clusters are introduced as localised sectoral agglomerations of symbiotic organisations that can achieve superior business performance because of their club-like interaction. "For us, innovative clusters are not a simple concentration of independent economic agents, but display at an inter-industrial level, underlying networks of interrelated co-operating businesses." (DeBresson, 1996, p. 161). Thus, coincidental agglomerations are excluded, e.g. those centred around raw materials and such, whose members do not show any particular interaction between each other. An innovative cluster is not likely to develop automatically (Steinle et al., 1998).

Without the intensive interaction that exceeds traditional market exchange, the advantages of proximity would basically confine to lower costs of transportation and lower switching costs for employees only. However, if actors have developed a milieu, which allows for an interaction that is simultaneously co-operative and competitive, a localised value-creating system can reach its full potential, i.e. transform into an "innovative cluster". Such an environment is characterised by a club-like atmosphere with intensive knowledge exchange (Weder and Grubel, 1993; Capello, 1999). The constituents are aware of their membership to the system. They adhere to a common set of rules. The "conventions" can be enforced because of the possibility to exclude members from the club and thus deprive them from the information flow only accessible to the constituents. Unlike in virtual communities, in geographically rooted clusters members cannot simply chose to join another group, as soon as they have ruined their reputation. Therefore, in a cluster the costs associated with bad reputation are so high that trustful collaboration becomes recommendable. Information about the opportunistic behaviour of an actor easily spreads within the cluster, so that breaking the conventions, an actor would not only put one single connection at risk, but the whole bundle of his relationships with other members of the system as well.

Now, in such a cluster, organisations can (1) be more efficient, i.e. work at lower costs, and (2) simultaneously be more innovative, i.e. upgrade products and performance continuously. The traditional "trade-off" between costs and differentiation can be overcome. The quintessential foundation of this advantage lies in the possibility of linking scale- and competence-adequate structural fragmentation with an efficient and flexible co-ordination of multiple actors. The costs of transaction are lower because in a cluster showing the character of a club, feedback due to intensive information exchange between its members reduces the success of opportunistic behaviour. Furthermore, the partners needed to transform ideas into business-innovations are available, well known and can be trusted (for details see Schiele, 2001). For these opportunities to translate into business benefits, however, several circumstances have to apply.

If all or most of the following conditions are met and the system can simultaneously develop its "soft side", a globally leading cluster can develop. The quality of interaction within a localised value-creating system can be improved by its members at will, even though the local culture may represent a limiting factor in a particular region (Fornahl, 2001; Pilon and DeBresson, 2001). The six conditions subsequently focussed upon, however, can hardly be influenced on a medium term and apply to an industry in general. Therefore, they can best serve to assess its potential for clustering. The conditions for clustering are not elaborated in order to predict where a cluster will develop, but to indicate which types of business are more likely to cluster-somewhere.

2. Conditions for clustering

2.1. Necessary conditions (NC): divisibility of process and transportability of product

First of all, specialisation into different activities along the value-chain must be possible and the final product of the system has to be transportable. These requirements can be considered as necessary conditions, for their absence in an industry makes clustering "technically" impossible. [1]

2.1.1. NC_1: divisibility of process as necessary condition for clustering

Only the possibility to divide a process of production into several distinct steps, considered to be the first necessary condition for clustering (NC_1), allows for specialisation, so that alternative forms of co-ordination can be considered (Brusco, 1982; Jarillo, 1995; Lazerson, 1995). Cupola production processes, for instance, are not divisible. For technical reasons several products appear simultaneously, so that an integrated form of organisation becomes largely predetermined. On one hand, the fragmentation of a process of production, therefore, depends on its technical divisibility (Piore and Sabel, 1984; Rosenkopf and Tushman, 1994). On the other hand, it can be considered to be a consensus that a critical mass, both in terms of actors and in terms of business volume, has to be present in order to enable the

[1] The following passage relies conceptionally on Schiele, 2001, p. 144 ff.

process of fragmentation, i.e. the forming of multiple specialised organisations. This aspect is further stressed by the observation that there should be several actors at each step of the value-chain to allow for healthy competition, as well as mutual learning.

2.1.2. NC_2: *transportability of the final product as necessary condition for clustering*

If a product cannot be relocated, the location of its providers is determined by the site of its consumers. It should be added that both, industrial products and services are transportable, e.g. if a service-provider moves with the equipment to the customer. Furthermore, a cluster without competitors or without the chance to enter in competition with these could not establish any boundaries and define an identity. Membership would not translate into any club-like business benefits. Therefore, the cluster's product has to be transportable (Lazerson, 1995; Genet, 1997). It is worth distinguishing between the final product and its components. If supplies are difficult to transport while the final product can easily be exported, the need for proximity during the production phase increases. As a consequence, the chances for clustering of the entire industry accentuate.

2.2. *Sufficient conditions (SC): flexible co-ordination of several distinct actors*

In addition to the necessary conditions, divisibility of process and transportability of product, a couple of sufficient conditions foster the process of clustering. The first two sufficient conditions for clustering (SC_1 and SC_2) draw upon Richardson's distinction between complementary activities, which succeed each other in the value-chain and similar activities, requiring similar competencies (Richardson, 1972). The more there is a need to co-ordinate complementary but dissimilar activities, the more clusters profit due to their reduced costs of co-ordination.

2.2.1. SC_1: *long value-chain: co-ordination of multiple components to form one final product*

The more specialised an organisation, the more its competitiveness depends on complementary actors and the more important the organisation of its environment becomes. The number of interfaces increases. Such a situation is typical for a very fragmented value-creating system. The challenge of co-ordination is accentuated, if diverse actors have to be co-ordinated not only after, but also before supplying their component, i.e. if instead of standard intermediate inputs, supplies have to be tailor-made (Wehrli and Jüttner, 1996).

When do long value-chains develop? On one hand, the length of a value-chain is technically determined. However, founders of specialised organisations find most fruitful ground if there is an incompatibility of scale, i.e. there are different optimal amounts of output per activity (Scott, 1988; Gertler, 1993). Regularly it is easier to source additional capacity externally than to sell own surplus on the market. Another reason for the segmentation of the value-chain resides in a different profitability of its segments (Jarillo, 1995).

2.2.2. SC_2: *diversity of competencies: multiple dissimilar, but complementary competencies*

The second part of Richardson's argumentation refers to the similarity between activities. The more distinct the competencies in one value-chain, the more challenging it gets for a single company to master them all. As a consequence, the problem of co-ordinating diverse partners, each focused on different competencies, arises. The presence of complementary, but dissimilar knowledge in one value-creating system, therefore, can be considered as another sufficient condition for clustering (Sternberg, 1995).

2.2.3. SC_3: *importance of innovation: network-innovations as a motor for agglomeration*

The more complementary actors contribute to a process of innovation and the shorter the time available for their co-ordination, the more the efficiency of their co-operation becomes a success-factor. If an industry does not honour innovation, however, advantages with the co-ordination of innovative actors do not provide any business benefit. This means that the importance of innovation in an industry can be considered as another sufficient condition fostering the formation of clusters.

While ample power of innovation is usually attributed to clusters (Baptista and Swann, 1998), there has been no consensus on which type of innovation would particularly promote clustering.

- One thesis states that radical innovations require a couple of complementary innovations in order to develop their potential. Therefore, the constitution of the proximate environment of an innovator would be of utmost importance with radical innovations, because the amount of organisations involved can be large (Sternberg, 1995; Moore, 1997). In such a simultaneous process of mutual adaptation, the exchange of implicit knowledge plays an important role. The presence of implicit knowledge as such is considered to foster clustering (Swann, 1993; DeBresson, 1996; Grant and Gregory, 1997; Malmberg and Maskell, 1997; Genet, 1997). Based on this, some authors expect clusters to be more important at the beginning of a product life cycle (Swann, 1993; Hobday, 1994; Audretsch and Feldman, 1995).

- A second and opposing thesis does not expect radical, but rather incremental innovations to lead to the formation of clusters, because for such "product-based learning" to take place, a constant and direct exchange of knowledge between diverse actors is necessary (Storper, 1992; Rallet, 1993; Malmberg et al., 1996). It would be the face-to-face interaction between incrementally innovating actors that allows for the exchange of implicit knowledge. As a consequence, the process of clustering would be expected to accelerate as an industry becomes mature and radical innovations are replaced by incremental ones (Grant and Gregory, 1997; Malmberg and Maskell, 1997).

From this short discussion we can conclude that it is difficult to deduce a clear-cut relationship. Common ground, however, is the expectancy of clustering in such industries that are characterised by the appearance of implicit knowledge, a high speed of transformation, and in particular, by the involvement of multiple actors with distinct competencies in the process of innovation. Independent of whether radical or incremental innovations are present or whether the industry is just forming or maturing, the distinction between "inventor-based innovations", "laboratory-type innovations" and "network-innovations" could be very fruitful for understanding an industry's innovation-induced propensity to cluster.

What are network-innovations? On the basis of extensive historical research, Freeman and Soete summarise the evolution of the process of innovation by distinguishing between three phases. In each of them, a distinct "design" shapes most of the innovations (Freeman and Soete, 1997; Gammage and Schiele, 2000).

1. Inventor–entrepreneur: the predominant actor in the 19th century. An invention leads to a radically new product, often commercialised by the inventor himself.
2. Large in-house research and development: beginning in the 1870s, in particular the chemical companies pioneered an organisational revolution. A considerable number of specialists were pooled in distinct departments. Most systematically improved already existing processes in their laboratories.
3. Network-innovation: currently a new design has the chance to become predominant. In the case of a network-innovation it is not a radically new invention that is brought into practical use as with the typical inventor–entrepreneur. Neither is it the improvement of one already existing competence as with the traditional laboratory. A network-innovation occurs when different actors with distinct competencies combine their skills, thus improving an existing product or process or even creating a new one. Unlike well organised research and development, network-innovations can happen without planning, but require the co-operation of several organisations.

It is worth remarking that the three types of innovation-design do not claim exclusivity for their period of time. In fact, it has been suggested that networks in innovation are not a new phenomenon, but have only been neglected in the past due to insufficient "intellectual lenses" (DeBresson and Amesse, 1991). What is crucial in the context of this paper is that an industry whose process of innovation is typically characterised by the appearance of network-innovations favours clustering.

In the case of the inventor–entrepreneur, innovations do have an exogenous character. Because inventor–innovations appear unexpectedly, they lead to sunk costs for all other actors in the field. The invention of the electric bulb, for example, reduced the value of the investments of all suppliers of kerosene lamps. Laboratory-type innovations, on the other hand, require the deepening of one competence and therefore grow best by continuous research and development in the stable environment of a large corporation. Network-innovations, finally, describe the

type of co-ordination / predominant type of innovation	market based co-ordination (single company)	club-like co-ordination (cluster)	hierachic co-ordination (integrated company)
inventor-innovations	+	o	−
laboratory-innovations	−	o	+
network-innovations	−	+	o

strategic fit of the form of co-ordination dependending on the type of innovation: + extensive, o medium, − reduced

Fig. 1. The predominance of network-innovations in a particular industry favours its clustering.

kind of innovations that evolve because of the combination of different competencies and tend to mature in a multi-organisational environment. In the case of network-innovations, the constitution of the whole value-creating system, the sum of all actors involved in the process of innovation, is of critical importance. In such a situation, the advantages of localised agglomerations become most relevant. The likelihood for complementors to exist or to form and to know from each other is higher. The need for efficient co-ordination becomes vital. If the development of an industry is driven by network-innovations we can, therefore, expect such an industry to favour clustering (Fig. 1).

2.2.4. SC_4: *volatility of market: rewarding flexible adaptation*

"In short, control over time allows an unusual form of control over space." (Schoenberger, 1994, p. 55). Put in other words, Schoenberger's conclusion means that the arbitrariness of spatial location is reduced as the control over timing of demand fades away. Advantages of co-ordination due to the proximity of the actors in a value-chain transform into competitive advantages if the speed of reaction is important, such as in volatile markets. Market-determined time-sensitivity, hence a reduction in producers' control of demand, therefore, fosters clustering (Scott, 1988; Sternberg, 1995; Wehrli and Jüttner, 1996).

In addition to that "... by the law of large numbers, the flow of economic activity through the system of subcontractors and input suppliers will always be more regular on average than the flow through any one vertically integrated channel." (Scott, 1988, p. 26). A multi-organisational system can adopt faster and at lower switching costs than an integrated company to market changes, which sums up to the assumption that dynamics and volatility foster clustering.

When can such time-sensitivity be expected? The imperative of speed does not apply to high-tech industries only or those whose process of production is characterised by just-in-time deliveries. In particular, in such value-creating systems whose final product is subject to fashion or suffers from cyclicity of demand, i.e. in systems which are active on volatile markets, time-based competition is likely to be present. Similarly, markets characterised by an outspoken heterogeneity of demand that requires individualisation of the products, honour the flexibility of the producers (Brusco, 1982; Lazerson, 1995; Malmberg and Maskell, 1997). Standardised products, on the other hand, allow for better control over time and therefore for control over space, with the result of a reduced likelihood for clustering.

The sufficient conditions for clustering add to each other: a long value-chain does not necessarily lead to the presence of multiple competencies, but increases its likelihood. Chains with multiple competencies do not necessarily lead to a higher speed of innovation, but are probably characterised by network-innovations. Finally, innovations can lead to market volatility, which, independent of the production process, can also be demand-side induced. Therefore, the conditions for clustering could be summarised using a picture of stairs (Fig. 2).

The necessary conditions for clustering are the foundations on which the stairs rely. Each of the sufficient conditions, then, is represented by a step. In this way it becomes possible to illustrate that more or less steps can be climbed. Because of the additivity of the sufficient conditions, it is even possible to "jump" over individual steps. The necessary conditions, the foundations of the stairs, however, always have to be present. For example, clustering would be expected even if an industry is not very innovative, as long as it is operating on a volatile market and relies on a long value-chain. Now, if the product or the service cannot be transported, i.e. one of the necessary conditions is missing, agglomerations hardly have a chance to develop (Box 1).

3. Summary and conclusion: sharpening the accuracy of cluster-based tools for management and policy

Not all industries are presently in the process of clustering at single regions or nations, but in the case the phenomenon applies to an industry, its players should be aware. In such industries, being outside the cluster leads to an acute "periphery discount" on returns. Here, a proposal on how to assess the potential of an individual industry to cluster has been developed.

In the following situation, a high probability of clustering would be expected: the product or service can be divided into several steps of production (divisibility of process), is universally applicable and has low costs of transportation (transportability of product). Its components are produced in different processes (long value-chain with distinct competencies). Demand is constantly being satisfied with new variations of

Box 1. Oilfield supply: an example of an industry with a high propensity to cluster.

The clustering-stairs (Fig. 2) offer a systematic way to analyse an industry's propensity to cluster in one region or nation. An illustrative example that may clarify the issue at hand refers to the upstream part of the oil and gas business, i.e. the exploration and production of oil and gas. While the natural reserves are spread over the globe, companies offering equipment and services to explore them are not. This is unlikely to be a coincidence. While in the early days one company alone would provide the whole service and just source some raw materials, presently the process of well construction has split up. No single company is contributing to it with more than one-third of the total costs (NC_1). Due to the nature of the business (remote location of reserves) all equipment is mobile (NC_2). The necessary conditions of clustering, divisibility of process and transportability of product, are clearly met.

During the construction of a well 20 different companies or more are employed, almost 10 of them usually contributing with more than 3% to the total costs (SC_1). Each has distinct competencies (SC_2). For example, the drilling fluid that is used to remove the cuttings and is being circulated by a special "mud company" contains a highly sophisticated mixture of chemicals. At the same time the drilling engineer, who works for the "drilling contractor" that runs a machinery that ultimately moves the drill bit, must emphasise on mechanical knowledge. Interestingly, it has been noticed that in certain geological situations the drill bit works better if a particular chemical additive is included into the drilling fluid (Gammage and Schiele, 2000). This pooling of competencies of different companies with the result of an improved overall service is an example of a network-innovation (SC_3). Finally, this highly innovative industry–with an average increase in productivity at around 10% per year—is also active in a volatile market (SC_4). In the US, for instance, certain years have been observed, in which a 1% change in demand for crude oil resulted in a 2% change in the demand for well construction services.

Summarising, it is fair to state that the conditions for clustering are met. Indeed, this industry is being dominated by US-based companies, who control more than 75% of the world market in the most important segments of the value-chain. A geographic centre could establish in Houston, TX. While the other "oil towns" lost employment in the last major cyclical downturn, Houston could even increase the payroll, in particular in knowledge-intensive jobs (for details see Schiele, 2001).

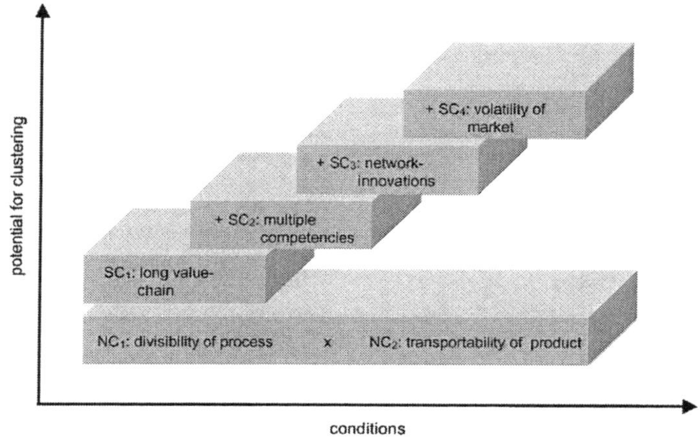

Fig. 2. The "clustering-stairs" indicate the potential for regional–sectoral agglomerations to form.

the product, which is manufactured by continuously improved methods (network-innovations). Finally, the market requires fast reactions to unpredictable changes in customers' demand (market volatility).

If all or most of these features apply to the industry a particular company belongs, it should closely monitor how far clustering has already taken place and how it can profit from this evolution. [2]

Additionally, the "clustering-stairs" (Fig. 2) allows industrial policy and regional planning to pre-select with greater accuracy the target-industry that should be supported by measures for regional networking or other forms of cluster-development.

Finally, deducing six conditions for clustering allows the testing of their precision by cross-checking these conditions with case studies and by broad empirical work. This could be a rewarding way forward for future research.

References

Amendoa, G., Guerrieri, P., Padoan, P.C., 1992. International patterns of technological accumulation and trade. Journal of International and Comparative Economics 1, 173–197

Archibugi, D., Pianta, M., 1992. The Technological Specialisation of Advanced Countries. A Report to the EEC on International Science and Technology Activities. Dordrecht, Boston, London.

Audretsch, D.B., Feldman, M., 1995. Innovative Clusters and the Industry Life Cycle. Centre for Economic Policy Research Discussion Paper Series No. 1161, London.

Aydalot, P., Keeble, D., 1988 High-technology and innovative environments in Europe: an overview. In: Aydalot, P., Keeble, D. (Eds.), High Technology Industry and Innovative Environments: The European Experience. London, New York, pp. 1–21.

Baptista, R., Swann, P., 1998. Do firms in clusters innovate more? Research Policy 27, 525–540.

Becattini, G., 1991. Italian industrial districts: problems and perspectives. International Studies of Management & Organisation 21, 83–90.

Brusco, S., 1982. The Emilian model: productive decentralisation and social integration. Cambridge Journal of Economics 6, 167–184.

Capello, R., 1999. Spatial transfer of knowledge in high technology milieux: learning versus collective learning processes. Regional Studies 33, 353–365.

Cooke, P., Uranga, M.G , Etxebarria, G., 1997. Regional innovation systems: institutional and organisational dimensions. Research Policy 26, 475–491.

DeBresson, C., 1996. Why innovative activities cluster. In: DeBresson, C. (Ed.), Economic Interdependence and Innovative Activity. An Input–Output Analysis. Cheltenham, Brookfield, pp. 149–164.

DeBresson, C., Amesse, F , 1991 Networks of innovators. a review and introduction to the issue Research Policy 20, 363–379.

DiMaggio, P.J., Powell, W.W., 1983. The iron cage revised: institutional isomorphism and collective rationality in

[2] A method for doing so is presented in Schiele (1999, 2001).

organısatıonal fields. Amerıcan Socıologıcal Revıew 48, 147–160.

Dunnıng, J.H., 1992. The competıtıve advantage of countrıes and the actıvıtıes of transnatıonal corporatıons. Transnatıonal Corporatıons 1, 135–168.

Enrıght, M.J., Tenti, P., 1990 How the dıamond works. The Italıan ceramıc tıle ındustry. Harvard Busıness Revıew 68, 90–91.

Fabıanı, S., Pellegrını, G., 1998. Un'analısı quantıtatıve della ımprese neı dıstrettı ındustrıalı ıtalıanı: reddıtvıtà, produttıvıtà e costo del lavoro, L'Industrıa. Rıvısta dı economıa e polıtıca ındustrıale 19, 811–831.

Fornahl, D., 2001. Regıonal economıc development: the ımpact of collective belıefs. In: Proceedıngs of the workshop on the Influence of Co-operatıons, Networks and Instıtutıons on Regıonal Innovatıon Systems, 8–10 February, Max-Planck-Instıtute for Research ınto Economıc Systems. Jena.

Freeman, C., 1995. The 'natıonal system of ınnovatıon' ın hıstorical perspectıve. Cambrıdge Journal of Economıcs 19, 5–24.

Freeman, C , Soete, L , 1997. The Economıcs of Industrıal Innovatıon. London, Washıngton.

Furtado, A., 1997. The French system of ınnovatıon ın the oıl ındustry: some lessons about the role of publıc polıcıes and sectoral patterns of technological change ın ınnovatıon networkıng. Research Polıcy 25, 1243–1259.

Gammage, J., Schıele, H., 2000. Netzwerkınnovationen verandern das Busıness Desıgn: Eın Fall aus der Olındustrıe. In: Steınle, C., Eggers, B., Thıem, H., Vogel, B. (Eds.), Vıtalısierung. Das Management der neuen Lebendıgkeıt. Frankfurt, a.m., pp. 122–141 (Englısh versıon avaılable wıth the authors).

Genet, C , 1997. Quelles condıtıons pour la formatıon des bıotechnopoles: Une analyse dynamıque. Revue d'Economıe Régıonale et Urbaıne, 1997, pp. 405–424.

Gertler, M.S., 1993. Implementıng advanced manufacturıng technologıes ın mature ındustrıal regıons: towards a socıal model of technology production. Regıonal Studıes 27, 665–680.

Gertler, M.S., 1995. Beıng there: proxımıty, organızatıon, and culture ın the development and adoptıon of advanced manufacturing technologıes Economıc Geography 71, 1–26.

Grant, E.B., Gregory, M.J , 1997. Tacıt Knowledge, the lıfe cycle and ınternatıonal manufacturıng transfer. Technology Analysıs & Strategıc Management 9, 149–161.

Hobday, M., 1994. The lımıts of Sılıcon valley. a crıtıque of network theory. Technology Analysıs & Strategıc Management 6, 231–244.

Hopkıns, T.K., Wallersteın, I., 1994. Conclusıon about commodıty chaıns. In: Gereffi, G , Konzenıewıcz, M. (Eds.), Commodıty Chaıns and Global Capıtalısm. Westport, London, pp. 48–50.

Isaksen, A., 1997. Regional clusters and competıtıveness. the Norwegıan case. European Plannıng Studıes 5, 65–76.

Jarıllo, J.C., 1995. Strategic Networks. Creatıng the Borderless Organızatıon. Oxford.

Judex, K., 1996. Der Eınfluß des technologıschen Wandels auf den Produktıonsstandort. Frankfurt., a.m , etc.

Lazerson, M., 1995. A new Phoenıx? Modern puttıng-out ın modena knıtwear ındustry. Admınıstratıve Scıence Quarterly 40, 34–59.

Lıst, F., 1910. Das natıonale System der Polıtıschen Oekonomıe, Neudruck nach der Ausgabe letzter Hand, eıngeleıtet von Prof. Dr. H. Waentıng, Zweıte Auflage. Jena.

Lundvall, B.-A., 1992. Introductıon. In: Lundvall, B. (Ed.), Natıonal systems of ınnovatıon: towards a theory of ınnovatıon and ınteractıve learnıng. London, pp. 1–19.

Malmberg, A., Maskell, P., 1997. Towards an explanatıon of regıonal specıalısatıon and ındustry agglomeratıon. European Plannıng Studies 5, 25–41.

Malmberg, A., Solvell, O., Zander, I., 1996. Spatıonal clusterıng, local accumulatıon of knowledge and firm competıtıveness. Geografiska Annaler 78B, 85–97.

Moore, J.F., 1997. The Death of Competıtıon: Leadershıp and Strategy ın the Age of Busıness Ecosystems. Chıchester.

Morvan, Y., 1991. Fondements d'Économıe Industrıelle. Parıs.

Patel, P., Pavıtt, K , 1991. Large firms ın the productıon of the world's technology: an ımportant case of non-globalısatıon. Journal of Internatıonal Busıness Studıes 22, 1–40.

Patel, P., Vega, M., 1999. Patterns of ınternatıonalısatıon of corporate technology: locatıon vs. home country advantages. Research Polıcy 28, 145–155.

Penttınen, R., 1994. Summary of the Crıtıque on Porter's Dıamond Model. Porter's Dıamond Model Modıefied to Suıt the Fınnısh Paper and Board Machıne Industry, Helsınkı: ETLA Elınkeınoelaman Tutkımuslaıtos. The Research Instıtute of the Fınnısh Economy Discussıon Papers No. 462 (ISSN 0781–6847).

Pılon, S., DeBresson, C., 2001. Local culture and regıonal ınnovatıve networks: New hypotheses and some proposıtıons. In: Proceedıngs of the Workshop on the Influence of Co-operatıons, Networks and Instıtutıons on Regıonal Innovatıon Systems, 8–10 February, Max-Planck-Instıtute for Research into Economıc Systems. Jena.

Pıore, M.J., Sabel, C F., 1984. The Second Industrıal Dıvıde. Possıbılıtıes for Prosperıty. New York.

Porter, M.E., 1990. The Competıtıve Advantage of Natıons. Houndmılls, London.

Porter, M E., 1998. Clusters and the new economıcs of competıtıon. Harvard Busıness Revıew 76, 77–90.

Rallet, A., 1993 Choıx de proxımıté et processus d'ınnovatıon technologıque. Revue d'Economıe Régıonale et Urbaıne, 1993, pp. 365–385.

Rıchardson, G.B., 1972. The organısation of ındustry. The Economıc Journal 82, 883–896.

Roelandt, T J.A., den Hertog, P., 1999 Cluster analysıs and cluster-based polıcy makıng in OECD countrıes. An ıntroductıon to the theme In: OECD (Ed.), Boostıng Innovatıon. The Cluster Approach. Parıs, pp. 9–26.

Rosenkopf, L., Tushman, M.L., 1994. The Coeveolutıon of Technology and Organızatıon. In. Baum, J.A.C., Sıngh, J.V. (Eds.), Evolutıonary Dynamıcs of Organızatıons. New York, Oxford, pp. 403–424.

Rugman, A.M., Hodgetts, R.M., 1995. Internatıonal Busıness: A Strategıc Management Approach. New York, etc.

Saxenıan, A., 1991. The orıgıns and dynamıcs of productıon networks ın Sılıcon Valley. Research Polıcy 20, 423–437.

Schiele, H., 1999. Partnering does not work everywhere. The advantages of belonging to and a method on how to detect an innovative cluster. In: Schruijer, S. (Ed.), Multi-Organizational Partnerships and Cooperative Strategy. Tilburg, pp. 179–184.

Schiele, H., 2001. Strategisches Management in Wertschopfungssystemen. Clusterbezogene Umweltanalyse, Gestaltungsempfehlungen, Anwendungsfall (with English summary). Wiesbaden.

Schoenberger, E., 1994. Competition, Time, and Space in Industrial Change. In: Gereffi, G., Korzeniewicz, M. (Eds.), Commodity Chains and Global Capitalism. Westport, London, pp. 51–66.

Scott, A.J., 1988. New Industrial Spaces. Flexible Production Organization and Regional Development in North America and Western Europe. London.

Steinle, C., Eickhoff, M., Schiele, H., 1998. Zur Perspektiveneweiterung des Strategischen Management: Der innovative Cluster als Wertschopfungssystem und die Entwicklung von Kernkompetenzen. Zeitschrift fur Planung 9, 367–390.

Sternberg, R., 1995. Technologiepolitik und High-Tech Regionen—ein internationaler Vergleich. Munster, Hamburg.

Stoffaes, C., 1980. Introduction: filières et stratégies industrielles. Annales des Mines 186, 9–20.

Storper, M., 1992. The limits to globalization: technology districts and international trade. Economic Geography 68, 60–93.

Swann, P., 1993. Clusters in high-technology industries. The Business Economist 25, 27–36.

Weder, R., Grubel, H.G., 1993. The new growth theory and coasean economics: institutions to capture externalities. Weltwirtschaftliches Archiv (Review of World Economics) 129, 488–513.

Wehrli, H.P, Juttner, U., 1996. Beziehungsmarketing in Wertsystemen. Marketing ZFP, 1986, pp. 33–43.

Yetton, P., Craig, J., Davis, J., Hilmer, F., 1992. Are diamonds a country's best friend? A critique of porter's theory of national competition as applied to Canada, New Zealand and Australia. Australian Journal of Management 17, 89–119.

Part IX
Competition and Public Policy

[25]

Economic Welfare and the Allocation of Resources for Invention

KENNETH J. ARROW

THE RAND CORPORATION

INVENTION is here interpreted broadly as the production of knowledge. From the viewpoint of welfare economics, the determination of optimal resource allocation for invention will depend on the technological characteristics of the invention process and the nature of the market for knowledge.

The classic question of welfare economics will be asked here: to what extent does perfect competition lead to an optimal allocation of resources? We know from years of patient refinement that competition insures the achievement of a Pareto optimum under certain hypotheses. The model usually assumes among other things, that (1) the utility functions of consumers and the transformation functions of producers are well-defined functions of the commodities in the economic system, and (2) the transformation functions do not display indivisibilities (more strictly, the transformation sets are convex). The second condition needs no comment. The first seems to be innocuous but in fact conceals two basic assumptions of the usual models. It prohibits uncertainty in the production relations and in the utility functions, and it requires that all the commodities relevant either to production or to the welfare of individuals be traded on the market. This will not be the case when a commodity for one reason or another cannot be made into private property.

We have then three of the classical reasons for the possible failure of perfect competition to achieve optimality in resource allocation: indivisibilities, inappropriability, and uncertainty. The first problem has been much studied in the literature under the heading of marginal-cost pricing and the second under that of divergence between social and private benefit (or cost), but the theory of optimal allocation of resources under uncertainty has had much less attention. I will summarize what formal theory exists and then point to the critical notion of information, which arises only in the context of uncertainty. The

NOTE: I have benefited greatly from the comments of my colleague, William Capron. I am also indebted to Richard R. Nelson, Edward Phelps, and Sidney Winter of The RAND Corporation for their helpful discussion.

economic characteristics of information as a commodity and, in particular, of invention as a process for the production of information are next examined. It is shown that all three of the reasons given above for a failure of the competitive system to achieve an optimal resource allocation hold in the case of invention. On theoretical grounds a number of considerations are adduced as to the likely biases in the misallocation and the implications for economic organization.[1]

Resource Allocation under Uncertainty

The role of the competitive system in allocating uncertainty seems to have received little systematic attention.[2] I will first sketch an ideal economy in which the allocation problem can be solved by competition and then indicate some of the devices in the real world which approximate this solution.

Suppose for simplicity that uncertainty occurs only in production relations. Producers have to make a decision on inputs at the present moment, but the outputs are not completely predictable from the inputs. We may formally describe the outputs as determined by the inputs and a "state of nature" which is unknown to the producers. Let us define a "commodity-option" as a commodity in the ordinary sense labeled with a state of nature. This definition is analogous to the differentiation of a given physical commodity according to date in capital theory or according to place in location theory. The production of a given commodity under uncertainty can then be described as the production of a vector of commodity-options.

This description can be most easily exemplified by reference to agricultural production. The state of nature may be identified with the weather. Then, to any given set of inputs there corresponds a number of bushels of wheat if the rainfall is good and a different number if rainfall is bad. We can introduce intermediate conditions of rainfall

[1] For other analyses with similar points of view, see R. R. Nelson, "The Simple Economics of Basic Scientific Research," *Journal of Political Economy*, 1959, pp. 297–306; and C. J. Hitch, "The Character of Research and Development in a Competitive Economy," The RAND Corporation, p. 1297, May 1958.

[2] The first studies I am aware of are the papers of M. Allais and myself, both presented in 1952 to the Colloque International sur le Risque in Paris; see M. Allais, "Généralisation des théories de l'équilibre économique général et du rendement social au cas du risque," and K. J. Arrow, "Rôle des valeurs bousières pour la répartition la meilleure des risques," both in *Economètrie*, Colloques Internationaux du Centre National de la Recherche Scientifique, Vol. XL, Paris, Centre National de la Recherche Scientifique, 1953. Allais' paper has also appeared in *Econometrica*, 1953, pp. 269–290. The theory has received a very elegant generalization by G. Debreu in *Theory of Values*, New York, Wiley, 1959, Chap. VII.

in any number as alternative states of nature; we can increase the number of relevant variables which enter into the description of the state of nature, for example by adding temperature. By extension of this procedure, we can give a formal description of any kind of uncertainty in production.

Suppose—and this is the critical idealization of the economy—we have a market for all commodity-options. What is traded on each market are contracts in which the buyers pay an agreed sum and the sellers agree to deliver prescribed quantities of a given commodity *if* a certain state of nature prevails and nothing if that state of nature does not occur. For any given set of inputs, the firm knows its output under each state of nature and sells a corresponding quantity of commodity-options; its revenue is then completely determined. It may choose its inputs so as to maximize profits.

The income of consumers is derived from their sale of supplies, including labor, to firms and their receipt of profits, which are assumed completely distributed. They purchase commodity-options so as to maximize their expected utility given the budget restraint imposed by their incomes. An equilibrium is reached on all commodity-option markets, and this equilibrium has precisely the same Pareto-optimality properties as competitive equilibrium under certainty.

In particular, the markets for commodity-options in this ideal model serve the function of achieving an optimal allocation of risk bearing among the members of the economy. This allocation takes account of differences in both resources and tastes for risk bearing. Among other implications, risk bearing and production are separated economic functions. The use of inputs, including human talents, in their most productive mode is not inhibited by unwillingness or inability to bear risks by either firms or productive agents.

But the real economic system does not possess markets for commodity-options. To see what substitutes exist, let us first consider a model economy at the other extreme, in that no provisions for reallocating risk bearing exist. Each firm makes its input decisions; then outputs are produced as determined by the inputs and the state of nature. Prices are then set to clear the market. The prices that finally prevail will be a function of the state of nature.

The firm and its owners cannot relieve themselves of risk bearing in this model. Hence any unwillingness or inability to bear risks will give rise to a nonoptimal allocation of resources, in that there will be discrimination against risky enterprises as compared with the optimum. A preference for risk might give rise to misallocation in the opposite direction, but the limitations of financial resources are likely to make underinvestment in risky enterprises more likely than the opposite. The inability of individuals to buy protection against uncertainty similarly gives rise to a loss of welfare.

In fact, a number of institutional arrangements have arisen to mitigate the problem of assumption of risk. Suppose that each firm and individual in the economy could forecast perfectly what prices would be under each state of nature. Suppose further there were a lottery on the states of nature, so that before the state of nature is known any individual or firm may place bets. Then it can be seen that the effect from the viewpoint of any given individual or firm is the same as if there were markets for commodity-options of all types, since any commodity-option can be achieved by a combination of a bet on the appropriate state of nature and an intention to purchase or sell the commodity in question if the state of nature occurs.

References to lotteries and bets may smack of frivolity, but we need only think of insurance to appreciate that the shifting of risks through what are in effect bets on the state of nature is a highly significant phenomenon. If insurance were available against any conceivable event, it follows from the preceding discussion that optimal allocation would be achieved. Of course, insurance as customarily defined covers only a small range of events relevant to the economic world; much more important in shifting risks are securities, particularly common stocks and money. By shifting freely their proprietary interests among different firms, individuals can to a large extent bet on the different states of nature which favor firms differentially. This freedom to insure against many contingencies is enhanced by the alternatives of holding cash and going short.

Unfortunately, it is only too clear that the shifting of risks in the real world is incomplete. The great predominance of internal over external equity financing in industry is one illustration of the fact that securities do not completely fulfill their allocative role with respect to risks. There are a number of reasons why this should be so, but I will confine myself to one, of special significance with regard to invention. In insurance practice, reference is made to the moral factor as a limit to the possibilities of insurance. For example, a fire insurance policy cannot exceed in amount the value of the goods insured. From the purely actuarial standpoint, there is no reason for this limitation;

the reason for the limit is that the insurance policy changes the incentives of the insured, in this case, creating an incentive for arson or at the very least for carelessness. The general principle is the difficulty of distinguishing between a state of nature and a decision by the insured. As a result, any insurance policy and in general any device for shifting risks can have the effect of dulling incentives. A fire insurance policy, even when limited in amount to the value of the goods covered, weakens the motivation for fire prevention. Thus, steps which improve the efficiency of the economy with respect to risk bearing may decrease its technical efficiency.

One device for mitigating the adverse incentive effects of insurance is coinsurance; the insurance extends only to part of the amount at risk for the insured. This device is used, for example, in coverage of medical risks. It clearly represents a compromise between incentive effects and allocation of risk bearing, sacrificing something in both directions.

Two exemplifications of the moral factor are of special relevance in regard to highly risky business activities, including invention. Success in such activities depends on an inextricable tangle of objective uncertainties and decisions of the entrepreneurs and is certainly uninsurable. On the other hand, such activities should be undertaken if the expected return exceeds the market rate of return, no matter what the variance is.[3] The existence of common stocks would seem to solve the allocation problem; any individual stockholder can reduce his risk by buying only a small part of the stock and diversifying his portfolio to achieve his own preferred risk level. But then again the actual managers no longer receive the full reward of their decisions; the shifting of risks is again accompanied by a weakening of incentives to efficiency. Substitute motivations whether pecuniary, such as executive compensation and profit sharing, or nonpecuniary, such as prestige, may be found, but the dilemma of the moral factor can never be completely resolved.

A second example is the cost-plus contract in one of its various forms. When production costs on military items are highly uncertain, the military establishment will pay, not a fixed unit price, but the cost of production plus an amount which today is usually a fixed fee. Such a contract could be regarded as a combination of a fixed-price contract with an insurance against costs. The insurance premium could be

[3] The validity of this statement depends on some unstated assumptions, but the point to be made is unaffected by minor qualifications.

regarded as the difference between the fixed price the government would be willing to pay and the fixed fee.

Cost-plus contracts are necessitated by the inability or unwillingness of firms to bear the risks. The government has superior risk bearing ability and so the burden is shifted to it. It is then enabled to buy from firms on the basis of their productive efficiency rather than their risk bearing ability, which may be only imperfectly correlated. But cost-plus contracts notoriously have their adverse allocative effects.[4]

This somewhat lengthy digression on the theory of risk bearing seemed necessitated by the paucity of literature on the subject. The main conclusions to be drawn are the following: (1) the economic system has devices for shifting risks, but they are limited and imperfect; hence, one would expect an underinvestment in risky activities; (2) it is undoubtedly worthwhile to enlarge the variety of such devices, but the moral factor creates a limit to their potential.

Information as a Commodity

Uncertainty usually creates a still more subtle problem in resource allocation; information becomes a commodity. Suppose that in one part of the economic system an observation has been made whose outcome, if known, would affect anyone's estimates of the probabilities of the different states of nature. Such observations arise out of research but they also arise in the daily course of economic life as a by-product of other economic activities. An entrepreneur will automatically acquire a knowledge of demand and production conditions in his field which is available to others only with special effort. Information will frequently have an economic value, in the sense that anyone possessing the information can make greater profits than would otherwise be the case.

It might be expected that information will be traded in, and of course to a considerable extent this is the case, as is illustrated by the numerous economic institutions for transmission of information, such as newspapers. But in many instances, the problem of an optimal allocation is sharply raised. The cost of transmitting a given body of information is frequently very low. If it were zero, then optimal allocation would obviously call for unlimited distribution of the informa-

[4] These remarks are not intended as a complete evaluation of cost-plus contracts. In particular, there are, to a certain extent, other incentives which mitigate the adverse effects on efficiency.

tion without cost. In fact, a given piece of information is by definition an indivisible commodity, and the classical problems of allocation in the presence of indivisibilities appear here. The owner of the information should not extract the economic value which is there, if optimal allocation is to be achieved; but he is a monopolist, to some small extent and will seek to take advantage of this fact.

In the absence of special legal protection, the owner cannot, however, simply sell information on the open market. Any one purchaser can destroy the monopoly, since he can reproduce the information at little or no cost. Thus the only effective monopoly would be the use of the information by the original possessor. This, however, will not only be socially inefficient, but also may not be of much use to the owner of the information either, since he may not be able to exploit it as effectively as others.

With suitable legal measures, information may become an appropriable commodity. Then the monopoly power can indeed be exerted. However, no amount of legal protection can make a thoroughly appropriable commodity of something so intangible as information. The very use of the information in any productive way is bound to reveal it, at least in part. Mobility of personnel among firms provides a way of spreading information. Legally imposed property rights can provide only a partial barrier, since there are obviously enormous difficulties in defining in any sharp way an item of information and differentiating it from other similar sounding items.

The demand for information also has uncomfortable properties. In the first place, the use of information is certainly subject to indivisibilities; the use of information about production possibilities, for example, need not depend on the rate of production. In the second place, there is a fundamental paradox in the determination of demand for information; its value for the purchaser is not known until he has the information, but then he has in effect acquired it without cost. Of course, if the seller can retain property rights in the use of the information, this would be no problem, but given incomplete appropriability, the potential buyer will base his decision to purchase information on less than optimal criteria. He may act, for example, on the average value of information in that class as revealed by past experience. If any particular item of information has differing values for different economic agents, this procedure will lead both to a nonoptimal purchase of information at any given price and also to a nonoptimal allocation of the information purchased.

It should be made clear that from the standpoint of efficiently distributing an existing stock of information, the difficulties of appropriation are an advantage, provided there are no costs of transmitting information, since then optimal allocation calls for free distribution. The chief point made here is the difficulty of creating a market for information if one should be desired for any reason.

It follows from the preceding discussion that costs of transmitting information create allocative difficulties which would be absent otherwise. Information should be transmitted at marginal cost, but then the demand difficulties raised above will exist. From the viewpoint of optimal allocation, the purchasing industry will be faced with the problems created by indivisibilities; and we still leave unsolved the problem of the purchaser's inability to judge in advance the value of the information he buys. There is a strong case for centralized decision making under these circumstances.

Invention as the Production of Information

The central economic fact about the processes of invention and research is that they are devoted to the production of information. By the very definition of information, invention must be a risky process, in that the output (information obtained) can never be predicted perfectly from the inputs. We can now apply the discussion of the preceding two sections.

Since it is a risky process, there is bound to be some discrimination against investment in inventive and research activities. In this field, especially, the moral factor will weigh heavily against any kind of insurance or equivalent form of risk bearing. Insurance against failure to develop a desired new product or process would surely very greatly weaken the incentives to succeed. The only way, within the private enterprise system, to minimize this problem is the conduct of research by large corporations with many projects going on, each small in scale compared with the net revenue of the corporation. Then the corporation acts as its own insurance company. But clearly this is only an imperfect solution.

The deeper problems of misallocation arise from the nature of the product. As we have seen, information is a commodity with peculiar attributes, particularly embarrassing for the achievement of optimal allocation. In the first place, any information obtained, say a new method of production, should, from the welfare point of view, be available free of charge (apart from the cost of transmitting information). This insures optimal utilization of the information but of course provides no incentive for investment in research. In an ideal socialist economy, the reward for invention would be completely separated from any charge to the users of the information.[5] In a free enterprise economy, inventive activity is supported by using the invention to create property rights; precisely to the extent that it is successful, there is an underutilization of the information. The property rights may be in the information itself, through patents and similar legal devices, or in the intangible assets of the firm if the information is retained by the firm and used only to increase its profits.

The first problem, then, is that in a free enterprise economy the profitability of invention requires a nonoptimal allocation of resources. But it may still be asked whether or not the allocation of resources to inventive activity is optimal. The discussion of the preceding section makes it clear that we would not expect this to be so; that, in fact, a downward bias in the amount of resources devoted to inventive activity is very likely. Whatever the price, the demand for information is less than optimal for two reasons: (1) since the price is positive and not at its optimal value of zero, the demand is bound to be below the optimal; (2) as seen before, at any given price, the very nature of information will lead to a lower demand than would be optimal.

As already remarked, the inventor will in any case have considerable difficulty in appropriating the information produced. Patent laws would have to be unimaginably complex and subtle to permit such appropriation on a large scale. Suppose, as the result of elaborate tests, some metal is discovered to have a desirable property, say resistance to high heat. Then of course every use of the metal for which this property is relevant would also use this information, and the user would be made to pay for it. But, even more, if another inventor is stimulated to examine chemically related metals for heat resistance, he is using the information already discovered and should pay for it in some measure; and any beneficiary of his discoveries should also pay. One would have to have elaborate distinctions of partial property rights of all degrees to make the system at all tolerable. In the interests of the possibility of enforcement, actual patent laws sharply restrict the range of appropriable information and thereby reduce the incentives to engage in inventive and research activities.

[5] This separation exists in the Soviet Union, according to N. M. Kaplan and R. H. Moorsteen of The RAND Corporation (verbal communication).

WELFARE ECONOMICS AND INVENTIVE ACTIVITY

These last considerations bring into focus the interdependence of inventive activities, which reinforces the difficulties in achieving an optimal allocation of the results. Information is not only the product of inventive activity, it is also an input—in some sense, the major input apart from the talent of the inventor. The school of thought that emphasizes the determination of invention by the social climate as demonstrated by the simultaneity of inventions in effect emphasizes strongly the productive role of previous information in the creation of new information. While these interrelations do not create any new difficulties in principle, they intensify the previously established ones. To appropriate information for use as a basis for further research is much more difficult than to appropriate it for use in producing commodities; and the value of information for use in developing further information is much more conjectural than the value of its use in production and therefore much more likely to be underestimated. Consequently, if a price is charged for the information, the demand is even more likely to be suboptimal.

Thus basic research, the output of which is only used as an informational input into other inventive activities, is especially unlikely to be rewarded. In fact, it is likely to be of commercial value to the firm undertaking it only if other firms are prevented from using the information obtained. But such restriction on the transmittal of information will reduce the efficiency of inventive activity in general and will therefore reduce its quantity also. We may put the matter in terms of sequential decision making. The a priori probability distribution of the true state of nature is relatively flat to begin with. On the other hand, the successive a posteriori distributions after more and more studies have been conducted are more and more sharply peaked or concentrated in a more limited range, and we therefore have better and better information for deciding what the next step in research shall be. This implies that, at the beginning, the preferences among alternative possible lines of investigation are much less sharply defined than they are apt to be later on and suggests, at least, the importance of having a wide variety of studies to begin with, the less promising being gradually eliminated as information is accumulated.[6] At each stage the decisions about the next step should be based on all available information. This would require an unrestricted flow of informa-

[6] The importance of parallel research developments in the case of uncertainty has been especially stressed by Burton H. Klein; see his, "A Radical Proposal for R. and D.," *Fortune*, May 1958, p. 112 ff.; and Klein and W. H. Meckling, "Application of Operations Research to Development Decisions," *Operations Research*, 1958, pp. 352–363.

tion among different projects which is incompatible with the complete decentralization of an ideal free enterprise system. When the production of information is important, the classic economic case in which the price system replaces the detailed spread of information is no longer completely applicable.

To sum up, we expect a free enterprise economy to underinvest in invention and research (as compared with an ideal) because it is risky, because the product can be appropriated only to a limited extent, and because of increasing returns in use. This underinvestment will be greater for more basic research. Further, to the extent that a firm succeeds in engrossing the economic value of its inventive activity, there will be an underutilization of that information as compared with an ideal allocation.

Competition, Monopoly, and the Incentive to Innovate

It may be useful to remark that an incentive to invent can exist even under perfect competition in the product markets though not, of course, in the "market" for the information contained in the invention. This is especially clear in the case of a cost reducing invention. Provided only that suitable royalty payments can be demanded, an inventor can profit without disturbing the competitive nature of the industry. The situation for a new product invention is not very different; by charging a suitable royalty to a competitive industry, the inventor can receive a return equal to the monopoly profits.

I will examine here the incentives to invent for monopolistic and competitive markets, that is, I will compare the potential profits from an invention with the costs. The difficulty of appropriating the information will be ignored; the remaining problem is that of indivisibility in use, an inherent property of information. A competitive situation here will mean one in which the industry produces under competitive conditions, while the inventor can set an arbitrary royalty for the use of his invention. In the monopolistic situation, it will be assumed that only the monopoly itself can invent. Thus a monopoly is understood here to mean barriers to entry; a situation of temporary monopoly, due perhaps to a previous innovation, which does not prevent the entrance of new firms with innovations of their own, is to be regarded as more nearly competitive than monopolistic for the purpose of this analysis. It will be argued that the incentive to invent is less under monopolistic than under competitive conditions but even in the latter case it will be less than is socially desirable.

WELFARE ECONOMICS AND INVENTIVE ACTIVITY

We will assume constant costs both before and after the invention, the unit costs being c before the invention and $c' < c$ afterward. The competitive price before invention will therefore be c. Let the corresponding demand be x_c. If r is the level of unit royalties, the competitive price after the invention will be $c' + r$, but this cannot of course be higher than c, since firms are always free to produce with the old methods.

It is assumed that both the demand and the marginal revenue curves are decreasing. Let $R(x)$ be the marginal revenue curve. Then the monopoly output before invention, x_m, would be defined by the equation,

$$R(x_m) = c \, .$$

Similarly, the monopoly output after invention is defined by,

$$R(x'_m) = c' \, .$$

Let the monopoly prices corresponding to outputs x_m and x'_m, respectively, be p_m and p'_m. Finally, let P and P' be the monopolist's profits before and after invention, respectively.

What is the optimal royalty level for the inventor in the competitive case? Let us suppose that he calculates p'_m, the optimal monopoly price which would obtain in the postinvention situation. If the cost reduction is sufficiently drastic that $p'_m < c$, then his most profitable policy is to set r so that the competitive price is p'_m, i.e. let,

$$r = p'_m - c' \, .$$

In this case, the inventor's royalties are equal to the profits a monopolist would make under the same conditions, i.e. his incentive to invent will be P'.

Suppose, however, it turns out that $p'_m > c$. Since the sales price cannot exceed c, the inventor will set his royalties at,

$$r = c - c' \, .$$

The competitive price will then be c, and the sales will remain at x_c. The inventor's incentive will then be, $x_c(c - c')$.

The monopolist's incentive, on the other hand, is clearly $P' - P$. In the first of the two cases cited, the monopolist's incentive is obviously less than the inventor's incentive under competition, which is P', not $P' - P$. The preinvention monopoly power acts as a strong disincentive to further innovation.

ALLOCATION OF RESOURCES FOR INVENTION

The analysis is slightly more complicated in the second case. The monopolist's incentive, $P' - P$, is the change in revenue less the change in total cost of production, i.e.,

$$P' - P = \int_{x_m}^{x'_m} R(x)\, dx - c'\, x'_m + c\, x_m .$$

Since the marginal revenue $R(x)$ is diminishing, it must always be less than $R(x_m) = c$ as x increases from x_m to x'_m, so that,

$$\int_{x_m}^{x'_m} R(x)\, dx < c\,(x'_m - x_m),$$

and,

$$P' - P < c\,(x'_m - x_m) - c'\,x'_m + c\,x_m = (c - c')\,x'_m .$$

In the case being considered, the postinvention monopoly price, p'_m, is greater than c. Hence, with a declining demand curve, $x'_m < x_c$. The above inequality shows that the monopolist's incentive is always less than the cost reduction on the postinvention monopoly output, which in this case is, in turn, less than the competitive output (both before and after invention). Since the inventor's incentive under competition is the cost reduction on the competitive output, it will again always exceed the monopolist's incentive.

It can be shown that, if we consider differing values of c', the difference between the two incentives increases as c' decreases, reaching its maximum of P (preinvention monopoly profits) for c' sufficiently large for the first case to hold. The ratio of the incentive under competition to that under monopoly, on the other hand, though always greater than 1, decreases steadily with c'. For c' very close to c (i.e., very minor inventions), the ratio of the two incentives is approximately x_c/x_m, i.e., the ratio of monopoly to competitive output.[7]

[7] To sketch the proof of these statements quickly, note that, as c' varies, P is a constant. Hence, from the formula for $P' - P$, we see that,

$$d(P' - P)/dc' = dP'/dc' = R(x'_m)(dx'_m/dc') - c'(dx'_m/dc') - x'_m = -x'_m ,$$

since $R(x'_m) = c'$. Let $F(c')$ be the difference between the incentives to invent under competitive and under monopolistic conditions. In the case where $p'_m < c$, this difference is the constant P. Otherwise,

$$F(c') = x_c\,(c - c') - (P' - P),$$

so that

$$dF/dc' = x'_m - x_c .$$

For the case considered, we must have $x'_m < x_c$, as seen in the text. Hence, $dF/dc' \leq 0$, so

WELFARE ECONOMICS AND INVENTIVE ACTIVITY

The only ground for arguing that monopoly may create superior incentives to invent is that appropriability may be greater under monopoly than under competition. Whatever differences may exist in this direction must, of course, still be offset against the monopolist's disincentive created by his preinvention monopoly profits.

The incentive to invent in competitive circumstances may also be compared with the social benefit. It is necessary to distinguish between the realized social benefit and the potential social benefit, the latter being the benefit which would accrue under ideal conditions, which, in this case, means the sale of the product at postinvention cost, c'. Clearly, the potential social benefit always exceeds the realized social benefit. I will show that the realized social benefit, in turn, always equals or exceeds the competitive incentive to invent and, a fortiori, the monopolist's incentive.

Consider again the two cases discussed above. If the invention is sufficiently cost reducing so that $p'_m < c$, then there is a consumers' benefit, due to the lowering of price, which has not been appropriated by the inventor. If not, then the price is unchanged, so that the consumers' position is unchanged, and all benefits do go to the inventor. Since by assumption all the producers are making zero profits both before and after the invention, we see that the inventor obtains the entire realized social benefit of moderately cost reducing inventions but not of more radical inventions. Tentatively, this suggests a bias against major inventions, in the sense that an invention, part of whose costs could be paid for by lump-sum payments by consumers without making them worse off than before, may not be profitable at the maximum royalty payments that can be extracted by the inventor.

---

that $F(c')$ increases as c' decreases.

Let $G(c')$ be the ratio of the incentive under competition to that under monopoly. If $p'_m < c$, then,

$$G(c') = P'/(P' - P),$$

which clearly decreases as c' decreases. For $p'_m > c$, we have,

$$G(c') = x_c(c - c')/(P' - P).$$

Then,

$$dG/dc' = [-(P' - P)x_c + x_c(c - c')x'_m]/(P' - P)^2.$$

Because of the upper bound for $P' - P$ established in the text, the numerator must be positive; the ratio decreases as c' decreases.

Finally, if we consider c' very close to c, $G(c')$ will be approximately equal to the ratio of the derivatives of the numerator and denominator (L'Hopital's rule), which is, x_c/x'_m, and which approaches x_c/x_m as c' approaches c.

Alternative Forms of Economic Organization in Invention

The previous discussion leads to the conclusion that for optimal allocation to invention it would be necessary for the government or some other agency not governed by profit-and-loss criteria to finance research and invention. In fact, of course, this has always happened to a certain extent. The bulk of basic research has been carried on outside the industrial system, in universities, in the government, and by private individuals. One must recognize here the importance of nonpecuniary incentives, both on the part of the investigators and on the part of the private individuals and governments that have supported research organizations and universities. In the latter, the complementarity between teaching and research is, from the point of view of the economy, something of a lucky accident. Research in some more applied fields, such as agriculture, medicine, and aeronautics, has consistently been regarded as an appropriate subject for government participation, and its role has been of great importance.

If the government and other nonprofit institutions are to compensate for the underallocation of resources to invention by private enterprise, two problems arise: how shall the amount of resources devoted to invention be determined, and how shall efficiency in their use be encouraged? These problems arise whenever the government finds it necessary to engage in economic activities because indivisibilities prevent the private economy from performing adequately (highways, bridges, reclamation projects, for example), but the determination of the relative magnitudes is even more difficult here. Formally, of course, resources should be devoted to invention until the expected marginal social benefit there equals the marginal social benefit in alternative uses, but in view of the presence of uncertainty, such calculations are even more difficult and tenuous than those for public works. Probably all that could be hoped for is the estimation of future rates of return from those in the past, with investment in invention being increased or decreased accordingly as some average rate of return over the past exceeded or fell short of the general rate of return. The difficulties of even ex post calculation of rates of return are formidable though possibly not insuperable.[8]

The problem of efficiency in the use of funds devoted to research

[8] For an encouraging study of this type, see Z. Griliches, "Research Costs and Social Returns: Hybrid Corn and Related Innovations," *Journal of Political Economy*, 1958, pp. 419–431.

is one that has been faced internally by firms in dealing with their own research departments. The rapid growth of military research and development has led to a large-scale development of contractual relations between producers and a buyer of invention and research. The problems encountered in assuring efficiency here are the same as those that would be met if the government were to enter upon the financing of invention and research in civilian fields. The form of economic relation is very different from that in the usual markets. Payment is independent of product; it is governed by costs, though the net reward (the fixed fee) is independent of both. This arrangement seems to fly in the face of the principles for encouraging efficiency, and doubtless it does lead to abuses, but closer examination shows both mitigating factors and some explanation of its inevitability. In the first place, the awarding of new contracts will depend in part on past performance, so that incentives for efficiency are not completely lacking. In the second place, the relation between the two parties to the contract is something closer than a purely market relation. It is more like the sale of professional services, where the seller contracts to supply not so much a specific result as his best judgment. (The demand for such services also arises from uncertainty and the value of information.) In the third place, payment by results would involve great risks for the inventor, risks against which, as we have seen, he could hedge only in part.

There is clear need for further study of alternative methods of compensation. For example, some part of the contractual payment might depend on the degree of success in invention. But a more serious problem is the decision as to which contracts to let. One would need to examine the motivation underlying government decision making in this area. Hitch has argued that there are biases in governmental allocation, particularly against risky invention processes, and an excessive centralization, though the latter could be remedied by better policies.[9]

One can go further. There is really no need for the firm to be the fundamental unit of organization in invention; there is plenty of reason to suppose that individual talents count for a good deal more than the firm as an organization. If provision is made for the rental of necessary equipment, a much wider variety of research contracts with individuals as well as firms and with varying modes of payment, including incentives, could be arranged. Still other forms of organiza-

[9] *Op. cit.*

tion, such as research institutes financed by industries, the government, and private philanthropy, could be made to play an even livelier role than they now do.

COMMENT

C. J. HITCH, The RAND Corporation

There is, I think, one important aspect of government policy with respect to research and development that Markham failed to discuss—the government's contracting and management policies in this area. The problems here are related to Arrow's discussion of incentives, particularly his remarks concerning risk bearing and the adverse allocative effects of cost-plus contracts.

Let me illustrate one set of such problems by contrasting the management policies used by the Army Air Corps before World War II in developing new combat aircraft with the policies now used by the Air Force in developing aircraft and missiles.

Before the war the Air Corps would typically announce a competition for a new aircraft type, the desired characteristics of which would be described in general terms. Any aircraft company that wanted to enter the competition would develop and build, with its own funds, the required number of prototypes (usually one to three). The competition consisted in flying and testing, at Dayton, the prototypes entered by the companies. The prize for the winner was a production contract. The losers lost their stake.

Now, typically, the competition occurs at the design stage. The competing companies enter drawings, with their estimates of performance (and time and cost of development). The winner of the design competition is awarded a development contract which, if things go favorably, is transformed later into a production contract. Occasionally, if the program is considered very important, two competing designs may be approved for development or even production (cf. the Atlas and Titan). But usually the development of more than one model is considered too expensive.

In fact, the greatly increased cost of development is the reason given for this change in policy. When the cost of developing a new vehicle is from $0.5 billion to $2.0 billion, it seems evident that no aircraft company can assume the risk, and perhaps also that even the government can not finance multiple developments.

WELFARE ECONOMICS AND INVENTIVE ACTIVITY

Nevertheless, the new policy appears far from ideal. The aircraft companies risk nothing (even the cost of preparing designs is usually reimbursed). Because of the great uncertainties involved in any major development, it is hard to make a wise selection at the design stage. Companies have a natural tendency to be optimistic in estimating performance, cost, and availability at this stage—sometimes much more so than at other stages as Klein and Marshall and Meckling have shown. And the Air Force has a natural tendency to favor the more optimistic proposals. As a result, the specifications in the development contract are sometimes unrealistic to the point of causing inordinate delay.

Moreover, once the winner has been selected, there is no more competition. The company may put its best team on the development for the sake of patriotism or its own long term reputation, but the powerful incentive of competition is lacking. Partly for this reason the Air Force, locked to a sole source with a cost-plus contract, has to exercise a kind and degree of control during the development process that is inconsistent with management prerogatives as they are understood and practiced in other parts of the free enterprise economy.

I believe the inefficiencies resulting from these policies and procedures are serious, and worthy of much more attention by economists. It does no good simply to inveigh against the iniquities of cost-plus contracting and government risk bearing when we are unable to propose a practical alternative. Perhaps part of the answer lies in some form of risk sharing. What is badly needed here is an economics invention or, more probably, several of them. The government is going to be in the business of supporting research and development on a large scale for a long time, and it is important that it use policies that take advantage of the incentives present in the economy.

[26]

Paradoxically, the enduring competitive advantages in a global economy lie increasingly in local things – knowledge, relationships, and motivation that distant rivals cannot match.

CLUSTERS AND THE NEW ECONOMICS OF COMPETITION

BY MICHAEL E. PORTER

NOW THAT COMPANIES can source capital, goods, information, and technology from around the world, often with the click of a mouse, much of the conventional wisdom about how companies and nations compete needs to be overhauled. In theory, more open global markets and faster transportation and communication should diminish the role of location in competition. After all, anything that can be efficiently sourced from a distance through global markets and corporate networks is available to any company and therefore is essentially nullified as a source of competitive advantage.

Michael E. Porter *is the C. Roland Christensen Professor of Business Administration at the Harvard Business School in Boston, Massachusetts. Further discussion of clusters can be found in two new essays – "Clusters and Competition" and "Competing Across Locations" – in his new collection titled* On Competition *(Harvard Business School Press, 1998).*

But if location matters less, why, then, is it true that the odds of finding a world-class mutual-fund company in Boston are much higher than in most any other place? Why could the same be said of textile-related companies in North Carolina and South Carolina, of high-performance auto companies in southern Germany, or of fashion shoe companies in northern Italy?

Today's economic map of the world is dominated by what I call *clusters:* critical masses – in one place – of unusual competitive success in particular fields. Clusters are a striking feature of virtually every national, regional, state, and even metropolitan economy, especially in more economically advanced nations. Silicon Valley and Hollywood may be the world's best-known clusters. Clusters are not unique, however; they are highly typical – and therein lies a paradox: the enduring competitive advantages in a global economy lie increasingly in local things – knowledge, relationships, motivation – that distant rivals cannot match.

Although location remains fundamental to competition, its role today differs vastly from a generation ago. In an era when competition was driven heavily by input costs, locations with some important endowment – a natural harbor, for example, or a supply of cheap labor – often enjoyed a *comparative advantage* that was both competitively decisive and persistent over time.

Competition in today's economy is far more dynamic. Companies can mitigate many input-cost disadvantages through global sourcing, rendering the old notion of comparative advantage less relevant. Instead, competitive advantage rests on making more productive use of inputs, which requires continual innovation.

Untangling the paradox of location in a global economy reveals a number of key insights about how companies continually create competitive advantage. What happens *inside* companies is important, but clusters reveal that the immediate business environment *outside* companies plays a vital role as well. This role of locations has been long overlooked, despite striking evidence that innovation and competitive success in so many fields are geographically concentrated – whether it's entertainment in Hollywood, finance on Wall Street, or consumer electronics in Japan.

Clusters affect competitiveness within countries as well as across national borders. Therefore, they lead to new agendas for all business executives – not just those who compete globally. More broadly, clusters represent a new way of thinking about location, challenging much of the conventional wisdom about how companies should be configured, how institutions such as universities can contribute to competitive success, and how governments can promote economic development and prosperity.

What Is a Cluster?

Clusters are geographic concentrations of interconnected companies and institutions in a particular field. Clusters encompass an array of linked industries and other entities important to competition. They include, for example, suppliers of specialized inputs such as components, machinery, and services, and providers of specialized infrastructure. Clusters also often extend downstream to channels and customers and laterally to manufacturers of complementary products and to companies in industries related by skills, technologies, or common inputs. Finally, many clusters include governmental and other institutions – such as universities, standards-setting agencies, think tanks, vocational training providers, and trade associations – that provide specialized training, education, information, research, and technical support.

> Untangling the paradox of location in a global economy offers insights into how companies continually create competitive advantage.

The California wine cluster is a good example. It includes 680 commercial wineries as well as several thousand independent wine grape growers. (See the exhibit "Anatomy of the California Wine Cluster.") An extensive complement of industries supporting both wine making and grape growing exists, including suppliers of grape stock, irrigation and harvesting equipment, barrels, and labels; specialized public relations and advertising firms; and numerous wine publications aimed at consumer and trade audiences. A host of local institutions is involved with wine, such as the world-renowned viticulture and enology program at the University of California at Davis, the Wine Institute, and special committees of the California senate and assembly. The cluster also enjoys weaker linkages to other California clusters in agriculture, food and restaurants, and wine-country tourism.

ANATOMY OF THE CALIFORNIA WINE CLUSTER

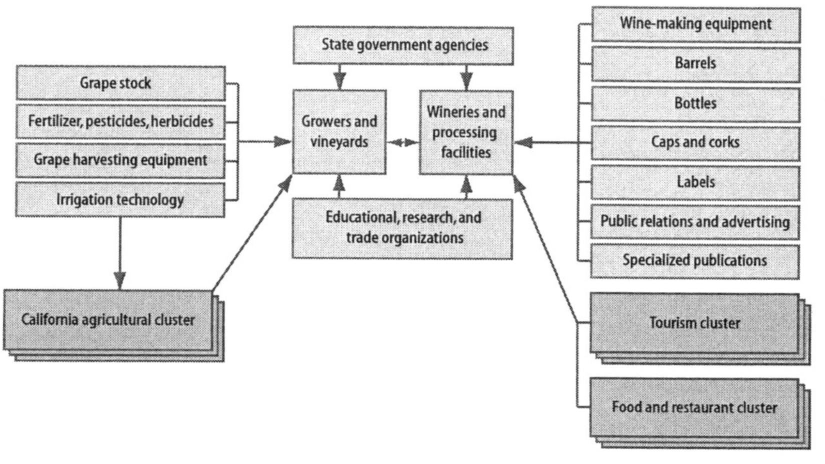

Consider also the Italian leather fashion cluster, which contains well-known shoe companies such as Ferragamo and Gucci as well as a host of specialized suppliers of footwear components, machinery, molds, design services, and tanned leather. (See the exhibit "Mapping the Italian Leather Fashion Cluster.") It also consists of several chains of related industries, including those producing different types of leather goods (linked by common inputs and technologies) and different types of footwear (linked by overlapping channels and technologies). These industries employ common marketing media and compete with similar images in similar customer segments. A related Italian cluster in textile fashion, including clothing, scarves, and accessories, produces complementary products that often employ common channels. The extraordinary strength of the Italian leather fashion cluster can be attributed, at least in part, to the multiple linkages and synergies that participating Italian businesses enjoy.

A cluster's boundaries are defined by the linkages and complementarities across industries and institutions that are most important to competition. Although clusters often fit within political boundaries, they may cross state or even national borders. In the United States, for example, a pharmaceuticals cluster straddles New Jersey and Pennsylvania near Philadelphia. Similarly, a chemicals cluster in Germany crosses over into German-speaking Switzerland.

Clusters rarely conform to standard industrial classification systems, which fail to capture many important actors and relationships in competition. Thus significant clusters may be obscured or even go unrecognized. In Massachusetts, for example, more than 400 companies, representing at least 39,000 high-paying jobs, are involved in medical devices in some way. The cluster long remained all but invisible, however, buried within larger and overlapping industry categories such as electronic equipment and plastic products. Executives in the medical devices cluster have only recently come together to work on issues that will benefit them all.

Clusters promote both competition and cooperation. Rivals compete intensely to win and retain customers. Without vigorous competition, a cluster will fail. Yet there is also cooperation, much of it vertical, involving companies in related industries and local institutions. Competition can coexist with cooperation because they occur on different dimensions and among different players.

Clusters represent a kind of new spatial organizational form in between arm's-length markets on the one hand and hierarchies, or vertical integration, on the other. A cluster, then, is an alternative

MAPPING THE ITALIAN LEATHER FASHION CLUSTER

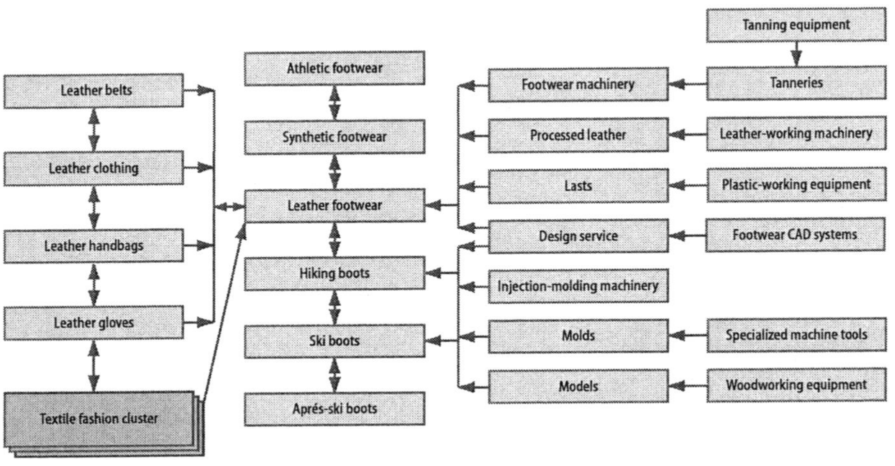

way of organizing the value chain. Compared with market transactions among dispersed and random buyers and sellers, the proximity of companies and institutions in one location – and the repeated exchanges among them – fosters better coordination and trust. Thus clusters mitigate the problems inherent in arm's-length relationships without imposing the inflexibilities of vertical integration or the management challenges of creating and maintaining formal linkages such as networks, alliances, and partnerships. A cluster of independent and informally linked companies and institutions represents a robust organizational form that offers advantages in efficiency, effectiveness, and flexibility.

Why Clusters Are Critical to Competition

Modern competition depends on productivity, not on access to inputs or the scale of individual enterprises. Productivity rests on *how* companies compete, not on the particular fields they compete in. Companies can be highly productive in any industry – shoes, agriculture, or semiconductors – if they employ sophisticated methods, use advanced technology, and offer unique products and services. All industries can employ advanced technology; all industries can be knowledge intensive.

The sophistication with which companies compete in a particular location, however, is strongly influenced by the quality of the local business environment.[1] Companies cannot employ advanced logistical techniques, for example, without a high-quality transportation infrastructure. Nor can companies effectively compete on sophisticated service without well-educated employees. Businesses cannot operate efficiently under onerous regulatory red tape or under a court system that fails to resolve disputes quickly and fairly. Some aspects of the business environment, such as the legal system, for example, or corporate tax rates, affect all industries. In advanced economies, however, the more decisive aspects of the business environment are often cluster specific; these constitute some of the most important microeconomic foundations for competition.

Clusters affect competition in three broad ways: first, by increasing the productivity of companies based in the area; second, by driving the direction and pace of innovation, which underpins future productivity growth; and third, by stimulating the formation of new businesses, which expands and strengthens the cluster itself. A cluster allows each member to benefit *as if* it had greater scale or *as if* it had joined with others formally – without requiring it to sacrifice its flexibility.

Clusters and Productivity. Being part of a cluster allows companies to operate more productively in sourcing inputs; accessing information, technology, and needed institutions; coordinating with related companies; and measuring and motivating improvement.

Better Access to Employees and Suppliers. Companies in vibrant clusters can tap into an existing pool of specialized and experienced employees, thereby lowering their search and transaction costs in recruiting. Because a cluster signals opportunity and reduces the risk of relocation for employees, it can also be easier to attract talented people from other locations, a decisive advantage in some industries.

A well-developed cluster also provides an efficient means of obtaining other important inputs. Such a cluster offers a deep and specialized supplier base. Sourcing locally instead of from distant suppliers lowers transaction costs. It minimizes the need for inventory, eliminates importing costs and delays, and – because local reputation is important – lowers the risk that suppliers will overprice or renege on commitments. Proximity improves communications and makes it easier for suppliers to provide ancillary or support services such as installation and debugging. Other things being equal, then, local outsourcing is a better solution than distant outsourcing, especially for advanced and specialized inputs involving embedded technology, information, and service content.

Formal alliances with distant suppliers can mitigate some of the disadvantages of distant outsourcing. But all formal alliances involve their own complex bargaining and governance problems and can inhibit a company's flexibility. The close, informal relationships possible among companies in a cluster are often a superior arrangement.

In many cases, clusters are also a better alternative to vertical integration. Compared with in-house units, outside specialists are often more cost effective and responsive, not only in component production but also in services such as training. Although extensive vertical integration may have once been the norm, a fast-changing environment can render vertical integration inefficient, ineffective, and inflexible.

Even when some inputs are best sourced from a distance, clusters offer advantages. Suppliers trying to penetrate a large, concentrated market will price more aggressively, knowing that as they do so they can realize efficiencies in marketing and in service.

Working against a cluster's advantages in assembling resources is the possibility that competition will render them more expensive and scarce. But companies do have the alternative of outsourcing many inputs from other locations, which tends to limit potential cost penalties. More important, clusters increase not only the demand for specialized inputs but also their supply.

Access to Specialized Information. Extensive market, technical, and competitive information accumulates within a cluster, and members have preferred access to it. In addition, personal relationships and community ties foster trust and facilitate the flow of information. These conditions make information more transferable.

Complementarities. A host of linkages among cluster members results in a whole greater than the sum of its parts. In a typical tourism cluster, for example, the quality of a visitor's experience depends not only on the appeal of the primary attraction but also on the quality and efficiency of complementary businesses such as hotels, restaurants, shopping outlets, and transportation facilities. Because members of the cluster are mutually dependent, good performance by one can boost the success of the others.

Complementarities come in many forms. The most obvious is when products complement one another in meeting customers' needs, as the tourism example illustrates. Another form is the coordination of activities across companies to optimize their collective productivity. In wood products, for instance, the efficiency of sawmills depends on a reliable supply of high-quality timber and the ability to put all the timber to use – in furniture (highest quality), pallets and boxes (lower quality), or wood chips (lowest quality). In the early 1990s, Portuguese sawmills suffered from poor timber quality because local landowners did not invest in timber management. Hence most timber was processed for use in pallets and boxes, a lower-value use that limited the price paid to landowners. Substantial improvement in productivity was possible, but only if several parts of the cluster changed simultaneously.

A cluster allows each member to benefit *as if* it had greater scale or *as if* it had joined with others without sacrificing its flexibility.

Logging operations, for example, had to modify cutting and sorting procedures, while sawmills had to develop the capacity to process wood in more sophisticated ways. Coordination to develop standard wood classifications and measures was an important enabling step. Geographically dispersed companies are less likely to recognize and capture such linkages.

Other complementarities arise in marketing. A cluster frequently enhances the reputation of a lo-

MAPPING SELECTED U.S. CLUSTERS

Here are just some of the clusters in the United States. A few – Hollywood's entertainment cluster and High Point, North Carolina's household-furniture cluster – are well known. Others are less familiar, such as golf equipment in Carlsbad, California, and optics in Phoenix, Arizona. A relatively small number of clusters usually account for a major share of the economy within a geographic area as well as for an overwhelming share of its economic activity that is "exported" to other locations. *Exporting clusters* – those that export products or make investments to compete outside the local area – are the primary source of an area's economic growth and prosperity over the long run. The demand for local industries is inherently limited by the size of the local market, but exporting clusters can grow far beyond that limit.

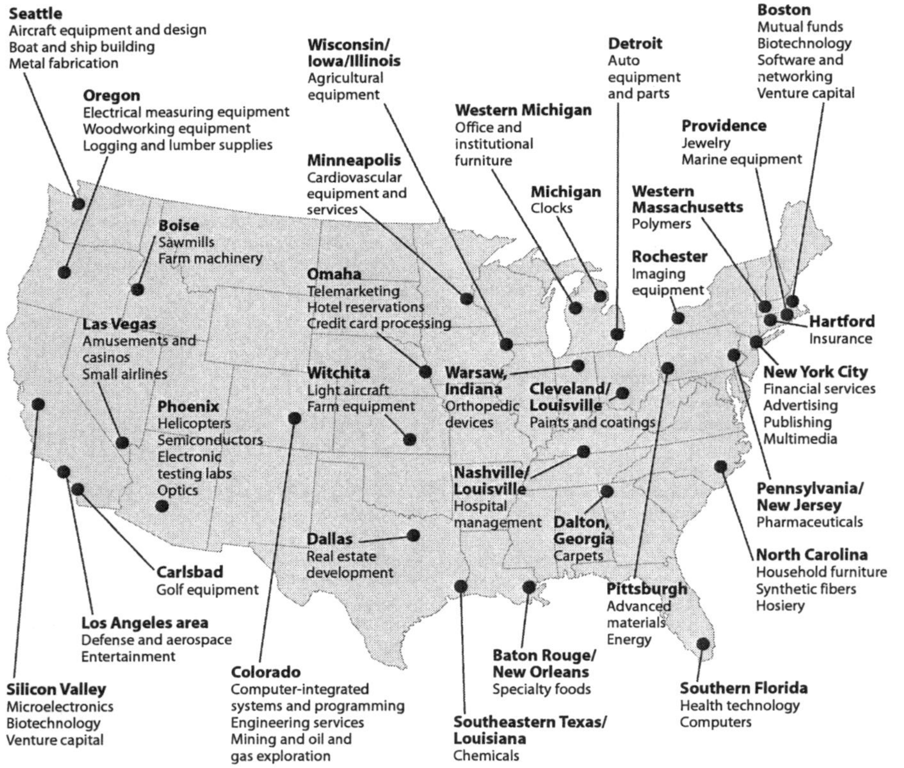

cation in a particular field, making it more likely that buyers will turn to a vendor based there. Italy's strong reputation for fashion and design, for example, benefits companies involved in leather goods, footwear, apparel, and accessories. Beyond reputation, cluster members often profit from a variety of joint marketing mechanisms, such as company referrals, trade fairs, trade magazines, and marketing delegations.

Finally, complementarities can make buying from a cluster more attractive for customers. Visiting buyers can see many vendors in a single trip. They also may perceive their buying risk to be lower because one location provides alternative suppliers. That allows them to multisource or to switch vendors if the need arises. Hong Kong thrives as a source of fashion apparel in part for this reason.

Access to Institutions and Public Goods. Investments made by government or other public institutions – such as public spending for specialized infrastructure or educational programs – can enhance a company's productivity. The ability to recruit employees trained at local programs, for example, lowers the cost of internal training. Other quasi-public goods, such as the cluster's information and technology pools and its reputation, arise as natural by-products of competition.

It is not just governments that create public goods that enhance productivity in the private sector. Investments by companies – in training programs, infrastructure, quality centers, testing laboratories, and so on – also contribute to increased productivity. Such private investments are often made collectively because cluster participants recognize the potential for collective benefits.

Better Motivation and Measurement. Local rivalry is highly motivating. Peer pressure amplifies competitive pressure within a cluster, even among noncompeting or indirectly competing companies. Pride and the desire to look good in the local community spur executives to attempt to outdo one another.

Clusters also often make it easier to measure and compare performances because local rivals share general circumstances – for example, labor costs and local market access – and they perform similar activities. Companies within clusters typically have intimate knowledge of their suppliers' costs. Managers are able to compare costs and employees' performance with other local companies. Additionally, financial institutions can accumulate knowledge about the cluster that can be used to monitor performance.

Clusters and Innovation. In addition to enhancing productivity, clusters play a vital role in a company's ongoing ability to innovate. Some of the same characteristics that enhance current productivity have an even more dramatic effect on innovation and productivity growth.

Because sophisticated buyers are often part of a cluster, companies inside clusters usually have a better window on the market than isolated competitors do. Computer companies based in Silicon Valley and Austin, Texas, for example, plug into customer needs and trends with a speed difficult to match by companies located elsewhere. The ongoing relationships with other entities within the cluster also help companies to learn early about evolving technology, component and machinery availability, service and marketing concepts, and so on. Such learning is facilitated by the ease of making site visits and frequent face-to-face contact.

Clusters do more than make opportunities for innovation more visible. They also provide the capacity and the flexibility to act rapidly. A company within a cluster often can source what it needs to implement innovations more quickly. Local suppliers and partners can and do get closely involved in the innovation process, thus ensuring a better match with customers' requirements.

Companies within a cluster can experiment at lower cost and can delay large commitments until they are more assured that a given innovation will pan out. In contrast, a company relying on distant suppliers faces greater challenges in every activity it coordinates with other organizations – in contracting, for example, or securing delivery or obtaining associated technical and service support. Innovation can be even harder in vertically integrated companies, especially in those that face difficult trade-offs if the innovation erodes the value of in-house assets or if current products or processes must be maintained while new ones are developed.

Peer pressure, pride, and the desire to look good in the community spur executives to outdo one another.

Reinforcing the other advantages for innovation is the sheer pressure – competitive pressure, peer pressure, constant comparison – that occurs in a cluster. Executives vie with one another to set their companies apart. For all these reasons, clusters can remain centers of innovation for decades.

Clusters and New Business Formation. It is not surprising, then, that many new companies grow up within an existing cluster rather than at isolated locations. New suppliers, for example, proliferate within a cluster because a concentrated customer base lowers their risks and makes it easier for them to spot market opportunities. Moreover, because developed clusters comprise related industries that normally draw on common or very similar inputs, suppliers enjoy expanded opportunities.

Clusters are conducive to new business formation for a variety of reasons. Individuals working within a cluster can more easily perceive gaps in products or services around which they can build businesses. Beyond that, barriers to entry are lower than elsewhere. Needed assets, skills, inputs, and staff are often readily available at the cluster location, waiting to be assembled into a new enterprise. Local financial institutions and investors, already familiar with the cluster, may require a lower risk premium on capital. In addition, the cluster often presents a significant local market, and an entrepreneur may benefit from established relationships. All of these factors reduce the perceived risks of entry – and of exit, should the enterprise fail.

The formation of new businesses within a cluster is part of a positive feedback loop. An expanded cluster amplifies all the benefits I have described – it increases the collective pool of competitive resources, which benefits all the cluster's members. The net result is that companies in the cluster advance relative to rivals at other locations.

At the intersection of clusters, insights and skills from various fields merge, sparking new businesses.

Birth, Evolution, and Decline

A cluster's roots can often be traced to historical circumstances. In Massachusetts, for example, several clusters had their beginnings in research done at MIT or Harvard. The Dutch transportation cluster owes much to Holland's central location within Europe, an extensive network of waterways, the efficiency of the port of Rotterdam, and the skills accumulated by the Dutch through Holland's long maritime history.

Clusters may also arise from unusual, sophisticated, or stringent local demand. Israel's cluster in irrigation equipment and other advanced agricultural technologies reflects that nation's strong desire for self-sufficiency in food together with a scarcity of water and hot, arid growing conditions. The environmental cluster in Finland emerged as a result of pollution problems created by local process industries such as metals, forestry, chemicals, and energy.

Prior existence of supplier industries, related industries, or even entire related clusters provides yet another seed for new clusters. The golf equipment cluster near San Diego, for example, has its roots in southern California's aerospace cluster. That cluster created a pool of suppliers for castings and advanced materials as well as engineers with the requisite experience in those technologies.

New clusters may also arise from one or two innovative companies that stimulate the growth of many others. Medtronic played this role in helping to create the Minneapolis medical-device cluster. Similarly, MCI and America Online have been hubs for growing new businesses in the telecommunications cluster in the Washington, D.C., metropolitan area.

Sometimes a chance event creates some advantageous factor that, in turn, fosters cluster development – although chance rarely provides the sole explanation for a cluster's success in a location. The telemarketing cluster in Omaha, Nebraska, for example, owes much to the decision by the United States Air Force to locate the Strategic Air Command (SAC) there. Charged with a key role in the country's nuclear deterrence strategy, SAC was the site of the first installation of fiber-optic telecommunications cables in the United States. The local Bell operating company (now U.S. West) developed unusual capabilities through its dealings with such a demanding customer. The extraordinary telecommunications capability and infrastructure that consequently developed in Omaha, coupled with less unique attributes such as its central-time-zone location and easily understandable local accent, provided the underpinnings of the area's telemarketing cluster.

Once a cluster begins to form, a self-reinforcing cycle promotes its growth, especially when local institutions are supportive and local competition is vigorous. As the cluster expands, so does its influence with government and with public and private institutions.

A growing cluster signals opportunity, and its success stories help attract the best talent. Entrepreneurs take notice, and individuals with ideas or relevant skills migrate in from other locations. Specialized suppliers emerge; information accumulates; local institutions develop specialized training, research, and infrastructure; and the cluster's strength and visibility grow. Eventually, the cluster broadens to encompass related industries. Numerous case studies suggest that clusters require a decade or more to develop depth and real competitive advantage.[2]

Cluster development is often particularly vibrant at the intersection of clusters, where insights, skills, and technologies from various fields merge, sparking innovation and new businesses. An example from Germany illustrates this point. The country has distinct clusters in both home appliances and household furniture, each based on different technologies and inputs. At the intersection of the two, though, is a cluster of built-in kitchens and appliances, an area in which Germany commands a higher share of world exports than in either appliances or furniture.

Clusters continually evolve as new companies and industries emerge or decline and as local institutions develop and change. They can maintain vibrancy as competitive locations for centuries; most successful clusters prosper for decades at least. However, they can and do lose their competitive edge due to both external and internal forces. Technological discontinuities are perhaps the most significant of the external threats because they can neutralize many advantages simultaneously. A cluster's assets – market information, employees' skills, scientific and technical expertise, and supplier bases – may all become irrelevant. New England's loss of market share in golf equipment is a good example. The New England cluster was based on steel shafts, steel irons, and wooden-headed woods. When companies in California began making golf clubs with advanced materials, East Coast producers had difficulty competing. A number of them were acquired or went out of business.

A shift in buyers' needs, creating a divergence between local needs and needs elsewhere, constitutes another external threat. U.S. companies in a variety of clusters, for example, suffered when energy efficiency grew in importance in most parts of the world while the United States maintained low energy prices. Lacking both pressure to improve and insight into customer needs, U.S. companies were slow to innovate, and they lost ground to European and Japanese competitors.

Clusters are at least as vulnerable to internal rigidities as they are to external threats. Overconsolidation, mutual understandings, cartels, and other restraints to competition undermine local rivalry. Regulatory inflexibility or the introduction of restrictive union rules slows productivity improvement. The quality of institutions such as schools and universities can stagnate.

Groupthink among cluster participants – Detroit's attachment to gas-guzzling autos in the 1970s is one example – can be another powerful form of rigidity. If companies in a cluster are too inward looking, the whole cluster suffers from a collective inertia, making it harder for individual companies to embrace new ideas, much less perceive the need for radical innovation.

Such rigidities tend to arise when government suspends or intervenes in competition or when companies persist in old behaviors and relationships that no longer contribute to competitive advantage. Increases in the cost of doing business begin to outrun the ability to upgrade. Rigidities of this nature currently work against a variety of clusters in Switzerland and Germany.

As long as rivalry remains sufficiently vigorous, companies can partially compensate for some decline in the cluster's competitiveness by outsourcing to distant suppliers or moving part or all of production elsewhere to offset local wages that rise ahead of productivity. German companies in the 1990s, for example, have been doing just that. Technology can be licensed or sourced from other locations, and product development can be moved. Over time, however, a location will decline if it fails to build capabilities in major new technologies or needed supporting firms and institutions.

> The term *high tech* has created the misconception that only a handful of businesses compete in sophisticated ways.

Implications for Companies

In the new economics of competition, what matters most is not inputs and scale, but productivity – and that is true in all industries. The term *high tech*, normally used to refer to fields such as information technology and biotechnology, has distorted

CLUSTERS, GEOGRAPHY, AND ECONOMIC DEVELOPMENT

Poor countries lack well-developed clusters; they compete in the world market with cheap labor and natural resources. To move beyond this stage, the development of well-functioning clusters is essential. Clusters become an especially controlling factor for countries moving from a middle-income to an advanced economy. Even in high-wage economies, however, the need for cluster upgrading is constant. The wealthier the economy, the more it will require innovation to support rising wages and to replace jobs eliminated by improvements in efficiency and the migration of standard production to low-cost areas.

Promoting cluster formation in developing economies means starting at the most basic level. Policymakers must first address the foundations: improving education and skill levels, building capacity in technology, opening access to capital markets, and improving institutions. Over time, additional investment in more cluster-specific assets is necessary.

Government policies in developing economies often unwittingly work against cluster formation. Restrictions on industrial location and subsidies to invest in distressed areas, for example, can disperse companies artificially. Protecting local companies from competition leads to excessive vertical integration and blunted pressure for innovation, retarding cluster development.

In the early stages of economic development, countries should expand internal trade among cities and states and trade with neighboring countries as important stepping stones to building the skills to compete globally. Such trade greatly enhances cluster development. Instead, attention is typically riveted on the large, advanced markets, an orientation that has often been reinforced by protectionist policies restricting trade with nearby markets. However, the kinds of goods developing countries can trade with advanced economies are limited to commodities and to activities sensitive to labor costs.

While it is essential that clusters form, *where* they form also matters. In developing economies, a large proportion of economic activity tends to concentrate around capital cities such as Bangkok and Bogotá. That is usually because outlying areas lack infrastructure, institutions, and suppliers. It may also reflect an intrusive role by the central government in controlling competition, leading companies to locate near the seat of power and the agencies whose approval they require to do business.

This pattern of economic geography inflicts high costs on productivity. Congestion, bottlenecks, and inflexibility lead to high administrative costs and major inefficiencies, not to mention a diminished quality of life. Companies cannot easily move out from the center, however, because neither infrastructure nor rudimentary clusters exist in the smaller cities and towns. (The building of a tourism cluster in developing economies can be a positive force in improving the outlying infrastructure and in dispersing economic activity.)

Even in advanced economies, however, economic activity may be geographically concentrated. Japan offers a particularly striking case, with nearly 50% of total manufacturing shipments located around Tokyo and Osaka. This is due less to inadequacies in infrastructure in outlying areas than to a powerful and intrusive central government, with its centralizing bias in policies and institutions. The Japanese case vividly illustrates the major inefficiencies and productivity costs resulting from such a pattern of economic geography, even for advanced nations. It is a major policy issue facing Japan.

An economic geography characterized by specialization and dispersion – that is, a number of metropolitan areas, each specializing in an array of clusters – appears to be a far more productive industrial organization than one based on one or two huge, diversified cities. In nations such as Germany, Italy, Switzerland, and the United States, this kind of internal specialization and trade – and internal competition among locations – fuels productivity growth and hones the ability of companies to compete effectively in the global arena.

thinking about competition, creating the misconception that only a handful of businesses compete in sophisticated ways.

In fact, there is no such thing as a low-tech industry. There are only low-tech companies – that is, companies that fail to use world-class technology and practices to enhance productivity and innovation. A vibrant cluster can help any company in any industry compete in the most sophisticated ways, using the most advanced, relevant skills and technologies.

Thus executives must extend their thinking beyond what goes on inside their own organizations and within their own industries. Strategy must also

MAPPING PORTUGAL'S CLUSTERS

In a middle-income economy like Portugal, exporting clusters tend to be more natural-resource or labor intensive.

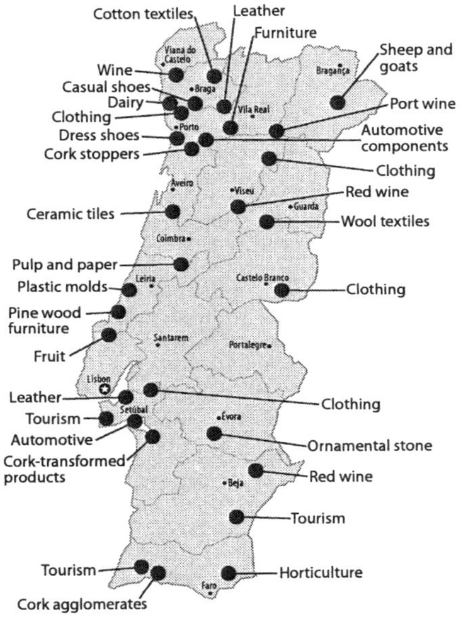

address what goes on outside. Extensive vertical integration may once have been appropriate, but companies today must forge close linkages with buyers, suppliers, and other institutions.

Specifically, understanding clusters adds the following four issues to the strategic agenda.

1. Choosing Locations. Globalization and the ease of transportation and communication have led many companies to move some or all of their operations to locations with low wages, taxes, and utility costs. What we know about clusters suggests, first, that some of those cost advantages may well turn out to be illusory. Locations with those advantages often lack efficient infrastructure, sophisticated suppliers, and other cluster benefits that can more than offset any savings from lower input costs. Savings in wages, utilities, and taxes may be highly visible and easy to measure up front, but productivity penalties remain hidden and unanticipated.

More important to ongoing competitiveness is the role of location in innovation. Yes, companies have to spread activities globally to source inputs and gain access to markets. Failure to do so will lead to a competitive *disadvantage.* And for stable, labor-intensive activities such as assembly and software translation, low factor costs are often decisive in driving locational choices.

For a company's "home base" for each product line, however, clusters are critical. Home base activities – strategy development, core product and process R&D, a critical mass of the most sophisticated production or service provision – create and renew the company's product, processes, and services. Therefore locational decisions must be based on both total systems costs and innovation potential, not on input costs alone. Cluster thinking suggests that every product line needs a home base, and the most vibrant cluster will offer the best location. Within the United States, for example, Hewlett-Packard has chosen cluster locations for the home bases of its major product lines: California, where almost all of the world's leading personal computer and workstation businesses are located, is home to personal computers and workstations; Massachusetts, which has an extraordinary concentration of world-renowned research hospitals and leading medical instrument companies, is home to medical instruments.

As global competition nullifies traditional comparative advantages and exposes companies to the best rivals from around the world, a growing number of multinationals are shifting their home bases to more vibrant clusters – often using acquisitions as a means of establishing themselves as insiders in a new location. Nestlé, for example, after acquiring Rowntree Mackintosh, relocated its confectionary business to York, England, where Rowntree was originally based, because a vibrant food cluster thrives there. England, with its sweet-toothed consumers, sophisticated retailers, advanced advertising agencies, and highly competitive media companies, constitutes a more dynamic environment for competing in mass-market candy than Switzerland did. Similarly, Nestlé has moved its headquarters for bottled water to France, the most competitive location in that industry. Northern Telecom has relocated its home base for central office switching from Canada to the United States–drawn by the vibrancy of the U.S. telecommunications-equipment cluster.

Cluster thinking also suggests that it is better to move groups of linked activities to the same place than to spread them across numerous locations. Colocating R&D, component fabrication, assembly, marketing, customer support, and even related businesses can facilitate internal efficiencies in sourcing and in sharing technology and information. Grouping activities into campuses also allows companies to extend deeper roots into local clusters, improving their ability to capture potential benefits.

2. *Engaging Locally.* The social glue that binds clusters together also facilitates access to important resources and information. Tapping into the competitively valuable assets within a cluster requires personal relationships, face-to-face contact, a sense of common interest, and "insider" status. The mere colocation of companies, suppliers, and institutions creates the *potential* for economic value; it does not necessarily ensure its realization.

To maximize the benefits of cluster involvement, companies must participate actively and establish a significant local presence. They must have a substantial local investment even if the parent company is headquartered elsewhere. And they must foster ongoing relationships with government bodies and local institutions such as utilities, schools, and research groups.

Companies have much to gain by engaging beyond their narrow confines as single entities. Yet managers tend to be wary, at least initially. They fear that a growing cluster will attract competition, drive up costs, or cause them to lose valued employees to rivals or spin-offs. As their understanding of the cluster concept grows, however, managers realize that many participants in the cluster do not compete directly and that the offsetting benefits, such as a greater supply of better trained people, for example, can outweigh any increase in competition.

Tapping into the competitively valuable assets within a cluster requires personal relationships and "insider" status.

3. *Upgrading the Cluster.* Because the health of the local business environment is important to the health of the company, upgrading the cluster should be part of management's agenda. Companies upgrade their clusters in a variety of ways.

Consider Genzyme. Massachusetts is home to a vibrant biotechnology cluster, which draws on the region's strong universities, medical centers, and venture capital firms. Once Genzyme reached the stage in its development when it needed a manufacturing facility, CEO Henri Termeer initially considered the pharmaceuticals cluster in the New Jersey and Philadelphia area because it had what Massachusetts lacked: established expertise in drug manufacturing. Upon further reflection, however, Termeer decided to influence the process of creating a manufacturing capability in Genzyme's home base, reasoning that if his plans were successful, the company could become more competitive.

Thus Genzyme deliberately chose to work with contractors committed to the Boston area, bypassing the many specialized engineering firms located near Philadelphia. In addition, it undertook a number of initiatives, with the help of city and state government, to improve the labor force, such as offering scholarships and internships to local youth. More broadly, Genzyme has worked to build critical mass for its cluster. Termeer believes that Genzyme's success is linked to the cluster's–and that all members will benefit from a strong base of supporting functions and institutions.

4. *Working Collectively.* The way clusters operate suggests a new agenda of collective action in the private sector. Investing in public goods is normally seen as a function of government, yet cluster thinking clearly demonstrates how companies benefit from local assets and institutions.

In the past, collective action in the private sector has focused on seeking government subsidies and special favors that often distort competition. But executives' long-term interests would be better served by working to promote a higher plane of competition. They can begin by rethinking the role of trade associations, which often do little more than lobby government, compile some statistics, and host social functions. The associations are missing an important opportunity.

Trade associations can provide a forum for the exchange of ideas and a focal point for collective action in overcoming obstacles to productivity and growth. Associations can take the lead in such activities as establishing university-based testing facilities and training or research programs; collecting cluster-related information; offering forums on common managerial problems; investigating solutions to environmental issues; organizing trade fairs and delegations; and managing purchasing consortia.

For clusters consisting of many small and midsize companies – such as tourism, apparel, and agriculture – the need is particularly great for collective bodies to assume scale-sensitive functions. In the Netherlands, for instance, grower cooperatives built the specialized auction and handling facilities that constitute one of the Dutch flower cluster's greatest competitive advantages. The Dutch Flower Council and the Association of Dutch Flower Growers Research Groups, in which most growers participate, have taken on other functions as well, such as applied research and marketing.

Most existing trade associations are too narrow; they represent industries, not clusters. In addition, because their role is defined as lobbying the federal government, their scope is national rather than local. National associations, however, are rarely sufficient to address the local issues that are most important to cluster productivity.

By revealing how business and government together create the conditions that promote growth, clusters offer a constructive way to change the nature of the dialogue between the public and private sectors. With a better understanding of what fosters true competitiveness, executives can start asking government for the right things. The example of MassMEDIC, an association formed in 1996 by the Massachusetts medical-devices cluster, illustrates this point. It recently worked successfully with the U.S. Food and Drug Administration to streamline the approval process for medical devices. Such a step clearly benefits cluster members and enhances competition at the same time.

What's Wrong with Industrial Policy

Productivity, not exports or natural resources, determines the prosperity of any state or nation. Recognizing this, governments should strive to create an environment that supports rising productivity. Sound macroeconomic policy is necessary but not sufficient. The microeconomic foundations for competition will ultimately determine productivity and competitiveness.

Governments – both national and local – have new roles to play. They must ensure the supply of high-quality inputs such as educated citizens and physical infrastructure. They must set the rules of competition – by protecting intellectual property and enforcing antitrust laws, for example – so that productivity and innovation will govern success in the economy. Finally, governments should promote cluster formation and upgrading and the buildup of public or quasi-public goods that have a significant impact on many linked businesses.

This sort of role for government is a far cry from industrial policy. In industrial policy, governments target "desirable" industries and intervene – through subsidies or restrictions on investments by foreign companies, for example – to favor local companies. In contrast, the aim of cluster policy is to reinforce the development of *all* clusters. This means that a traditional cluster such as agriculture should not be abandoned; it should be upgraded. Governments should not choose among clusters, because each one offers opportunities to improve productivity and support rising wages. Every cluster not only contributes directly to national productivity but also affects the productivity of *other* clusters. Not all clusters will succeed, of course, but market forces – not government decisions – should determine the outcomes.

Clusters offer a constructive way to change the nature of the dialogue between the public and private sectors.

Government, working with the private sector, should reinforce and build on existing and emerging clusters rather than attempt to create entirely new ones. Successful new industries and clusters often grow out of established ones. Businesses involving advanced technology succeed not in a vacuum but where there is already a base of related activities in the field. In fact, most clusters form independently of government action – and sometimes in spite of it. They form where a foundation of locational advantages exists. To justify cluster development efforts, some seeds of a cluster should have already passed a market test.

Cluster development initiatives should embrace the pursuit of competitive advantage and specialization rather than simply imitate successful clusters in other locations. This requires building on local sources of uniqueness. Finding areas of specialization normally proves more effective than head-on competition with well-established rival locations.

New Public-Private Responsibilities

Economic geography in an era of global competition, then, poses a paradox. In a global economy – which boasts rapid transportation, high-speed communication, and accessible markets – one would expect location to diminish in importance. But the opposite is true. The enduring competitive advantages in a global economy are often heavily local, arising from concentrations of highly specialized skills and knowledge, institutions, rivals, related businesses, and sophisticated customers. Geographic, cultural, and institutional proximity leads to special access, closer relationships, better information, powerful incentives, and other advantages in productivity and innovation that are difficult to tap from a distance. The more the world economy becomes complex, knowledge based, and dynamic, the more this is true.

Leaders of businesses, government, and institutions all have a stake – and a role to play – in the new economics of competition. Clusters reveal the mutual dependence and collective responsibility of all these entities for creating the conditions for productive competition. This task will require fresh thinking on the part of leaders and the willingness to abandon the traditional categories that drive our thinking about who does what in the economy. The lines between public and private investment blur. Companies, no less than governments and universities, have a stake in education. Universities have a stake in the competitiveness of local businesses. By revealing the process by which wealth is actually created in an economy, clusters open new public-private avenues for constructive action.

1 I first made this argument in *The Competitive Advantage of Nations* (New York Free Press, 1990) I modeled the effect of the local business environment on competition in terms of four interrelated influences, graphically depicted in a diamond. factor conditions (the cost and quality of inputs); demand conditions (the sophistication of local customers), the context for firm strategy and rivalry (the nature and intensity of local competition); and related and supporting industries (the local extent and sophistication of suppliers and related industries). Diamond theory stresses how these elements combine to produce a dynamic, stimulating, and intensely competitive business environment.

A cluster is the manifestation of the diamond at work Proximity – the colocation of companies, customers, and suppliers – amplifies all of the pressures to innovate and upgrade.

2 Selected case studies are described in "Clusters and Competition" in my book *On Competition* (Boston Harvard Business School Press, 1998), which also includes citations of the published output of a number of cluster initiatives. Readers can also find a full treatment of the intellectual roots of cluster thinking, along with an extensive bibliography

Reprint 98609 To place an order, call 1-800-988-0886.

[27]

Entrepreneurial Enterprises, Large Established Firms and Other Components of the Free-Market Growth Machine*

William J. Baumol

ABSTRACT. The paper studies the principal influences accounting for the unprecedented growth and innovation performance of the free-market economies. It indicates that vigorous oligopolistic competition, particularly in high-tech industries, forces firms to keep innovating in order to survive. This leads them to internalize innovative activities rather than leaving them to independent inventors, and turns invention into an assembly-line process. The bulk of private R&D spending is shown to come from a tiny number of very large firms. Yet the revolutionary breakthroughs continue to come predominantly from small entrepreneurial enterprises, with large industry providing streams of incremental improvements that also add up to major contributions. Moreover, these firms voluntarily disseminate much of their innovative technology widely and rapidly, both as a major revenue source and in exchange for complementary technological property of other firms, including direct competitors. This helps to internalize the externalities of innovation and speeds elimination of obsolete technology. Some policy implications for industrialized and developing countries are also discussed.

1. Introduction

Entrepreneurship has long been valued as a key contributor to the growth of an economy. This view often refers to the entrepreneurial activity entailed in the original usage of the term – the mere organization or establishment of just any new firm, even one that merely duplicates something that has been done many times before. In the following discussion, it is natural and appropriate to emphasize instead the meaning that was assigned to the term by Joseph Schumpeter (indisputably the 20th century's prime contributor to the economic analysis of entrepreneurship and innovation), taking the entrepreneur as the partner of the inventor – as the businessperson who recognizes the value of the invention, determines how to adapt it to the preferences of prospective users and whose tasks include bringing the invention to market and promoting its utilization. It is widely believed that economies that are abundantly supplied with entrepreneurs will tend to grow far more rapidly than those in which entrepreneurial talent is scarce. Yet Schumpeter himself was led to conclude that the day of the entrepreneur was waning, that the expanding role of routinized innovation by big business was threatening to make the entrepreneur obsolete. I will argue here that part of the pertinent mechanism has been correctly discerned both by those who continue to have faith in the individual entrepreneur's critical role in economic growth and by any who follow Schumpeter in concluding that routinized innovation by giant enterprises is assuming a primary role. Yet each side here is telling only part of the story and, as a result, overlooks much of its essence. The entrepreneur continues to play a critical part in the growth process, and there is no reason to expect that role to disappear. But in the modern economy the entrepreneur, working alone in the marketplace, cannot carry out the task most effectively. Fortunately, the market mechanism has provided the partners that the entrepreneur needs for the purpose.

Final version accepted on September 19, 2003

C.V. Starr Center for Applied Economics
New York University
269 Mercer Street
NY 1003
New York
U.S.A.
E-mail: william.baumol@nyu.edu

Small Business Economics **23** 9–21, 2004.

In seeking to explain the unprecedented and, indeed, miraculous growth performance of the free-market economies, this paper focuses upon the difference in, but complementary relationship between, the characteristic innovative contributions of large and small entrepreneurial firms, pointing out that these two groups have tended to specialize in different components of society's innovation process. The major breakthroughs have tended to come from small new enterprises, while the invaluable incremental contributions that multiply capacity and speed, and increase reliability and user-friendliness have been the domain of the larger firms. Together, the two have contributed far more than either would have by itself.

In addition, important innovations continue to flow from two groups outside the market sector: the government and the universities. I will point out some of the truly astonishing contributions that have come from each of the four sectors. The implication is that to ensure that the pertinent arrangements and institutions are really effective in the promotion of economic growth, it is essential that each is provided with the appropriate incentives to undertake its role in the process. For any modern economy concerned with this issue, understanding of the roles of the four key sectors and of the requisites for effectiveness of their participation constitutes a road map for public sector growth policy.

2. Market pressures for an enhanced large-firm role in technical progress

Free competition – that is, competition not handicapped by severe government regulations or tightly enforced customary rules, like those of the medieval guilds that prevented gloves-off combat among rival firms – has arguably played a critical role in the growth of the capitalist economies. Of particular significance here is rivalry among oligopolistic firms – those large firms in markets dominated by a small number of sellers. And crucial here is the fact that in today's economy many rival oligopolistic firms use *innovation* as their main battle weapon, with which they protect themselves from competitors and with which they seek to beat those competitors out. The result is precisely analogous to an arms race – to the case of two countries, each of which fears that the other will attack it militarily and, therefore, feels it necessary always at least to match the other country's military spending. Similarly, either of two competing firms will feel it to be foolhardy to let its competitor outspend it on the development and acquisition of *its* battle weapons. Each firm is driven to conclude that its very existence depends, at the least, on matching its rivals' efforts and spending on the innovation process. In an economy in which this is so, a constant stream of innovations can be expected to appear, because the giant warring firms to whom the story pertains do not dare relax their innovation activities.

The entrepreneur is naturally associated with the small, startup firm; indeed, as we know, widespread and long-employed usage simply *defines* entrepreneurs as the creators of new enterprises. This is relevant to investigation of their role in the economy's innovation process, because for the reasons just indicated, the apportionment of the task of supplying the resources invested in innovation has been changing materially. Increasingly, at least in the United States, the funding for innovation has been supplied by large oligopolistic enterprises, hardly the sort of firms that one associates with the entrepreneur.

Today some 70 percent of R&D expenditure in the U.S. is carried out by private business, and most of this is provided by the larger firms.[1] In these enterprises, innovative activities are carefully designed to prevent unwelcome surprises and to keep risks to a minimum. As a result, there is little of the free-wheeling, imaginative, and risk-taking approach that characterizes the entrepreneur. Instead, the large firm's top management often keeps a tight rein on the activities of the company's laboratories, with budgets determined by the upper strata of control within the firm, who also may determine how many persons and what sort of specialists at what levels will be employed on R&D endeavors. It is not even unusual for persons untrained or inexperienced in research to determine what new products and processes the laboratories should next seek to discover. Sometimes, large firms try to unleash their employees engaged in innovative activity by organizing a subsidiary operation that is more inviting to the free exercise of entrepreneurship, but often without much success.

The natural incentive system for a bureaucratically governed enterprise is to run research and development in accord with bureaucratic rules and procedures. All of this leads to the conjecture, voiced by Schumpeter, that the work responsibilities the economy assigns to the entrepreneur are narrowing and are destined to shrink even further. One can easily surmise what prompted Schumpeter to foresee a limited future for the entrepreneur where industry and its innovation processes are widely characterized in the manner just described. Yet, I will argue next that this is fundamentally a mischaracterization. Rather than being condemned to obsolescence, a vital role continues to be played by independent entrepreneurs.

3. Revolutionary breakthroughs: A small-firm specialty

It is convenient here to divide up inventions with the aid of two polar categories: revolutionary breakthroughs and cumulative incremental improvements. Of course, many new products and processes fall into neither extreme category, but are somewhere in-between. Still, it will become clear that the distinction is useful. Moreover, there are many examples that clearly fit into one of these categories or the other quite easily. For instance, the electric light, alternating electric current, the internal combustion engine, and a host of other advances must surely be deemed revolutionary, while successive models of washing machines and refrigerators – with each new model a bit longer-lasting, a bit less susceptible to breakdown, and a bit easier to use – arguably constitute a sequence of incremental improvements.

The relevance of the distinction should be evident, given what has been said about the working and organization of R&D in the large business organization. The inherent conservatism of the process naturally leads to the expectation that these firms will tend to specialize in the incremental improvements and tend to avoid the risks of the unknown that the revolutionary breakthrough entails. The latter, rather, is left most often to the small or newly founded enterprise, guided by its enterprising entrepreneur. Though that is to be expected, the degree of asymmetry in the apportionment of this specialized activity between large and small firms in reality is striking. The U.S. Small Business Administration has prepared a chart listing breakthrough innovations of the twentieth century for which small firms are responsible (reproduced here in Table I), and as will be seen, its menu of inventions literally spans the range from A to Z, from the airplane to the zipper.

This remarkable list includes a strikingly substantial share of the technical breakthroughs of the twentieth century. Besides the airplane, it lists FM radio, the helicopter, the personal computer, and the pacemaker, among a host of others, many of enormous significance for our economy.

A very recent study, also sponsored by the U.S. Small Business Administration (2003), provides more-systematic and powerful evidence to similar effect.[2] This report examines technical change through patenting and defines "small firms" as "businesses with fewer than 500 employees." Perhaps most notably, the study finds that ". . . a small firm patent is more likely than a large firm patent to be among the top 1 percent of most frequently cited patents." Among other conclusions, in the words of its authors, this study reports that,

> Small firms represent one-third of the most prolific patenting companies that have 15 or more U.S. patents.
>
> Small firm innovation is twice as closely linked to scientific research as large firm innovation on average, and so is substantially more high-tech or leading edge.
>
> Small firms are more effective in producing high-value innovations – the citation index for small firm patents averaged 1.53 compared to 1.19 for large firms.
>
> Small patenting firms are roughly 13 times more innovative per employee than large patenting firms.
>
> A small firm patent is at least twice as likely to be found among the top 1 percent of highest-impact patents as a patent from a large firm (p. 2).

One is, then, led to the plausible conjecture that most of the revolutionary new ideas of the past two centuries have been, and are likely to continue to be, provided more heavily by independent innovators who, essentially, operate small business enterprises. Evidently, the small entrepreneurial firms have come close to monopolizing the portion of R&D activity that is engaged in the search for revolutionary breakthroughs.

But now we seem to have leapt to the opposite conclusion, that rather than likely disappearance of the innovative role of the entrepreneur and the

TABLE I
Some important innovations by U.S. small firms in the twentieth century

Air Conditioning	Heart Valve	Prestressed Concrete
Air Passenger Service	Heat Sensor	Prefabricated Housing
Airplane	Helicopter	Pressure Sensitive Tape
Articulated Tractor Chassis	High Resolution CAT Scanner	Programmable Computer
Cellophane Artificial Skin	High Resolution Digital X-Ray	Quick-Frozen Food
Assembly Line	High Resolution X-Ray Microscope	Reading Machine
Audio Tape Recorder	Human Growth Hormone	Rotary Oil Drilling Bit
Bakelite	Hydraulic Brake	Safety Razor
Biomagnetic Imaging	Integrated Circuit	Six-Axis Robot Arm
Biosynthetic Insulin	Kidney Stone Laser	Soft Contact Lens
Catalytic Petroleum Cracking	Large Computer	Solid Fuel Rocket Engine
Computerized Blood Pressure Controller	Link Trainer	Stereoscopic Map Scanner
Continuous Casting	Microprocessor	Strain Gauge
Cotton Picker	Nuclear Magnetic Resonance Scanner	Strobe Lights
Defibrillator	Optical Scanner	Supercomputer
DNA Fingerprinting	Oral Contraceptives	Two-Armed Mobile Robot
Double-Knit Fabric	Outboard Engine	Vacuum Tube
Electronic Spreadsheet	Overnight National Delivery	Variable Output Transformer
Freewing Aircraft	Pacemaker	Vascular Lesion Laser
FM Radio	Personal Computer	Xerography
Front-End Loader	Photo Typesetting	X-Ray Telescope
Geodesic Dome	Polaroid Camera	Zipper
Gyrocompass	Portable Computer	

Source: U.S. Small Business Administration, 1995, p. 114.

small firm, little would appear to be left for the large enterprise to do. That, as we will see next, is also a very misleading conclusion.

4. Revolutionary consequences of aggregated *incremental* improvements

As we have seen, the type of innovation in which the giant enterprises tend to specialize is primarily devoted to product improvement, increased reliability and enhanced user-friendliness of products and the finding of new uses for those products. The approach tends to be conservative, seeking results whose applicability is clear and whose markets are relatively unspeculative. As already noted, the bureaucratic control typical of innovative activity in the large firm serves to ensure that the resulting changes will be modest, predictable and incremental. These firms are not predisposed to welcome the romantic flights of the imagination, the entrepreneurial leaps of faith and plunges into the unknown that often lead only to disaster, but which alone are likely to open up new worlds.

However, having recognized the critical role of the smaller enterprises, one should not go to the other extreme and undervalue the incremental contribution of the routine activity that at least sometimes arguably adds even more to growth than do the more revolutionary prototype innovations. Though each such small improvement may be relatively unspectacular, added together they can become very significant indeed. Thus, consider how little computing power the first clumsy and enormously expensive computers provided, and what huge multiples of such power have been added by the many subsequent incremental improvements.

Table I provides a set of extreme examples of the contributions of the small, entrepreneurial firms. But one can easily obtain equally startling examples of the magnitude of the innovative contributions of the large companies, whose incremental contributions can add up and compound to results of enormous magnitude. One such illustration is the progress in computer chip manufacture by the Intel Corporation, which is the leading manufacturer of these devices and has brought to market successive generations of chips and transistors, on which the performance of computers is so heavily dependent. According to a recent report,[3] over the period 1971–2003, the "clock speed" of Intel's microprocessor chips – that is, the number of instructions each chip can carry out per second – has increased by some *3 million percent*, reaching about 3 billion computations per second today. During the period 1968–2003, the number of transistors embedded in a single chip has expanded more than *10 million percent*, and the number of transistors that can be purchased for a dollar has grown by *five billion percent*. These are evidently no minor contributions. Added up, they surely contribute far more computing capacity than was provided by the original revolutionary breakthrough of the invention of the electronic computer. Of course, that initial invention was an indispensable necessity for all of the later improvements. But it is only the combined work of the two together that made possible the powerful and inexpensive apparatus that serves us so effectively today.

5. On the role of government and the university in innovation

A revised analysis of the forces making for economic growth also must not overlook the public sector's role in promotion of economic output and its expansion. My writing on the free-market growth mechanism emphasizes the importance of oligopolistic competition and independent entrepreneurship, and has given less attention to the contribution of the public sector. Yet in that story the government plays two critical roles, one active and the other passive. The passive contribution is provided primarily through the legal infrastructure that encourages entrepreneurship, the formation of new firms and investment in the innovation process by larger competing enterprises. That entails well-recognized provisions such as property rights and enforceability of contracts. It also entails absence of government acts of interference in the exchange of technical information and access to patented intellectual property, as well as avoidance of rules on employment and rental that inhibit the formation of new firms. On the active side, government support of basic research has proven to be invaluable, since with its uncertainties and unpredictable beneficiaries, such research is not highly attractive to private enterprise, though it can be critical for innovation and growth in the long run.

There are, as a matter of fact, two key players that are not guided directly by market forces, the universities and the pertinent government agencies, which have also made direct contributions to technological progress themselves. Here, one need only mention the electronic computer once again, as well as the Internet. But the contributions of these institutions have also tended to be rather specialized and different from those discussed above. It is to them that we must look primarily for the results provided by basic research as distinguished from applied research. The reasons for this division of labor with private industry are well understood, and only a few words need be said on the subject here.

From the point of view of the unthinking market mechanism, expenditure on basic research is a "wasteful" expenditure, because the outlay promises no addition to the profits of the firm. By its very nature, it is nearly impossible to predict whether basic research will yield any financial benefit at all and, if so, who will ultimately be the beneficiary. Certainly, it need not be the enterprise that carried it out. That is why governments and universities have had to step in, if basic research of any magnitude was to be carried out. And as we know, it is important for growth in the long run that this be done, for so much of applied innovation is made possible or is at least stimulated by its results.

The importance for technological progress of this conjunction of academia, business, and public sector was dramatically confirmed in an American Philosophical Society symposium (2003) on recent and projected biomedical advances.[4] There, it was strikingly demonstrated that the contributions of universities, government (notably the military) and (apparently small) private firms have produced a truly mind-boggling array of medical breakthroughs, aptly summed up by one speaker in a quotation from Hollywood's Steven Spielberg: "There is no more science *fiction*." The outlook is, indeed, that there will be no break in the acceleration of innovation, and that the innovations in prospect will be as difficult for us to comprehend as those now thoroughly familiar to us would have been for our ancestors. Just a small sample indicates the nature of what was described at the symposium:

- Surgery carried out by computer-guided robots, with immediate and automatic restocking (without reordering or human intervention) of surgical equipment and medication (this is partly already in use), and remote surgery in which the operating surgeon (who guides the computer) may be thousands of miles from the patient during the procedure (this is already done successfully);
- Growth of replacement bodily organs in specially maintained groups of animals (pigs apparently favored currently) for use in human transplantation, promising to make shortages of transplanted organs a thing of the past (this is predicted to become available within one to three years);
- Artificially induced hibernation and what can only be described as "reversible pseudo-death" of patients as a substitute for anesthesia, eliminating the latter's undesirable side effects and other perils;
- Inducement of growth of new organs within patients, as a substitute for transplants, eliminating the difficult and dangerous problem of transplant rejection and the need for post-procedure anti-rejection medications, with their undesirable side effects, perhaps for the remainder of the patient's lifetime;
- Use of certain insects (notably butterflies, cockroaches, and sphinx moths with their very specialized and powerful senses) to transmit information about the presence of dangers such as buried land mines or anthrax spores (or to locate earthquake victims), with no human presence, making it possible to take countermeasures remotely.

This list, surely, must suffice to stir the imagination to overload, to indicate that the end of innovation is nowhere in sight, and to confirm that the large corporations cannot do it alone.

6. Dissemination of invention and rapid termination of the obsolete

There is another key activity for growth that is elicited by market forces but that has not yet been mentioned – the incentive for rapid dissemination and widespread utilization of new or improved products and processes. This would appear simultaneously both to help and to hinder growth, but as will be argued next, current practices by industry have tended to contain the hindering effects.

One of the attributes of an effective economic arrangement for the encouragement of beneficial technological change is the innovator's financial gain derived from the temporary acquisition of monopoly power through the improved product or process in his possession. However, encouragement of growth also requires rapid *dissemination* of any improved techniques and products and their widespread adoption by others beside the innovator. These two desiderata would, however, appear to be in conflict. After all, rapidity and ease of dissemination can threaten the innovator's reward. While the free market has hardly eliminated this conflict, it has nevertheless ameliorated the problem to a considerable degree. Though it is something of a digression, my discussion will turn briefly to this issue because it will play a substantial role in the possible programs that merit consideration for facilitation of an economy's growth.

As is to be expected, many business firms do guard their proprietary technology and strive with the aid of patents, secrecy and other means to prevent other firms, notably rivals, from using the new products and processes. This is unfortunate for economic progress because it means that consumers who purchase from other firms are forced to accept obsolete features in the items they buy.[5] Moreover, two firms that deny one another access to their proprietary improvements in the firms' common product can evidently both survive, marketing their somewhat differentiated outputs, each of which is rendered inferior in terms of what is currently possible technologically by the obsolete features that it is forced to provide. Happily, however, that is hardly the norm. On the contrary, voluntary licensing of access to proprietary technology is widespread in the economy. Many firms derive substantial incomes from the sale of such licenses. The logic is straightforward. Suppose firm A invents a new widget and expects to make a net profit of X dollars per widget of the new type that it produces. Then if rival firm B offers firm A a license fee of Y dollars ($Y > X$) for each unit of the new widget it is able to sell, then A obviously can be better off letting B do so, even if every widget sold by B means one less sale for A. Of course, B will generally be able to afford so high a fee only if it is a more efficient *producer* of widgets than A, even though it may be an inferior inventor. In this way the price mechanism will not only encourage licensing, but will, as usual, elicit efficient specialization: inventive activity will be undertaken primarily by the more effective inventor, while production of the resulting products will be undertaken predominantly by the more efficient producer. This sort of unreciprocated licensing does take place in practice, but it seems most frequently to entail the sale of licenses by large firms that are in a position to undertake extensive R&D activity, the licensees being smaller enterprises that cannot afford to carry out such activity and do not possess personnel qualified to do so.

There are a number of other incentives for such profitable and voluntary exchanges in the free market. For example, the most straightforward reason, and the one that seems most frequently offered by businesspersons, is the very high cost of R&D activity. By entering into some sort of sharing consortium this burden can obviously be divided and reduced for each participant. Given the public-good attribute of the resulting information, it is far less expensive (per user) to provide such information to several firms than only to supply it to one. Another reason is reduction of risk. In any given year, a single firm's R&D division may fail to come up with any significant breakthroughs. The fear by management of firm A that this will happen to it in a year when its rival, B, manages a significant breakthrough is a fear that is replicated in firm B. Since, as already emphasized, product and process improvement are a matter of life and death in the high-tech industries characterized by vigorous oligopolistic competition, technology-sharing agreements serve as effective insurance policies, protecting each participant from such catastrophes.

A further, and less obvious, reason for voluntary dissemination also entails trading of technology, but it is undertaken because it protects the firm from entry. To see how this works, consider, for example, an industry with 10 firms of identical size, each with an R&D division with similar staffing and similar funding to those of the others. Each firm in such a consortium will then have available to it not only the discoveries of its own R&D establishment, but those of nine other firms in addition. Now suppose an eleventh firm wants to enter the market, but is not invited to join the technology-sharing consortium. Having only the products of its own R&D division at its disposal, while the other firms each obtain the outputs of ten R&D establishments, the entrant can find itself at a severe competitive disadvantage.

This type of arrangement evidently has its pros and cons. It can be shown to stimulate innovative effort (provided that anticompetitive conspiracy is absent). For it helps to internalize the externalities generated by the innovative efforts of each firm. Indeed, if as happens in practice, in such an exchange each firm undertakes compensation equalization payments to any other member of the consortium if the latter's innovations are of market value significantly superior to its own, then the firm has a direct incentive to come to the contract bargaining table with a menu of valuable innovations to offer. It can also be shown that the formation of such a consortium tends to be welfare-enhancing (Baumol, 2002, Chapter 7).

Yet there are evidently dangers against which the authorities must be vigilant. Such consortia *can* serve as vehicles or as camouflage for anticompetitive behavior. For example, the contract discussions can conceivably serve as a disguise for price fixing by the competitors. Or they can enter into an agreement for mutual restriction of their R&D expenditures, each firm knowing that it can safely limit its innovative efforts if it can rely on its rivals to do the same. Or the contracts can be offered in a discriminatory manner that limit the benefits offered to entrants or denies them access altogether.

Similar perils for the public interest arise in the last of the reasons for voluntary technology sharing – the problem of "patent thickets" and the widespread patent pools that have been formed to deal with the thicket problem. A complex piece of equipment, such as a computer, characteristically is made up of components each of which is covered by a surprisingly large number of patents, and the patents pertinent for such an item are often owned by a considerable number of different firms, many of them direct competitors in the final-product market. For example, Peter N. Detkin, vice president and assistant general counsel at Intel Corporation, estimates that there were more than 90,000 patents generally related to microprocessors held by more than 10,000 parties in 2002 (Federal Trade Commission, 2002, p. 667). This puts many of these firms in a legal position that can enable each to bring the manufacturing process of the others to a halt. The most effective way to prevent the catastrophic consequences this threatens for each of them is the formation of a patent pool in which each makes use of its patents available to the other members of the pool, and even to outsiders (as a step to avoid intervention by the anti-monopoly authorities). There are many such pools in the U.S., with widely varying membership rules, license fee arrangements and other differences that are not germane here.

It is of some interest that in the U.S. the Department of Justice and the Federal Trade Commission have recognized the two sides of the issue, the benefits of coordination in the arena and the attendant danger of anticompetitive behavior. Their 2000 *Guidelines for the Licensing of Intellectual Property* very explicitly discuss the substantial pro-competitive benefits of licensing, coordination of research efforts and trading of proprietary technology. What is significant for us here is that licensing as the prime instrument for technology dissemination has become sufficiently important to merit this sort of attention by the antitrust agencies.

7. Indicators of the magnitude of the free-enterprise growth miracle

But what has actually been accomplished by the components of the growth mechanism that have been reviewed here? The growth record of the free-market economies is generally recognized, though its magnitude seems often to be overlooked. It is seems worthwhile to pause and attempt to suggest the size of the accomplishment. The improvement in the growth performance of the industrial economies is, in fact, so enormous that it is difficult to comprehend. Average growth rates for about one and a half millennia before the Industrial Revolution are estimated to have been approximately zero, and while there was undoubtedly some growth starting around the tenth century, it proceeded at a snail's pace by modern standards. But in the 18th century, with the advent of capitalism in Great Britain, GDP per capita is estimated to have grown some 20 to 30 percent. In the 19th century, this figure rose, perhaps, tenfold, to some 200 percent. In the 20th century, growth in the U.S. has conservatively been estimated at about 700 percent, with some qualified observers arguing, with the aid of striking data, that this is a substantial *underestimate*. Elsewhere, the twentieth century growth rate was even more spectacular. The British record is described in Figure 1, which reports Angus Maddison's (2001) estimates of per capita income since the 16th century. The negligible rate of increase during the first three centuries reported in the graph is clear. The explosive growth path since that time must also be striking. It is also worth noting how early in the rising trajectory it was that Marx and Engels made their cogent observations: "The Bourgeoisie [i.e., capitalism] cannot exist without constantly revolutionizing the instruments of production. . . . Conservation of the old modes of production in unaltered form was, on the contrary, the first condition of existence for all earlier industrial classes. . . . The bourgeoisie, during its rule of scarce one hundred years has created more massive and more colossal productive forces than have all preceding generations together" (the arrow labeled 1848 marks the date of publication of these words).

Even the most well-off consumers before the Industrial Revolution had virtually no goods that were unavailable in ancient Rome. Clocks, hunting guns, window glass and paper virtually exhaust the list. Moreover, many choices available to affluent Romans – like hot baths – had long disappeared by the time of the Industrial Revolution. Since then, the pace of innovation has evidently grown into a flood. It has been said that the 18th century was the period when the process first took off; that only in the 19th century did the change achieve widespread recognition and did the belief in continuing technical progress take hold; while in the 20th century, the arrival of new products and processes became so frequent and commonplace that it began to be taken for granted and hardly worth noticing. It is surely arguable that during the

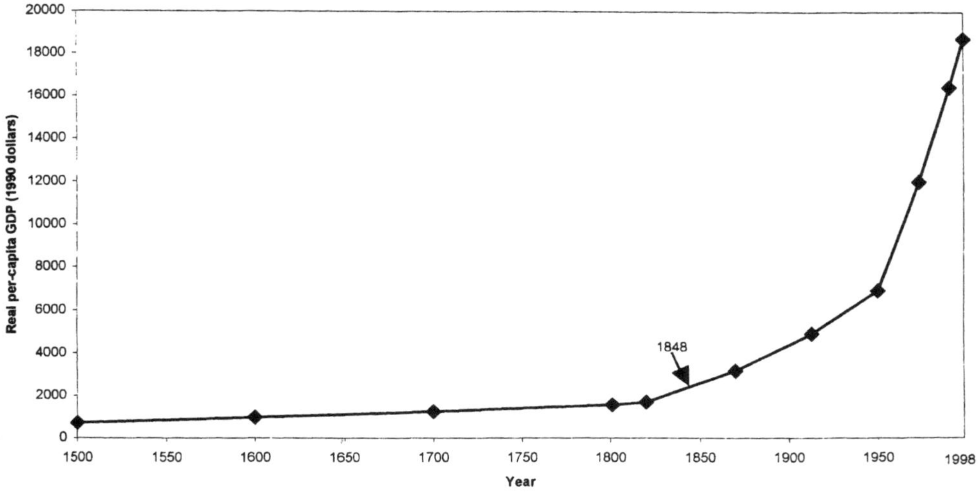

Source: Maddison, 2001.

Figure 1. Real per-capital GDP in the United Kingdom, 1500–1998.

collapse of many of the communist regimes in the late 1980s and early 1990s, when even China turned toward capitalist enterprise, what the public wanted was to participate in the capitalist growth miracle just described.

Outlays on research and invention, like GDP per capita, have also been exploding. Figure 2 shows real private (business) expenditures on research and development activity in the U.S. for nearly half a century after the Second World War. Here again, the near-exponential growth path is immediately observable. It is also worth observing how small was the effect of the recessions of the postwar period in holding back the growth in real expenditures on the invention process. The intervals of decline in R&D outlays are merely small deviations from what appears to be an inexorable rising path.

8. The invaluable contribution of "mere imitation"

It is predictable that most of the innovation that a relatively small industrial economy can expect to introduce will not have been contributed by the country's own R&D activities, but by those of other countries. This is not to be regarded as a deficiency. In a world in which almost all major technological development takes place in some 25 countries, and in which technology licensing and trading is increasingly common, it is a tautology that if none of the countries falls significantly behind, then the average country should expect some 24/25ths of its new technology to come from abroad.

The imitation process that is evidently so important is the source of a significant misapprehension. The notion is that the imitation process has little or none of the attributes of a truly innovative activity. But that is simply incorrect. History is replete with examples of substantial improvements that were contributed by the imitators. In part, these improvements are elicited by the need to adapt the technology to local conditions, including differences in size of the market, in the nature of consumer preferences, in climatic conditions and in the character of available complementary inputs. Thus, there is nothing inherently inferior about a process of organized imitation of foreign technology. Indeed, as one

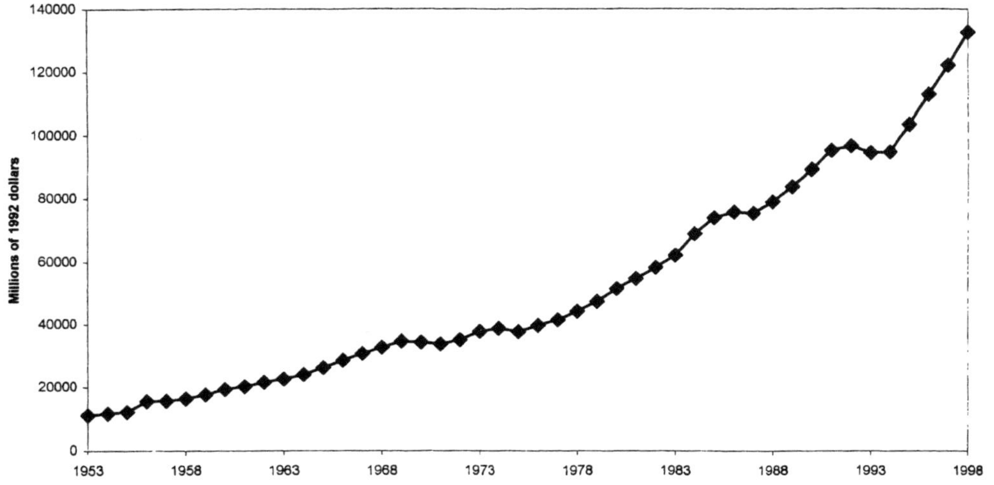

Source: National Science Board, 2000; and *Economic Report of the President*, 2001; expenditures are deflated using the GDP implicit price deflator.

Figure 2. Real U.S. Private R&D expenditures, 1953–1998.

historian who specialized in the history of innovation has observed, "... every invention contains some borrowing and every borrowing some invention" (De Camp, 1963).

Moreover, as just noted, every economically advanced nation can be expected to do it and to run the risk of falling behind if it does not. Even the U.S. and Japan, the two leading contributors to the world's stock of new products and processes, derive a substantial proportion of their latest technology from elsewhere. For every advanced economy, innovation will continue to be of prime importance for economic growth. But one may well expect that a substantial proportion of that innovation will be obtained from foreign sources. And to be an effective user of such foreign technology, it is important for the country to ensure that it is a skilled imitator as well as an effective innovator.

9. On governmental policy for promotion of innovation and growth

Given the four contributory sources that play critical roles in expanding an economy's innovation and growth – entrepreneurs and small firms, large firms with internal R&D capacity, universities, and government – one is driven to conclude that effective programs for facilitation and stimulation of entrepreneurship are important, but that there is more that can be done than is currently being done for this purpose. In this section, I offer a few illustrative suggestions that seem to merit consideration by the designers of policy. Of course, since the subject here is public policy, what will be said relates to the role of government – not the role of government as innovator on its own, but, rather, as facilitator of the innovative work of others. In particular, I will focus on one illustrative subject, the facilitation of the acquisition of foreign technology.

Funding and execution of basic research

As already noted, the universities and the pertinent government agencies have also made major contributions to technological progress, the electronic computer and the Internet providing striking examples. However, the contributions of these institutions have also tended, appropriately, to be rather specialized and different from those of private industry. I begin with just a few words about the character of those contributions.

It is to the public sector and universities that we must look primarily for the results provided by *basic research* as distinguished from applied research. The reasons for this division of labor with private industry are well understood and were laid out earlier in this paper. Basic research can contribute materially to the economy's growth, at least in the long run, but its questionable returns to the private investor makes it unattractive to business firms. What is clearly called for is governmental funding of basic research, to be carried out by its own agencies, or some appropriate outside agencies, most notably the universities, with the outstanding scientists and engineers to be found in their faculties. And indeed, specialization of the research activities funded by government have tended to focus on such basic research because that is the most effective and reliable way to elicit such activity that can be so important for long-run growth.

A government role in acquisition of foreign technology

Economists generally agree that government can play a useful role in the provision of certain socially valuable goods and services because private enterprises lack the incentive to supply optimal quantities of such outputs. Basic research, as just noted, is one important example. The encouragement of technology transfer from abroad is another significant case in point. In particular, it is important for an economy of smaller or medium size to recognize the contribution to its growth offered by rapid acquisition and absorption of technological information from elsewhere. But the transfer process has significant attributes that invite a role for the public sector. For example, the work of monitoring foreign technical journals and of providing translations of pertinent articles can be carried out nearly as cheaply for a multiplicity of firms, or even for a considerable number of industries, as it can on behalf of any single business enterprise.

Countries appear to have differed substantially in the quantity of resources they devote to this purpose. Edwin Mansfield (1990, p. 343) reports, on the basis of a survey of 100 American firms in 13 industries, that these respondents believed only 29 percent of U.S. firms spend as large a percentage of their sales on the monitoring of foreign technology as the average amount spent by the Japanese, only 47 percent as much as the Germans do, only 51 percent as much as the French do, and only 70 percent spend as much as the average British enterprises in the corresponding industries. Such disparities may well constitute an opportunity for a country to gain a differential advantage in its monitoring and adoption of foreign technology. For example, it may prove to be profitable socially for a government to establish a special Office of Technology Transfer, with a staff of specialists qualified to monitor, translate and disseminate pertinent materials in foreign publications. This, surely, constitutes a form of industrial policy that should make sense even to those economists who are most suspicious of public-sector intervention. More specifically, it is easy to describe some illustrative steps that can be taken in carrying out such a program:

a. *Education and training*. The government can establish a set of well-funded scholarships for the study of engineering and other pertinent subjects by a smaller country's students studying in the U.S., Japan, Germany, and several other countries that are leading producers of innovation. This program would include funding of an intensive set of language courses that would prepare the students for their studies abroad. The students can be obligated, upon completion of their courses, to take suitable jobs in their home country's industry or government for a period of (say) five years after completion of their studies.

b. *Immigration of foreign technicians and related personnel*. A fund can be established to provide subsidies for the immigration, permanent or temporary, of foreign scientists, engineers and technicians who can provide knowledge to their home-country industry about current foreign technological developments. It may be desirable to require such a prospective immigrant to be sponsored by a home-country firm, university or government agency, with the sponsor obligated to provide employment to the immigrant, and to make a case that the immigration of the individual in question will make a substantial contribution to absorption of useful technological information by native industry.

c. *Establishment of observer staff in the country's embassies*. At several countries that are leaders in innovation, the embassy in such a leader country of any growth-seeking economy should be provided with a special technology monitoring staff of a size that can be deemed adequate to keep track of technical journals, company newsletters and other available published materials, and arrange for translations where desirable, making those translations available to home-country firms, and to provide such information from other legitimate sources. These specialists can also help to facilitate technology transfer agreements between those firms and home-country enterprises. Frequent contact between these embassy staff members and suitable representatives of home-country industry can ensure that the efforts of the embassy observer staff are directed in a manner that is as useful as possible.

d. *Study of measures taken by governments in other countries to facilitate absorption of foreign technology by their industry*. A suitable group in the universities or elsewhere can be given funds to support an extensive and systematic study of the programs of other governments to encourage the acquisition of foreign technology, with the study describing the programs, their workings and their degree of success. Interviews with businesspersons in the countries in question can be used to provide analysis of the effectiveness and shortcomings of each of the programs. The study can provide a report and a set of recommendations based on foreign experience to the appropriate committees of the home government.

The preceding remarks are merely offered as examples of what sorts of measures are at least worthy of consideration. Reading of this article, and of the book upon which it draws, should readily suggest other promising avenues. Stimulation of such investigation is all that the preceding brief remarks on policy are meant to accomplish.

10. Concluding comment

The primary lesson that follows from the discussion here is that the future prosperity of any economy depends to a considerable extent on the its success in promoting entrepreneurship, innovation, and the effective and prompt importation of technological advance from abroad. It also follows that the incentives for these developments should not be left exclusively to chance and the natural spirit of enterprise to be found within a nation's population. There are measures that can be adopted to stimulate and facilitate them. This article's analysis of the mechanism underlying the free-market growth miracle should suggest promising directions for growth policy as well as actions that should be avoided because they are likely to impede progress in the economy.

Acknowledgement

As always, I must express my gratitude to my colleague, Sue Anne Batey Blackman, for her editing and contribution of sense to this paper. The discussion here is based upon, but goes beyond, the materials in my recent book, *The Free-Market Innovation Machine: Analyzing the Growth Miracle of Capitalism* (Princeton, 2002).

Notes

* Acceptance Address for the 2003 International Award for Entrepreneurship and Small Business Research by the Swedish Foundation for Small Business Research.

[1] According to data gathered by the National Science Foundation (National Science Board, 2000, Chapter 2, p. 24), in the year 2000, 46 percent of total U.S. industrial R&D funds were spent by 167 companies with 25,000 or more employees; 60 percent of these funds were spent by 366 companies with at least 10,000 employees, and 80 percent was spent by 1,990 firms of 1,000 or more employees. At the other end of the spectrum, about 15 percent of total U.S. industrial R&D funds were spent by 32,000 companies with fewer than 500 employees each.

[2] Quoting the release describing the study, "A total of 1,071 firms with 15 or more patents issued between 1996 and 2000 were examined. A total of 193,976 patents were analyzed. CHI [the firm that carried out the study] created a data-base of these firms and their patents. This list excluded foreign-owned firms, universities, government laboratories, and nonprofit institutions" (p. 2).

[3] John Markoff, "Technology; Is There Life After Silicon Valley's Fast Lane?," *New York Times*, Business/Financial Desk, Section C, April 9, 2003, p. 1.

[4] For more on this, see also "A Futurist's View: An Interview with Richard M. Satava, M.D., *Yale Medicine*, Winter 2003, available at www.med.yale.edu.

[5] It is, however, not always recognized that patents are not designed to *prevent* the spread of information about novel technology. On the contrary, patent holders are required to make full information on their inventions public so that others can profit from the ideas even if they cannot replicate the patented products themselves without the patent holder's permission.

References

American Philosophical Society, 2003, Session on "Biomedical Advances, Experienced and Projected, During One Surgeon's Seven Decade Career: In Honor of Jonathan E. Rhodes, MD, DSc (1907–2002; APS 1958)" April 25 Meetings (Speakers: J.C. Thompson, University of Texas at Galveston, T. E. Starzl, University of Pittsburgh, and R. M. Satava, University of Washington), Philadelphia, Pennsylvania, proceedings volume forthcoming in 2003.

Baumol, William J., 2002, *The Free-Market Innovation Machine: Analyzing the Growth Miracle of Capitalism*, Princeton, New Jersey: Princeton University Press.

De Camp, L. Sprague, 1993, *The Ancient Engineers: Technology and Invention From the Earliest Times to the Renaissance*, New York: Barnes and Noble.

Economic Report of the President, January 2001, Washington, D.C.: U.S. Government Printing Office.

Maddison, Angus, 2001, *The World Economy: A Millennial Perspective*, Paris: Organization for Economic Cooperation and Development.

Mansfield, Edwin, 1990, 'Comment', in C. R. Hulten (ed.), *Productivity Growth in Japan and the United States*, Chicago, Illinois: Chicago University Press, pp. 341–346.

Marx, Karl and Friedrich Engels, 1848, *Manifesto of the Communist Party*.

National Science Board, *Science and Engineering Indicators-2000*, Arlington, Virginia: National Science Foundation.

Ordover, Janusz A., 1991, 'A Patent System for Both Diffusion and Exclusion', *Journal of Economic Perspectives* **5** (Winter), 43–60.

The Patent Office [United Kingdom], *The Patent Office Annual Facts and Figures 2000–2001*, available at www.patent.gov.uk.

Reynolds, Paul D. et al., *Global Entrepreneurship Monitor [GEM], 2001 Executive Report, 2001*, Wellesley, Massachusetts, Kansas City, Missouri, and London, UK: Babson College, Kauffman Center for Industrial Leadership and London Business School.

Rosenberg, Nathan, 1976, *Perspectives on Technology*, Cambridge, UK: Cambridge University Press.

Schumpeter, Joseph A., 1947, *Capitalism, Socialism and Democracy*, 2nd ed., New York, New York: Harper and Brothers.

U.S. Small Business Administration, 1995, *The State of Small Business: A Report of the President, 1994*, Washington, D.C.: U.S. Government Printing Office.

Name Index